World Resources
1990-91

World Resources
1990-91

**A Report by
The World Resources Institute**

in collaboration with

**The United Nations
Environment Programme**

and

**The United Nations
Development Programme**

**New York   Oxford
Oxford University Press
1990**

The cover shows a view of the Greater Himalayas looking into India. The valley of the Indus River in Pakistan can be seen at the upper right. The photo was taken with a handheld camera from a U.S. space shuttle flight. Photo by the National Aeronautics and Space Administration, courtesy of Kevin Kelley.

The World Resources Institute, The United Nations Environment Programme and the United Nations Development Programme gratefully acknowledge permission to reprint from the following sources:

Part II: Figure 2.1, *Nature*; Table 2.4, American Geophysical Union; Box 2.2 Figure 1, Goddard Institute for Space Studies and Geophysical Fluid Dynamics Laboratory.

Part III: Figure 5.4, CDM and Associates; Table 5.4, The Johns Hopkins University Press; Table 6.4 and Figure 6.6, The World Bank; Figure 6.7, International Center for Tropical Agriculture; Figure 7.3, Smithsonian Institution Traveling Exhibition Service; Table 8.4, National Academy Press; Tables 9.1 and 9.2, British Petroleum Company; Figure 9.2, Motor Vehicle Manufacturers' Association of the United States; Table 9.3 and Figure 9.3, Greenwood Publishing Group; Figures 10.3, 10.4, and 10.5, Basil Blackwell; Table 11.2, Office of Technology Assessment; Box 11.1 Figure 1, Witherby & Co. Ltd.; Table 11.3, National Academy Press; Figure 11.1, James Dobbin Associates; Figure 11.2, Cambridge University Press; Figures 12.1 and 12.2 and Table 12.3, Japanese External Trade Organization; Figures 13.1, 13.2, Box 13.1, Table 1, and Tables 13.1, 13.2, and 13.3, World Meteorological Organization. World Map from Goode's World Atlas © 1988 by Rand McNally, R.L. 88-S-153-Renewal.

Oxford University Press
Oxford   New York   Toronto
Delhi   Bombay   Calcutta   Madras   Karachi
Petaling Jaya   Singapore   Hong Kong   Tokyo
Nairobi   Dar es Salaam   Cape Town
Melbourne   Auckland
and associated companies in
Berlin   Ibadan

Contents

The tabs along the sides of pages connect text chapters in Parts II and III with corresponding data tables in Part IV.

Preface

The *World Resources* series is intended to meet the critical need for accessible, accurate information on some of the most pressing issues of our time. Wise management of natural resources and protection of the global environment are essential to achieve sustainable economic development and hence to alleviate poverty, improve the human condition, and preserve the biological systems on which all life depends.

Publication of *World Resources 1990-91*, the fourth in the series, marks a number of new dimensions. With this volume, the United Nations Development Programme (UNDP) joins with the United Nations Environment Programme (UNEP) and the World Resources Institute (WRI) in an ongoing collaborative effort to provide the most objective and up-to-date report of conditions and trends in the world's natural resource base and in the global environment.

This volume of *World Resources* contains a wealth of information on resource use and environmental issues around the world. Part I gives an overview of topics that stand out as having a pervasive impact on the human condition, exerting unique leverage on the future of the planet, or requiring urgent action. Part II includes a chapter on global climate change, a timely and crucial topic; the chapter includes the first publication of the Greenhouse Index, which tracks each country's current responsibility for greenhouse gas emissions. Part II also includes an overview of resource use and environmental problems in Latin America—from deforestation to urban pollution—continuing and expanding on a tradition of examining in each volume a particular geographical region in more detail. Both the potential impact of climate change and Latin America resource issues are discussed in additional sections throughout the volume. Part III reports on basic conditions and trends, major problems and efforts to resolve them, and recent developments in each of the major resource categories, from population and health to energy and freshwater. Supporting data, as well as the core data tables from the World Resources Database, are found in Part IV.

Additional information and data can be found in the *Environmental Data Report*, published every other year by UNEP in cooperation with WRI and the U.K. Department of the Environment. The *Environmental Data Report* and *World Resources* are published in alternating years.

The audience for the *World Resources* series is steadily expanding, with English, Spanish, Arabic, German, Japanese, and Chinese editions now in print and a Russian edition in preparation. Final distribution for this volume of the series is expected to exceed 100,000 copies in all languages—to policymakers, resource managers, scholars, teachers and students, the media, and other concerned individuals.

WRI, UNEP, and UNDP share the conviction that the *World Resources* series can best contribute to management of the world's natural resources and to a broadened awareness of environmental concerns by providing an independent perspective on these critical global issues. Therefore, while both UNEP and UNDP have provided essential information and invaluable critical advice, *final responsibility for substance and editorial content of the series remains with WRI.*

The effort required to put together *World Resources 1990-91* was enormous. The *World Resources* staff has assembled and analyzed a massive and unique collection of information on conditions and trends in global natural resources and the environment. We commend staff members for their accomplishment. The Editorial Advisory Board, chaired by Dr. M.S. Swaminathan, provided active advice and support at all stages of the project.

We wish to thank the John D. and Catherine T. MacArthur Foundation for its support in this endeavor. Its financial commitment to the continuation of the *World Resources* series and its distribution throughout the developing world help to make this all possible.

James Gustave Speth
President
World Resources Institute

Mostafa K. Tolba	William H. Draper III
Executive Director	Administrator
United Nations	United Nations
Environment Programme	Development Programme

Acknowledgments

World Resources 1990–91 is the product of a unique international collaboration involving many institutions and individuals. Without their advice, support, and hard work this volume could not have been produced.

We are especially grateful for the advice and assistance of our many colleagues at the World Resources Institute (WRI), the United Nations Environment Programme (UNEP), and the United Nations Development Programme (UNDP). Their advice on the selection of material to be covered and their diligent review of manuscript drafts, often under time pressure, have been invaluable.

Institutions

We wish to recognize and thank the many other institutions that have contributed data, reviews, and encouragement to this project. They include:

The Food and Agriculture Organization of the United Nations

The Global Environment Monitoring System (GEMS) of UNEP

The World Bank

The Institute of Geography of the Soviet Academy of Sciences

The GEMS Monitoring and Assessment Research Center of UNEP

The United Nations Statistical Office

The World Health Organization

The World Conservation Union

The World Conservation Monitoring Center

Individuals

Many individuals contributed to the development of this volume by providing expert advice, data, or careful review of manuscripts. While final responsibility for the chapters rests with the *World Resources* staff, the contributions and help of these colleagues are reflected throughout the book. Special thanks go to several UNEP colleages who aided us with logistics and administration: Joan Martin-Brown, special advisor to the executive director, was always ready to help as well as to review chapters; Mitchell Loeb and Wittold Sartorius coordinated access to pertinent UNEP experts in Nairobi. Special thanks also go to Dan Tunstall of WRI, who advised on and reviewed data chapters. We also thank our authors, who performed diligently and then endured patiently our numerous queries and often substantial editorial changes. The primary authors are also listed at the end of each chapter. Authors, reviewers, consultants, and major sources include:

Atmosphere and Climate Kathy Kaufman, U.S. Environmental Protection Agency; Mick Kelly, Climate Research Unit, University of East Anglia; Nancy Kete, U.S. Environmental Protection Agency; Alan Lloyd, South Coast Air Quality Management District, California; Elaine Matthews, Goddard Institute for Space Studies; Curtis Moore.

Basic Economic Indicators Aysel Baschi, the World Bank; Betty Dow, the World Bank; Paul Emerson, United Nations Statistical Office; John O'Connor, the World Bank; Reza Farivari, the World Bank; Robert Repetto, WRI; Richard Roberts, United Nations Statistical Office; Michael Ward, the World Bank.

Climate Change Dean Abrahamson, Humphrey Institute of Public Affairs; Thomas Boden, Carbon Dioxide Information Analysis Center, Oak Ridge National Laboratory; Paul J. Crutzen, Max Planck Institute for Chemistry; John Firor, National Center for Atmospheric Research; Erik Helland-Hansen, UNDP; R.A. Houghton, Woods Hole Research Center; C.D. Keeling, Scripps Institute of Oceanography; M.A.K. Khalil, Oregon Graduate Center; Jean Lerner, Goddard Institute for Space Studies; Gordon MacDonald, MITRE Corporation; Greg Marland, Carbon Dioxide Information Analysis Center, Oak Ridge National Laboratory; Jim MacKenzie, WRI; Mack McFarland, E.I. duPont de Nemours & Company, Inc.; Gregory A. Mock; Bill Moomaw, Tufts University; Rafe Pomerance, WRI; V. Ramanathan, University of Chicago; Susan Subak, Stockholm Environmental Institute; Mark Trexler, WRI; P. Usher, UNEP; C.C. Wallen, UNEP; George Woodwell, Woods Hole Research Center.

Energy Deborah Bleviss, International Institute for Energy Conservation; Tom Burroughs; John Christensen, UNEP; William Clive, United Nations Statistical Office; Charles Komanoff, Komanoff Energy Associates; Bill Liggett, U.S. Department of Energy; Edwin Moore, the World Bank; R.G. Muranaka, International Atomic Energy Agency; Robert Pollard, Union of Concerned Scientists; G. Samdani, International Atomic Energy Agency; William Zajac, U.S. Bureau of Mines.

Food and Agriculture A. Balasubramaniam, Pesticides Board of Malaysia, Department of Agriculture; Ed Barbier, International Institute for Environment and Development; Bill Barclay, WRI; Stephen Batt, Monitoring and Research Centre, UNEP; Kurt Becker, Food and Agriculture Organization of the United Nations; Robert Blake, WRI; Robert Brinkman, Food and Agriculture Organization of the United Nations; Lester Brown, Worldwatch Institute; Janet Brown, WRI; Walter Couto; Pierre Crossen, Resources for the Future; Axel Dourojeani, Economic Commission for Latin America; H. Hassan, Ministry of Science, Technology, and Environment, Malaysia; Mickey Glantz, National Center for Atmospheric Research; Abbas Kesseba, International Fund for Agricultural Development; Marian Mitchell, Tulane University; Ed Overton, U.S. Department of Agriculture; Francesco Pariboni, Food and Agriculture Organization of the United Nations; János Pógany, United Nations Industrial Development Organization; Martin Parry, Atmospheric Impacts Research Group, University of Birmingham; Eric Rodenburg, WRI; Pedro Sanchez, Department of Soil Science, North Carolina State University; M.S. Swaminathan, International Union for the Conservation of Nature and Natural Resources; B. Waiyaki, UNEP; Phyllis Windle, Office of Technology Assessment, U.S. Congress; John Young, Worldwatch Institute; Garth Youngberg, Institute for Alternative Agriculture; Monty Yudelman.

Forests and Rangelands Utz Bahm, International Livestock Center for Africa; Burton Barnes, University of Michigan; Robert Bruck, North Carolina State University; Bill Burley, Programme for Belize; A.H. Contreras, the World Bank; Bert Drake; Dennis Child, Winrock International; James Detling; Robert Goodland, the World Bank; Peter Hazlewood, CARE; Jean Gorse; David Harmon; J. Kauffman, Oregon State University; Harold Heady; J.P. Lanly, Food and Agriculture Organization of the United Nations; Liberty Mhlanga, Agricultural and Rural Development Authority, Zimbabwe; Judith Moore; Helen Morris, Monitoring and Assessment Research Centre, UNEP; Dan Nepstad; Theodore Panayotou, Harvard Institute for International Development; Duncan Poore; Marco Romero; Stephen Sanford, International Livestock Center for Africa; Carlos Sere, Center for Tropical Agriculture; E.A.S. Serrao; Dr. Eneas Pinheiro; Thomas Stone,

Woods Hole Research Center; Boyd Strain; José Toledo, Center for Tropical Agriculture; Victor Toledo; P. Wardle, Food and Agriculture Organization of the United Nations; Robert Winterbottom, WRI; Richard Woodward, WRI.

Freshwater J. Balek, UNEP; Silvio Barabas, Global Environmental Monitoring System/World Health Organization, Canada Centre for Inland Waters; Alexander Belyaev, Institute of Geography, National Academy of Sciences, U.S.S.R.; Debbie Chapman, Monitoring and Assessment Research Centre, UNEP; R.J. Daley, Global Environmental Monitoring System/World Health Organization, Canada Centre for Inland Waters; Malin Falkenmark; Erik Helland-Hansen, UNDP; Richard Helmer, World Health Organization; Maynard Hufschmidt, East-West Center; J. Lisa Jorgenson; Allen Kneese, Resources for the Future; Michel Meybeck, Universite Pierre et Marie Curie Laboratoire de Geologie Applique; David Moody, U.S. Geological Survey, National Center; Helen Morris, Monitoring and Assessment Research Centre, UNEP; Peter Rodgers, Harvard University; Roger Sedjo, Resources for the Future; V. Vandeveerd, Global Environmental Monitoring System, UNEP; Gilbert White, Natural Hazards Information Center; Angelo Zingaro, Global Environmental Monitoring System/World Health Organization, Canada Centre for Inland Waters.

Global Systems and Cycles Frances Bretherton, University of Wisconsin; Tom Malone, St. Joseph College; Gregory Mock; Jim Neilon, National Weather Service, National Oceanic and Atmospheric Administration; Ralph Peterson, National Weather Service, National Oceanic and Atmospheric Administration; James Rasmussen, World Weather Watch; Charles Redmond, National Aeronautics and Space Administration; Joseph Tribbia, Global Dynamics Section, National Center for Atmospheric Research; Paul Uhlir, Space Science Board, National Research Council; P. Usher, UNEP.

Human Settlements Walter Arensberg, WRI; Carl Bartone, the World Bank; Alice Clague, United Nations Statistical Office; Robert Fox; Dennis Gallagher, Refugee Policy Group; Harvey Garn, the World Bank; N. Gebremedhin, UNEP; Virginia Hamilton, U.S. Committee for Refugees; James Listorti, the World Bank; Robert Livernash, WRI; Leslie Pean, the World Bank; Eduardo Rojas, Inter-American Development Bank; William Seltzer, United Nations Statistical Office; Albert Wright, the World Bank; Guillermo Yepes, the World Bank.

Latin America Sheldon Anis, Overseas Development Council; Patricia Ardila, Panos Institute; Rafael Asenjo, American University, International Development Program; J. Balek, UNEP; Nina Bohlen, World Wildlife Fund/Conservation Foundation; Gerardo Budowski, Universidad para la Paz; Marc Dourojeani, Inter-American Development Bank; Edward Farnsworth, Inter-American Development Bank; Jorge Ferraris,

Inter-American Development Bank; Curtis Freese, World Wildlife Fund/Conservation Foundation; Nicolo Gligo, Unidad Conjunta CEPAL/PNUMA de Desarrollo y Medio Ambiente; José Goldemberg, Universade de São Paulo; Yolanda Kakabadse, Fundacion Ecuatoriana para la Conservacion de la Naturalenz; Gabinete do Reitor; Arsenio Rodriguez, UNEP; Peter Rogers, Harvard University; Lane Krahl; Carlos Lopez-Ocana, Universidad Nacional Agraria la Molina; Gus Medina, World Wildlife Fund/ Conservation Foundation; Michael Nelson; Paulo Nogueira-Neto, University of São Paulo; Diana Page, WRI; Eric Rodenburg, WRI; Kirk Rodgers, Organization of American States; Eneas Salati; Richard Saunier, Organization of American States; W.G. Sombroek, UNEP/ISRIC Project on Global Assessment of Soil Degradation; Pablo Stone; Isabel Valencia, University of Arizona; Ray Victurine, United States Agency for International Development; Richard Woodward, WRI; P. Zentilli, UNEP.

Oceans and Coasts Salvano Briceño, Caribbean Action Plan, UNEP; Clif Curtis, Oceanic Society; A. Dahl, UNEP; Bruce Davis, Murdock University; Paul Dingwall, International Union for the Conservation of Nature and Natural Resources; James Dobbin, James Dobbin Associates Inc.; Charles Ehler, National Oceanic and Atmospheric Administration; Daniel Elder, International Union for the Conservation of Nature and Natural Resources; John Gulland, Imperial College; Donald Heisel, Food and Agriculture Organization of the United Nations; Robert Hofman, U.S. Marine Mammal Commission; Maurice Jorgens, United Nations Office of Ocean Affairs and the Law of the Sea; Graeme Kelleher, Great Barrier Reef Marine Park Authority; Lee Kimball, Council on Ocean Law; Stephen Leatherman, University of Maryland; A.D. McIntyre, Department of Agriculture & Fisheries for Scotland; Beverly Miller, Caribbean Action Plan, UNEP; Mary Paden, WRI; Polly Penhale, National Science Foundation; Walter Reid, WRI; Hope Robertson; Michael Robinson, Food and Agriculture Organization of the United Nations; Arsenio Rodriquez, UNEP; Kenneth Sherman, Northeast Fisheries Center, Natural Oceanic and Atmospheric Administration; Steve Tibbitt, National Oceanic and Atmospheric Administration; Edward Towle, Island Resources Foundation; Manuel Vegas; Michael Weber, Center for Marine Conservation; Miranda Wecker, Council on Ocean Law; Harold Weeks, North Pacific Fishery Management Council; Douglas Wolf, National Oceanic and Atmospheric Administration.

Policies and Institutions Yusuf Ahmad, UNEP; Peter Bartelmus, United Nations Statistical Office; Clare Billington, World Conservation Monitoring Centre; Mary Brandt, U.S. State Department; Wan-Li Cheng, UNEP; Uttam Dabholkar, UNEP; Arthur Dahl, UNEP; Shantayanan Devarajan, Kennedy School of Government, Harvard University; Mark Halle, International Union for the Conservation of Nature and Natural Resources; Stjepan Keckes, UNEP; Ernst Lutz, the World Bank; Henry Peskin; Inez Price, United Nations

Treaty Section; Robert Repetto, WRI; John Rigby; Iwona Rummel-Bulska, UNEP; David Runnalls, Institute for Research on Public Policy.

Population and Health C.K. Campbell, Centers for Disease Control; Alice Clague, United Nations Statistical Office; P. Cornu, International Labor Office; Robert Fox; H.R. Hapsara, World Health Organization; Donald F. Heisel, United Nations Statistical Office; Shunichi Inoue, United Nations Population Division; Ann Marie Kimball, Pan American Health Organization; Alan Lopez, World Health Organization; Tom Merrick, Population Reference Bureau; Fred T. Sai, the World Bank; K.L. Gnana Sekaran, United Nations Statistical Office; G. Watters, World Health Organization; Tensie Whelan; G. David Williamson, Centers for Disease Control.

Wildlife and Habitat Janet Abramovitz, WRI; Jonathan Barsdo, World Conservation Monitoring Centre; Richard Block, World Wildlife Fund/Conservation Foundation; William Burnham, M.A.S. Burton, Monitoring and Assessment Research Centre/UNEP; J.R. Caldwell, World Conservation Monitoring Centre; Nathan Flesness, International Species Inventory System, Minnesota Zoological Gardens; David Harmon; Jeremy Harrison, World Conservation Monitoring Centre; Gary Hartshorn, World Wildlife Fund/Conservation Foundation; V.H. Heywood, International Union for the Conservation of Nature and Natural Resources; Nels Johnson, WRI; Christian Leon; Richard Luxmoore, Wildlife Trade Monitoring Unit; John MacKinnon; P.J.S. Olney, Zoological Society of London; Robin Pellew, World Conservation Monitoring Centre; Rob Peters, World Wildlife Fund/Conservation Foundation; Walter Reid, WRI; Scott Swengel, International Crane Foundation; Jane Thornback, World Conservation Monitoring Centre; A. Willcocks, Monitoring and Assessment Research Centre, UNEP.

World Environment Outlook C. Boelcke, UNEP; Mohamed El-Ashry, WRI; Andrea Matte-Baker, UNEP.

Production Staff

A talented team of copyeditors, factcheckers, proofreaders, desktop publishing experts, and layout artists accomplished the enormous task of making this volume ready for the printer in record time. We thank them for their dedication, hard work, long hours, and high professional standards. In addition to the *World Resources* staff, they include:

Editors Rosemarie Philips, Hugh McIntosh, Seth Beckerman, Jim Singer, Tensie Whelan.

Copyeditors Sheila Mulvihill, Audrey Pendergast, Helena Hallden, Carolyn Pool.

Factcheckers Tony Zamparutti, Joe Dever, Nina Burns, Kara Page.

Data Entry Maurice Allen.

Proofreader Evelyn Harris.

Manuscript Production Suraiya Mosley, Renee Rochester.

Desktop Production Robert Llewellyn (Idea Tech Associates, Inc., Alexandria, Virginia.)

Mechanical Production The Forte Group, Inc., Lanham, Maryland.

We are especially grateful to WRI Librarian Sue Terry for assisting us with research and materials.

Translation and Distribution

We would also like to acknowledge the help of the Instituto PanAmericano de Geografia e Historica in translating *World Resources 1987* into Spanish and the support of the Rockefeller Brothers Fund for distributing the 1988–89 edition around the world.

It has been a privilege to work with so many outstanding individuals throughout the world in producing *World Resources 1990–91*.

Allen L. Hammond
Editor-in-Chief

1. World Environment Outlook

When the 20th Century began, the world was home to about 1.6 billion of the human species. Although pollution and environmental degradation were common—some cities lived under a pall of smoke, soot, and ash—the problems were local. The world as a whole seemed vast, with huge regions virtually untouched by its human inhabitants.

By midcentury, airplanes and radio broadcasts had begun to shrink distances and bring the peoples of the world, now 2.5 billion, into greater contact. Industrial growth had multiplied per capita consumption of natural resources—and per capita pollution—in many countries. As that growth continued, air and water pollution became more widespread, as did concern about the cumulative impact of toxic industrial products on living species.

As the 1990s begin, world population has more than doubled since 1950, to 5.2 billion people, and world economic activity has nearly quadrupled (1). To local concerns about environmental degradation have been added new, global worries. The pressures of agricultural and industrial development have begun to crowd out and extinguish other species at a rapid rate, to visibly erode the carrying capacity of the planet's soils, forests, estuaries, and oceans, and to alter its atmosphere. These pressures can only increase if the human population doubles again and economic activity continues its explosive growth.

The scale of the changes under way are spelled out in the following passage from *Our Common Future*:

> The planet is passing through a period of dramatic growth and fundamental change. Our human world of 5 billion must make room in a finite environment for another human world. The population could stabilize at between 8 billion and 14 billion sometime next century, according to UN projections.... Economic activity has multiplied to create a $13 trillion world economy, and this could grow five- or tenfold in the coming half century (2).

The world's environmental dilemma is that the scale on which natural resources are being consumed and wastes are being produced is already immense, yet many poor countries still lack and desperately need the benefits of industrialization and economic development. How this dilemma is resolved is likely to dictate our planet's prospects for the coming century.

Of the hundreds of topics addressed in this volume, a few stand out as having a pervasive impact on the human condition, exerting unique leverage on the future of the planet, or requiring urgent action. Together, they highlight how seemingly isolated environmental problems are, in fact, bound in a contin-

---

Datapoint: Population Growth Will Add 959 Million People In 1990s

The world's population passed 5 billion in 1987 and continues to rise: an increase of 842 million during the 1980s and a projected increase of 959 million during the 1990s. Even though the average number of births per woman is declining in most countries, the excess of births over deaths means that world population is not yet approaching stability.

Most of the growth is occurring in developing countries. (See Figure 1.1.)

Rapid population growth in these countries recently has created populations in which young people predominate. As these young people reach childbearing age, some additional population growth is virtually certain. This demographic momentum makes it more difficult to stabilize populations; it also means that stresses on natural resources and food supplies and the demand for jobs are likely to intensify in the next

few decades.

About 40 percent of the world's people now live in urban areas, and urban population is growing rapidly—again, predominantly in the developing world (see Figure 1.2.)—while the population in rural areas is growing very slowly, if at all. This trend will intensify the need for and the difficulty of supplying basic services and maintaining infrastructure in large urban areas.

Figure 1.1 World Population Growth[a]

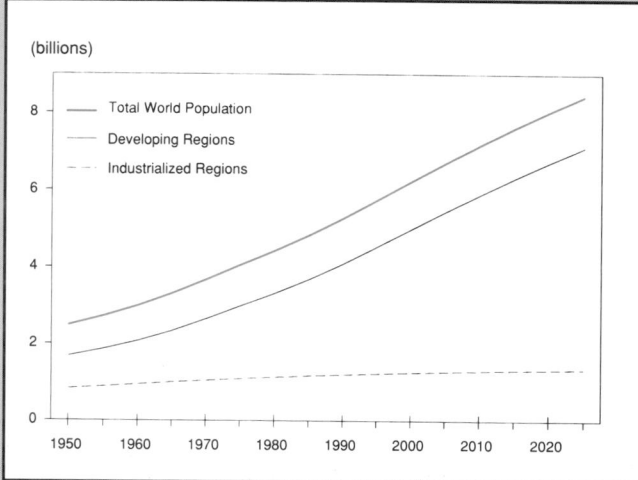

Source: United Nations Population Division, *World Population Prospects 1988* (United Nations, New York, 1989).
Note: a. Based on United Nations medium variant projection, the middle projection of possible high, medium, and low projections.

Figure 1.2 Urban Population Growth, 1950–2025

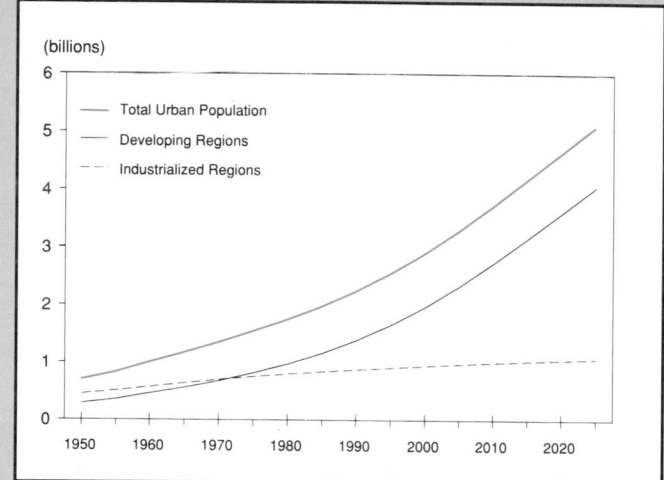

Source: United Nations Population Division, *World Population Prospects 1988* (United Nations, New York, 1989).

---

uum of cause and effect, and suggest the scope of the problems that must be addressed if our civilization is to be ultimately sustained and the dignity and potential of human life enhanced in the decades to come.

ENDANGERED RESOURCES

Air—The Global Commons

From the ground, Earth's atmosphere is nearly invisible and easy to take for granted. From space, it is perceived more readily as a thin blanket of gases, shielding our planet from the harshest of the sun's ultraviolet radiation and trapping the sun's warmth to keep Earth's rivers and oceans from freezing. Now, human activity is releasing gases into the atmosphere on a scale that threatens to alter both of these atmospheric functions significantly, degrading the ultraviolet-shielding ozone layer and intensifying the heat-trapping properties of the atmosphere as a whole.

Because the atmosphere thoroughly mixes the pollutants that cause these changes, it does not matter where the gases are emitted. The global effect of a billion metric tons of carbon dioxide is the same, whether it is released from a northern industrial country or a southern developing nation, whether it comes from burning coal in a power plant, gasoline in a car, or trees in a forest. Preventing atmospheric degradation is, therefore, a task that must be shared globally.

The need is becoming increasingly urgent. In 1987, human activity added the equivalent of 5.9 billion metric tons of carbon—in the form of carbon dioxide or other heat-trapping "greenhouse" gases—to the atmosphere, more than a metric ton for every person on the planet (3). The total would be 14.6 billion metric tons of carbon equivalent—the amount actually emitted—were it not for natural processes that remove greenhouse gases from the atmosphere. The most rapidly rising levels are of chlorofluorocarbons, which in the lower atmosphere trap heat but in the stratosphere are the source of chemicals that degrade the ozone layer.

Datapoint: The Greenhouse Index: Each Country's Contribution To Global Warming Potential

The Greenhouse Index combines net annual emissions of three major greenhouse gases, weighting each gas according to its heat-trapping capability. When calculated for a given country, the Index is a measure of that country's responsibility for the increase in the atmosphere's warming potential. The annual increase for the world as a whole is rising rapidly, having nearly tripled in three decades. (See Figure 1.3.) Thus, in 1987, the world is adding to the heating potential of the atmosphere at three

times the rate of 1957.

The responsibility for these increases is widely shared. Worldwide, 30 industrialized countries account for 55 percent of total emissions. But of the 10 largest emitters in 1987 (counting the European Community as a single entity), half are industrialized countries and half are developing countries. (See Figure 1.4.) The top 50 emitting countries, which account for an estimated 92 percent of global emissions, span every region of the world, including the developing portions of

Africa, Latin America, and Asia, and include both capitalist and planned economies. Clearly, the problem is already a global one, and only common global efforts can hope to stabilize or reduce greenhouse gas emissions. However, solutions must also address the need for equity and for industrial development: Industrial countries with less than 20 percent of the world's population are responsible for over 50 percent of global emissions. (See Chapter 2, "Climate Change," Greenhouse Gas Emissions.)

Figure 1.3 Annual Additions of Three Major Greenhouse Gases to the Atmosphere 1957–87

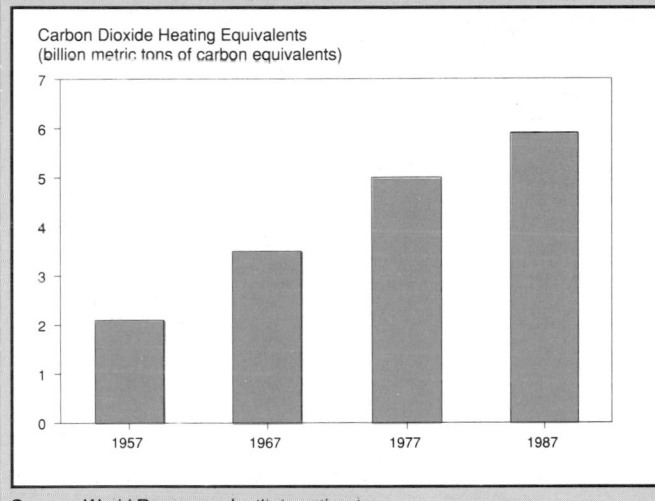

Source: World Resources Institute estimates.

Figure 1.4 Greenhouse Index: 10 Countries with the Most Greenhouse Emissions, 1987

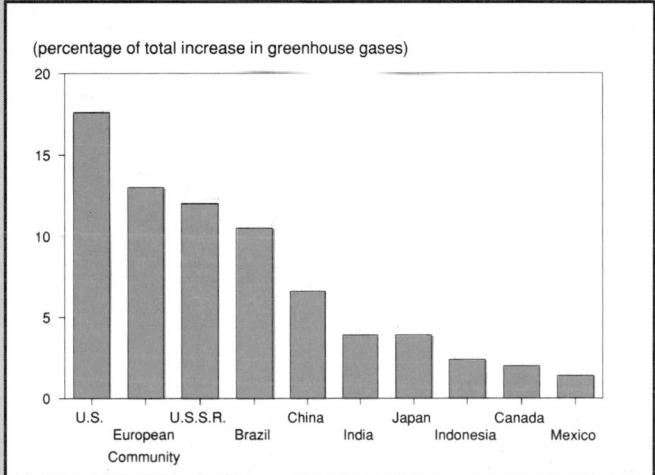

Source: Chapter 2, "Climate Change: A Global Concern," Table 2.2.

The Uncertain Climate

There is no doubt that degrading the protective layer of stratospheric ozone will increase the amount of biologically damaging ultraviolet radiation that reaches the surface of the Earth. How the climate will change as greenhouse gases accumulate in the lower atmosphere and alter its heat-trapping properties is less certain. If present trends continue, greenhouse gases will reach the equivalent of twice the preindustrial levels of carbon dioxide within 40 years and will double again before the end of the 21st Century. If current models of the Earth's climate system are correct, even one doubling will raise global temperatures a few degrees centigrade. The best current consensus appears to be that temperatures would rise an average of 1.5-4.5° C, with smaller increases in the tropics and larger increases at high latitudes.

The prospective shift in temperatures may seem small, but it is nearly comparable to the change between the extreme climate of the last ice age, 18,000 years ago, and today's climate (4). It is enough to alter significantly rainfall patterns and temperature regimes in much of the world (see Chapter 10, "Fresh-

water"), with impacts on agriculture, forestry, and virtually all living things. (See Chapter 6, "Food and Agriculture," Chapter 7, "Forests and Rangelands," and Chapter 8, "Wildlife and Habitat.") Even a 2° increase would take temperatures higher than human societies have ever experienced.

Global climatic feedbacks—involving clouds, ocean currents, the biota, or several other, as yet, poorly understood mechanisms—might decrease or amplify such temperature changes. These uncertainties—what is known and what is not known about climate change—are discussed in Chapter 2, "Climate Change: A Global Concern"; efforts to monitor and model the climate are discussed in Chapter 13, "Global Systems and Cycles."

If the present patterns of industrial and agricultural activities that are the sources of these atmosphere-altering gases are not themselves altered, there is a distinct possibility of large-scale climate change, forcing what could be very difficult adjustments. In fact, the world may be committed to some degree of global warming already. Whether a warming is detectable now is controversial; what is certain is that the 1980s was the warmest decade of the past

Datapoint: Energy Consumption: Is Continued Growth Sustainable?

Energy, in one form or another, is essential to industrial civilization. But, in addition to its beneficial role, energy production and use are major causes of environmental degradation. Production and use of coal, oil, and natural gas—fossil fuels—are the major sources of local air pollution, regional problems such as acid precipitation, and greenhouse gases that are changing the composition of the atmosphere and that may lead to changes in the global climate. Nonfossil energy sources, such as nuclear power and large-scale hydroelectric power, also raise environmental concerns. Together, these sources constitute virtually all commercial energy production. (In develop-

ing countries, however, biomass—largely wood—is an important source of energy and one that, with proper management, is renewable.) Thus, energy occupies a special place in any discussion of global environmental and resource problems and is central to strategies aimed at reducing air pollution and preventing global warming.

Energy use had long been closely associated with economic growth. During the late 1970s and early 1980s, however, energy use stabilized and even declined, while economic growth moved ahead rapidly. This important development was largely the result of shifts to more efficient use of energy in cars and trucks, in

homes, and in industrial processes. In recent years, however, as energy prices have declined, the world's use of energy has again begun to climb. (See Figure 1.5.)

If economic growth is to continue without exacerbating the environmental problems caused by energy, then fundamental changes in energy production and use are likely to be necessary in the coming decades. The easiest gains will come from still higher energy efficiency, the technology for which needs to be shared widely with all nations. However, in view of the long periods required to phase in new energy technology on a global scale, the development of nonfossil energy sources should also be encouraged.

Figure 1.5 Consumption of Primary Energy

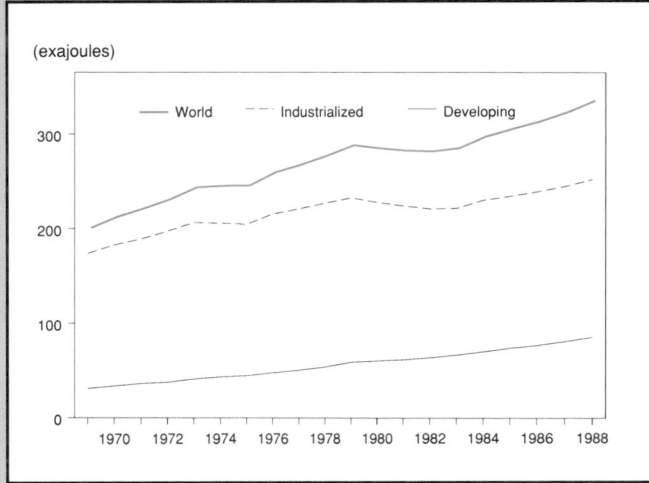

Source: British Petroleum (BP), *BP Statistical Review of World Energy* (BP, London, 1989).

Figure 1.6 Global Emissions of Nitrogen and Sulfur Oxides from Fossil Fuel Combustion

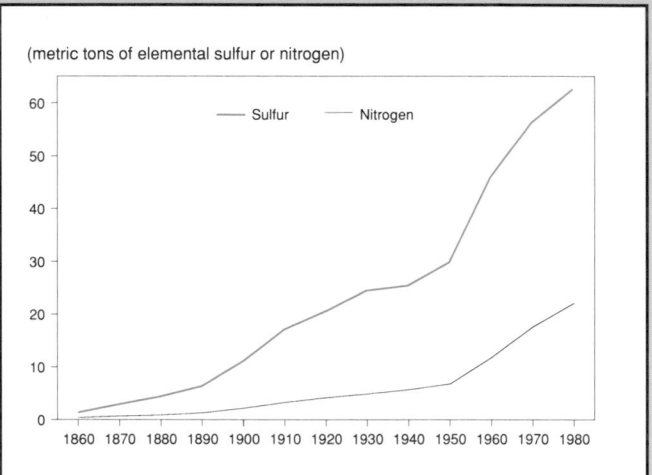

Source: Jane Dignon, Sultan Hameed, "Global Emissions of Nitrogen and Sulfur Oxides from 1860 to 1980," in *Air Pollution Control Association Journal,* Vol. 39, No. 2, p. 153.

century and included 6 of the 10 warmest years on record. If the predicted levels of warming do occur, the changes would be extraordinarily rapid compared with most climatic shifts in the geological record. The changes may occur more rapidly than trees and other biota can adjust to new temperature regimes or migrate to new ranges; the result may be disrupted ecosystems and accelerated extinction for many plant and animal species.

The largest source of greenhouse gas emissions intimately linked with human activity is the use of fossil fuels as a source of energy, but other major sources include the clearing of forested land for agriculture, industrial and consumer use of chlorofluorocarbons, the growing of rice in flooded paddies, and the raising of domestic livestock. Potential alternatives to fossil fuels as sources of energy exist, but they are not available on the scale needed, not yet

economically competitive, or, in some cases, not without environmental problems of their own. (See Chapter 9, "Energy," Nonfossil Sources: A Look Ahead.) Yet energy is so central to economic activity and its current sources are so embedded in our industrial infrastructure that promising efforts to develop and spread the use of nonfossil sources should be intensified.

Dealing with Greenhouse Gases

The Montreal Protocol on Substances that Deplete the Ozone Layer, agreed in 1987, was an important step toward limiting the production and release of the chlorofluorocarbon chemicals that cause the destruction of ozone and that are also long-lived greenhouse gases. But there is already widespread consensus that additional steps will be needed if ozone depletion is not to grow, intensifying ultraviolet rays that can harm many forms of life. In late 1989, the Antarctic ozone hole was again large (see Chapter 12, "Atmosphere," Severe Ozone Depletion Returns in 1989), and scientists reported that conditions were ripe for a hole to form in the Arctic as well. The urgency is underscored by the fact that some chlorofluorocarbons remain in the atmosphere for up to 100 years. On the positive side, however, many of the industrial companies that manufacture chlorofluorocarbons were moving to develop less-damaging substitutes, and other companies announced plans to phase out use of the chemicals or to recapture them before their release into the atmosphere.

Preparations are under way for a similar, but more difficult, step toward controlling greenhouse gas emissions. An Intergovernmental Panel on Climate Change, under the sponsorship of the United Nations Environment Programme and the World Meteorological Organization, has been studying the science and the impacts of climate change and policy responses and is scheduled to report its findings in the summer of 1990. In the fall of 1990, the Second World Climate Conference will take place in Geneva to consider those findings. The United States has offered to host an international conference in 1991, at which governments might negotiate the foundations of a treaty for regulating greenhouse gas emissions. If successful, such an agreement would be followed by the development of a specific protocol analogous to that established in Montreal. International agreements will be critical, but they also must be supplemented by national actions to share technologies that can create a more sustainable industrial system and to provide the resources to acquire and use them.

Industrial Air Pollution

At regional levels, meanwhile, urban and industrial air pollution remains a serious problem. Concern about the impact of air pollution on human health and on forests has led many industrial countries to tighten regulations and reduce emissions of lead, sulfur dioxide, and oxides of nitrogen in recent years. (See Chapter 12, "Atmosphere," Approaches to Air Pollution Control.) In most centrally planned economies in the developing world, and in the emerging market economies of Eastern Europe, these concerns are just beginning to be addressed; often, not even the extent of the problem is known. Since most of the growth in populations will occur in the urban areas of the developing world, urban air quality will become increasingly crucial to human health.

Worldwide, emissions of sulfur dioxide and oxides of nitrogen continue to climb. (See Figure 1.6.) These pollutants cause economic, environmental, and aesthetic harm, as well as damage to human health. The oxides of sulfur form particulates that reduce atmospheric visibility, degrading recreational values. Oxides of nitrogen help to form ozone, a principal

constituent of urban smog. Concentrated in urban areas, sulfur dioxide and smog cause lung damage and raise the incidence of many respiratory diseases. But even when these pollutants are widely dispersed, they pose a significant environmental threat through the formation of acids that precipitate as rain, snow, or in other forms.

Acid precipitation has altered the chemistry of tens of thousands of lakes and streams to the point where they can no longer support life. It is implicated in damage to crops and forests in many areas of the world, among them China, North and South America, Africa, and nearly 50 million hectares of forest in Europe alone. It can also help to mobilize toxic metals from soil, sediments, and even drinking water pipes. It is, in short, a major indirect cost attributable to the use of fossil fuels.

Land and Sea—The Fertile Earth

Life on Earth depends very directly on the living soil and the aquatic ecosystems of rivers, lakes, estuaries, and oceans. Without fertile soil and the microbial fauna that inhabit it, food would not grow, dead things would not decay, and nutrients would not be recycled. Without the plankton and algae that form the base of the aquatic food web, there would be no fish to harvest; even the ocean's ability to remove carbon dioxide from the atmosphere—and thus moderate global warming—would be greatly decreased.

Yet the Earth's soils are being stripped away, rendered sterile, or contaminated with toxic chemicals at a rate that cannot be sustained. By some estimates, 10 percent of the land surface of the planet has been transformed by human activities from forest and rangeland into desert (5); as much as 25 percent more is at risk (6). Cropland is already scarce in much of the developing world; it is getting scarcer as urban areas expand and high dams flood fertile valleys. A country is often described as "land scarce" when more than 70 percent of the potentially arable land is under cultivation. In Asia, an estimated 82 percent of all potential cropland is under production. In Latin America and Africa, the percentages are lower, but misleading. Latin America has the highest concentration of land ownership—10 percent of the population owns 95 percent of the arable land—but subsistence farmers must seek land on steep hillsides or in tropical forests, hastening erosion and deforestation (7). Large areas in Africa are plagued by drought or by the tsetse fly. As a result, continuing the rate of expansion of cropland in these regions— by 14.1 percent in South America and 4.6 percent in Africa, from the mid-1970s to the mid-1980s—may prove unsustainable.

Expanding Agriculture

Most areas of potential expansion for agriculture either lack water or have physical or chemical constraints such as steep slopes, easily eroded or

Figure 1.7 Per Capita Cereal Production 1967–87

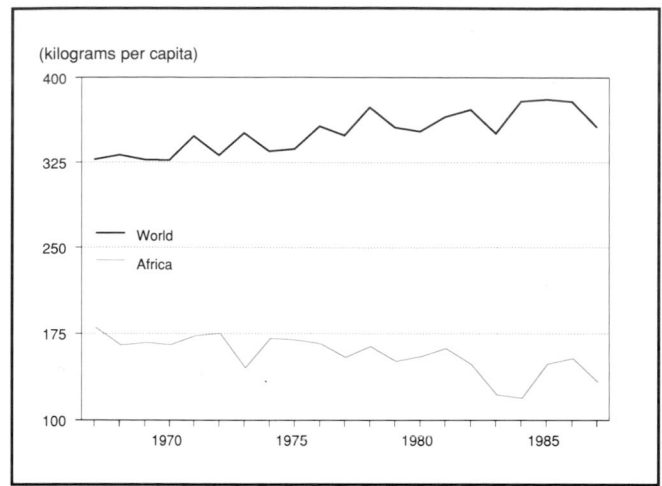

(kilograms per capita)

World
Africa

Source: Food and Agriculture Organization of the United Nations, *1948-1985 World Crop and Livestock Statistics* (FAO, Rome, 1987), pp. 8-9. *Production Yearbook*, 1987, Vol. 41 (FAO, Rome, 1988), p. 113. (Population figures derived from the World Bank Database.)

poorly drained soils, alkaline or other conditions toxic to plants. New agricultural lands often prove illusory, as evidenced by the failure of an estimated 20 percent of new ranches in the Amazon within a few years. Indeed, the loss of existing agricultural land through erosion is estimated at 6–7 million hectares per year, with an additional loss of productivity to 1.5 million hectares from waterlogging, salinization, and alkalinization (8). Modern irrigation systems can bring water from long distances, careful terracing can make steep slopes usable, and imaginative new technical solutions such as those being used in Latin America's tropical savannas (see Chapter 6, "Food and Agriculture," Sustainable Land Use for South America's Tropical Savannas) may overcome soil limitations, but these are not solutions for subsistence agriculture. All too often, expansion into new and marginal agricultural land is, out of basic necessity, led by those least able to overcome its difficulties or to farm it in a sustainable manner.

Asia has increased its per capita food production over the past decade through a massive increase in agricultural inputs such as fertilizers, pesticides, and mechanical equipment. Over the same period, Latin America barely maintained its per capita food production (despite substantial increases in inputs and in land used for agriculture), while Africa's per capita food production declined. For grains, specifically, world production has continued to rise, even on a per capita basis, except during the North American drought years of 1987 and 1988. The major exception is Africa, which continued a 20-year trend of declining per capita grain production. (See Figure 1.7.) Meanwhile, the world's margin of error for food declined in 1988, as grain stocks fell by nearly a quarter, to 305 million metric tons, in large part because of droughts in North America.

At present, the world's cropland averages about 0.28 hectare per capita, but varies widely. Latin America has nearly double the world average (but, as noted above, concentrated in the hands of a few); Asia, in contrast, has only 0.15 hectare per capita. If world food production is to increase 60 percent by the year 2025—as it must to maintain current nutritional levels for a population projected at 8.5 billion—then either croplands must expand or crop yields must increase. The latter is likely to require more inputs of fertilizer, pesticides, and irrigation—with the risk of worsening water pollution—or novel means of increasing productivity, such as higher yielding or pest-resistant strains developed through biotechnology. The extension of plant breeding methods typical of the Green Revolution to dryland crops—if that proves possible—could also increase yields and reduce losses to pests.

Growing Demands on Freshwater Supply

Freshwater for irrigation is increasingly limited in many regions of the world. Withdrawals of water for municipal and industrial use are rising. Projections based on a detailed model of water use prepared by the Institute of Geography of the Soviet Academy of Sciences suggest that withdrawals of freshwater will continue to rise rapidly in the near future, approaching 20 percent of the total runoff for both Asia and Europe and much higher percentages in specific river basins. (See Figure 1.8.) The environmental effects of excessive withdrawals are demonstrated dramatically by the devastation of the Soviet Union's Aral Sea, which has lost 40 percent of its area, 60 percent of its volume, and virtually all of its once productive fishery over the past 30 years. (See Figure 1.9.)

The first global assessment of water quality, completed in 1989 under the auspices of the United Nations Global Environmental Monitoring System, found growing contamination of freshwater. (See Chapter 10, "Freshwater," Freshwater Quality: A Global Perspective.) Runoff from urban and agricultural areas is a major continuing source of pollution. In developing countries and in many industrialized countries with centrally planned economies, domestic sewage and industrial effluents are often dumped untreated into rivers and lakes; the problem is severe in Eastern Europe and the U.S.S.R. and growing in the concentrated urban areas of the developing world where perhaps 500 million people do not have access to treated water for drinking, bathing, and preparing food.

Polluting the Oceans

The oceans, vast as they are, are beginning to show the effects of pollution too, particularly in coastal areas. This was demonstrated graphically in Alaska when the tanker *Exxon Valdez* went aground and spilled 38.5 million liters of oil into the cold waters of Prince William Sound. Oil spills are isolated, if recur-

Figure 1.8 Freshwater Withdrawals By Region, 1980s and projections for 2000.

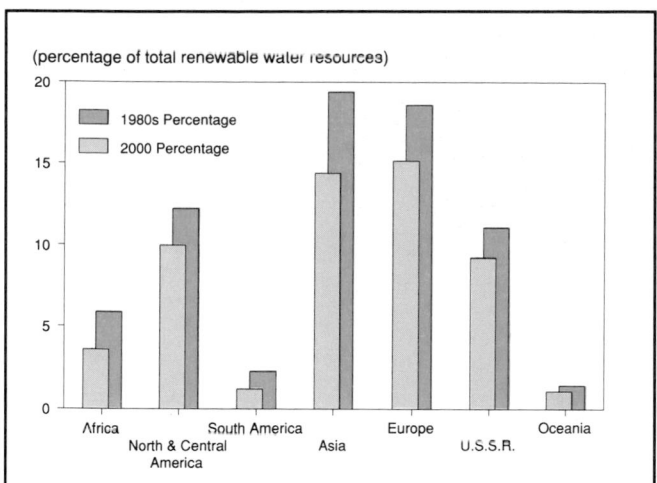

(percentage of total renewable water resources)

■ 1980s Percentage
□ 2000 Percentage

Africa / North & Central America / South America / Asia / Europe / U.S.S.R. / Oceania

Source: Chapter 10, "Freshwater," Tables 10.4, 10.5, and 10.6.

Figure 1.9 The Rising Consumption of Irrigation Water and the Declining Volume of the Aral Sea, 1960–87

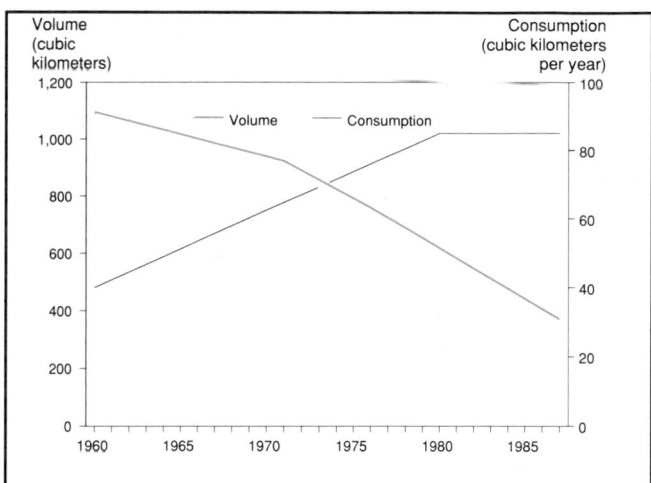

Volume (cubic kilometers) / Consumption (cubic kilometers per year)

Volume — Consumption

Source: Philip P. Micklin, "Desiccation of the Aral Sea: A Water Management Disaster in the Soviet Union," *Science*, Vol. 241 (Sept. 2, 1988), pp. 1170-1176.

ring, incidents, but the effects of urban sewage, silt, plastics, pesticides and other organic substances are pervasive and compound the damage. A recent study by the United Nations Group of Experts on the Scientific Aspects of Marine Pollution concluded that "chemical contamination and litter can be observed from the poles to the tropics and from beaches to abyssal depths....We fear, especially in view of the continuing growth of human populations, that the marine environment could deteriorate significantly in the next decade unless strong, coordinated national and international action is taken now." (See Chapter 11, "Oceans and Coasts," New Assessment of Ocean Pollution.)

Coastal waters, where pollution is at its worst, shelter the vast majority of the ocean's life. Although the world's fish catch continues to climb, some fisheries in heavily fished and polluted areas are declining. And although the global fish catch is just below its estimated sustainable yield, fishermen exceed the estimated sustainable yield in 4 of 16 major fisheries—the Pacific Northwest, the Mediterranean and Black Sea, the Eastern Indian Ocean, and the Southeast Pacific—and are close to it in many others.

The Earth cannot produce its bounty if we continue to damage its productive capacity. Human action has increased the global rate of soil erosion threefold since prehistoric times (9), based on the evidence of the sediment loads of the world's rivers compared with the history of earlier sediment loads written in the muds of the ocean floor. Toxic materials abound—oil spilling into the oceans at 10 times the rate of natural seeps (10), lead being deposited on soils and in waterways at 100 times the prehistoric levels (11), cadmium being released into the environment at 40 times the natural rates (12), radioactive

materials contaminating soils at many sites, and the acidity of precipitation over millions of square kilometers increasing 10-fold (13).

Living Resources

The living resources of the planet surely include the few dozen species of plants and animals grown for food and fiber. They also include the human resources—people themselves. But it is the vast diversity of other species that constitute the biological systems on which life itself depends. Now, that diversity, and the genetic heritage that it represents, is increasingly at risk.

The largest density of those species is to be found in the tropical forests of the world, which are under intense pressure. Over the past decade, land conversion and deforestation in these regions have accelerated dramatically. An estimated 17 million hectares of closed forest were deforested and converted to other uses in 1987, up from 7.5 million hectares in 1981. (See Figure 1.10.) Estimates of deforestation vary substantially and the true extent is not known with any great accuracy, but the trend is not in doubt and it is alarming. The best current estimate for 1987, in what may have been a peak year, is that the worldwide loss of both closed and open tropical forest amounted to 20.4 million hectares. (See Chapter 7, "Forests and Rangelands," New Estimates of Tropical Deforestation.) In many areas of the world, the loss is not only the trees and the habitat for an enormous variety of species that the forest provides, but also the relatively poor soils beneath it, which often compact, laterize, or erode when the tree cover has gone.

Figure 1.10 World Total of Closed Tropical Forest Area Lost Per Year

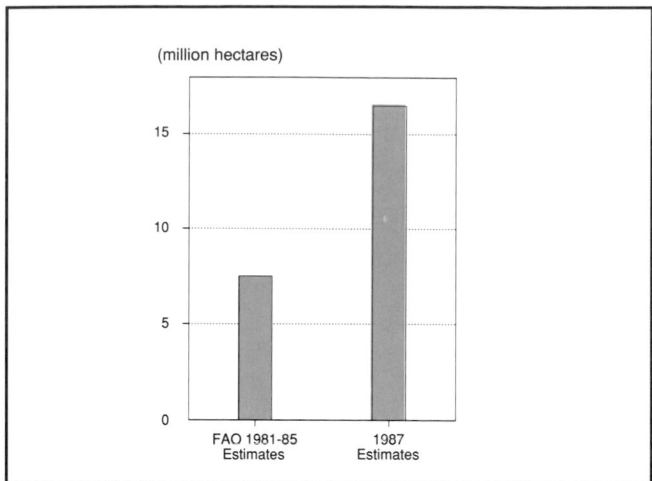

Source: Chapter 7, "Forests and Rangelands," Table 7.1 and Food and Agriculture Organization of the United Nations (FAO), Forest Resources Division, *An Interim Report of the State of the Forest Resources in the Developing Countries* (FAO, Rome, 1988).

Figure 1.11 Elephant Populations in Africa, 1979[a] and 1989

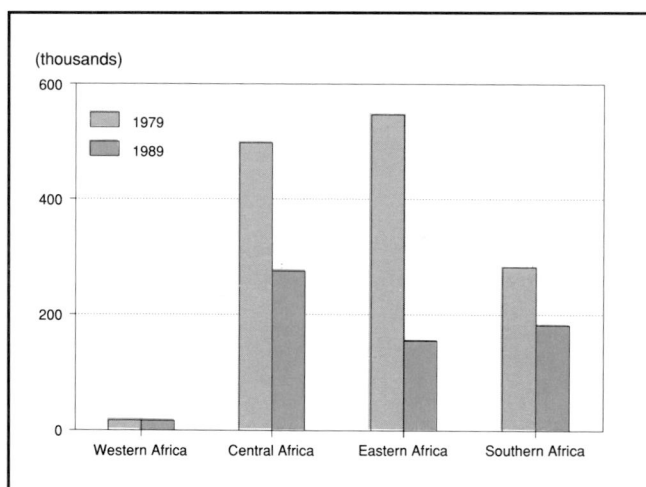

Source: TRAFFIC USA, World Wildlife Fund, unpublished data (Washington, D.C., October 1989).
Note: a. Estimated in 1979.

Clearing land for agriculture and environmental degradation of other kinds have accelerated the rate of habitat loss and species extinction to levels not seen on Earth for 60 million years. A precise measure is not possible because scientists have not yet identified many of the life forms that are threatened and they do not know, with any precision, the number of species that inhabit the Earth, but careful estimates are that we are losing 100 species per day (14). The

trends are all too clear—from declining American songbirds whose winter habitat is the disappearing Central American forests to the African elephant whose numbers have been sharply reduced over the past decade by ivory poachers. (See Figure 1.11.)

Human Poverty and Ill Health

The human species is not declining in numbers, but some 350 million to 1 billion people remain seriously undernourished. Estimates are that 1 billion people live in households too poor to obtain the food they need in order to be able to work, that one child in three is underweight by age 5, and that hundreds of millions of people suffer from anemia, goiter, and impaired sight—conditions associated with a lack of iron, iodine, or vitamin A (15). Beyond outright hunger, many people lead lives trapped in subsistence agriculture or in urban poverty.

Conditions in urban environments are deteriorating in many developing countries: untreated sewage, uncontrolled industrial effluents, and rapidly growing automobile fleets pollute air, water, and nearby cropland. By the year 2000, a billion people are expected to be without satisfactory sanitation and half a billion without adequate drinking water. (See Chapter 5, "Human Settlements," Conditions and Trends.) Competition for housing in already overcrowded cities is expected to increase. The strong trend to urbanization in developing countries means that more and more of the world's population must contend with these conditions.

The consequences for human health are severe. An estimated 60 percent of Calcutta's population suffer from respiratory diseases related to air pollution; lung cancer in Chinese cities (where coal is the main fuel) is 4 to 7 times higher than in the nation as a whole; and lead from auto exhausts is found in the blood of children in Mexico City at levels high enough to cause brain damage (16).

Access to health care has improved in recent decades, but the health care systems of many developing countries cannot cope with many serious health problems. Diseases related to inadequate sanitation and absence of clean water are widespread, leaving hundreds of millions of people debilitated and unable to reach their full potential. Malaria is resurgent—the total number of cases may reach 100 million per year—and AIDS is spreading steadily in Africa, Latin America, and the Caribbean, as well as in many industrialized countries. (See Chapter 4, "Population and Health," Focus on a Major Killer: Malaria and AIDS in the Americas.)

With the expected increase in the world's population rising by about 96 million additional people per year during the decade of the 1990s (17), it will take extraordinary efforts to supply the world's population with adequate food and health care and to slow

Datapoint: Latin American Debt

The context for Latin American environmental issues during the 1980s—and much else—has been the economic stagnation of the region. A factor in that stagnation has been the region's mounting external debt, which in 1987 totaled $387 billion. By comparison, the external debt for all of Asia was $342 billion and for Africa $212 billion. To pay the interest on that debt, countries struggled to increase exports, with varying success. For the largest countries in Latin America, debt service consumed between 25 and 45 percent of export earnings. (See Figure 1.12.) To make matters worse, prices for agricultural commodities (one of Latin America's major exports) dropped on world markets. The result was a decline in per capita gross domestic product of almost 1 percent per year between 1982 and 1989. Under such conditions, it is hard to make protection of the environment the highest priority.

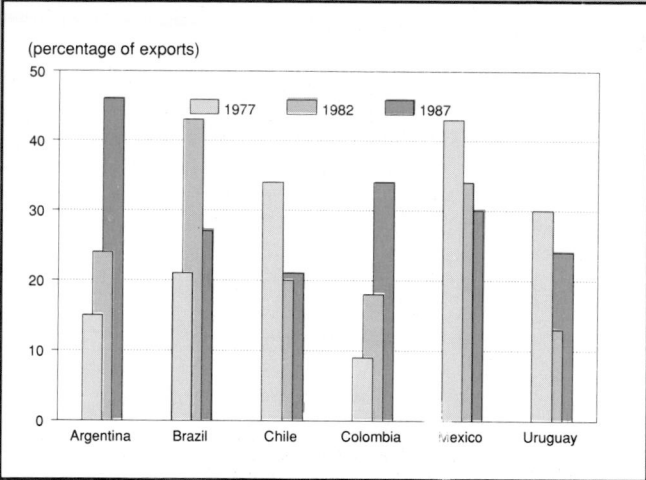

Figure 1.12 Debt Service as a Percentage of Exports of Goods and Services in Six Latin American Countries

Source: Chapter 15, "Basic Economic Indicators," Table 15.2.

or halt the deterioration of the biological systems on which all life depends.

A CHANGING WORLD

The year 1989 saw unprecedented political change in Eastern Europe, as country after country adopted democratic reforms. Equally dramatic transitions to new or strengthened democratic governments took place in Latin America, and there were strong currents of political change in the Soviet Union and South Africa. This sweeping tide is a reminder that attitudes and human institutions can and do change, and change quickly, given appropriate conditions. That is a source of great hope, not only for a freer and more peaceful world, but also for a world less threatened by continuing environmental degradation.

Where will the financial resources come from to build an enduring economic base and to combat poverty in the developing countries, to clean up and rebuild the pollution-prone industrial base of Eastern Europe and the Soviet Union, and to develop new technologies everywhere that can provide the foundation for economic growth without degrading the environment? One potential source, if the world turns from confrontation to peaceful cooperation, is the huge resources now devoted to military budgets—an estimated $1 trillion per year worldwide and significant fractions of the industrial workforce: 11 percent in the United States, 9.7 percent in the Soviet Union [18]. One rationale for such a refocusing of resources is the growing concern that local and global environmental problems can be as serious a

threat to national security and prosperity as military conflicts [19].

In order for sustainable economic development to proceed, however, the environmental consequences of development must be considered from the start, in developing countries and industrial countries alike. It is a hopeful sign that major funding organizations such as the World Bank and a growing number of large international corporations have made commitments to environmental standards [20]. It may also prove essential to measure economic progress more realistically. At present, such economic indicators as gross national product give misleading indications. For example, when applied to countries that rely heavily on their natural resources, gross national product fails to take into account the depletion or destruction of those assets in the process of economic development. (See Chapter 14, "Policies and Institutions.")

Public Opinion on the Environment

As democratic governments spread, support for rechanneling resources to environmental problems may come more and more from ordinary citizens. Widespread concern about the environment is becoming a potent political force. Environmental issues have played a strong role in the wave of democracy sweeping over Eastern Europe and have increasingly become an election issue in the new democracies of Latin America. Environmental questions are being raised repeatedly in the Soviet Union and are already a recurring issue in elections in Western Europe and North America. Increasingly, the environment is becoming a global political issue, occupying a major place on the agenda at the "Group of Seven" (Canada, France, Federal Republic of Germany, Italy, Japan, the United Kingdom, and the United States) economic summit meeting in 1989.

The reasons for the emergence of the environment as a political concern are not hard to find. In 1988 and 1989, the Harris polling organization surveyed public opinion and leadership attitudes around the globe on behalf of the United Nations Environment Programme (21). The survey encompassed 15 countries, including the developing regions of Africa, Asia, and Latin America as well as North America, Western Europe, and Eastern Europe, and found widespread concern about the quality of the environment. Most of those surveyed in all but one country, Saudi Arabia, believed that their environments had become worse in the past decade; large majorities in every country surveyed believed that environmental protection should have a high priority and favored stronger actions by both governments and international organizations. Women surveyed were more aware and concerned about environmental degradation than were men; younger people were more aware and concerned than older people.

The greatest concern of those polled was pollution of drinking water, but large majorities also thought that air and water pollution, the dumping of toxic chemicals on the land, loss of agricultural or farm land, loss of forests, and overuse of pesticides and herbicides were serious problems.

Perhaps most striking was the finding that, by large margins in all the countries surveyed, the public put a higher priority on reducing environmental health risks than they did on increasing living standards. Most people in all 15 countries surveyed (except Argentina) declared that they would be willing to pay higher taxes to see an improvement in the environment, and substantial majorities everywhere (except Japan) would be willing to spend some of their time or money to help improve the environment. Overwhelming majorities of those polled wanted governments to pass and enforce stronger laws and to make protection of the environment a higher priority.

A Common Future

A more peaceful and cooperative world, governments that are more responsive to their citizens and capable of redirecting resources to deal with environmental problems and assist the development process, widespread public concern and growing knowledge about the environment—these are the hopeful signs for the coming decade. The growing role of nongovernmental organizations and of grassroots participation in environmental issues is also a positive trend. The 1992 United Nations Conference on Environment and Development, for which preparation is now under way, ought to be regarded as a unique and crucial opportunity for industrial and developing countries to work together to mobilize world attention on environmental issues and to further sustainable development. These and other initiatives will need widespread support if the world's environmental problems are to be successfully confronted and Earth is to be sustained as a fertile, bountiful planet, the common home of all living things.

References and Notes

1. World Resources Institute estimate, based on growth in world energy production; this method of estimating understates the real growth in economic activity, including agricultural activity in the developing world that uses human, rather than commercial, energy.
2. The World Commission on Environment and Development, *Our Common Future*, (Oxford University Press, New York, 1987), p. 4.
3. See Chapter 24, "Atmosphere and Climate," Tables 24.1 and 24.2.
4. Stephen H. Schneider, "The Greenhouse Effect: Science and Policy," *Science*, Vol. 243, No. 4892 (1988), p. 774.
5. Paul R. Ehrlich, Anne H. Ehrlich, and John P. Holdren, *Ecoscience: Population, Resources, Environment* (W.H. Freeman, San Francisco, 1977), p. 149.
6. World Resources Institute and International Institute for Environment and Development, *World Resources 1986*, (Basic Books, New York, 1986), p. 5.
7. World Resources Institute and International Institute for Environment and Development in collaboration with the United Nations Environment Programme, *World Resources 1988-89*, (Basic Books, New York, 1988), Table 17.3, p. 276-277.
8. *Ibid.*, p. 217.

9. *Op. cit.* 5, p. 257.
10. National Research Council, *Oil in the Sea: Inputs, Fates and Effects* (National Academy Press, Washington, D.C., 1985), Table 2-22, p. 82.
11. National Research Council, *Lead in the Human Environment* (National Academy Press, Washington, D.C., 1980), p. 184.
12. National Research Council, *Atmospheric-Biosphere Interactions: Toward a Better Understanding of the Ecological Consequences of Fossil Fuel Combustion* (National Academy Press, Washington, D.C., 1981).
13. John P. Holdren, "Global Environmental Issues Related to Energy Supply: The Environmental Case for Increased Efficiency of End Use," *Energy*, Vol. 12, Nos. 10-11 (1987), p. 982.
14. Walter V. Reid and Kenton R. Miller, *Keeping Options Alive* (World Resources Institute, Washington, D.C., 1989).
15. *The Bellagio Declaration: Overcoming Hunger in the 1990's*, adopted by the participants in the Bellagio Conference, November, 1913-17, 1989, Bellagio, Italy.
16. *Op. cit.* 2, p. 240.
17. United Nations Department of International Economic and Social Affairs, *World Population Prospects 1988* (U.N., New York, 1989), p. 34.

18. Michael Renner, "Converting to a Peaceful Economy," in Worldwatch Institute, *State of the World 1990: A Worldwatch Institute Report on Progress Toward a Sustainable Society* (W.W. Norton, New York, 1990), p. 155 and Table 9-1, p. 158.
19. Jessica T. Mathews, "Redefining Security," *Foreign Affairs*, Vol. 68, No. 2 (The Council on Foreign Relations, New York, Spring 1989), pp. 162-177.
20. Barber B. Conable, President, The World Bank and International Finance Corporation, in a speech made to the World Resources Institute, Washington, D.C., May 5, 1987; E.S. Woolard, Chairman, E.I. du Pont de Nemours and Co., in a speech to the American Chamber of Commerce, London, U.K., May 4, 1989.
21. Louis Harris and Associates, *Public and Leadership Attitudes to the Environment in Four Continents: A Report of a Survey in 14 Countries* (Louis Harris and Associates, New York, 1988), describes the surveys in Kenya, Nigeria, Senegal, Zimbabwe, China, India, Japan, Saudi Arabia, Argentina, Jamaica, Mexico, Hungary, Norway and West Germany. The same survey taken in the United States is described in Louis Harris and Associates, *American Results of the UNEP Survey* (Louis Harris and Associates, New York, 1989).

2. Climate Change: A Global Concern

We tend to take for granted the moderate nature and relative stability of the Earth's climate. Modern human societies have experienced little else, because of the prevailing interglacial period. But the Earth's climate has changed dramatically in the past. Early human societies experienced a colder climate, a glacial period or ice age, extending until about 10,000 years ago. In still earlier periods, the Earth was ice-free and far warmer than today. To these natural variations in climate must now be added a new concern: human activity is altering the composition of the atmosphere rapidly in ways that could bring on rapid and profound changes in climate.

The temperature of the Earth is determined by the balance between the rate at which sunlight reaches the Earth's surface and the rate at which the warmed Earth sends infrared radiation back into space. It is well established that the warm temperatures (which make life on Earth possible) are the direct result of the trapping of part of the Earth's radiant heat by traces of atmospheric water vapor, carbon dioxide (CO_2), methane (CH_4), and other infrared-absorbing (greenhouse) gases (1). Now, however, human activities are not only increasing the atmospheric concentrations of the naturally occurring greenhouse gases, but also they are adding new, and very powerful, infrared-absorbing gases—such as the industrial chemicals known as chlorofluorocarbons (CFCs)—to the mix.

The rate at which the composition of the atmosphere is being altered through human actions has accelerated spectacularly in recent decades. Levels of carbon dioxide, methane, chlorofluorocarbons, nitrous oxide (N_2O), ozone (O_3), and other greenhouse gases released by human industrial, agricultural, and forestry activities are building up in the lower atmosphere, known as the troposphere. There is strong scientific consensus that, if current trends continue, the buildup of these gases is likely to cause significant warming of the global climate; some warming may already be inevitable because of past emissions. This global climate change could destabilize the natural and societal systems upon which we have come to depend.

In recent years, the link between growing atmospheric concentrations of greenhouse gases and eventual global warming has become broadly accepted by scientists. Nevertheless, both the magnitude and the timing of the warming remain uncertain and many related determinants of future climate change are still inadequately understood. This chapter reports on what is known and what is not known about the factors that may influence changes in the global climate. The chapter also reports on past and current emissions of the major greenhouse gases and on the relative contributions of the countries of the world to these emissions. Each country is assigned a rank based on the combined warming effect

Table 2.1 Summary of Atmospheric Growth of Greenhouse Gases

Gas	Concentration in Air Preindustrial	1986	Present Annual Rate of Increase (percent)
Carbon dioxide	275 ppm	346 ppm	1.4 ppm (0.4)
Methane	0.75 ppm	1.65 ppm	17 ppb (1.0)
Chlorofluorocarbon-12	0	400 ppt	19 ppt (5.0)
Chlorofluorocarbon-11	0	230 ppt	11 ppt (5.0)
Nitrous oxide	280 ppb	305 ppb	0.6 ppb (0.2)
Tropospheric ozone	unknown	35 ppb	unknown

Source: Office of Policy, Planning and Evaluation, Office of Research and Development, U.S. Environmental Protection Agency, *Potential Effects of Global Climate Change on the United States,* Joel B. Smith, Dennis A. Tirpak, eds. (U.S. Environmental Protection Agency, Washington, D.C., draft, 1988).

ppm = parts per million.
ppb = parts per billion.
ppt = parts per trillion.

of the country's estimated emissions of the major greenhouse gases. The chapter discusses the scientific knowledge base and some of the major uncertainties involved in our knowledge of climate change. And it reviews policy options that governments can pursue to reduce the risks or counter the effects of climate change. Other chapters in this volume discuss the potential impacts of climate change on public health (Chapter 4), agriculture (Chapter 6), forests and rangelands (Chapter 7), wildlife (Chapter 8), precipitation patterns and water availability (Chapter 10), and oceans and coastal areas (Chapter 11). Chapter 13, "Global Systems and Cycles," examines present and future means of monitoring global changes in the climate. Chapter 24, "Atmosphere and Climate," contains the primary data, estimates of greenhouse gas emissions by country, and a discussion of how these estimates were reached.

STATUS OF CURRENT KNOWLEDGE

The status of current knowledge can be straightforwardly summarized.

SCIENTIFIC FACT: Infrared-Absorbing Gases Are Responsible for Keeping the Earth Warmer Than It Would Otherwise Be

It has been proved conclusively that, in recent millennia, the "greenhouse effect"—the insulating effect of the atmosphere—has kept the planet at an average temperature of about 13° C, or 33° C warmer than it otherwise would be. Earth's greenhouse warming turns out to be consistent with the temperatures found on the planets of Mars and Venus, once varying distance from the sun and varying atmospheric composition are taken into account. Mars has a very thin atmosphere and maintains an average temperature of -53° C, while Venus, with its CO_2-dominated atmosphere, suffers from a "runaway greenhouse effect" and experiences an average temperature of 447° C (2).

On Earth, carbon dioxide and water vapor are the most important greenhouse gases in creating the insulating effect of the atmosphere, but methane and chlorofluorocarbons also play a major and growing role (3). The heat-absorbing potential of these gases varies widely and depends on such factors as their relative concentration in the atmosphere and their infrared absorption profile. The addition of an incremental molecule of methane, for example, will result in the trapping of 20–30 times as much heat as the addition of an incremental CO_2 molecule. An incremental CFC molecule will result in the trapping of 20,000 times as much heat as would a CO_2 molecule (4).

SCIENTIFIC FACT: Concentrations of Infrared-Absorbing Gases Are Increasing at Unprecedentedly Rapid Rates

Table 2.1 illustrates what is known about the growing atmospheric concentrations of the major greenhouse gases. Based on direct measurements as well as on ice core analyses, a record of carbon dioxide concentrations 160,000 years into the past has been constructed (5). While recent growth rates in the concentrations of individual gases vary, they all have increased significantly since preindustrial times. Continuous sampling done at the Mauna Loa Observatory in Hawaii since 1958 has shown an inexorable rise in atmospheric CO_2 concentrations. The CO_2 concentration of 351 parts per million (ppm) in 1988 is 20–25 percent higher than at any time in the past 160,000 years, notwithstanding wide swings during that period as climate shifted. The very warm interglacial period of 130,000 years ago, for example, was accompanied by CO_2 levels of just under 300 ppm, while concentrations during the previous great ice age dropped to around 200 ppm. By the beginning of the Industrial Revolution, CO_2 concentrations had climbed back up to 280 ppm (6). Figure 2.1 shows the parallel trends in CO_2 and temperature records over the past 160,000 years.

Within the past 100 years, however, atmospheric CO_2 concentrations have risen by another 70 ppm. More than half of this 25 percent increase has occurred in the past 30 years alone. If the current growth rate of 0.4 percent per year continues, atmospheric concentrations of CO_2 will double from preindustrial levels by around 2075.

Atmospheric methane concentrations have more than doubled from their preindustrial levels, and are growing at a rate of about 1 percent annually (7). Trends in methane levels have tracked those of CO_2 closely during the past 160,000 years. Today, methane levels are much higher than ever before (8). Table 2.1 shows that concentrations of other significant greenhouse gases, namely, nitrous oxide and CFCs, also are increasing. Chlorofluorocarbon concentrations, for example, are growing at about 5 percent annually.

Figure 2.1 Long-Term Variations of Global Temperature and Atmospheric Carbon Dioxide

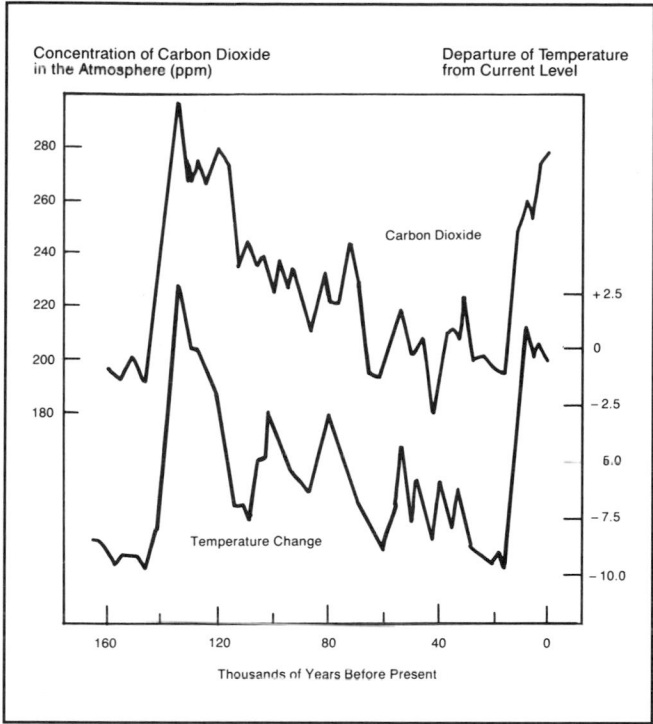

Source: J.M. Barnola *et al.*, "Vostok Ice Core Provides 160,000-year Record of Atmospheric CO₂," *Nature*, Vol. 329, No. 6138 (1987), p. 410.

SCIENTIFIC THEORY: Continued Greenhouse Gas Emissions Will Lead to Global Warming

As the levels of greenhouse gases rise, they will trap more heat in the lower atmosphere, raising the temperature at the Earth's surface and changing the climate—unless other climatic mechanisms counteract the warming. Extrapolating projected growth rates of greenhouse gases into the future and combining this information with the known, heat-absorbing properties of the individual gases, scientists have modeled the amount of warming from past emissions as well as from current and future emissions that each gas is expected to contribute to a new atmospheric equilibrium temperature (9). Deriving actual short- and mid-term temperature projections from these models is made difficult by several key facts. First, it is far easier to model the temperatures that are likely to prevail once a new climatic equilibrium has been reached than it is to model transient temperature changes. Second, it is believed that the thermal inertia of the oceans can delay a new temperature equilibrium for several decades, although the precise mechanism and duration of this delay are not well understood (10) (11). Therefore, it is not now possible to accurately predict the rate of temperature rise as greenhouse gas concentrations accumulate. (See State of the Science, below.)

If CO_2 (or its equivalent in other greenhouse gas concentrations) reaches twice the preindustrial level, the consensus within the scientific community is that we can expect increases of a few degrees Celsius (12). When such a doubling is modeled on general circulation models (GCMs), which make projections of climate changes and which can only be run on the most powerful computers available, the results suggest that average global temperatures would rise by 1.5° to 4.5° C, depending on the particular model (13). (For additional discussion of GCMs and the data on which they depend, see The Limits of Climate Models: Key Uncertainties, below, and Chapter 13, "Global Systems and Cycles.")

While atmospheric concentrations of CO_2 are not expected to double until the third quarter of the 21st Century, modeling suggests that, if current trends continue, the combined effects of all greenhouse gases could cause a warming commitment equivalent to that of a doubled-CO_2 environment by about 2030 (14). Because of delays in the climate's response, exactly when warming would occur is not known.

Impacts of Climate Change

The impacts of a global warming are still uncertain in many respects, not least because the computer models of climate change cannot predict regional changes with any reliability. Nonetheless, the effects of a global warming over the next half century are likely to include a rise in sea level of perhaps 30 centimeters, along with changes in winds, ocean currents, accumulations of ice and snow in polar ice caps, and the frequency of severe storms (see Chapter 11, "Oceans and Coasts"); variations in the range of disease-bearing organisms and other impacts on public health (see Chapter 4, "Population and Health"); alterations in precipitation patterns that will affect water availability and agriculture (see Chapter 6, "Food and Agriculture," and Chapter 10, "Freshwater"); and changes in wetlands, forests, and other natural ecosystems, possibly leading to the increased extinctions of plant and animal species (see Chapter 7, "Forests and Rangelands," and Chapter 8, "Wildlife and Habitat").

GREENHOUSE GAS EMISSIONS

Sources of greenhouse gases are distributed widely around the world, with both developed and developing countries sharing major responsibility for emissions. Nonetheless, in 1987, just five countries contributed 50 percent of the warming potential added to the atmosphere that year.

Past Emissions

The key greenhouse gases responsible for most of the projected warming emanate from a wide variety of human activities, ranging from fossil fuel combustion to wet rice cultivation. The most important, carbon dioxide, arises primarily from the burning of

Figure 2.2 Cumulative Emissions of Carbon Dioxide from Fossil Fuels for the 26 Countries with the Highest Emissions, 1950—87

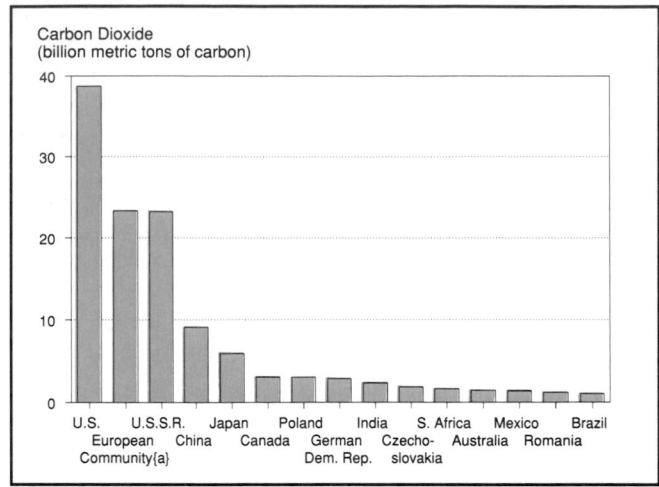

Source: Carbon Dioxide Information Analysis Center (CDIAC), Oak Ridge National Laboratory, Oak Ridge, Tennessee, unpublished data, August 1989.
Note: a. The European Community comprises 12 countries: Belgium, Denmark, France, Federal Republic of Germany, Greece, Ireland, Italy, the Netherlands, Luxembourg, Portugal, Spain, United Kingdom.

fossil fuels, the manufacture of cement, and changes in land use through large-scale deforestation, including burning and clearing land for agricultural purposes. An enormous quantity of this gas has been released through human activities since the onset of the Industrial Revolution. Worldwide consumption of fossil fuels in the period 1860 to 1949 is estimated to have released 51 billion metric tons of carbon (15). (By convention, carbon dioxide releases are measured in terms of the weight of carbon contained in the gas.) In the past four decades, fossil fuel use has accelerated and carbon dioxide emissions from fossil fuel consumption between 1950 and 1987 totaled an additional 130 billion metric tons (16). Figure 2.2 shows the relative contributions for the period 1950–87 of the 26 countries that have been the largest sources of fossil-fuel-based carbon dioxide emissions (emissions from the 12 countries of the European Community are combined). Land use change, including deforestation for agricultural purposes, is estimated to have released another 60 billion metric tons of carbon dioxide since 1860 (17). Thus, in the period 1860–1987, the total release of carbon dioxide from human activities equals an estimated 241 billion metric tons of carbon.

The Greenhouse Index

While carbon dioxide has received the most attention, it accounts for about only half of the warming potential attributable to human activity. An estimate for the contributions of the major greenhouse gases, based on the atmospheric concentrations of the gases during the mid-1980s and their relative heat-

trapping potential, yields the following picture: carbon dioxide, 50 percent; chlorofluorocarbons, 20 percent; methane, 16 percent; tropospheric ozone, 8 percent; nitrous oxide, 6 percent (18). Thus, estimates of the relative contributions by country based solely on CO_2 (as in Figure 2.2) distort the true picture.

A preferable approach would be to calculate a Greenhouse Index that represents the contribution of each gas, appropriately weighted by its heat-trapping potential, based on annual emissions for each country. For CO_2, CFCs, and CH_4, it is possible to make estimates of annual emissions by country in a systematic way. In 1987, for example, human activities released about 8.5 billion metric tons of carbon as carbon dioxide, 255 million metric tons of methane, and more than 770,000 metric tons of CFCs (CFC-11 and CFC-12). (See Chapter 24, "Atmosphere and Climate," Table 24.1.) These gases account for about 86 percent of the current global warming potential attributable to human activity.

Estimates of tropospheric ozone contributions by country are more problematic, and even the global sources of nitrous oxide are not well understood. Because of such uncertainties, these and other minor greenhouse gases—and natural sources of all greenhouse gases not directly attributable to human activity—are omitted from calculations of a Greenhouse Index.

For the three major greenhouse gases, 1987 estimates of current annual additions to the atmosphere are weighted appropriately by their infrared effectiveness. (See Box 2.1.) These weighted contributions are combined to create a single Greenhouse Index, given as carbon dioxide heating equivalents and expressed in metric tons of carbon, for each of 146 countries. The results are summarized below (see National Rankings). The Greenhouse Index provides a more balanced measure of each country's responsibility for potential global warming than CO_2 alone; the Index also provides a means of tracking increases or decreases in each country's share of greenhouse gas emissions over time.

National Rankings

Table 2.2 gives the carbon dioxide heating equivalents of annual atmospheric increases of CO_2, CH_4, and CFCs—and their total contribution—for 50 countries, ranked by their Greenhouse Index. (Complete data for 146 countries can be found in Chapter 24, "Atmosphere and Climate," Table 24.2.) Table 2.2 also gives the percentage contributed by each country to the global total; thus, in 1987, the United States with a Greenhouse Index of 1 billion metric tons of carbon equivalent, contributed the largest share, 17.6 percent, of the world's increase in warming potential represented by the three major greenhouse gases; the U.S.S.R. contributed 12 percent, and Brazil, 10.5 percent. Altogether, the 50 countries in Table 2.2 accounted for 92 percent of the world's

Table 2.2 The Greenhouse Index: 50 Countries with the Highest Greenhouse Gas Net Emissions, 1987

(Carbon Dioxide Heating Equivalents, 000 metric tons of carbon)

| Country | Greenhouse Index Rank | Greenhouse Gases | | | | |
		Carbon Dioxide	Methane	CFCs{a}	Total	Percent of Total
United States	1	540,000	130,000	350,000	1,000,000	17.6
U.S.S.R.	2	450,000	60,000	180,000	690,000	12.0
Brazil	3	560,000	28,000	16,000	610,000	10.5
China	4	260,000	90,000	32,000	380,000	6.6
India	5	130,000	98,000	700	230,000	3.9
Japan	6	110,000	12,000	100,000	220,000	3.9
Germany, Fed. Rep.	7	79,000	8,000	75,000	160,000	2.8
United Kingdom	8	69,000	14,000	71,000	150,000	2.7
Indonesia	9	110,000	19,000	9,500	140,000	2.4
France	10	41,000	13,000	69,000	120,000	2.1
Italy	11	45,000	5,800	71,000	120,000	2.1
Canada	12	48,000	33,000	36,000	120,000	2.0
Mexico	13	49,000	20,000	9,100	78,000	1.4
Myanmar	14	68,000	9,000	0	77,000	1.3
Poland	15	56,000	7,400	13,000	76,000	1.3
Spain	16	21,000	4,200	48,000	73,000	1.3
Colombia	17	60,000	4,100	5,200	69,000	1.2
Thailand	18	48,000	16,000	3,500	67,000	1.2
Australia	19	28,000	14,000	21,000	63,000	1.1
German Dem. Rep.	20	39,000	2,100	20,000	62,000	1.1
Nigeria	21	32,000	3,100	18,000	53,000	0.9
South Africa	22	34,000	7,800	5,800	47,000	0.8
Côte d'Ivoire	23	44,000	550	2,000	47,000	0.8
Netherlands	24	16,000	8,800	18,000	43,000	0.7
Saudi Arabia	25	20,000	15,000	6,600	42,000	0.7
Philippines	26	34,000	6,700	0	40,000	0.7
Lao People's Dem. Rep.	27	37,000	1,000	0	38,000	0.7
Viet Nam	28	28,000	10,000	0	38,000	0.7
Czechoslovakia	29	29,000	2,200	2,700	33,000	0.6
Iran	30	17,000	6,400	9,000	33,000	0.6
Argentina	31	13,000	12,000	5,500	31,000	0.5
Korea, Rep.	32	21,000	2,900	5,400	29,000	0.5
Turkey	33	16,000	3,600	9,200	29,000	0.5
Romania	34	25,000	3,100	0	28,000	0.5
Venezuela	35	19,000	4,700	3,200	27,000	0.5
Yugoslavia	36	15,000	2,800	8,200	26,000	0.4
Malaysia	37	22,000	1,400	2,500	26,000	0.4
Belgium	38	12,000	1,200	12,000	25,000	0.4
Algeria	39	8,400	12,000	4,100	25,000	0.4
Peru	40	22,000	870	0	23,000	0.4
Bangladesh	41	2,300	20,000	0	22,000	0.4
Ecuador	42	19,000	570	1,700	21,000	0.4
Greece	43	7,000	1,100	12,000	20,000	0.4
Korea, Dem. People's Rep.	44	18,000	2,300	0	20,000	0.3
Portugal	45	3,700	1,000	13,000	17,000	0.3
Egypt	46	9,000	3,100	5,100	17,000	0.3
Bulgaria	47	15,000	660	1,600	17,000	0.3
Austria	48	6,500	960	9,100	17,000	0.3
Zaire	49	16,000	790	0	16,000	0.3
Cameroon	50	16,000	580	0	16,000	0.3

Source: Chapter 24, "Atmosphere and Climate," Table 24.2.
Note: a. Chlorofluorocarbons

total increase in warming potential in 1987. The countries that are the largest contributors are displayed in slightly different form—with the European Community as a single source–in Figure 2.3. When all countries are considered, Asia (excluding the Soviet Union) is the largest contributor among the major regions of the world, followed by North and Central America, and by Europe. (See Chapter 24, Table 24.2.)

What is evident is that responsibility for greenhouse emissions is spread widely around the world.

Three of the six countries that are the largest contributors to the atmosphere's warming potential—the United States, the U.S.S.R., Brazil, China, India, and Japan—have heavily industrialized economies; three do not. The same conclusion holds if the European Community is considered as one unit. Ranked by Greenhouse Index, every major region of the world and every continent are represented in the top 50 countries; all except Africa are represented in the top 20. Such widespread responsibility for significant greenhouse gas emissions means that any effective

Box 2.1 Estimating Responsibility for Potential Climate Change

Greenhouse gases are emitted into the atmosphere by a wide variety of sources, both natural and attributable to human activities. These gases participate in complex chemical, biological, and even geological cycles and are removed from the atmosphere ultimately by processes that differ for each gas. The processes proceed at different rates that must be taken into account in order to estimate the relative contributions of each gas—and each source—to the atmosphere's warming potential.

Carbon dioxide (CO_2) molecules are very stable and are capable of remaining in the atmosphere almost indefinitely. But CO_2 is absorbed in the oceans and the soils and is taken up by plants during photosynthesis—the Earth itself is its sink. So active are the removal processes and so close are they to ground-level sources of CO_2 that less than half of the gas emitted appears in the atmosphere as an increase. Exactly where the rest goes is still uncertain. Methane is decomposed rapidly by chemical processes in the atmosphere, principally oxidation by the hydroxyl radical molecule, so that a given methane molecule may survive only about a decade. Thus, a large portion of the methane emissions merely replaces methane that has decomposed; less than a fifth of the methane emitted manifests itself in the atmosphere as an increase.

Chlorofluorocarbon molecules are stable in the lower atmosphere, but gradually make their way into the stratosphere where they are slowly destroyed by ultraviolet radiation. Since there are no natural sources and no sinks in the lower atmosphere, all of the CFCs emitted increase the atmospheric levels of these gases.

CALCULATING THE GREENHOUSE INDEX

Measurements of concentrations of these gases indicate that atmospheric levels are rising, indicating that—whatever the processes involved—human activity is emitting more of these gases than the natural processes can remove. One approach to estimating their potential warming effects is to focus on the increase in atmospheric levels and to allocate responsibility for that increase based on share of total emissions. A comparison of the increase of these gases in the atmosphere with estimated anthropogenic emissions finds that about 44 percent of the CO_2, 17 percent of the CH_4, and 100 percent of the CFCs represent net additions to the atmosphere. (See Chapter 24, "Atmosphere and Climate," for a more detailed discussion of these percentages.)

The amount of each greenhouse gas added to the atmosphere can be weighted by its relative effectiveness in absorbing infrared radiation and, for convenience, is expressed as an equivalent CO_2 heating effect. This allows the direct comparison of the current warming contributions of methane and CFCs with those of carbon dioxide.

In 1987, for example, human activities released an estimated 8.5 billion metric tons of carbon in the form of CO_2, 255 million metric tons of CH_4, and 771,500 metric tons of CFCs. (See Chapter 24, "Atmosphere and Climate," Table 24.1, and the discussion below.) The net additions to the atmosphere of these gases, however, were 3.7 billion metric tons of carbon as CO_2, 43 million metric tons of CH_4, and 771,500 metric tons of CFCs. (See Chapter 24, "Atmosphere and Climate," Table 24.2.) Weighted by their heating effectiveness and converted to common units, these numbers imply that human activities added the equivalent of 5.9 billion metric tons of carbon to the atmosphere. Of this, 63 percent was from carbon dioxide, 14 percent from methane, and 24 percent from CFCs.

COUNTRY-LEVEL GREENHOUSE DATA

This procedure can be repeated on a country-by-country basis to obtain a measure of each country's yearly contribution to the greenhouse warming potential of the atmosphere. The resulting measure, a Greenhouse Index, provides a means of tracking increases or reductions in each country's contributions.

Table 24.1 of this report contains estimates of current CO_2, CH_4, and CFC emissions for the major anthropogenic sources by country. Table 2.2 (and in more complete form, Table 24.2) presents net additions of these gases to the atmosphere and their equivalent CO_2 heating effect by country. These estimates and calculations are the basis for the national rankings in this chapter. The major sources of carbon dioxide emissions used in these rankings are combustion of fossil fuels, flaring of natural gas, and land clearing. There is considerable uncertainty, controversy—and probably year-to-year variation—in land clearing figures. The base year for most data on which the country rankings are determined is 1987. In Brazil, however, 1987 was probably an anomalously high year for land use clearing in the Amazon region because immigration of poor people to that region peaked in that year and because of the impending suspension of tax credits for wealthy landowners that could be secured only by "improving" land—a requirement usually satisfied by

clearing that land through burning. Comparable estimates for land use clearing in 1988 give a figure 35 percent lower—which still would not change Brazil's rank in Table 2.2. (See Chapter 7, "Forests and Rangelands," New Estimates of Tropical Deforestation.) An alternative and controversial estimate [1] of annual average deforestation between 1977 and 1987—which does not take into account the acceleration of deforestation in recent years—is 52 percent lower still, and would change Brazil's rank in Table 2.2 from third to fourth. Despite such uncertainties, carbon dioxide emissions are dominated by fossil fuel use, which is known with reasonable reliability.

SOURCES OF METHANE AND CFCs

Much less is known about the sources of atmospheric methane. The major sources used in these rankings are estimates for anaerobic fermentation in wet rice paddies, in cattle and other livestock, and in landfills and sewage treatment plants. (See the technical notes for Table 24.1 for specific assumptions.) Additional sources are estimated leakages from natural gas pipelines and estimates of methane released during the mining of hard coals.

Estimates of methane emissions are further complicated by carbon monoxide, which is released during the fossil fuel combustion and biomass burning and which, in turn, scavenges OH radicals from the atmosphere. This prolongs the average lifetime of methane molecules in the atmosphere and helps to increase atmospheric concentrations. Exactly what proportion of the observed increase for methane is due to increased emissions and what proportion is due to decreases in OH radicals is still uncertain.

A third major set of greenhouse gases are the CFCs, industrial products, that obviously did not exist in the atmosphere until they were created 60 years ago. They are the most potent greenhouse gases. Table 24.1 reports estimates of current national per capita consumption of the two major CFC gases, from which national consumption can be calculated. Consumption approximates the annual release of these materials. (See Table 24.2.)

References and Notes

1. Roberto Pereira da Cunha, "Deforestation Estimates Through Remote Sensing: The State of the Art in the Legal Amazonia," National Space Research Institute of Brazil (INPE), São José dos Campos, São Paulo, Brazil, 1989.

Figure 2.3 The Greenhouse Index: 26 Countries with the Highest Greenhouse Gas Net Emissions, 1987.

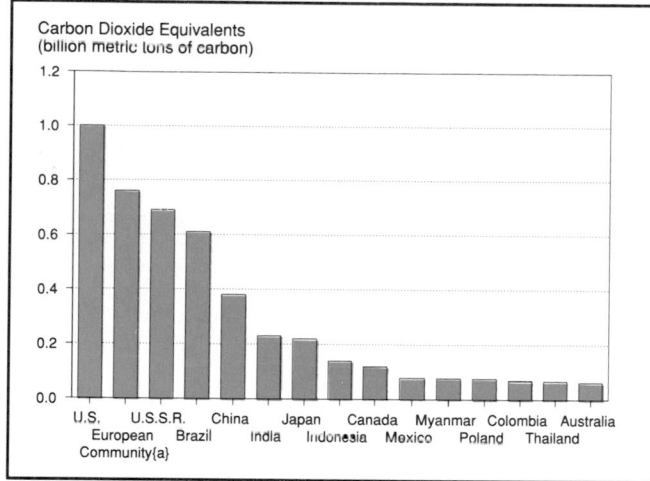

Source: Chapter 24, "Atmosphere and Climate," Table 24.2

Note: a. The European Community comprises 12 countries: Belgium, Denmark, France, Federal Republic of Germany, Greece, Ireland, Italy, Luxembourg, the Netherlands, Portugal, Spain, United Kingdom.

Table 2.3 Per Capita Greenhouse Index: 50 Countries with the Highest Per Capita Greenhouse Gas Net Emissions, 1987

(Carbon Dioxide Heating Equivalents, metric tons of carbon)

Country	Rank	Metric Tons per Capita
Lao People's Dem Rep	1	10.0
Qatar	2	8.8
United Arab Emirates	3	5.8
Bahrain	4	4.9
Canada	5	4.5
Luxembourg	6	4.3
Brazil	7	4.3
Côte d'Ivoire	8	4.2
United States	9	4.2
Kuwait	10	4.1
Australia	11	3.9
German Dem Rep	12	3.7
Oman	13	3.5
Saudi Arabia	14	3.3
New Zealand	15	3.2
Netherlands	16	2.9
Denmark	17	2.8
Costa Rica	18	2.8
Singapore	19	2.7
United Kingdom	20	2.7
Germany, Fed Rep	21	2.7
Finland	22	2.6
Ireland	23	2.5
Belgium	24	2.5
U.S.S.R.	25	2.5
Switzerland	26	2.4
Nicaragua	27	2.4
Colombia	28	2.3
Trinidad and Tobago	29	2.3
France	30	2.2
Austria	31	2.2
Czechoslovakia	32	2.1
Israel	33	2.1
Ecuador	34	2.1
Italy	35	2.1
Norway	36	2.1
Greece	37	2.1
Poland	38	2.0
Myanmar	39	2.0
Bulgaria	40	1.9
Spain	41	1.9
Japan	42	1.8
Iceland	43	1.8
Liberia	44	1.7
Portugal	45	1.7
Sweden	46	1.7
Guinea-Bissau	47	1.6
Malaysia	48	1.6
Cameroon	49	1.6
Venezuela	50	1.5

Source: World Resources Institute calculation based on Chapter 24, "Atmosphere and Climate," Table24.2.

agreement to stabilize or reduce these emissions will have to be equally widely based. Global warming is truly a global phenomenon, in both cause and potential effect.

Per Capita Greenhouse Index

National rankings reflect the contributions of individual countries to the atmosphere's growing store of greenhouse gases. But it is also of interest to compare countries by per capita contributions of these gases, since the citizens of our planet vary widely in their relative responsibility for these emissions. Table 2.3 gives per capita contributions to the atmosphere's warming potential, calculated by apportioning each country's share equally among its citizens, for the 50 countries with the highest per capita Greenhouse Index. High rates of deforestation account for elevated rankings in countries such as the Lao People's Democratic Republic and the Côte d'Ivoire; energy consumption and gas flaring account for high rankings in energy-producing countries such as Qatar, the United Arab Emirates, and Kuwait. Industrialized countries with elevated rates of energy consumption such as Canada, Australia, and the United States also tend to have high per capita rankings.

Per capita figures also can be looked at as an indication of how much greenhouse gas emissions and global warming potential might grow if developing countries, in the process of development, significantly increase their per capita emissions. If just China and India, for example, raised their per capita emissions to the current world average, 1.1 metric

tons of carbon dioxide heating equivalent per annum, total worldwide additions to the atmosphere would increase 28 percent; if they raised per capita emissions to the level of France, worldwide additions to the atmosphere would increase 68 percent.

Figure 2.4A compares the per capita emissions for the 10 countries of the Organisation for Economic Cooperation and Development (OECD) with the highest *national* Greenhouse Index rank. Similarly, Figures 2.4B and 2.4C compare per capita emissions for the

Figure 2.4 Per Capita Net Greenhouse Gas Emissions, 1987

A. OECD Countries

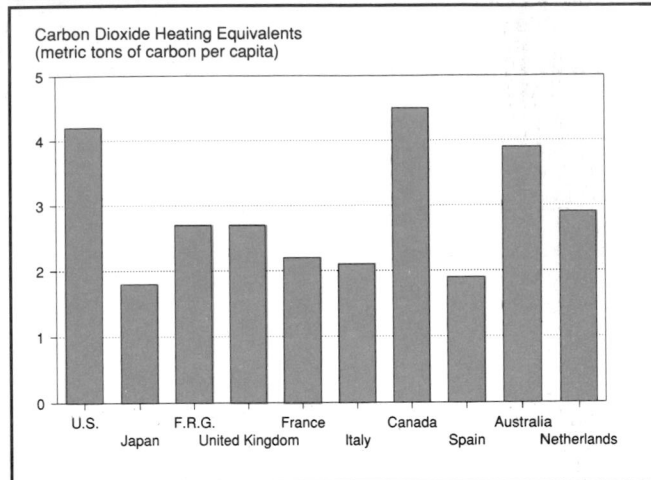

Carbon Dioxide Heating Equivalents
(metric tons of carbon per capita)

U.S. Japan F.R.G. United Kingdom France Italy Canada Spain Australia Netherlands

B. Planned Economies

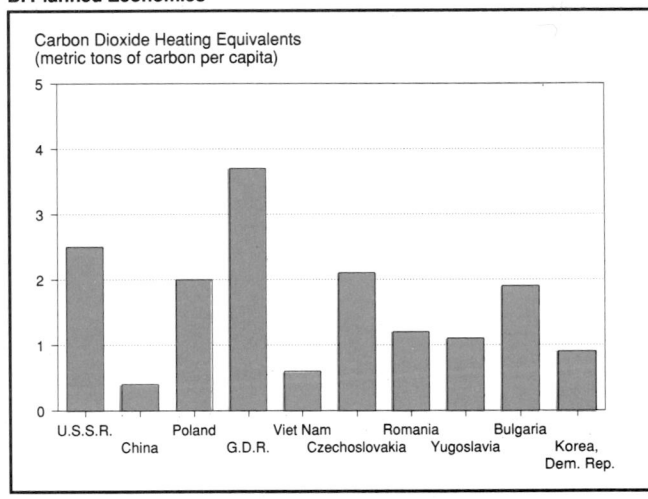

Carbon Dioxide Heating Equivalents
(metric tons of carbon per capita)

U.S.S.R. China Poland G.D.R. Viet Nam Czechoslovakia Romania Yugoslavia Bulgaria Korea, Dem. Rep.

C. Developing Countries

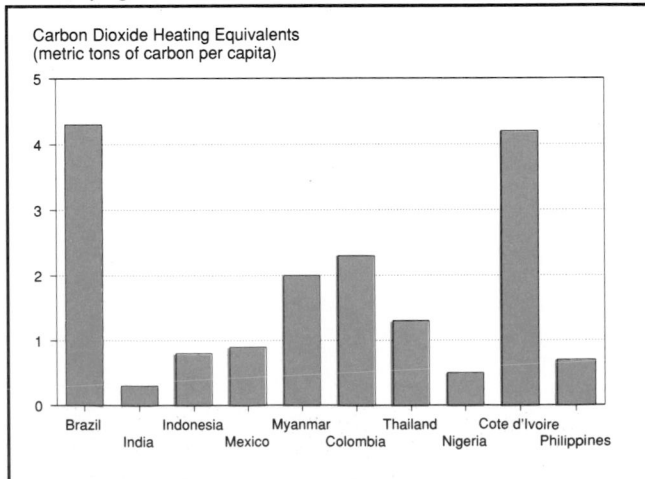

Carbon Dioxide Heating Equivalents
(metric tons of carbon per capita)

Brazil India Indonesia Mexico Myanmar Colombia Thailand Nigeria Cote d'Ivoire Philippines

Source: Table 2.3

10 planned economies and 10 developing countries, respectively, with the highest national emissions. The large ranges among per capita emissions in each economic group suggest that national policies can make a significant difference in a country's contributions to atmospheric warming potential, although, clearly, geography also plays a role. Both the United States and Canada, for example, emit more than twice the per capita greenhouse gases that Japan does, a difference attributable to both geography and policy. Mexico emits about half the per capita emissions of Colombia and about one quarter those of Brazil. At the same time, differences between economic groups are a reminder that any agreement to limit greenhouse gas emissions will have to take into account questions of individual equity and the legitimate aspirations of nations to develop.

Greenhouse Emissions and Gross National Product

A third index of comparison is greenhouse gas emissions scaled by gross national product (GNP). The United States emits the heating equivalent of 200 grams of carbon for each dollar of its gross national product. Indonesia emits nearly nine times as much for each dollar of its gross national product. But while developed countries generally emit far smaller amounts of greenhouse gases per unit of economic output than do developing countries—and, in that sense, constitute a model for other countries to strive toward—their emissions are high in an absolute sense. Figure 2.5 compares this index for those OECD, planned economies, and developing countries with the highest national emissions, as was done previously for the per capita comparisons.

Both per capita emissions and per GNP emissions bring huge national emissions into a more human—and perhaps more easily comprehended—scale. One metric ton of carbon or its heating equivalent is added to the atmosphere each year for each person on the planet. One metric ton of carbon equivalent also is added to the atmosphere for every $5,000 of economic activity in developed countries and for roughly every $1,000 of economic activity in the developing world. As the world's population doubles to a projected 10 billion persons sometime in the next century—and as economic activity multiplies several-fold to feed, house, employ, and fulfill the rising expectations of these people—greenhouse gas emissions are likely to more than double, absent major efforts to reduce emissions. And, in any such efforts, both per capita and per GNP emissions will need to be taken into consideration in establishing a fair international basis for limiting greenhouse gas emissions.

STATE OF THE SCIENCE

The current view of global climate change is marked by surprising scientific consensus, but clouded by great scientific uncertainty as well. Though scientists generally regard the evidence for substantial climate change induced by greenhouse warming as quite convincing in its outlines, there is much less understanding or agreement about the details of such global change. The timing, magnitude, and precise mechanisms of both the warming and its global consequences remain the subject of intense scrutiny and lively debate.

What, then, can we say with relative certainty today about the greenhouse phenomenon and its potential effects upon global climate? Equally important, what are the key uncertainties–those unknowns preventing an accurate depiction of the progress and processes of the greenhouse effect and its consequences?

Greenhouse Knowns

The present list of generally accepted greenhouse facts is meager, but important. First, although there is much argument about the exact climatic effects of adding greenhouse gases to the atmosphere rapidly, there is no dispute that the greenhouse phenomenon itself—the trapping of heat by atmospheric gases—is a reality (19). Without the substantial greenhouse effect routinely conferred by the Earth's atmosphere, surface temperatures would plunge about 33° C (20) (21), resulting in a world of frozen seas and little life as we know it. Satellite measurements of the Earth's energy budget—the solar energy the Earth receives and the energy it reflects or reemits in turn—confirm the trapping of infrared radiation by the atmosphere and the surface warming it causes (22).

A second point of consensus is that, as shown in the previous section, the atmospheric concentrations of greenhouse gases are rising at unprecedented rates—rates which, in many cases, show signs of accelerating. No one disagrees that we are altering our atmosphere very rapidly. In fact, such a fast rise in heat-absorbing gases is without parallel in history, at least within the period for which we have measurements (160,000 years).

Another undisputed fact is that, historically, changes in greenhouse gas concentrations are closely correlated with changes in the Earth's surface temperature. Analysis of gas bubbles trapped in ice deposited over the past 160,000 years reveals that a warming of the global climate and a rise in CO_2 and CH_4 levels coincide. Conversely, lower atmospheric CO_2 and CH_4 levels coincide with global cooling (23). The implication for today's climate is that higher global temperatures will track the higher levels of greenhouse gases now in the atmosphere. Even if the climate changes of the past were not initiated by the historic fluctuation in CO_2 and CH_4 levels, this fluctuation undoubtedly played a significant role in determining the intensity of the warming or cooling that took place (24) (25).

These three knowns—that the greenhouse effect is real, that levels of greenhouse gases are rising at un-

Figure 2.5 Net Greenhouse Gas Emissions per U.S. Dollar of Gross National Product, 1987

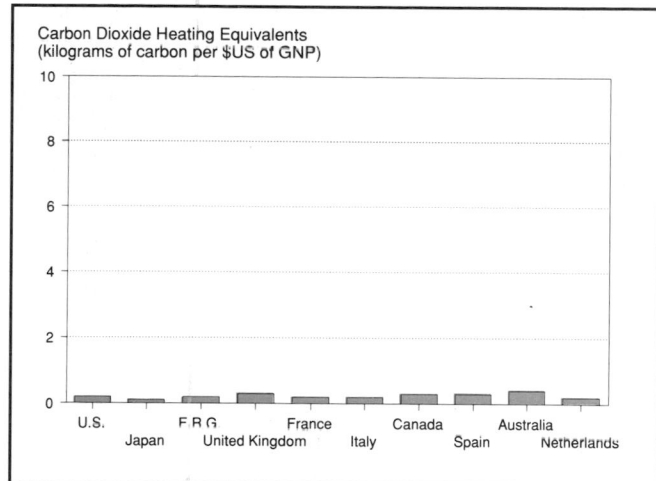

A. OECD Countries

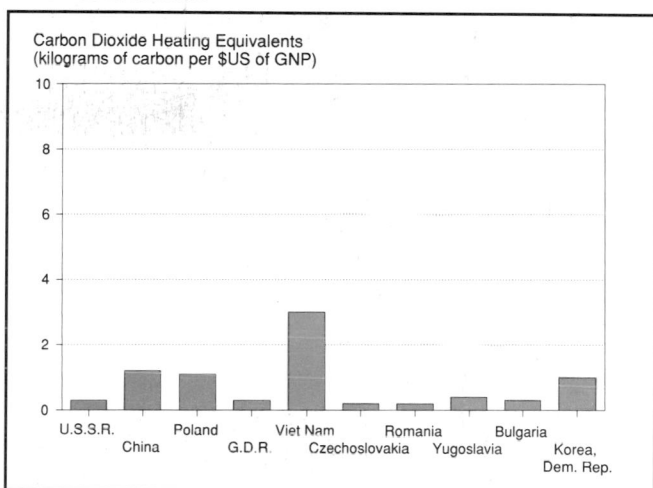

B. Planned Economies

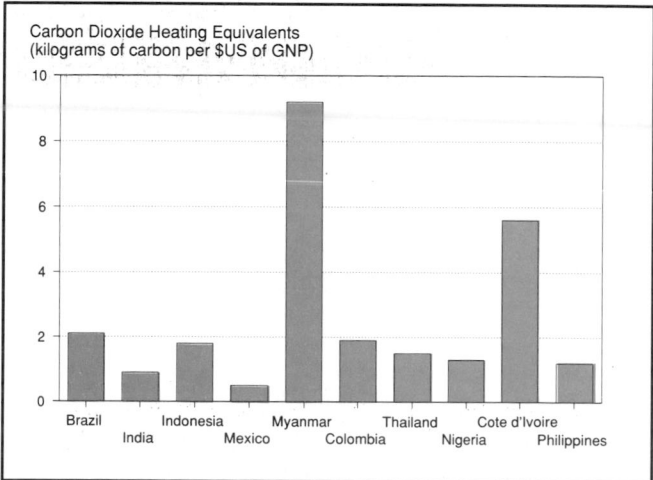

C. Developing Countries

Source: World Resources Institute calculation based on Chapter 24, "Atmosphere and Climate," Table 24.2.

precedented rates, and that greenhouse gas concentrations have tracked global climate change closely in the past—when combined with the predictions of the latest climate models have resulted in the consensus that some global warming is likely within the next century.

Some evidence exists that the predicted warming has begun to manifest itself already, but this evidence is controversial. Two separate evaluations of temperature records taken around the world since 1860 suggest that a global warming of 0.5°–0.7° C (26) (27) (28) (29) has occurred since that year, with nearly half this rise occurring since 1965. Some question the reliability of these temperature readings, but corrections made to satisfy such criticisms still leave a warming of about 0.5° C (30). This correlates well with temperature measurements made from weather balloons since 1958, which show a rise in surface temperatures of about 0.08° C per decade (31) (32).

Moreover, these temperature records show that six of the warmest years on record have occurred this decade; in descending order, they are 1988, 1987, 1983, 1981, 1989, and 1980. (33). Many scientists believe that such an aggregation of warm years goes beyond the normal variability expected of the climate and is in line with the predictions for warming based on the present level of greenhouse gases (34). Others contend that the evidence linking this record warmth directly to the greenhouse effect is still insufficient. They point to an unexplained cooling trend from 1940–1965 to demonstrate that climate records do not unambiguously bear the fingerprint of the greenhouse effect (35) (36).

The Limits of Climate Models: Key Uncertainties

Even if we were certain of future emission levels for greenhouse gases, we still could not predict in accurate detail how the global climate would respond. This is largely because the computer models—the general circulation models—responsible for such predictions cannot account adequately yet for the complex interactions between the Earth's systems that routinely influence climate (37). (See Box 2.2.)

The biosphere and the geosphere, the Earth's water cycles and reflective ice surfaces, and the changing flux of solar energy reaching the Earth all affect atmospheric processes and contribute to world climate. Climate models must catalog these interactions, model them separately, and then integrate them into the larger model. If these interactions are poorly understood, as is the case in several key areas, they are omitted from the models or represented only crudely (38) (39).

Current climate models perform well in many important respects; the results they give are generally consistent with theory and experience (40) (41). For instance, provided initial data on Venus and Mars,

they can predict successfully the climatic conditions that we know have developed there.

Global climate models also can reproduce the Earth's seasonal cycle—its summer and winter temperature extremes—reasonably well and the temperatures they calculate for known climatic events of the past, such as ice ages, are consistent with scientific evidence (42).

In spite of these successful tests of their validity, present-day models suffer many limitations that can make their predictions inexact. For example, their spatial resolution is poor. That is, they cannot "see" climate-making processes that occur on small scales (43) (44). When several hundred kilometers are shrunk into a single data point, mountain ranges disappear, storm fronts cannot be seen, and Hawaii is invisible.

Such limitations arise in the course of simplifying the overwhelming details of the chemical and physical interactions that go into determining climate. Modelers divide the Earth's surface and the atmosphere above it into a three-dimensional grid of up to 11,000 boxes, each typically several hundred kilometers on a side and a kilometer or more in height. The atmospheric processes within these boxes are then approximated and the temperature, humidity, wind speed, and air pressure are calculated within each layer.

As adjacent boxes are allowed to interact over time, weather patterns develop. But significant detail—mountains and storm fronts, for example—is lost because of the size of the grid. This makes climate models poor predictors of regional climate patterns and explains in part why they are most useful on continental scales or larger (45) (46) (47).

The question of spatial resolution aside, current climate models do not account adequately for several factors that could be crucial in determining the magnitude and rate of greenhouse warming and impacts of climate change such as variations in rainfall, especially on a regional level (48). These factors include the role of the oceans, the behavior of clouds, and the response of living systems, especially plants and soil organisms, to increased CO_2 levels and global warming. A related factor is the prospect of increased natural releases of methane and carbon dioxide as global warming occurs (49) (50).

These factors fall into the category of climate feedbacks—interconnected processes that can amplify or lessen a primary climatic effect, such as the heating caused by adding greenhouse gases to the atmosphere. In some cases, these feedback mechanisms can act as buffers, helping the climate system to remain relatively stable; in other instances, their cyclical action may help drive the climate system to a new equilibrium that may be substantially different from the current state. In any case, they cannot be ignored, but at the present time they also cannot be incorporated into the models satisfactorily.

Box 2.2 Modeling the Effects of Climate Change

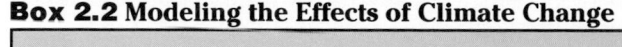

Figure 1 Comparison of Climate Change Models

A Summer Temperature Differences[a]   GFDL Model   GISS Model

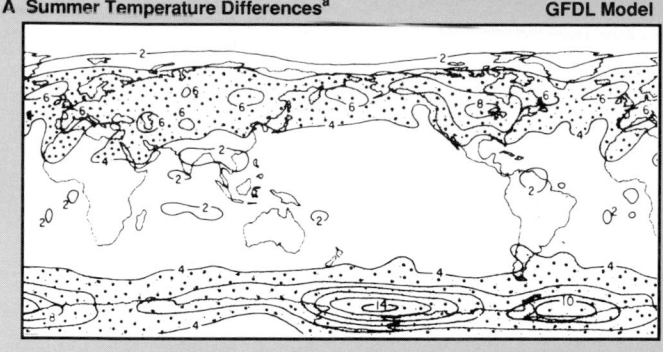
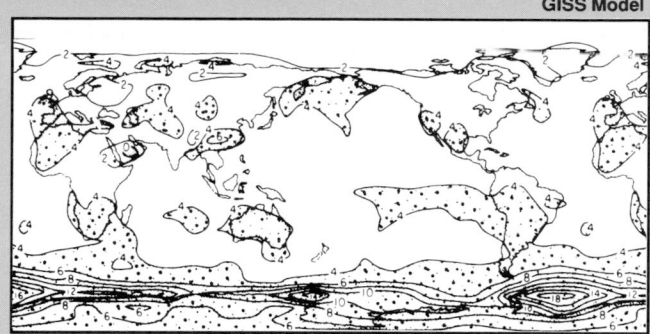

B Summer Soil Moisture Differences[b]   GFDL Model   GISS Model

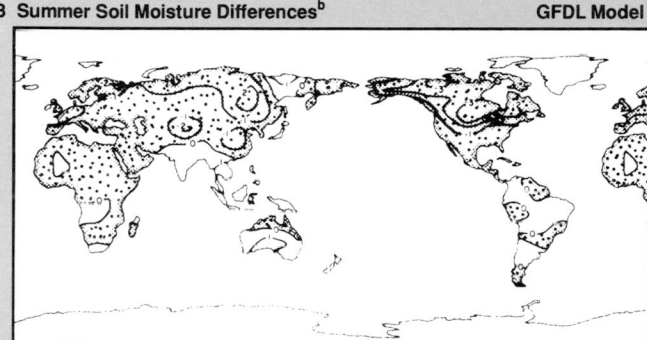
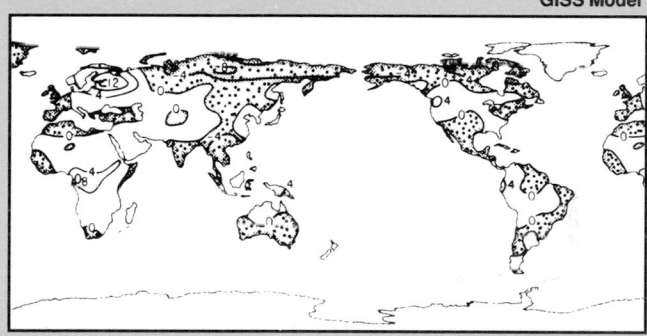

Sources:
1. Geophysical Fluid Dynamics Laboratory, Princeton, New Jersey.
2. Goddard Institute for Space Studies, New York.

Notes:
a. Stippled areas indicate a temperature increase greater than 4° C.
b. Stippled areas indicate a decrease in soil moisture.

Studies of the effects of climate change on wildlife, agriculture, forests, fisheries, and other natural resource systems must start with projections of how temperature, rainfall, soil moisture, and other aspects of climate will change. A common source of these projections is a general circulation model of the atmosphere, which despite its limitations provides a useful starting point for considering what climate change might mean. At least five general circulation models are referred to in the recent scientific literature. Two such models—or variations of them simplified for use in climate studies—are used most often in studies cited in subsequent chapters of this volume: that developed by the Geophysical Fluid Dynamics Laboratory in Princeton, New Jersey (the GFDL model), and that developed by the Goddard Institute for Space Studies in New York (the GISS model).

Although climate projections from these models give roughly similar global results, they differ significantly on a regional scale. Projections of soil moisture under conditions of doubled CO_2 levels, for example, differ significantly in the GFDL and GISS models in many parts of the world. (See Figure 1.) These differences arise because the two models use different means of representing the climatic effects of phenomena such as clouds, which are too small to be modeled individually, or of the oceans, which still are not understood sufficiently.

Thus, regional projections from climate models, while they serve as a useful starting point for impact studies, should be understood more as a set of assumed conditions and not as predictions. Projections from the GISS model, for example, show that summer temperatures would be warmer in parts of western Europe, northern and southern Africa, Australia, Central Asia, central South America, Mexico, and the western United States. Rainfall would be reduced

in many of these areas, leading to drier summer soils in western Europe, northern Africa, parts of South America, Mexico, the northeast and southwest United States, northern Canada, the western Soviet Union, and most of China.

Changes in temperature, rainfall, and soil moisture such as those indicated in climate projections could alter food production (see Chapter 6, "Food and Agriculture," The Effects of Climate Change on Agriculture), shift the natural ranges of forests and wildlife species (see Chapter 7, "Forests and Rangelands," The Effects of Global Warming on Forests and Rangelands and Chapter 8, "Wildlife and Habitat," Focus on Greenhouse Warming and Biodiversity), raise sea levels (see Chapter 11, "Oceans and Coasts," Effects of Climate Change on Oceans and Coasts), and both increase and decrease precipitation. (See Chapter 10, "Freshwater," Focus on Global Climate Change.)

Ocean Dynamics

The ocean system, which covers about three fourths of the planet, is the greatest single climate-feedback system and, perhaps, the greatest source of uncertainty in present models because relatively little is known about ocean dynamics and the intricacies of ocean-atmosphere interactions (51).

The ocean can act initially as a buffer for global climate change because it acts as both a heat sink and, possibly, as a sink for some CO_2 emissions (52) (53). As the climate initially warms, some of the excess heat produced by greenhouse processes is siphoned off to warm the enormous bulk of ocean waters. Thus, the full effect of greenhouse warming is not manifest at first in the atmosphere. Thermal inertia of the oceans could explain why the global warming observed to date is only about half of the 1.0° C warming that climate models predict should have occurred already on the basis of present levels of greenhouse gases (54).

The mixing of ocean waters and their global circulation by ocean currents are key processes governing how the oceans warm and transport heat (55) (56) (57). Heat transport by the oceans plays a crucial role in shaping details of the world's climate. The Earth is heated unevenly by the sun, with the tropics absorbing more solar energy than the polar regions. Ocean currents provide an important means to redistribute this heat; for example, without them, much of Europe would likely be considerably cooler. Indeed, estimates are that the oceans account for 40 percent of the total heat transported from low to high latitudes in the Northern Hemisphere (the rest being transported through the atmosphere by weather systems) (58). Yet, existing climate models do not account for the complexity of ocean-mixing or for the details of deep ocean circulation.

Deep Ocean Currents

The disruption of deep ocean currents by global warming could have profound consequences on the global climate. These currents, which act as a great conveyor belt carrying roughly 20 times the volume of water flowing through all the world's rivers, depend on a delicate balance of salinity and temperature to drive them—a balance that may be shifted by warmer temperatures in the polar regions and consequent changes in the amount of polar ice (59).

Evaporation from prevailing winds across the Atlantic brings rain to Europe and Asia, but raises the Atlantic's salinity. Chilled by arctic cold in the North Atlantic, this salty water increases in density and sinks to the bottom of the ocean, where it travels south in a huge circuit that takes it around Africa and through the Indian Ocean to the Pacific Ocean, there to upwell. The Pacific's less saline waters make their way in a shallower current back along a roughly similar route that skirts several continents, ultimately balancing the oceans' salt budget and warming the European land mass in the process (60).

An infusion of less salty water, as might be brought on by the melting of arctic ice, could cause the deep ocean currents to stall since surface waters could not gain the density needed to sink. Some suggest that this stalling of the ocean current system has happened in the past, leading to dramatic effects on the climate. According to this theory, the rapid changes in climatic conditions that have signaled the end of recent ice ages may have been precipitated by shifts in deep currents (61).

The role of the oceans in taking up or supplying atmospheric CO_2 is another key uncertainty in the attempt to model greenhouse effects. Of the roughly 8.5 billion metric tons of CO_2 emitted yearly by humans from fossil-fuel burning and deforestation, only about half remain in the atmosphere (62). The rest is taken up by the various links in the planet's carbon cycle, including land-based ecosystems and the oceans. The terrestrial biosphere—including all plants alive and dead—contains about 2.5 times as much CO_2 as the atmosphere. But the world's oceans contain about 50 times as much CO_2 as the atmosphere and, each year, over 200 billion metric tons cycle in and out of the oceans through gas exchange at the sea surface (63) (64).

Oceans: Source or Sink?

Currently, the ocean system is considered likely to be one of the primary sinks for CO_2; that is, the oceans absorb more CO_2 than they release. However, that may change in unforeseen ways as global warming proceeds. Even minor changes in ocean temperature, chemistry, or circulation patterns could cause the ocean system to alter its uptake of CO_2, increasing its role as a sink or possibly changing it to a significant source of CO_2—a source with the potential to dwarf human-caused emissions (65). For instance, a change in ocean-mixing patterns, which comprise the limiting factor in ocean CO_2 uptake by transporting CO_2-saturated surface waters to deeper regions, could alter easily the present balance between uptake and release, providing either a positive or negative feedback, depending on how the mixing pattern changed.

Clearly, accurate forecasts of greenhouse-induced climate change demand a more complete understanding of ocean dynamics than we now possess.

Cloud Behavior

The behavior of clouds in future greenhouse scenarios is also a major source of uncertainty in global climate models. Clouds play an important role in regulating the Earth's energy budget. They routinely cover about half the Earth and account for a good deal of the 30 percent of incident sunlight that the Earth reflects into space (66). But clouds do not merely reflect solar radiation from the Earth; they also absorb some of the heat radiated by the Earth and reflect it back to the surface below. In other words, they also exert a greenhouse effect of their own (67). Currently, averaged over the globe, the reflective property of clouds dominates their greenhouse effect and, thus, they exert a substantial, net cooling effect on the Earth.

Both the reflective effect and the greenhouse effect of clouds are large compared with the greenhouse effect exercised by CO_2 and other trace gases in the atmosphere. Recent data from a satellite-borne experiment show that, on a scale where the warming effect from CO_2 doubling is 1, the cloud cooling effect is currently 11 and the cloud warming effect is currently 7. Therefore, the result is a net cooling effect four times larger than the expected warming that would be caused by a doubling of CO_2 (68).

High Clouds Trap Heat, Low Clouds Reflect It

But will the cooling effect that clouds now exercise persist in a warmer world? That depends on the amount and kinds of clouds produced. Cloud generation is governed by atmospheric and ocean circulations and they may well change with global warming (69). Cloud areas, altitudes, and water content all may be affected (70). This is important because cloud altitude (the height of the cloud tops) governs their heat-trapping ability, while their water and ice content determine their reflectivity (71). Thus, high cirrus clouds tend to trap heat well, while low- to mid-level stratus and cumulus clouds tend to be more reflective (72).

If global warming escalates the proportion of high clouds, the current cooling effect might be diminished or shifted to a warming effect, adding to the overall global warming. Conversely, increasing the percentage of low- and midlevel clouds might intensify the cloud cooling effect and oppose the global warming (73). Since the cloud effect—both cooling and warming—is so large relative to the atmospheric greenhouse effect, small changes from current cloud behavior could have profound climatic repercussions. For example, a shift in cloud dynamics during the previous ice age caused by a migration of cooler ocean waters toward the equator may have amplified the global cooling significantly (74). Unfortunately, cloud modeling is not sufficiently advanced to predict with confidence which scenario—heating or cooling—will prevail in a greenhouse world, let alone the magnitude of the effect (75).

Biosphere Response

Predicting how living systems will respond to global warming and what climate feedbacks they will exercise may be just as difficult as solving the riddles of ocean dynamics or cloud behavior and provide still another major source of uncertainty in climate modeling. Terrestrial ecosystems play an important role in the global carbon cycle, storing about 100 billion metric tons of carbon each year in plant materials through the process of photosynthesis and releasing about the same amount through the processes of respiration and decay. This annual flux is equivalent to almost 30 percent of all the carbon held in the atmosphere, so even minor changes in the balance between photosynthesis, respiration, and decay could affect the level of atmospheric CO_2 quickly (76).

Fertilization Effect of Higher CO_2 Levels

Anticipating how this balance will change is complicated by the many competing influences that could pertain under greenhouse conditions. For instance, photosynthesis rates are known to be sensitive to atmospheric CO_2 levels. The effect known as CO_2 fertilization can cause an average 30 percent increase in growth rates in many, but not all, species when CO_2 concentration is doubled (77) (78). Recent data suggest that this effect is temperature-dependent and could be increased up to threefold by a 3° C rise in air surface temperatures (79).

Further, there is evidence that CO_2 enrichment helps some plants minimize water loss, presumably making them more drought-tolerant—an important consideration if the drier conditions predicted by climate models for many midcontinent areas come to pass (80). If all these favorable effects of greenhouse conditions were realized, a net increase in plant matter might increase the carbon uptake of the biosphere and act as a negative climate feedback, helping to alleviate further CO_2 buildup. Indeed, there is evidence that the annual carbon uptake in the Northern Hemisphere has increased during the growing season in recent years, perhaps indicating greater carbon storage (81).

But many believe that such an optimistic scenario is unlikely. Plants may not be able to take full advantage of the CO_2 fertilization; in the wild, lack of nutrients and sufficient water may well provide the limiting factors for plant growth, not the level of CO_2. In addition, respiration and decay rates respond strongly to temperature changes, with a 1° C change influencing the respiration rate up to 30 percent. Indeed, the increased carbon uptake in the Northern Hemisphere may be matched by an equal increase in respiration, leading to no net gain in storage (82).

In fact, there is deep concern that greater respiration and decay will overwhelm any increases in photosynthesis. Not only will higher temperatures mean greater CO_2 production, but also they will mean greater releases of CH_4. Dead plant matter, particularly in tundra regions, represents a large reservoir of carbon. Cold temperatures have kept this matter from rotting, leading to its accumulation as peat beneath the tundra vegetation. As temperatures rise, this matter will begin to decay at much faster rates, possibly leading to a surge in the release of methane, which is produced as organic matter breaks down in the boggy, oxygen-poor, tundra soils (83).

This potential rise in methane production is particularly worrisome. Methane is 20–30 times more efficient than carbon dioxide at trapping heat, and its concentration in the atmosphere has been increasing more than 1 percent per year. Within 50 years, it may be the most significant greenhouse gas (84) (85). Large methane releases from a warmer tundra—or from other sources sensitive to warming, such as deep ocean deposits—could create a major climate feedback.

Table 2.4 Contributions to Global Warming by Greenhouse Gases and Human Activity

(percent of global warming)

Sector	Carbon Dioxide	Methane	Ozone	Nitrous Oxide	Chlorofluorocarbons (and others)	Percent Warming by Sector
Energy						
Direct	35	3	X	4	X	49
Indirect	X	1	6	X	X	
Deforestation	10	4	X	X	X	14
Agriculture	3	8	X	2	X	13
Industry	2	X	2	X	20	24
Percent Warming by Gas	50	16	8	6	20	100

Sources:
1. V. Ramanathan, R.J. Cicerone, H.B. Singh, *et al.*, "Trace Gas Trends and their Potential Role in Climate Change," *Journal of Geophysical Research*, Vol. 90, No.. D3 (1985) pp. 5547-5566.
2. William Moomaw, Director, Center for Environmental Management, Tufts University (personal communication).
x = not available.

Species Would Migrate Poleward

Perhaps the most severe disruption of the terrestrial carbon cycle could come as result of the rapidity of the expected global warming. One consequence of this warming will be a movement toward the poles of the range of many plant species. But will plants be able to migrate poleward as fast as their ranges shift? Species migration is usually a rather slow affair, especially for many long-lived tree species. However, a rapid rise of 2°-5° C in global temperatures in the next 100 years could well translate into a 500-1,000-kilometer range shift poleward—a distance that many species will not be able to travel. This may result in the gradual extinction of many forest communities and, perhaps, a concomitant decrease in carbon storage rates (86) (87). (See also Chapter 7, "Forests and Rangelands," Possible Effects of Global Warming on Forests and Rangelands and Chapter 8, "Wildlife and Habitat," Focus on Greenhouse Warming and Biodiversity.)

Other Uncertainties

In addition to the questions surrounding ocean, cloud, and plant responses to global warming, there are other scientific uncertainties as well. Climate models do not describe sufficiently such important factors as the variation in solar output, the influence of volcanic and human-caused aerosols, the Earth's reflectivity, and the influence of soil chemistry on the carbon cycle, as well as several other minor factors (88).

In spite of these many deficiencies, today's climate models provide a first estimation of global climate change with enough internal consistency and apparent relation to reality that they cannot be ignored, though their detailed predictions will require years of refinement.

POLICY OPTIONS AVAILABLE TO ADDRESS GLOBAL WARMING

The responses of human societies to global warming can take several approaches. First, prevention strategies can be employed to reduce the quantities of greenhouse gases being emitted. Mitigation mechanisms strategies can attempt to compensate for emissions that do occur; for example, through reforestation policies that increase the uptake of CO_2 from the air by trees and other plants. Lastly, strategies can be employed that help communities and nations adapt to changes in climate and their consequences. In practice, all three types of policies are likely to be important. Here, we discuss a range of prevention and mitigation policies.

Table 2.4 translates our knowledge of the science of greenhouse gas emissions into the categories of human activities through which policy could be brought to bear on the problem of global warming. The relative contributions may well shift in the future if the provisions in the Montreal Protocol to Control Substances that Deplete the Ozone Layer are implemented successfully or if more efficient and alternative energy options are adopted.

Although there are substantial uncertainties underlying several of the estimates in Table 2.4, the data provide a clear sense of where policy leverage is likely to be the greatest. The highest potential lies with policies targeted at the energy and industrial sectors. Nevertheless, virtually all elements of human activity contribute to greenhouse gas emissions. The fact that nearly one quarter of the potential warming arises from the forestry and agricultural sectors suggests that each of these sectors should be examined on a country-by-country basis to determine which strategies are most readily implementable at local and national levels. By the nature of the patterns of greenhouse gas emissions, reducing these emissions will be an incremental process.

It is worth noting that the problem of global warming is intimately linked to other, serious environmental problems involving the atmosphere, in particular those of acid precipitation, urban smog, and depletion of the stratospheric ozone layer. The problems are linked chemically because, once released, many of the pollutants that cause trouble interact in complex and synergistic ways within the atmosphere or play a role in more than one problem. They are linked economically because it is often the same human activities that release the pollutants responsible for all three problems. And they are linked in a policy sense because policies designed to attack one problem—by altering an economic activity to reduce emission of the pollutant responsible, for example—inevitably will affect other problems as well (89). A policy designed to produce automobiles that use less fuel, for example, would reduce not only CO_2 emissions, but also may reduce the oxides of nitrogen (NOx) emissions that contribute to acid precipitation, urban smog, and global warming.

Prevention: Reducing the Risk of Global Warming

While work should proceed simultaneously on mitigation and adaptation strategies, prevention deserves the highest priority. Preventing the emission of greenhouse gases that would be released into the atmosphere not only delays the onset of significant global warming, but also slows its advance and reduces its ultimate magnitude.

To reduce greenhouse gas emissions or even to slow their rate of growth on a global scale will require an extraordinary degree of political consensus. As the greenhouse gas indicators discussed earlier in this chapter make clear, no one country or even one region can prevent the buildup of greenhouse gases by itself, although leadership by individual countries will be important in achieving global consensus. To be broadly acceptable, ideally, prevention policies should confer other local benefits on their adopting countries in addition to the worldwide benefit of reducing the risk of a global warming.

Many policies have been proposed and are under study. It seems clear, however, that any successful prevention strategy will include five key elements:
■ increasing the efficiency of energy production and use,
■ switching from carbon-intensive fuels such as coal to hydrogen-intensive fuels such as natural gas, where possible,
■ encouraging the rapid development and use of solar and other carbon-free energy sources,
■ eliminating the production of most CFCs and developing the means to recapture those now in use,
■ reducing the rate of deforestation (90).

Energy Efficiency

Increased energy efficiency appears to be the single most promising option for reducing the risk of global warming (91). Higher efficiency in energy production and in a wide range of energy uses is the fastest and most cost-effective method for reducing emissions of carbon dioxide and other energy-related greenhouse gases. The replacement of a single 75-watt incandescent bulb with an 18-watt fluorescent bulb, for example, provides the same light but prevents the emission of over 100 kilograms of carbon as CO_2 over the life of the new bulb. Replacing a car that burns 10 liters (l) of gasoline per 100 kilometers (km) with one that is twice as efficient would prevent the emission of more than five metric tons of carbon over a 10-year period. Experience during the past 15 years in the United States, Western Europe, Japan, the People's Republic of China, Korea, the Republic of China and many other countries has shown that large improvements in energy efficiency are possible without a detrimental effect on the rate of economic growth. The evidence suggests that economic growth is not tied inexorably to increased energy consumption, as was once thought. A recent study published by the World Resources Institute suggests that an energy strategy based on increased efficiency can sustain economic growth in both developing and industrialized countries (92).

One way to characterize the prevention potential is to compare the energy-efficiency of different national economies, which vary markedly. The United States, for example, is only half as energy efficient as Western Europe or Japan in generating a unit of gross domestic product. On the other hand, it is twice as efficient as the People's Republic of China (93) and more efficient than the Soviet Union and most of Eastern Europe. Many newly industrializing countries are the least efficient of all. Although a large and geographically heterogeneous country is unlikely to match the energy efficiency achievable in some smaller countries, the current variation is still remarkable and suggests the scope of the energy and CO_2 savings that are possible.

Policies that promote higher efficiency are particularly timely for consideration by international development agencies because many developing countries are planning major expansions and modernizations of their energy sectors. A 1985 study by the Companhia Energetica de São Paulo, Brazil (the world's largest electric utility) suggested that a $2 billion investment in cost-effective improvements in electricity efficiency could offset the need for 22 gigawatts of new generating capacity estimated to cost at least $44 billion by the year 2000 (94). Such investments have many arguments in their favor, quite apart from countering the threat of global warming.

More Efficient Lights, Stoves and Cars

End-use efficiency options range from the introduction of more efficient cooking stoves, the installation of more efficient light bulbs in homes and commercial buildings, the construction of better-insulated buildings, and the manufacture of more fuel-efficient vehicles. In each area, the best technology available today consumes at least 50 percent less energy (and, in most instances, produces at least 50 percent less CO_2) than the technology most commonly used (95). For example, new jet engines and composite materials for aircraft construction offer the promise of advanced planes that can fly longer distances while burning half the fuel of current models. New car designs by various Japanese and European manufacturers show that cars that use less than 4 l/100 km can offer safe, attractive, and efficient transportation for four or five passengers—compared with the 9 l/100 km efficiency of the average, new U.S. automobile (96). Considering the rapid growth of the world's automobile fleet and its growing role as a source of carbon dioxide and urban air pollution, higher efficiency in cars and light trucks is particularly important. (See the sections on the automobile in Chapter 9, "Energy" and Chapter 12, "Atmosphere.")

More Efficient Power Generation

On the supply side, new power generation technologies offer significantly higher energy conversion efficiencies than have been available in the past. The spread of cogeneration offers some of the greatest potential savings. Cogeneration systems convert primary energy in the fuel to electricity and then use the otherwise-wasted heat to make steam or hot water for local processes requiring thermal energy. Such systems can deliver between 60–80 percent of the heat in the fuel to useful purposes, compared with about 33 percent for conventional steam-electricity technologies. Cogeneration has many important applications in industry. In the U.S. paper industry, for example, almost all process steam and heat requirements are met through the use of cogeneration systems. In Brazil, cogeneration systems in alcohol distilleries provide revenue from electricity sales as well as heat for distillation and purification processes. In Germany and Scandinavia, cogeneration systems are used widely to supply steam or hot water to heat nearby buildings and provide power.

Other Greenhouse Gases

Improvements in energy efficiency also would have significant effects on the emissions of other greenhouse gases, in addition to carbon dioxide. Combustion of fossil fuels is a major source of NOx, which contributes to the formation of tropospheric ozone (97). Combustion of fossil fuels also produces carbon monoxide (CO), which reacts with and depletes the atmosphere of a reactive chemical known as the hydroxyl radical, and which otherwise would break down and remove methane molecules. Thus, carbon monoxide emissions lengthen the atmospheric lifetime of methane, an extremely potent greenhouse gas (98) (99). Reductions in fuel use through more effi-

cient energy technologies may reduce emissions of N_2O, NOx, and CO, resulting in less tropospheric ozone and slower growth of methane levels. Greater care and efficiency in the extraction and transport of natural gas (which is more than 90 percent methane) to reduce leaks into the atmosphere also would slow the growth of methane levels.

By pursuing the efficiency opportunities discussed above, it appears possible to continue economic growth and to maintain high standards of living while stabilizing or reducing per capita energy use in industrialized countries. Such a strategy would allow economic growth in developing countries to proceed vigorously without significantly increasing global greenhouse gas emissions, provided these countries also adopt efficient, low-emitting technologies.

Fuel-Switching

Because the carbon content of fossil fuels varies over a wide range, one option for reducing CO_2 emissions is to emphasize reliance on the less carbon-intensive fuels. Coal, for example, produces nearly twice as much CO_2 per unit of energy as natural gas and 1.5 times as much as oil (100). If gasified prior to combustion, coal emissions of CO_2 can be three times as high as natural gas because of the energy requirements of the gasification process.

If combined with a move to the more efficient technologies described above, fuel-switching can provide major benefits in reducing CO_2 emissions. By shifting electricity production from conventional, coal-fired, steam turbines to advanced gas turbines, CO_2 emissions per electrical kilowatt-hour can be reduced by 60 percent (50 percent from the coal-to-gas conversion and an additional increment from the higher conversion efficiencies offered by advanced gas turbines). Care must be taken to reduce leakage of natural gas from production facilities and pipelines, however, because the benefits of fuel-switching can be negated by a methane leakage rate as seemingly modest as 3–4 percent. This is because methane is 20–30 times as powerful a greenhouse gas as CO_2 on a molecule-per-molecule basis.

Switching from oil-based fuels such as petrol to methanol made from natural gas or ethanol made from biomass also can reduce CO_2 emissions. Such a policy has been pursued in Brazil, where the majority of cars burn alcohol-based fuels made from sugar cane, and has been proposed on a limited basis in the United States. Fuel-switching policies must be evaluated carefully, however, to ensure that sufficient, long-term supplies of the appropriate fuels exist or can be provided without creating additional environmental problems. If a significant portion of the U.S. automobile fleet were to be converted to the use of methanol fuels, it is likely that the methanol would be made from coal eventually, rather than from natural gas, in order to ensure long-term domestic supplies. However, coal-based fuel would increase CO_2 emissions significantly compared with present fuel sources (101).

Fuel-switching is a useful, near-term strategy. But the world's resources of natural gas and oil are small compared with coal deposits and may be largely consumed before the end of the next century. That would leave coal—the most damaging fuel from a global warming perspective—as the primary fossil energy source. For countries with large coal deposits and little natural gas or oil—such as China, Poland, and India—the option of switching to other fossil fuels might not be very feasible. For these countries in the near future and for virtually all countries when readily available sources of oil and natural gas have been consumed, the only realistic alternative to coal will be biomass or carbon-free energy sources.

Carbon-Free Energy Sources

Energy sources and fuels that produce no CO_2 are ideally suited to reduce the risk of global warming. To the extent that biomass energy is produced in a renewable manner—with growth of new biomass equaling that which is burned—no net CO_2 is added to the atmosphere. Such technologies are in operation already in many applications around the world, although in many cases they are not yet economically competitive with fossil fuels. Some carbon-free energy sources—such as hydroelectric dams and conventional nuclear power stations—raise serious environmental questions of their own. Nonetheless, hydroelectric, wind, solar thermal, photovoltaic, and nuclear technologies deserve additional attention. As the eventual alternative to coal-based energy, carbon-free energy sources occupy a unique strategic role.

Encouragingly, significant progress is being made in developing these carbon-free energy sources. Hydropower is a major energy source now and can be expanded further, although the environmental costs of large dams are receiving more careful scrutiny. Wind turbines, both large and small, are in use in Denmark, the United States, Australia, and the United Kingdom. Prototype, solar thermal plants to provide electricity are in operation in a number of countries, and solar thermal installations for heating water are widespread. Geothermal energy is finding applications in a number of countries. As the costs of production decline and the costs of conventional fuels rise, these technologies can make an ever larger contribution to global energy. Solar thermal electricity and wind machines are already cost-competitive in some parts of the United States. These developments should be encouraged.

Solar photovoltaic systems are finding applications in remote areas, but are still too costly for central station applications. Technical progress is rapid, however, and may lead to photovoltaic power becoming economically competitive with conventional electricity sources within a decade (102). (See also Chapter 9, "Energy," Photovoltaic Power.) If sufficient investments in research and development are made in the

next decade, these small modular systems could become a major source of energy in the early part of the next century. Therefore, policies that promote research and early commercialization deserve wide support.

Nuclear power's future in the global energy picture remains quite uncertain. Commercial nuclear power plants produce a substantial portion (although still less than 10 percent) of the world's electricity. But, because of public concern about safety and radioactive waste disposal and commercial concern about high capital costs, long construction times, and financial risks, construction of nuclear plants has slowed or halted in all major markets except France. There is growing interest, however, in a new generation of "passively safe" nuclear technologies, some of which also offer the benefits of small-scale, modular units of standardized design that can be factory-produced under controlled conditions. In view of the long-term relevance of nuclear power as a potential alternative to coal for baseload power generation, these new nuclear technologies deserve careful attention and increased research and development efforts. (See Chapter 9, "Energy," Nonfossil Energy Sources: A Look Ahead.)

Eventually, hydrogen may prove to be an important carbon-free fuel for transportation use. Hydrogen is an energy carrier and must be produced by a primary energy source through, for example, the electrolysis of water. Hydrogen has the advantage that its combustion products are primarily water vapor, but as yet it is expensive to produce and difficult to store. Prototype automobiles that run on hydrogen have been demonstrated, however, and a recent study proposes a strategy by which photovoltaic power facilities could produce hydrogen for use in automobiles in some areas of the United States within a couple of decades (103). This option deserves further study.

Modifying Industrial Practices to Reduce CFC Emissions

Although the industrial sector can be seen as encompassing many of the activities described under the energy sector, it has one important unique component: chlorofluorocarbons, which, molecule-for-molecule, are the most potent greenhouse gases known. Reducing production and controlling release of these substances in the immediate future is critical for reducing the risk of global warming.

International cooperation to limit CFC emissions is well under way in response to the role played by CFCs in depleting the stratospheric ozone layer. The *Montreal Protocol to Control Substances that Deplete the Ozone Layer*, agreed in 1987, is intended to reduce the production and use of CFCs in the industrialized countries by 50 percent from 1986 levels before the year 2000. However, developing countries are permitted to increase CFC use to 0.3 kilograms per person. While it is hoped that CFC substitutes can be deployed in the developing world as well, unless management practices, substitutes, and new technologies are implemented universally, the anticipated 50-percent cut in CFC production easily could become a 50-percent increase, even if all nations comply with the provisions of the Protocol. On the other hand, reductions to essentially zero by the year 2000 are now being proposed by many countries in order to protect the ozone layer; such cuts would prevent CFCs and related compounds from becoming an even larger component of the global warming problem. Unfortunately, these chemicals are among the longest-lived of all the greenhouse gases, residing in the atmosphere for a century or more.

A number of options exist for reducing rapidly the most widely used CFCs to the level specified in the Montreal Protocol. The first strategy involves incentives (or penalties) to improve equipment maintenance and to encourage the more efficient use of CFCs in applications such as residential refrigeration. In the process of manufacturing foam products, substantial potential exists for recapture, rather than release to the atmosphere, of the blowing agents. Enclosing process lines, aging the product before shipment, and controlling spray heads carefully can reduce emissions in manufacturing substantially and may allow recapture of the dangerous gases. Several recent advances have occurred in the use of CFC-based solvents in the electronics industry. Some corporations have been highly successful in minimizing solvent losses by enclosing their production lines and passing the captured gases through a bed of absorbent carbon. Similar opportunities in solvent applications exist throughout the electronics industry.

The second option involves policies that encourage recapture and recycling of these compounds rather than their release and periodic replacement. It would be simple and cheap to require the installation of service valves in the cooling loops of most devices so that the refrigerant could be removed and captured either during routine servicing or at the point of disposal. In the case of automobile air conditioners, higher quality hoses and fittings would reduce leakage significantly, and it is now possible to capture and recycle the air conditioning cooling fluid rather than replacing it whenever the unit is serviced. For those applications where the CFCs can be recaptured but for which recycling is not feasible or is prohibitively expensive, careful disposal techniques are essential.

Slowing Deforestation and Altering Agricultural Practices

As shown in Table 2.4, activities in the agricultural and forestry sectors contribute significantly to greenhouse gas emissions. The largest emissions arise from land use changes, principally deforestation, which release large quantities of CO_2 and methane to the atmosphere as well as some N_2O and ozone. Therefore, reducing and ending deforestation ulti-

mately is an immediate priority that would reduce emissions of two principal greenhouse gases dramatically. By suspending tax credits that encouraged deforestation as a means of establishing title to land in the Amazon region, Brazil has decreased the rate of deforestation sharply in the past two years, although the rate remains high. (See Chapter 7, "Forests and Rangelands," Deforestation in Brazil.) Overall, however, deforestation rates in many key, tropical forest countries have accelerated during the 1980s. Consequently, there exist opportunities for significant, near-term reductions in the risk of global warming.

Some agricultural practices release CO_2 and N_2O through energy inputs, fertilizer inputs, and soil erosion. Enclosing manure piles or feedlots would allow the capture and use of methane as a commercial-quality fuel, rather than allowing it to become an additional atmospheric burden. New strains of rice may not require the flooding now needed to control weed growth, thus reducing methane production significantly. Revising agricultural practices to reduce fertilizer requirements would reduce N_2O emissions, no-till agriculture can stabilize carbon in the soil rather than releasing it to the atmosphere through erosion, and improved productivity reduces pressure for deforestation, which currently emanates from the need to rotate continually among poorly performing agricultural plots.

Each of these policy options has local benefits besides slowing global warming. There is some evidence, for example, that tropical forests in regions of poor soils have greater economic value for natural products harvested on a sustained basis than as farmland (104). Indeed, significant resources are being dedicated to making progress in these areas (see Chapter 7, "Forests and Rangelands," Deforestation and Development), although such progress is woefully slow. The threat posed by global warming should encourage a rededication of effort and resources to these areas.

Mitigating the Effects of Greenhouse Gas Buildup

The agricultural and forestry sectors offer one of the few opportunities for actively mitigating and slowing global warming. This option consists of using biotic processes to remove carbon from the atmosphere and to store it over the long-term in vegetation, in soils, or in long-term products such as building materials. Forests and soils play a key role in the global carbon cycle, serving as a large sink for carbon that otherwise would have to be distributed between the oceans and the atmosphere. Policy options such as reforestation and soil carbon enhancement can increase the net size of the biotic carbon sink, and can help to stabilize atmospheric carbon dioxide concentrations.

Just how much of an offset to carbon emissions is possible through reforestation is not certain. The size of the biotic carbon sink varies widely by the type of forestry, the rotation times of the trees being planted, and the ultimate disposition of the wood being grown. The area of new forest required merely to offset current CO_2 emissions for the next 40 to 50 years probably exceeds 1 billion hectares. Clearly, net reforestation of this magnitude is unlikely to occur, and global warming itself may harm some of the world's forests seriously. Whether the potential for increased uptake of CO_2 by green plants is large enough to offset significantly the rate of emissions growth from fossil fuel combustion, or to reverse the historical trend of increasing concentrations, remains to be determined. Some analysts are skeptical that significant additions to the world's forest stocks are possible and suggest that neither more intensive forest management nor reduced industrial harvests would have more than a very modest impact on the atmospheric carbon buildup (105). Other analysts are more optimistic, pointing to the possibility of expansion of northern boreal forests as northern temperatures and rainfall increase and higher CO_2 levels stimulate more rapid growth rates (106). It is most likely that such approaches can help to slow the buildup of CO_2 during a transition period in which fossil-fuel-related, greenhouse gas emissions are reduced. Estimating the potential of individual forestry projects will require the evaluation of a number of strategic factors, among them land availability and suitability, the growth rates likely to be achieved, and the many implementation uncertainties that can interfere with such projects. Clearly, more study of these policy options warrants a high priority. In addition to mitigating the effects on global warming, reforestation would confer local benefits in many developing countries, such as reducing erosion and increasing supplies of firewood. (See Chapter 7, "Forests and Rangelands," Possible Effects of Global Warming on Forests and Rangelands and Box 7.5.)

Institutional Issues and Policy Instruments

Several conclusions can be drawn from the preceding analysis of the scientific and political issues surrounding global warming:

■ The threat posed by global climate change to human and natural systems is unparalleled in recorded history and means that the community of nations making up the planet must confront the likelihood of rapid and significant global warming.
■ Technological and policy decisions made by nations in the relatively near term will affect substantially the timing and severity of any global warming that occurs.
■ The technologies needed to prevent, reduce, or mitigate substantially further human impacts on the atmospheric concentrations of the gases contributing to global warming exist or are likely to exist soon, although they are not disseminated widely.

If these conclusions are accepted, it becomes evident that the challenge facing nations is fundamentally of an institutional, rather than technical, character. Institutional issues that must be addressed range from the structure of east-west and north-south trade, patterns of developing country debt, and facilitating the transfer of new technologies, to restructuring the taxation and subsidy policies that steer energy investment in many countries, agricultural and forestry policy, development, and consumption patterns. (See Chapter 14, "Policies and Institutions.") Ranging across these issues are many institutional impediments to addressing global warming.

While the mandates of many national and international institutions are related tangentially to the threat posed by global warming, few are charged with responding to the threat in any way. Energy departments and ministries around the world, for example, may have extensive responsibilities in the areas of energy exploration, development, and consumption without being charged with evaluating and responding to broader threats such as that of climate change. This lack of a mandate to deal comprehensively with energy or other issues often results in policy decisions that actively aggravate, rather than offset, these broader threats.

In many countries, tax incentives that underwrite with public money the exploration and development of fossil fuel resources, for example, subsidize the artificially low prices paid for fossil fuels. International trade policies frequently have the unintended impact of compounding, rather than ameliorating, the severity of human impacts on environmental systems. Until recently, international institutions charged with promoting economic development in developing countries often have ignored the environmental implications of their actions, promoting indirectly activities such as tropical deforestation that contribute to global warming.

This suggests that a significant restructuring of organizational mandates will be required at all levels of government. Responsibility for evaluating and responding holistically to the threat of climate change, or at least coordinating the evaluation and response processes, must be assigned to specific institutions. In turn, these institutions must seek to rationalize the policy processes operating at other levels of government and in other agencies. This will require comprehensive education not only of governmental bureaucracies, but also of individual citizens.

Need for National Action

The realities of international cooperation and action suggest that the institutional infrastructures most prepared and most able to address and regulate the emission of greenhouse gases are agencies at the national level. Notwithstanding the international ramifications of global warming and the fact that solving the problem over the long term cannot be achieved without organized international cooperation, national action remains the obvious point for immediate policy intervention. International coordination and cooperation will follow much more easily once national actions reflect the existence of consensus on the need for action and a political willingness to take action.

Therefore, national governments should move to adopt policies of global warming prevention, mitigation, and adaptation on a continuing basis, without waiting for formal international agreements.

International action in the short term should focus on the activities under way through the Intergovernmental Panel on Climate Change, sponsored by the United Nations Environment Programme and the World Meteorological Organization. Efforts to understand better both the effects of and possible responses to global warming are pivotal to developing the type of international consensus that will have to precede formal international cooperation. Within this category, multilateral research efforts deserve a high priority.

An additional vehicle for action with a clear international flavor is available immediately: namely, bilateral and multilateral development banks and other development assistance institutions that can assess the contributions their lending and foreign assistance programs are making to these problems.

A Global Bargain

In the longer term, considerable attention should be devoted to the components of a "global bargain" that will allow an equitable sharing of responsibilities for addressing the problem. Global warming is not the only policy problem facing human societies that is characterized by massive uncertainty and equally massive risk. In evaluating national defense needs, for example, countries also are confronted with the possibility of future destruction, with the timing and likelihood of such destruction remaining uncertain. Nevertheless, countries invest huge resources in national defense programs, even though many mechanisms exist to help prevent war, ranging from a fear of mutually assured destruction to memories of the horrors of past wars. Such mechanisms are not available to help avoid global warming. Climate change will result from societal actions taken incrementally, largely in ignorance, and with no past history from which to learn.

If climate change is to be avoided, the world cannot wait to begin responding to the threat of global warming until the precise character and timing of the threat are known.

There is no single solution to the complex challenge posed by global warming. To one degree or another, virtually all elements of human societies are involved in creating the problem. All must play a role in bringing it under control.

The introductory section on Status of Current Knowledge and parts of the concluding section on Policy Options Available to Address Global Warming were originally drafted by Mark Trexler, an associate with the Program in Climate, Energy, and Pollution of the World Resources Institute; the section on Greenhouse Gas Emissions was written by Allen L. Hammond, editor-in-chief of World Resources 1990-91; and the section on the State of the Science was written by Gregory Mock, a science writer based in Ben Lomond, California. William Moomaw, director of the Center for Environmental Management, Tufts University, served as a principal consultant for the chapter and assisted in the construction of the Greenhouse Index.

References and Notes

1. A. John Arnfield, "Greenhouse Effect," *The Encyclopedia of Climatology*, Encyclopedia of Earth Sciences Series, Vol. 11, John E. Oliver and Rhodes W. Fairbridge, eds. (Van Nostrand Reinhold Company, New York, 1987), pp. 463-464.
2. James I. Hansen, Fung, A. Lacis, *et al.*, "Prediction of Near-Term Climate Evolution: Can We Tell Decision-Makers Now?" in the Climate Institute, *Preparing for Climate Change: Proceedings of the First North American Conference on Preparing for Climate Change* (Government Institutes, Inc., Rockville, Maryland, 1988), Figure 1, pp. 35-36.
3. V. Ramanathan, R.J. Cicerone, H.B. Singh, *et al.*, "Trace Gas Trends and Their Potential Role in Climate Change," *Journal of Geophysical Research*, Vol. 90, No. D3 (1985), pp. 5547-5566.
4. L. Donner and V. Ramanathan, "Methane and Nitrous Oxide: Their Effects on the Terrestrial Climate," *Journal of Atmospheric Science*, Vol. 37 (1980), pp. 119-124.
5. J.M. Barnola, D. Raymond, Y.S. Korotkevich, *et al.*, "Vostok Ice Core Provides 160,000-Year Record of Atmospheric CO_2," *Nature*, Vol. 329, No. 6188 (1987), pp. 408-414.
6. *Ibid.*, Table 1, p. 409.
7. Richard A. Houghton and George M. Woodwell, "Global Climatic Change," *Scientific American*, Vol. 260, No. 4 (1989), p. 39.
8. *Ibid.*, pp. 39-40.
9. *Op. cit. 3.*
10. T.M.L. Wigley and M.E. Schlesinger, "Analytical Solution for the Effect of Increasing CO_2 on Global Mean Temperature," *Nature*, Vol. 315, No. 6021 (1985), p. 649.
11. R.E. Dickinson, "How Will Climate Change? The Climate System and Modelling of Future Climate," in Bert Bolm Bolin, *et al.*, eds., *The Greenhouse Effect, Climate Change, and Ecosystems: SCOPE 29* (New York, John Wiley and Sons, 1986), pp. 221-251, 256-262.
12. Because different greenhouse gases have radically different temperature-forcing potentials, the equivalent of doubling CO_2 concentrations is not a doubling of all greenhouse gases. It is possible, however, to formulate an atmospheric composition in which the incremental warming effect of all greenhouse gases combined is equivalent to the warming effect of doubling CO_2 concentrations alone.
13. National Research Council, *Changing Climate: Report of the Carbon Dioxide Assessment Committee* (National Academy Press, Washington, D.C., 1983), Figure 113, p. 35.
14. World Meteorological Organization, *Report of the International Conference on the Assessment of the Role of Carbon Dioxide and of Other Greenhouse Gases in Climate Variations and Associated Impacts* (World Meteorological Organization, Geneva, 1986).
15. Susan Subak, Harvard University, "Accountability for Climate Change," unpublished paper, 1989.
16. Greg Marland, T.A. Boden, R.C. Griffin, *et al.*, "Estimates of CO_2 Emissions from Fossil Fuel Burning and Cement Manufacturing Using the United Nations Energy Statistics and the U.S. Bureau of Mines Cement Manufacturing Data," Oak Ridge Na-

tional Laboratory, Oak Ridge, Tennessee, 1988.

17. J.F. Richards *et al.*, "Development of a Data Base for Carbon Dioxide Releases Resulting from Conversion of Land" (Institute for Energy Analysis, Oak Ridge, Tennessee, 1983). Note: This estimate of land clearing does not take into account the acceleration of deforestation in many tropical forest countries in recent years.

18. Estimates based on radiative forcing estimates from V. Ramanathan, R.J. Cicerone, H.B. Singh, *et al.*, "Trace Gas Trends and Their Potential Role in Climate Change," *Journal of Geophysical Research*, Vol. 90, No. D3 (1985), pp. 5547-5566; and concentrations from Office of Policy, Planning, and Evaluation, Office of Research and Development, U.S. Environmental Protection Agency, *Potential Effects of Global Climate Change on the United States*, Joel B. Smith and Dennis A. Tirpak, eds. (U.S. Environmental Protection Agency, Washington, D.C., draft, 1988).

19. Stephen H. Schneider, "The Greenhouse Effect: Science and Policy," *Science*, Vol. 243, No. 4892 (1989), p. 771.

20. Bette Hileman, "Global Warming," *Chemical and Engineering News* (March 13, 1989), p. 25.

21. *Op. cit.* 2., p. 35.

22. V. Ramanathan, Bruce R. Barkstrom, and Edwin F. Harrison, "Climate and the Earth's Radiation Budget," *Physics Today* (May 1989), p. 23, 28.

23. *Op. cit.* 7, pp. 39-40.

24. *Op. cit.* 7, p. 39.

25. Richard A. Houghton, Senior Scientist, Woods Hole Research Center, Woods Hole, Massachusetts, July 1989 (personal communication).

26. James Hansen and Sergej Lebedeff, "Global Surface Air Temperatures: Update Through 1987," *Geophysical Research Letters*, Vol. 15, No. 4 (1988), p. 323.

27. P.D. Jones, T.M.L. Wigley, C.K. Folland, *et al.*, "Evidence for Global Warming in the Last Decade," *Nature*, Vol. 323, No. 6167 (1988), Scientific Correspondence, p. 790.

28. James Hansen and Sergej Lebedeff, "Global Trends of Measured Surface Air Temperature," *Journal of Geophysical Research*, Vol. 92, No. D11 (1987), pp. 13,345-13,372.

29. *Op. cit.* 7, p. 37.

30. *Op. cit.* 19, pp. 771-772.

31. *Op. cit.* 28, p. 13,371-13,372.

32. J.K. Angell and J. Korshover, "Global Temperature Variations in the Troposphere and Stratosphere, 1958-1982," *Monthly Weather Review*, Vol. 111, No. 5 (1983), pp. 901-921.

33. Richard A. Houghton and George M. Woodwell, "Global Climatic Change," *Scientific American*, Vol. 260, No. 4 (1989), p. 38. Note: More recent data from the British Meteorological Office adds 1989 to the list as the fifth warmest year on record.

34. R. Monastersky, "Scientist Says Greenhouse Warming Is Here," *Science News*, Vol. 134 (July 2, 1988), p. 4.

35. *Op. cit.* 7, p. 38.

36. *Op. cit.* 20, p. 27.

37. *Op. cit.* 20, pp. 30-31.

38. *Op. cit.* 20, pp. 30-31.

39. Stephen H. Schneider, "Climate Modeling," *Scientific American*, Vol. 256, No. 5 (1987), pp. 72-74.

40. *Op. cit.* 20, p. 31.

41. *Op. cit.* 7, p. 38.

42. *Op. cit.* 19, pp. 175-176.

43. *Op. cit.* 39, p. 74.

44. *Op. cit.* 20, p. 30.

45. *Op. cit.* 20, pp. 30-31.

46. William Booth, "Computers and 'Greenhouse Effect': The Genesis of Understanding," *The Washington Post* (June 12, 1989), p. A-3.

47. *Op. cit.* 39, p. 74.

48. *Op. cit.* 20, p. 31.

49. *Op. cit.* 20, pp. 31-32.

50. *Op. cit.* 7, pp. 40-41.

51. *Op. cit.* 20, p. 34.

52. *Op. cit.* 20, p. 34.

53. Berrien Moore and Bert Bolm Bolin, "Ocean Models in the Global Carbon Cycle," *CDIAC Communications* (Carbon Dioxide Information Analysis Center, Oak Ridge Laboratory, Oak Ridge, Tennessee, Winter 1989), pp. 4-6.

54. *Op. cit.* 10, pp. 649-652.

55. Taro Takahashi, "Seasonal Variation of CO_2 Chemistry in the North Pacific Ocean: Implications for the CO_2-Induced Climate Change," *CDIAC Communications* (Carbon Dioxide Information Analysis Center, Oak Ridge National Laboratory, Oak Ridge, Tennessee, Summer 1987), p. 3.

56. Jorge L. Sarmiento, "Models of Carbon Cycling in the Ocean," *CDIAC Communications* (Carbon Dioxide Information Analysis Center, Oak Ridge National Laboratory, Oak Ridge, Tennessee, Winter 1988), p. 6.

57. *Op. cit.* 20, p. 34.

58. *Op. cit.* 22, p. 24.

59. *Op. cit.* 20, pp. 34-35.

60. Wallace Broecker, Professor of Geochemistry, Lamont-Doherty Geological Observatory of Columbia University, Palisades, New York, July 14, 1989 (personal communication).

61. *Op. cit.* 20, p. 35.

62. *Op. cit.* 7, p. 38.

63. *Op. cit.* 55.

64. *Op. cit.* 7, p. 38.

65. *Op. cit.* 55.

66. Richard A. Kerr, "How to Fix the Clouds in Greenhouse Models," *Science*, Vol. 243, No. 4887 (1989), p. 28.

67. V. Ramanathan, R.D. Cess, E.F. Harrison, *et al.*, "Cloud-Radiative Forcing and Climate: Results from the Earth Radiation Budget Experiment," *Science*, Vol. 243, No. 4887 (1989), p. 57.

68. *Op. cit.* 22, p. 32.

69. *Op. cit.* 22, p. 32.

70. *Op. cit.* 66.

71. *Op. cit.* 67, p. 62.

72. *Op. cit.* 22, Figure 7, p. 31.

73. *Op. cit.* 20, p. 36.

74. *Op. cit.* 67, p. 62.

75. *Op. cit.* 66, p. 29.

76. *Op. cit.* 7, p. 41.

77. Sherwood B. Idso, Bruce. A. Kimball, Michael G. Anderson, *et al.*, "Effects of Atmospheric CO_2 Enrichment on Plant Growth: The Interactive Role of Air Temperature," *Agriculture, Ecosystems and Environment*, Vol. 20 (1987), p. 1.

78. Bruce A. Kimball, "Adaptation of Vegetation and Management Practices to a Higher Carbon Dioxide World," in *Direct Effects of Increasing Carbon Dioxide on Vegetation*, Boyd R. Strain and Jennifer D. Cure, eds. (U.S. Department of Energy, Office of Energy Research, Office of Basic Energy Sciences, Carbon Dioxide Research Division, Washington, D.C., 1985), pp. 187-188.

79. *Op. cit.* 77, p. 7.

80. *Op. cit.* 78, p. 193.

81. *Op. cit.* 20, p. 39.

82. *Op. cit.* 7, pp. 41-42.

83. *Op. cit.* 7, p. 41.

84. *Op. cit.* 7, p. 41.

85. Fred Pearce, "Methane: The Hidden Greenhouse Gas," *New Scientist*, No. 1663 (1989), p. 37.

86. Leslie Roberts, "How Fast Can Trees Migrate?" *Science*, Vol. 243, No. 4892 (1989), p. 735.

87. *Op. cit.* 7, p. 42.

88. *Op. cit.* 20, p. 40.

89. Irving Mintzer, William R. Moomaw, and Mark C. Trexler, "Reducing the Risks: A Preliminary Investigation of Strategies to Reduce the Risk of Rapid Climate Change," in *The Full Range of Responses to Anticipated Climatic Change* (United Nations Environment Programme and The Beyer Institute, Nairobi, Kenya, April 1989), pp. 39, 41.

90. *Ibid.*, p. 41.

91. D. Fisher and M. Oppenheimer, "Analysis of Limitation Strategies," in *The Full Range of Responses to Anticipated Climatic Change* (United Nations Environment Programme and The Beyer Institute, Nairobi, Kenya, April 1989), p. 128.

92. José Goldemberg, Thomas B. Johansson, Amulya K.N. Reddy, *et al.*, *Energy for a Sustainable World* (World Resources Institute, Washington, D.C., 1987), pp. 70-86.

93. World Resources Institute and International Institute for Environment and Development in collaboration with the United Nations Environment Programme, *World Resources 1988-89* (Basic Books, New York, 1988), Table 7.3, p. 114.

94. H.S Geller, "The Potential for Electricity Conservation in Brazil" (Companhia Energetica de São Paulo, São Paulo, Brazil, 1985), cited in José Goldemberg, Thomas B. Johansson, Amulya K.N. Reddy, *et al.*, *Energy for a Sustainable World* (World Resources Institute, Washington, D.C., 1987), p. 58.

95. *Op. cit.* 92, pp. 70-74.

96. Deborah L. Bleviss, *The New Oil Crisis and Fuel Economy Technologies: Preparing the Light Transportation Industry for the Nineteen Nineties* (Quorum Books, Greenwood, 1988).

97. John H. Seinfeld, "Urban Air Pollution: State of the Science," *Science*, Vol. 243, No. 4892 (1989), pp. 745-752.

98. Ralph J. Cicerone and R.S. Oremland, "Biogeochemical Aspects of Atmospheric Methane," *Global Biogeochemical Cycles*, Vol. 2, No. 4 (1988), pp. 229-327.

99. Ralph J. Cicerone, "Methane Linked to Global Warming," *Nature*, Vol. 334, No. 6179 (1988), p. 198.

100. *The Long-Term Impacts of Increasing Atmospheric Carbon Dioxide Levels*, Gordon J. MacDonald, ed. (Ballinger, New York, 1982).

101. James Mackenzie, Senior Associate, Program in Climate, Energy, and Pollution, World Resources Institute, February 1990 (personal communication.)

102. H.M. Hubbard, "Photovoltaics Today and Tomorrow," *Science*, Vol. 244, No. 4902 (1989), pp. 297-304.

103. Joan M. Ogden and Robert H. Williams, *Solar Hydrogen: Moving Beyond Fossil Fuels* (World Resources Institute, Washington, D.C., 1989).

104. Charles M. Peters, Alwyn H. Gentry, and Robert O. Mendelsohn, "Valuation of an Amazonian Rainforest," *Nature*, Vol. 339, No. 6227 (1989), pp. 655-656.

105. Roger A. Sedjo and Allen M. Solomon, "Climate and Forests," in *Greenhouse Warming: Abatement and Adaptation*, Norman J. Rosenberg, *et al.*, eds. (Resources for the Future, Washington, D.C., 1989), p. 105.

106. Roger R. Revelle, "Thoughts on Abatement and Adaptation," in *Greenhouse Warming: Abatement and Adaptation*, Norman J. Rosenberg, *et al.*, eds. (Resources for the Future, Washington, D.C., 1989), pp. 169-170.

3. Latin America: Resource And Environment Overview

Latin America is a land of promise and paradox. The region as a whole is rich in natural resources, including oil and minerals, fertile soils and forests, and abundant sources of water. But these resources are not distributed equally, and many countries within the region face severe resource constraints. Likewise, the human resources of Latin America are a potential source of strength, yet much of the population lives in chronic poverty. Increasingly, pollution and environmental degradation are blighting the natural resources of the region, decreasing its productive potential for current and future generations, and threatening human health and the very existence of countless plant and animal species.

Ecuador, as an example, is a medium-size country that is neither a symbol of environmental problems nor known as a leader in environmental preservation. When flying into Quito, the capital, one's first impression is a country of incredible beauty—soaring snow-covered volcanoes rising above the clouds. But coming down from the peaks, the view is bleaker. The trees that once covered the mountain re-

gion are nearly gone, cleared by small farmers desperate for land, and the bare slopes show the ravages of erosion. Ecuador is also facing rapid destruction of the tropical rainforests that cover its eastern regions, increasing air pollution in its major cities, damage to coastal estuaries, and loss of protective coastal mangrove forests. Increasing the pressure on natural resources is a population that is expected to double before the year 2020 and an economic crisis that includes a large foreign debt and rapid inflation. These environmental and resource problems do not occur in isolation but are closely interrelated as discussed more fully below. (See Box 3.1.)

Thus Ecuador faces environmental problems on all fronts—in the cities, in the rainforests, along the coast, and in the mountains. Despite these problems, Ecuador also has a decade's experience with a democratic government and a growing number of governmental agencies, nongovernmental organizations, and grassroots movements devoted to protecting the environment. Indeed, Ecuador's problems, while

Figure 3.1 **Map of Latin America**

Source: The World Bank.

illustrative of the Latin American situation, are far from the worst in the region, economically or environmentally.

A HUGE AND DIVERSE REGION

From the Rio Grande at Mexico's northern border down to Chile's southernmost outposts, Latin America encompasses a huge and diverse region. (See Figure 3.1.) The countries of Latin America share a history of Iberian colonialism, the Spanish and Portuguese languages, and membership in a continental-scale culture of art, music, literature, and religion. Shared languages also give the nations of Latin America similar access to information, common interest in events throughout the region, and a cultural similarity.

Cultural similarities aside, however, Latin America is a region of great diversity. Its nations range from huge, industrializing Brazil to tiny Central American nations largely dependent on forestry, fishing, and agriculture, and include Andean nations such as Peru and Colombia that host both ancient cultures and modern drug-war violence. With diverse native cultures, each country adds a unique history of immigration, resulting in 20 distinct nations. Latin America extends through 90 degrees of latitude, from the heat of the Sonoran desert to the cold and gale-lashed shores of Tierra del Fuego. (As used in this chapter, Latin America includes the continent of South America and the southern part of North America—Mexico and Central America; it does not include Caribbean countries.)

The countries of South America are dominated by the Andean Cordillera, a mountain chain that shapes continental (and global) climate, transportation, the geographic focus of political action, and economic development. On its slopes and plateaus are ancient cities and modern capitals, dense populations, bare mountain peaks, intensive cultivation, and massive erosion, both natural and that caused by human activities. The continent also contains the huge watershed of the Amazon River basin and its namesake tropical forest, the largest in the world. There are substantial areas of tropical forest in Central America as well, and large unforested areas potentially suitable for farming or rangeland on the Mexican plateau and in Brazil and Argentina.

Health and Services

The population of the region is 414 million, 8 percent of the world's total, and grows at a rate of 2.1 percent annually—faster than Asia (1.6 percent), but slower than Africa (3.0 percent). Despite serious health problems, people in Latin America have the longest life expectancy and the lowest child mortality of any developing region. (See Chapters 4 and 16 on "Population and Health" for further data and analysis.) The provision of services and the maintenance of economic opportunity for nearly 8.7 million more

Figure 3.2 Urban and Rural Share of Population in Eight Latin American Countries, 1960 and 1990

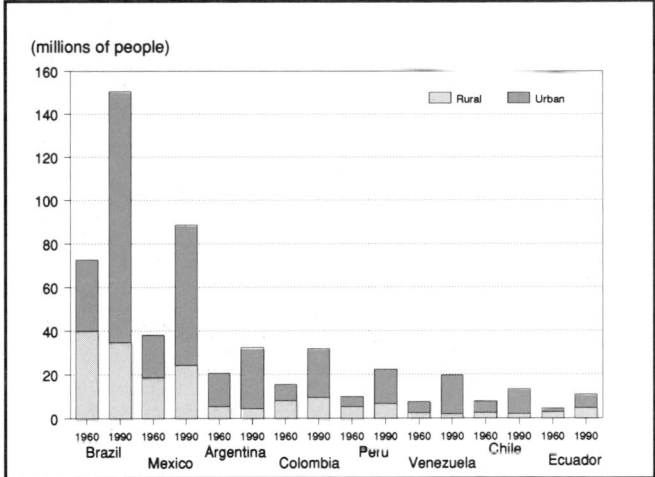

Source: Chapter 16, "Population and Health," Table 16.1 and Chapter 17, "Land Cover and Settlements," Table 17.2.

citizens a year strain the infrastructure and political fabric of Latin American societies.

Most of this growth is in urban centers. Since 1960 the urban population has grown at an average annual rate of 3.8 percent while rural growth has been almost stagnant at only 0.4 percent. Most of the population—72 percent—in Latin America now live in urban settings, a higher percentage than in Europe. (See Figure 3.2.) Air and water quality suffer from this rapid urban growth through the insults of unregulated emissions from vehicles and factories and of waste dumping and inadequate sewage treatment.

Figure 3.3 External Debt of Latin American Countries as a Percentage of Gross National Product (GNP), 1977, 1982, and 1987

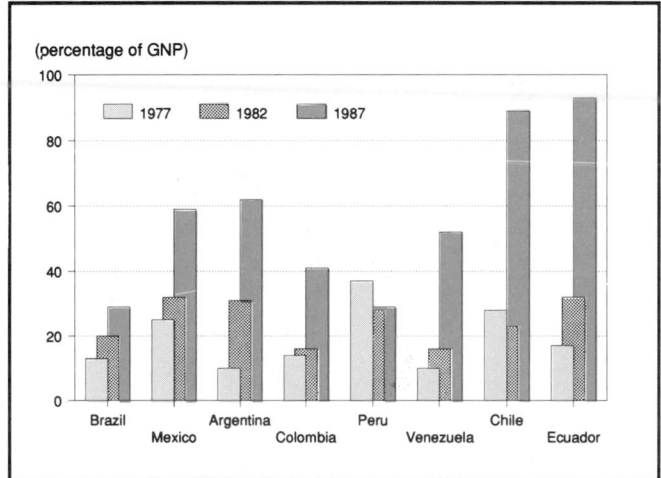

Source: Chapter 15, "Basic Economic Indicators," Table 15.2.

Box 3.1 Ecuador: A Case Study

Ecuador provides an instructive example of the interrelationship of economic, social, resource, and environmental problems in a Latin American country. In Ecuador, despite past attempts at agrarian reform, 1.2 percent of the landholders control 66 percent of the arable land, and 90 percent of farmers own fewer than 10 hectares each, often on steep slopes that scarcely seem arable (1). Soil erosion is severe, partly as a result of farming on steep slopes. Estimates of average soil loss run as high as two hundred tons of soil per hectare per year (2). The small farmers, whose lots are usually the areas with the steepest slopes, feel the effects of land degradation first.

Farms have a tendency to get smaller with divisions among each succeeding generation, and the population in Ecuador is growing at a rate of over 2.6 percent annually. This growth will be hard to slow in the near future, since, in 1990, 40 percent of the population was under the age of 14, and the entire population is expected to double by the year 2025 (3).

For many small landowners, the crops they can grow do not support their families. As a temporary solution, the men migrate to cities or provide seasonal labor to harvest the crops of large landowners. Thus, over 50 percent of family farm work is done by women.

MIGRATING TO THE AMAZON

Another common solution is for the whole family to migrate to the Amazon region, where new roads are opening up virgin tropical forests. Estimates based on the extent of roads and studies of the forestry practices lead experts to conclude that at least 100,000 hectares (4) —and perhaps as many as 340,000 hectares—of Ecuador's forests are cut down each year.Thus, Ecuador is losing its forests at the rate of perhaps 2.4 percent per year—the second highest rate in South America. (See Chapter 19, "Forests and Rangelands," Table 19.1.)

The migrant colonists clear land wherever roads penetrate the forest, usually in the wake of oil companies exploring new sites to drill. In 1989, 32 oil companies were drilling in an area of 3.5 million hectares, 80 percent of which is in the Amazon region. The Ecuadoran government itself estimated that these companies observed only half of the relevant environmental regulations; oil slicks have oozed out of leaky pipes to float down the rivers (5).

GOLD MINING A NATIONAL PARK

Oil is not the only business lure in the forests, but it is the one that is officially sanctioned. Ecuador has not entirely escaped from the attentions of drug traffickers who dominate parts of all its neighboring countries, but cocaine is not a major source of income for the country—yet. The effect of the drug trade is similar to another phenomenon that draws desperate people into the wilderness to seek a fast route out of poverty: gold. In southeastern Ecuador, a gold rush has begun to devour Podocarpus National Park, an area containing cloud forests rich in endangered species. Approximately 500 miners are currently panning for gold there while more than 90 percent of the park has been sold under concessionary rights to national and international mining companies. The mountain streams that flow from the park and provide water for nearby towns are now threatened with mercury contamination owing to on-site gold processing. In northern Ecuador, the ecosystems of Sangay National Park and Cayambe-Coca Reserves face similar destruction under gold mining activity (6).

Mining in the Podocarpus Park is not an isolated example of the weak environmental protection offered by Ecuador's park system. The borders of the Cuyabeno Wildlife Reserve in the Amazon, home of the Cofan and Siona Indians, have been changed to accommodate African palm oil plantations and petroleum companies, both of which continue to challenge the borders of this and other parks (7).

Ecuador's foreign debt is more than $11 billion. Oil used to provide about 50 percent of Ecuador's export earnings, but when the debt crisis hit simultaneously with the drop in oil prices, Ecuador turned to other natural resources. Patterns of agriculture changed over the 12-year period ending in 1986, with the land dedicated to basic foods—corn,rice, beans and potatoes—dropping by 25.7 percent. The planting of soybeans, African palm for oil, sugar cane, and feed corn expanded by 171 percent in the same period. Ecuador's cattle pastures expanded by 31 percent nationwide between 1977 and 1983, most cleared from forests in the Amazon region (8). Despite these efforts, however, agricultural exports for Ecuador as a whole, measured

(See Chapter 5, "Human Settlements," The Urbanization Trend.) Despite these significant hazards to human health, Latin America has better access to safe drinking water than do other developing regions.

Economic Crisis

Economic problems, especially those arising from external debt, have played a major role over the past decade in constraining the ability of Latin American governments to accelerate development or to respond to environmental concerns. The debt crisis of the 1980s is expressed most purely in the experience of Latin American countries. Together their external debt totaled $387 billion in 1987, compared with $342 billion in Asia and $212 billion in Africa. (See Chapter 15, "Basic Economic Indicators," Table 15.2.) Brazil's long-term debt in 1987 was equal to 29 percent of its GNP; Mexico's was 59 percent, and Ecuador's was a staggering 93 percent. (See Figure 3.3.) This relatively huge debt, coupled with economic stagnation or worse, a decline in international commodity prices, and lack of confidence among domestic investors, has limited the economic potential of the region. Even oil-producing countries find they are part of a growing miasma of despair. (See The Economic Crisis, below, and Chapter 15, "Basic Economic Indicators.")

Rich in Resources

For all its economic problems and demographic realities, Latin America is rich in natural resources and potential. It is the most forested of all developing regions, with 966 million hectares of forest covering 48 percent of its land area. (See Chapter 17, "Land Cover and Settlements," Table 17.1.) Pasture comprises another 28 percent of the region, and agricultural land about 8.7 percent.

Between 1977 and 1988, the region's agricultural production increased 25 percent. Total food production increased 27 percent during the same period, which exactly paced population growth (1). (See Figure 3.4.) By 1998, however, food production must increase another 26 percent to maintain per capita

Box 3.1

in dollars, have not grown appreciably over the past decade.

The soils and the forests are not the only resources that are under pressure to produce exports. The shrimp business grew until exports reached $360 million, enough to pay one third of the service costs of the nation's debt. To make way for the shrimp-growing pools, however, 100,000 hectares of mangroves were cleared (9).

URBAN SERVICES

Ecuador is finding it difficult to provide sanitation and services for its residents at a time of austerity imposed by the economic crisis. More than half of all Ecuadorians—52 percent—live in urban areas. Guayaquil and Quito are both growing at an annual rate of 4.5 percent. The government estimated the housing shortage at 840,000 homes in 1985, and found that 400,000 existing homes lacked the basics of plumbing and water, or were considered uninhabitable (10). Guayaquil, Ecuador's second-largest city, with a population of 1.2 million, is located on the estuary of two rivers that have been fouled by untreated sewage, garbage, and industrial wastes. Air pollution, like water pollution, is not monitored carefully in Ecuador, although the smog is clearly visible in Guayaquil and Quito. One indication of the health threat from pollution is the amount of lead allowed in the gasoline used by Ecuador's motor vehicles; the level is more than 20 times that permitted by the U.S. Environmental Protection Agency (11).

Ecuador is thus facing a conjunction of environmental and economic crises—

and yet is better off than many other countries of the region. Ecuador's per capita gross domestic product fell only 1 percent in total form 1981 to 1989, much better than in Mexico (-9 percent) or Argentina (-23 percent). Ecuador's inflation rate fell to 59 percent in 1989, far below that of Brazil, Peru, Argentina, or Nicaragua (12).

Trying to rationalize the use of natural resources under these economic conditions requires a strong political commitment on the part of Latin Americans. In the 1970s, environmental problems were not even discussed; today there are environmental agencies within the governments, a dozen organizations outside advocating better protection of resources, and hundreds of articles in the press on these issues. Even the constitutions, like those recently written in Brazil and Chile, give citizens the right to a healthy environment.

SIGNS OF HOPE

Today in Ecuador there are instances of peasant cooperatives reforesting hillsides, ecology being taught in grade schools, and indigenous people's organizations drawing up their own forest management plans. While these initiatives are still on a small scale, they point toward the solutions. As Ecuadoran environmentalists themselves put it:

"The fundamental prerequisite for any action is to recognize the problems...to identify and analyze their roots, and above all, to understand that these are not simply technical problems—on the contrary, they are strongly woven into

our idiosyncrasies, our behavior, and the way that we conceive our rights" (13).

References and Notes

1. James D. Nations, *Sociocultural Factors of Watershed Management in Ecuador*, U.S. Agency for International Development (U.S. AID), Project 518-0023 (U.S. AID and Ecuadorian Institute of Electrification, Ministry of Energy and Mines, Quito, Ecuador, October 1985), pp. 3- 4.
2. Barbara d'Achille, "Peru, Ecuador y Colombia: Los Mismos Errores," *El Comercio* (June, 18, 1988, Lima, Peru), p. D1.
3. Department of International Economic and Social Affairs, United Nations, *World Population Prospects, 1988* (United Nations, New York, 1989), Table 17-B, p. 356.
4. Fernando Ortiz, Natural Resources Specialist, USAID/Ecuador (personal communication).
5. "Ecuador: La Industria Petrolera Dana el Medio Ambiente," *Interciencia*, Vol. 14, No. 3 (May-June 1989), pp. 146-147.
6. Lisa Naughton, Wildlife Conservation International, New York, February, 1990 (personal communication).
7. Theodore MacDonald, Cultural Survival, Cambridge, Massachusetts, June 1989 (personal communication).
8. Helena Landazuri and Carolina Jijon, *El Medio Ambiente en el Ecuador* (Instituto Latino Americano de Investigaciones Sociales (ILDIS), Quito, Ecuador, 1988), pp. 64, 66.
9. *Op. cit.* 2.
10. *Op. cit.* 8, p. 50.
11. *Op. cit.* 8, pp. 51-55.
12. United Nations Economic Commission for Latin America and the Caribbean (ECLAC), *Preliminary Overview of the Economy of Latin America and the Caribbean 1989* (ECLAC, Santiago, Chile, December 1989), Table 3, p. 19 and Table 5, p. 20.
13. *Op. cit.* 8, p. 181.

food production at current levels. The rural population, which is not expected to grow appreciably, will be called upon to expand agricultural productivity. It is likely that this increase will be met by more mechanization, the application of agricultural inputs, and the expansion of the area under cultivation.

Currently, 8.7 percent of Latin America's land area is under cultivation. Although there are no inherent physical or chemical constraints on agriculture on fully 12.3 percent of the land area, availability of water supplies in arid regions limit potential agricultural expansion. In Mexico, for example, much of the land with no other constraints for agriculture lies in the arid zone. Opening land with physical or chemical constraints to agriculture will require additional inputs. (See Chapter 6, "Food and Agriculture," Sustainable Agriculture for South America's Tropical Savannas, and Chapter 18, "Food and Agriculture" Sources and Technical Notes, Tables 18.5 and 18.6.)

Latin America, as a region, has no lack of freshwater; 26.4 percent of the world's renewable water re-

sources exist there. (See Chapter 22, "Freshwater," Table 22.1.) This wealth of water, while unevenly distributed, helps support large areas of diverse habitat throughout the region, some of which is at risk from the seemingly inevitable expansion of agricultural lands.

Desert, cloud forest, alpine tundra, temperate forest, extensive savanna, rain forest, temperate grassland, tropical dry forest, mangroves, and swamps cover the region. Latin America has the world's largest river and driest desert. This environmental diversity has led to genetic diversity, which in itself is a valuable resource. (See Chapter 8, "Wildlife and Habitat," Box 8.1.) Latin America has given the world the staple crops of maize, manioc (cassava), potato, sweet potato, peanut (groundnuts), and many beans as well as squash, peppers, tomatoes, cacao, cashews, and brazil nuts. Local crops throughout the region may someday also have a global distribution; they, and their wild cousins, provide a genetic bank account to tap when necessary.

Figure 3.4 Food Production in Latin America, 1965–85

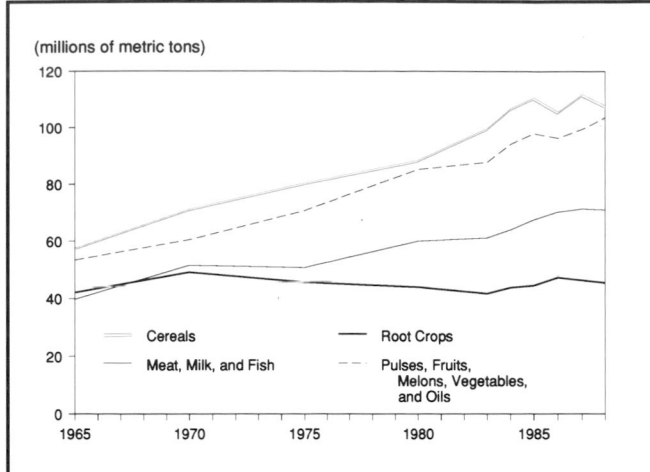

Source: Chapter 6, "Food and Agriculture," Table 6.1.

Figure 3.5 Population Density by Region, 1989

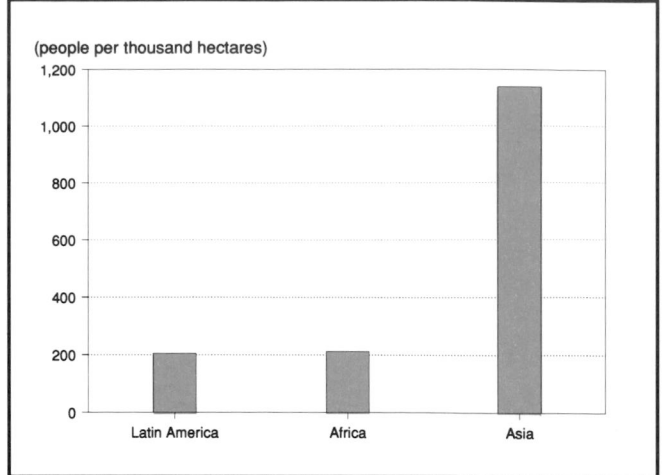

Source: Chapter 17, "Land Cover and Settlements," Table 17.1.

Accelerating Land Clearing

Part of this diversity are the large numbers of species that are integral parts of each ecosystem. Indigenous people have exploited a few of these species over the past 10,000–15,000 years. Nonetheless, millennia of heavy development on the plateau of Mexico, the Yucatan, the Andes, and along the Pacific coast of South America have transformed the landscape, flora, fauna, micro-climate, and ecosystemic potential of the region. This transformation has increased in recent years as population pressure has led to the intensification of agriculture, the opening of marginal lands, and the diversion of river and ground water.

One of the most visible examples of this habitat destruction is the accelerating clearing of moist tropical forest. It appears to have reached a crescendo in 1987 with the loss of as much as 9 million hectares in Brazil through massive land clearing. (See Chapter 7, "Forests and Rangelands," New Estimates of Tropical Deforestation.) Still another potential cause of habitat destruction is the creation of large dams for hydroelectric power production.

Energy is an important resource for industrial development. At present, Latin America consumes only 4 percent of the world's commercial energy supplies, and a relatively large proportion of those are from sources other than fossil fuels. Hydroelectric power supplies 63 percent of total electricity, and the exploitable hydroelectric potential at large-scale sites is over six times the current capacity. Over half of Latin America's hydroelectric capacity is installed in Brazil, which also has the region's greatest unexploited potential. Brazil is also unique in its use of renewable energy resources to meet a large part of its liquid fuel needs, producing ethanol from biomass (primarily from sugarcane) equivalent to 27 bil-

lion liters of petrol per year (2). Some of the countries of the region produce oil for export, while others are net energy importers.

In comparison with the rest of the world, the amount of forested area and the bounteous supplies of water in Latin America are striking. With 414 million people in a region spanning more than 20 million square kilometers, the ratio of land to inhabitants is still enviable (see Figure 3.5), but no other region exhibits such concentration of land ownership in a tiny segment of the population. Cities' sizes are expanding quickly, but the basic sanitation and social services are reaching ever smaller proportions of the urban mass.

Critical Resource Problems

The most critical environmental and resource problems of the region include: air, water, and other forms of pollution in crowded urban areas; deforestation and the concomitant loss of habitat and biodiversity; soil erosion and other forms of land degradation that threaten agricultural productivity; river pollution from untreated sewage and from mining and other industrial operations; and damage to coastal estuaries and other marine resources through pollution, clearing of mangrove swamps, and overfishing. In addition, Latin America is a significant contributor to global environmental problems through emissions of carbon dioxide and other greenhouse gases. (See Chapter 2, "Climate Change: A Global Concern," and Chapter 24, "Atmosphere and Climate.")

Many of the regional and local environmental problems are the result of ill-planned development, lack of emission controls, or other forms of private and public mismanagement; others arise from the combination of population pressure and land tenure problems. This overview chapter discusses a few of these problems in more detail, together with the economic crisis that has afflicted Latin America during the 1980s; other concerns are discussed or documented with data in the remaining 22 chapters of this volume. But the overall picture is of a region that, despite some encouraging trends, has not made much recent progress toward a better life for most of its citizens. Indeed, the danger exists that many countries in Latin America are seriously damaging the resource base on which future economic growth must rely. A growing trend toward democratic governments, increasing popular concern over environmental problems (see Chapter 1, "World Environment Outlook"), and signs of renewed governmental interest in environmental reforms are positive changes.

THE ECONOMIC CRISIS

The context for the discouraging statistics of the 1980s is what Latin Americans refer to simply as "la crisis." For the United States and Europe, there have been recessions and "slowdowns," but what Latin America has experienced is a one-step-forward, two-steps-backward slide into a depression, with bursts of hyperinflation. In 1989, for example, annual inflation in Nicaragua was more than 3,400 percent (but one tenth the year's before); in Argentina inflation reached 3,700 percent, in Brazil almost 1,500 percent, and in Peru nearly 3,000 percent. Ecuador, with only 60 percent inflation, did comparatively well (3). The external debt of Latin America, nearly $400 billion in 1987 (see Chapter 15, "Basic Economic Indicators," Table 15.2), is large—nearly $1,000 for every man, woman, and child in the region. In 1985, Brazil, Argentina, and Mexico each made interest payments twice as large in proportion to their GNPs as those of Weimar Germany (4). Contributing to the crisis are such factors as a drop in agricultural commodity prices, capital flight, rapid population growth, concentration of land ownership, income disparities, and inefficient—if not corrupt—governments. Whatever the exact mix of causes, the result is a region with deep and pervasive economic problems—and equally pervasive environmental dilemmas.

Some observers see a link between the fiscal indebtedness and the ecological impoverishment that is all too widespread in Latin America. That may well be the case, but evidence for such a direct impact is scanty. It is also true that most of Latin America's environmental problems predate the debt crisis. What seems likely is that the economic crisis has prevented some environmentally unsound investments or development projects, as well as needed environmental improvements, from taking place.

Past Decade

The statistics show the following: economic growth came to a halt in the 1980s. Between 1982 and 1989, the increase in gross domestic product (GDP)—a measure of economic performance that excludes interest paid to or received from abroad—averaged 1.4 percent per year in real terms. When population growth is considered, however, real per capita GDP *fell by almost 1 percent per year* (5). Enrique Iglesias, president of the Inter-American Development Bank, told his board of governors in March 1989, "In many of our countries, per capita income when the decade ends will be less than when it began. In some cases, per capita income has fallen back not 10 years but 20. It is not just one lost decade, but two."

Normally, to increase exports and expand the economy, large amounts of capital are invested. Yet so few new loans or investments were going to Latin America in the 1980s that the net capital flow reversed, and the developing countries were sending money north. In 1989, the net transfer flowing out of Latin America was $24.6 billion (6). What was not being paid on loans was flowing out of the Latin American countries as capital flight, $166 billion by 1988 (7), since the economic crisis discouraged Latin Americans from investing their own money at home. The overall 1987 level of gross domestic investment in Latin America was over 25 percent less than the 1980 level, whereas in the two previous decades the level of investment had doubled (8).

Austerity imposed by the debt crisis can affect the ability of governments to undertake protective measures for the environment in several ways. For example, the economic crisis may have prevented imports of pollution control equipment, halted investment in many badly needed cleanup efforts, and contributed to the lack of maintenance and resulting decay of water and sewage systems and other basic infrastructure. Likewise, austerity measures in many countries are reported to have left natural resource agencies without enough money to carry out their basic tasks: agronomists confined to offices, parks without rangers, and environmental regulators without paychecks. In Brazil, for example, 6,500 employees of the government's new environmental institute went on strike three months after it was created to protest the lack of routine funding (9). So long as the economic crisis in Latin America continues, investment in environmental improvements—such as cleaning up the heavily polluted urban areas where most Latin Americans live—will be difficult.

THE URBAN ENVIRONMENT

The cities of Latin America are increasingly polluted, exposing their populations to a wide variety of environmental hazards. One fundamental cause of a deteriorating urban environment is the pressure of rapidly growing populations. The region's major cities have gone beyond merely large—in 1990, Mexico City had 19 million people in the metropolitan area and is expected to have 24.4 million by the year 2000. São Paulo's population is projected to jump from 18.4 million in 1990 to 23.6 million by 2000, and Rio de Janeiro's and Buenos Aires' are projected to exceed 13 million. Overall, population in Latin America's cities is projected to increase by over 90 million people between 1990 and 2000 (10).

Urban growth rates in Latin America appear to have slowed somewhat in the 1980s to 3.2 percent per year, compared with 4.4 percent for the 1960s. In Venezuela, for example, urban growth dropped from 4.1 to 3.3 percent, and in Brazil from 4.9 to 3.3 percent (11). Nevertheless, this still-rapid growth will inevitably put further stress on the abilities of governments to provide basic environmental services and on what has become a delicate balance between human ecology and natural systems.

Many of Latin America's major colonial cities were founded in the 16th and 17th Centuries in the shadows of mountains or surrounded by mountain ranges, often at high elevations. These locations took advantage of cool climates, the proximity of mountain forests that could provide fuel and lumber, and an important public health consideration—they were above the "mosquito line." To facilitate agriculture, most sites also were chosen for the presence of excellent soils, deposited in alluvial fans over thousands of years by the rivers that spilled from the mountains and coursed their way through the city centers. But many of these advantages have turned to disadvantages in the industrialized and heavily populated cities of today.

City Air Is Trapped by Mountains

The same mountains that once brought cool evening breezes now help trap polluted air—the product of car, truck, and bus exhausts and factory smokestacks—in the thermal inversions that form above the cities. As urban sprawl covers once-rich agricultural land, farmers push farther up the slopes of the mountains, removing the trees and often accelerating erosion. Acid rain from the city's pollutants also damages the forests. The city's rivers carry garbage, human wastes, and industrial effluents including heavy metals, depositing some of them on agricultural lands downstream where they contaminate the city's food supply.

The extreme example of these problems is Mexico City, where conditions of life for the city's residents have become steadily worse. Its geographical location is partly responsible—the city is situated in a high basin that traps smog and other pollutants during the dry season. But unregulated sources of pollution are the main cause of the city's misery. An estimated 36 thousand factories—half of Mexico's industry—is located in the city, along with some 3 million cars, trucks, and buses. Together they emit 5.5 million metric tons of contaminants each year. Approximately 80 percent of the emissions that make up ozone, recognized to be one of the most difficult pollutants to control, come from motor vehicles. An astounding 30 percent of these contaminants could be eliminated from the atmosphere if this fleet were given a tune-up (12).

In the past, vehicle emission standards have rarely been enforced. Governments in Latin America are more likely to ignore industrial pollution than decrease manufacturing and risk aggravating unemployment (13). These policies of neglect cannot be continued without grave threats to human health. In Chile, for example, the winter of 1989 brought a level of smog so dangerous that Santiago hospitals reported 30 to 60 percent increases in the number of patients treated for respiratory ailments. The government then decided to temporarily close down 50 percent of the 132 polluting industries and to stop half the cars and buses from circulating (14). The private bus companies initially went on strike for several days rather than cut their service, throwing the city into chaos (15).

One of the greatest health threats from automobiles, buses and trucks in Latin America is the amount of lead in the petrol; in Ecuador, for example, petrol contains more than 20 times as much lead as allowed under U.S. standards (16). The result is often lead levels high enough to damage the health of children. For example, in Mexico City, a high proportion of newborns tested had lead in their bloodstreams at levels believed to cause intelligence and psychomotor deficits and other nervous system damage (17). The average lead levels in the blood of Mexico City residents, for example, is nearly four times that of a Tokyo resident (18). In Mexico and in some other Latin American countries, these concerns are leading to change. After environmentalists in Mexico publicized the dangers of the lead in the air, the state oil company Pemex introduced a new petrol in 1986 with lower lead content. This new petrol, however, used a mix of hydrocarbons to replace the lead; as a consequence, Mexico City's ozone level has shot up since 1988 (19). Median ozone levels tripled between 1986 and 1988, reaching 60 percent above the World Health Organization (WHO) safety standard (20). Ozone irritates the eyes and throat, also causing respiratory problems. For children in Mexico City, breathing the air can be as bad as smoking cigarettes from birth (21), and can cause chronic lung effects (22). Although few data exist, emissions of toxic chemicals from industrial plants, largely unregulated in most of Latin America, might also be adding to the health threat. Mexico City is beginning mandatory testing of exhaust emissions for automobiles and hopes to begin a crackdown on industrial emissions as well. In 1990, drivers were banned from using their cars on a rotating basis one day a week during a two-month test program (23). And beginning in 1993, all new automobiles produced in Mexico will be required to have catalytic converters. (See Chapter 12, "Atmosphere and Climate.")

Wastes and Water

Besides chemical pollutants, air is polluted by uncollected or abandoned solid wastes and fecal material that dries into dust. In Mexico City, winds carrying

Box 3.2 Cubatao: New Life in the Valley of Death

Called the "valley of death" and the "most polluted place on earth," Cubatao—located on the Atlantic underbelly of Brazil's industrial potentate, São Paulo—had a reputation that no city would want to claim (1).

Before the 1970s, Cubatao might have been considered a pleasant, well-situated town, looking upon the bay, with tree-covered mountain slopes rising on three sides around it. The valley where the town is located was traversed by rivers that flowed into the sea, lined with mangroves.

When a river was first dammed up to generate energy, industries began to crowd into this town near the major port of Santos. A steel plant, a huge oil refinery, and fertilizer and chemical producers squeezed into the valley, while workers and job seekers set up shacks on swamplands and hillsides around the industrial nucleus. By 1985, Cubatao was producing 3 percent of Brazil's gross national product (2).

By then, the mangroves were gone, the waters were the color of mud, and silt backed up in the rivers, which overflowed regularly. The changes in the rivers also signaled the losses on the mountain slopes as the smog-filled air left trees like skeletons with threadbare leaves. Loss of vegetation led to erosion, and hillsides crumbled, collapsing around the town.

The 110,000 residents were under siege; poisonous wastes from industries were scattered around, dumped in the rivers, and pumped into the air where winter weather would trap the gases in the valley for long periods (3). Accidental spills and chemical accidents claimed hundreds of lives and led to emergency evacuations on a yearly basis.

This is no longer the situation. The town of Cubatao has changed, and today the state of São Paulo points with pride to the results of a cleanup effort that re-

quired industries to control pollutants. Out of 320 sources of pollution, 249 had been cleaned up and controlled by late 1988. Particulate pollution had been reduced 92 percent, ammonia was down by 97 percent, and the hydrocarbons that cause ozone were cut 78 percent. Sulfur dioxide levels fell by 84 percent, but nitrogen oxides dropped only 22 percent (4). No major emergencies from dangerous levels of air pollution were reported in 1987.

Water quality in the three main rivers of the valley has improved notably. The obvious sign is the return of fish, after a 30-year absence, to the Cubatao River (5). The industrial effluents being dumped into the rivers were cut from 64 metric tons each day to 6 metric tons, with corresponding reductions in the amounts of metal in the water (6). Parts of the rivers were also dredged of 780,000 cubic meters of material to improve the flow (7).

What did it take to restore the environment of Cubatao to livable levels? First came the restoration of democracy to Brazil, allowing complaints from residents in Cubatao to be publicized in the press. Gubernatorial candidate Franco Montoro made Cubatao an issue in his political campaign, then committed his government to reducing the pollution. During the Montoro administration from 1983 to 1987, the state worked with the private sector, sharing the costs of the cleanup. The state government's plan had an estimated cost of $98 million, and the total investment in pollution devices by industries was $220 million (8). The program was backed by a $100 million loan from the World Bank (9).

The government also launched a program to reforest the mountain slopes with pollution-resistant trees bred from species native to the area, to contain the erosion and mudslides that were threatening to bury the town. The Forestry and Botanical Institutes developed seed pel-

lets that could be sown on the mountainsides from helicopters, and millions of seeds were spread over an area of 60 square kilometers.

The change in Cubatao is heartening, but not yet complete. The principal industry that is still polluting the valley is the government-owned steel plant, which has not been able to invest in pollution control because of the restrictions imposed on the government by the debt crisis.

References and Notes

1. "Cubatao: Brazil's Ecological Success," [The] Financial Times (London, June 10, 1988).
2. Jorge Wilheim, "Air Pollution in the Third World Context: Controlling the Cubatao Region," a paper presented to the 4th World Congress on Conservation of the Built and Natural Environments: The Siting of Industry and Its Effects on the Environment, Toronto, May 1989, pp. 10-11.
3. Governo do Estado de São Paulo, A Batalha do Meio Ambiente no Governo Montoro (Governo do Estado de São Paulo, São Paulo, Brasil, March 1987), pp. 83-99.
4. Op. cit. 2.
5. Edison Martinez Alsono, "Peixes votam Rio Cubatao revive," O Estado de São Paulo (May 8, 1988), p. 34.
6. "Acao da Cetesb em Cubatao," Quarterly Progress Report of the Company for Environmental Sanitation Technology (Cetesb) of the state of São Paulo, August-December 1987.
7. Op. cit. 3, p. 95.
8. Op. cit. 2.
9. "A Volta dos Bons Ars na Velha Capital da Fumaca," Veja (February 11, 1987), quoted in Governo do Estado de São Paulo, Brasil, A Batalha do Meio Ambiente no Governo Montoro (Governo do Estado de São Paulo, São Paulo, Brasil, March 1987), pp. 96-97.

dust from the body wastes of the 6 million people without indoor plumbing and excrement from the city's 2 million dogs are the likely source of many illnesses (24). Similar problems on a smaller scale can be found in most major Latin American cities.

Another source of disease or other health hazards is water, particularly untreated waste water dumped into rivers, then used to irrigate vegetables grown for urban food supplies. This is the case for Mexico City (25), at least, and also Bogotá, where the Tunjuelito river has high levels of heavy metals such as cadmium and lead (26).

Access to clean drinking water and to sewage treatment is still a significant problem in urban areas. (See Chapter 5, "Human Settlements," Progress to Date; Water and Sanitation.) Despite recent improvements in Central America, for example, in 1988 only

59 percent—of the population was served by adequate water facilities; only 58 percent had sanitation service (27). In Mexico City, the portion of the population with access to clean drinking water has dropped slightly from 1980 to 1986; so did the portion of homes connected to the sewage system (28). Almost one half of the wastewater treatment plants surveyed recently in Mexico were not in service. (See Chapter 12, "Human Settlements.") With the growth in population, the amount of untreated sewage entering the rivers or dumped on land is rising steadily.

Although most Latin American cities have abundant water supplies, population growth can in fact outstrip water supplies. Again, Mexico City is perhaps the extreme example. Built on what was once a lake, Mexico City is pumping ground water far faster than it is replenished, an unsustainable situation. Ad-

Table 3.1 Deforestation in Latin America, 1980s

	Average Annual Deforestation 1980s					
	Closed Forest		Open Forest		Total Forest	
	Extent (000 ha)	Percent	Extent (000 ha)	Percent	Extent (000 ha)	Percent
Central America	1,070	1.6	20	0.0	1,090	1.6
Belize	9	0.7	X	X	9	0.7
Costa Rica	124	7.6	X	X	124	6.9
El Salvador	5	3.2	X	X	5	3.2
Guatemala	90	2.0	X	X	90	2.0
Honduras	90	2.3	X	X	90	2.3
Mexico	595	1.3	20	1.0	615	1.3
Nicaragua	121	2.7	X	X	121	2.7
Panama	36	0.9	X	X	36	0.9
South America	9,837	1.5	1,293	0.6	11,180	1.3
Argentina	X	X	X	X	X	X
Bolivia	87	0.2	30	0.1	117	0.2
Brazil	8,000	2.2	1,050	0.7	9,050	1.8
Chile	X	X	X	X	50	0.7
Colombia	820	1.8	70	1.3	890	1.7
Ecuador	340	2.4	0	X	340	2.3
Guyana	2	0.0	1	0.2	3	0.0
Paraguay	190	4.7	22	0.1	212	1.1
Peru	270	0.4	0	X	270	0.4
Suriname	3	0.0	X	X	3	0.0
Uruguay	X	X	X	X	X	X
Venezuela	125	0.4	120	6.0	245	0.7

Sources:
1. Food and Agriculture Organization of the United Nations Forest Resources Division, An Interim Report on the State of the Forest Resources in the Developing Countries, (FAO, Rome, 1988).
2. Alberto Waingort Setzer, Marcos da Costa Pereira, Alfredo da Costa Pereira, Jr., and Sergio Alberto de Oliveira Almeida, Relatorio de Atividades do Projeto IBDF-INPE "SEQE"—Ano 1987," unpublished paper (Instituto Nacional de Pesquisas Espaciais (INPE), Sao Jose dos Campos, Sao Paulo, May 1988).
3. Steven A. Sader and Armond T. Joyce, "Deforestation Rates and Trends in Costa Rica, 1940 to 1983," Biotropica, Vol. 20, No. 1, (1988), p. 14.
Note: X = not available.

ditional water is brought from other areas at great expense, about $50 million annually. Water brought into Lima, Peru, by tanker now supplies 2 million people. Despite its cost and growing scarcity, water is still wasted. About 25 to 35 percent of the water consumed by Latin American cities is lost through leaks and breaks in the water system. What Mexico City loses alone would serve the needs of Rome (29).

While urban environmental problems are great, there are ways of rethinking the needs, changing priorities, and finding funds to reverse the trends. A striking example is the city of Cubatao, once the most polluted in Brazil, but now in the process of recovering. (See Box 3.2.) For the recovery to become more widespread, governments must find the political will to enforce environmental regulations. More efficient use of natural resources and the elimination of harmful economic policies will also be necessary.

DEFORESTATION

Tropical forests are a unique and endangered resource. Those in Latin America merit special concern, both because the region contains more than half—57 percent—of the remaining tropical forests (30) and because the rate of deforestation is high. In

fact, the rate of deforestation in Latin America is the highest in the developing world: about 1.3 percent of the existing forests are lost annually, compared with 0.9 percent in Asia and 0.6 percent in Africa. (See Figure 3.6.) In part, these disparities in rates reflect the fact that many Asian and African countries have cleared most of their forests.

Brazil alone contains 30 percent of the world's tropical forests, more than the rest of Latin America combined, and tends to dominate discussions of tropical deforestation to the point that other areas are overlooked. Brazil also dominates the loss figures: an estimated 9 million hectares in 1987, which may have been an anomalously high year. The amount of deforestation is not in fact known with a high degree of accuracy: estimates for Brazil vary by more than 100 percent (see, Chapter 2, "Climate Change: A Global Concern," and Chapter 7, "Forests and Rangelands," Deforestation in Brazil); estimates for Ecuador vary by as much as 300 percent. (See Box 3.1.) Regardless of the uncertainties, there is no doubt that deforestation rates in many countries are extremely high. (See Table 3.1.) Besides Brazil, other Latin American countries facing large-scale deforestation are Colombia (890,000 hectares per year), Mexico (615,000 hectares), and Ecuador (340,000 hectares).

Deforestation in Central America

Central American countries, although their forests are smaller, are also experiencing very high rates of loss. Costa Rica, for example, has had the highest rate of deforestation in Latin America: 6.6 percent annually, although this is now being reduced. In 1984, forests covered only 18 percent of Costa Rica, and the rate of deforestation was so high that Costa Rican officials predicted that forests outside the national parks and protected areas would be completely gone in another decade (31). According to a Costa Rican analysis, the reasons include: increasing population and the concentration of the best agricultural lands in a few hands; the expanding agricultural frontier, with tax credits to stimulate forest cutting and cattle ranches; the demand for high-value timber, which is typically logged by laying waste to the rest; and finally, the fact that the government has not enforced its own laws to protect the forests (32).

The long-term costs to Costa Rica will be high. The country depends on hydroelectric facilities for over 80 percent of its electricity, and deforestation of watersheds has led to high sedimentation rates. The lost revenue to date at the Cachi Dam is estimated to be at least $133 million, and other dams are silting up as well (33). Learning from its mistakes, Costa Rica has rallied to save its natural resources, both through official and private programs drawing on the country's impressive human resources in higher education and public service.

Other countries in Central America lack the human and institutional resources of Costa Rica, but find the same problems festering. In El Salvador, only 3

Figure 3.6 Annual Deforestation in Three Regions

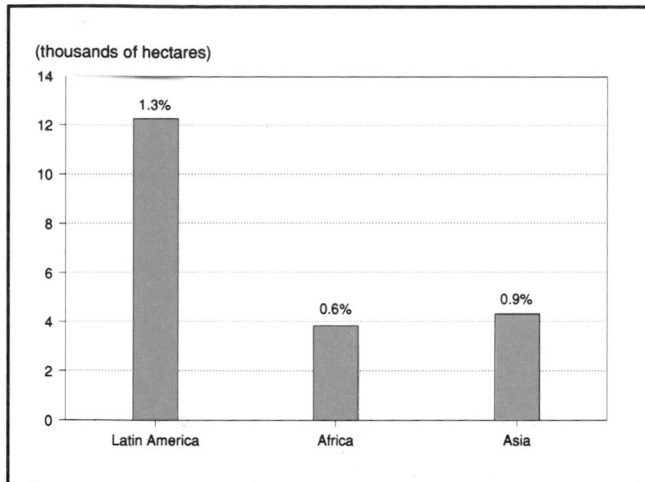

Source: Chapter 19, "Forests and Rangelands," Table 19.1.
Note: The percentages indicate the area of each region deforested annually.

percent of the country is covered with tracts of woodland large enough to be considered forests (34). Ravaged by a decade of war, the country has been unable to react to the deteriorating environment, even though it suffers from severe soil erosion and siltation of reservoirs. In Nicaragua, as if the destruction brought by war were not enough, the country's largest tract of rainforest north of the Rio San Juan was hit by Hurricane Joan in October 1988. An estimated 80 percent of the trees were blown over or otherwise damaged over an area of Southeastern Nicaragua (35).

Settling the Amazon

Thus it is not only in Brazil that deforestation is a problem, but throughout Latin America. By its sheer size, the problems of Brazil are writ larger. All of Central America is smaller than the 410,000-square-kilometer area of western Amazonia that became the focus of Brazil's controversial Northwest Brazil Integrated Development Program, called Polonoroeste.

Polonoroeste was intended to promote sustainable farming systems based on tree crops and to include environmental protection. Instead, unguided colonization took place without regard to land-use studies, ecological reserves, or the rights of native people. Settlers poured into the state of Rondonia, increasing its population at an average rate of 14 percent a year since 1980. Many of the settlers came from southern Brazil, where small coffee farms were being turned into large soybean-producing enterprises. Over a quarter of Rondonian immigrants came from Parana state, where the soybean acreage went from 400,000 hectares to 2 million in the 1970s, with the massive increase taking place primarily on large

mechanized farms. Displaced small farmers sought new land on the Brazilian frontier (36).

The wave of settlers in Rondonia cleared and burned the forest. Deforested areas that in 1980 accounted for 3 percent of Rondonia had by 1988 increased to 24 percent. (See Chapter 7, "Forests and Rangelands," Figure 7.3.) The deforestation was in part a direct result of existing government policies. First, clearing the land was required as evidence of "land improvement," which confers on settlers the right to claim title to an amount of land in direct proportion to the area of forest cleared. Second, land values soared as roads improved and more migrants came. Brazil's high inflation makes investment in land seem safer, and income tax exemptions for large-scale agriculture such as ranching make it a tax shelter. Poor settlers, who do not benefit from the tax incentives, gained more by clearing land, selling it to large ranchers, and moving on than they could from farming the land (37).

In addition to the deforestation caused by small farmers, large tracts of the Amazon have been cleared for cattle ranches by wealthy investors and corporations. These ventures have not fared well economically and would not be viable without major government subsidies and tax incentives. Livestock investors benefited from $730 million in tax exemptions and subsidized rural credit between 1966 and 1983 (38). More than 527 livestock ranches gained from the regional tax credits offered: their average size was 23,600 hectares (39)—not small farms even by Amazonian standards.

A more fundamental problem, however, is that experience has shown that livestock activities in Amazon regions with poor soils are not sustainable. A large fraction of these ranches—at least 20 percent of the 4 million hectares of Amazon ranchland, by one estimate—have been abandoned (40). Once abandoned, the pastures are often taken over by weeds that grow on the poor, now compacted soil. Some abandoned pastures do return to forests, in 100–500 years, but the secondary growth may never achieve the complexity of the original forest ecosystem (41). (See Chapter 7, "Forests and Rangelands," Box 7.3.)

Soil erosion and loss of species are the obvious results from deforesting the Amazon. Less obvious, but beginning to be a source of concern, are local climate changes. The Amazon basin is a major source of water vapor—forests evapotranspire much more than nearby savannas—not only for itself, but also for the central plateau to the south (42) where major agricultural investments are being made. Lower amounts of rainfall could harm the farmers of the savannas and the Amazon ecosystem. Besides the micro climate changes, burning the forest is a significant source of carbon dioxide, a gas that is the major contributor to the greenhouse effect. In 1987, deforestation in Brazil accounted for more than 15 percent of the world's total carbon dioxide emissions. (See Chapter 24, "Atmosphere and Climate," Table 24.3.)

Box 3.3 Drugs in the Wilderness

Coca plants—hardy, drought-resistant, providing at least three harvests each year—appear to be a poor farmer's dream. The payoff from a hectare of coca plants can be 18 times more than any other crop would earn, and more cash than most small farmers have seen in their lives (1).

The trade in cocaine, marijuana, and heroin is having a major effect in many parts of Latin America. Production of drugs affects the environment directly through deforestation, soil erosion, and stream pollution. More important, it is also distorting local and national economies, creating social problems including addiction and corruption locally, and violently undermining governments' ability to govern their own territories.

THE BOOM CROP OF THE 1980s

Cocaine, the boom crop of the 1980s, has led to the planting of more and more coca bushes. Peru is the largest producer of the leaf in Latin America, with about 60 percent of the crop, followed by Bolivia with 28 to 30 percent, and the rest grown in Colombia, with a few plantings in Ecuador and Brazil (2).

Traditionally, coca plants grow on hillsides in subtropical moist forests at 1,000 to 1,200 meters above sea level, but one variety, "epadu," has been found suitable for tropical lowlands such as the Amazon region. To produce cocaine, the coca leaves are converted into paste, then refined into cocaine hydrochloride, using chemicals, such as sulfuric acid, kerosene, lime, and carbides. It takes about half a metric ton of dry leaves to produce a kilo of cocaine.

The increase in land devoted to coca reflects both growing demand and the enormous profits to be reaped from this illegal trade. Bolivia went from 22,788 hectares in 1980 to 70,900 hectares in 1986, according to its own government (3), and Peru went from approximately 18,000 hectares in 1979 to 200,000 in 1987 (4).

Most coca farms are found on newly deforested land in some of the most biologically rich areas of the continent, including parks and forest reserves. The amount of deforestation attributed to the cocaine business is much larger than the actual area planted, because the growers are constantly moving to clear new areas either to escape detection or to increase yields as the soil wears out. In Peru, cocaine growers are blamed for about 10 percent of all deforestation (5).

Coca is not a sustainable crop unless it is cultivated in the traditional Andean way, on terraced hillsides, interspersed with other crops. Growers trying to maximize profits while avoiding detection do not worry about soil erosion on the steep slopes where they plant coca. Because each trimesterly harvest strips the plant of leaves, the soil is more exposed to the erosive effects of heavy rains. Migrant growers, with little knowledge of either traditional or modern agriculture, often apply excessive amounts of pesticides to the plants; most of the pesticides wash downstream with the soil polluting the rivers. A number of disasters such as mud slides and flooding along Peru's rivers are the result of deforestation and erosion in the coca regions.

COCA'S TOXIC WASTES

Another environmental concern raised by the cocaine industry is the amount of chemicals dumped into the rivers and abandoned by the cocaine laboratories. In 1986 alone, the amount of chemicals used in Peru's upper Huallaga valley to convert the average crop to cocaine base was estimated at 57 million liters of kerosene, 32 million liters of sulfuric acid, 6.4 million liters each of acetone and toluene, 3,200 metric tons of carbide, 16,000 metric tons of lime and 16,000 metric tons of toilet paper that was used to filter the paste (6). Not only the narcotics dealers dump the toxic wastes; the police who confiscate the chemicals also pour them into rivers (7). The people who fish along these rivers are at risk,

and only the lack of inhabitants in some areas keeps the pollution from being perceived as a fatal side effect of the drug trade.

Plans to combat the drug trade by spraying herbicides raise additional environmental concerns. The herbicide Tebuthiuron or "Spike," intended for use in Peru, is scattered in pellet form to keep it from drifting into surrounding rainforests. But the herbicide destroys more than just coca plants, and the U.S. Environmental Protection Agency labels Spike as not intended for cropland and dangerous for wetlands, because it can remain in rivers and farmlands for up to five years (8). A similar problem is reported in Guatemala, where opium production is spreading from Mexico, and farmers complain that the herbicide sprayed from planes kills honeybees and affects legal crops as well as poppy fields (9). There also has been controversy about the chemicals used to spray marijuana fields in the rainforest regions of Guatemala and Belize. In environmental terms, the use of herbicides only increases the extent of land damaged. Once one coca plantation is destroyed, not only is that land degraded, but the growers have probably moved on to clear another area of forest.

ERADICATION EFFORTS INEFFECTIVE

Eradication efforts, with or without herbicides, have been ineffective in controlling drug production. When Peru had its "best" year in 1985 by eliminating 5,000 hectares of coca plantations, about 12,000 new hectares went into cultivation (10). In 1988, when Bolivia passed tougher narcotics laws, coca production increased by over 20 percent (11).

Prices paid for coca leaves are increasingly unstable and the drug business is notoriously violent, even for those who grow the leaves. Thus if farmers are offered the alternative of an assured market for legal crops, with credit and technical support for sustainable farming techniques, there is at least a chance

World Resources 1990–91

Economics of the Tropical Forest

The economic costs of destroying tropical forests are not easy to calculate. These complex ecosystems are believed to contain half of all plant and animal species on Earth. Some medicines and foods are still to be found in the future, and new woods and timber products may be adapted to future industrial uses; their value cannot be estimated. Massive deforestation may fundamentally alter the hydrological characteristics of the Amazon basin, and may cause climatic changes on a regional or even global scale. Yet, even the known values are not being taken into account. A study of the annual market value of edible fruits, cocoa, and rubber from one hectare of Peru's Amazon forest showed that, over time, the revenue was approximately six times the amount that could be earned from harvesting all the timber in a single year or twice the value of converting the land to cattle pasture (43). This study supports the advocates of "extractive reserves" where forests are set aside as a source of natural products that can be harvested indefinitely without clearing the land.

Meanwhile, even the current value of cut timber is ignored. The vast majority of the trees cleared in the Amazon region are burned or left to rot (44). For example, of the Amazon ranches supported by government tax breaks, only 20 percent ever marketed

Box 3.3

they may choose security over the risks associated with windfall drug profits. A successful drive against the drug trade in Latin America, however, would be a major shock to the economy. In Bolivia, about 500,000 people are working at jobs derived from coca production; coca provides almost one out of every three jobs to a work force of 1.7 million (12). Bolivia's coca crop grossed at least $750 million in 1987 and about one half of that stayed in the country (13)—a sum larger than all of Bolivia's legal exports combined. Peru's cocaine trade was estimated to bring between $750 million and $1.5 billion, while its legal exports were worth $2–4 billion in 1989 (14). Colombia is involved in exporting more than half the world's cocaine, but its economy is large enough that the revenues of $2 billion are less than 40 percent of legal exports (15).

LATIN YOUTHS ADDICTED

If governments only considered the economic benefits of drug trafficking, developing countries would be tempted to rebel against the demand of consuming nations for controls on supply. But there are disincentives. One is the destruction of human minds and bodies, which Latin Americans can see in their own countries. Colombians see the danger in Bogotá, where studies estimate that 130,000 youth between the ages of 12 and 30 are addicted to cocaine or smoking the coca base, basuco, made from byproducts of cocaine production (16). In the slums of Lima, pastillos, cigarettes laced with coca paste, are all too common. In one survey of peasants who migrated to coca-growing areas to work in the laboratories, 80 percent had used the drug regularly (17).

The corruption and violence have hurt the producer nations most, as well as the intermediate countries webbed into the narcotics transport network. The Colombian drug cartel has financed rightist paramilitary groups and set up schools for assassins. In the war on drugs, Colom-

bian casualties have included not only supreme court justices, an attorney general, the minister of justice, two dozen journalists, more than fifty judges, and hundreds of police officers (18), but also, in August 1989, the leading presidential candidate. In Peru, the narcotics dealers formed an alliance with the guerrillas of Sendero Luminoso (Shining Path), which gives extreme leftists slush funds to buy arms, and leaves out-gunned law enforcement officers turning to military counterinsurgency forces for help.

POLITICAL COURAGE NEEDED TO FIGHT DRUGS

As the corruptive power of drug money and the threats of violence overwhelm governments, what happens to programs for natural resource management? National parks have been taken over by cocaine producers in Peru and Colombia. As repression increases in one area, drug producers move to another area, usually farther into the wilderness. The ecological side effects of the drug trade are all undesirable, but the greatest threat is to the ability of Latin American countries to control their own political systems and to govern the territory where the resources lie. Ecological zoning in the Amazon will have no effect if it cannot be enforced because drug traffickers and guerrillas determine what happens there. Yet to oppose the narcotics trade requires political courage. Colombia's struggle since the assassination of the leading presidential candidate in August 1989 shows the difficulty of this decision.

Colombian President Virgilio Barco told the United Nations General Assembly a year earlier that "nuclear arms are a distant threat to Colombians compared to terror, crime, and drug dealers." He called the worldwide demand for narcotics "the principal threat to democracy in Latin America." Without democracy, the chances for Latin Americans to tackle their own environmental problems will diminish.

References and Notes

1. United Nations Environment Programme (UNEP), "El Medio Ambiente y el Desarrollo en America Latina y el Caribe: Una Vision Evolutiva," UNEP Regional Office for Latin America, discussion paper, Mexico City, May 1989, p. 50.
2. Robert Graham, "Coca cultivation encouraged by deep Andean traditions," *Financial Times* (London, Nov. 28, 1988), p. 10.
3. *Op. cit.* 1, p. 53.
4. Marc Dourojeanni, Senior Environmental Advisor, the World Bank, October 1989 (personal communication).
5. *Ibid.*
6. *Ibid.*
7. Mark Mardon, "The Big Push," *Sierra Magazine* (Sierra Club, San Francisco, November/December 1988), pp. 68-75.
8. Cynthia McClintock, "The War on Drugs: The Peruvian Case," *Journal of Interamerican Studies and World Affairs*, Vol. 30, Nos. 2 and 3 (Institute of Interamerican Studies, Miami, Florida, Summer/Fall 1988), p. 134.
9. Wilson Ring, "Opium Production Rises in Guatemala Mountains," *The Washington Post* (June 30, 1989), p. A25.
10. *Op. cit.* 8, p. 130.
11. Joseph B. Treaster, "In Bolivia, U.S. Pumps Money into the Cocaine War, but Victory is Elusive," *The New York Times* (June 11, 1989), p. 18.
12. "The Cocaine Economies," *The Economist* (October 8, 1988), p. 22.
13. *Ibid.*, pp. 21-22.
14. Gustavo Gorriti, "How to Fight the Drug War," *The Atlantic Monthly* (July 1989), p. 71.
15. *Ibid.*
16. Alcaldia Mayor de Bogotá, "Bogotá y el Consumo de Substancias Psicoactivos," Alcaldia Mayor de Bogatá, March 1989, pp. 58, 127.
17. E. Morales, "Coca and the Cocaine Economy and Social Change in the Andes of Peru," *Economic Development and Cultural Change*, Vol. 35, No. 1 (October 1986), cited in Cynthia McClintock, "The War on Drugs: The Peruvian Case," *Journal of Interamerican Studies and World Affairs*, Vol. 30, Nos. 2 and 3 (Institute of Interamerican Studies, Miami, Florida, Summer/Fall 1988), p. 136.
18. Michael Massing, "The War on Cocaine," *New York Review of Books*, Vol. 35, No. 20 (December 22, 1988).

some of their timber, wasting an estimated $100 million to $250 million in commercially valuable wood (45). More effective government efforts to support and encourage sustainable uses of the forest—and to enforce more strictly the environmental laws already in place—are needed.

In Rondonia, however, fiscal incentives played only a small role in the slash-and-burn destruction—only 20 of Brazil's 631 government-subsidized livestock projects were located in the state. Here the human pressure of settlers desperate for free land was the main source of destruction. Government austerity measures had drastically reduced the available

credit that was needed to buy fertilizers and other necessary inputs to grow the planned tree crops, and the absence of government extension services left unprepared settlers on their own (46).

Broken Dreams

A large number of Amazon settlers fail. On average, 60 percent surrender (47) to disease, violence, crop failure, or the temptation of quick cash from selling out. Disease is a major problem where health facilities are rare, and, in 1987, an estimated 20 percent of the settlers were infected with malaria (48). Violence is the standard way of settling rival claims, personal vendettas, and political challenges. One victim, Francisco "Chico" Mendes, leader of the rubber tappers' union, brought to the world's consciousness the violent way that land-use disputes are settled in the Amazon region. (See Chapter 7, "Forests and Rangelands," Box 7.4.) The Catholic Church in Brazil recorded 533 assassinations over land disputes between 1984 and 1988, the majority in the Amazon region (49).

The waste of biological, fiscal, and human resources in the deforestation of the Amazon region has begun to cause national and international concern. A dramatic demonstration of the extent of the problem came in 1987, when 170,000 fires were photographed from satellites, 7,600 in a single day, according to Brazil's National Space Research Institute (Instituto Nacional de Pesquisas Espaciais) (50).

In diverting the pressure of the landless poor from the tropical forests, one alternative can be just as destructive, that is, pushing subsistence farmers farther up steep slopes in the hills and mountains. About 25 percent of Latin America's land is found on hills or plateaus vulnerable to erosion and land degradation, and 50 to 75 percent of the small farms are concentrated in these areas (51).

Land Reform and Other Options

To slow the rush of landless settlers to endangered frontiers such as the tropical forests will require that other areas be made available for farming, that urban job opportunities be improved, and that agricultural production on existing farmland be intensified. Resolving land tenure questions is likely to be a necessary part of reducing not just deforestation but the erosion and soil degradation that occurs when steep slopes or other marginal lands are farmed with unsophisticated methods. In Brazil, for example, 2 percent of the landowners control 60 percent of all arable land (52). In 1985, then President José Sarney launched an agrarian reform program. Four years later, he canceled the program after only 10 percent of the targeted land had been distributed. Out of about 12 million landless rural workers (53), the original goal was to help 1.4 million families. Only 77,351 families—5 percent of the goal—eventually received land (54). Land reform programs have been initiated in most Andean countries—Chile, Peru, Ecuador, and Bolivia—but, over time, they have not had the impact intended, and in some cases, they have been reversed (55).

In Peru, agrarian reform in the 1970s led to large-scale cooperative farms nearly all of which have since been broken up after internal management disputes and labor problems. These problems went hand in hand with a severe deterioration of the economy. Facing economic problems and growing poverty, Peru also decided that the poor should be

offered new land in the Amazon. In 1989 the Peruvian legislature passed, over extensive opposition, a law to promote development in the Amazon region. The law appeared to ignore all recent experiences in the region, such as the Polonoroeste project, and called for migration from the Andes to the rainforests (56).

DEMOCRACY

While most indicators of economic and social progress in Latin America have not been encouraging, the trend toward democratic governments has been positive. Democratic governments appear to be surviving even the severe challenge of the drug trade. (See Box 3.3.) With the end of military regimes and gradual transitions from one elected government to another in the 1980s, environmental activism in Latin America became a tiny voice growing louder.

Another manifestation of democracy in Latin America is the proliferation of private, nonprofit environmental organizations in the 1980s. These organizations have begun to contribute in important ways. Some like Fundacion Natura in Ecuador have become highly professional, capable of mounting education campaigns and running nature reserves. Others remain largely volunteer groups rallying around a single issue, such as a Mexican group fighting air pollution by organizing drivers to leave their cars at home one day a week.

In Ecuador, Fundacion Natura spearheaded a hopeful development—an innovative debt-for-nature swap that converted $10 million in foreign debt into local conservation bonds. In the first year alone, interest payments on those bonds equaled the government's budget for national parks (57). Similar swaps took place earlier in Bolivia and Costa Rica. The Honduran Ecological Association collaborated with the government in helping to manage national parks and has received as much as 20 percent of its income in government grants for this purpose. The Peruvian Association for the Conservation of Nature establishes and operates environmental education programs for primary school teachers throughout the country (58).

As the nongovernmental organizations have grown, some have decided to increase their strength by banding together. Thus, Bolivia's environmental groups formed a coalition, LIDEMA, League in Defense of the Environment. A similar coalition is the Conservation Federation of Guatemala. Cooperation is also spreading across national boundaries, with several groups forming networks to exchange information and support across Latin America, including REDES (the Regional Network of Nongovernmental Organizations for the Sustainable Development of Central America) a coalition of conservation and development organizations from seven Central American countries, and a Tropical Rain Forest Network being established in the countries bordering the Amazon basin (59).

A New Constituency

Recently, these environmental groups began to be seen as representing a new political constituency. Some Latin American leaders took notice. Former Costa Rican President Oscar Arias committed his government to conservation and set up a strong ministry of natural resources. Guatemalan President Vinicio Cerezo pushed for a regional commission on environment as part of the Central American peace plan. In Brazil, Fabio Feldman, a São Paulo lawyer known for his defense of communities affected by pollution, was elected to the national congress that wrote a new constitution. He helped create a new chapter on the environment in that constitution, ratified in 1988. Venezuela's presidential candidates both made environmental statements part of their 1988 platforms. As Chile returned to democracy in 1989, the political parties pounded on the doors of environmental groups asking for proposals on how to deal with Santiago's smog. The two leading presidential candidates appeared before an environmental conference to outline their positions on the pollution and resource issues. In Costa Rica, the two principal presidential candidates for the 1990 elections set forth environmental programs.

Two factors are apparent in the new political activism on environmental issues. One is the importance of a democratic climate. When the channels of political expression are controlled and sealed off, there is little debate of these issues. If there is no way to change laws by electing committed candidates or by using a free press to alert citizens to problems, those with environmental concerns can only solicit favors from the governing power, without any other recourse.

Another characteristic of the environmental politics is the urban base of the most influential organizations, those closest to the government elite. Deforestation and soil erosion do not get so much attention from the politicians as air and water pollution, unless the economic bills from rural land degradation turn up in the form of increased flooding of downstream urban areas or decreased hydroelectric capacity for industry.

Latin Americans: Concerned and Pessimistic

A public opinion survey conducted for the United Nations Environment Programme on four continents in 1988 revealed that Latin Americans were more concerned about the environment and more pessimistic about its future than the average global citizen. (See Chapter 1, "World Environment Outlook.") From those surveyed around the world, 87 percent said protecting the environment should be a major priority for governments—94 percent from Latin America

insisted on this. In this region, 74 percent said the governments' failure to adopt effective policies was a major cause of environmental damage, and 85 percent said that nongovernmental organizations can play a major part in solving environmental problems.

Latin Americans—92 percent of those surveyed—thought the government should have the major responsibility for environmental protection, but they were less willing to pay higher taxes for this purpose. Only 28 percent were "very willing" to accept a tax increase, but 84 percent would be willing to work two hours a week or contribute financially to environmental projects. The fact that Latin Americans are less accustomed to paying taxes and more cynical about the misuse of public funds might have influenced their preference for volunteering or supporting nongovernmental organizations in order to improve the environment. Nonetheless, 85 percent said they would choose a lower standard of living with fewer health risks than a higher standard of living with more risks, if given the alternative.

Despite the impact of the economic crisis, it is clear that the fundamental causes of many of Latin America's environmental problems are poor planning, mismanagement, and neglect of the environment by both government and private industry. All too often, dams have been built without watershed protection, irrigation projects have not considered salinization of the soil, and environmental regulations have been either nonexistent or not enforced. Industrial plants commonly have been built and operated without emission controls, as have cars and trucks. The result has been that government policies intended to promote growth and development have often had disastrous effects on the environment, leading to threats to human health, lost revenues, or greater costs at a later date (60). In Costa Rica, for example, eroding topsoil from a deforested hillside that might have been protected for $5 million has drastically shortened the life of an expensive hydroelectric dam (61). In the Carajas region of Brazil, dozens of pig-iron smelters depend for their economic viability on consuming 72,000 hectares of the surrounding forest each year as "free" fuel (62).

To offset this kind of ill-conceived development and the need for expensive cleanup efforts, it may be essential to improve the capacity for environmental planning and regulation in Latin American countries. Such efforts will require significant resources, but the costs of planning and prevention are likely to be significantly less than the cost of rehabilitation and restoration.

The overview section, A Huge and Diverse Region, was written by Eric Rodenburg, Research Director of World Resources 1990-91; *the rest of the chapter was written by Diana Page, a specialist on Latin America and special liaison for nongovernmental organizations with the Center for International Development and Environment of the World Resources Institute.*

References and Notes

1. Food and Agricultural Organization of the United Nations (FAO), *FAO Quarterly Bulletin of Statistics*, Vol. 2, No. 2, 1989, pp. 12, 16.
2. World Energy Conference, 1989 Survey of Energy Resources, (World Energy Conference, London, 1989), pp. 101-104, 133.
3. United Nations Economic Commission for Latin America and the Caribbean (ECLAC), *Preliminary Overview of the Economy of Latin America and the Caribbean 1989* (ECLAC, Santiago, Chile, December 1989), Table 5, p. 20.
4. Carol Graham, "The Latin American Quagmire," *The Brookings Review* (The Brookings Institution, Washington, D.C., Spring 1989), p. 43.
5. *Op. cit.* 3, Table 2, p. 18 and Table 3, p. 19.
6. *Op. cit.* 3, Table 3, p. 19.
7. International Monetary Fund (IMF), *International Financial Statistics* (IMF), Washington, D.C., May 1989), Vol. XLII, No. 5.
8. Inter-American Development Bank (IDB), *Economic and Social Progress in Latin America 1988 Report* (IDB, Washington, D.C., 1988), p. 2.
9. Mac Margolis, "Amazon Rain Forest Faces New Season of Controversy," *The Washington Post* (June 24, 1989), p. A24.
10. Department of International Economic and Social Affairs, United Nations Population division, *Prospects of World Urbanization, 1988* (United Nations, New York, 1989), Table A-10, pp. 82, 89, 152, and Table A-1, p. 31.
11. *Op. cit.* 8, Table A-2, p. 535.
12. United Nations Environment Programme (UNEP), "El Medio Ambiente y el Desarrollo en America Latina y el Caribe: Una Vision Evolutiva," UNEP Regional Office for Latin America, discussion paper, Mexico, June 1989, pp. 34- 35.
13. *Ibid.*, p. 36.
14. "Brusco Aumento de las consultas respiratorias," *La Epoca*, (Santiago, Chile, June 21, 1989), p. 16.
15. Ricardo Katz, Comision Nacional de Energía, Santiago, Chile, February 1990 (personal communication).
16. Helena Landazuri and Caroline Jijon, *El Medio Ambiente en el Ecuador* (Instituto Latina Americano de Investigaciones Sociales, Quito, Ecuador, 1988), pp. 51-52.
17. Stephen J. Rothenberg, Lourdes Schnaus-Arrietu, Irving A. Perez-Guerrero, *et al.*, "Evoluación del Riesgo Potencial de la Exposicion Perinatal al Plonou en el Valle de Mexico, *Perinatologia y Reproduccion Humans*, Vol. 3, No. 1 (1989), p. 49.
18. Larry Rohter, "Mexico City's Filthy Air, World's Worst, Worsens," *The New York Times* (April 12, 1989), p. A-1.
19. William Branigan, "Bracing for Pollution Disaster: Mexico City Smog Seen Eventual Death Trap," *The Washington Post* (November 28, 1988) p. A14.
20. *Op. cit.* 18.
21. *Op. cit.* 19.
22. *Op. cit.* 12, p. 33.
23. Brook Larmer, "Mexico Fights Pollution: New Program Thins Smog, Street," *The Christian Science Monitor* (Boston, Massachusetts, January 31, 1990).

24. *Op. cit.* 18.
25. *Op. cit.* 12, p. 30.
26. Fernando Casas Castaneda, "The Risks of Environmental Degradation in Bogota, Colombia," *Environment and Urbanization*, Vol. 1, No. 1 (International Institute for Environment and Development, London, U.K., April 1989), p. 17.
27. Frederick S. Mattson, *Planning for Central America Water Supply and Sanitation Programs: Update* (CDM and Associates, Arlington, Virginia, May 1989), Table 1, p. 4, and Table 2, p. 5.
28. *Op. cit.* 12, p. 29.
29. *Op. cit.* 12, pp. 29-30.
30. Judith Gradwohl and Russell Greenberg, *Saving the Tropical Forests* (Earthscan Publications, London, U.K., 1988), p. 33.
31. Alvaro Umana, "Costa Rica Swaps Debt for Trees," *Wall Street Journal* (March 6, 1987).
32. D.A. Matamoros, "Los Recursos Forestales: Borrador de Trabajo, estudio del estado del ambiente," Fundación Neotrópica San José, Costa Rica, 1987, cited in Fundación Neotrópica, Desarrollo Socioeconomico y el Ambiente Natural de Costa Rica: Situacion Actual y Perspectivas (Editorial Heliconia, San José, Costa Rica, May 1988), pp. 61-62.
33. U.S. Agency for International Development (AID), *Regional Tropical Watershed Management* (AID), Washington, D.C., 1983) cited in H. Jeffrey Leonard, *Natural Resources and Economic Development in Central America* (International Institute for Environment and Development/Transaction Books, New Brunswick, New Jersey, 1987), p. 135.
34. S. Hilty, *Environmental Profile of El Salvador* (Arid Lands Information Center, University of Arizona, Tucson, 1982), cited in H. Jeffrey Leonard, *Natural Resources and Economic Development in Central America* (International Institute for Environment and Development/Transaction Books, New Brunswick, New Jersey, 1987), p. 119.
35. Doug Boucher, "Nicaraguan Rainforest Regenerating Well After Hurricane," unpublished report, Rockville, Maryland, March 1989.
36. Dennis J. Mahar, *Government Policies and Deforestation in Brazil's Amazon Region* (The World Bank, Washington, D.C., 1989), pp. 30-31, 34-35.
37. *Ibid.*, pp. 34, 37-39.
38. Robert Repetto, *The Forest for the Trees? Government Policies and the Misuse of Forest Resources* (World Resources Institute, Washington, D.C., May 1988), p. 76.
39. Hans P. Binswanger, "Brazilian Policies that Encourage Deforestation in the Amazon" (The World Bank, Washington, D.C., 1989), p. 13.
40. João Meirelles Filho, *Amazonia: O Que Fazer Por Ela?* (Companhia Editora Nacional, São Paulo, 1986), p. 77.
41. C. Uhl, R. Buschbacher, and E.A.S. Serrao, "Abandoned Pasture in Eastern Amazonia, I. Patterns of Plant Succession," *Journal of Ecology*, Vol. 76 (1988) pp. 676, 678.
42. Eneas Salati, R.L. Victoria, L.A. Martinelli, *et al.*, "Deforestation and Its Role in Possible Changes in the Brazilian Amazon," Es-

cola Superior de Agricultura Luiz Queiroz, Piracicaba, Brasil, April 26, 1989.
43. Charles M. Peters, Alwyn Gentry, and Robert O. Mendelsohn, "Valuation of an Amazonian Rainforest," *Nature*, Vol. 339, No. 6227 (June 29, 1989), p. 655-656.
44. *Op. cit.* 36, p. 9.
45. *Op. cit.* 38, p. 80.
46. *Op. cit.* 36, pp. 36, 37.
47. Alain de Janvry and Raúl Garcia, "Rural Poverty and Environmental Degradation in Latin America: Causes, Effects, and Alternative Solutions," paper presented at the International Consultation on Environment, Sustainable Development and the Role of Small Farmers, International Fund for Agricultural Development (IFAD), Rome, October 1988.
48. *Op. cit.* 12, p. 64.
49. "Nova Republica Enterra 533 Agricultores em Quatro Anos," *Jornal do Brasil* (Brasilia, March 5, 1989), p. 8.
50. Marlise Simons, "Vast Amazon Fires, Man-Made, Linked to Global Warming," *The New York Times* (August 12, 1988), p. A6.
51. *Op. cit.* 47, p. 3.
52. Alan Riding, "The Struggle for Land in Latin America," *The New York Times* (March 26, 1989).
53. Ricardo Reboucas, "Adjustment Policies and the Poor in Brazil," Instituto Brasileiro de Analises Sociais e Economicas, Rio de Janeiro, September 1987, p. 7.
54. "Reforma Agraria de Sarney Fica Apenas nas Promessas," *Jornal do Brasil* (Brasilia, March 5, 1989), p. 8.
55. See William Thiesenhusen (Ed.), *Searching for Agrarian Reform in Latin America* (Unwin Hyman, Boston, Massachusetts, 1989).
56. Pablo de la Cruz, lawyer, Centro de Investigacion e Promocion Amazonica, Lima, Peru, February 1990 (personal communication).
57. Roque Sevilla L., "El Canje de Deuda por Conservacion en America Latina y el Caribe," paper presented at the United Nations Environment Programme Meeting of High- Level Governmental Experts on Regional Co-operation in Environmental Matters in Latin America and the Caribbean, Brasilia, March 1989, p. 14.
58. Lisa Fernandez, "Private Conservation Groups on the Rise in Latin American and the Caribbean," *World Wildlife Fund Letter*, 1989, No. 1, World Wildlife Fund, Washington, D.C., p. 4.
59. *Ibid.*, pp. 4-5.
60. Robert Repetto, and Malcolm Gillis, eds., *Public Policy and the Misuse of Forest Resources* (Cambridge University Press, New York, 1988).
61. Mostafa Tolba, address to the Sixth Ministerial Meeting of the Environment in Latin American and the Caribbean, Brasilia, Brazil, March 30, 1989, p. 4.
62. Philip Fearnside, "The Charcoal of Carajas: A Threat to the Forests of Brazil's Eastern Amazon Region," *Ambio*, Vol. 18, No. 2 (1989), p. 142.

Part III

4. Population and Health

Much of the discussion of population issues in recent years has focused on falling birth and fertility rates. And, in fact, population growth *rates*— that is, the percentage change over time—have peaked everywhere except in Africa. In absolute terms, however, the world's population is increasing rapidly and will continue to do so well into the 21st Century.

The cause of this continued rapid growth is quite simple: Even though the number of children born per woman has decreased, the number of women of childbearing age has increased more rapidly. This phenomenon—the huge bulge in numbers of younger people in the age structure of populations in most of the developing world—is itself a consequence of past rapid growth in these populations. And although the details vary by country and by region, the fact of youthful populations and the virtually inevitable consequence of rapid population growth in coming decades will have enormous impact on the need for food, water, housing, and jobs, and on the natural resources needed to provide them.

On a national scale, the age structure of a country's population is crucial to development planning efforts; on an international scale, the implications of this rapid population growth are enormous for regional and global environmental issues such as climate change. Much of the predicted population growth is expected to occur in urban areas, where,

for example, use of fossil fuels for heating and for transport is more common than in rural areas.

This chapter also examines two health problems—an old one and a new one. Malaria has long been a major cause of death and ill health in the tropical areas of the world. Despite partially successful efforts to combat this disease in recent decades, malaria is now on the upswing again. Attempts to eradicate the disease are, in consequence, giving way to health strategies that attempt to manage its human impact.

AIDS—acquired immune deficiency syndrome—is still a relatively new disease, whose full impact on death rates, health costs, and social disruption has yet to be realized. Moreover, evidence shows that the disease is continuing to spread rapidly in many regions. This chapter reports on the growing presence of AIDS in the Americas and particularly in Latin America and the Caribbean, where the disease, without aggressive efforts to counter its spread, may well become a significant health problem.

CONDITIONS AND TRENDS

WORLD POPULATION TRENDS

Between 1990 and 2025, world population will increase by 3.2 billion people, according to projections of the United Nations Population Division (1). Of this growth, 3 billion will occur in the developing regions of Africa, Asia, and Latin America. The now developed regions of the world will experience population growth of only about 166 million. (See Table 4.1.)

The developed regions' share of total world population has dropped continuously since 1950, when it was 33 percent. By 2025, it will be only 16 percent. During that time, the developing regions' share will increase from 67 percent in 1950 to 84 percent in 2025. Africa's large increase in projected population is the most dramatic, with its share of total world population increasing from 9 percent in 1950 to 19 percent in 2025.

While population projections are not infallible and forecasts have changed from time to time, particularly for individual countries, overall the projections have proved a remarkably reliable guide to future population growth. (See Box 4.1.)

Historically, population growth was low because most societies had both high birth rates *and* high death rates. With only a small gap between the number born and the number who died, population grew slowly. With the introduction in many developing countries after World War II of ideas and technologies affecting health, sanitation, medicine, and education, death rates plummeted quickly in those countries. At the same time, birth rates remained high, in accordance with traditional cultural norms, values, and attitudes (2).

For the first time, there was a large discrepancy between birth and death rates, producing the high population growth rates that are now associated with developing countries. Today, death rates are near or at their anticipated minimum, and birth rates have declined or are starting to decline in most countries. If these trends continue, population growth will eventually return to traditional low rates sometime in the 21st Century—this time with births and deaths stabilized at low rather than high levels (3).

But birth and death rates are far from the whole story. A very youthful population such as in most de-

veloping countries or an aging population like in many industrial countries has different implications for future population growth and for social needs. Thus the age structure of a population and how it

Figure 4.1 Comparision of Population Age Profiles, More Developed Regions and Latin America, 1989

A. More Developed Regions

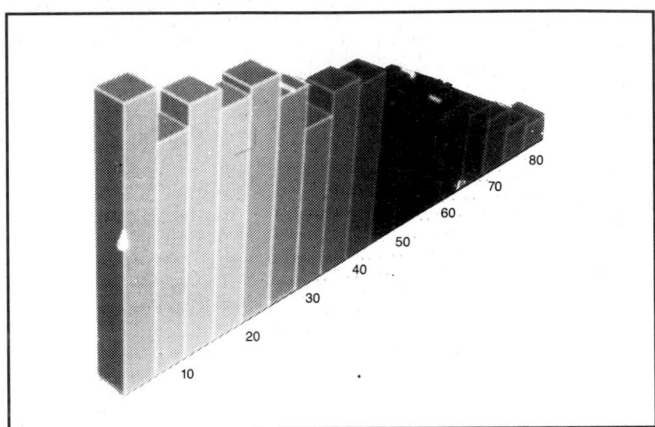

B. Latin America

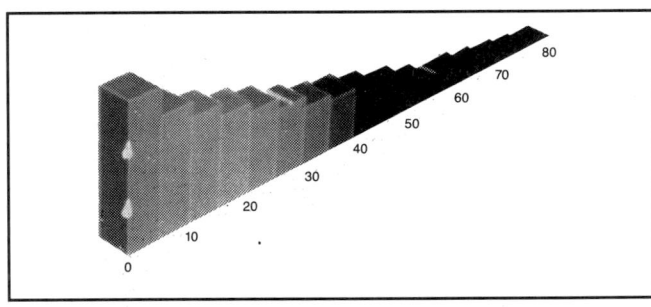

Source: Graphics by Robert Fox, Allen Carroll, and Melvin L. Prueitt, Population Images, Bethesda, Maryland, with data from Department of International Economic and Social Affairs, United Nations, *Global Estimates and Projections of Population by Sex and Age: The 1988 Revision* (United Nations, New York, 1989), p. 10.

Table 4.1 Population Size and Projections for Major World Regions, 1950–2025

	Population (in millions)					Percentage Share of the World				
	1950	1975	1990	2000	2025	1950	1975	1987	2000	2025
World	2,515.3	4,079.8	5,292.2	6,251.1	8,466.5	100.0	100.0	100.0	100.0	100.0
More Developed Regions	832.4	1,095.6	1,205.2	1,262.5	1,352.1	33.1	26.9	23.6	20.2	16.0
Less Developed Regions	1,682.9	2,984.1	4,087.0	4,988.6	7,114.4	66.9	73.1	76.4	79.8	84.0
Africa	224.1	415.1	647.5	872.2	1,581.0	8.9	10.2	11.8	14.0	18.7
Latin America	165.4	322.7	448.1	539.7	760.4	6.6	7.9	8.4	8.6	9.0
China	554.8	927.3	1,135.5	1,285.9	1,492.6	22.1	22.7	21.7	20.6	17.6
Other East Asia	33.0	57.2	75.1	86.3	107.4	1.3	1.4	1.4	1.4	1.3
Southeastern Asia	182.0	323.5	440.8	523.8	700.5	7.2	7.9	8.3	8.4	8.3
Southern Asia	478.7	848.6	1,202.9	1,502.3	2,173.8	19.0	20.8	22.3	24.0	25.7
Western Asia	42.4	85.2	130.8	170.4	286.5	1.7	2.1	2.4	2.7	3.4
Oceania	2.5	4.4	6.4	7.9	12.2	0.1	0.1	0.1	0.1	0.1

Source: United Nations Department of International Economic and Social Affairs, *World Population Prospects* (United Nations, New York, 1989), Part 2, Table 1, pp. 74–75 and Table 2, pp. 82–83.
Note: Some totals may not add because of computer rounding.

Box 4.1 Changes in International Population Forecasts

Global population forecasts have remained remarkably consistent over time, varying little in their population projections, despite the constant accumulation of new data and surveys. In its most recent population assessment, referred to as the "1988 Assessment," the United Nations projects a global world population of 8.467 billion by the year 2025 (1). The previous assessment—the "1986 Assessment"—had indicated a world population of 8.206 billion in 2025, only 261 million fewer people (2).

This consistency at the global level, however, hides some dramatic changes in the projected population levels of individual countries. Between the 1986 and the 1988 Assessments, 25 countries showed a larger projected population size of 10 percent or more for the year 2025. Eighteen countries showed a smaller projection in their 2025 population size of 10 percent or more.

DEVELOPED COUNTRIES GROW SLOWLY

The developed countries, for example, have been forecast for some time to show only very small increases in population in the coming years. But the 1988 Assessment shows that increase to be 44 million fewer than had been anticipated as little as two years ago (3) (4). This is primarily because total fertility rates

(TFRs) in many countries are proving to be even lower than anticipated. TFRs represent the number of children projected to be born to each woman upon completion of her childbearing years. Some countries have now dropped well below replacement level fertility (2.1 children). Italy, for example, now has the world's lowest fertility rate, at 1.3. Spain is 1.5, Greece 1.6, the United States 1.9, and Sweden 1.9 (5).

Other countries whose populations are projected to be smaller than had previously been thought include the Soviet Union, Ethiopia, Nigeria, South Africa, and Zimbabwe. In three of these countries—Ethiopia, Nigeria, and Zimbabwe—fertility rates, while still very high, have been readjusted downward, producing by 2025 significantly lower populations. The Soviet Union, with its already low fertility rate, and South Africa, with a moderately high fertility rate, are dropping faster than had been anticipated.

MANY DEVELOPING COUNTRIES GROWING FASTER

In contrast, a number of developing countries are now projected to have considerably larger populations by 2025 than had been thought at the time of the 1986 Assessment. Four countries in Asia—Bangladesh, India, Iran and Pakistan—together are projected to have 319 mil-

lion more people by the year 2025 than was thought in 1986. This includes 16 million for Bangladesh, 217 million for India, 25 million for Iran, and 57 million for Pakistan. In all of these countries, fertility is now believed to be higher than had previously been projected.

Another country for which revisions were made is China, which now shows an expected 17 million more people in 2025 than was expected in the 1986 Assessment. Although large in absolute terms, this represents only minor changes between the Assessments in underlying fertility and mortality assumptions.

References and Notes

1. United Nations Department of International Economic and Social Affairs, *World Population Prospects, 1988*, Population Studies No. 106 (United Nations, New York, 1988), Part I, p. 28, Table 2.1.
2. United Nations Department of International Economic and Social Affairs, *World Population Prospects: Estimates and Projections as Assessed in 1984*, Population Studies No. 98 (United Nations, New York, 1986), Part II, p. 19, Table 2.
3. *Op. cit.* 1, Part II, p. 74, Table 1.
4. *Op. cit.* 2, Annex 1, p. 49, Table A-2.
5. Carl Haub, Population Reference Bureau, Washington, D.C., June 1989 (personal communication).

---

changes over time is also a major factor in understanding and planning for the impact of population on society. In particular, the youthful age structure of the population in much of the developing world—where large proportions of young people have not reached their childbearing and working years—means that the absolute number of births and the total population will continue to rise rapidly for the next 20 to 30 years, if present trends continue.

Age Structure of the Population

The age structure of a population can be defined as the pattern created when people are grouped by age, so that the size of each group represents the number of people in a given age range, such as 0–5, 5–10, 10–15, etc. The pattern is, in effect, a snapshot of the distribution of the population by age at a given time. In a western industrial country, for example, such a distribution for the year 1985 would show a relatively flat pattern, indicating roughly equal numbers of persons in each age range, except for the very old and for a bulge in the 30–35 and 35–40 age groups, reflecting those persons born in the post-war baby boom. (See Figure 4.1.) The 1985 age structure for a developing region such as Latin America, however, would show a pattern that is sharply tilted to the

younger ages, reflecting the fact that there are far more people in young age groups than in older groups. (See Figure 4.1.)

If such a pattern is combined with past and likely future patterns for a given region, the result is a three-dimensional chart that summarizes both earlier population events (in terms of the net impact of births and deaths as well as of migration) and the probable direction of future trends for each age group. While every country's age structure is distinct, two patterns are common around the globe—one in the industrialized countries and one in developing countries. The chart for Africa, for example, shows the tilt to younger age groups characteristic of most developing regions, but in extreme form. (See Figure 4.2.) Moreover, the youthful tilt gets more pronounced in each succeeding time period or row in the chart, reflecting the continuing acceleration of population growth in Africa (and the projection of that acceleration continuing in future years, as the growing number of women of childbearing age have children). The chart can also be read along the other axis, marked in years, reflecting the number of people of a given age group in each successive time period. The trend in the 0–5 age group, for example, documents the rising number of births in Africa. Thus the chart illustrates four sets of relationships (also documented in Tables 4.2, 4.3, 4.4, and 4.5):

Figure 4.2. Africa: Population Size by Age, 1950—2025

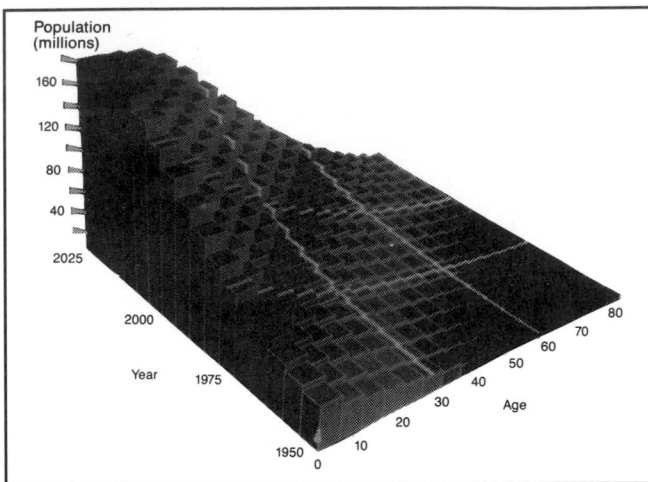

Source: Graphics by Robert Fox, Allen Carroll, and Melvin L. Prueitt, Population Images, Bethesda, Maryland, with data from Department of International Economic and Social Affairs, United Nations, *Global Estimates and Projections of Population by Sex and Age: The 1988 Revision* (United Nations, New York, 1989), p. 10.

■ The size of one age group relative to any other in a given year (the age structure of the population read along the "age" axis);
■ The increase (or decrease) in the size of each age group over time read along the "year" axis;

■ Overall trends for the entire mix of age groups over the 75-year period; and
■ The trends for each birth group, or cohort, as it ages, seen by following a diagonal axis (i.e., those 0–4 in one interval become 5–9 in the next 5-year interval, 10–14 in the next, and so on).

Africa

Africa represents an extreme case. Throughout the 75-year period under review (1950–2025), the age composition of the African population is dominated by a very high proportion of youth. (See Table 4.2.) This results primarily from a sustained high birth rate combined with a rapidly falling death rate. (See Table 4.3.) The gap between the two rates—births and deaths—is the rate of natural increase or, absent migration, population growth. In Africa, the gap is projected to continue to widen until at least the year 2000, because the death rate is falling faster than the birth rate. As a result, the African population increased at an average annual rate of 2.2 percent in the period 1950–55, rising to 2.7 percent in 1970–75 and 3.0 percent in 1985–90. It is projected to fall only slightly, to an average annual rate of 2.95, for 1995–2000 (4).

By the period 2020–2025, Africa is projected to have reached its anticipated minimum death rate of seven deaths per thousand population (declining from 27 deaths per thousand population in 1950–55). Additional moderate drops in the birth rate will then cut more deeply into the pace of population growth.

Table 4.2 Distribution of the Population in Major Age Groups, 1950–2025

	1950 Age Groups			1975 Age Groups			2000 Age Groups			2025 Age Groups		
	0-14	15-16	65+	0-14	15-64	65+	0-14	15-64	65+	0-14	15-64	65+
World	34	61	5	37	57	6	31	62	7	24	66	10
Developed Regions	28	64	8	25	64	11	20	66	14	18	63	19
Less Developed Regions	38	58	4	41	55	4	34	61	5	26	66	8
Africa	42	54	4	45	52	3	44	53	3	34	62	4
China	34	62	4	40	56	4	26	67	7	18	69	13
Latin America	41	56	3	41	55	4	33	62	5	26	66	8

Source: United Nations Department of International Economic and Social Affairs, *World Population Prospects* (United Nations, New York, 1988), Part 2, Table 17-A, pp. 200, 202, 204, 206, 218 and Table 17–13, p. 326

Table 4.3 Birth and Death Rates, 1950–2025
(per 1,000 population)

	Birth Rates				Death Rates			
	1950-55	1970-75	1995-2000	2020-25	1950-55	1970-75	1995-2000	2020-25
World	37.4	31.5	24.8	17.4	19.7	12.2	8.7	7.7
Developed Regions	22.6	16.7	13.5	11.9	10.1	9.3	9.5	10.6
Less Developed Regions	44.6	37.1	27.7	81.5	24.3	13.3	8.5	7.1
Africa	48.9	46.7	41.4	25.5	27.0	19.4	12.0	7.0
China	43.6	30.6	18.0	12.6	25.0	8.7	6.6	8.2
Latin America	42.5	35.3	24.9	18.5	15.3	9.7	6.7	7.0

Source: United Nations Department of International Economic and Social Affairs, *World Population Prospects* (United Nations, New York, 1988), Part 2, Table 6, pp. 118, 120 and Table 9, pp. 134, 136.

Table 4.4 Number of Women of Reproductive Age 15–49

(millions)

	1950	1975	2000	2025
World	623.8	950.6	1,569.0	2,099.7
Developed Regions	223.9	276.1	310.6	290.7
Less Developed Regions	398.8	677.4	1,257.1	1,807.1
Africa	52.7	93.8	200.6	419.0
China	134.3	205.9	345.9	335.8
Latin America	39.5	74.9	141.4	193.9

Source: United Nations Department of International Economic and Social Affairs, *World Population Prospects* (United Nations, New York, 1988), Part 2, Table 17-A, pp. 200, 202, 206, 218 and Table 17-B, p. 326.
Note: Some totals may not add because of computer rounding.

Table 4.5 Average Number of Annual Births

(millions)

	1950-55	1970-75	1995-2000	2020-25
World	98.6	122.4	148.8	143.9
Developed Regions	19.5	17.9	16.8	16.1
Less Developed Regions	79.1	104.5	132.0	127.8
Africa	11.6	18.2	33.6	38.5
China	25.4	26.9	22.5	18.6
Latin America	7.5	10.7	12.9	13.7

Source: United Nations Department of International Economic and Social Affairs, *World Population Prospects* (United Nations, New York, 1988).
Note: Some totals may not add because of computer rounding.

Yet, even after 2025, Africa's population will continue to increase rapidly, if present trends continue. This paradox of falling birth rates alongside a continuation in the sharp rise in the African population is a direct result of the massive increase in the number of women of reproductive age, many of whom are "survivors" from the earlier era of falling death rates. The number of African women between the ages of 15 and 49 is projected to double between 2000 and 2025, rising from 201 to 418 million. Thus, the trend of a redoubling of those in this age group will continue, as it has doubled every 25 years since 1950. (See Table 4.4.)

Accordingly, although the birth rate in Africa will fall, the absolute number of births is expected to rise. (See Table 4.5.) Each woman, in other words, will have fewer children on average, but there will be many more women aged 15 to 49 who will have children.

Some 34 million births annually are projected for Africa by the year 2000, compared with 18 million in 1975 and 11 million in 1950. (See Table 4.5.) Because of the demographic momentum already built into the age structure, this number is projected to continue to rise to about 38 million births annually by 2025. Figure 4.2 shows the resulting disproportionately large size of Africa's young population for the years ahead. The median age of Africa's population (that is, the point at which half is above and half is below)

was 17 years in 1975, and is projected to be 17 in 2000, and 22 in 2025.

The More Developed Regions

The "more developed regions," as defined by the United Nations, include the United States and Canada, Japan, Europe, Australia, New Zealand, and the Soviet Union. Together, these countries share a population pattern that is at the opposite extreme of Africa's. (See Figure 4.3.) This pattern is characterized by a much flatter and generally well-balanced age distribution pattern that will continue to flatten out over time. It reflects the trend of low and continually falling birth rates and low death rates that have combined to produce an older and continuously aging population in the more developed countries. (See Table 4.3.) The median age of these countries' populations was 30 years in 1975, and is projected to be 36 years in 2000 and 41 years in 2025 (5).

According to the projected age distribution trends for the more developed regions, a near 1 to 1 ratio of those 65 and older to those under age 15 will prevail by 2025, when the oldest group will comprise 19 percent and the youngest group 18 percent of the total population. (See Table 4.2.) This is evident in the relatively flat, even distribution of population across age ranges in Figure 4.3. In Africa, in contrast, the ratio of those 65 and older to those under age 15 will be 1 to 8.5. (See Table 4.2.) This means that even by 2025, the largest population groups will only be entering their childbearing years and beginning the period when the need for jobs and social services will be greatest.

Figure 4.3 More Developed Regions, Population Size by Age, 1950–2025

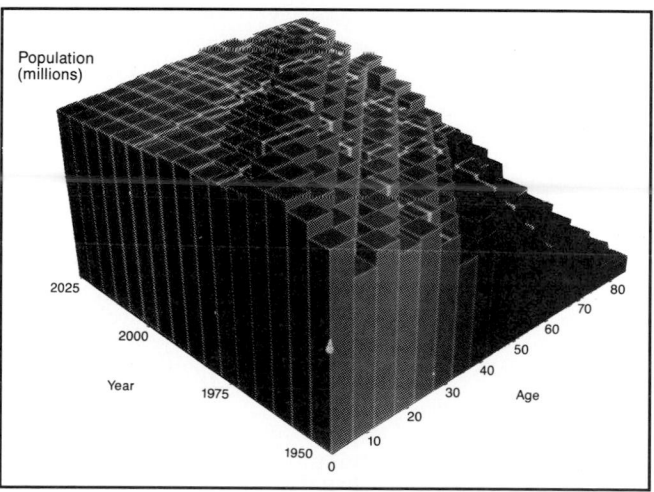

Source: Graphics by Robert Fox, Allen Carroll, and Melvin L. Prueitt, Population Images, Bethesda, Maryland, with data from Department of International Economic and Social Affairs, United Nations, *Global Estimates and Projections of Population by Sex and Age: The 1988 Revision* (United Nations, New York, 1989), p. 6.

Figure 4.4 Latin America: Population Size by Age, 1950–2025

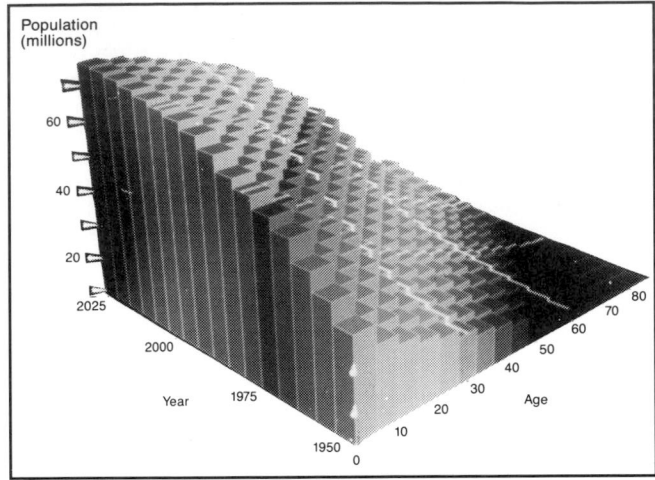

Source: Graphics by Robert Fox, Allen Carroll, and Melvin L. Prueitt, Population Images, Bethesda, Maryland, with data from Department of International Economic and Social Affairs, United Nations, *Global Estimates and Projections of Population by Sex and Age: The 1988 Revision* (United Nations, New York, 1989), p. 22.

Latin America

The age distribution pattern in Latin America for the period 1950 to 1975 is similar to that of Africa, although less extreme. (See Figure 4.4.) The population under age 15 represented a very high share of the total (41 percent in both 1950 and 1975), and climbed rapidly from 67 to 133 million during that time. (See Table 4.2.) As in Africa a quarter century later, the sharp increase is attributable to high fertility rates and rapidly falling mortality rates.

Since the early 1980s, fertility rates—that is, the number of births per woman—have fallen more sharply than earlier anticipated. But because there is continuously a large number of young women just entering their childbearing years, there will continue to be significant—if slowing—increases in population for some time to come. (See Table 4.1.) Between 1950 and 1975, the region had the world's highest population growth rate, increasing at an average annual rate of 2.7 percent in the period 1950–55, rising to 2.8 percent in the period 1960–65, before falling slightly to a 2.5 percent in 1970–75 (6).

This average, however, disguises sharp variations among the more than 25 nations of the region—variations that are now narrowing. Argentina and Mexico are at the extremes among the major countries. Argentina's population increased moderately, at 1.5 percent per year during 1965–70, while Mexico's population increased at an average annual rate of 3.3 percent (equaling the highest in the region) during that time. Since then, Argentina's population growth rate has dropped to 1.3 percent in 1985–90, and Mexico's to 2.2 percent (7).

After 1975, Latin America as a whole began to experience a marked decline in its pace of population growth. The 1985–90 rate of 2.1 percent per year is

projected to fall to 1.6 percent by 2000–05 and to 1.1 percent by 2020–25. Figure 4.4 shows the leveling out of the increases in the youngest age groups.

After the year 2000, the growth momentum will begin to drain out of the system. As the population under age 15 begins to approach stability in Latin America, the adult population claims larger and larger shares of the total. By 2025, those under age 15 will comprise only 26 percent of the total population. (See Table 4.2.)

A close examination of the data shows that this tendency actually began to develop during 1970–75 and will persist through 2025 and beyond. The population between the ages of 15 and 64, which represented 55 percent of the Latin American total in 1975, will rise to 62 percent in 2000 and to 66 percent in 2025. (See Table 4.2.) On the one hand, this means that there will be far fewer dependents per worker in the labor force. On the other hand, there will be sharp increases in the number of people seeking employment in already crowded labor markets.

China

Excepting China, Asia resembles the Latin American patterns. China, however, represents a unique age distribution pattern. (See Figure 4.5.) This pattern is characterized by enormous change both in births and deaths in a short time—first in the form of extremely rapid declines in mortality, followed immediately by the forced implementation of a policy that allowed only one child per couple.

Historically, such abrupt changes have only been found in countries that suffered major trauma through disease, warfare or famine. Such events, however, have been one-time occurrences that left relatively little impact on population size over the

Figure 4.5 China: Population Size by Age, 1950–2025

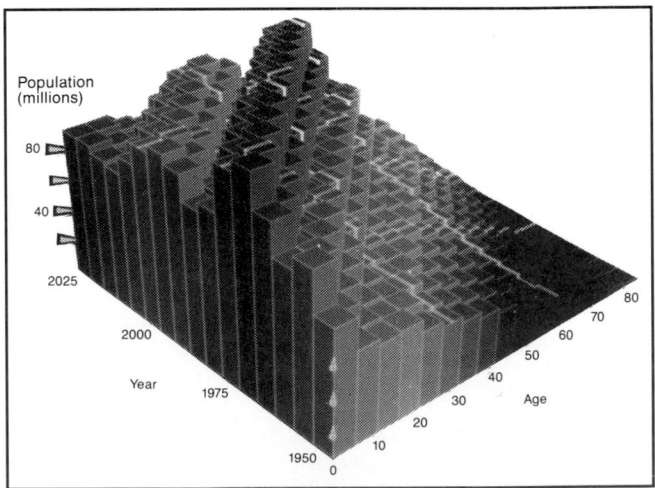

Source: Graphics by Robert Fox, Allen Carroll, and Melvin L. Prueitt, Population Images, Bethesda, Maryland, with data from Department of International Economic and Social Affairs, United Nations, *Global Estimates and Projections of Population by Sex and Age: The 1988 Revision* (United Nations, New York, 1989), p. 130.

long term. China's abrupt declines in birth and death rates, in contrast, resulted from deliberate policy decisions to bring births and deaths into balance in a short period of time, and have had significant consequences for the age structure of the population. Moreover, because China alone constitutes one quarter of the world's population, its unique pattern merits special attention.

China's initially high birth rate in the period 1950–55 (44 per thousand population) dropped moderately to 31 by 1970–75. During that period, its death rate experienced a radical drop—falling sharply from 25 to 9 (see Table 4.3)—and the rate of natural increase rose from 1.9 to 2.2 percent between 1950 and 1975.

As in many developing countries, high mortality in China in the years 1950 to 1975 was concentrated in the 0–4 year age bracket. With major improvements in health, sanitation, education and medical conditions, child mortality levels plummeted from 266 per thousand live births in the period 1950–55 to 83 in 1970–75 (8). The patterns in Figure 4.5 reflect the large number of children born in the 1950–75 period (see also Table 4.5), along with their increased probabilities of surviving.

Today, those who were born between 1965 and 1970—some 132 million—comprise the single largest population group in China. The slightly smaller group of 126 million that were born between 1970 and 1975 is the next largest group. Together, these two groups—now ages 15 to 24—comprise 22 percent, or nearly one quarter, of the total Chinese population (9). These two groups can be seen as the high "wall" extending diagonally across Figure 4.5.

With the initiation of the "one child" population policy in the late 1970s, numerous and profound changes to the Chinese age structure occurred that are clearly seen in the chart. Several are particularly noteworthy:

■ The number of children in the next birth group—those born between 1975 and 1980—fell dramatically to 98 million. Moreover, the number of births in the period from 1980 to 1985 continued to stay near this level. Since then, however, the number of births has risen slightly and is projected to rise still further, as the "one child" policy and bans on early marriage have been relaxed somewhat, particularly in rural areas, and as the large number of women born in the earlier population boom reach childbearing age.
■ The population groups currently (in 1990) in the age groups 15 to 19 and 20 to 24 will remain the largest segment of the population throughout the period ending in 2025.
■ The "aging" of the Chinese population is projected to produce a situation by 2025 in which those 65 years and over (194 million) will be closing in on the population under age 15 (272 million) at a ratio of 1 to 1.4. This is in marked contrast to earlier times when there were many more children; the ratio of over 64 to those under 15 was 1 to 7.5 in 1950, and is 1 to 4.5 in 1990. (See Table 4.2.) In a society that has traditionally venerated old age precisely because it was so rarely attained, this kind of shift in the numeric re-

lationship among generations could bring about radical change in Chinese social, economic, cultural and political systems.
■ The reduction in the proportion of Chinese children to total population in coming years means that a proportionately larger segment of the population will be in the productive years (see Table 4.2), thus lowering the dependency ratio, that is, the ratio of those between 15 and 64 to those either under 15 or over 64 years of age. Sixty-nine percent of the Chinese population will be in their productive years and in need of jobs by 2025, compared with only 56 percent in 1975.

Age Structures in the World Context

For some regions, including the more developed countries and the African continent, age patterns will continue in today's directions for some time. The Chinese age distribution pattern shown in Figure 4.5, however, is new to the world's demographic experience. It is evidence that deliberate policy can alter a country's demographic future. And while the Chinese pattern is still an anomaly on the global scene, it may also start to emerge in modified form in other developing countries, including Mexico, where family planning targets and programs are making strong headway.

From another perspective, however, the entire post-1945 Third World "population explosion" typified by the Africa model is also a demographic anomaly. The initial explosion came about through the combination of continued high birth rates and an extremely rapid and sharp first-time drop in death rates, resulting in suddenly large populations of children.

For most of the world, the "inflection point"—that is, the point at which the rate of population growth begins to decline—has been reached. But it will take generations before birth and death rates are brought back into equilibrium— this time at low birth and low death rates. As this occurs, societies will experience dramatic shifts in age structure.

More attention is needed to the implications of age structure within populations for resource and environmental management and for economic and social development. For too long, the focus has been on population growth rates, with an implicit assumption that, if these fell, the population "problem" would be resolved. When the United Nations announced in the late 1970s, for example, that the world's annual population growth rate had reached its high point and was starting to diminish, press reports interpreted this to mean that the "explosion" phase was over. In terms of absolute numbers, however, earlier United Nations population projections for the year 2000 were unaltered—both then and now. They have remained consistent since the early 1960s. In 1963, the United Nations projected a global population total of 6.13 billion in the year 2000 (10). In 1988, the projection for the year 2000 is 6.25 billion. Thus population growth at the global level continues on the same track as before.

Today, however, planners and development officials need a more sophisticated appreciation of the implications of population growth of this magnitude. The consequences of rapid population growth can no longer be perceived only in terms of the prospect of food shortages and the shortages of land available for agriculture. Since much of the growth—particularly in Latin America and to a lesser extent in Asia—is projected to occur in urban areas, it will mean the continued development of megacities and the need to manage the particular development problems associated with urban concentrations. The maturing of youthful populations will mean massive growth of the labor force and the need for jobs—again, particularly in urban areas. In economies heavily dependent upon the exploitation of natural resources, rapid population growth means growing pressure for rapid exploitation of those resources in unsustainable ways—and consequent damage to forests, soils, and other resources. And even with more careful development, economic growth rapid enough to absorb and fulfill the basic needs of the projected numbers of the world's people will require significant increases in energy and basic materials flows. Continued rapid population growth means rising pollution pressure on rivers and estuaries and the potential for worsening air quality in urban areas. In addition to local and regional pollution, population growth is likely to mean greater worldwide emissions of greenhouse gases capable of altering the global climate. Changing age distributions within populations may also disrupt social and cultural patterns and alter health care needs.

The projected growth of the world's population is thus likely to be a major force affecting virtually every aspect of human and social development, natural resource and environmental management, and economic progress, particularly in the developing world, over the next half century.

FOCUS ON A MAJOR KILLER: MALARIA

Around the world, there has been a resurgence of malaria since the early 1970s when the disease was nearly beaten. Every year it is estimated that some 100 million people contract the disease. Despite massive efforts in the 1950s and 1960s to eradicate it, malaria today, according to the World Health Organization, "remains a major public health problem and continues to be an obstacle to development" (11). The actual number of reported cases worldwide is about 10 million per year, but it is widely recognized that underreporting and poor surveillance are prevalent in many of the countries most affected (12).

Approximately 80 percent of reported cases worldwide occur in Africa south of the Sahara, where the prevalence of infection is often over 50 percent (13). Among young children, it can approach 100 percent (14). Over 90 percent of the people living in sub-Saharan Africa live in malarious areas (15). This combination of factors means that there is high infant and childhood mortality, but those who survive childhood often acquire immunity against the most severe manifestations of the infection.

In central and southeast Asia, over 96 percent of the population is exposed to malarial risks. In almost all countries in this region, the drug chloroquine is used to treat malarial fevers, with the result that the most prevalent strain of malaria parasite today, the deadly *P. falciparum*, is increasingly chloroquine-resistant (16).

Latin America enjoyed perhaps the greatest success in its antimalaria efforts in the 1950s and 1960s. By 1970, deaths from malaria had all but disappeared in most Latin American countries. Today, however, at least 17 countries are again seriously affected by malaria; it is estimated that in 1984, some 2.1 million people in the region were affected by the disease (17).

Malaria is caused by one or more of four species of the parasite of the genus *Plasmodium*, with *P. falciparum* being both the most deadly and the most common in tropical areas. Other species are *P. vivax*, which has the broadest range; *P. ovale*, found almost entirely in Africa; and *P. malariae*, present where *P. falciparum* occurs, but less prevalent (18).

Malarial infections most commonly begin with mosquito bites, although transfusions of infected blood and shared needles can also be responsible. When a female mosquito injects infective forms of the *Plasmodium* parasite (or "sporozoites") into the human body, they find their way to the liver. There they multiply into thousands of new forms of *Plasmodium* (or "merozoites"), although some *P. vivax* and *P. ovale* sporozoites may remain dormant for up to three years. Once the liver-based, or hepatic, parasites mature (in six to 16 days, depending on the species), they enter the bloodstream and invade the red blood cells, where they reproduce asexually. This cycle repeats itself until immunity develops, treatment is given, or the victim dies. If immunity or treatment is not complete, relapses may occur (19).

Some merozoites, instead of reproducing asexually, differentiate into male and female sex "gametocytes," which can be picked up by the next mosquito to bite. It then takes 8–35 days, depending on the species and the climate, to produce sporozoites in the mosquito's salivary glands. As the insect bites more people, new rounds of infection start (20).

Epidemiologic Factors

Five major epidemiologic factors contribute to the transmission of malaria: human immunity (or the lack thereof), the habits and biology of the vector mosquitoes, environmental conditions, climate, and the presence of high-risk groups.

Immunity

Where malaria is endemic, particularly in tropical Africa, people can acquire immunity to malaria either in utero or in a series of infections. Very young children frequently have passive immunity acquired in utero from infected mothers. After this initial period of immunity passes, children gradually build up their own antibodies, provided they do not die from their first attacks from the malaria parasite. As they acquire immunity, they are less vulnerable to subsequent attacks. In areas where a large portion of the population has built up immunity in this way, malaria will remain endemic, but is not likely to become epidemic, until there is a change in the balance between the human and the mosquito population (21).

There are also genetic factors in some populations that either prevent them from getting malaria or alter the severity of the infection. Perhaps the best known, but not the only, example is the presence of the sickle-cell gene in African populations. The presence of a pair of sickle-cell genes causes sickle-cell anemia, in itself highly dangerous and deadly. But the presence of just a single sickle-cell gene provides immunity against the *P. falciparum* parasite. Although the parasite still reaches the bloodstream, the sickle hemoglobin does not provide adequate nutrients to the parasite, causing individuals with this trait to have less severe malarial disease or to be asymptomatic. As many as 20 to 30 percent of Africans have this genetic protection.

The Vector Mosquito

A number of factors affect the ability of mosquitoes to carry the malaria infection. Mosquitoes vary in their ability to support all stages of the development of the parasite and thus in their ability to inoculate sporozoites into new victims. Long-lived mosquitoes are the best vectors, whereas not even one generation of sporozoite can develop in mosquitoes that survive less than 10 days. Mosquitoes also differ in their preference for feeding and resting inside houses (22).

The extent to which malaria is endemic in an area is largely determined by the habits and longevity of the vector mosquitoes and the species of malaria. For example, in warm, humid climates, where mosquitoes have a long life (four weeks or more) and feed by choice on humans, malaria is likely to be endemic. Under these circumstances, a small number of mosquitoes continually reinfects the human population, ensuring that immunity against at least the most severe forms of malaria is gradually built up (23).

Even in endemic areas, however, periodic epidemics may occur, usually because of such factors as a temporarily improved opportunity for vectors to transmit the disease (for example, because of increased breeding sites or the removal of cattle, causing mosquitoes to increase feeding on humans); the introduction of a different and highly efficient vector; the arrival of a large group of infected people; or an influx of nonimmune people.

Environmental Conditions

Environmental factors contributing to the spread of malaria include the location of human settlements near mosquito breeding sites, overcrowding and unsanitary living conditions, and housing construction that facilitates mosquito entry. Additionally, failure to remove collections of fresh water in and around the home can foster spread of the disease, as does the digging of pits for construction of houses or roads, as these soon fill with pools of water that breed mosquitoes. At the same time, the proximity of nonhuman feeding sources for mosquitoes (for example, cattle) can aid in reducing the spread.

Climate

Temperature, humidity, and rainfall play an essential part in the ability of the mosquito to develop and transmit malarial infection. Temperatures between 20°C and 30°C and humidity in excess of 60 percent are optimal, allowing the mosquito to survive long enough to transmit infection to several people. Unless rain is excessive enough to flush out the mosquito larvae, it, too, contributes to further spread, by creating more pools of water in which the mosquitoes will breed (24).

The Presence of High-Risk Groups

High-risk groups are those with no previous exposure, and therefore no immunity, to malaria. These include young children, immigrants, travelers, and expatriates. When there is a large influx of nonimmune people into a previously endemic area, full-blown epidemics can occur quickly.

In the Amazon, where malaria has been endemic for centuries, there has been a virtual explosion of malaria—from 51,000 reported cases in 1970 to 1 million expected cases in 1990. Annual deaths are estimated at 5,000. The onset of the epidemic follows the opening up of roads into the western Amazon state of Rondonia, which has drawn large numbers of landless migrants from the cooler parts of southern Brazil, where there is no malaria and therefore no resistance to the disease (25).

Where malaria is endemic, there is also heightened risk to pregnant women and their unborn children (26). It is less certain what the impact is on pregnancy in areas where malaria is endemic and adult women have acquired significant immunity. Placental malaria is estimated to occur among 20–34 percent of the women living in tropical Africa. Studies have found that the mean birth weight of a child drops with the presence of placental malaria, but have yet to determine a correlation with spontaneous abortion or stillbirth. No relationship has been established with maternal morbidity (27). The infant is unlikely to acquire a malaria infection from the mother, and in fact during the first weeks of life, enjoys the protection offered by antibodies acquired transplacentally.

Box 4.2 The Africa Child Survival Initiative—Combatting Childhood Communicable Diseases

The Africa Child Survival Initiative–Combatting Childhood Communicable Diseases (CCCD)–is a major international effort to reduce childhood mortality due to malaria, diarrhea, and vaccine-preventable diseases. The program is funded by the U.S. Agency for International Development and implemented by the U.S. Centers for Disease Control. Started in 1982, it was by 1988 operating in 13 sub-Saharan African countries, with a total population of 170 million, of which 12 have endemic malaria.

With respect to malaria, CCCD strategies include ensuring prompt, effective treatment for high-risk groups; encouraging prophylactic drug use for pregnant women; monitoring treatment and prevention practices as well as patterns of illness and death; and developing national malaria control strategies (1).

CCCD researchers have found that where malaria reinfection is likely to occur after treatment, it is not necessary to eliminate parasites completely in order to reduce malaria morbidity. For this reason, CCCD efforts have concentrated on improving malaria diagnosis and therapy as an initial step in developing effective and locally appropriate delivery systems for antimalarial drugs.

In 1988, 11 of the CCCD countries had malaria control units and national plans. In 1982, when the CCCD project began, only four of these countries had such units—all concerned primarily with vector control issues. Since then, the CCCD units have concentrated on 1) disseminating information on malaria, 2) training health workers in the diagnosis, treatment and prevention of malaria, and 3) conducting operational research on the

many still unresolved issues related to malaria. This approach has enjoyed some significant early successes. In Togo, for example, unnecessary quinine injections as a front-line treatment for all fever episodes in children went from 56 percent at the outset of the program to 7 percent in 1987, after widespread dissemination of the results of operational research that showed that a single dose of chloroquine was very effective in eliminating malarial infection (2).

References and Notes

1. J.G. Breman and C.C. Campbell, "Combatting Severe Malaria in African Children," *Bulletin of the World Health Organization*, Vol. 66, No. 5 (1988), p. 612.
2. *Ibid.*, p. 618.

In sub-Saharan Africa an estimated 100,000 children under the age of one and 575,000 between one and four years of age die annually from malaria (28).

Controlling the Disease

Malaria control has been an important priority both for the countries afflicted and the international community since the 1950s. Early efforts concentrated on malaria eradication. It was assumed that active intervention to control or interrupt malaria transmission in large areas on a sustained basis would eventually completely eliminate the disease. The 1950s and 1960s saw war declared on malaria worldwide, with scientists and policymakers optimistic about their chances of eradicating the killer infection. The two-pronged strategy consisted of spraying infected mosquitoes and to destroy malarial parasites in human bloodstreams using antimalarial drugs.

Initially, the effort seemed promising. Significant progress was achieved in some areas of the world, especially in Latin America and some parts of Asia. By 1970, in most countries of Latin America, deaths from malaria had all but disappeared. Over time, however, this strategy became increasingly less effective. Inadequate financial and human resources led to a relaxation of surveillance, and the prolonged application of the strategy led to vector resistance to insecticides and parasite resistance to drugs (29). This resulted in increased transmission of malaria after an initial downswing, with a peak of approximately 20 million cases reported in 1976, finally stabilizing at about 10 million in 1984 (30).

Today, the World Health Organization estimates that there are some 100 million clinical cases of malaria in the world each year (31). Estimates by a Swiss researcher suggest that in 1986 the number in fact may have been as high as 489 million clinical cases of malaria (32).

The resurgence of malaria has been most pronounced in areas where the introduction of wide-scale antimalaria tactics preceded the development of a primary health care system. Among other problems, this led to uncontrolled availability and unmonitored consumption of antimalaria drugs, which resulted in overuse or, alternatively, insufficient doses that relieved symptoms but gradually increased parasite resistance (33). *P. falciparum* developed a resistance to chloroquinine, which is the drug of choice because it is inexpensive, readily available, easy to use, and has few side effects. This resistance was first observed in Thailand in the 1950s, showed up in Kenya in 1978, and is now widespread.

Additionally, the massive use of insecticides for agricultural purposes affected mosquitoes at all stages of their life cycle and caused some 60 species to become resistant to DDT, as well as other insecticides (34). Moreover, the cost of insecticides has increased greatly, and there is a growing understanding of their toxicity to humans and to the environment. A related problem is that mosquitoes developing resistance to insecticides change breeding sites and feeding habits, sometimes contributing to a new epidemic outbreak.

Social and economic conditions prevalent in most developing countries also had a major impact on the continued spread and virulence of malaria (35). These include:

■ Population growth and civil instability causing movement of nonimmune people to infected areas, and infected people to previously malaria-free areas; and

■ Increases in uncontrolled breeding sites due to a) intensive agricultural development requiring irrigation or flooding, b) more industrial, hydroelectric and communications projects requiring or resulting in impoundment of water, and c) changes in ecosystems caused by widespread deforestation, soil erosion and flooding.

Box 4.3 A Community Effort to Control Malaria

An Indian village in the malaria belt has succeeded in controlling malaria with a community action program that does not use pesticides or high technology.

Before the program began three years ago, one of every four people had malaria in Pudukkupam, an Indian fishing village on India's east coast about 15 kilometers south of Pondicherry. Malaria was common in the area despite years of spraying with DDT (1).

The Vector Control Research Center in Pondicherry, part of the Indian Council of Medical Research, selected Pudukkuppam and three other villages to test the idea that malaria-carrying mosquitoes could be eliminated through a community participation project that focused on closing mosquito-breeding sites or preventing the development of mosquito larvae. To get the villagers interested in the program, center officials first spent about a year solving some of the village's basic problems, such as the provision of a borewell for drinking water (2).

The village had several sources of mosquito breeding: water-filled earthen pots, which were used in homes to soak coconut husks for production of coir ropes; about 600 pits dug along the coast that irrigated casuarina trees, the village's source of firewood; and large expanses of lagoons infested with aquatic algae that mosquitoes used to lay their eggs (3).

The community replaced the earthen pots with a single large shallow pit for soaking coconut husks; to prevent egg-laying by mosquitoes, the pit was covered with a lid made of palm leaves. To control egg-laying in the casuarina pits, the center persuaded villages to introduce larvae-eating fish, *Gambusia affinis*. Fish also were successfully introduced into the ponds used to irrigate rice paddies; income from the sale of mature fish was used for clearing weeds from the ponds (4).

Larvae-eating fish were not the answer in the lagoons, because the algae sheltered the larvae. Center scientists did find, however, that the algae, when mixed with cotton waste, could be made into writing paper. Regular clearing of the algae from the lagoons helped control mosquito-breeding and paper-making turned out to be profitable for the village (5).

Malaria in the village was mainly spread by a type of mosquito, *Anopheles subpictus*, that normally bites cattle rather than people. Center officials arranged loans for the village to expand its cattle herd, so mosquitoes would feed more on cattle rather than on people (6).

So far, the results have been very impressive: by early 1989, the village had not had a single malaria case for three years (7).

References and Notes

1. K.S. Jayaraman, "Profiting from Malaria Control," *Panoscope* No. 12 (May 1989), pp. 3-4.
2. *Ibid.*, p. 4.
3. *Ibid.*
4. *Ibid.*, p. 5.
5. *Ibid.*
6. *Ibid.*
7. *Ibid.*, pp. 3, 5.

The strategy adopted by the twenty-second World Health Assembly in 1969 proposed the following prerequisites for successful malaria efforts: (36)

■ There must be a long-term governmental commitment supporting antimalarial activities;

■ Malaria control must be an integral part of the country's health program;

■ There must be evidence that the target objectives are feasible and practical;

■ There must be extensive community participation; and

■ Malaria control must be an integral part of other development programs (for example, irrigation, drainage, highways, hydroelectric development).

Today, however, the WHO acknowledges that eradication of malaria is not feasible in many countries and many areas. (Eradication remains, of course, the long-range goal.) In 1985, the WHO Expert Committee on Malaria proposed two contrasting approaches to malaria control (37):

■ A modest strategy of concentrating on improving local health services (see Box 4.2);

■ An active strategy to control and reduce transmission of the disease.

Four main kinds of measures are used to control malaria today. (See Box 4.3.) These are: 1) reducing contact between people and mosquitoes (e.g., through the use of bednets and coils, protective clothing, and building houses away from breeding areas); 2) controlling mosquito breeding (e.g., through draining swamps, training streams, applying chemical larvicides on certain water surfaces, cultivating larvivorious fish, and clearing vegetation); 3) destroying adult mosquitoes through spraying interior walls and other areas every three to six months;

and 4) destroying the malarial parasite using chemotherapy and, in the future, vaccines (38).

On the Horizon: Malaria Vaccines

Malaria vaccines are in varying stages of development ranging from analysis of chemical makeup to cloning to chemical synthesis. These vaccines are just beginning to be tested but are still many years away from use.

Three vaccine strategies are being pursued, each mimicking one or another form of the parasite by stimulating protective immune responses at specific stages in the parasite's development. Each type of vaccine will work at only one of the stages.

The first type of vaccine copies the essential parts of the sporozoite. It prevents sporozoites from reaching the liver and establishing themselves in the host, thereby inducing sterile immunity. In other words, as parasitemia does not occur, no clinical illness will emerge and the immunized person can no longer infect mosquitoes. Unfortunately, that person will continue to be susceptible to transmission at later stages of the parasite development (for example, by receiving a transfusion of infected blood).

The second type of vaccine copies the more numerous merozoites and stimulates the immune system to intercept them before they infect the red blood cells. This reduces morbidity and mortality, without necessarily inducing sterile immunity. The immunized person remains susceptible to first-stage sporozoite infection and therefore is capable of infecting mosquitoes. Despite this drawback, the vaccine is important for young children living in endemic areas be-

Figure 4.6 AIDS Cases in the Americas, 1986–88

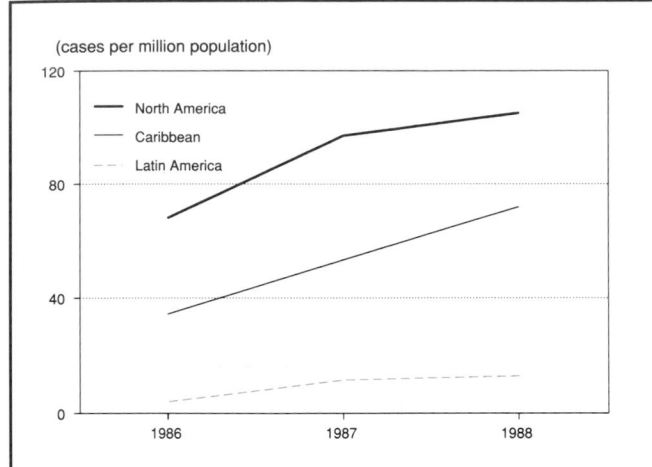

(cases per million population)

Legend:
— North America
— Caribbean
-- Latin America

Source: National Program Support–AIDS Program, "AIDS Surveillance in the Americas" (Pan American Health Organization, Washington, D.C., 1989).

AIDS IN THE AMERICAS

The worldwide AIDS epidemic continues to take a major toll in human lives and a significant claim on health care resources. As of December 1989, more than 198,000 cases had been reported worldwide, with 66 percent of these occurring in the Americas.

Although the great majority of AIDS cases occurring in the Western Hemisphere continues to be reported by the United States, the epidemic is increasingly affecting Latin America and the Caribbean as well. The number of reported cases, and a review of the limited existing published scientific reports, suggest that:
■ The number of AIDS cases in Latin America and the Caribbean is increasing rapidly,
■ The Caribbean area has relatively high rates of infection with the AIDS virus, and
■ The pattern of cases in much of the region may be shifting from known risk groups such as homosexual and bisexual males to a growing number of cases occurring in women and children.

AIDS in these areas is concentrated in cities, and transmission is primarily through sexual contact.

Throughout the Americas, the number of AIDS cases is increasing. As Figure 4.6 shows, the increase in reported rates was most dramatic between 1986 and 1987, and appears to have leveled off slightly in the past year. However, this apparent leveling off is probably due to a delays in reporting.

The illness AIDS is caused by infection with the HIV virus (42) and is characterized by a long latency between infection and the clinical onset of illness. Recent information suggests that on average this period is between seven and eight years (43). Thus the cases being diagnosed now could be the result of viral transmission through sexual contact or use of infected blood or blood products that took place up to a decade ago.

This kind of case reporting is important for maintaining public awareness of AIDS and allowing some planning for the care of AIDS patients. But it does not provide good information for targeting public health programs for the prevention of AIDS. For this reason, international emphasis is shifting gradually from the mere reporting of new cases of AIDS to gathering information about the likely incidence of AIDS through antibody testing of selected populations, known as serologic surveillance. Exactly which populations are tested varies from country to country, depending on which populations are thought to be most at risk. As access to the technology of screening tests increases even in relatively poor countries, serologic surveillance will eventually become the major system for tracking the epidemic.

In the meantime, the World Health Organization and its regional affiliates continue to gather data and report on new AIDS cases on a regular basis. The number of cases being reported in the Caribbean suggests that the level of infection in this area is approaching that of North America. In 1988, there were 72 reported cases of AIDS per million population for

cause it can raise their immunity enough to make a serious malaria-related illness unlikely. Any reinfection would only boost the vaccine-induced immunity.

The third type of vaccine, the so-called transmission-blocking vaccine, copies the gametocytes, stimulating an immune reaction against them. Serum antibodies block fertilization of females by male gametes in the mosquito (39), thus interrupting the development of the parasite within the mosquito and preventing further transmission.

Over the years, malaria control methodology has shifted emphasis from eradication, which proved impossible, to control and abatement. Financial and human resources are increasingly allocated to the improvement of immunity in the human host, and only secondarily to controlling the vector or the parasite. But however promising, vaccines are not a cure-all. They are likely to be costly and require more than a single inoculation. The poorest countries currently spend less than $2 per citizen on public health, and with the population in the developing countries expected to increase by 1.2 billion between now and the year 2000, this ratio can only worsen.

Even if unlimited funds were available, vaccine coverage for the *P. falciparum* parasite would need to be 99 percent by age three to six months for transmission to cease, an unattainable goal at present (40). In addition, the vaccine would change the pattern of immunity in a specific age group. Once the effect of the infantile vaccine wore off, that group might be subject to an epidemic (41).

While vaccines represent an exciting new opportunity for malaria control and management, they will be most effective when combined with other interventions. Both now and over the long term, a multiprong approach attacking all links in the malaria chain is required.

Table 4.6 Reported AIDS Cases by Sex Ratio and Occurrence in Children, Selected Countries in the Americas

Country	1988 AIDS Rate (per million)	Cumulative Sex Ratio (M/F)	Cases in Children[a] (percent)
Argentina	5.6	17.9	1
Brazil	21.1	8.8	3
Haiti	116.7	1.9	2
Honduras	26.9	1.5	2
United States	118.6	9.2	1

Source: National Program Support–AIDS Program, "AIDS Surveillance in the Americas" (Pan American Health Organization, Washington, D.C., 1989).
Note: a. Age less than 15 years.

the Caribbean compared with 105 cases for North America. (See Figure 4.6.) In fact, case rates—measured in the number of cases per million population—in the Bahamas, Bermuda, and Haiti now are equal to or exceed those in the United States. AIDS cases have been reported from all countries in the Western Hemisphere. In Central America and South America, the prevalence of AIDS is variable but substantially lower than for North America or the Caribbean.

Shifting Patterns of AIDS Infection

In the Caribbean and some Central American countries, the pattern of AIDS infection appears to be shifting from an illness initially confined to homosexual and bisexual males to an illness that affects heterosexual males, women, and their children (through perinatal transmission of the virus or transfusion of infected blood). Public health authorities monitor this shift by recording the sex ratio of cases and by classifying cases according to alleged risk factors.

Table 4.6 shows data for five countries. The incidence of reported AIDS cases in these countries ranges from 5.6 per million population in Argentina to 117 per million population for Haiti and the United States. In some countries, the ratio of male to female cases has dropped since the onset of the epidemic, with a growing number of cases now being reported in women. In Haiti and Honduras, for example, this ratio is now 1.9 and 1.5 to one, respectively. This shift has not occurred in the United States, where the ratio of men to women is approximately 9 to one. In Argentina and Brazil, the sex ratios remain high (17.9 and 8.8 to 1, respectively), but the level of infection has not yet reached that seen in other countries. The percentage of cases in children remains quite low throughout the region.

Thus, it appears that in some countries, at least, the sex ratio of reported AIDS cases has shifted, with the illness occurring increasingly in women. Although it is not completely known what factors in a given country cause the infection to spread to the general population, certainly the persistence of unsafe blood banking, using blood unscreened for AIDS, plays a role in many countries. Women are particularly at risk from blood transfusions, since peripartum hemorrhage is more common in countries where maternal care programs are less developed.

In addition, there is increasing evidence that untreated sexually transmitted diseases such as gonorrhea and syphilis play an important role as cofactors in the transmission of AIDS (44)(45). Current programming efforts to control AIDS in the Caribbean and Latin America include strengthening treatment programs. The reported rate of sexually transmitted diseases is several times higher for the Caribbean than for other areas of the Americas.

Potential Impact of AIDS on the Americas

The potential impact of the AIDS epidemic on the countries of Latin America and the Caribbean is significant. Many countries have health programs that give priority to the survival of women and children through the provision of oral rehydration, immunizations, antenatal care and nutrition education. Projections based on data from Central Africa and other parts of the world suggest that the presence of AIDS infection can eliminate gains in childhood and maternal survival that could be expected from these programs (46). In addition, tuberculosis remains highly endemic in Latin America, and recent reports from New York City suggest that AIDS can also exacerbate this threat to health (47). Because AIDS affects immunity, AIDS patients are both more likely to contract TB and to remain sick longer, thus also potentially spreading it more.

Efforts to stem the AIDS epidemic in the Americas are coordinated by the Global Program on AIDS of the Pan American Health Organization (PAHO) and the World Health Organization (WHO), with the support of many international donors. Every country in the Americas now has begun national activities to combat AIDS with financing and technical collaboration from PAHO/WHO and other international agencies. Programs focus on the promotion of safe sexual practice (including the use of condoms) and the systematic screening of all blood products for evidence of AIDS infection.

As in other regions of the world, increasing effort is being given to ongoing serologic surveillance as an additional—and potentially more useful—means of analyzing the course of the disease than mere case reporting.

RECENT DEVELOPMENTS

BRAZIL'S FERTILITY RATE DECLINES

Brazilian women are having far fewer children than demographers predicted just a decade ago, but the increasing number of women entering childbearing age will continue to rapidly increase the country's absolute population.

Brazil's population in 1985 was estimated at 135.6 million, and that total is expected to increase to 179.5 million by the year 2000 and to 245.8 million by 2025 (48). Nearly all of that growth will occur in Brazil's urban areas. Sao Paulo, which in 1985 totaled 15.54 million people, is estimated to grow to 23.6 million by 2000, which is expected to make it the second largest city in the world after Mexico City. Rio de Janeiro is expected to jump from 10 million to 13 million by the end of the century (49).

Those totals still represent a substantial improvement over earlier predictions, because the nation's fertility rate is dropping sharply. The Brazil Demographic and Health Survey, which was conducted by the Society for the Welfare of the Family in Brazil (BEMFAM), estimated fertility at 3.5 children per women in the 1983–86 period. That compares with estimates of 4.2 children per woman in the 1975–80 period and 4.7 children in the 1970–75 period (50). The survey also found that women living in urban areas had a substantially lower fertility rate than women in rural areas (3.0 compared with 5.0) and that fertility rates declined sharply as educational levels increased (6.5 for women with no education; 2.5 for women with more than primary education) (51).

A few other Latin American countries, notably Colombia and Mexico, have had similar declines in fertility with the help of strong government-supported family planning programs. Surprisingly, the decline in Brazil has been largely self- induced; the Brazilian government has not played an active role (52).

Nearly two thirds (65.8 percent) of the women surveyed were using some sort of contraception, generally either the pill (25.2 percent) or female sterilization (26.9 percent). Nearly all women (93 percent) taking birth control pills were buying them from pharmacies (53). Slum dwellers appear to be no exception to this trend: a separate study of contraceptive use among low-income women living in three favelas in Rio de Janeiro found 62.9 percent of women aged 15–44 were current users (70.1 percent of non-pregnant women) (54). Illegal abortions are estimated at about 3 million each year (55).

Aside from the rapid increase in the use of contraceptive devices, most experts attribute the decline to economic stagnation in the 1980s, which has made it difficult for low-income couples to afford children. The Brazilian government in 1988 also passed a law requiring a four-month maternity leave for female employees. That law may be inadvertently penalizing fertile women, because some employers reportedly are insisting that young women seeking jobs show certificates of sterility (56).

Social scientists also think that television, which is now widespread in Brazil and constantly portrays images of small, affluent, consumer-oriented families, also is having an effect (57).

THE IMPACT OF CLIMATE CHANGE ON HUMAN HEALTH

Climate change and stratospheric ozone depletion could affect human health in a number of ways according to researchers who have recently looked into a wide range of possibilities (58). A thinned ozone layer would allow more skin-cancer-causing ultraviolet rays to strike the earth, causing a projected increase in skin cancer rates. Ultraviolet light can also damage the eye. An increase in the number of hot summer days caused by global warming is likely to cause an increase in deaths from heat stroke. And, as some temperate areas become warmer, tropical diseases carried by insects or parasites might invade new areas.

Studies generally fall into two categories: the relatively direct health effects caused by additional ultraviolet light reaching the earth's surface as a result of stratospheric ozone depletion; and the somewhat more uncertain, often indirect effects of global warming.

Considerable work is under way in the United States and elsewhere. For example, in 1988, the U.S. Environmental Protection Agency prepared a draft report to the U.S. Congress on the effect of global climate change that included a chapter on human health, and the Second North American Conference on Preparing for Climate Change, which was held in December 1988, included several papers on the subject. The U.S. National Institute of Environmental Health Sciences held a conference on the issue in November 1989 in Research Triangle Park, North Carolina, which was followed by a similar conference by the Center for Environmental Information in December 1989 in Washington, D.C. (59).

Ozone Depletion and Ultraviolet Light

The stratospheric ozone layer helps prevent much of the sun's ultraviolet light (UVR) from reaching the earth's surface. The protection varies somewhat, however: virtually all of the shortest wavelength (UV-C, 200 to 290 nanometers) ultraviolet light is absorbed, whereas medium wavelength ultraviolet light (UV-B, 290 to 320 nanometers) is partially absorbed and the longer wavelengths (UV-A, 320 to 400 nanometers) are not absorbed at all (60).

As the ozone layer deteriorates, more UV-B is expected to reach the earth's surface. Additional UV-B is fairly certain to cause additional cases of skin cancer. Nonmelanoma skin cancer, which is quite common and benign, seems certain to increase, with the greatest risk among individuals with the lightest skin and highest sun exposure and individuals living closer to the equator and therefore exposed to more UVR. It is estimated that for every 1 percent decrease in stratospheric ozone, nonmelanoma skin cancers will increase by about 3 percent (61).

Cutaneous (malignant) melanoma, which results from the transformation of pigment-producing skin cells (melanocytes), accounts for about 7,800 deaths in the United States annually. Incidence of the disease in the United States over the past 35 years has increased by some 200 percent; mortality has increased by 150 percent (62). The evidence linking malignant melanoma and ozone depletion is less clearcut, but suggests that a 1 percent decrease in ozone could cause roughly a 1 percent increase in both incidence and mortality among light-skinned populations (63).

UV-B reduces the ability of the body's immune system to fight foreign substances that enter the body through the skin. For example, UVR irradiation during a first infection of leishmaniasis (64) or the *Herpes simplex* virus (65) may weaken the body's ability to fight subsequent infections.

Ultraviolet radiation also has been associated with various diseases of the eye such as cataracts (66) and deterioration of the cornea and retina.

Health Effects of Global Warming

Global warming probably will affect mortality due to heat stress and increase the incidence of chronic and infectious respiratory diseases, allergic reactions, and reproductive illnesses. It may also affect the geographic range of vector-borne diseases.

Warming is likely to have a direct impact on mortality, but the extent of the mortality is uncertain because people will partially adjust to hotter temperatures. One study of 15 cities in the United States estimated that about 1,100 deaths would occur in an average summer with no increase in warming, but that an increase of 3.9°C would cause about 4,500 deaths assuming partial acclimatization and 9,000 deaths assuming no acclimatization (67). The elderly are disproportionately affected by hot spells. Heat waves that occur early in the summer, or

that occur in areas where hot weather is uncommon, appear to be particularly dangerous (68).

Temperature extremes also tend to worsen the effects of cardiovascular, cerebrovascular, and respiratory diseases, particularly on older people. Studies of selected American cities have found sharp increases in the incidence of heart disease and stroke when temperatures rise above 25°C (69).

Temperature and humidity changes could change the range and life cycles of plants, animals, insects, bacteria, and viruses. Changes in the geographic range of vector-borne diseases— diseases transmitted to humans by mosquitoes, for example— might be significant (70). Five mosquito-borne diseases—malaria, dengue fever, and arbovirus-induced encephalitis, and to a lesser extent yellow fever and Rift Valley fever—are thought to pose a potential risk to U.S. populations if significant warming does occur (71). Warming also could increase the risk of a northward spread of diseases caused by parasites, such as Ascariasis and Chagas' disease, which are common in Central and South America (72).

Some studies also have found statistically significant increases in preterm births and in perinatal mortality (death just before, during, or just after birth) in the summer months, suggesting that this may be temperature-related (73).

The section on World Population Trends was authored by Robert Fox, a Washington, D.C. consultant. The section on Malaria was written by Tensie Whelan, a New York City environmental writer, based on material supplied by Carlos Kent Campbell, Chief of the Malaria Branch of the Centers for Disease Control, Atlanta, who served as principal consultant. The section on AIDS in the Americas was authored by Ann Marie Kimball and Fernando Zacarias of the Pan American Health Organization, Washington, D.C., for which Dr. Kimball is Regional Advisor on AIDS.

References and Notes

1. United Nations Department of International Economic and Social Affairs, United Nations, *World Population Prospects, 1988,*Population Studies No. 106 (United Nations, New York, 1989). All statistics are from this publication unless otherwise noted.
   Also called "The 1988 Assessment," this biannual publication is the United Nations Population Division's major statistical report on global population trends. Along with the corresponding computer tape, it serves as the principal international source of demographic information on regional and country-specific trends. It includes data for fertility and mortality, life expectancy, distribution by age and sex and by urban and rural place of residence, in addition to 29 estimates of population size and projections for the period 1950–2025. Population projections used here are based on the U.N.'s *medium variant* projection, as opposed to its high or low population projections.

The statistical series are supported by numerous data collection and evaluation efforts at the national level. These include censuses, vital registries and continuous household sample surveys. Results filter up to the international level where technical adjustments are often made in keeping with international data standards. Contributing to the assessment are national census and statistics organizations, the Statistics Division of the United Nations and its five regional economic commissions, as well as specialized agencies such as the Latin American Demographic Center in Santiago, Chile. The International Statistical Programs Center of the United States Bureau of the Census also makes significant contributions.
   The internal consistency among the numerous data series in the assessment is of fundamental importance. Each set of computerized statistics is linked to the next. A change to a national level data base has a ripple effect, creating changes for other series at national, regional and

global levels. Survey data, for example, reporting new fertility trends will affect birth rate statistics and the composition of the population by age. Affecting forecasts of future population size, revisions also mean changes back to 1950, modifying earlier estimates of population size and composition. Backcasting is particularly important for Third World countries without censuses prior to the 1960s or 1970s.
2. Thomas Merrick, "World Population in Transition," *Population Bulletin*, Vol. 41, No. 2 (April 1986), pp. 8-11.
3. Based on Projects in Department of International Economic and Social Affairs, United Nations, *World Population Prospects, 1988*, Population Studies, No. 106 (United Nations, New York, 1989).
4. Department of International Economic and Social Affairs, United Nations, *World Population Prospects, 1988*, Population Studies No. 106 (United Nations, New York, 1989), Part 2, p. 75, Table 1.
5. *Ibid.*, Part 2, p. 202, Table 17.A.
6. *Ibid.*, Part 2, p. 75, Table 1.

7. *Ibid.*, Part 2, p. 88, Table 2.
8. Department of International Economic and Social Affairs, United Nations, *Mortality of Children Under Age 5: World Estimates and Projections, 1950-2025*, Population Studies No. 105 (United Nations, New York, 1988), p. 13.
9. Department of International Economic and Social Affairs, United Nations *Global Estimates and Projections of Population by Sex and Age: The 1988 Revision* (United Nations, New York, 1989), pp. 130-131.
10. Department of Economic and Social Affairs, United Nations, *World Population Prospects as Assessed in 1963*, Population Studies No. 41 (United Nations, New York, 1966) p. 15, Table 4.3.
11. D. Sturchler, cited in "Focus on Malaria—Global Eradication: The Unrealistic Goal," *Panoscope*, No. 12, May 1989, p. 3.
12. David F. Clyde, "Recent Trends in the Epidemiology and Control of Malaria," *Epidemiologic Reviews*, Vol. 9 (1987), p. 219.
13. J.G. Breman and C.C. Campbell, "Combatting Severe Malaria in African Children," *Bulletin of the World Health Organization*, Vol. 66, No. 5 (1988), p. 611.
14. WHO Expert Committee on Malaria, *Eighteenth Report* WHO Technical Report Series No. 735 (World Health Organization, Geneva, 1986), p. 13.
15. *Op. cit.* 13.
16. *Op. cit.* 12, p. 222.
17. *Op. cit.* 12, p. 220.
18. *Op. cit.* 12, p. 222.
19. *Op. cit.* 12, p. 223.
20. *Op. cit.* 12, p. 223.
21. *Op. cit.* 12, pp. 224-25.
22. *Op. cit.* 12, p. 225.
23. *Op. cit.* 12, p. 225.
24. *Op. cit.* 12, p. 226.
25. Richard House, "Malaria Spreads in Brazil as Development Opens Up the Amazon," *Washington Post*, Health Section (July 18, 1989), p. 5.
26. Ian A. McGregor, "Epidemiology, Malaria and Pregnancy," *Tropical Medicine and Hygiene*, Vol. 33, No. 4 (1984), p. 517.
27. *Ibid.*, p. 519.
28. *Op. cit.* 13, p. 617. A precise figure is, however, impossible to obtain, both because cause of death is determined in few cases in sub-Saharan Africa, and because many children die from multiple symptoms, including malaria.
29. *Op. cit.* 14, p. 12.
30. *Op. cit.* 12.
31. Malaria Action Programme, World Health Organization, "World Malaria Situation 1985," *World Health Statistics Quarterly*, Vol. 40, No. 2 (1987), p. 142.
32. *Op. cit.* 11.
33. *Op. cit.* 14.
34. *Op. cit.* 12, p. 230.
35. *Op. cit.* 12, pp. 228-231.
36. *Op. cit.* 12, p. 233.
37. *Op. cit.* 14, pp. 48-49.
38. *Op. cit.* 12, pp. 235-239.
39. Statement on the Development of Malaria Vaccines, Document WHO/MAP/TDR, 1985. Reprinted in the Epidemiologic Bulletin (n.d.).
40. G.H. Mitchell, "An Update on Candidate Malaria Vaccines," *Parasitology*, Vol. 98 (1989), p. S30.
41. *Ibid.*
42. Robert C. Gallo and Luc Montagnier "AIDS in 1988," *Scientific American*, Vol. 259, No. 4 (October 1988), p. 41.
43. G.F. Medley, R.M. Anderson, D.R. Cox, L. Billard, "Estimating the incubation period for AIDS patients," *Nature*, Vol. 333 (June 9, 1989), p. 505.

44. Jacques Pepin, Frances Plummer, *et al.*, "The Interaction of HIV Infection and Other Sexually Transmitted Diseases: An Opportunity for Intervention," *AIDS*, Vol. 3 (1989), pp. 3-9.
45. Global Programme on AIDS and Programme of STD, World Health Organization, "Consensus Statement from Consultation on Sexually Transmitted Diseases as a Risk Factor for HIV Transmission" (World Health Organization, Geneva, January 1989), pp. 4-6.
46. James Chin, Gopal Sankaran, and Jonathan Mann, *Mother to Infant Transmission of HIV: An Increasing Global Problem* (World Health Organization, Geneva, 1988).
47. "Leads from MMWR: Tuberculosis and the Acquired Immunodeficiency Syndrome," *JAMA*, Vol. 259, No. 3 (1988), pp. 338-339.
48. *Op. cit.* 4, p. 304.
49. United Nations (U.N.), *World Population Trends and Policies: 1989 Monitoring Report*, unpublished data (United Nations, New York, February 1989), p. 326.
50. "Brazil 1986: Results from the Demographic and Health Survey," *Studies in Family Planning*, Vol. 19, No. 1 (1988), p. 61, Table 2.1.
51. *Ibid.*, Table 2.2.
52. James Brooke, "Births in Brazil are on Decline, Easing Worries," *The New York Times* (August 8, 1989), p. A9.
53. *Op. cit.* 50, p. 61, Tables 4.3, 5.1.
54. Maria J. Wawer, Karen Johnson Lassner, and Beatriz B. Collere Hanff, "Contraceptive Prevalence in the Slums of Rio de Janeiro," *Studies in Family Planning*, Vol. 17, No. 1 (1986), p. 47.
55. *Op. cit.* 52.
56. *Op. cit.* 52.
57. *Op. cit.* 52.
58. Alexander Leaf, "Potential Health Effects of Global Climatic and Environmental Changes," *The New England Journal of Medicine*, Vol. 231, No. 23 (December 7, 1989), pp. 1577-1583.
59. U.S. Environmental Protection Agency, *The Potential Effects of Global Climate Change on the United States*, J. Smith and D. Tirpak, eds. (Environmental Protection Agency, Washington, D.C., 1988). The Climate Institute, *Coping with Climate Change: Proceedings of the Second North American Conference on Preparing for Climate Change: A Cooperative Approach*, John C. Topping, Jr., ed. (The Climate Institute, Washington, D.C., 1989).
60. Janice Longstreth, "Global Climate Change: Potential Impacts on Public Health," paper presented at the Global Climate Change and Life on Earth Conference, Albany, New York, April 1989, p. 1.
61. Frederick Urbach, "Potential Health Effects of Climatic Change: Effects of Increased Ultraviolet Radiation on Man," paper presented at the Conference on Global Atmospheric Change and Human Health, National Institute of Environmental Health Sciences, Research Triangle Park, North Carolina, November 6-7, 1989, pp. 3-4.
62. National Cancer Institute, *1987 Annual Cancer Statistics Review, Including Cancer Trends: 1950-85* (National Cancer Institute, Bethesda, Maryland, 1988), as cited in Janice Longstreth, "Global Climate Change: Potential Impacts on Public Health," paper presented at the Global Climate Change and Life on Earth Conference, New York State Museum, Albany, New York, April 1989, pp. 1-2.
63. U.S. Environmental Protection Agency, *Ultraviolet Radiation and Melanoma with a Special Focus on Assessing the Risks of Stratospheric Ozone Depletion* (Govern-

ment Printing Office, Washington, D.C., 1987), as cited in Janice Longstreth, "Global Climate Change: Potential Impacts on Public Health," paper prepared for the Global Climate Change and Life on Earth Conference, New York State Museum, Albany, New York, April 1989, p. 2.
64. Suzanne Holmes Giannini, "Effects of UVB on Infectious Diseases," paper presented at the Conference on Global Atmospheric Change and Public Health, Center for Environmental Information, Washington, D.C., December 5, 1989.
65. J. J. Perna, M. L. Mannix, J. E. Rooney, *et al.*, "Reactivation of latent herpes simplex virus infection by ultraviolet light," *Journal of the American Academy of Dermatology*, Vol. 17 (1987), pp. 473—478, as cited in Janice Longstreth, "Global Climate Change: Potential Impacts on Human Health," paper presented at the Global Climate Change and Life on Earth Conference, New York State Museum, Albany, New York, April 1989, p. 4.
66. A. Kornhauser, "Implication of Ultraviolet Light in Skin Cancer and Eye Disorders," in *Coping with Climate Change: Proceedings of the Second North American Conference on Preparing for Climate Change* (The Climate Institute, Washington, D.C., 1989), p. 178.
67. Laurence S. Kalkstein, "Potential Impact of Global Warming: Changes in Mortality from Extreme Heat and Cold," in *Coping with Climate Change: Proceedings of the Second North American Conference on Preparing for Climate Change* (The Climate Institute, Washington, D.C., 1989), p. 156.
68. Laurence S. Kalkstein and Robert E. Davis "Weather and Human Mortality: An Evaluation of Demographic and Interregional Responses in the United States," *Annals of the Association of American Geographers*, Vol. 79, No. 1 (1989), pp. 44, 54-56.
69. E. Rogot and S. J. Padgett, "Associations of coronary and stroke mortality with temperature and snowfall in selected areas of the United States 1962-66," *American Journal of Epidemiology*, No. 103 (1976), pp. 565-575, as cited in U.S. Environmental Protection Agency, *The Potential Effects of Global Climate Change on the United States* (Government Printing Office, Joel B. Smith and Dennis A. Tirpak, eds. Office of Policy, Planning and Evaluation, Washington, D.C., 1988), pp. 12-7, 12-8.
70. Robert E. Shope, "Global Climate Change and Infectious Diseases," paper presented at the Conference on Global Atmospheric Change and Human Health, National Institute of Environmental Health Sciences, Research Triangle Park, North Carolina, November 6-7, 1989, p. 4.
71. U.S. Environmental Protection Agency, *The Potential Effects of Global Climate Change on the United States* (Government Printing Office, Washington, D.C., 1988), p. 12-11.
72. Andrew Dobson, "Climate change and parasitic diseases of man and domestic livestock in the United States," in *Coping with Climate Change: Proceedings of the Second North American Conference on Preparing for Climate Change* (The Climate Institute, Washington, D.C., 1988), p. 149.
73. C. A. Keller and R. P. Nugent, "Seasonal patterns in perinatal mortality and preterm delivery," *American Journal of Epidemiology*, No. 118 (1983), pp. 689-698, as cited in U.S. Environmental Protection Agency (EPA) Office of Policy, Planning and Evaluation, Joel B. Smith and Dennis A. Tirpak, eds., *The Potential Effects of Global Climate Change on the United States* (Government Printing Office, Washington, D.C., 1988), p. 12-4.

5. Human Settlements

In 1990, we are just past the midpoint of a major trend in population movement that will span the 75-year period between 1950 and 2025. The trend toward urbanization that is engulfing the developing world is still gathering momentum. By 2025, the world's population is expected to reach about 8.5 billion; much of this growth will be in Third World cities. In 1950, less than 30 percent of the world's population was urban; by 2025, it is estimated, more than 60 percent will be urban (1).

As urban populations grow and as rural people continue to migrate to urban areas, city service agencies will be hard pressed to keep up with the basic requirements of new residents. Shelter and services such as water and sanitation are particularly important because of their impact on the health of the urban poor. In both of these areas, rapid population growth and limited financial resources are forcing managers—with the support of international institutions such as the World Bank—to look for low-cost options, improve the efficiency and cost-effectiveness of services and shelter projects, and stretch budgets further with better recovery of costs.

This chapter looks at a few services such as water and sanitation that have particularly important links to natural resources and environmental quality.

Most Third World cities are working under difficult financial constraints in the provision of basic services. Development banks, foreign aid programs, and other external sources have been contributing increasing amounts to fund water and sanitation development projects for Third World cities, but in recent years, aid has been eroded by inflation. (See Financing Water and Sanitation Services, below.)

Where city governments have been unwilling or unable to cope with rapid population growth, sometimes residents have taken the provision of shelter and services into their own hands. In Lima, Montevideo, Nairobi, Karachi, and elsewhere, ad hoc groups of residents—often with the help of governmental agencies or nongovernmental organizations (NGOs)—have helped to build housing and infrastructure projects. These groups—loosely referred to as the "informal sector"—are filling the voids where official agencies have failed. (See The Informal Sector, below.) As successes of the informal sector become obvious, development agencies have studied, discussed, and made loans to support informal operations. The informal sector should not be expected to carry the burden of municipal services that a city government ought to provide. But the enterprise and resourcefulness that residents show in creating a better urban environment for themelves—at times in the face of overwhelming odds—is an important, and largely untapped, resource.

Figure 5.1 Actual and Projected Urban Population Growth, 1950–2025

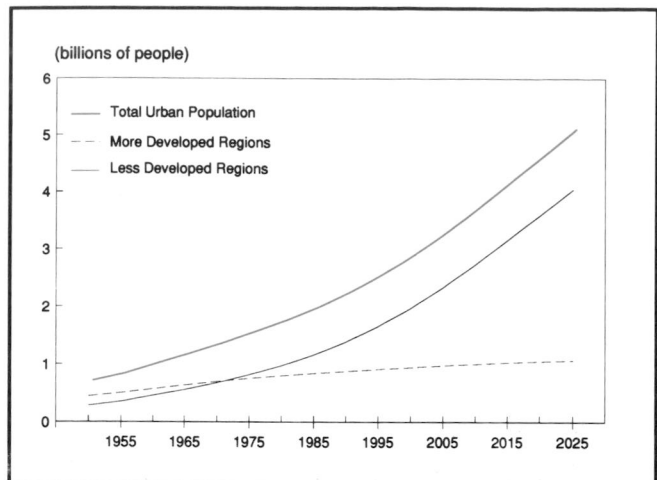

Source: Department of International Economic and Social Affairs, United Nations (U.N.), *Prospects of World Urbanization, 1988* (U.N., New York, 1989), Table A-1, pp. 28-29.

Figure 5.2 Actual and Projected Percentages of Urban Population in Less Developed Regions, 1950–2025

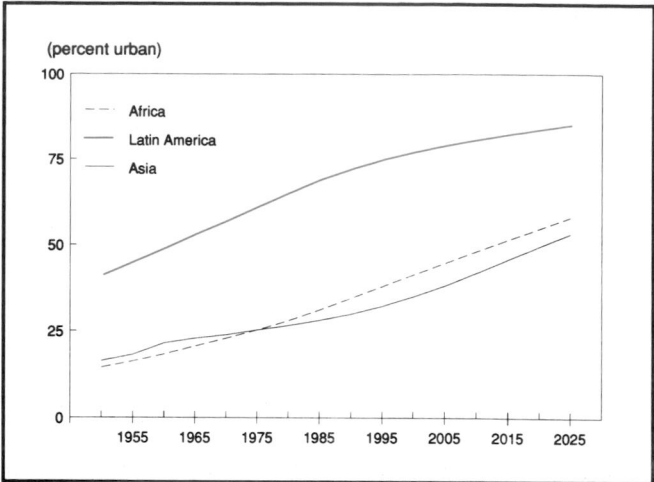

Source: Department of International Economic and Social Affairs, United Nations (U.N.), *Prospects of World Urbanization, 1988* (U.N., New York, 1989), Table A-3, pp. 40-41.

CONDITIONS AND TRENDS

THE URBANIZATION TREND

In the mid-1700s, only 3 percent of the world's inhabitants lived in urban areas (2). By 1950, urban areas held 29 percent of the population, and just 35 years later, they contained 41 percent. In 2025, 60 percent of the world's population is expected to be living in and around cities (3).

Nearly all of this future growth is taking place in the cities of developing countries. Figure 5.1 also indicates the rapid progression toward urbanization in the developing world. By comparison, urban growth in the developed world has been relatively smooth and slow. Only 16.9 percent urban in 1950, developing countries by 1990 had about 34 percent of their population living in cities and are projected to be about 57 percent urban by the year 2025. The developed world, in contrast, already 54 percent urban in 1950, was 73 percent urban in 1990, and is projected to increase only marginally—to 79 percent—by the year 2025 (4).

The level of urbanization varies significantly among the regions of the developing world. As shown in Figure 5.2, in 1990, Latin America was 72 percent urbanized—comparable to the developed world—and is projected to be 85 percent urban in 2025, a figure that exceeds projections for the developed world. In Asia and Africa, much smaller proportions of the populations live in urban areas, but these regions are urbanizing faster than Latin America (5).

While Latin America has the largest percentage of urban dwellers, Asia has the largest number of urban dwellers. As shown in Table 5.1, in 1990 Asia had 931

million urban dwellers, far more than Latin America's 324 million and Africa's 223 million urban inhabitants and fully 41 percent of the world's total urban population. By 2025, Asia's urban population is expected to rise to nearly 2.6 billion people, which will be just over 50 percent of the world's total.

Third World cities have been straining under the weight of population increases since 1950, but the worst is yet to come. Of the total increase of nearly 3.8 billion projected over the 1950–2025 period, roughly 2.66 billion (71 percent) will occur after 1990. (See Table 5.1.)

Natural increase has contributed more to the growth in urban population than rural-to-urban migration (though there are numerous exceptions). Many urban migrants are young and will soon have children, so the high rate of natural increase in a city's population is partly due to the in-migration of young people in previous years (6).

In Latin America, the nations with the strongest growth in their economies and in manufacturing output in the 1960s and 1970s—Mexico, Colombia, and Brazil—tended to have the largest increase in urbanization. Chile, Uruguay, and Argentina had much slower rates of urban population growth, in part because of slow economic growth and in part because they are heavily urbanized already. Several core regions have emerged, particularly those centered on Mexico City and Buenos Aires and the São Paulo/Rio de Janeiro/Belo Horizonte triangle. The regions include examples of the centers of cities growing faster than their suburban rings and of suburban areas outpacing the city centers (7).

In most nations, there is an increasing concentration of population and economic activity in one or

Table 5.1 Actual and Projected Urban Population by Region, 1950–2025

(millions)

	1950	1960	1970	1980	1990	2000	2010	2025
World Total	733	1,030	1,374	1,770	2,260	2,917	3,737	5,119
More Developed Regions	448	571	699	798	876	945	1,004	1,068
Less Developed Regions	285	459	675	972	1,385	1,972	2,733	4,051
Africa	33	51	83	135	223	361	552	914
Latin America	69	107	163	237	324	417	509	645
Asia	226	359	503	688	931	1,292	1,772	2,589
Europe	221	259	307	340	364	387	405	422
Oceania	8	10	14	16	19	21	24	29
U.S.S.R.	71	105	138	167	195	217	237	260

Source: Department of International Economic and Social Affairs, United Nations (U.N.), *Prospects of World Urbanization, 1988* (U.N., New York, 1989), Table A1, pp. 28-29.

two cities or core regions. Megacities are growing rapidly in number and size in the Third World. Mexico City will leap from 17 million inhabitants in 1985 to more than 24 million by 2000; São Paulo will experience a similar jump from 15 million to 24 million. Of the 24 cities expected to exceed 10 million inhabitants by 2000, 18 will be in developing countries (8). Small and medium-size cities have also grown in developing countries. (See *World Resources 1988-89*, pp. 42-44.)

The Urbanization-Environmental Quality Gap

In most Third World countries, the effort to deliver essential services—getting water supplies to new neighborhoods, for example—has far outweighed the commitment to treat and safely dispose of waste.

The resulting environmental quality problems, according to many accounts, appear to have grown much worse in Third World urban areas during the 1980s. Industrial, human, and solid wastes piled up in lagoons in cities such as Abidjan, Côte d'Ivoire (9). In cities throughout the Third World, rivers, estuaries, and coastal zones are polluted by sewage; groundwater resources are threatened by uncontrolled dump sites; and forests around some cities are denuded by the urban demand for timber and domestic fuels (10).

In the past, investment in water and sanitation services tended to favor water supply over sewerage, other sanitation systems, and wastewater treatment. The bias also may have been an unintentional result of the United Nations International Drinking Water Supply and Sanitation Decade, which set ambitious goals for the global extension of water and sanitation services in the 1980s. The surge in water and sanitation services in that decade far outpaced gains in wastewater treatment and disposal. For example, it is estimated that less than 2 percent of total urban sewage flows in Latin America receive treatment (11). Data for 24 rivers in Central and South America suggest that the situation in these regions is worse than in most other parts of the world; over half the rivers sampled showed fecal coliform counts in excess of 1,000 per 100 milliliters (12). (Safe drinking water has zero fecal coliforms, according to World Health Organization (WHO) guidelines (13).)

These environmental problems put added financial pressures on city governments. Water and sanitation agencies that face limited local water supplies are forced to extend water lines to more distant sources, pump and store more water, and treat water from increasingly polluted sources. For example, in Shanghai, pollution forced water intakes to be moved upstream more than 40 kilometers (14). In Lima, upstream pollution has increased treatment costs by about 30 percent (15).

There is growing recognition that lagging investment in sanitation takes a heavy toll on human health and living conditions (16). Although it is difficult to quantify, most studies show that improvements in water supplies, sanitation, excreta disposal, and water quality can reduce illnesses such as diarrhea among children in poor neighborhoods (17). Improved rainwater drainage in poor neighborhoods also can have important benefits.

Figure 5.3 Actual and Projected Percentages of Urban Population in Developing Countries with Access to Water and Sanitation Services, 1980–2000

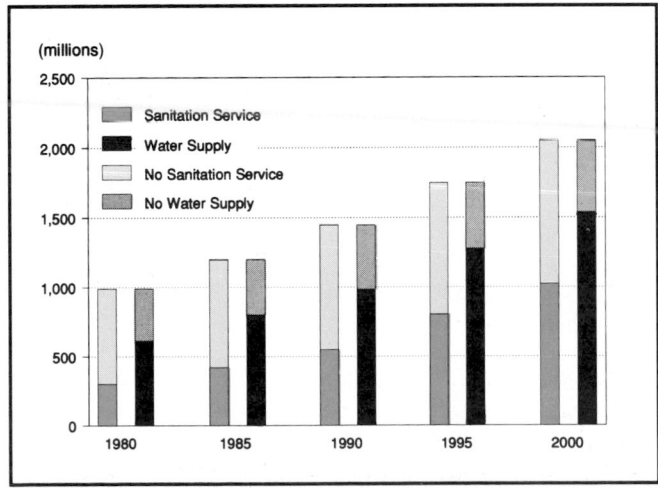

Source: Policy, Planning, and Research Staff, The World Bank, *FY88 Annual Sector Review: Water Supply and Sanitation* (The World Bank, Washington, D.C., November 1988), Figure 2, p. 2.

Figure 5.4 Actual and Projected Water Supply and Sanitation Coverage in 20 African Countries, 1980–2000

A. Water Supply Coverage

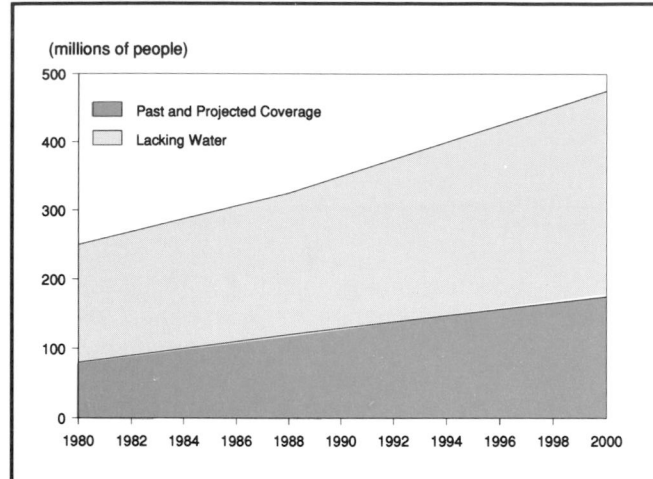

B. Sanitation Coverage

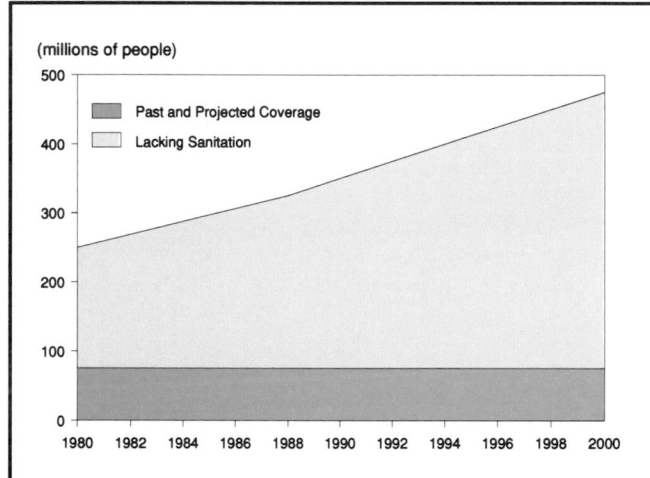

Source: Water and Sanitation for Health Project (WASH), *Water and Sanitation Sector Profiles of Twenty African Countries* (CDM and Associates, Arlington, Virginia, 1989), Figures 1, 2, p. 16.

Notes:

a. Based on 1980–88 coverage rates. Projected sanitation coverage appears to show no improvement in sanitation over the 1980–88 period because the data for the initial years were poor and inflated and, for the later years, somewhat less inflated. The deceptively high baseline in 1980 masks what was probably a slight improvement in sanitation coverage over the period.

b. The 20 African countries profiled are Benin, Burkina Faso, Burundi, Cameroon, Côte d'Ivoire, Guinea, Kenya, Liberia, Malawi, Mali, Niger, Nigeria, Rwanda, Senegal, Sudan, Swaziland, Tanzania, Togo, Uganda, and Zaire.

URBANIZATION: AN OPPORTUNITY AND A CHALLENGE

Urbanization is both an opportunity and a challenge. Cities produce jobs and economic activity; they transform resources into productive goods and services (18). Contrary to many early assertions that migrants to cities are poor backward peasants who contribute nothing to economic growth, many studies now argue that, for the most part, urban migrants are energetic and ambitious and that the urban poor make significant contributions to urban economies.

The challenge for city managers is to absorb new residents into the city fabric efficiently. Under the weight of rapid population growth and a generally dismal economic performance in the 1980s, many Third World cities struggled just to keep up, and some fell behind. A host of problems confront city managers, among them:

■ Effective planning to ensure some order in the growth of cities and to coordinate development strategies. Good planning can help build relationships among city agencies, between the public and private sectors, and between city officials and community members (19). Local participation can be particularly important; most of the neighborhoods in Cali, Colombia, for example, participate in the assessment of problems and needs. This involvement has helped the city cope successfully with rapid population growth (20).

■ Building a strong institutional framework, especially giving local authorities the money, manpower, and training to be effective (21).

■ Mobilizing financial resources to support the growth of settlements; particularly troublesome issues include the public financing of infrastructure and services, savings institutions for the urban poor, and the financing of low-income shelter. (See Financing Water and Sanitation Services and The Informal Sector and the Development Process, below.)

■ Managing land; often, information about land is unreliable and out of date, land speculation is rampant in many cities (22), and the urban poor are forced to build settlements on marginal lands that they do not own. (See Housing, below.)

Two other issues—shelter and services—present equally daunting challenges for city officials. Local governments are partly or fully responsible for providing basic physical infrastructure such as roads, safe water supplies, sanitation and sewerage systems, and drainage works. Usually, local government units are fully responsible for many other services such as markets, street lighting, garbage collection, fire protection, public open spaces, libraries, and cemeteries. Third World governments also have made a major commitment to provide shelter for the urban poor.

This chapter focuses on water and sanitation services because of their important links with environmental quality.

Progress to Date: Water and Sanitation Services

In December 1985, at the midpoint of the International Drinking Water Supply and Sanitation Decade, 75 percent of the urban inhabitants in developing countries had access to safe drinking water. That figure represented an increase of 133 million people from the start of the decade, but in percentage terms was unchanged from 1980. Access to sanitation services went up from 50 percent at the beginning of the decade (350 million people) to 59 percent (512 mil-

Figure 5.5 Comparison of Urban Water Supply and Sanitation Coverage in Central America, 1980 and 1988

A. Urban Water Supply

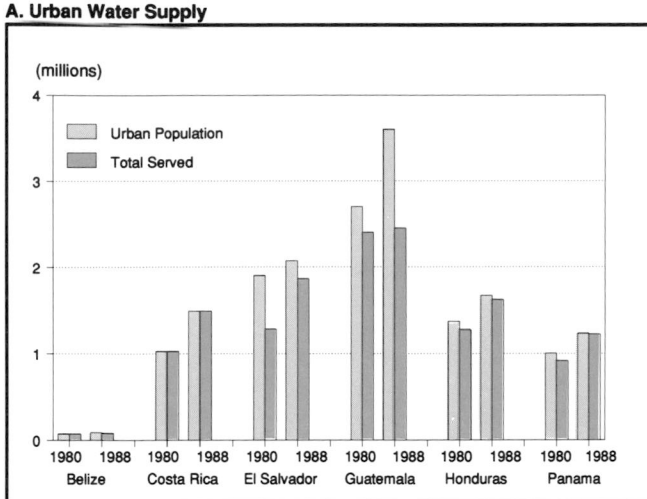

B. Sanitation Coverage

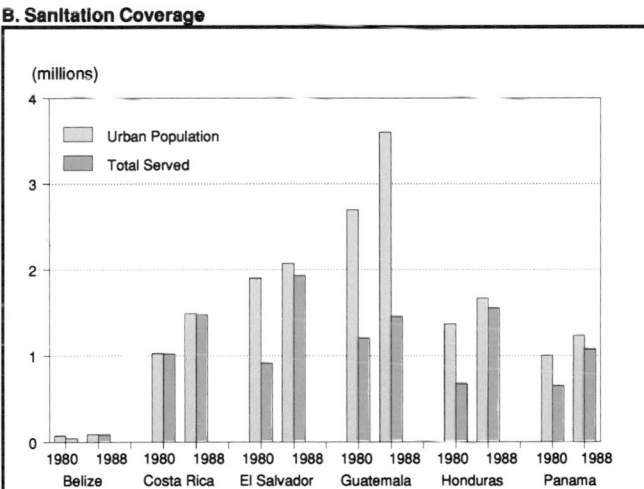

Source: Frederick S. Mattson, *Planning for Central America Water Supply and Sanitation Programs: Update* (CDM and Associates, Arlington, Virginia, 1989). Figure 5.5A: Table A-1, p. 25; Table B-1, p. 36; Table C-1, p. 46; Table D-1, p. 55; Table E1, p. 65; Table F-1, p. 74. Figure 5.5B: Table A-2, p. 26; Table B-2, p. 37; Table C-2, p. 47; Table D-2, p. 56; Table E-2, p. 66; Table F-2, p. 75.

Figure 5.6 Access to Safe Urban Drinking Water and Sanitation in Five Asian Countries, 1980–87

A. Access to Safe Urban Drinking Water

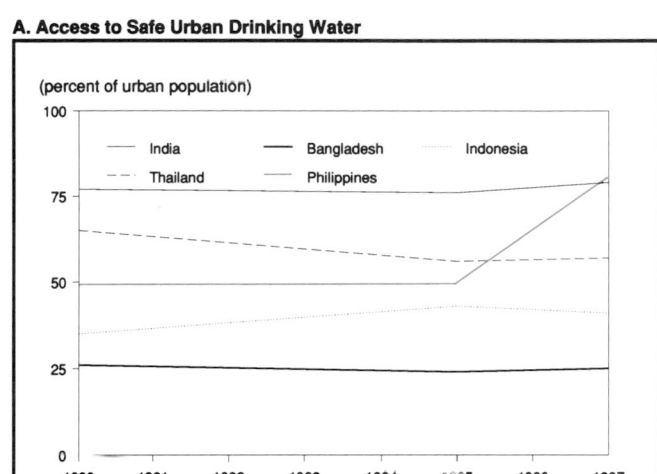

B. Sanitation Coverage

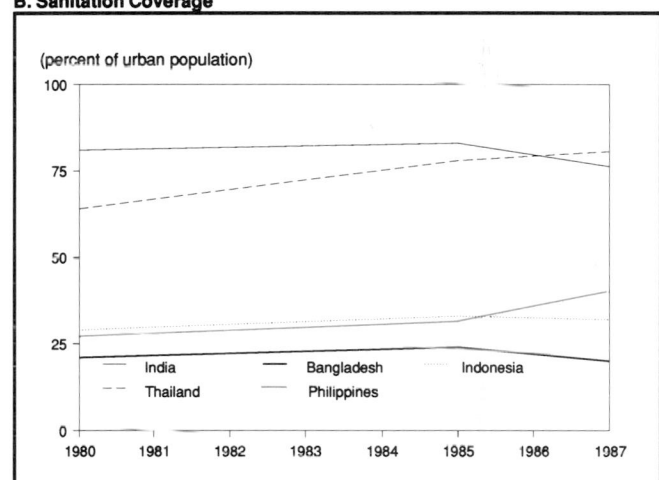

Sources:
1. World Health Organization (WHO), *The International Drinking Water Supply and Sanitation Decade: Review of National Progress (as at December 1983)* (WHO, Geneva, 1986), Table 5.3, p. 154 and Table 3.3, p. 156.
2. World Health Organization (WHO) *The International Water Supply and Sanitation Decade: Review of Mid-Decade Progress (as at December 1985)* (WHO, Geneva, 1987), Table 3.3, p. 102 and Table 6.3, p. 174.
3. World Health Organization, (WHO), "Global Strategy for Health for All: Monitoring 1988-89," WHO, Geneva, 1989, Table 2, pp. 97-98.

lion people) by 1985 (23). WHO defines reasonable access to safe drinking water in urban areas as access to piped water or to a public standpipe within 200 meters of a dwelling or housing unit. Access to sanitation is defined as waste disposal via connections to public sewers or household systems such as pit latrines, pour-flush latrines, septic tanks, and communal toilets (24).

The uncertainty of the data and the broadness of the definitions probably disguise a situation that is worse than the numbers suggest. In Africa, for example, WHO's estimate of urban sanitation coverage indicated that only about 12 percent of urban residents were served by sewer connections; 59 percent were served by "other" household systems (25). Furthermore, the existence of sewers does not guar-

antee the proper disposal of waste because sewers often run into rivers or lakes from which the urban poor may draw their water. In addition, people who must rely on piped water from a communal tap are likely to use minimal amounts of water for washing, cooking, or personal hygiene (26). Where communal taps are provided within a walking distance of 200 meters, each person commonly consumes 20 to 40 liters per day; with house connections provided by a single yard tap, individual consumption rises to 40–60 liters daily (27).

After factoring in independent World Bank studies and estimates for China and using more stringent definitions (28), World Bank officials estimated that, in

1985, 66 percent of Third World urban residents had access to a satisfactory water source and 35 percent had satisfactory sanitation services. World Bank officials estimate that urban drinking water coverage may reach 68 percent by 1990 and 75 percent by 2000, and that sanitation coverage could rise to 38 percent by 1990 and 50 percent by 2000 (29).

Interpretation of the World Bank's estimates shown in Figure 5.3 is a case of whether "the cup is "half empty or half full." On one hand, the figures project impressive gains both in absolute numbers and in the percentages of the population served. On the other hand, because of urban population growth by the year 2000, more people will be without water and sanitation service. By that year, more than a billion people might not have satisfactory sanitation, compared with about 780 million people in 1985. About 500 million people may not have adequate drinking water, compared with about 400 million in 1985.

The U.S. Agency for International Development's (U.S. AID's) Water and Sanitation for Health Project (WASH) has studied water and sanitation in Africa and Central America recently. WASH bases its estimates mainly on reports from AID bureaus; WHO uses estimates submitted by governments for the International Drinking Water Supply and Sanitation Decade. WASH estimates often are more conservative than the country reports used by WHO. (For the figures reported by WHO, see Chapter 16 "Population and Health," Table 16.4.)

Africa

A WASH study of the 325 million people in 20 African countries in 1988 showed that 201 million lacked access to safe drinking water supplies and 257 million lacked adequate sanitation facilities (30). The study found that $20 billion would be required over the next 12 years to meet the International Water Supply and Sanitation Decade's goals of extending water service to 240 million additional people and sanitation service to an additional 154 million people by the year 2000. Comprehensive data on past investments are not available, but $20 billion is far larger than past investments by these countries (31). Figure 5.4 (showing total population, both urban and rural) suggests that even if coverage continues to increase as it did from 1980 to 1988, it will not satisfy the demands of growing populations by the end of the century.

Central America

WASH's review of water supply and sanitation programs in Central America suggests that the region as a whole is keeping pace with population growth. From 1986 to 1988, the combined population of Belize, Honduras, Guatemala, El Salvador, Costa Rica, and Panama increased by 757,000; the number served by adequate water facilities increased by

860,000; and the number served by adequate sanitation facilities increased by 1.23 million (32).

Among these countries, as shown by Figure 5.5, the picture is mixed. Costa Rica, blessed with a stable political system and a relatively prosperous economy, has maintained essentially complete coverage for water and sanitation throughout the period. In contrast, provision of safe water in urban areas did not keep up with population growth in Guatemala, where coverage dropped from 89 percent in 1980 to 68 percent by 1988 (33).

El Salvador made significant progress on urban water supply and sanitation projects through 1988, although rural water supply fell dramatically and rural sanitation improved only slightly (34).

Few data are available on the performance of most South American countries since 1985. From 1980 to 1985, most countries reported slight increases in the percentage of the urban population with safe drinking water. For sanitation, the pattern was much less consistent. (See Chapter 16, "Population and Health," Table 16.4.)

Asia

WHO data for Asian countries also present a mixed picture. India reported a slight increase in access to safe urban drinking water and a more substantial increase in sanitation services; Bangladesh appeared to be struggling to keep up with population growth; Thailand's urban inhabitants experienced progress in sanitation services but a decline in safe drinking water; the Philippines reported a decline in sanitation and a startling increase in drinking water coverage to 80.5 percent in 1985, up from 49 percent in 1980. (See Figure 5.6.)

Overall, the outlook for Asia is considerably brighter than for Latin America and Africa because of strong economic growth. Asia's per capita gross domestic product increased at an average annual rate of 4.7 percent from 1980 to 1987 (35).

FINANCING WATER AND SANITATION SERVICES

The countries of the developing world are in a race with population growth in their efforts to extend water and sanitation services. The evidence is sparse, but suggests that many developing countries were falling behind in the mid-1980s.

Much of the decline is attributable to a difficult economic climate for developing countries during the 1980s. Extensive external borrowing and high inflation contributed to a generally dismal economic performance; between 1980 and 1987, more than half of them experienced declines in per capita income. By comparison, from 1965 to 1980, 9 out of 10 developing countries showed gains in per capita income (36).

Low-income groups probably have fared worse economically than other groups in the 1980s (37), sug-

Figure 5.7 Donor Agency Funding of Water and Sanitation Projects, 1981–88

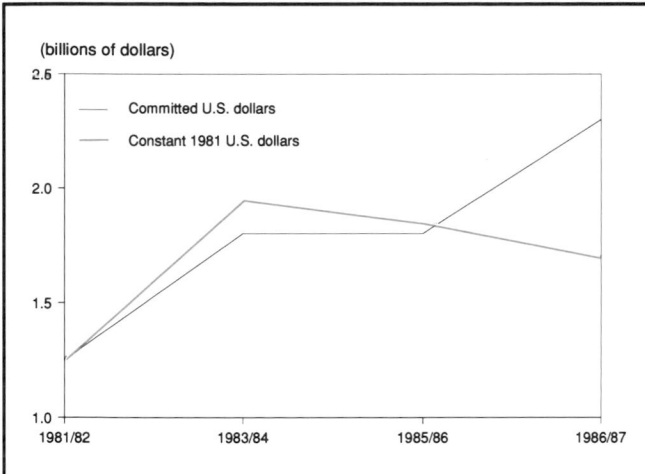

Source: Division of Information, United Nations Development Programme (UNDP), "Donor Dollars Go Further as Agencies Push Cost Effectiveness," *Source,* Vol. 1, No. 2 (UNDP, New York, September 1989), p. 21.

gesting a reduction in their ability to pay for water and sanitation services.

In that bleak context, it is surprising that spending on water and sanitation appears to have stayed about the same during the period 1985–88. It is estimated that spending for water and sanitation (in constant 1985 dollars) in all developing countries was about $8.9 billion in 1985, rose to $9.2 billion in 1986, and then returned to $8.9 billion in 1988 (38). Although they have held their investments fairly constant during hard economic times, developing countries have not been able to provide adequate water and sanitation for current populations or to extend coverage significantly or to keep up with population growth. For example, it is estimated that providing each additional person with low-cost water and sanitation services costs about $200 (39) ($270 in urban areas and $60 in rural areas), yet investment per additional person has declined from $122 in 1985 to $110 in 1988 (40).

Third World governments are not expected to boost spending significantly on water supply and sanitation in the 1990s (41). Consequently, water and sewer agencies will face severe spending constraints in the coming decade; however, improved overall economic performance could help relieve those constraints somewhat (42).

External aid agencies, such as foreign aid programs, development banks, and nongovernmental organizations, contribute about $3 billion to water and sanitation projects in 1989, according to World Health Organization officials. A WHO survey of 24 of these agencies showed that their annual support rose to $2.3 billion during 1986–87, a substantial increase over the $1.25 billion annual level of 1981–82. Computed in constant dollars, however, aid has been declining since 1983. (See Figure 5.7.) WHO also surveyed 11 of the 24 countries belonging to the Or-

ganisation for Economic Co-operation and Development (OECD) and found that they contributed about 4 percent of their aid to water and sanitation projects in 1985–87, roughly a doubling of the share devoted to these areas in 1980 (43).

The WHO study noted that aid agencies are increasingly interested in water and sanitation projects that emphasize human resources development, management improvement, community participation, the involvement of women, health education, public information, and institution building. According to the survey, projects with at least one of these elements accounted for 34 percent of the total funded by aid agencies in 1985–87, up from 24 percent in 1981–82 (44).

Loan support from the World Bank has averaged about $715 million a year, or about 4 percent of the Bank's total loans. Recipient governments contributed an average of 15 percent of the costs to the 10 projects supported by the Bank in 1989; 7 of those projects were cofinanced with other multilateral banks and development agencies (45). That represents a substantial decline in local participation from previous years. In 1985, the average contribution by local participants was 53 percent by Southeast Asian countries, 58 percent by eastern Mediterranean countries, and 60 percent by Latin American nations (46).

Project Performance

The financial performance of most projects has not been very good. A review of 54 projects begun before 1981 and supported by the World Bank found that most projects encountered financial problems. The problems—fairly typical compared with the Bank's experience in other sectors—included construction cost overruns that averaged about 23 percent and higher-than-expected operating and maintenance costs; in half of the projects, the ratio of revenue to operating and maintenance costs was 68 percent less than anticipated (47). Most projects also had fewer customer connections, disappointingly high levels of unaccounted-for water, and lower sales volumes than expected. As a result, six years after project initiation, only slightly more than 10 percent of these projects were able to cover operation and maintenance costs (plus depreciation and interest) from their revenues (48).

What caused this poor performance? The study found that project planners consistently overestimated consumer demand and willingness or ability to pay for water and sanitation services. In addition, weak institutions often did not provide strong incentives for managers to operate efficiently (49).

Are there solutions to these recurring problems? Possibilities include decentralizing responsibility to more autonomous public utilities or community-based efforts or, perhaps, privatizing water and sanitation services, which could help put managers in closer touch with consumers and improve their ability to deliver services that consumers can afford (50).

Table 5.2 Cost Per Household of Different Sanitation Systems

(1978 U.S. dollars)

	Total Investment Cost	Monthly Operational Cost	Monthly Water Cost	Total Monthly Cost[a]	Percentage of Income of Average Low-Income Household[b]
Low-cost					
Pour-flush toilet	70.7	-0.2-	0.3	2.0	2
Pit latrine	123.0	x	x	2.6	3
Communal toilet[c]	355.2	0.3	0.6	8.3	9
Vacuum truck cartage	107.3	1.6	x	3.8	4
Septic tank	204.5	0.4	0.5	5.2	6
Composting toilet	397.7	0.4	x	8.7	10
Bucket cartage[c]	192.2	2.3	x	5.0	6
Medium-cost					
Sewer aqua privy	570.4	2.0	0.9	10.0	11
Aqua privy	1,100.4	0.3	0.2	14.2	16
Truck cartage	709.9	5.0	x	13.8	15
High-cost					
Septic tanks	1,645.0	5.0	5.9	46.2	51
Sewerage	1,478.6	5.1	5.7	41.7	46

Source: John M. Kalbermatten, DeAnne S. Julius, and Charles G. Gunnerson, *Appropriate Sanitation Alternatives: A Technical and Economic Appraisal*, World Bank Studies in Water Supply and Sanitation (The Johns Hopkins University Press, Baltimore and London, 1982), Table 3-11, p. 64.

Notes:

a. Assumes that the investment cost is financed by loans at 8 percent over 5 years for the low-cost systems, over 10 years for the medium-cost systems, and over 20 years for the high-cost systems.

b. Assumes that average annual income is $180 per capita, with six persons in a household.

c. Based on per capita costs scaled up to household costs to account for multiple household use in some of the case studies.

x = not available.

The opportunities to reduce costs appear numerous. For example, reductions could be made through better personnel management; improved training; more efficient metering, billing, and collection; and rigorous programs to reduce unaccounted-for water (51).

Low-Cost Sanitation Systems

Lower cost sanitation systems offer another important prospect for cost-cutting. There are several options in the sanitation area. (See Table 5.2.)

For example, the Indian government promotes a pour-flush latrine with a water seal and twin leach pits. The latrine is flushed by hand using a container that holds 1.5–2 liters of water. Each pit is designed to last about three years; excreta then are diverted to the second pit. After about two years, the contents of the filled pit turn into organic humus, which can be removed and used as manure; the pit is then ready for reuse (52).

Unlike conventional sewerage, the success of low-cost sanitation often depends on persuading consumers to buy a latrine. Financing is being handled in a variety of ways in India; one state provides 50 percent of the funds as a loan and the rest as a grant; others provide 100 percent subsidies for some low-income groups and partial subsidies for others. The costs of latrines vary, but are roughly $150–$200 in urban areas (53). (For other examples, see Improving Housing and Infrastructure, below.)

The Institutional Gap

An important part of the challenge confronting most developing countries is to strengthen the institutional effectiveness of services.

The basic urban service systems that do exist in the Third World often are plagued with inefficiencies. For example, a 1989 study of Mexico City's solid waste collection system found that 38 percent of the vehicles in the Federal District and 19 percent of the vehicles in the contiguous State of Mexico were out of service for maintenance and repair; 10–15 percent is considered an acceptable number (54). A 1989 survey of 223 wastewater treatment plants in Mexico found 45 percent of them out of service and 35 percent with severe operational problems (55). Water supply systems surveyed in 40 countries routinely were unable to account for 30 percent or more of the water they deliver (15 to 20 percent is considered a good figure in developed countries (56)); usually, the unaccounted-for water is lost because of leaks in the system, illegal connections, or poor metering (57). The city of Lima cannot account for more than half of its water (58).

In Latin America, many water and sewerage enterprises face high inflation, difficult access to capital markets, rapid urban population growth, a large proportion of low-income residents in the population to be served, politically appointed managers, controlled rates, high turnover in technical and managerial positions, low salaries, and an untrained labor force (59). Nevertheless, a 1989 World Bank study

identified five water and sewerage companies in
Latin America—COPASA of Minas Gerais, Brazil;
ACUAVALLE of Valle de Cauca, Colombia; E.P.M. of
Medellin, Colombia; EMOS of Santiago, Chile; and
C.A.D. of Monterrey, Mexico—that have succeeded
despite these problems. All five are public enterprises; two are regional (COPASA and ACUAVALLE)
and three are municipal. They are all controlled by
national and local government agencies; some are su-
pervised by central government ministries (60).

The study found that all five companies had a reasonable amount of institutional autonomy and had
been able to resist pressures for abrupt shifts in policies. Each had sound personnel management, with
highly esteemed top managers and a relatively stable
group of middle managers and professionals. As a result, they had high productivity, using an average of
5.4 employees per 1,000 water connections; most
companies in the region typically use 10–20 employees per 1,000 connections. The five companies also
had been able to implement reasonable rates, which
gave them enough financial strength to cover operating and maintenance costs (61).

The companies' unaccounted-for water averaged
34 percent, a lower rate than those at most other
companies in the region. As part of its effort to reduce lost water, COPASA has supported the development of small meters that read the low-flow rates of
most domestic consumers more accurately (62). In addition, three of the five companies were carrying out
watershed protection programs, which focused on
forest conservation and environmental monitoring
and control (63).

The World Bank study also identified important
economies of scale in the operation of water and sewerage systems. In the Latin American context, the
study found that water and sewerage services for cities with populations of fewer than 150,000 can be provided more efficiently and economically by large
regional companies that each serve several small cities. Large companies typically require substantially
fewer employees per 1,000 connections than small
water and sewer companies, can afford to pay their
staff higher salaries, and are in a better position to
construct large facilities that can be shared by several municipalities (64).

The Self-Financing Option

In view of the limited outlook for government spending on water and sanitation, some governments are
trying new approaches that provide low-cost infrastructure as part of housing projects that are largely
self-financed.

One successful infrastructure project improved
two squatter settlements—Hai el Salam and Abu
Atwa—and adjacent land on the outskirts of Ismailia,
Egypt. The project was financed largely through the
sale of plots in the existing areas and in the new
land; all the land was government-owned. To help
low-income residents participate, the least expensive

plots were set at a low price while corner plots with
commercial potential were priced higher. Some particularly valuable commercial plots were sold at auc-
tion. Funds from land sales provided enough money
for the installation and maintenance of the new
settlement's infrastructure (65).

To ensure that low-income residents could participate, the project agency—a small local agency that
received technical help from the British and Egyptian governments—decided against installing a waterborne sewerage system, instead installing cesspits or
pit latrines, standpipes at 150-meter intervals, street
lighting, and services such as solid waste collection.
Once the project was under way, residents improved
their own homes and plots, investing some 6.25 million Egyptian pounds (66) over the first three-and-a-
half years of the project, which began in 1978 (67).

Another self-financing project that includes installation of infrastructure is Batikent, a planned, 45,000-
unit settlement of the Kent Koop cooperative in
Ankara, Turkey. Kent Koop is an umbrella group of
mini-cooperatives of common-interest groups such
as civil servants and taxi drivers. Kent Koop uses
payments from its members plus other external
grants and loans to pay the municipal government
for the land and for the installation of infrastructure.
The municipal government has sold the land to the
cooperative at below-market prices; prospective
homeowners also have been able to obtain below-
market credit (68).

FOCUS ON THE INFORMAL SECTOR

The burgeoning growth of Third World cities creates
extraordinary demands for housing, services, and
jobs—demands that are far beyond the capacity of
most cities to fulfill.

Lacking affordable housing or conventional jobs,
the urban poor use whatever means are available to
find shelter and work. They build settlements on va-
cant land and subsist through a wide variety of small-
scale family businesses; the most visible examples
are street vendors, the hawkers of fruits, vegetables,
and innumerable other goods, who are a common
sight in the streets of Third World cities.

For years, the "informal sector," as it is now frequently called in development literature, was dispar-
aged by governments and ignored as economically
marginal by planners (69) (70). In that view, members
of the informal sector weakened the economy by not
paying taxes and undermined city services by, for example, illegally tapping into water mains. Furthermore, they aggravated urban environmental
problems by building shelters that did not conform
to official building standards and by providing services such as solid waste collection that were not
subject to governmental controls. Governments perceived informal-sector activities as haphazard and
unorganized, its members with no political voice.

A second view emerged in the mid-1970s and 1980s
as planners began to grasp the sheer size and vigor

5 Human Settlements

Box 5.1 Interview with a Squatter Settlement Leader

One large squatter settlement stands out among the many surrounding the capital of a Latin American country. The housing units are among the poorest, often constructed of little more than cardboard and metal scraps. The alleys separating the houses are only about 3 meters wide, but they are perfectly straight, and the lots are rectangular and of standard sizes. Public water taps have been installed at regular intervals throughout the community. Many households have made drains in front of their dwellings to remove rainwater. In excerpts from a taped interview, the "presidente" of this community describes the planning and organization of the settlement. His words have been translated and edited slightly. Financial data have been converted to U.S. dollars. This version of the interview is quoted from sections of a World Bank report (1).

COMMUNITY ORGANIZATION

Our residents come from many areas of the country. Some came straight from rural areas, others moved to the city many years ago. Those of us who lived in the city did well enough until recent conditions brought drastic increases in rent. We felt forced to leave the city to occupy this land.

Some neighbors say that about 30 to 50 families had begun to arrive by midwinter. By spring, people had already occupied an area about 3 kilometers long by a few hundred meters wide. As more and more came, they started creating different communities within the settlement.

About 30,000 people live in this settlement.

It was generally recognized that we needed some physical order in our community. We thought that we had to include sidewalks and lanes for the circulation of vehicles, considering that this was going to be our permanent home. We now have alleys 3 meters wide. Initially, we had planned for lots of 7 by 6 meters, 6 by 5 meters, and 4 by 5 meters. We now believe that the ideal size is about 72 square meters. At first, because there was plenty of land, we took all the land we needed. But, as many more came, we all had to divide our lots into smaller plots.

Some residents saw that I had an honest interest in helping, and they asked me to join a new board of directors. The first thing was to organize a community council for each of the communities in the settlement. My own community, the oldest, has 38 sectors, each with 50 to 100 families; each sector has two representatives, sometimes three. All the community councils together became what is today the United Neighborhood Association. After a year, the association had obtained a degree of legal recognition. Once we had obtained a legal status, we started working by organizing residents in each of the communities.

With the help of an international group of doctors called Médicos sin Fronteras (Doctors Without Boundaries), we established our first medical and health center. We have also created what we call integrated health commissions, run by

groups of women from each community. We also have a good group of midwives who can be called to attend at any time of the day or night, saving the patients a trip to the hospital.

In addition to the health center, we have already built our school facilities and a gymnasium. We started with some jerry-built sheds for classrooms. But, later, with the help of a private school in the city, from which we received some construction materials, we built the school with our own hands. The gymnasium was conceived to have dual functions: first, to enable our children to receive some physical education and, second, to serve as a wrestling arena for the entertainment of our community. We have also formed play groups for young people and football teams that compete toward an all-settlements championship. We have sponsored some beauty contests.

There are 12 community leaders on our community board of directors (all men so far, no women). We are very close and share our problems so that we can make ourselves available to one another in times of serious need. Because our people know who we are, we can walk freely around the settlement without having to carry arms. The people themselves act as police, keeping order and applying the law to thieves regardless of whether they live here or not. Neighbors here have even recovered things that have been stolen and returned them to the authorities.

---

of the informal sector. New research in the 1980s showed that the informal sector often did pay taxes (71) and that informal workers had considerable energy and dynamism (72). Now, many planners see the informal sector as probably the greatest source of new urban jobs in the next few decades and as an important safety valve to fill the gap created by inadequate city services. Planners also are beginning to realize the potentially important role that informal groups can play in building city infrastructures and services. Often lacking outside help or facing the active discouragement of city governments, informal local groups nevertheless provide services such as transportation, the delivery of drinking water, arranging cooperative loans, and recycling materials. With effective governmental help or community organization, the informal sector can play a role in other services such as garbage collection or the upgrading of sanitation and sewerage systems.

Organization and leadership—in the form of neighborhood organizations, NGO support, or simply the energy and imagination of a single informal leader—can be particularly important. (For an example of an informal sector leader, see Box 5.1.)

Settling on a definition of the informal sector has been a vexatious problem for researchers. One of the first definitions, given in a study of employment in Kenya, described the informal sector by its characteristics: ease of entry, reliance on indigenous resources, family ownership of enterprises, small scale of operation, labor-intensive and adapted technology, skills acquired outside the formal school system, and unregulated and competitive markets (73). Other definitions describe the informal sector in comparison with the formal sector. They point out that, for example, the informal sector is characterized by self-employed workers, whereas the formal sector is typified by wage earners; the informal sector does not accept the legal and administrative rules imposed by a government, whereas the formal sector does; or the informal sector cannot be measured easily by conventional statistics, whereas the formal sector can (74).

The debate continues over the proper definition of the informal sector. But, in the meantime, according to one critic of the term, at least the research has helped give standing to a variety of economic activities that policymakers might have ignored otherwise

Box 5.1

OBTAINING THE WATER SUPPLY

The first year the settlement was begun, we heard a rumor that typhoid fever was around and that some tanker trucks were selling contaminated water that was making our people sick. This turned out to be true. We had to stand in line for hours to buy water from the tanker trucks. Then the vendors started raising their prices, until finally they were incredibly high. Yet we had to drink this water or buy clean water from the neighboring town.

We tried every means to persuade the government water authority to install a water system in our community. But they kept telling us that, because we were squatters, we had no right to receive water services. So we called a meeting at which we decided that everyone would install a pipeline in his own area at the same time. Well, we did it, and we also connected our pipelines with a major line that supplies water to a town nearby. With this improvised connection, we were able to secure some water supply for families in 24 of the 38 sectors. We installed plenty of taps all over the community, about 150 of them. It turned out that, once we connected with the major line, only about 25 to 30 percent of the taps really worked, but that was good enough for us.

Once the water authority found out that we were stealing water, it called us to a meeting. The water authority proposed a project to provide our community with potable water. We were promised a 3,000-meter, piped water system that would cost our community about $11,000. We started by asking each household to give $20. We soon realized that this amount was unrealistic for many poor families, so we lowered it to $6 to cover the cost of basic construction materials. Even though we were unable to collect the $11,000 originally planned, we did gather about $5,000, and we did put into the project our own labor as well as some construction materials, including pipelines and water meters.

Our need for potable water is only part of an overall problem. We also need to plan for a sewer system. So far, by working on Saturdays and Sundays, we have been able to construct a series of drainage systems that have solved the problem of waste disposal temporarily.

OBTAINING ELECTRIC POWER

Some people became interested in having electricity in their homes and managed to raise about $600 toward having meters installed. But, after receiving several messages from the owners of the land we were occupying, the power company decided to stop the connections. Several times, we went to request connections, at least for our school and meeting-room buildings, but the power company always turned us down. So together with other friends, we decided to steal energy just as we had stolen water.

With the help of an electrician, we connected to the public power lines without public authorization. We now provide electricity to some 70 families. Not long afterward, the power company came to our community and destroyed our connections to the power lines. As soon as they left, we reconnected them. The power company's officials came back to disconnect us from the system again. This time, we armed ourselves with sticks and did not allow them to carry out their intentions.

I immediately went to speak with the power company's executive manager and explained to him our frustration with his company for cutting off our power after it had installed meters in most of our homes. Throughout our interview, the manager insisted that the company could not provide our community with electricity because we were squatters and because of the uncertain nature of our settlement. But, in the end, he decided to be indulgent with us, in exchange for our promise that we would not abuse our arrangement by "going too far" with domestic connections. The fact is that we do not pay for electricity now. It is free, and we have been able to light our streets, our school, and our homes. If there were anything else that we needed and if the government were unwilling to give it to us, then we would just take it without authorization. It is the hardest thing to do, but, in the end, it has always worked.

References and Notes

1. Donald T. Lauria and Dale Whittington, *Planning in Squatter Settlements: An Interview with a Community Leader* (The World Bank, Washington, D.C., October 1989). (In this report, details of time and location have been altered or omitted to protect the privacy of the community.)

and has converted these activities from the sphere of social welfare to the sphere of economic planning (75). Furthermore, the informal sector is an indisputable fact in Third World cities (as well as in many developed countries), and there is little doubt that governmental policies can have an important impact on the informal sector's prospects.

The Informal Economy

What kinds of jobs typify the informal sector? The range is broad. A study of the informal sector in the Sahel in Africa found, for example, that half the informal businesses in Niamey, Niger, a city of over 500,000 people, were engaged in commerce with food, clothing, or other items; about one quarter were in production, working in milling, tailoring, garment-making, embroidery, leather work, carpentry, furniture-making, blacksmithing, or jewelry-crafting; and one quarter were in services, in jobs such as mechanics, radio and refrigerator repairs, hotel and restaurant services, laundering, hairdressing, and photography (however, most of these "businesses" had only one or two workers) (76). Energy production also can be an important component of the informal economy. For instance, a survey of the informal economy in Penang, Malaysia, found that 14.8 percent of the informal firms (defined as firms with fewer than 30 workers) were engaged in the production of charcoal and fuelwood (77). Much of this activity occurs at informal markets. In Lima, for example, 274 (83 percent) of the city's 331 markets are informal; that is, they were built by vendors who decided to move off the streets (78).

How large is the informal sector? Estimates for most developing countries show it ranging from 30 to 70 percent of the economically active population. The quality of the data is uneven, and the figures may vary greatly depending on definitions. For instance, in Latin America about 42 percent of the workers are in the informal sector. Many wage earners, however, are exempted from labor legislation and could be counted as part of the informal sector. Including informal wage earners, the total number of informal workers in Latin America rises to about 60 percent (79).

5 Human Settlements

Table 5.3 Percentage of Urban Populations in Informal Settlements, 1980

	Total Population	Population in Informal Settlements	
	(thousands)	(thousands)	Percent
Addis Ababa, Ethiopia	1,668	1,418	85
Luanda, Angola	959	671	70
Dar es Salaam, Tanzania	1,075	645	60
Bogota, Colombia	5,493	3,241	59
Ankara, Turkey	2,164	1,104	51
Lusaka, Zambia	791	396	50
Tunis, Tunisia	1,046	471	45
Manila, Philippines	5,664	2,666	40
Mexico City, Mexico	15,032	6,013	40
Karachi, Pakistan	5,005	1,852	37
Caracas, Venezuela	3,093	1,052	34
Nairobi, Kenya	1,275	421	33
Lima, Peru	4,682	1,545	33
São Paulo, Brazil	13,541	4,333	32

Sources:
1. United Nations (U.N.), *Patterns of Urban and Rural Population Growth* (U.N., New York, 1980), Table 48, pp. 125-154.
2. Other sources cited in United Nations Centre for Human Settlements (Habitat), *Global Report on Human Settlements, 1986* (Oxford University Press, New York, 1987), Table 5.18, p. 77.

From 1950 to 1980, the size of the urban informal sector in Latin America (excluding informal wage earners) stayed at about 30 percent, even as the continent's industrial economy grew about four times larger (80). This trend contrasts with the experience of the developed countries. In the United States, for example, informal employment (including self-employed workers) during an equal number of years (1900–30) declined from about 50 percent of the labor force to about 30 percent (81).

Some studies suggest that informal-sector employment is growing considerably faster than formal-sector employment. An illustration is Abidjan, Côte d'Ivoire, where urban employment was estimated to be growing at 6 percent annually in the informal sector, compared with 4.6 percent in the formal sector (82). The Inter-American Development Bank has estimated that, from 1980 to 1985, in Latin America as a whole, the informal sector grew at an annual rate of 6.8 percent, while the formal sector increased by only 2 percent over the entire five-year period (83).

Informal enterprises rely on informal financing to get started and keep going. Family and friends provide a large percentage of the informal loans; other sources include professional moneylenders, pawnbrokers, tradespeople, and associations of acquaintances (84). Rotating savings and credit associations (ROSCAs), found throughout the developing world, are another popular source of informal finance. In a ROSCA, typically from 6 to 40 individuals form a group, select a leader, and contribute a given amount per member. The money is lent in rotation to each member. In Bolivia, one study found that one third to one half of all urban adults participated in ROSCAs, and that their ROSCA payments averaged about one sixth of their salaries (85). Such associations are found throughout Africa also, where they are used to finance equipment purchases and other

small business investments, consumer purchases, and even international trade (86).

Housing

Running roughly parallel to the demand for jobs is the demand for housing. In 14 cities studied in the Third World, from 32 to 85 percent of the residents live in informal settlements. (See Table 5.3.)

Informal housing often is built on economically or ecologically undesirable land such as flood-prone, low-lying areas or unstable hillsides. Generally, residents do not own the land and pay no property taxes. But they often upgrade their neighborhoods and increase their legal rights to the land incrementally, as the settlements grow older.

Hernando de Soto, president of the Instituto Libertad y Democracia (ILD) in Lima and author of an influential study of Lima entitled *The Other Path: The Invisible Revolution in the Third World*, found that many informal settlements do not begin haphazardly, but are actually the result of careful preparations designed to reduce the risk of rapid eviction by governmental authorities. In Peru, a group usually chooses unoccupied government-owned land, attracts a "critical mass" of settlers to reduce the risk of repression, draws up a settlement plan, and prepares to negotiate with the authorities and resist eviction. The group then "invades" the land, usually at night or in the early morning on a civic anniversary to reduce the risk of a rapid reaction by the police. Initially constructed of temporary materials, informal housing gradually becomes more permanent as the threat of eviction recedes (87).

De Soto estimates that, by 1982, informal settlements accounted for 42.6 percent of all dwellings in Lima, housed 47 percent of the city's population, and had a total value of $8.3 billion (in 1984 prices) (88). The numbers have continued to rise. During 1985, 69 percent of the houses built in Lima were constructed without regard to formal building codes, while 31 percent conformed to them (89).

Recycling, Transportation, and Water Supply Services

Members of the informal sector provide services—recycling, transportation, water supply, food vending, and the provision of traditional medicines, to name but a few—that present modest economic opportunities.

Recycling is driven by industry demand for recycled materials to use as feedstocks in making new paper, steel, glass, and plastic products. This practice is widespread in many Third World cities. In Mexico City, it is estimated that 25 percent of the municipal waste collected is recycled, mostly by the estimated 10,000 scavengers working at official dump sites (90). In many Third World cities, recycling also is accomplished by door-to-door buyers, who purchase materials directly from households, or by collection-crew workers, who recycle the most valuable and easily retrievable items (91). In Montevideo, Uruguay, many dwellers in the peripheral shantytowns known as "cantegriles" engage in recycling. Cantegril dwellers collect food waste, paper, plastics, and metal. Some collectors also work in the separation of various types of waste and the packaging or tying up of materials (92).

Recycling also can create opportunities for informal enterprises to process materials prior to their reuse. For example, plastics must be washed before they are reused. In Montevideo, companies in the formal sector have found that washing is too labor-intensive and have subcontracted the work to small, informal-sector enterprises, which can do the job at lower cost by using family labor (93).

Recycling has, in one instance, contributed to the overall savings generated by energy conservation. In Niamey, Niger, tinsmiths and blacksmiths made an estimated 30,000 energy-efficient stoves from recycled scrap metal over a two-year period; these stoves have cut household consumption of fuelwood. Throughout sub-Saharan Africa, artisans use recycled metal, glass, tires, and other materials (94).

Reusing materials that otherwise would have to be purchased abroad can bring substantial savings in foreign exchange (95). It is estimated that recycling for the paper, steel, glass, and plastic industries in Lima in 1983 saved about $19–26 million in foreign exchange (96). Also, scavenging by the informal sector before waste is picked up reduces the volume of municipal waste, which helps reduce governmental costs.

In at least one case, the informal sector's expertise in recycling and waste collection has been put to formal use by a government. When Cairo devised a new solid waste collection system in late 1986, it turned to the Wahis and Zabbaleen, two informal groups that have long supplemented the municipal sanitation force's waste collection and recycling work. City officials—recognizing that the Wahi-Zabbaleen enterprise was experienced, efficient, inexpensive, and had established client relationships—supported their proposal to create a formal company using modern organizational and management methods. In January 1988, a successful pilot program was started in two Cairo neighborhoods (97).

Informal transportation enterprises using cars, vans, or minibuses also are widespread in Third World cities. Some of the most common examples include colectivos in Argentina and Chile, jeepneys in the Philippines, peseros in Mexico, por puestos in Venezuela, dolmus in Turkey, matatus in Kenya, and trotros in Ghana. Some of these informal networks are estimated to provide more than one half of all public transport trips (98).

Informal transportation operators have had an uneasy relationship with local governments and public transportation systems. Critics charge that informal transport may undermine the efficiency of conventional bus service (99), is difficult to regulate, is generally indifferent to public safety, and does not provide uniform high-quality service. Supporters assert that

Table 5.4 Prices Charged By Water Vendors, mid-1970s—1980

City	Country	Multiples of Price Charged by Public Water Utility
Kampala	Uganda	4-9
Lagos	Nigeria	4 10
Abidjan	Côte d'Ivoire	5
Lomé	Togo	7-10
Nairobi	Kenya	7-11
Istanbul	Turkey	10
Dacca	Bangladesh	12-25
Tegucigalpa	Honduras	16-34
Lima	Peru	17
Port-au-Prince	Haiti	17-100
Surabaya	Indonesia	20-60
Karachi	Pakistan	28-83

Source: Urban Development Division, The World Bank, "Urban Strategy Paper," draft, The World Bank, Washington, D.C., May 1989, Table 3.2, p. 70.

informal transport is responsive, flexible, and provides extensive coverage (100); serves a large section of the population; is a significant source of employment; helps to discourage the use of private automobiles; and can operate without public subsidies (101).

Informal transport can meet the need created by inadequate, formal transport systems. Nairobi's matatus (private minibuses), legalized after 1973, are a good example. Ostensibly in response to this "unfair competition," the city's bus service (run by a foreign-owned company under a franchise with the city) decided not to increase its fleet. While the public system stayed about the same size, the number of matatus expanded to accommodate Nairobi's rapidly growing population. Managers of the public system have acknowledged that without the matatus they would have had to undertake an expansion that probably would have cost more than the likely returns (102).

The informal sector also is active in providing water for the urban poor. In Onitsha, Nigeria, an important market town of 700,000 people in West Africa, the vast majority of households obtain water from a private, water-vending system. Roughly 275 tanker operators purchase water from about 20 private boreholes, then sell it to households and businesses equipped with water storage facilities. Many of these households resell the water to individuals or to water vendors, who carry water in two four-gallon tins suspended from a shoulder yoke and deliver it directly to customers. In the dry season, the private vending system provides about two thirds of household water to city residents (103). The study of Onitsha and other studies of water-vending have found that urban residents are paying a relatively high price for water bought from vendors. As indicated by Table 5.4, consumers pay much more than they would if public water were available.

The Informal Sector and the Development Process

Third World governments, development banks, and nongovernmental organizations have been wrestling for years with two key problems whose resolution could unlock even more productive activity by the informal sector: how to nurture the growth of informal enterprises and how to improve the informal sector's housing and infrastructure. Most studies and projects deal with one subject or the other; few consider both.

What needs to be done to unleash the potential of informal enterprises? De Soto argues that the key problem for informal enterprises is government red tape. He created an imaginary small business in Lima and found that it took 289 days and cost the equivalent of 32 months of minimum monthly wage to register the business legally (104). Furthermore, the cost of formal regulations tends to be fixed, which imposes a disproportionate penalty on small enterprises.

Legalizing informal enterprises, in de Soto's view, would enhance their access to credit greatly, enabling them to enter into legally binding contracts and to join partnerships and associations.

Critics of de Soto's view point out that legality has obligations, such as compliance with labor laws, which may be costly for informal enterprises. Instead of legality, others emphasize the need for access to credit, skills, and markets. They draw attention to success stories like the Grameen Bank in Bangladesh, which has about 400,000 borrowers—most of them women—and makes small loans averaging about $67. The bank's recovery rate is around 98 percent, a fact that has been instrumental in convincing banking agencies that the poor will repay loans (105). The bank requires that borrowers organize into five-member groups, that each group member establish a pattern of regular weekly saving before applying for a loan, and that the first two members of a group make several repayments on their loans before other group members can apply for loans. The bank also requires weekly repayments and makes future borrowing contingent on repayment of loans. Most loans—for example, to finance trading or the purchase of livestock—help to set up or to expand businesses. Aside from these factors, the bank's success has been attributed to a close relationship with its borrowers, careful supervision of field operations, and a dedicated staff (106).

The World Bank has supported several similar banking projects in Calcutta, India; Manila, the Philippines; El Salvador; and Guatemala that rely on formal groups for loan repayments; require frequent loan collections; do not provide any technical assistance; and use grass-roots organizations such as cooperatives or neighborhood associations as a bridge between borrowers and financing agencies (107).

These new banking projects forgo the training and technical assistance that traditionally go hand in hand with development credit. The Carvajal Foundation of Cali, Colombia, arguing against the new approach, believes that training and technical assistance are the key elements and that credit will be used better by those who have passed through training programs on managing their enterprises.

This sort of "integrated" approach is still favored by most informal-enterprise assistance programs (108).

The preference for integrated programs may be changing, however, due in part to successes like the Grameen Bank. A recent study of successful assistance programs found that they usually focused on one activity, such as credit, or on a particular trade or income-earning enterprise, such as garbage collection or food preparation (109).

Nongovernmental organizations are viewed widely as a key source of support for informal-sector development. NGOs are generally either membership organizations, such as women's and self-employed persons' associations, or technical or financial assistance organizations, such as ACCION International. They understand that project success depends on community involvement and have dedicated staff members who live in the project communities. NGOs develop the capacity to understand what informal enterprises need, and tend to be flexible and willing to take risks (110). NGOs also may be better suited than development banks to handle projects that deal with low-income customers and require much painstaking attention and time.

Some of these strengths also can be weaknesses. Generally, NGO projects are small and reach limited numbers of people. And research suggests that it can be difficult to replicate many NGO projects on a larger scale (111) (112).

Improving Housing and Infrastructure

What needs to be done to improve living conditions—housing and services—in informal settlements? Many studies suggest that self-help can be a powerful tool and that the poor are willing to undertake many projects themselves, under the right conditions. (See Box 5.1.)

One critical condition is the legal right to own their homes. Many studies have found that even the poorest people will spend money to upgrade their homes if they are given security of tenure (113). In Lima, de Soto studied two adjacent neighborhoods of similar informal origins, one legally classified as permanent and the other as temporary. He found that the value of a typical building in the legally secure settlement was 41 times greater than in the other settlement. Using a larger sample, he found that the average value of buildings whose owners had received title was nine times that of buildings whose owners had not (114).

Usually, municipal governments now recognize that it makes more sense to accommodate informal housing than to provide public housing, which has high construction standards, is much too expensive for the poor, and has fallen far short of demand. With encouragement from the World Bank and other international organizations, planners in the past devised "sites-and-services" projects, which simply provided a rudimentary infrastructure and let the owners build the rest at their own pace. These projects were intended to be affordable for the poor. But project planners often overestimated how much the poor were willing or able to spend on a house. As a result, many projects did not benefit the poorest residents (115). Studies of five projects indicated that subsidies ranged from 55 to 72 percent of their costs (116).

The general experience to date suggests that slum upgrading—that is, narrowly targeting projects in existing settlements—tends to reach the poorest of the poor more effectively than sites-and-services ventures. However, experience also suggests that upgrading usually requires a larger subsidy than do sites-and-services projects (117). Choosing between more effective help to the poor and better recovery of costs poses a difficult political problem for Third World governments.

Community Participation in Projects

Community participation in upgrading projects is frequently high, though planners and residents are commonly at odds over the pace of construction and the type of building materials. As a rule, planners favor a rapid pace and new materials, while residents often do not care about pace and prefer used or discarded materials (118).

Community-based upgrading and housing programs also have the additional benefit of providing employment for informal enterprises. For example, the production of building materials has been an important feature of many self-help projects (119).

Communal work on projects such as digging ditches for sewerage and drainage has had a mixed history (120), but there are some success stories. Two are in Brazil's Pernambuco state. In Olinda, slum dwellers are working with a team of professionals from the municipal government to improve infrastructure and housing with materials provided by community workshops (121). The Olinda project includes road resurfacing, road drainage installation, and solid waste collection; local enterprises were set up to make latrine components and drain tiles (122). Residents have controlled rainwater drainage successfully by installing locally made, precast cement drainage channels. Each household is responsible for cleaning the debris trapped in nearby drains (123). In Passira, residents participate in a lower-cost, "condominium" sewer system (so-called because connecting lines pass through several homes on their way to a main line). The project uses locally made ceramic pipes, which cost about one thirtieth as much as standard pipes, and terracotta inspection chamber rings, which cost about one fourth as much as concrete rings (124).

Another success story is unfolding in Orangi, a settlement of 800,000 people in Karachi, Pakistan, where residents are active participants in the construction of a low-cost, urban sewerage system. The pilot project area includes about 43,000 of Orangi's 90,000 housing units, along 3,181 lanes. Project leaders stressed that residents of each lane must organize and elect lane managers, who would formally apply to participate in the project. After technical staff surveyed the lane and prepared plans, the lane manager collected money from the residents, received tools from project leaders, and made arrangements to get the work done. So far, sewerage systems have been installed in more than 2,200 lanes, with most of the cost borne by residents (125). Project organizers estimate that the locally planned and implemented work and simplified system cost 80 percent less than a conventional sewerage system (126). Planners may be looking increasingly at projects like Orangi that do not put much pressure on public funds.

Institutions like the World Bank are seeking ways to promote the construction of affordable sanitation facilities in urban settlements. For some time, Bank officials have been working on procedures to help cities decide which conventional, intermediate, or lower cost systems—small-bore sewers, condominium sewers, on-site latrine systems, or upgraded existing facilities, for example—make sense in particular urban areas. To a greater extent, both donors and governments have realized that users must pay at least a share of the cost of these services. However, this depends on developing a close match between the cost of the system and the residents' willingness to pay for it, thus ensuring that residents can pay for part (but not necessarily all) of the cost of the services they receive. Recovering more costs will enable municipalities to stretch their budget dollars farther (127).

Banks May Encourage Self-Help

In the future, aid agencies and development banks are likely to encourage self-help projects such as the Orangi sewerage project as well as projects with potentially large-scale impacts, such as grassroots banking organizations like the Grameen bank. To help reduce the cost of regulation, they probably will encourage Third World governments to simplify building codes, design standards, and permit processes (128). Support also is building for programs to improve access to credit for the informal sector, for removal of formal-sector subsidies that hurt the competitiveness of the informal sector, and for more long-term help to develop the institutional skills of NGOs and strengthen their ability to assist the informal sector (129).

Planners and municipal officials also will be more likely to encourage the informal sector to play a greater role in housing development and in services such as transport and solid waste removal. For example, in Ecatepec, Mexico, officials have modified a sanitary landfill so that scavengers can recover materials safely and efficiently. The site now has two levels. Trucks dump their loads on the upper level for sorting by scavengers, who push unwanted waste down to the lower level. There, the unwanted waste is spread and graded. The site is fenced and restricted to adult scavengers only; trucks with potentially infectious or hazardous waste are diverted to a section of the landfill where scavenging is prohibited (130).

These sorts of efforts to accommodate and channel the energies of the informal sector could prove to be important strategies for Third World city managers in the challenge to handle the rapid urbanization of the 1990s and beyond.

RECENT DEVELOPMENTS

HEAT ISLANDS

The rapidly growing cities of the developing world can expect a double dose of warming: the expected increase in temperatures caused by global climate change plus an additional increase caused by urbanization.

It has been known for some time that urban areas tend to be hotter than surrounding rural areas. A variety of factors contributes to this effect, including the release of heat from buildings and automobiles; intensified heat storage during the daytime due to the thermal properties of buildings; decreased evaporation caused by the replacement of trees and other vegetation with building materials; the effects of urban "canyon geometry," which tends to reduce wind speeds and raise short-wave radiation absorption; and increased long- and short-wave radiation absorption by air pollutants (131).

Arthur Viterito, a geography professor at The George Washington University in Washington, D.C., completed a study in 1988 of the differences between urban and rural temperatures in 30 cities of the United States. Viterito compared urban and rural temperatures in 1980, when accurate census data were available. He found the average yearly difference between urban and rural temperatures to be 1.13° C. The New York City urban area, with a population of more than 15 million in 1980, led the list with an urban/rural temperature difference of 3.0° C. Minneapolis, Minnesota, with a much smaller population of 1.7 million, nevertheless averaged 2.11° C warmer than adjacent rural areas (132).

Another study of Kuala Lumpur, Malaysia, measured on the same evening temperatures of 24° C on the outskirts of the city and 28.5° C near the city's commercial center (133).

Little research has been done yet on the impact of this "heat island" effect on rapidly growing cities of the developing world. For example, what will happen to temperatures in Mexico City and São Paulo, both of which are projected to have populations of roughly 24 million by the year 2000?

Using a conservative extrapolation of the U.S. data, Viterito estimates that cities with a population of 25 million can expect an urban/rural temperature difference of 3.13° C; cities with 30 million inhabitants will tend to be 3.24° C hotter than adjacent rural areas

(134). The jump in temperature tends to become smaller with each additional million inhabitants.

The potential increases in urban temperatures could have important effects on human health. Increased mortality from heat stress can be expected. (See Chapter 4, "Population and Health.") Higher energy costs for air conditioning also are likely, though in colder climates this may be offset by energy savings during warmer winter months.

USING BAMBOO AS A BUILDING MATERIAL

The government of Costa Rica is encouraging the use of bamboo as a building material to help relieve the country's housing shortage and to stem deforestation by providing an alternative to wood.

Costa Rica's National Bamboo Project is the outgrowth of a 1980 college thesis by Ana Cecilia Chaves, a Costa Rican architecture student. Bamboo has numerous advantages; it can be harvested five years after planting, does not require much labor, and costs only about one third as much as cement block. In addition, homes built with bamboo can withstand earthquakes better. Bamboo plantations can help reforestation and can control soil erosion (135).

By 1988, international donors—including the government of the Netherlands, the Central American Bank for Economic Integration, and the United Nations Development Programme—had provided $7 million to support the project. Consultants from the Netherlands and Colombia, which have had experience with bamboo as a building material, served as advisers (136).

The project's 12 construction technicians are going to rural communities throughout Costa Rica to teach local people how to build bamboo homes and to assist in their construction; 220 of the homebuilders will set up 10 small enterprises, five to plant, harvest and prepare bamboo, and five to build bamboo houses. The houses have wood frames onto which lengths of bamboo are fixed to make walls; these are covered with stucco. Each house, about 48 square meters in size, costs about $2,000 (137) (138). The project's goal is to build 30,000 homes in four years (139).

Project leaders also plan to sow 700 hectares of *Bambusa guadua*, a stronger species of bamboo from Colombia and Brazil that can grow to 33 meters (140).

Project leaders in Costa Rica hope to take what they have learned to other countries in the region. In addition to Colombia, Peru and Ecuador have had some experience with bamboo as a building material. Bamboo has been used for centuries for construction in China, Japan, and other parts of Asia.

This Chapter was written by Robert Livernash, a writer and consultant in Washington, D.C. James Listorti of the World Bank's Environment Department was principal consultant on the Informal Sector section.

References and Notes

1. Department of International Economic and Social Affairs, United Nations (U.N.), *Prospects of World Urbanization, 1988* (U.N., New York, 1989), Table 2, p. 7, and Table A-3, pp. 40-41.
2. United Nations Centre for Human Settlements (Habitat), *Global Report on Human Settlements, 1986* (Oxford University Press, New York, 1987), p. 23.
3. (Most of the data used here rely on the United Nations *Prospects of World Urbanization, 1988*, which accepts the various definitions of "urban" used by national governments. Most countries define an urban area as having a minimum of 2,000 to 5,000 persons, but some countries have minimum sizes as small as 200 persons or as large as 30,000.)
4. *Op. cit.* 1, Table A-3, pp. 40-41.
5. *Op. cit.* 1, Table 1, p. 5.
6. Jorge E. Hardoy and David Satterthwaite, *Squatter Citizen: Life in the Urban Third World* (Earthscan Publications, London, 1989), p. 229.
7. *Op. cit.* 6, pp. 232-236.
8. *Op. cit.* 1, Table A-9, p. 76.
9. The World Bank, "Deepening Pollution Demands Solutions," *The Urban Edge*, Vol. 13, No. 5 (The World Bank, Washington, D.C., 1989), p. 1.
10. Carl Bartone, "Environmental Issues in Urban Management," The World Bank, Washington, D.C., 1989, p. 2.
11. United Nations Economic Commission for Latin America and the Caribbean (ECLAC), "The Water Resources of Latin America and the Caribbean: Water Pollution," ECLAC, Santiago, Chile, May 1989, p. 5.
12. *Ibid.*, pp. 5-9.
13. World Health Organization (WHO), *Guidelines for Drinking-Water Recommendations Quality: Volume 1* (WHO, Geneva, 1984), p. 19, Table 6.
14. Harvey A. Garn, Senior Economist, The World Bank, Washington, D.C., January 1990 (personal communication).
15. Policy, Planning, and Research Staff, The World Bank, *FY88 Annual Sector Review: Water Supply and Sanitation* (The World Bank, Washington, D.C., November 1988), p. 10.
16. *Ibid.*, p. 2.
17. S.A. Esrey, R.G. Feachem, and J.M. Hughes, "Interventions for the Control of Diarrhoeal Diseases Among Young Children: Improving Water Supplies and Excreta Disposal Facilities," *Bulletin of the World Health Organization*, Vol. 63, No. 4 (1985), p. 763.
18. World Resources Institute and International Institute for Environment and Development in collaboration with the United Nations Environment Programme, *World Resources 1988-89* (Basic Books, New York, 1988), pp. 41-42.
19. *Op. cit.* 2, pp. 98-101.
20. *Op. cit.* 2, p. 99, Box 71.
21. *Op. cit.* 2, pp. 114-116.
22. *Op. cit.* 2, pp. 133-135.
23. World Health Organization (WHO), *The International Drinking Water Supply and Sanitation Decade: Review of Mid-Decade Progress as at December 1985)* (WHO, Geneva, 1987), pp. 8-9.
24. *Op. cit.* 18, p. 261.
25. *Op. cit.* 23, Table A.3.2, p. 22.
26. World Health Organization (WHO), *Urbanization and Its Implications for Child Health* (WHO, Geneva, 1988), pp. 15-17.
27. *Op. cit.* 2, p. 148.

28. Alfonso Zavala, Advisor, Water and Sanitation Division, The World Bank, Washington, D.C., January 1990 (personal communication).
29. *Op. cit.* 15, p. 8.
30. Water and Sanitation for Health Project (WASH), *Water and Sanitation Sector Profiles of Twenty African Countries* (CDM and Associates, Arlington, Virginia, 1989), Table 3, p. 11.
31. *Ibid.*, p. 8.
32. Frederick S. Mattson, *Planning for Central America Water Supply and Sanitation Programs: Update* (CDM and Associates, Arlington, Virginia, 1989), p. 3.
33. *Ibid.*, Table 13-1, p. 36.
34. *Ibid.*, Table D-2, p. 56.
35. *Op. cit.* 15, p. 11.
36. Harvey A. Garn, "Financing Water Supply and Sanitation Services," paper presented at the Collaborative Council Meeting in Sophia Antipolis, France, November-December 1989, The World Bank, Washington, D.C., p. 3.
37. *Ibid.*
38. *Ibid.*, Figure 3, p. 18.
39. *Op. cit.* 14.
40. *Op. cit.* 36, p. 8.
41. The World Bank, *FY88 Annual Sector Review: Water Supply and Sanitation*, The World Bank, Washington, D.C. November 1988, p. 2.
42. *Op. cit.* 36, Figure 3, p. 28.
43. Ingvar Ahman, quoted in Division of Information, United Nations Development Programme (UNDP), "Donor Dollars Go Further as Agencies Push Cost Effectiveness," *Source*, Vol. 1, No. 2 (UNDP, New York, September 1989), p. 21.
44. *Ibid.*
45. *Op. cit.* 36, p. 8.
46. World Health Organization (WHO), *International Drinking Water Supply and Sanitation Decade: Review of National Progress (as at December 1983)* (WHO, Geneva, Switzerland, 1986), p. 17.
47. Harvey A. Garn, "Patterns in the Data Reported on Completed Water Supply Projects, The World Bank, Washington, D.C., 1987, pp. 6-7.
48. *Op. cit.* 36, pp. 5.
49. *Op. cit.* 36, pp. 5 6.
50. *Op. cit.* 36, pp. 11-12.
51. *Op. cit.* 36, p. 17.
52. The World Bank, "Low-Cost Options for Urban Sanitation," *The Urban Edge*, Vol. 11, No. 10 (The World Bank, Washington, D.C., December 1987), p. 2.
53. *Ibid.*, p. 3.
54. Carl R. Bartone, "Institutional and Management Approaches to Solid Waste Disposal in Large Metropolitan Areas," The World Bank, Washington, D.C., 1989, p. 4.
55. Carl R. Bartone, "Urban Wastewater Disposal and Pollution Control," The World Bank, Washington, D.C., 1989, p. 2.
56. Carl R. Bartone, Senior Environmental Specialist, The World Bank, Washington, D.C., January 1990 (personal communication).
57. The World Bank, "Tracking the Problem of Lost Water," *The Urban Edge*, Vol. 10, No. 6 (The World Bank, Washington, D.C., June-July 1986), p. 1.
58. Robin Wiseman, "Latin Crisis Needs Political Priority," *World Water*, Vol. 11, No. 10 (November 1988), p. 16.
59. The World Bank, "Management and Operational Practices of Municipal and Regional Water and Sewerage Companies in Latin America and the Caribbean," The World

Bank, Washington, D.C., 1989, restricted distribution, pp. 2- 3.
60. *Ibid.*, pp. 2-3, 6.
61. *Ibid.*, pp. 5, 9.
62. *Ibid.*, pp. 7, 9.
63. *Ibid.*, p. 6.
64. *Ibid.*, pp. 13-14.
65. The World Bank, "Annual Awards Focus on Innovative Housing," *The Urban Edge*, Vol. 13, No. 9 (The World Bank, Washington, D.C. 1989), pp. 1-2.
66. Project prices are given in 1977 Egyptian pounds, when the exchange rate was approximately E1=$1; in January 1990, the rate was about E2.6=$1.
67. *Op. cit.* 65, pp. 2-3.
68. *Op. cit.* 65, pp. 4-5.
69. *Op. cit.* 2, p. 172.
70. Janice Perlman, "Six Misconceptions About Squatter Settlements," *Development: Seeds of Change. Village Through Global Order*, Vol. 4, No. 4 (1986), p. 40.
71. Leslie Péan, "Working Paper on the Informal Sector in the Sahel," The World Bank, Washington, D.C., February 1989, summary, p. 6.
72. *Op. cit.* 81.
73. International Labour Office, International Labour Organisation, *Employment, Incomes and Equality: A Strategy for Increasing Productive Employment in Kenya* (International Labour Organisation, Geneva, 1970), p. 6, cited in Lisa Peattie, "An Idea in Good Currency and How it Grew: The Informal Sector," *World Development*, Vol. 15, No. 7 (1987), p. 854.
74. Lisa Peattie, "An Idea in Good Currency and How It Grow: The Informal Sector," *World Development*, Vol. 15, No. 7, 1987, p. 856.
75. *Ibid.*, p. 855.
76. Ministére du Plan, Republique de Niger, "Enquête Nationale sur le Secteur Informel et la Petite Entreprise, Premiers Résultats," Niamey, Niger, October 1987, cited in Leslie Péan, "Working Paper on the Informal Sector in the Sahel," the World Bank, Washington, D.C., 1989, p. 57-61.
77. Kamal Salih, Mei Ling Young, Lean Heng Chan, *et al.*, *Young Workers and Urban Services: A Case Study of Penang, Malaysia* (Universite Sains Malaysia, Penang, 1985), cited in T.G. McGee, Kamal Salih, Mei Ling Young, *et al.*, "Industrial Development, Ethnic Cleavages and Employment Patterns: Penang State, Malaysia," in *The Informal Economy: Studies in Advanced and Less Developed Countries*, Alejandro Portes, Manuel Castells, and Lauren A. Benton, eds. (The Johns Hopkins University Press, Baltimore and London, 1989), pp. 269, 272.
78. Hernando de Soto, *The Other Path: The Invisible Revolution in the Third World* (Harper and Row, New York, 1989), pp. 59, 61.
79. Manuel Castells and Alejandro Portes, "World Underneath: The Origins, Dynamics and Effects of the Informal Economy," in *The Informal Economy: Studies in Advanced and Less Developed Countries*, Alejandro Portes, Manuel Castells, and Lauren A. Benton, eds. (The Johns Hopkins University Press, Baltimore and London, 1989), Table 1.1, p. 17.
80. *Ibid.*, Table 1.1, p. 16.
81. *Ibid.*, Table 1.2, p. 19.
82. J.P. Lachaud, *Pauvretve' et Marche du Travail Urbain: le Cas d'Abidjan* (Institut International d'Etudes Sociales, Geneva, 1988), cited in Leslie Péan, "Working Paper on the

Informal Sector in the Sahel," The World Bank, Washington, D.C., February 1989), summary, p. 4.

83. Jacob Levitsky, "Summary Report" of the Washington, D.C., June 1988, International Conference on Microenterprise Development, Committee of Donor Agencies for Small Enterprise Development, Washington, D.C., August 1988, p. 2.

84. The World Bank, *World Development Report 1989* (The World Bank, Washington, D.C., 1989), p. 112.

85. *Ibid.*, Box 8.2, p. 114.

86. The World Bank, *Sub-Saharan Africa: From Crisis to Sustainable Growth* (The World Bank, Washington, D.C., 1989), pp. 140-141.

87. *Op. cit.* 78, pp. 20-26.

88. *Op. cit.* 78, p. 18.

89. *Op. cit.* 78, p. 23.

90. Sandra J. Cointreau, "Solid Waste Recycling: Case Studies in Developing Countries," paper prepared for the United Nations Development Programme/World Bank Resource Recovery Project, The World Bank, Washington, D.C., October 1987, p. 39.

91. *Ibid.*, p. 69.

92. Juan Carlos Fortuna and Suzana Prates, "Informal Sector versus Informalized Labor Relations in Uruguay," in *The Informal Economy: Studies in Advanced and Less Developed Countries*, Alejandro Portes, Manuel Castells, and Lauren A. Benton, eds. (The Johns Hopkins University Press, Baltimore and London, 1989), pp. 86-87.

93. *Ibid.*, pp. 90-91.

94. *Op. cit.* 71, p. 63.

95. The World Bank, "Recycling from Municipal Waste," *The Urban Edge*, Vol. 11, No. 5 (The World Bank, Washington, D.C., June 1987), p. 3.

96. *Op. cit.* 90, p. 66.

97. Environmental Quality International, "Extension of Technical and Advisory Services to the Zabbaleen Gameya—Phase II," report to the Ford Foundation, Cairo, December 1988, pp. 3-5.

98. *Op. cit.* 2, Box 11.7, p. 158.

99. *Op. cit.* 2, Box 11.7, p. 158.

100. Romeo B. Ocampo, *Low-Cost Transport in Asia: A Comparative Report on Five Cities* (International Development Research Centre, Ottawa, Canada, 1982), p. 69.

101. *Op. cit.* 2, Box 11.7, p. 158.

102. Elliot Berg, "The Potentials of the Private Sector in Sub-Saharan Africa," in *Crisis and Recovery in Sub-Saharan Africa*, Tore Rose, ed. (Organisation for Economic Co-operation and Development Development Centre, Paris, 1985), p. 140.

103. Dale Whittington, Donald T. Lauria, and Xinming Mu, "Paying for Urban Services," The World Bank, Washington, D.C., March 1989, pp. 3-6, 10.

104. *Op. cit.* 78, pp. 133-134.

105. Muhammad Yunus, "Grameen Bank: Organization and Operations," paper prepared for the International Conference on Microenterprise Development, Washington, D.C., June 1988, p. 2.

106. The World Bank, "Issues in Informal Finance," in *World Development Report* (The World Bank, Washington, D.C., 1989), p. 117.

107. Friedrich Kahnert, Economist, Asia Regional Office, "Micro-Enterprise Lendings," The World Bank, Washington, D.C., June 1988 (memorandum).

108. *Op. cit.* 83, pp. 18-19.

109. Judith Tendler, "What Ever Happened to Poverty Alleviation?" paper prepared for the International Conference on Microenterprise Development, Washington, D.C., June 1988, p. 9.

110. *Op. cit.* 83, p. 23.

111. *Op. cit.* 109, pp. iv-v.

112. Eduardo Rojas, Sanitation Specialist, Inter-American Development Bank, Washington, D.C., 1989 (personal communication).

113. Stephen K. Mayo, Stephen Malpezzi, and David J. Gross, "Shelter Strategies for the Urban Poor in Developing Countries," *Research Observer*, Vol. 1, No. 2 (The World Bank, Washington, D.C., July 1986), p. 199.

114. *Op. cit.* 78, p. 24.

115. *Op. cit.* 113, p. 190.

116. Stephen K. Mayo and David J. Gross, "Sites and Services—and Subsidies: The Economics of Low-Cost Housing in Developing Countries," draft paper, The World Bank, Washington, D.C., 1985, cited in The World Bank, "Making Shelter Projects Replicable," *The Urban Edge*, Vol. 9, No. 10 (The World Bank, Washington, D.C., December 1985), p. 2.

117. *Op. cit.* 2, p. 175.

118. *Op. cit.* 2, p. 176.

119. United Nations Centre for Human Settlements (Habitat), *Supporting the Informal Sector in Low-Income Settlements* (Habitat, Nairobi, 1986), p. 19.

120. *Op. cit.* 2, p. 176.

121. *Op. cit.* 58, p. 17.

122. Carl R. Bartone, Senior Environmental specialist, The World Bank, Washington, D.C., December 1989 (personal communication).

123. Trudy Harpham, Tim Lusty, and Patrick Vaughan, *In the Shadow of the City: Community Health and the Urban Poor* (Oxford University Press, Oxford, U.K., 1988), p. 118.

124. *Op. cit.* 58, p. 17.

125. *Op. cit.* 123, pp. 126-131.

126. The World Book, *Rural Water Supply and Sanitation in Pakistan: Lessons from Experience*, Hafiz A. Pasha and Michael G. McGarry, eds. (The World Bank, Washington, D.C., 1989), pp. 31-34.

127. Albert Wright, Senior Sanitary Engineer, The World Bank, Washington, D.C., December 1989 (personal communication).

128. Urban Development Division, The World Bank, "Urban Strategy Paper," The World Bank, Washington, D.C., May 1989, p. 109.

129. *Op. cit.* 83, p. 25.

130. Sandra J. Cointreau, "Solid Waste Recycling: Case Studies in Developing Countries," paper prepared for the United Nations Development Programme/World Bank Resource Recovery Project, The World Bank, Washington, D.C., June 1987), p. 13.

131. T.R. Oke, *Review of Urban Climatology 1973-76* (World Meterological Organization Technical Note No. 169, Geneva, 1979), cited in Arthur Viterito, "Implications of Urbanization for Local and Regional Temperatures in the United States," in *Coping with Climate Change: Proceedings of the Second North American Conference on Preparing for Climate Change: A Cooperative Approach*, John C. Topping, Jr., ed. (The Climate Institute, Washington, D.C., 1989), p. 115.

132. Arthur Viterito, "Implications of Urbanization for Local and Regional Temperatures in the United States," in *Coping with Climate Change: Proceedings of the Second North American Conference on Preparing for Climate Change: A Cooperative Approach*, John C. Topping, Jr., ed. (The Climate Institute, Washington, D.C., 1989), p. 116.

133. Sham Sani, "The Urban Heat Island—Its Concept and Application to Kuala Lumpur," in *Urbanization and the Atmospheric Environment in the Low Tropics*, Figure 3, p. 249, Sham Sani, ed. (1987), (Universiti Kebangsaan Malaysia, Banki, Malaysia, cited by P.J. Peterson, Director, GEMS Monitoring and Assessment Research Centre, London, January 1990 (personal communication).

134. Arthur Viterito, Assistant Professor of geography, The George Washington University, Washington, D.C., December 1989 (personal communication).

135. Lisa Swenarski, "Bamboo to the Rescue in Costa Rica," *The Christian Science Monitor* (December 9, 1988, Boston), p. 27.

136. *Ibid.*

137. Projecto Nacional de Bamb, "Lestas las Premeras Casas," *Bambusetem* (Projecto Nacional de Bamb, San José, Costa Rica, September 1989), pp. 3, 6.

138. Ana Cecilia Chaves, Director, Projecto Nacional de Bamb, San José, Costa Rica, January 1990 (personal communication).

139. *Op. cit.* 135.

140. Projecto Nacional de Bamb, "Projecto Nacional de Bamb," brochure, San José, Costa Rica, n.d., pp. 10, 22.

6. Food and Agriculture

The increase in food production in developing countries over the past 23 years is remarkable. It is also contrary to the Malthusian vision that prevailed in the 1960s. The success was due to a variety of factors—expanded croplands, new high-yielding seeds, and heavier applications of fertilizers and pesticides. Enlargement of the area under cultivation and application of increasing amounts of chemicals have exacted significant environmental costs that cannot be paid indefinitely. The challenge facing farmers in developing countries today is to increase production while pursuing methods that are both economically and environmentally sustainable.

World production of cereals was expected to continue on an upward path in 1989. Since 1983, however, increases have not been dramatic. Stocks of world cereals, meanwhile, have fallen to their lowest level in the 1980s. (See Conditions and Trends, below.)

The evidence suggests a relatively grim outlook for Africa (see Focus on Sub-Saharan Africa: A Region in Crisis, below)—and an uncertain but generally brighter future for other parts of the developing world.

Efforts to boost food production to outpace population growth have relied on expanding the land area under cultivation and increasing productivity. Some regions have already expanded nearly to the limits of their arable land; farmers in other regions are clearing land with marginal agricultural potential.

Research under way in Latin America could open the way for sustainable agriculture in both existing and new areas of the tropical savannas. For example, the International Center for Tropical Agriculture (CIAT) in Cali, Colombia, is testing a form of intensified agriculture appropriate to the vast tropical savannas of Latin America. It incorporates crops and livestock and uses cultivars especially adapted to the savannas' low-nutrient, acidic soil. (See Sustainable Land Use for South America's Tropical Savannas, below.)

Global climate change is casting a cloud of uncertainty over the future of agriculture. Some northern areas may benefit from a longer growing season; many others could suffer from less rainfall. (See The Effects of Climate Change on Agriculture, below.)

CONDITIONS AND TRENDS

A MIXED PICTURE

Global food and agriculture present a mixed picture that is not easily amenable to an unqualified progress report:

■ Global production of cereals is increasing, but the rate of increase has slowed since 1983. The easiest gains of the Green Revolution may have occurred. Although further gains are expected, they are unlikely to match the increase of the past two decades.

■ In per capita terms, Asia and some other regions of the developing world are doing better; still others, especially Africa, are doing worse.

■ World cereal stocks have declined to the lowest level of the 1980s.

■ Many regions have little potential for adding cropland area; cropland in per capita terms will decline in all regions of the developing world if population projections are accurate. Feeding increasing numbers is stressing agricultural lands and degrading soils and watersheds.

■ The number of undernourished people has increased in absolute terms but has probably declined in percentage terms in most regions, with the exception of Africa. Access to food remains a serious problem in some regions; uneven food distribution and the inability of the very poor to afford sufficient food contribute to this problem.

Global Food Production: 1988–89

The world's production of cereals continued to increase during the 1980s and was projected to set a new record of roughly 1.88 billion metric tons in 1989, up 6.8 percent from the 1.76 billion metric tons estimated for 1988. Nearly all the increase was in the United States and Canada, which were bouncing back from a severe drought in 1988, and in the Soviet Union (1).

World production of root crops and tubers declined in 1988, largely because of declines in developed countries, but increased in the developing world. Modest increases were seen in the production of meat, milk, fish, pulses, fruits, melons, vegetables, and oils. (See Table 6.1.)

Long-Term Trends

Since 1965, world cereal production increased over 70 percent and more than doubled in developing countries, with Asia leading. Cereals provide about half the calories in the human diet, so cereal production is the single best indicator of the health of agriculture and the nutritional status of whole populations. (See Figure 6.1.)

Gains in root crops and tubers were not so dramatic in absolute numbers over the 1965–88 period, although Asia again experienced significant percentage increases, as did Africa until 1987.

Production of concentrated protein in the forms of meat, milk, and fish in the developing countries has more than doubled since 1965. Aggregate production of pulses, vegetables, melons, fruits, oils and other food crops increased over 95 percent throughout the developing world. (See Table 6.1.)

Trends since 1984

Since 1984, the rate of increase in cereal production has not matched the spectacular growth of the previ-

Figure 6.1   World Production of Selected Food Crops, 1965–88.

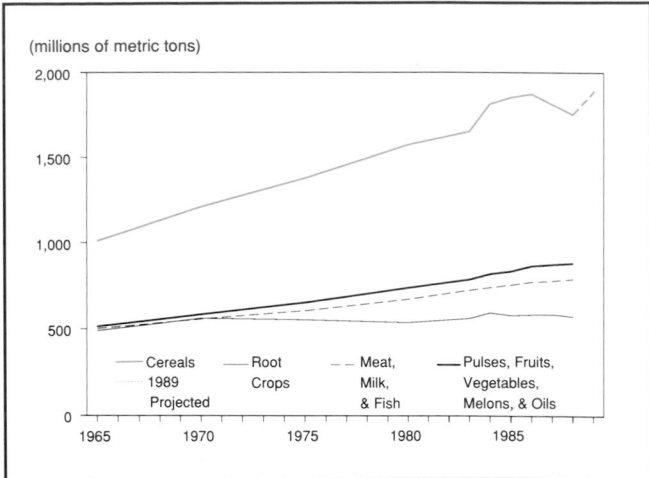

Source: Table 6.1.

ous two decades. Production since 1984 has generally hovered around 1.8–1.85 billion metric tons. For example, the record 1989 crop of 1.88 billion metric tons is only 4 percent higher than the 1984 crop of 1.80 billion metric tons. By comparison, the 1984 crop was 15 percent higher than the 1980 crop. Alternatively, as Figure 6.1 suggests, 1984–86 could be interpreted as anomalously high years that disguise a continuing upward trend.

Global production of root crops actually peaked in 1984 and has declined since then. Production of meat, milk, and fish has made fairly steady gains, rising 66 million tons from 1980 to 1984 and 46 million tons from 1984 to 1988. Production of fruits and vegetables made steady gains over the entire 1980–88 period, with nearly all the gain in the developing countries. (See Table 6.1 and Figure 6.1.)

The Green Revolution

For countries such as India, much of this growth is clearly attributable to the continuing gains of the Green Revolution, especially the development of new rice and wheat varieties suited to regional conditions. Increases in world food production also have followed increases in the use of fertilizers, pesticides, and agricultural machinery. (See Chapter 18, "Food and Agriculture," Table 18.2.) For example, average annual fertilizer use increased in Asia from 42 kilograms per hectare of cropland in 1975–77 to 93 kilograms per hectare in 1985–87.

Whether these gains can continue is unclear. The fact that many high-yielding varieties are in relatively widespread use suggests that the biggest gains of the Green Revolution may already have been realized. Nevertheless, other varieties may have significant potential for marginal lands. For example, triticale, a

Table 6.1 Production of Selected Food Crops, 1965–88
(thousand metric tons)

Cereals

	1965	1970	1975	1980	1983	1984	1985	1986	1987	1988
World	1,005,926	1,205,128	1,372,727	1,567,472	1,642,704	1,805,319	1,843,757	1,863,877	1,803,118	1,742,985
Developing Countries	470,248	587,418	683,263	770,799	892,483	922,126	926,231	943,504	933,639	969,099
Africa	37,877	43,087	47,679	49,957	46,876	45,857	58,846	62,524	55,370	63,053
Far East	157,652	212,254	239,075	273,652	316,883	318,756	323,485	325,800	309,355	339,693
Latin America	57,640	71,307	80,545	88,498	99,647	106,740	110,573	105,552	111,724	107,730
Near East	37,821	39,962	51,689	55,536	55,830	55,791	63,028	66,748	65,207	75,632
Asian Centrally Planned Economies	179,240	220,779	264,245	303,114	373,213	394,945	370,257	382,845	391,946	382,945
Developed Countries	535,678	617,710	689,463	796,673	750,221	883,193	917,527	920,373	869,480	773,886
North America	215,893	215,389	286,543	311,336	255,259	357,704	395,590	373,099	331,133	241,815
Western Europe	116,285	128,230	146,628	177,513	173,603	211,614	195,704	191,151	186,525	196,126
Oceania	10,106	13,482	18,422	17,159	31,940	29,717	26,363	25,181	20,994	22,990
Eastern Europe & U.S.S.R.	168,412	234,742	208,405	264,130	268,617	260,496	273,624	303,837	304,611	287,847

Root Crops

	1965	1970	1975	1980	1983	1984	1985	1986	1987	1988
World	489,283	561,774	553,230	538,214	562,032	593,971	578,890	582,496	583,838	571,182
Developing Countries	246,432	302,483	330,014	353,652	357,837	368,018	362,109	355,292	369,720	375,543
Africa	58,502	68,228	79,713	84,779	88,404	93,242	97,935	99,616	94,257	94,783
Far East	33,261	36,713	45,684	58,011	59,480	65,564	65,218	57,020	62,948	70,305
Latin America	42,089	49,184	45,743	44,047	41,654	43,791	44,525	47,299	46,385	45,491
Near East	3,417	3,869	4,901	7,234	7,747	8,158	9,533	9,611	10,069	10,251
Asian Centrally Planned Economies	107,952	143,173	152,601	158,121	159,003	155,674	143,286	140,146	154,449	153,090
Developed Countries	242,851	259,290	223,216	184,563	204,195	225,953	216,781	227,204	214,118	195,639
North America	16,004	17,886	17,398	16,715	18,253	19,833	22,137	19,737	21,067	19,199
Western Europe	63,637	63,254	47,519	49,186	42,526	50,517	51,573	47,988	47,863	45,651
Oceania	812	1,068	977	1,091	1,127	1,327	1,277	1,252	1,310	1,364
Eastern Europe & U.S.S.R.	152,143	169,291	151,145	111,251	135,629	147,334	134,596	150,729	136,563	122,165

Meat, Milk, and Fish

	1965	1970	1975	1980	1983	1984	1985	1986	1987	1988{a}
World	501,748	557,500	606,277	673,377	720,603	735,017	749,856	765,297	771,382	781,045
Developing Countries	114,547	136,140	152,288	183,957	205,320	214,320	225,882	237,652	245,312	250,829
Africa	10,389	12,435	13,190	15,105	17,929	17,353	17,840	19,084	19,579	19,979
Far East	37,162	42,188	51,442	61,860	73,751	77,942	80,363	84,541	87,477	91,273
Latin America	39,731	51,606	50,870	60,147	61,169	64,053	67,543	70,334	71,388	71,072
Near East	12,890	14,291	16,530	20,294	21,861	21,464	22,789	22,904	23,021	23,116
Asian Centrally Planned Economies	14,208	15,215	19,817	25,960	30,610	33,508	37,347	40,759	43,847	45,389
Developed Countries	387,200	421,362	453,988	489,421	515,283	520,697	523,974	527,645	526,070	530,216
North America	88,497	88,811	87,802	98,056	104,538	104,629	107,626	108,421	109,620	111,907
Western Europe	143,113	150,917	162,914	182,459	192,233	191,953	189,507	185,723	181,797	179,258
Oceania	15,756	16,997	16,471	16,334	16,967	17,756	18,318	18,786	18,291	18,937
Eastern Europe & U.S.S.R.	124,178	143,932	164,168	167,188	174,637	179,461	180,739	186,202	187,601	191,042

Pulses, Fruits, Melons, Vegetables, and Oils

	1965	1970	1975	1980	1983	1984	1985	1986	1987	1988
World	514,161	584,298	653,611	738,782	786,016	817,501	832,337	860,466	870,289	876,620
Developing Countries	284,179	322,986	373,709	435,070	469,886	494,932	516,868	532,313	542,250	554,535
Africa	36,700	44,237	51,176	51,232	53,006	53,399	55,726	58,838	58,854	60,130
Far East	84,073	95,380	110,279	130,854	143,363	151,492	154,417	159,418	155,637	158,209
Latin America	53,435	60,534	70,841	85,392	87,804	94,205	97,941	96,388	99,515	103,587
Near East	32,584	37,859	45,714	55,521	61,720	62,793	67,287	68,911	70,055	71,820
Asian Centrally Planned Economies	76,014	83,483	94,041	110,246	122,072	131,024	139,425	146,684	156,078	158,646
Developed Countries	229,982	261,313	279,903	303,713	316,128	322,570	315,468	328,153	328,038	322,085
North America	46,734	52,368	62,953	69,765	65,536	67,281	68,212	67,285	72,037	69,201
Western Europe	99,210	110,172	111,183	118,437	121,641	121,778	120,055	126,543	124,232	120,997
Oceania	3,201	3,794	3,813	4,334	4,520	5,082	5,776	5,931	6,487	6,595
Eastern Europe & U.S.S.R.	58,631	68,798	73,319	81,621	94,869	99,595	92,216	99,114	95,379	95,510

Sources:

1. For 1965-80 data, Food and Agricultural Organization of the United Nations (FAO), *1987 Country Tables: Basic Data on the Agricultural Sector* (FAO, Rome, 1987), pp. 310-336.
2. For 1983-88 data, Food and Agricultural Organization of the United Nations (FAO), *Quarterly Bulletin of Statistics,* Vol. 2 (FAO, Rome, 1989), pp. 19-23.
3. Food and Agricultural Organization of the United Nations (FAO), *Fishery Statistics: Catches and Landings*, Vol. 64, 1987 (FAO, Rome, 1989).

Notes:

a. 1988 fish catch data unavailable; 1987 data repeated in their place.
b. Some small countries not appearing in regional totals are included in the total for their development status and in the world totals.

Figure 6.2 Index of Per Capita Food Production

A. Developing Regions

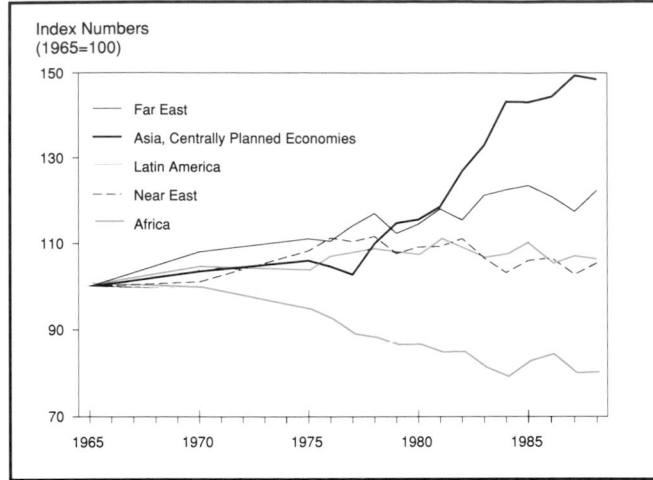

B. Developed Regions

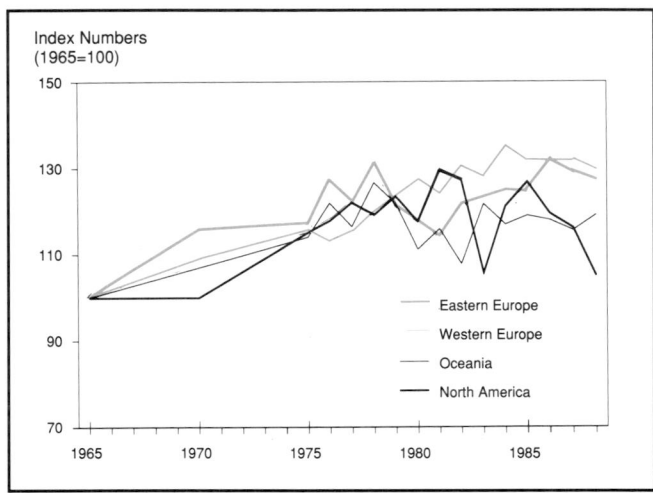

Sources: For 1965 and 1970, United Nations Food and Agriculture Organization (FAO), *1987 Country Tables* (FAO, Rome, 1987), pp. 312-336; for 1975-76, FAO, *1986 Production Yearbook* (FAO, Rome, 1987), p. 48; for 1977-88, FAO, *FAO Quarterly Bulletin of Statistics*, Vol. 2 (1989), pp. 17-18.

Figure 6.3 World Cereal Carryover Stocks, 1961–89

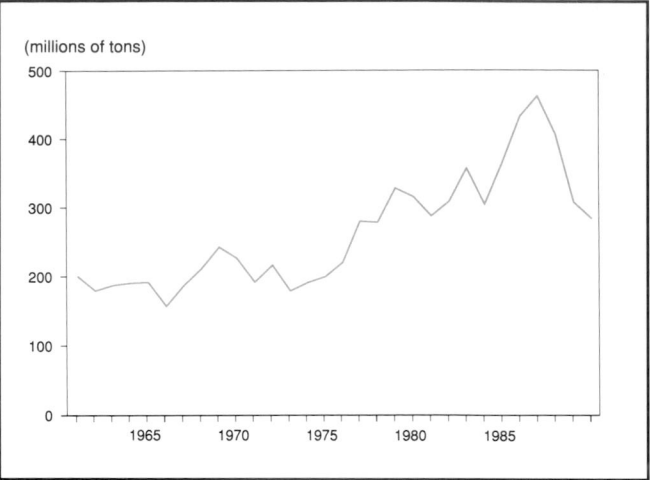

Source: U.S. Department of Agriculture, Economic Research Service, PS&D View Database on cereal production created by Karl Gudmunds and Alan Webb, September 1989.
Note: Stocks shown are those held at the beginning of each calendar year.

wheat-rye hybrid, shows promise in difficult growing conditions. (See Recent Developments, below.)

Per Capita Production

The relationship between food production and population in the developing countries is illustrated in Figure 6.2A. Using 1965 as a base year, Figure 6.2 shows increases or declines in per capita food production through 1987 for developing and developed regions. China and the other centrally planned economies of Asia have shown the largest gains in per capita food production; other Asian nations are slightly ahead of the population curve. In Latin America and the countries of the Near East, food production appears to be staying about even with population, and Africa is con-

tinuing a long-term decline in per capita food production. (See Figure 6.2A.)

In Africa, total food production has increased significantly, 23 percent between the 1976–78 average and the 1986–88 average. Despite this increase, population growth outpaced agricultural growth, thus per capita food production declined 8 percent.

Figure 6.2B shows that food production per capita in developed countries has been extremely variable but has shown an upward trend.

World Cereal Carryover Stocks

For the third successive year, estimated global consumption in 1989 exceeded production, leading to a further decline in world cereal carryover stocks (2). As shown in Figure 6.3, carryover stocks have fallen to the lowest point of the 1980s.

Global cereal stocks fell 100 million metric tons in 1988 to 305 million metric tons (3) (4). Preliminary figures for 1989 suggest that total cereal stocks will again fall, to about 287 million metric tons by the beginning of 1990 (5). These stocks are now slightly below stocks in 1983, the time of the Sahelian drought and famine.

Total world production, and therefore total cereal carryover stocks, declined in 1988, primarily because of drought in the United States and secondarily because of federal policies designed to slow production. Total U.S. cereal reserves dropped from a historic high of 204 million metric tons at the beginning of 1987 to 84 million metric tons at the beginning of 1989; the U.S. reserve was still the largest in

Figure 6.4 Cropland Per Capita, 1989 and 2025 (projected)

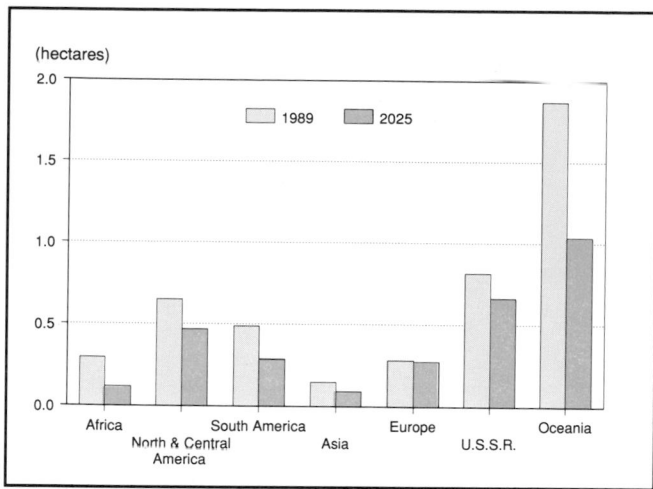

Source: Chapter 16, "Population and Health," Table 16.1 and Chapter 18, "Food and Agriculture," Table 18.2.

absolute size. Stocks in other regions of the world remained relatively stable (6).

Policies and Prices Affect Stocks

Stocks in specific countries can vary widely from year to year. Poor government management can destabilize prices, which in turn can amplify the effects of climatic variability on production and stocks. Such a swing in cereal stocks occurred in the Sudan, where the 1984 drought depleted most cereal stocks (down to 14,000 metric tons) and led to high prices for sorghum and millet at planting time in 1985. Responding to these prices, Sudanese farmers planted sorghum and millet on every available hectare. A record 1985 harvest and huge carryover stock (1.4 million metric tons) led to a price collapse and a lower (but still good) harvest in 1986. Farmers responded to continuing low prices and continuing high carryover stocks in 1987 by planting even less area to

grain. Poor rainfall affected the 1987 harvest, and stocks were depleted to 224,000 metric tons at the beginning of 1988. At planting time in 1988, high prices motivated farmers to increase the area planted in cereals; they produced a record harvest and bumper cereal reserves, but prices fell. Preliminary data suggest that the low prices once again contributed to a reduction in the area planted to grain and limited potential 1989 production (7). In a multiyear drought, these swings could lead to inadequate stocks at a time of greatest need. Aside from prices, access to markets, inputs, and credit can be important factors.

Food Aid

The Food and Agriculture Organization of the United Nations (FAO) forecast that food aid in 1989–90 would be about 11 million metric tons, up from an estimated 9.8 million metric tons in 1988–89. FAO predicted much larger shipments of food aid to Eastern Europe and no change in deliveries to low-income food-deficit countries (8).

During 1985–87, Africa received an average of only 6.5 million metric tons of food aid per year. Of that amount, 2.6 million metric tons (40 percent) went to countries in the Sahel and the Horn of Africa as part of the worldwide effort to mitigate famine there. (See Chapter 18, "Food and Agriculture," Table 18.4.)

Cropland Area

Without a major expansion of arable land, the world average of 0.28 hectares of cropland per capita (2,800 square meters) is expected to decline to 0.17 hectares by the year 2025 if current population projections are accurate. In Asia, cropland per capita would decline to 0.09 hectares. (See Figure 6.4.)

In Asia, an estimated 82 percent of potential cropland is already under production. There are large reserves of potential cropland in Latin America and sub-Saharan Africa, but for much of these reserves the soil is marginal, it is suitable only for perennial tree crops, or rainfall is unreliable (9). New agricul-

Table 6.2 Estimated Number of Undernourished People in the World, 1969–71, 1979–81, and 1983–85
(millions of people)

Region	1969–70 Number	1969–70 Share of Population (percent)	1979–81 Number	1979–81 Share of Population (percent)	1983–85 Number	1983–85 Share of Population (percent)
Africa	92	32	110	29	140	32
Far East	281	29	288	24	291	22
Latin America	51	18	52	15	55	14
Near East	35	22	24	12	26	11
Total	460		475		512	

Source: United Nations World Food Council (WFC), "The Global State of Hunger and Malnutrition and the Impact of Economic Adjustment on Food and Hunger," World Food Council, Thirteenth Ministerial Session, Report by the Secretariat, Beijing, China, 1987, p. 16.

6 Food and Agriculture

Table 6.3 Food Available for Direct Human Consumption

(calories per capita per day)

	1961-63	1969-71	1979-81	1984-86
World Total	2,300	2,440	2,600	2,690
Developing Countries				
94 developing countries	1,940	2,120	2,330	2,460
Africa (sub-Saharan)	2,040	2,100	2,140	2,060
Near East/North Africa	2,240	2,390	2,870	3,050
Asia	1,830	2,030	2,260	2,430
Asia[a]	1,970	2,070	2,200	2,280
Latin America	2,380	2,520	2,670	2,700
Low-income countries	1,850	2,020	2,200	2,360
Middle-income countries	2,160	2,340	2,620	2,680
Developed Countries	3,060	3,230	3,340	3,380
North America	3,180	3,380	3,510	3,620
Western Europe	3,090	3,230	3,370	3,380
Other developed market economies	2,590	2,810	2,900	2,930
European centrally planned economies	3,140	3,330	3,390	3,410

Source:
1. Food and Agricultural Organization of the United Nations (FAO), *World Agriculture: Toward 2000*, Nikos Alexandratos, ed. (Belhaven Press, London, 1988), p. 27.
2. Nikos Alexandratos, 1990 (personal communication).
Note: a. Does not include China.

tural lands also tend to be constrained by physical restrictions (e.g., soil structure, slope), chemical limits (e.g., acidity, alkalinity), or water availability.

Cropland expansion will most often come at the expense of rangeland, forests, wetlands, and other areas that are both economically important and ecologically fragile.

For existing cropland, urbanization and industrial development are a threat. No comprehensive data are available, but there is little doubt that desertification, salinization, waterlogging, and erosion degrade significant amounts of cropland. They may remove as much land from production as is added each year (10).

World Hunger, Access to Food

No comprehensive data have been collected on the number of people who are hungry or undernourished. Estimates range from about 500 million to about 1 billion people; part of the discrepancy is due to different definitions of hunger (11).

FAO estimated the number of undernourished people at 512 million in 1983–85, as shown in Table 6.2. The total was up slightly from previous years, but the share of the population that was undernourished declined in all regions except Africa. FAO defined "undernourished" as those with a daily caloric intake below 1.4 times the basal metabolic rate (the estimated energy requirements in a state of fasting and at complete rest), which in India is equal to 1,600 calories and in Egypt 1,700 calories (12).

Estimates of food available for direct human consumption show a generally upward trend. (See Table 6.3.) The number of calories available per person in 94 developing countries rose from 1,940 per day in

1961–63 to 2,460 per day in 1984–86. Sub-Saharan Africa was a striking exception, with a small decline from 1969–71 to 1984–86 (13)(14).

A wide range of reasons is attributed to the continuing problem of hunger and malnourishment in some parts of the world. Meager and unreliable rainfall and periodic droughts are serious problems in the Sahel and Ethiopia. Other sub-Saharan nations may have sharply skewed income distribution, with a part of the population that is very poor (Kenya), may be contending with civil wars (Angola, Ethiopia, and Mozambique), may have poor infrastructure (Uganda and Zaire), may have large, poor urban populations (Zambia), or may have pursued economic policies that do not support equitable distribution of benefits or discouraged production (15). (See Focus On Sub-Saharan Africa: A Region in Crisis, below.)

The ability to meet food requirements is also affected by food production policies within countries. Ethiopia, in addition to contending with a civil war and periodic droughts, remains a food-deficit nation in part because it discourages small farmers through poor access to inputs and denial of land tenure. Inadequate planning is another factor; for example, the villagazation, a program that brought together isolated households, which distanced people from their farmsteads (16).

FOCUS ON SUB-SAHARAN AFRICA: A REGION IN CRISIS

Sub-Saharan Africa is experiencing a profound economic and environmental crisis. It is the only region in the developing world with declining per capita food production. Of all developing regions, it relies most heavily on imported food, and it has the high-

Box 6.1 Root Causes: Rapid Population Growth, Slow Economic Growth

At the base of Africa's crisis are two factors—rapid population growth and slow economic growth.

RAPID POPULATION GROWTH

In 1985, Africa's total population stood at 555 million, including 450 million in sub-Saharan Africa. With an annual growth of 3 percent between 1980 and the year 2000, the continent will add more people than the present population of the United States. If present trends continue, the population will triple by 2025. With 45 percent of the people under the age of 15, Africa's near-term demographic history is already partially written.

Since 1965, African nations have brought infant mortality rates down one third and raised life expectancy about 7 years [1]. Yet even with these and related health indicators, the gulf between Africa and the rest of the world remains shocking. In parts of the continent, children under the age of 5 make up 20–25 percent of the population but account for 50–80 percent of the deaths; in Europe, for example, the total mortality for the same age group is 3 percent [2].

Africa's fertility rate has risen. It now stands at 6.6 (i.e., the average woman will have 6.6 children), twice the world average [3]. Government efforts to reduce fertility have been generally weak, but the prospects for rapid fertility decline are not wholly bleak. Since the first World Population Conference in 1974, the number of African countries with an active government population program has grown from 1 (Mauritius) to 24. Most have had little impact; nonetheless, their number is growing steadily. Recent surveys across the continent indicate changing attitudes among African women—interest in smaller families is growing, for example. (See *World Resources 1986*, pp. 15–23.)

SLOW ECONOMIC GROWTH

In Africa, a region dependent on the export of mineral and agricultural commodities for one quarter of its GNP, the terms of trade fell 35 percent between 1980 and 1986 [4]. Total export earnings fell almost as much, from $50 billion in 1980 to $35 billion in 1986 [5]. Africa's foreign debt continued the meteoric rise that began in the 1970s, tripling in the 1980s and reaching 81 percent of GNP in 1987 [6]. Today, scheduled payments on the debt amount to two thirds of expected export earnings and, in many countries, exceed such earnings [7]. Most ominous of all, average per capita GNP, which had grown slowly in the 1960s and remained nearly flat throughout the 1970s, began to decline. As the economic devastation deepened and population growth continued, per capita GNP slid faster; in 1980–87, per capita GNP declined -2.8 percent per year on average [8].

References and Notes

1. World Resources Institute and International Institute for Environment and Development in collaboration with the United Nations Environment Programme, *World Resources 1988-89* (Basic Books, New York, 1988), p. 250.
2. Fred T. Sai, "Population and Health: Africa's Most Basic Resource and Development Problem," *Strategies for African Development*, Robert J. Berg and Jennifer S. Whitaker, eds. (University of California Press, Berkeley, California, 1986), p. 139.
3. Overseas Development Council (ODC), *Africa's Recovery and Development: Reassessing U.S. Policy* (ODC, Washington, D.C. 1988), p. 3.
4. The World Bank, *World Development Report, 1989* (The World Bank, Washington, D.C., 1989), p. 217.
5. *Ibid.*, p. 151.
6. *Ibid.*, p. 211.
7. *Op. cit.* 3, p. 4.
8. The World Bank, *Sub-Saharan Africa: From Crisis to Sustainable Growth* (The World Bank, Washington, D.C., 1989), Table 1, p. 221.

est proportion of land area losing its fertility and the highest percentage of population suffering from severe malnourishment. Land degradation, caused largely by the demand for fuelwood and livestock grazing, is a serious problem. GNP, population growth, infant mortality, and other indicators of the regions' health are equally alarming.

Rapid population growth and slow economic growth (see Box 6.1) go far to explain the crisis in Africa. Beyond these broad causes, however, is the unaccommodating physical environment for agriculture and a complex policy environment that has often undermined progress in domestic food production.

Added to the problems are armed conflict and civil strife. In the late 1980s, wars raged in nine sub-Saharan countries [17].

Though the crisis is grave, it is not without hope. Policy changes in a few countries have had important positive impacts on food production. In addition, intercropping, a new farming technique, shows great promise.

Unless otherwise noted, the use of Africa in this section refers to "sub-Saharan Africa," which includes all of Africa except the belt of nations along the northern coast, South Africa, and Namibia. (See Figure 6.5.)

Dimensions of the Food Crisis

Figure 6.2A tells the story of Africa's agriculture as well as any words could. Overall food production is growing, but per capita production has been falling ever since 1970, despite a substantial expansion of land under cultivation. A food exporter in the early 1960s, Africa was transformed—because of a rapid extension of life expectancy in newly independent countries—into an importer during the 1970s. By the 1980s, cereal imports had quadrupled (their costs increased 10-fold), and as the bill grew, the region became heavily dependent on food aid.

Between 1970 and 1985, the estimated number of undernourished people increased by over 50 percent. While in other regions the share of the population lacking sufficient food declined, Africa's remained the same. (See Table 6.2.)

The Physical Environment

Sub-Saharan Africa includes a vast and diverse area of 22,245 square kilometers. Many countries are large, but have little arable land. An estimated 30 percent of the area can sustain production of rainfed crops; only about one fourth of that amount is used

Figure 6.5 Sub-Saharan Africa

Source: The World Bank, *Sub-Saharan Africa: From Crisis to Sustainable Growth* (The World Bank, Washington, D.C., 1989), p. ii.

for crops. FAO estimates that cropland acreage expanded 0.7 percent annually over the past 20 years (18).

Agricultural potential varies widely. Central Africa, humid West Africa, and Southern Africa have relatively large areas of arable land and relatively low population densities. Most of the Sahel, parts of mountainous East Africa, and the dry region from Angola through Botswana, Lesotho, and southern Mozambique all have large populations that import food (19).

Where land scarcity is becoming common, the challenge African farmers face is to increase productivity. In other regions, African farmers have more land for food production. In Rwanda, for example, cropland increased from 26 percent to 45 percent of the total land area during 1965–87 (20).

African farmers work in a uniquely inhospitable environment. One fifth of the entire continent is desert, and in another 10 percent, the soils are too sandy. An area larger than the continental United States is rendered largely unusable by the tsetse fly. Only 19 percent of the soils have no fertility limitations. Most arable soils are coarse; the clay content is low so that they cannot absorb and hold moisture and therefore are highly susceptible to erosion (21).

As in all tropical regions, Africa's rainfall is seasonal, and when it comes, it is heavy. As a general rule, rain begins to cause erosion when it falls more rapidly than 25 millimeters per hour. Only 5 percent of rainfall is in this category in the temperate zones, but in the tropics, it is 40 percent. Vegetation is crucial to preventing erosion in these areas; leaves soften the rain's impact and roots hold the soil. In areas barren of vegetation, the results can be devastating. Not only is heavy rain erosive, it also leaches out vital soil nutrients (22).

Large parts of Africa are arid or semiarid, and low rainfall is the norm but highly variable. For much of the region, the interannual variability is 30–40 percent. Thus, normal rainfall, as many nations experience it, has a completely different meaning for sound agricultural planning in Africa than it does for temperate conditions prevalent in most developed regions. Also, multiyear droughts are normal for Africa (23).

Some 5 million hectares are irrigated, mostly in Madagascar, Nigeria, and Sudan. About 20 million hectares may be suitable for irrigation, most of it in these three countries plus Chad, Ethiopia, Malawi, Mali, Mauritania, Senegal, and Uganda (24).

Traditionally, African farmers have adapted to their environment by practicing shifting or fallow cultivation. They clear the land, plant crops for a few years, and as nutrient levels decline and weeds take over, abandon the land and allow it to return to its natural state. Then, given time, trees, shrubs, and grasses naturally refurbish the soils. Shifting cultivation can sustain agriculture indefinitely, even in harsh conditions, when fallow periods are long enough. But when there are too many people and fallow periods are shortened or disappear completely, land productivity declines rapidly.

Fuelwood and Livestock

Added to the pressures on the land caused by agriculture are the people's needs to gather fuelwood and graze their livestock.

Fuelwood supplies an estimated 80 percent of energy needs in sub-Saharan Africa. It is without question in very short supply, and the situation is likely to worsen (25).

As fuelwood supplies become scarcer, farmers have to burn animal dung and crop residues instead of using them to enrich and sustain the soil. As more land loses nutrients, vegetative cover decreases, soil erosion grows, and the moisture-holding capacity of soils declines. Vulnerability to naturally variable rainfall grows, and in a prolonged dry spell, the entire system may collapse.

The estimated 160 million head of cattle in Africa include many that are not well suited to arid conditions, as are mixed herds of sheep, goats, and camels. Livestock provide both manure and draft power, but overgrazing is an acute problem that is contributing to land degradation in many regions (26).

Figure 6.6 Projected Food Gap: Three Scenarios for Sub-Saharan Africa, 1990–2020

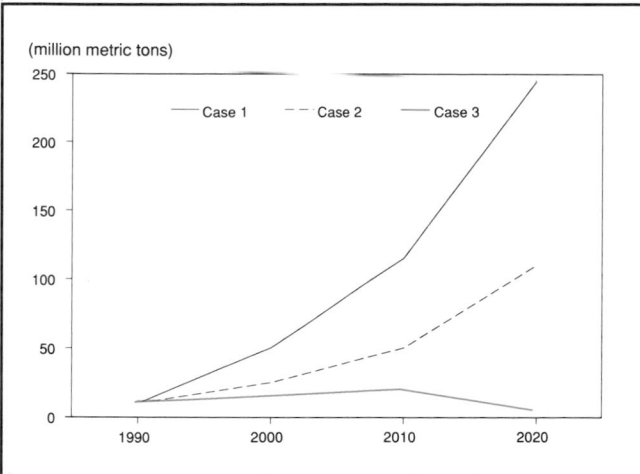

Source: The World Bank, *Sub-Saharan Africa: From Crises to Sustainable Growth* (The World Bank, Washington, D.C., 1989), p. 74.

The end result is that more than one quarter of sub-Saharan Africa's land area—750 million hectares—is probably moderately to very severely desertified. Desertification refers broadly to the loss of the land's biological productivity. It encompasses erosion, compaction, salinization, waterlogging, and other forms of land degradation. Arid and semiarid lands are especially vulnerable. Vast areas of Africa have lost agricultural potential, some of them, where the damage is most severe, permanently. (See *World Resources 1988-89*, pp. 215–231.)

The Policy Environment

Governments in Africa have often adopted agricultural policies that are counterproductive. Most African governments have, for example:

■ Imposed controls on the marketing of key crops, often banned private trade, and frequently set up state-owned food-processing enterprises that were often costly and inefficient and slowed agricultural development;
■ Set producer prices so low, in order to keep consumer prices low to satisfy rapidly growing urban populations, that there was little incentive for farmers to market through official channels (27);
■ Allowed overvalued exchange rates, resulting in cheaper imported foods than local products, so that urban populations developed a taste for imported wheat and rice, neither of which is well suited to local growing conditions (28); and
■ Neglected rural roads and rural infrastructure, in general, and the development of rural and secondary towns.

Aid Experience

Foreign assistance programs indirectly measure the deeply rooted policy and institutional problems that encumber many African governments and slow agricultural progress.

In 1985, the World Bank reviewed the short-term results of more than 1,000 of the projects it had funded. The failure rate in Africa was the worst of any region, and its agricultural failure rate was the worst of any sector. The failure rate for Asia's agriculture projects was 5 percent; by comparison, it was 33 percent in West Africa and 50 percent in East Africa. The worst sector within Africa's agriculture was livestock projects, which averaged a shocking -2 percent economic return. FAO concluded that the more than $1 billion spent on rangeland management projects over the past 15 years had been "largely wasted" (29) (30). (See *World Resources 1988-89*, pp. 82–84.)

Among 17 African agricultural projects judged over the long term, all showed an economic return of 10 percent or more at the end of the project, but 5–10 years later, 13 averaged only 3 percent and 2 had negative returns. Reforestation projects may not have

Table 6.4 Population and Food Security in Sub-Saharan Africa, 1990—2020

	1990	2000	2010	2020
Case I				
Population (millions of people) (with constant fertility rate)	500	700	1,010	1,500
Food production (mtme) (at current-trend growth rate of 2 percent per year)	90	110	135	165
Food requirement (mtme for universal food security by 2020)	100	160	250	410
Food gap (mtme)	10	50	115	245
Case II				
Population (as in Case I)	500	700	1,010	1,500
Food production (at 4 percent annual growth)	90	135	200	300
Food requirement (as in Case I)	100	160	250	410
Food gap (as in Case I)	10	25	50	110
Case III				
Population (millions of people) (with total fertility rate declining 50 percent to 3.3 by 2020)	500	680	890	1,110
Food production (mtme at 4 percent annual growth)	90	135	200	300
Food requirement (mtme)	100	150	220	305
Food gap (mtme)	10	15	20	5

Source: The World Bank, *Sub-Saharan Africa: From Crises to Sustainable Growth* (The World Bank, Washington, D.C., 1989), p. 73.
Note: mtme = millions of tons of maize equivalent.

Table 6.5 Percentage Increase in Soil Nitrogen and Organic Carbon in Agroforestry Plantings

Country	Nitrogen	Carbon
Niger	231	269
Senegal	33	40
Senegal	194	192
Senegal	110	91
Sudan	600	200

Source: Robert Winterbottom and Peter T. Hazlewood, "Agro-forestry and Sustainable Development: Making the Connection," *Ambio*, Vol. 16, No. 2-3 (1987), p. 103.

fared much better; a 1984 U.S. Agency for International Development review found no large-scale plantation or small village woodlot project in Africa that could be judged successful (31).

Aside from policy and institutional problems, cultural differences among donors and recipients may help explain project failure. A World Bank report concluded that misreading economic or social conditions was a factor in virtually every project in which new technology was rejected. The introduction of technology that was unfamiliar or required major lifestyle changes was a major cause of rejection (32).

Donors note better project outcomes when official funds are disbursed through indigenous or expatriate nongovernmental organizations (NGOs). This result is especially true for agriculture, forestry, and conservation projects. NGO projects are usually small, however, and do not generally handle large development funding amounts.

The Role of Women

African women produce roughly 70 percent of staple food. Women in many cultures are allocated fields from their fathers' or husbands' land, are responsible for specific crops and operations, and may receive independent income from marketing certain crops. As more men work in cities, women's responsibilities for farm management increase; in many areas, women manage one half the farms (33). In parts of Africa, women provide 60–90 percent of the subsistence agricultural labor force (34). Most land registration and land settlement schemes result in husbands' registering as sole owners. Lacking land titles or security of tenure, women are often unable to buy fertilizer and other inputs on credit (35).

Formalizing women's rights to land, increasing their representation in agricultural training, encouraging extension workers to include women farmers among their contacts, improving their access to technology and tools, and fostering access to credit for women's groups are among the many initiatives that could help African agriculture (36).

The Challenge Ahead

The challenge to Africa's agriculture is illustrated in a recent World Bank report that presented three scenarios for population growth and food production. (See Figure 6.6.) If present trends continue (case 1)—population grows 3.5 percent a year and food production 2 percent—the food gap (the amount required to provide 110 percent of the recommended intake of calories for the entire population (37)) would grow from 10 million metric tons in 1990 to 245 million metric tons 30 years later, in 2020. Even if population growth remains the same and food production grows 4 percent annually, the food gap would be 110 million metric tons by 2020 (case 2). Only if population growth declines to 2.75 percent and food production grows 4 percent annually (case 3) will the food gap stabilize and fall to 5 million metric tons by 2020. The World Bank report concludes that Africa must curb population growth and maintain a 4 percent growth in food over the next three decades in order to come close to providing the recommended amount of calories for its population (38). (See Table 6.4 and Figure 6.6.)

Can Africa meet this challenge? Several countries met the 4 percent target for extended periods between 1965 and 1987: Botswana, Cameroon, Côte d'Ivoire, Kenya, Malawi, Mauritius, and Rwanda (39).

Policies that encourage food production can make a significant difference. In Rwanda, for example, food production grew 4.7 percent a year between 1966 and 1982, well ahead of population growth. Market forces fixed the level of food prices, and the government remained sensitive to the needs of the farming community in determining pricing policies, exchange rates, and fiscal priorities, and promoting effective rural institutions (40).

In Zimbabwe, per capita cereal production increased 80 percent and maize production doubled between 1979 and 1985. Smallholders, who were responsible for most of the maize production gain, benefited from an increase in the availability of credit, a rise in producer prices, better access to credit, and the introduction of new hybrid seeds (41).

New Approaches: Intercropping

Among the most promising new approaches are the many forms of intercropping, especially agroforestry, in which trees and shrubs are planted alongside crop plants. In these mixed systems, trees improve soil fertility by protecting against soil erosion. Nitrogen-fixing species of trees and crops add nitrogen directly to the soil, and the deep roots of trees pump up nutrients that have leached down to deep soil where they are inaccessible to crops. Fallen tree leaves add organic material to the soil and improve its moisture-holding capacity, making more water available to crops. Tree branches also provide fuelwood, allowing dung or crop residues to be returned to the soil. Table 6.5 shows the improvements that can result from intercropping.

Alleycropping is being developed by the International Institute for Tropical Agriculture in Nigeria, one of the members of the Consultative Group for International Agricultural Research. In alleycropping, rows of shrubs and crop plants are sowed alternately; the shrubs are vigorously pruned as mulch for the crops and they provide soil cover, thereby reducing erosion. Mulch reduces soil temperatures, helps fix nitrogen in the soil, and reduces weed growth. The mixed system not only increases the yield of the crop over what it would have been planted alone but also provides enough fuelwood from each hectare to meet the needs of four people (42). The International Fund for Agricultural Development is supporting alleycropping tests under different conditions in 18 sub-Saharan countries (43).

Intercropping schemes are but one of many new strategies that could provide African farmers with productive, sustainable agriculture. Others include improved strains of native crops (especially drought-resistant species), small-scale, no-cost irrigation schemes, soil or rock barriers to soil erosion, no-tillage planting systems, village fuelwood lots and other types of reforestation, and natural pest control.

Agricultural Research

Agricultural research centers have produced impressive results, including improved strains of cassava, sorghum, maize and rice. But they are also criticized for their being slow to develop new agricultural technology, an effect of weak government commitment and poor management. Most basic research will be conducted at international agriculture research centers (IARCs) and at universities and research centers in industrial countries. Close IARC cooperation with national agricultural research programs and local farmers is necessary for new technologies to be accepted (44).

Agricultural extension services, which are fragmented in most African countries, need to be consolidated into national systems that include regular training sessions for fieldworkers and links with research institutions (45).

Prospects

What must be done to prevent the suffering that may lie ahead for Africa? A recent study listed six key agricultural needs:
■ Reform agricultural economies to increase the role of the private sector, provide for the free marketing of agricultural products, allow prices to reflect supply and demand, promote private investment in production, agricultural processing, and farm input supplies, and encourage easier access to credit in rural areas;
■ Strengthen agricultural research management at the national level and boost international research efforts;
■ Develop and maintain rural infrastructure;
■ Develop environmental protection plans for each country to address soil erosion, deforestation, and watershed management issues;

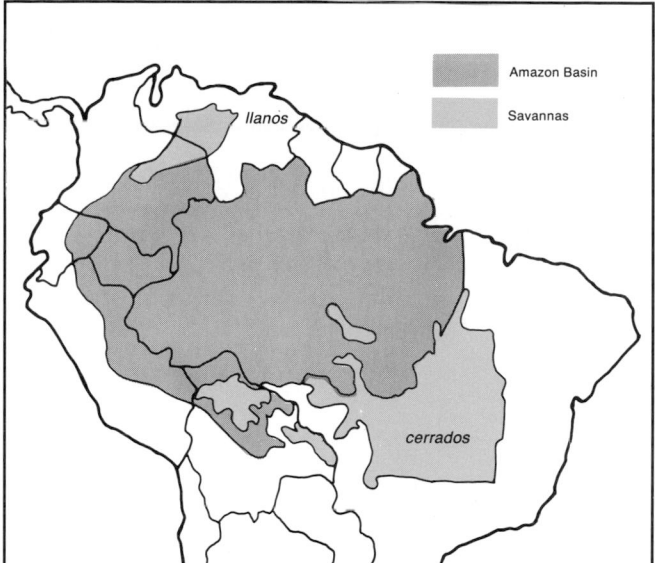

Figure 6.7 Amazon Basin and Savannas

Source: Adapted from T.T. Cochrane et al., Land in Tropical America, Vol. 1 (International Center for Tropical Agriculture, Cali, Colombia, 1985), p. 59.

■ Strengthen programs to assist women as farmers and traders and foster women's farm, credit, and marketing groups; and
■ Provide legal and administrative mechanisms to facilitate land tenure (46).

These macroeconomic reforms will require an increase in spending on agriculture. The Organization of African Unity has recommended 20–25 percent of government spending go to agriculture. Above all, a greatly expanded program of agricultural research, focused on the small farmer and dryland farming, closely linked to forestry, and building on traditional systems tailored to Africa's environment, is sorely needed (47).

SUSTAINABLE LAND USE FOR SOUTH AMERICA'S TROPICAL SAVANNAS

The settlement of the Amazon basin—a complex process affected by population, government policies, patterns of land ownership, and other factors—generally has not provided migrants with access to viable farmland. Soils there quickly lose their nutrients when they are cleared of forest.

As an alternative to farming the Amazon rainforest, intensified agriculture and ranching in the less fragile savannas adjacent to the Amazon are being studied. A shift could help reduce pressure for further deforestation of the Amazon.

Tropical Savannas

Tropical savannas are found in four of the six Amazon countries. They include the cerrados of Brazil, the llanos of Venezuela, and the llanos orientales of Colombia—a total of 140 million hectares.

Box 6.2 Annual Crops of the Savannas

Pasture grasses and crop plants with potential for increased yields under a crop-pasture system include the grass llanero, the legume Vichada, and cultivars of rice, cassava, soybeans, and sorghum.

In the Colombian llanos, combining cultivations of the recently released grass llanero (Andropon gayanas) and the legume Vichada (Centrosema acutifolum) for cattle fodder produced twice the liveweight gain per head of cattle and more than 15 times the weight gain per hectare that could be achieved on traditional, less-managed pastures (1).

UPLAND RICE

For more than 10 years, upland rice was often planted as a pioneer crop on some 4 million hectares in the Brazilian cerrados, 2.2 million hectares of it in moisture-favored areas. The improved tall varieties have a low yield potential (an average 1.1 metric tons per hectare) and poor grain quality, and they are susceptible to rice blast disease and lodging. The International Center for Tropical Agriculture in Cali, Colombia, has developed some promising cultivars for dry upland based on native upland plants from Africa and Asia. With their well-developed root systems, these cultivars are more drought tolerant, their yield potential is high, they are disease resistant, and the grain quality is excellent. Under low-input conditions in the llanos of Colombia, yields were 3.6 metric tons per hectare, and in the cerrados of Brazil, they exceeded 4 metric tons per hectare (2).

CASSAVA

Cassava, which is naturally adapted to acid, infertile soils, has great potential as both a food staple and cattle fodder. The national agricultural program in Colombia is expected to release four high-yielding hybrids in 1990 and is developing two others that are well adapted to the infertile soil of the llanos and can resist several diseases and insects in the area. The hybrids had average yields of more than 20 metric tons per hectare in unfavorable conditions (3).

SOYBEANS

Soybeans have been planted in the cerrados under high-input conditions since the late 1970s. Yields of varieties tolerant to acid soils are low (1.5 metric tons per hectare), and although higher yielding varieties yield an additional ton per hectare, their yields may fall after five or six years of production without good soil management practices. The Center for Agricultural Research for the Semi-Arid Tropics in Brazil has developed several lines adapted to these soils. Average yields are about 3 metric tons per hectare, with a potential for 4.5–5 metric tons per hectare with adequate fertilization (4).

SORGHUM

Sorghum is another potential component of savanna systems, but it is particularly sensitive to high levels of aluminum in the soil. The Oxisol and Ultisol soils of the llanos in Colombia have an average aluminum saturation of 88 percent. The new cultivars being developed by the International Sorghum and Millet Program perform well at aluminum levels of up to 60 percent, and they can efficiently extract water from the subsoil during the dry season. These sorghums have yielded 2 metric tons per hectare when lime was applied to reduce aluminum saturation to 60 percent, along with application of nitrogen, phosphorus, and potassium (5).

MAIZE

In collaboration with national programs in Latin America and Asia, the International Maize and Wheat Improvement Center has been breeding germplasm for different stress environments, including acid soils. Each year progeny of five maize populations with different grain colors and textures are evaluated under a range of aluminum and phosphorus concentrations, and the superior ones are selected and recombined. Thus far, 10 cultivars have been developed that perform up to 200 percent better than intolerant plants under medium to high levels of soil acidity. Given the importance of maize as a subsistence and cash crop in many parts of the developing world, these trials are especially encouraging (6).

References and Notes

1. Carlos E. Lascano and Jos M. Toledo, "Desarrollo y Potencial de Pasturas Mejoradas para America Tropical," paper presented at VI Encuentro Nacional de Zootecnia, Segunda Conferencia Nacional de Produccion y Utilizacion de Pastos y Forrajes Tropicales, Cali, Colombia, October 28-31, 1987, p. 91.
2. R.S. Zeigler, Leader, Rice Program, International Center for Tropical Agriculture, personal communication.
3. Clair H. Hershey, "Clones de yuca promisorios de CIAT/ICA con posibilidades de liberacion para Colombia," Instituto Colombiano Agropecuario, Bogota, Colombia.
4. Plineo de Souza, Brazilian Agricultural Research Corporation/Center for Agricultural Research for the Semi-Arid Tropics, Planaltina, Brazil (1989) (personal communication).
5. Carlos Ser and Ruben D. Estrada, "Potential Role of Grain Sorghum in the Agricultural Systems of Regions with Acid Soils in Tropical Latin America," in Sorghum for Acid Soils: Proceedings of a Workshop on Evaluating Sorghum for Tolerance to Al-Toxic Tropical Soils in Latin America, Cali, Colombia, 28 May to 2 June 1984 (International Sorghum Millet Program/International Crop Research Institute for the Semi-Arid Tropics/International Center for Tropical Agriculture, Cali, Colombia, 1987), pp. 152-159.
6. Shivaji Pandey, Head, International Maize and Wheat Improvement Center, International Center for Tropical Agriculture, Andean Region Maize Project, 1989 (personal communication).

(See Figure 6.7.) A savanna is a treeless plain, but the term refers to most tropical grasslands, even the cerrados of Brazil, in which trees of medium height form a closed canopy. Savannas do not compare with the ecological treasurehouse of the rainforest; yet, their flora and fauna include anteaters, armadillos, the large quara wolf, and the rhea, an ostrich-like bird. Brazil has several national parks in the cerrados. Brazilian conservationists think erosion-prone savanna areas, those that are hilly or have thin soil, should not be farmed, but they agree that most savannas can be farmed (48).

Since 1940, Brazil's cerrados, the largest of the savannas, have attracted increasingly more ranchers. By 1985, one third of Brazil's cattle herd was located on the cerrado of the states of Goias and Mato Grosso, up from 15 percent in 1940.

Ranches are generally left in native grasses, on which cattle graze. Because grass production is fairly low, ranches must be large to be economically viable; they average 500 hectares. In Brazil, most savanna farms raise soybeans for export. Because soybeans require costly harvesting machinery, most of the farms are fairly large.

Savanna Soils

Savanna soils have three characteristics that restrict farming: high acidity, low nutrient content, and a variable erosion potential, depending mostly on topogra-

Table 6.6 Major Net Cereal Importers and Exporters, 1987

Importers		Exporters	
Country	Million Metric Tons	Country	Million Metric Tons
U.S.S.R.	29	United States	83
Japan	27	Canada	28
China	16	France	26
Egypt	9	Australia	18
Korea	9	Argentina	9
Saudi Arabia	8	Thailand	6
Iran	6	United Kingdom	4
Italy	5	South Africa	2
Mexico	5	Denmark	1
Iraq	4	New Zealand	0.2

Source: Food and Agriculture Organization of the United Nations (FAO), *Trade Yearbook 1987*, Vol. 41 (FAO, Rome, 1988), pp. 114-116.

phy. Annual cropping systems, particularly in the cerrados of Brazil and the llanos of Venezuela, are profitable in the short term, but they eventually degrade the soil (49). As a result of this degradation, the medium- to long-term productivity of the savannas' annual cropping systems is expected to decline, even with a high level of chemical inputs.

Research on Crop-Pasture Associations

Several research programs are considering ways to develop the savannas on a low-input, sustainable basis. One such program is being conducted by CIAT in Cali, Colombia. CIAT is an international agricultural research institution working with national agricultural research and development systems on pasture and rice development.

CIAT research focuses on combining crop and pasture systems by alternating production of forage grasses with annual crops specially bred for suitability to the low-nutrient, acidic savanna soil. Because the land is always covered, there is less erosion.

To address the problem of high soil acidity, CIAT developed acid-tolerant rice, grass, and legume cultivars. To address the low level of soil nutrients, CIAT researchers selected cultivars known for their efficient use of nutrients (50). Crop varieties of rice, cassava, soybeans, and sorghum have been selected for their ability to produce under savanna conditions with limited chemicals and fertilizer. (See Box 6.2.)

Associations of agricultural crops and pasture grasses can be both ecologically and economically complementary, particularly when crops and grasses are planted in sequence. CIAT field tests show that, if properly managed, grazing systems based on mixtures of acid-tolerant grasses and legumes can ameliorate the chemical and biological deficiencies of savanna soils. The systems also reduce the need for chemical inputs because the leguminous grasses fix nitrogen in the soil; when plowed under, they become fertilizer for the annual crops. Subsequent crops benefit from the improved soils and require fewer chemical inputs, reducing production costs.

CIAT's savanna agriculture system is characterized by a low-intensity use of inputs. Land preparation, estimated at about $47 per hectare, does not require deep plowing or other intensive practices. The CIAT system assumes a six-year rotation, with the fields allowed to lie fallow the sixth year. Fertilizer use is about one third of normal use under intensive practices such as in irrigated plots. Expenditures for weed, insect, and pest control, estimated at $8–16 per hectare, are quite low.

CIAT estimates that the system will obtain rice yields of about 1.8 metric tons per hectare, 44 percent better than current yields of about 1.25 metric tons per hectare (51). Pastures used for beef production are projected to obtain yields of 200 kilograms per hectare, a marked advance from current yields of 12 kilograms per hectare (52).

Adoption of crop-pasture practices could provide a sustainable agricultural system for tropical savannas that might boost production substantially. The savanna area in South America suitable for crop-pasture technology is estimated at 76 million hectares—44 percent of the total savanna area (53). There are also similar agroecological zones in Africa and Asia with potential for crop-pasture associations.

The savannas are relatively accessible, requiring little additional infrastructure. Farmers in the Amazon rainforest, by comparison, must deal with a relatively undeveloped infrastructure, long distances to markets, and low-nutrient soils that require heavy doses of chemical fertilizers and pesticides (54).

THE EFFECTS OF CLIMATE CHANGE ON AGRICULTURE

Predictions by global climate modelers that additional atmospheric carbon dioxide will warm the planet are subject to considerable uncertainty because the models are in a relatively early stage of development. For this reason, the possible effects on agriculture should be viewed cautiously—as more or less educated guesswork—at this stage.

Not much is known about how different regions would be affected and how various crops would respond. Although much research still needs to focus on whether climate change will enhance or diminish a particular region's potential for food production, it is clear that climate change would have a major effect on the world food system.

The impacts on agriculture could be double-edged; by altering production in the main food-producing areas, climate change could weaken our ability to manage food crises, and by making growing conditions worse in food-deficit nations, it could increase the risk of famine.

Climate changes could alter the location of the main food-producing regions. For example, warming at high latitudes is likely to benefit agriculture where temperature now limits crops, whereas decreased rainfall in Mediterranean climates could affect agriculture adversely.

If the major food exporters have a much reduced production potential because of climate change, prices might go up and importers would be affected. Table 6.6 lists the major net cereal importers and exporters.

Regions with a large imbalance between population and agricultural potential may be particularly vulnerable to climate change. These countries occupy 22 percent of the global land area and include 11 percent of the population in 117 countries (55). These areas include the Andean region, the Maghreb in Northwest Africa, the Sahel, the Horn of Africa, the mountain regions of Southwest Asia, the Indian subcontinent, and parts of mainland and insular Southeast Asia.

Other changes would occur in regions where agriculture is already vulnerable and where the capability to absorb the potentially large-scale shock of climate change or to invest in mitigation measures is limited.

The assessments made here, unless otherwise noted, are based on the climate model of the Goddard Institute of Space Studies (GISS). (See Chapter 2, "Climate Change: A Global Concern," Box 2.2.)

Changing Climate Patterns

Global warming would change both temperature and rainfall patterns around the world. GISS and other climate models broadly predict that if greenhouse gases reach a level equivalent to the doubling of carbon dioxide (CO_2), temperature increases would be greater at higher latitudes than at lower ones. Warming may be slightly more pronounced in winter than in summer. Possible rainfall changes are more uncertain. In the higher midlatitudes (45–60°), precipitation could increase 5 percent in summer and as much as 15 percent in winter. In the lower midlatitudes (e.g., the semiarid Mediterranean climates), summer rainfall could be limited and winter rainfall could decrease 5–10 percent. In the low latitudes (0–30°), precipitation is expected to increase about 5–10 percent. The increase in precipitation may be large in equatorial regions but would diminish toward higher latitudes where weather patterns over arid and semiarid regions are likely to continue to operate (56) (57).

Assessing Climate Change and Agriculture

Late in 1989, GISS and the University of Birmingham (United Kingdom) began a three-year study of the impacts of climate change on global agricultural output and food trade. To date, only three comprehensive regional or national assessments of the consequences of climate change for agriculture have been completed. National assessments have begun in two other countries and the International Institute for Applied Systems Analysis has conducted several case studies for the United Nations Environment Programme (58) (59) (60) (61) (62). Several countries including the United Kingdom, Australia, and New Zealand have completed literature reviews (63) (64) (65). Other countries, including Japan and Finland, have well-developed plans for national impact assessments.

Probable Effects on Agriculture

A higher level of greenhouse gases would have two broad effects on agriculture; increased atmospheric CO_2 would enhance growth rates of certain types of crop plants, and climate changes would affect livestock, crops, pests, weeds, and soils.

The Fertilizing Effect of Increased CO_2

Increased CO_2 in the atmosphere can stimulate plant growth by increasing the rate of photosynthesis, though shortages of water and other nutrients may limit yields. Some crop plants respond more to CO_2 enhancement than others, usually depending on how they photosynthesize. Scientists classify some plants as either C_3 or C_4, depending on whether they have a 3-carbon or 4-carbon path for photosynthesis (66). C_3 crops (e.g., wheat, barley, rice, potatoes) respond quite vigorously to CO_2 enhancement, but C_4 crops (e.g., maize, sorghum, millet, sugar cane) do not. More research is needed before the size of yield increases in the field is known. Experiments in doubled CO_2 conditions have shown yield increases of about 35 percent for wheat and 70 percent for barley (67). For maize and sorghum, the increase is minimal (68). Much depends on how plants respond to enhanced CO_2 under suboptimal temperature and rainfall conditions.

On balance, increasing CO_2 could benefit agriculture, particularly because C_3 crops provide a relatively large proportion of total human diet worldwide. Moreover, because greater yield improvements have been achieved recently in C_3 than C_4 crops, there has been a significant shift to C_3 crops in the past two decades. In India, the area under C_4 crops dropped from 40 percent to 23 percent of the total during this period (69).

Increased CO_2 has some negative effects. The food quality of plants tends to decline; leaves become richer in carbon and poorer in nitrogen, and insect pests may need to consume more to reach their required nitrogen nutrient levels (70). In addition, if plants grow more quickly, they may require more fertilizer. Furthermore, if climate change increases rainfall in areas that do not need additional moisture, runoff may erode more soil.

Regional Effects

Some predicted regional changes are described below:

■ In maritime climates above 45° north latitude, where temperature rather than rainfall is generally the constraint on agricultural potential, increased temperatures would likely increase productive potential. For example, mean annual temperatures in Finland are projected to increase more than 4° C under the GISS scenario, lengthening the growing season for spring wheat by two-three weeks (71). At similar latitudes in Iceland, on the northern edge of arable agriculture in the U.S.S.R. (near Leningrad), and in northern Japan, broadly similar lengthening of the growing season is estimated (72) (73) (74).

■ In cool temperate and cold regions, yields of most crops can be expected to increase with rising temperatures, except where moisture is a limiting factor. For example, in northern Japan, rice yields are estimated to increase by 3.8 percent under the 3.6° C increase in growing-season temperatures projected by the GISS model (75).

■ In the major cereal-exporting regions of the world, the lower midlatitudes (30–50°), much would depend on changes in moisture. In the United States, there are indications that yields of wheat, corn, and soybeans would decrease under a warmer climate. In some places, these decreases may be compensated for by the direct fertilizing effect of more CO_2, but overall, the productive potential of staple food grains could decrease (76) (77). In Australia, yields of winter wheat may be reduced if winter rainfall decreased, particularly in marginal growing areas such as southwestern Australia (78).

■ Where precipitation does not increase much or where moisture is now barely sufficient for agriculture, warming would shorten the growing season. It would do so because temperature increases elevate evapotranspiration rates at 2–3 percent per degree of warming. Thus plants would lose more moisture in a warmer climate. Consequently, in Saskatchewan, for example, where mean May-August temperatures would increase about 3.5° C under the GISS scenario, the overall growing season would be lengthened four–nine weeks (from onset of growth to the first fall freeze) but would be effectively split into two parts by a moisture shortage in July and August (79).

■ In tropical areas where rainfall decreases or increases only slightly, the growing season would shorten. How these rainfall changes would be distributed is not at all clear, but there is some evidence that total rainfall in the Sahel, India, and Southeast Asia could increase. The general circulation models agree somewhat that summer monsoon rainfall would increase in Central and South China, Southeast Asia, and India but that winter rainfall might decrease. Summer rainfall could also be more intense, so that flooding and erosion would increase, and despite the increased rainfall, less moisture would be available for agriculture (80) (81).

Geographic Changes

Climate change is likely to shift the geographic distribution of crops. Areas currently best suited to given crops may no longer meet the needs of those crops. At high latitudes, these shifts may change the area under cultivation. In northern Japan, for example, the area suitable for rice cultivation would be doubled (82). There may also be substantial opportunity for northward expansion of the arable area in the U.S.S.R., particularly western Siberia, but in North America, new arable areas would probably be limited to certain valley regions in Alaska because soils and terrain limit potential (83).

A wide range of adjustments may be required to match new land uses to new climates. For example, Iceland's climate is expected to become broadly similar to that of northeast Scotland today (84). Computer model experiments confirm that Iceland's hay yields may increase 50–66 percent, to approximate those in Scotland now (85). Would future Icelandic farmers switch from raising sheep (currently the major livestock) to rearing cattle, which are raised in northeast Scotland today? Nothing is likely to be so simple, but the extent of land use changes would be substantial.

Effects on Agricultural Systems

As a result of the effects of climate change on crops and livestock, extensive changes can be expected in farm output and profitability and in national food production (86).

Yield changes would affect the profitability of farming. Much will depend, of course, on changes in input costs, particularly fertilizers. Assuming no change in input costs, yield decreases would reduce farm incomes. In Saskatchewan, decreased wheat yields could lower average farm incomes an estimated 7 percent (87).

The aggregate effect on regional and national food production is not easy to evaluate because yields would be affected not only by climate changes but also by crop and soil types, input levels, and forms of management just as they are today. Only model-based estimates for single crops are available because of these variables. Under the GISS scenario for Saskatchewan that includes the effects of increased temperature and precipitation, total wheat production in the province would fall 18 percent (88). Total rice production on the northern Japanese island of Hokkaido would increase about 5 percent (89). Without any policy change, the national rice stock in Japan could double as a result of increased yields under the GISS climate (90).

Outside the northern countries that would benefit from warmer growing seasons, substantial adjustments may be needed to maintain production, largely because less water would be available for agriculture. Projections indicate that the changed climate would not threaten the U.S. food supply, but grain exports may decline significantly (91). Australia could be in a similar situation, although estimates are less certain (92).

Average agricultural production costs are projected to increase 10–20 percent worldwide, although this estimate has been disputed (93).

Adapting Agriculture to Climate Changes

Despite the uncertainty of new climate patterns over the next decades, farmers will undoubtedly continue to adapt to their physical and economic environments. Their most significant adaptation measure is likely to be the use of different cultivars and crops, particularly in high midlatitude countries. There, farmers would shift to crops with higher thermal requirements to take advantage of longer growing seasons (if water is available locally).

In continental midlatitude regions where summer drought may become more of a problem, a switch from spring- to winter-sown crops may be necessary (94).

Numerous management changes are likely to be necessary, such as matching farm operations to periods of rainfall. Changes in the amounts and use of water and chemicals would probably be needed with small changes in temperature and rainfall.

Farmers would make these adjustments as they perceive the need if they have the foresight, the know-how, and the money. But this process would require more accurate forecasting plus government help when technologies and capital are inadequate.

The scale of agricultural changes that could be caused by climate change leaves little room for complacency, especially where the capability to absorb the shock of climate change is limited and the cost of adapting may be great (e.g., areas of marginal agriculture in the semiarid tropics).

RECENT DEVELOPMENTS

LOW-INPUT FARMING WINNING NEW CONVERTS

Some large-scale farmers in the United States are cutting back on chemicals and adopting alternative farming practices that are both economically and environmentally beneficial. Though still only a small minority of all the big farmers, they are giving new legitimacy to the alternative agriculture movement.

These practices are called low-input agriculture, sustainable agriculture, and alternative agriculture, and their meanings differ. Most often they refer to incorporation of nutrient cycles, nitrogen fixation, and other natural processes in production; reduction or elimination of chemical fertilizers, pesticides, herbicides, insecticides, and fungicides, following integrated pest management practices; improving the match between cropping patterns and the productive potential and physical limitations of farm lands; and managing farms to maintain profitability while conserving soil, water, energy, and biological resources (95).

The new practices are a substantial departure from conventional farming in the United States. Since the 1950s, it has relied on heavy doses of chemical fertilizers and pesticides and frequent plowing with heavy machinery. The approach has improved productivity but has become increasingly costly and has raised many concerns about soil erosion, food safety, and surface and groundwater pollution (96).

Though alternative agriculture is practiced by only 5 percent of the farmers, some large-scale producers

are taking an interest (97). In the Salinas Valley in California, for example, companies such as the Superior Farming Company, one of the biggest table grape and fruit growers in the world, have begun to grow table grapes using alternative techniques. Superior uses hoes and mechanical cultivators to fight weeds, provides habitat for a wasp that feeds on grape leaf hoppers, uses natural bacterial and insecticidal compounds to fight other insects, and has built a prototype vacuum to test the practicality of sucking insects off the vines. The company expects yields of about 15,000 pounds per acre, similar to those when conventional techniques are used (98).

Steven Pavich and Sons, which primarily grows grapes on 580 hectares in south-central California and southern Arizona, uses no herbicides; it uses composted steer manure for soil fertility and for the most part is able to avoid the use of insecticides. The farm yields about 653 boxes of grapes per acre; the normal yield is 522 boxes (99).

The Pavich operation was one of 11 case studies included in a 1989 National Research Council (NRC) report. The other cases included two crop and livestock farms in Ohio; a diversified crop and livestock farm in Virginia; tree fruits, walnuts, and vegetables in California; fresh vegetable production in Florida; livestock farming in Colorado; and rice production in California (100).

The report found that innovative farmers have developed alternative farming methods suited to the specific needs, resource bases, and economic conditions of their farms. Those using alternative farming systems generally derived "significant sustained economic and environmental benefits" (101).

NEW CEREAL GRAIN SHOWS PROMISE

Agronomists have perfected a cereal grain that will flourish in various conditions, including the potentially hotter temperatures and more difficult growing conditions of the next century.

Another NRC report predicts that this grain, called triticale, "seems likely to play a role in sparing millions of the poor from the ravages of malnutrition in Africa, Asia, and Latin America" (102).

Triticale is produced by crossbreeding wheat (genus Triticum) and rye (genus Secale). After the pollen-carrying parts of the wheat flower is removed to prevent self-fertilization, pollen is transferred from a rye flower (the male parent) to the stigmata of a wheat flower (the female parent). The result is a grain that is generally larger than wheat and plumper than rye (103).

Triticale was promoted heavily in the 1960s, but there were numerous problems. It did not set seed well, the grains were often small and shriveled, yields were often disappointing, the stalks tended to fall over, and the plants were vulnerable to disease (104).

An accidental breakthrough occurred in 1967 at the International Maize and Wheat Improvement Center in Mexico, when windblown pollen from dwarf wheat plants accidentally fertilized a triticale plant. The result was Armadillo, a variety with improved fertility, a higher yield, stronger stalks, earlier maturation, and plumper grains. Further improvements have been made since then [105].

Triticale combines some of the best properties of rye and wheat. Rye, for example, does well in poor soils, in cold climates, at higher elevations, and in other unfavorable conditions, but it does not pollenate well, is susceptible to a fungal disease called ergot, and is far less popular than wheat. Wheat, which provides 20 percent of the calories and 45 percent of the protein in human diets, is not a good crop for marginal lands with highly acidic or alkaline soils or for regions subject to drought or climatic extremes [106].

Triticale's ruggedness helps it outperform wheat in areas with poor soils or extreme climates. Under marginal conditions, triticale yields usually exceed wheat by 20–30 percent. The plant does well in soils that are high in boron, acid, or salt or deficient in manganese. In dry and sandy soils or drought conditions, triticale's biomass production falls, but its relative advantage against wheat increases [107].

Some problems remain. Triticale dough is too sticky to roll out of the high-speed mixers used by industrial bakeries, though blending triticale with wheat flour works. The Armadillo variety, though essential to triticale's development, has a fairly narrow genetic base, which could threaten the plant's resilience, so additional genetic diversity is needed. Triticale is subject to some diseases, and the plant's resistance could break down if the crop is grown on larger areas [108].

Notwithstanding these problems, triticale appears to show great promise. It is already being grown on more than 1.5 million hectares in 32 nations, with substantial production in Poland, the U.S.S.R., France, and Australia [109]. It is easily used to make tortillas, chapatis, and other unleavened flat breads. Recently, CIMMYT researchers have developed strains of triticale for use in leavened breads comparable to breads using wheat flours [110].

The section on global food trends, A Mixed Picture, was written by Eric Rodenburg, World Resources 1990-91 Research Director and Robert Livernash, a Washington, D.C. environment writer and consultant. The focus on Sub-Saharan Africa was written by Livernash, based largely on work by Jessica Tuchman Mathews, World Resources Institute Vice President. Sustainable Land Use for South America's Tropical Savannas was authored by Filimon Torres, Deputy Director General, International Center for Tropical Agriculture, Cali, Colombia. The Effects of Climate Change on Food and Agriculture was authored by Martin Parry, Professor of Environmental Management, Atmospheric Impacts Research Group, The University of Birmingham, U.K.

References and Notes

1. Food and Agriculture Organization of the United Nations (FAO), *Food Outlook*, No. 12 (FAO, Rome, December 1989), p. 2.
2. *Ibid.*, pp. 1-2.
3. The stocks are the cereals in excess of consumption at the end of the 1988 or the beginning of the 1989 crop year. Crop years differ from calendar years and usually extend from one main harvest to the next. No attempt was made in these statistics to adjust all countries' crop years to a common calendar base.
4. U.S. Department of Agriculture, Economic Research Service, PS&D View Database on cereal production, created by Karl Gudmunds and Alan Webb, September 1989.
5. Ed Overton, Chief, Agricultural and Trade Indicators Branch, Economic Research Service, U.S. Department of Agriculture, Washington, D.C., 1990 (personal communication).
6. U.S. Department of Agriculture, Economic Research Service, PS&D View Database, September 1989.
7. Marian Mitchell, Analyst, Famine Early Warning System, Africa Bureau, U.S. Agency for International Development, Washington, D.C., 1989 (personal communication).
8. *Op. cit.* 1, pp. 1-2.
9. Nikos Alexandratos, *World Agriculture: Toward 2000* (Belhaven Press, London, 1988), pp. 128, 319.
10. Essam El-Hinnawi and Manzur H. Hashmi, *The State of the Environment* (United Nations Environment Programme, Butterworths, London, 1987), pp. 35-41.
11. Robert S. Chen, Assistant Professor, Alan Shawn Feinstein World Hunger Program, Brown University, Providence, Rhode Island, 1990 (personal communication).
12. Food and Agriculture Organization of the United Nations (FAO), *The Fifth World Food Survey* (FAO, Rome, 1987), pp. 51, 68.
13. *Op. cit.* 9, p. 27.
14. Nikos Alexandratos, 1990 (personal communication).
15. The World Bank, *Sub-Saharan Africa: From Crisis to Sustainable Growth* (The World Bank, Washington, D.C., 1989), pp. 72-73.
16. Office of Technical Resources, Africa Bureau, United States Agency for International Development (AID), *FEWS Monthly Country Report: Ethiopia, Famine Early Warning System* (U.S. AID, Washington, D.C. June 1986—June 1989).
17. Ruth Leger Sivard, *World Military and Social Expenditures, 1989* (World Priorities Inc., Washington, D.C., 1989), p. 22.
18. *Op. cit.* 15, p. 89.
19. *Op. cit.* 15, p. 89.
20. *Op. cit.* 15, Table 36, p. 279.
21. Paul Harrison for the International Institute for Environment and Development, *The Greening of Africa* (Penguin Books, Middlesex, U.K., 1987), p. 35.
22. *Ibid.*, pp. 36, 37.
23. World Resources Institute and International Institute for Environment and Development, *World Resources 1986* (Basic Books, New York, 1986), p. 126.
24. *Op. cit.* 15, p. 97.
25. *Op. cit.* 23, p. 68.
26. *Op. cit.* 15, p. 97.
27. *Op. cit.* 15, p. 91.
28. Food and Agriculture Organization of the United Nations (FAO), *African Agriculture: The Next Twenty-Five Years*, (FAO, Rome, 1986), p. 2.
29. *Op. cit.* 21, pp. 46-47.
30. *Op. cit.* 28, p. 5.
31. *Op. cit.* 21, pp. 46-47.
32. *Op. cit.* 21, p. 64.
33. *Op. cit.* 15, p. 103.
34. Dounia Loudiyi, Bill Nagle, and Waafus Ofosu-Amaah, *The African Women's Assembly: Women and Sustainable Development* (WorldWide, Washington, D.C., 1989), p. 3.
35. *Op. cit.* 15, p. 103.
36. *Op. cit.* 15, pp. 103-104.
37. The figure 110 percent is used in the World Bank study to allow for differences in purchasing power within the population and to account for the fact that the wealthy consume more calories than is necessary for good health.
38. *Op. cit.* 15, pp. 73-74.
39. *Op. cit.* 15, p. 105.
40. *Op. cit.* 15, p. 105.
41. *Op. cit.* 15, p. 106.
42. Robert Winterbottom and Peter Hazelwood, "Agroforestry and Sustainable Development: Making the Connection," *Ambio*, Vol. 16, No. 2-3 (1989), p. 102.
43. A. Kasseba, Coordinator, Technical Unit, International Fund for Cultural Development, Rome, 1990 (personal communication).

44. *Op. cit.* 15, p. 99.
45. *Op. cit.* 15, p. 100.
46. *Op. cit.* 15, pp. 106-107.
47. Idriss Jazairy, "How to Make Africa Self-Sufficient in Food," in *Development: Journal of the Society for International Development*, Vol. 2, No. 3 (1987), pp. 53-54.
48. Paulo Neguero-Neto, former Secretary of the Environment, Brazil, 1989 (personal communication).
49. L. Seguy *et al.*, *Perspectiva de Fixação da agricultura na Região Centro-norte do Mato Grosso (EMPA/EMBRAPA/CNPAF /CIRAD/IRAT*, Cuiaba, Mato Grosso, Brazil, 1988), pp. 10-11.
50. Filimon Torres, Deputy Director General, International Center for Tropical Agriculture, Cali, Colombia, 1989 (personal communication).
51. *Ibid.*
52. Carlos Seré, Economist, International Center for Tropical Agriculture, Cali, Colombia, 1990 (personal communication).
53. T.T. Cochrane *et al.*, *Land in Tropical America*, Vol. 1, International Center for Tropical Agriculture, Cali, Colombia, 1985), p. 54.
54. Robert J.A. Goodland, "Environmental Ranking of Amazonian Development Projects in Brazil," *Environmental Conservation*, No. 7 (1980), pp. 11, 17, 19.
55. Food and Agriculture Organization of the United Nations, *Food, Land and People* (FAO, Rome, 1985), p. 19.
56. Jill Jager, "Developing Policies for Responding to Climatic Change," World Climate Programme Impact Study (World Meteorological Organization, Geneva, 1988).
57. For a more detailed look at the climate change data generated by these models, refer to the article, Michael E. Schlesinger and John F.B. Mitchell, "Model Projections of the Equilibrium Climatic Response to Increased Carbon Dioxide," in *Projecting the Climatic Effects of Increasing Carbon Dioxide*, Michael G. MacCracken and Frederick M. Luther, eds. (U.S. Department of Energy (DOE), Washington, D.C., 1985).
58. Government of Canada, Environment Canada, *Canadian Climate Impacts Program* (Environment Canada, Downsview, Ontario, 1987).
59. U.S. Environmental Protection Agency (EPA), *Regional Study*, Vol. 1 of *The Potential Effects of Global Climate Change on the United States*, Draft Report to Congress, Washington, D.C., 1988).
60. U.S. Environmental Protection Agency (EPA), *National Study*, Vol. 2 of the *Potential Effects of Global Climate Change on the United States* (EPA, Washington, D.C., 1988).
61. Martin Parry, Timothy Carter, and Nicolaas Konijn, eds., Vol. 1 of *The Impact of Climatic Variations on Agriculture: Assessments in Cool Temperate and Cold Regions* (Kluwer Academic Publishers, Dordrecht, The Netherlands, 1988).
62. Martin Parry, Timothy Carter, and Nicholas Konijn, eds., *Assessment in Semi-Arid Regions*, Vol 2 of *The Impact of Climatic Variations on Agriculture* (Kluwer Academic Publishers, Dordrecht, the Netherlands, 1988).
63. U.K. Department of the Environment (DOE), *Possible Impacts of Climate Change on the Natural Environment in the United Kingdom* (DOE, London, 1988).
64. Graham I. Pearman, ed., *Greenhouse: Planning for Climatic Change* (E.J. Brill, Leiden, The Netherlands, 1988).
65. M.J. Salinger *et al.*, *Carbon Dioxide and Climate Change, Impacts on Agriculture* (New Zealand Meteorological Service, Wellington, New Zealand, 1989).
66. R.H.M. Langer and G.D. Hill, *Agricultural Plants* (Cambridge University Press, Cambridge, U.K., 1982), pp. 300-302.
67. Jennifer D. Cure, "Carbon Dioxide Doubling Responses: A Crop Survey" in *Direct Effects of Increasing Carbon Dioxide on Vegetation*, B.R. Strain and J.D. Cure, eds. (U.S. Department of Energy, Washington, D.C., 1985), pp. 103-105.
68. R.A. Warrick, R.M. Gifford, and M.L. Parry, "CO_2, Climate Change and Agriculture: Assessing the Response of Food Crops to the Direct Effects of Increased CO_2 and Climate Change," in Bert Bolin *et al.*, *The Greenhouse Effect, Climate Change and Ecosystems* (John Wiley & Sons, Chichester, U.K., 1986), pp. 397, 405.
69. Research Group, University of Birmingham, U.K., based on data from the Food and Agriculture Organization of the United Nations, Rome, 1989. Calculated by Dr. Martin Parry, Atmospheric Impacts.
70. W.C. Oechel and Boyd Strain, "Native Species Responses to Increased Carbon Dioxide Concentrations," in *Direct Effects of Increasing Carbon Dioxide on Vegetation*, B.R. Strain and J.D. Cure, eds. (U.S. Department of Energy, Washington, D.C., 1985), pp. 103-105.
71. Lauri Kettunen *et al.*, "The Effects of Climatic Variations on Agriculture in Finland," in *Assessments in Cool Temperate and Cold Regions*, Vol. 1 of *The Impact of Climatic Variations on Agriculture*, M.L. Parry, T.R. Carter, and N.T.Konijn, eds. (Kluwer Academic Publishers, Dordrecht, The Netherlands, 1988), pp. 520, 540-541.
72. Martin Parry and Timothy Carter, "The Assessment of Effects of Climatic Variations on Agriculture: Aims, Methods and Summary of Results," in *Assessments in Cool Temperate and Cold Regions*, Vol. 1 of *The Impact of Climatic Variations on Agriculture*, M.L. Parry, T.R. Carter, and N.T. Konijn, eds. (Kluwer Academic Publishers, Dordrecht, The Netherlands, 1988), p. 54.
73. Sergei E. Pitovranov *et al.*, "The Effects of Climatic Variations on Agriculture in the Subarctic Zone of the U.S.S.R.," in *Assessments in Cool Temperate and Cold Regions*, Vol. 1 of *The Impact of Climatic Variations on Agriculture*, M.L. Parry, T.R. Carter, and N.T. Konijn, eds. (Kluwer Academic Publishers, Dordrecht, The Netherlands, 1988).
74. Masatoshi Yoshino *et al.*, "The Effect of Climatic Variations on Agriculture in Japan," in *Assessments in Cool Temperate and Cold Regions*, Vol. 1 of *The Impact of Climatic Variations on Agriculture*, M.L. Parry, T.R. Carter, and N.T. Konijn, eds. (Kluwer Academic Publishers, Dordrecht, The Netherlands, 1988), pp. 816, 824.
75. *Ibid.*
76. *Op. cit.* 59.
77. *Op. cit.* 60, p. 10-11.
78. A. Barrie Pittock, Climate Impact Group, Commonwealth Scientific and Industrial Research Group (CGIRO), Victoria, Australia, 1990 (personal communication).
79. Dan Williams *et al.*, "Estimating Effects of Climatic Change on Agriculture in Saskatchewan, Canada," in *Assessments in Cool Temperate and Cold Regions*, Vol. 1 of *The Impact of Climatic Variations on Agriculture*,
   M.L. Parry, T.R. Carter, and N.T. Konijn, eds. (Kluwer Academic Publishers, Dordrecht, The Netherlands, 1988), pp. 248, 280, 309.
80. Stephen Schneider and Norman Rosenberg, "The Greenhouse Effect: Its Causes, Possible Impacts and Associated Uncertainties," in *Greenhouse Warming: Abatement and Adaption*, Norman J. Rosenberg *et al.*, eds. (Resources for the Future, Washington, D.C., 1989), pp. 10-11.
81. Michael E. Schlesinger and John F.B. Mitchell, "Model Projections of the Equilibrium Climatic Response to Increased Carbon Dioxide," in *Projecting the Climatic Effects of Increasing Carbon Dioxide*, M.C. MacCracken and F.M. Luther, eds. (U.S. Department of Energy, Washington, D.C., 1985), pp. 98, 118.
82. *Op. cit.* 74, p. 807.
83. *Op. cit.* 56.
84. *Op. cit.* 72, pp. 38-39.
85. Pal Bergthorsson *et al.*, "The Effect of Climatic Variations on Agriculture in Iceland," in *Assessments in Cool Temperate and Cold Regions*, Vol. 1 of *The Impact of Climatic Variations on Agriculture*, M.L. Parry, T.R. Carter, and N.T. Konijn, eds. (Kluwer Academic Publishers, Dordrecht, The Netherlands, 1988), p. 496.
86. *Op. cit.* 72, pp. 69-71.
87. *Op. cit.* 72, p. 85.
88. *Op. cit.* 72, p. 85.
89. *Op. cit.* 74, p. 805.
90. *Op. cit.* 74, p. 847.
91. U.S. Environmental Protection Agency, *The Potential Effects of Global Climate Change on the United States*, Executive Summary, draft report (Office of Policy, Planning and Evaluation, United States Environmental Protection Agency, Washington, D.C. 1988), p. 10-11.
92. *Op. cit.* 78.
93. Thomas Schelling, "Climate Change: Implications for Welfare and Policy," in National Research Council, *Changing Climate: Report of the Carbon Dioxide Assessment Committee* (National Academy Press, Washington, D.C., 1983), p. 475.
94. *Op. cit.* 79, pp. 272, 278.
95. National Research Council, *Alternative Agriculture* (National Academy Press, Washington, D.C., 1989), p. 4.
96. Arthur S. Brisbane, "Farmers Learning Lesson of the Till: Profit Can Grow as Chemicals Are Cut," *The Washington Post* (September 30, 1989), p. A3.
97. Kenneth R. Sheets, "Nature vs. Nurture on the Farm: A New Study Challenges the Efficacy of Fertilizers and Pesticides," *U.S. News and World Report* (September 18, 1989) p. 54.
98. Keith Schneider, "Big Farm Companies Try Hand at Organic Methods," *New York Times* (May 28, 1989) p. 1.
99. *Op. cit.* 95, pp. 351, 361.
100. *Op. cit.* 95.
101. *Op. cit.* 95, pp. 5-6.
102. National Research Council, *Triticale: A Promising Addition to the World's Cereal Grains* (National Academy Press, Washington, D.C., 1989), p. 28.
103. *Ibid.*, p. 5.
104. *Ibid.*, pp. 10-11.
105. *Ibid.*, pp. 12-13.
106. *Ibid.*, pp. 6-7.
107. *Ibid.*, pp. 28-29.
108. *Ibid.*, pp. 30-33.
109. *Ibid.*, pp. 14-15.
110. *Ibid.*, pp. 16.

7. Forests and Rangelands

Tropical deforestation has become a popular environmental issue over the past five years, but the extent of this deforestation remains unmeasured. Most publications still use national deforestation rates based on late 1970s data published by the Food and Agriculture Organization of the United Nations (FAO). However, a literature survey of more recent studies in eight countries reveals a startling increase in deforestation in most of them. In Brazil and India, internal debate over the figures leaves the course of action in doubt. Until a reasonable assessment of forest conditions and the rate of depletion is made, the best base for formulating forest policies is lacking. However, because evidence in these mid-1980s studies points to much higher rates of depletion than were measured in the late 1970s, policies affecting forests should reflect these changes.

In studying the impacts of climate change, this report examines how a warmer world could affect the location, health, and productivity of forests and rangelands. According to studies and theories based on computer projections of temperature and rainfall patterns, rangelands would expand and forests contract in area. Although local climates cannot be predicted, subtropical, temperate, and boreal tree species would probably shift northward. Many of the world's major grasslands would become drier, favoring the growth of forage plants that are tolerant of

fire and water stress, but these plants may not be the most nutritious for livestock grazing.

Each *World Resources* rangelands section surveys range conditions in a particular area. Previous volumes covered selected countries in Africa, West Asia, and Asia. Latin America is the regional focus this year. Most of this immense and diversified area—from the shrubland of northern Mexico to the Patagonian rangelands on the tip of Argentina to short-lived pastures newly carved from the Amazon rainforest—suffers some form of degradation.

CONDITIONS AND TRENDS

FOCUS ON NEW ESTIMATES OF TROPICAL DEFORESTATION

Recent studies covering several key countries suggest that deforestation in the tropics may be much worse than was previously thought. Until recently, the most authoritative estimate of annual deforestation in the tropics was 11.4 million hectares, based on a 1980 FAO assessment of tropical forestry research, literature, and surveys (1) (2). Several recent studies show that deforestation is much higher in Brazil, Costa Rica, India, Myanmar (formerly Burma), the Philippines and Viet Nam. Forest clearing also in-

Table 7.1 Deforestation Estimates for Closed Tropical Forests,[a] Selected Countries

(thousand hectares and annual deforestation rate)

Country	FAO Estimates 1981-85[1b]	Annual Rate of Loss (percent)	Recent Estimates	Annual Rate of Loss (percent)	Period of Recent Estimates
Brazil	1,480	0.4	8,000[2c]	2.2	1987
Cameroon[d]	80	0.4	100[3]	0.6	1976-86
Costa Rica	65	4.0	124[4]	7.6	1977-83
India[d]	147	0.3	1,500[5]	4.1	1975-82
Indonesia	600	0.5	900[6]	0.8	1979-84
Myanmar	105	0.3	677[7]	2.1	1975-81
Philippines	92	1.0	143[8]	1.5	1981-88
Thailand[e]	379	2.4	397[9]	2.5	1978-85
Viet Nam	65	0.7	173[10]	2.0	1976-81

Sources:
1. Food and Agriculture Organization of the United Nations (FAO), Forest Resources Division, *An Interim Report on the State of the Forest Resources in the Developing Countries* (FAO, Rome, 1988).
2. Alberto Waingort Setzer, *et al.*, "Relatório de atividades do projeto IBDF-INE 'SEQE'—ano 1987 National Space Research Institute of Brazil (INPE) São José dos Campos, São Paulo, Brazil, 1988.
3. Joint Interagency Planning and Review Mission (JIM) for the Forestry Sector, *Cameroon Tropical Forestry Action Plan* (JIM, Rome, 1988).
4. Steven A. Sader and Armond T. Joyce, "Deforestation Rates and Trends in Costa Rica, 1940 to 1983," *Biotropica*, Vol. 20, No. 1 (1988), p. 14.
5. B.B. Vohra, "Confusion on the Forestry Front," Advisory Board on Energy, New Delhi, 1987, based on reconciliation of figures in Natural Remote Sensing Agency, *Mapping of Forest Cover in India from Satellite Imagery 1972-75 and 1980-82* (Government of India, Hyderabad, 1983).
6. The World Bank, Asia Regional Office, *Indonesia: Forests, Land and Water: Issues in Sustainable Development*, August 1988, p. 2.
7. U.S. Kyaw, "National Report: Burma," in *Proceedings of Ad Hoc FAO/ECE/FINNIDA Meeting of Experts on Forest Resource Assessment, Kotka, Finland, 26-30 October* 1987 (Finnish International Development Agency, Helsinki, 1987).
8. Philippines Forest Management Bureau, Department of Environment and Natural Resources, "1987 Philippine Forestry Statistics" (Philippines, 1988).
9. Royal Forestry Department of Thailand, Planning Division, *Forestry Statistics of Thailand, 1986*, Center for Agricultural Statistics, Office of Agricultural Economics, Ministry of Agriculture and Cooperatives (Bangkok, Thailand) Table 2, p. 8.
10. Vo Quy, "Vietnam's Ecological Situation Today," *ESCAP Environment News*, Vol. 6, No. 4 (1988), pp. 4-5.

Notes:
a. Closed forests are forests in which trees cover a high proportion of the ground and in which grass does not form a continuous layer on the forest floor. Open forests are forests in which trees are interspersed with grazing lands. b. Unless otherwise noted, annual deforestation rate is calculated from FAO 1981 total stock estimates. c. For Legal Amazon only. Brazil also has a small amount of closed coastal forest remaining. d. Annual deforestation rate is calculated from stock numbers found in sources 3 and 5, above. e. Represents total forests, open and closed.

Figure 7.1 Percentage of Closed Forest Cleared Annually in Selected Tropical Countries, 1980s

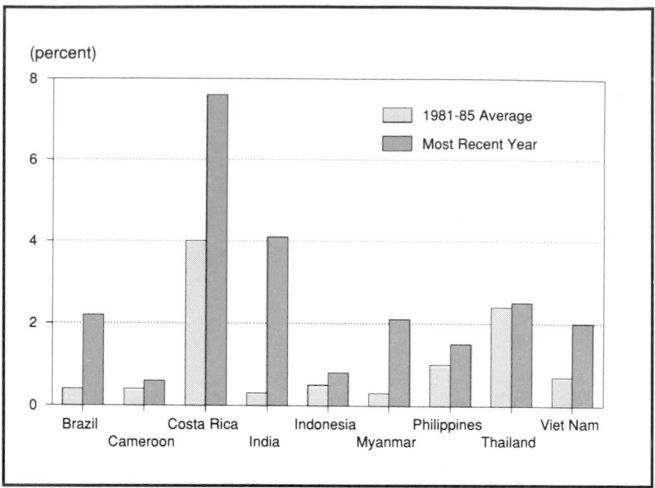

Source: Table 7.1.
Note: This figure shows the annual percentage of closed forest lost as calculated by the Food and Agriculture Organization of the United Nations (1981–85) and by more recent studies.

creased sharply in Cameroon, Indonesia, and Thailand. (See Table 7.1 and Figures 7.1 and 7.2.) The new studies examine loss of closed forest, defined by FAO as "land where trees cover a high proportion of the ground where grass does not form a continuous layer on the forest floor." They do not examine open forests, in which trees are widely spaced. If these new studies are accurate, the world is losing up to 20.4 million hectares of tropical forest annually—79 percent over FAO's 1980 estimate. The additional amount represents an annual forest loss the size of Panama.

Deforestation in Brazil

Brazil has the world's largest remaining tropical forest, and by far the largest area of annual deforestation. Forest outside the Amazon—mostly open forest—is being lost at the rate of 1.05 million hect-

ares per year. The Amazon forest covers 3.37 million square kilometers (337 million hectares) in Brazil. Brazil's Legal Amazon—a larger area of 5 million square kilometers that encompasses grasslands, wetlands, shrublands, lakes, and forests—is a political boundary encompassing six states and territories and parts of three others (3). It is the basis for several Brazilian deforestation studies. (See Figure 7.3.)

Two recent studies of deforestation in the Amazon have produced widely different results. The most up-to-date studies were conducted by Alberto Setzer of the National Space Research Institute of Brazil (INPE). In 1987, Setzer's team used satellite imagery to determine that 8 million hectares of virgin forest in the Legal Amazon were cleared that year (4). But 1987 may have been an anomalously high year for deforestation in the Brazilian Amazon; it was the last year tax credits were available to landholders who cleared their Amazon holdings, and many large landowners may have rushed to gain the advantage while they could. At the same time, Brazil's legislature was discussing taking "unimproved" land as part of a land-reform effort. Many who owned large forested tracts for mineral rights or future development burned and cleared as much of their land as possible in order to retain ownership (5).

In 1988 and 1989, tax credits were suspended and later canceled, and, in the face of international pressure, Brazil began a campaign to slow the burning. Using 5 helicopters and 60 trucks, forestry agents from Brazil's Institute of Environment and Renewable Natural Resources sought out illegal burning and issued millions of dollars in fines. Wetter weather also discouraged burning (6). Follow-up studies by Setzer in 1988 and 1989 showed a decline in the area cleared each year, a 40 percent drop to 4.8

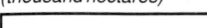

Figure 7.2 Area of Closed Forests Cleared Annually in Selected Tropical Countries, 1980s

(thousand hectares)

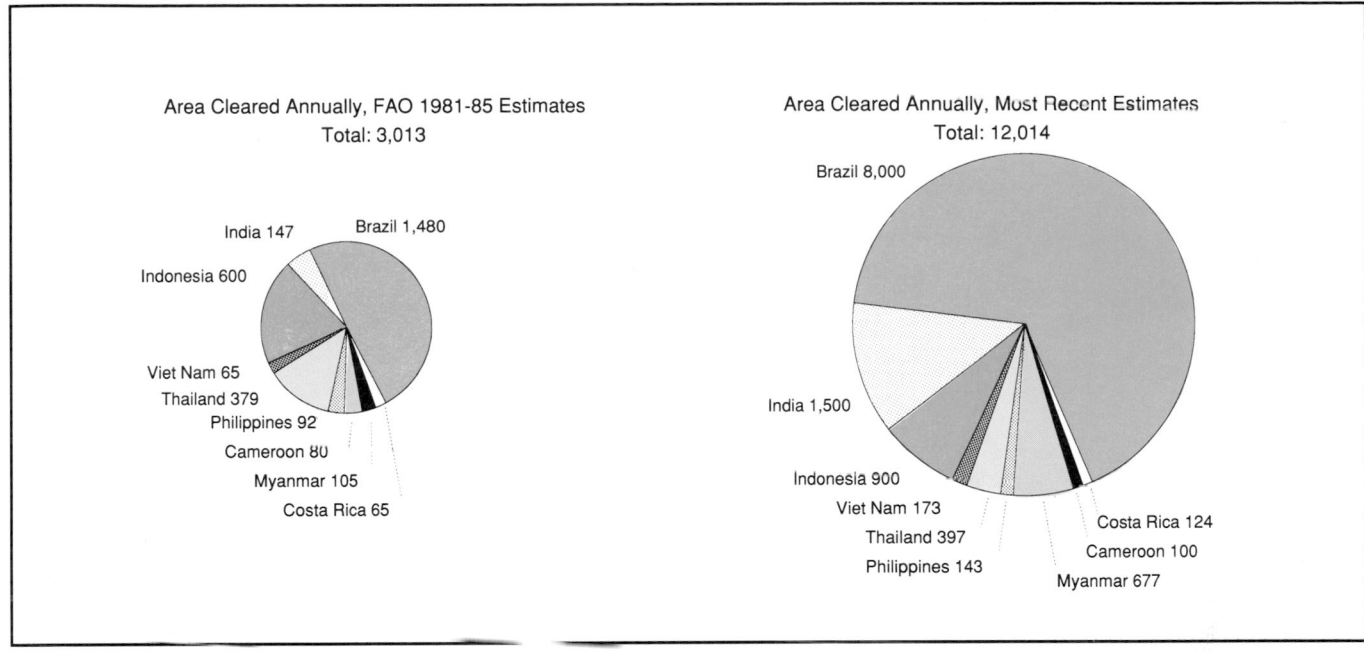

Area Cleared Annually, FAO 1981-85 Estimates
Total: 3,013

India 147
Brazil 1,480
Indonesia 600
Viet Nam 65
Thailand 379
Philippines 92
Cameroon 80
Myanmar 105
Costa Rica 65

Area Cleared Annually, Most Recent Estimates
Total: 12,014

Brazil 8,000
India 1,500
Indonesia 900
Viet Nam 173
Thailand 397
Philippines 143
Costa Rica 124
Cameroon 100
Myanmar 677

Source: Table 7.1.
Note: This figure shows the area lost to deforestation in nine tropical countries as calculated by the Food and Agriculture Organization of the United Nations (1981–85) and more recent estimates. Note that the country with the largest percentage loss (Costa Rica) in Figure 7.1 has the second smallest area lost.

million hectares in 1988 (7) and a further 40–50 percent drop in 1989, according to preliminary figures (8). The Setzer study methodology has been criticized because it uses space photos of smoke, which can extend far beyond the burned area. On the other hand, the study would not detect deforestation that occurs without burning (9).

A 10-year linear projection of deforestation by INPE scientist Philip Fearnside, based on a 1978 INPE survey, predicted that deforestation would reach a total of 35 million hectares in the Legal Amazon by 1989 (10). However, another INPE remote sensing survey in 1988 showed less deforestation over the 10-year period, 17 million hectares, at an average annual rate of 1.7 million hectares (11).

Thus, the range of annual deforestation rates for Brazil's Amazon is between 1.7 and 8 million hectares per year, numbers large enough to affect significantly the new global deforestation rate, which could be 13.9–20.4 million hectares per year, depending on Brazil's rates. Most likely, deforestation accelerated in the early 1980s, peaked in 1987, and declined somewhat in 1988–89 because of changed policies and wetter weather combined.

Also under debate is the extent of Amazon deforestation. According to the 1988 INPE study, 5.12 percent of the Legal Amazon has been deforested. Because this figure is much lower than previous projections, it is controversial. A recalculation of the INPE data, using only the forested portion of the Amazon as a base, showed that 7 percent of the forested

Amazon has been lost (12). (See Figure 7.3 and Box 7.1.)

New Studies in Other Countries

New studies of several other countries followed the 1980 FAO assessment. Many were based on more advanced survey methods. In Myanmar, for example, the 1980 assessment relied on mid-1970 data and the FAO computer model of rural population trends. The new estimate of 677,000 hectares lost per year—545 percent over the 1980 FAO figure—is based on recent fieldwork, a continuing analysis of aerial photographs and satellite imagery dating from the 1980s, and extrapolations from FAO data.

Likewise, the startlingly higher estimates of deforestation in India, the Philippines, Cameroon, and Costa Rica derive from newer satellite data. India's deforestation appears to have soared from 147,000 hectares to 1.5 million hectares per year. Its national Remote Sensing Agency, which calculated both the new and old deforestation rates based on satellite imagery, found that 10.4 million hectares were lost between 1975 and 1982 (13). Deforestation statistics in India are as contradictory as those in Brazil. For example, an Indian forestry expert argues that because of the population increase, the deforestation rate may have climbed to 2 million hectares per year since 1982, leaving the entire country with only 25 million hectares of forests (14). On the other hand, FAO believes India's deforestation may have peaked

Figure 7.3 **Tropical Deforestation in Latin America to 1988**

Tropical Rainforests

Deforested Areas

Source: Tropical Rainforests: A Disappearing Treasure (Smithsonian Institution, Washington, D.C., Traveling Exhibition Service, 1988).

at 1.5 million in the mid-1980s but fell to 150,000 since then (15).

The updated Indonesian estimate of 900,000 hectares—a 50 percent increase over 1980—was drawn from studies of forest clearing associated with the massive transmigration program and from measurements of forest burned in the 1982–83 El Niño fires on the island of Kalimantan (16). Only in Thailand is the new estimate near the older one. (Thailand is the only country where both old and new estimates in-

clude both open and closed forests.) FAO's assessment of 379,000 hectares deforested per year in Thailand was extrapolated from data obtained prior to 1975; the new number, 397,000 hectares per year, is derived from 1978 and 1985 satellite imagery. (See Table 7.1. For more detail on survey methods, see Chapter 19, "Forests and Rangelands," Table 19.1, Sources and Technical Notes.) A public outcry over the devastation of Thai forests led the government to ban all logging and revoke all timber concessions in January 1989 (17). (See Recent Developments, below.)

Counting the Trees

Most of the new studies benefit from the progress made over the past decade in collecting and interpreting remote-sensing data. Ground surveys of tropical forests are now being augmented by airborne sampling methods such as the use of low-altitude photography, satellite systems, and side-looking radar. (See *World Resources 1988–89*, pp. 72-73, for an explanation of these techniques.) For its 1980 assessment, FAO relied on remote-sensing data for only 31 countries and on ground-based information for 45 countries (18). Some of the ground-based surveys relied on interpretations of old maps (e.g., land-use planning and soil survey maps), literature surveys, and extrapolations from outdated inventories.

Remotely sensed data are not a panacea; even state-of-the-art satellite images have drawbacks. Some can distinguish only broad classifications of vegetation; others provide greater resolution but at a smaller scale, so the cost of obtaining and interpreting coverage over a meaningful area can be prohibitive (19). Any space or airborne system can be foiled by the frequent cloudiness of the tropics and, ironically, by the smoke from the fires used to clear the forests (20). Moreover, because deforestation is such a highly political issue, even the most impeccable data are subject to considerable differences of interpretation.

It is possible that the newer studies are not comparable to FAO's original analysis because of improved survey techniques, differences in the forest types included, and differences in defining deforestation and conversion (21). It has been suggested that what seems like a soaring deforestation rate is simply caused by the fact that the new surveys are more accurate and thus reveal old deforestation that had not previously been detected. It is also possible that the new studies show accelerating deforestation in these and other tropical countries. In any case, it seems certain that the world is losing much more of its vir-

Box 7.1 Forest Politics in Brazil

The Brazil Space Agency's 1988 report estimating that 5 percent of Brazil's Legal Amazon region had been deforested was controversial because it was seen as an attempt by the Brazilian government to minimize the extent of Amazonian deforestation. Previous estimates (made by projection rather than by direct measurement) have estimated that as much as 12 percent of the Amazon was deforested.

International pressure on Brazil to conserve its rainforest began to build in August 1988 when extensive burning to clear land showed up in satellite photos that were widely publicized. In December, union organizer Francisco Chico Mendez, who had protected Amazon forest used by rubber tree tappers from ranchers who wanted to burn it, was shot and killed. (Three ranchers were later convicted.) During a trip to Washington, D.C., Mendez had gained a considerable following among U.S. environmentalists and members of Congress as he helped successfully lobby the Inter-American Development Bank to stop funding a highway through Mendes' state of Acre (1).

Brazilian officials reacted to international pressure with angry remarks about outside interference. To statements that the Amazon should be left intact as a storehouse of biodiversity for the world or that its destruction could lead to global climatic chaos, they pointed out that Brazil intended to develop its resources like any temperate country had and that U.S. citizens use 15 times more energy than Brazilians, thus barely showing concern for the global ecology (2) (3).

At the same time, however, Brazil had joined with other Amazon nations in an Amazon Pact, which declared the "importance of genetic and biotic conservation" along with "sustainable use of natural resources." They established a Special Commission on the Amazon Environment charged with promoting research, methods for assessing environmental impacts, cooperation on projects, and studies of the compatibility of environmental legislation. The Commission is also charged with "preventing the deterioration of the Amazon's natural resources, particularly deforestation and soil erosion." The statement also emphasized respect for the "cultural identity of indigenous people in the region (4)."

Shortly afterward, former Brazilian President José Sarney announced his conservation master plan for the Amazon: Our Nature. As part of the plan, controversial tax credits for clearing land were suspended and later ended. Working with the Food and Agricultural Organization of the United Nations, Brazil is mapping the Amazon and designating land uses for each part (5).

In late summer 1989, an international symposium on the Amazon at the University of São Paulo concluded that given the fact that Amazon forest burning releases a substantial amount of the greenhouse gas carbon dioxide and destroys an unknown number of species, "the problem demands the recognition of co-responsibility and of the need for international cooperation in the management of solutions (6).

The symposium recommended more debt-for-nature swaps, more international research, and creation of an international fund to protect the Amazon (7).

Despite efforts to plan for a better use of the Amazon, the goal of large-scale de-velopment remains. For example, Brazil is seeking funding from Japan to complete a road that would connect the Amazon with the seaport of Lima, Peru, on the Pacific coast, opening a number of new markets for Amazon timber and other products (8). Another development that threatens a large area of rainforest is the $3.5 billion Grande Carajas Program, a mining project begun in 1985. Its smelters that convert ore into pig iron are powered by charcoal, and the easiest way to obtain charcoal is to chop down surrounding trees. Although Grande Caraja has not yet made a significant impact, environmentalists point to pig-iron production in the southeastern state of Ninas Gerais where it consumed nearly two thirds of the state's forests (9).

References and Notes

1. Eugene Linden, "Playing with Fire," *Time*, September 18, 1989, p. 80-82.
2. *Ibid.*, p. 77.
3. "It's Our Forest to Burn if We Want to," *The Economist*, March 11, 1989, p. 42.
4. Helena Landazuri T., *La Cuenca Amazonica Argumentos en favor de un manejo integrado*, Ediciones Abya- Yala, Quito, 1987.
5. Santiago Mourao, Head of Environmental Affairs, Brazilian Embassy, Washington, D.C. (personal communication) 1990.
6. José Goldemberg, *Amazonia: Facts, Problems and Solutions* (University of São Paulo and Institute for Space Research [INPE], São Paulo, Brazil, 1989), p. 29.
7. *Ibid.*
8. *Op. cit.* 3, pp. 42-43.
9. *Op. cit.* 1, p. 80.

Box 7.2 Searching for Sustainable Tropical Natural Forestry

In 1988, the International Tropical Timber Organization (ITTO), which is devoted to protecting the future of the tropical timber trade, enhancing nontimber products and services, and conserving forest ecosystems, commissioned a worldwide survey of how much forest was under sustainable management (1) (2). The survey found only a negligible amount of forest being managed for sustainable, long-term timber production. In the 17 ITTO member countries studied, only about 800,000 hectares were being managed sustainably (3) (4). In India (the 18th ITTO producer- member), another 3.6 million hectares are managed over successive rotations, bringing the ITTO total to 4.4 million hectares out of a global total of 828 million hectares of productive tropical forest (5) (6) (7). In other words, timber is produced sustainably on less than 1 percent of the exploitable tropical forests. Probably no more than 20 percent of the forested lands are subject to any harvesting control or silviculture, practices that are considered unsustainable (8) (9). (Sustainable timber production is predicated upon doing nothing to reduce irreversibly the forest's potential to produce marketable timber (10).)

While searching each tropical region for sustainably managed forests, ITTO researchers painted a discouraging picture. In Latin America and the Caribbean, they reported, the total area being sustainably managed at an operational level is limited to 75,000 hectares in Trinidad and Tobago, of which 16,000 have been declared as fully regenerated after logging (11). They found similar conditions in Africa: "One can safely say that there are no sustained yield management systems that are being practiced over large areas in the six countries studied. Management has been progressively abandoned, maybe with the partial exception of Ghana" (12). Asia was different in that all the forests under logging concessions are nominally under management. However, "there is a very great difference between theory and practice" (13). Tropical timber harvesting has the longest history in Asia, and examples of sustained timber management include the Mae Poong forest in Thailand and peninsular Malaysia's selective management system (14).

References and Notes

1. Mark Timm, "Timber Pact to Protect Tropical Forests," *D+C* (March 1987), p. 23.
2. Duncan Poore, "Executive Summary," in "Natural Forest Management for Sustainable Timber Production," International Institute for Environment and Development, London, 1988, p. 18.
3. *Ibid.*
4. International Tropical Timber Organization, "The Role of the International Tropi-
cal Timber Organization," in *The Future of the Tropical Rain Forest*, Melanie J. McDermott, ed. (Oxford Forestry Institute, Oxford, U.K., 1988), p. 52.
5. Theodore Panayotou and Peter Ashton, "The Case for Multiple-Use Management of Tropical Hardwood Forests," Harvard Institute for International Development, Cambridge, Massachusetts, 1988), p. 38.
6. The 18 ITTO producer-members are Bolivia, Brazil, Cameroon, the Congo, Côte d'Ivoire, Ecuador, Gabon, Ghana, Honduras, India, Indonesia, Liberia, Malaysia, Papua New Guinea, Peru, the Philippines, Thailand, and Trinidad and Tobago.
7. Duncan Poore, "Overview," in "Natural Forest Management for Sustainable Timber Development," International Institute for Environment and Development, London, 1988, p. 18.
8. *Ibid.*, p. 6.
9. *Op. cit.* 5, p. 44.
10. *Op. cit.* 7, p. 6, and 11-14.
11. *Op. cit.* 7, p. 16.
12. Simon Rietberger, quoted in Duncan Poore, "Overview," in "Natural Forest Management for Sustainable Timber Production," International Institute for Environment and Development, London, 1988, p. 17.
13. *Op. cit.* 7, p. 17.
14. Peter Burgess, "Asia," in "Natural Forest Management for Sustainable Timber Production," International Institute for Environment and Development, London, 1988, p. 24.

gin tropical forest than had been thought. This loss is occurring despite a surge of international concern to conserve the rainforests. A more complete picture of trends must await the next FAO global assessment, which is due to be published in 1992 (22). Until then, the information reported here indicates how dire the problem of tropical deforestation has become.

All the deforestation figures cited above refer to land that is cleared of forest and converted to some other use such as agriculture, shifting agriculture, or settlements. The surveys do not measure the amount of forest lands that are degraded by selective logging, overgrazing, livestock browsing, fires, or stripping for fuelwood collection. (See Chapter 19, "Forests and Rangelands," Table 19.1, Sources and Technical Notes.)

Causes of Deforestation

Tropical deforestation has three direct causes, which often act together in the same area. The first cause is permanent conversion of forest to agricultural land. This conversion can reflect a government policy to expand the agricultural base or resettle people. For example, the Indonesian government, in a full-page U.S. magazine advertisement, explained its forestry conservation policies, but noted that because its 170 million people have the same aspirations as anyone in the United States, 20 percent of its forests must be converted to plantations to produce teak, rubber, rice, coffee, and other agricultural crops (23). Deforestation in Brazil's state of Rondônia resulted from a government policy to resettle people from poor urban areas. Agricultural conversion also occurs in the absence of policy or as a result of inadequate or conflicting policies as settlers follow loggers into a new forest.

Policies conflicted in Côte d'Ivoire when the government, in an effort to increase foreign exchange earnings, allowed loggers to cut the remaining tracts of natural forest. Farmers, encouraged by government policies to promote the export of cacao, moved in on this newly logged land to plant cacao and other crops. Although the government forestry department has a policy of regenerating deforested areas, it has not been able to prevent deforestation of the remaining natural forest because of poor logging practices and agricultural policies that encourage immediate conversion of the land to agriculture (24).

Slash-and-burn agriculture, a traditional form of agriculture in which land is cleared, planted, and burned, then allowed to regrow into forest, has been an environmentally sound practice in the tropics.

But with increasing populations, plots are now often put back into agricultural use before they have had time to regenerate and replenish soil nutrients. Thus, slash-and-burn agriculture often degrades land (25).

Whether another cause of deforestation, logging, does in fact lead to deforestation depends on the methods used and on what follows the logging. In a few areas, it has been demonstrated that tropical hardwoods can be selectively logged and the forest left to regenerate. This situation is the exception. A survey published by the International Tropical Timber Organization in 1988 found the amount of sustainable forestry in tropical timber production to be "negligible." (See Box 7.2.) Logging is more often carried out in the tropics in a way that damages nontarget trees and opens new forest areas to settlement by farmers. In a typical logging operation, only 10–20 percent of the trees are cut, but another 30–50 percent of the trees are destroyed and the soil is sufficiently disturbed to impede regeneration, leaving a much degraded forest (26).

A third cause of deforestation is the demand for fuelwood, fodder, and other forest products where the resources cannot meet the demands. Inadequate supplies, a major problem in dry tropical areas such as the Sahel and the Aravalli region of India, can cause serious land degradation. (See Chapter 6, "Food and Agriculture," Agriculture in Africa.)

These direct causes as well as grazing, fire, and drought are often interrelated and are exacerbated by government economic policies, population pressures, and poverty. A wide array of government policies ranging from timber concession pricing to land tenure can have a major, if sometimes unintentional, impact on forests. For example, many countries convey title to parties that have "improved" the land by clearing the forest, charge inordinately low rents to timber companies that log government forests, or subsidize cattle ranching on cleared forest (27).

Historical Deforestation

Since preagricultural times, the world's forests have declined one fifth, from 5 to 4 billion hectares. Temperate forests have lost the highest percentage of their area (32–35 percent), followed by subtropical woody savannas and deciduous forests (24–25 percent), and old-growth tropical forests (15–20 percent). Tropical evergreen forests, now under the most pressure, have lost the least area (4–6 percent) because they were inaccessible and were sparsely populated (28). Today, just over one half the world's forests are in developing countries.

From 1850 (before the Industrial Revolution) to 1980, the greatest forest losses occurred in North Africa and the Middle East (-60 percent), South Asia (-43 percent), and China (-39 percent). The highest *rates* of deforestation (the percentage of forest lost *per year*) are now in South America (1.3 percent) and Asia (0.9 percent). (See Chapter 19, "Forests and Rangelands," Table 19.1.)

Deforestation and Development

An economic argument is made that mature forests should be harvested and the proceeds from sale of the lumber used to convert the land to higher yielding plantations and farms (29). In practice, however, tropical timber harvesting has not often brought prosperity to the local people or to the country as a whole. There are several reasons. First, agricultural yields based on cleared tropical forests have been overestimated. The soils underlying 95 percent of the remaining tropical forests are infertile and are degraded easily through erosion, laterization, or other processes once the vegetative cover is removed (30). Many tropical rainforest ecosystems survive on poor soils by quickly recycling the nutrients leached from dead leaves, plants, and other matter before they can accumulate and decay in the top layer of soil, as they do in most temperate forests. Because an agricultural system cannot duplicate the rapid and complex recycling ability of the rainforest, its limited nutrients are lost to leaching and erosion within a few years. (See Box 7.3.)

Second, the economic benefits of timber harvests rarely enrich the community or even the government. Most of the profits have gone to domestic and foreign timber concessionaires and the politicians and military officers who are often their silent partners (31). The pervasive practice of undercharging logging companies for the right to harvest public timber in numerous countries is documented in several World Resources Institute reports (32) (33) (34) (35) (36).

Third, berries, nuts, game, fish, honey, and other forest products have substantial local value even if they never see foreign markets; resins, essential oils, medical substances, rattans, flowers, and a variety of other products have commercial value for the local people who collect and trade them. Indonesia annually exported $125 million worth of the latter group of products in the early 1980s (37). (See Box 7.4.) Although most rainforest species have not even been identified, they may well offer considerable opportunity for development of pharmaceutical and industrial products. Once a species is identified for certain products, infrastructures could be set up for collection, delivery, and manufacturers' contracts. For example, the National Biodiversity Institute in Costa Rica (INBIO) and North American pharmaceutical companies are discussing contractual arrangements for chemical testing of animals and plants. The goal is to protect the interests of the local people as well as these natural resources (38).

Attempts to Slow Deforestation

Growing concern among professional foresters during the 1970s and 1980s about the rate of tropical deforestation has now been overtaken by public concern about rainforest destruction. Early on, FAO and others worked largely within the forestry com-

Box 7.3 Cattle Pastures in the Brazilian Amazon

Over the past two decades, more than 20 million hectares (3 percent) of humid tropical forests in Latin America have been converted to cattle pastures. At least one half this conversion has taken place in Brazil's Amazon, one fourth in Mexico, and the rest in Colombia, Peru, Venezuela, and Central America (1) (2).

Pasture expansion in eastern Amazonia began in the late 1960s, stimulated by a growing network of paved roads, investment tax credits, government-subsidized credit, (3) (4) and land speculation opportunities (5). Forests were cut and burned during the dry season and planted with African forage grasses. These grasses grew well during the first two–three years, perhaps because of the pulse of phosphorous and other nutrients released into the soil when the rainforest was burned (6). However, within 10 years, phosphorus levels fell to those of the original forest soil. As phosphorus availability dropped, the nutrient-demanding forage grasses were gradually overtaken by shrubs and nonforage grasses.

These pastures were stocked heavily with cattle (more than one animal unit per hectare is considered heavy on this type of soil), which accelerated the degradation (7). By the late 1970s in the Paragominas region of northeastern Amazonia, more than one half the forest-derived pastures were degraded (8).

"REFORMING" PASTURES

To recuperate a degraded Amazonian pasture periodically, a rancher would first hire teams of field laborers to cut weedy shrubs and sprouting trees manually, then burn the cut vegetation. This practice triggers a flush of grass, but the response diminishes with each repetition. Some ranchers with access to capital (e.g., those with forested lands that can be exploited for lumber) "reform" their degraded pastures by scraping the soil surface with bulldozers, fertilizing, and replanting with forage grasses that have relatively low nutrient demands. A small portion of Amazon pastures have been reformed in this way (9).

The government of Brazil no longer subsidizes pasture formation in Amazonia, but pastures continue to be formed as a means of securing and retaining title to land, thereby capturing escalating land values (10). Cattle are valued as an inflation-free investment and as a source of milk products and status, even when cattle production is not profitable

(11). As much as 90 percent of Amazonian pasturelands may have been formed without government subsidies (12). Unlike government-supported pastures, unsubsidized pastures are typically formed on sites that have been cultivated in annual crops following forest cutting and burning.

REGROWTH OF ABANDONED PASTURES

Amazonian pastures are usually invaded by regrowth forests within a few years of abandonment. Estimates differ on how long an abandoned pasture takes to regenerate into a mature forest; a best case estimate of regeneration of a forest from a pasture that was lightly used and not subjected to fire is slightly less than 100 years (13). Where pastures were moderately used (e.g., six-eight years of grazing) and then burned, regeneration is slower; a pasture in Venezuela's Amazon that was repeatedly burned to maintain the grasses has regenerated only 50 percent of its forest biomass 80 years after abandonment (14).

The potential for forest regrowth is greatly diminished when pastures are reformed with bulldozers. Eight years after abandonment, a pasture near Paragonimus that had been bulldozed was dominated by weedy grasses and shrubs and supported no tree species native to the original forest (15). Although abandoned pastures are rare in Amazonia, they demonstrate the potential for ranching practices to produce degraded ecosystems that can persist for many years.

THE THREAT OF FIRE

Pasture formation in Amazonia also introduces the threat of fire to neighboring forests. During the dry season, degraded pastures can burn within one day of a rain storm, but primary forests cannot be ignited even after months of drought (16). As forests that lie adjacent to pastures and were logged dry out and become flammable, the stage is set for extensive forest fires (17). In 1987, 20 million hectares in Amazonia burned (18), a large portion of which had supported forests exploited for timber.

References and Notes

1. E.A.S. Serrao and J.M. Toledo, "Sustaining Pasture- Based Production Systems for the Humid Tropics," in T. Downing, S. Hecht, and H. Pearson, *Development or Destruction of the Livestock Sector in Latin America*, Westview, Boulder, Colorado, 1990.
2. Instituto Brasileiro de Geografia e Estatística (IBGE), "Sinopse preliminar do censo agropecuario—1985," Rio de Janeiro, 1983.
3. J. Browder, "The Social Costs of Rain Forest Destruction: A Critique and Economic Analysis of the 'Hamburger debate,'" *Interciencia*, Vol. 13, (1988) pp. 115-120.
4. Dennis J. Mahar, *Government Policies and Deforestation in Brazil's Amazon Region* (World Bank World Wildlife Fund, Conservation Foundation, Washington, D.C., 1989), p. 20.
5. P.M. Fearnside, "A Prescription for Slowing Deforestation in Amazonia," *Environment*, Vol 31, No. 4 (1989), pp. 18-20.
6. E.A.S. Serrao, et al., "Productivity of Cultivated Pastures on Low Fertility Soils in the Amazon of Brazil," in *Pasture Production in Acid Soils of the Tropics: Proceedings of a Seminar Held at CIAT, Cali, Colombia, April 17-21, 1978* (Centro Internacional de Agricultura Tropical, Cali, Colombia), pp. 195-225.
7. *Ibid.*
8. A.P. dos Santos, E.M.L. de Moraes Novo, and V. Duarte, *Relatoria final do Projeto INPE/SUDAM* (Space Research Institute of Brazil (INPE), São Paulo, 1979).
9. *Op. cit.* 1.
10. *Op. cit.* 5, pp. 18-40.
11. Brent Hayes Millikan, The Dialectics of Devastation: Tropical Deforestation, Land Degradation and Society in Rondonia, Brazil," Master's thesis, University of California at Berkeley, p. 131.
12. *Op. cit.* 4.
13. C. Uhl, R. Buschbacher, and E.A.S. Serrao, "Abandoned Pastures in Eastern Amazonia. I. Patterns of Plant Succession," *Journal of Ecology*, Vol. 76 (1988), pp. 663-681.
14. Juan G. Saldariaga, "Recovery Following Shifting Cultivation, a Century of Succession in the Upper Rio Negra," in *Amazonian Rain Forest: Ecosystem Disturbance and Recovery*. C.F. Jordon, ed. (Springer-Verlag, New York, 1987), pp. 24-33.
15. *Op. cit.* 13.
16. C. Uhl and J.B. Kauffman, "Deforestation Effects on Fire Susceptibility and the Potential Response of Tree Species to Fire in the Rainforest of Eastern Amazonia," *Ecology*, Vol. 71, No. 2, 1990, pp. 437-449.
17. C. Uhl and R. Buschbacher, "A Disturbing Synergism between Cattle Ranch Burning Practices and Selective Tree Harvesting in the Eastern Amazon," *Biotropica*, Vol. 17, No. 4 (1985), pp. 265-268.
18. Alberto Waingrot Seltzer, et al., *Relatorio de atividades do projeto IBDF-INPE 'SEQE' — Ano 1987*, (Space Research Institute of Brazil (INPE) São Paulo, 1988).

munity to find new and better ways to manage forests. With introduction of the Tropical Forestry Action Plan (TFAP) in 1985, FAO, the United Nations Development Programme, several bilateral development agencies, the World Bank and other international lending banks, several tropical country governments, and several nongovernmental organizations including the World Resources Institute hammered out a new strategy. They issued a statement recognizing that efforts to improve forest management had failed; because many causes of deforestation lie outside the domain of national forestry departments, the problem must be addressed at higher political and economic levels (39). TFAP was designed to deal not only with deforestation but also with forest degradation, with emphasis on agroforestry, fuelwood, industrial forestry, and other methods by which forestry can provide goods and services to rural people and national economies (40).

Since its beginning, TFAP has faced criticism and internal debate. At the ninth meeting of the TFAP advisors in Washington, D.C. in 1989, workshop participants criticized the program for being too government oriented and not seeking the participation of nongovernmental organizations or forest dwellers.

More than 60 countries have decided to prepare national forestry action plans to guide management of their forests. This TFAP planning process has helped stimulate an increase in investment in tropical forestry from $500 million in 1985 to $1 billion in 1989—70 percent of the amount originally called for by TFAP. The program has made many governments aware of the economic costs of deforestation and has mobilized many people and organizations to discuss how to address the causes of deforestation and forest degradation (41). Progress under the TFAP framework is being evaluated by an independent team commissioned by FAO.

Attitudes on Deforestation

During the 1980s, conserving the tropical rainforests was a popular issue in North America and Europe. As the role of tropical forests in tempering global warming and harboring a diverse range of species has become clear, environmental groups have pressured international development agencies to halt projects that cause deforestation. Environmental groups in northern countries have also formed links with groups of indigenous people and forest dwellers such as the rubber tappers of Brazil and the Penan of Malaysia. (See Box 7.4.) A global poll of environmental attitudes commissioned by the United Nations Environment Programme and conducted in 1988 found considerable concern over deforestation among Third World people. In Latin America and the Caribbean, 78 percent of the interviewees expressed concern; in Asia, 73 percent; and in Africa, 77 percent (42).

Some governments initially reacted with irritation at having northern politicians and nongovernmental organizations tell them they should preserve forests rather than develop as temperate countries did. But a number of decisionmakers in developing countries are now declaring that they hope both to conserve their forests and to increase forest productivity in an effort to increase the economic benefits from forests on a sustainable basis (43).

POSSIBLE EFFECTS OF GLOBAL WARMING ON FORESTS AND RANGELANDS

Effects on Forests

Climate is paramount in determining the broad distribution of forests. The vast differences among the tropical, subtropical, temperate, and boreal forests can be explained in no other way, though soils and topography are also important. It is evident, too, that forests have a reciprocal if not as dominant an effect on climate. Their presence usually serves to temper extremes of global climatic variables such as precipitation and albedo, the reflectiveness of Earth's surface. Forests also affect local climates; they alter the surface energy balance by reflecting sunlight, evaporating water, and drawing groundwater up through tree roots. The rate of these processes depends on the species present and their abundance (44).

Perhaps the most vital connection between climate and forests is the latter's role in the global carbon cycle. The world's forests store 450 billion metric tons of carbon, which is 20–100 times more carbon per area unit than croplands (45) (46). When forests are cleared, their capacity to withhold carbon from the atmosphere is lost. Once the trees are cut, the sequestered carbon oxidizes and is released, quickly if the trees are burned and slowly if they are left to decay or harvested and converted into longlasting products such as housing or furniture (47).

Deforestation and the Greenhouse Effect

Deforestation is second only to the burning of fossil fuels as a human source of atmospheric carbon dioxide. Almost all carbon releases from deforestation originate in the tropics; temperate forests no longer add to the total because their growth and cutting rates cancel each other. Of course, the massive deforestation of Europe and North America in the past contributed heavily to current global carbon levels (48).

Our global estimates of the amount of carbon given off annually by deforestation is 2.8 billion metric tons. (See Chapter 24, "Atmosphere and Climate," Table 24.1.) Previous estimates range from 0.4 to 2.5 billion metric tons (49). Deforestation (or land-use change) accounts for about 33 percent of the annual emissions of carbon dioxide caused by humans. (See Table 24.1.) In 1987, 11 countries were responsible for 82 percent of this net carbon release: Brazil, Indonesia, Colombia, Côte d'Ivoire, Thailand, Laos, Nigeria, Viet Nam, Philippines, Myanmar, and India (50). This figure suggests that much more carbon is given off because of deforestation than was previously thought. During 1987, a year of intense land clearing by fire in Brazil's Amazon, more than 1.2 million metric tons of carbon are believed to have been re-

Box 7.4 Social Justice and Deforestation: Forest Dwellers Unite to Press the Connection

In all three tropical regions—Asia, Africa, and Latin America—forest dwellers recently organized to demand changes in the treatment of their homelands. As befits people whose sustenance and culture are inseparable from the natural surroundings and who feel their way of life is threatened, their concerns are fundamentally social or political. Often their goal is to redress underlying inequities in forest use and land tenure. The solutions they propose, not coincidentally, are environmentally sound.

The connection between tropical deforestation and social justice was brought forcefully to international attention in December 1988, when the charismatic Brazilian labor leader Franciso Mendes Filho was assassinated outside his home in the state of Acre. Chico Mendes, as he was widely known, had helped build a union of peasant rubber tappers into a potent national political power. The rubber tappers, who live by gathering Brazil nuts and native latex, are fighting to escape a history of debt peonage and to thwart attempts by cattle ranchers and politicians to evict them from lands they have occupied since early in the century. Under Mendes, the union won the creation of four reserves in Acre to be managed by rubber-tapper communities. Mendes made several overseas trips and was considered instrumental in persuading the U.S. Congress to pressure the Inter-American Development Bank into making its leading Amazon policies more environmentally sensitive. Over the years, Mendes's union often confronted employees of land speculators who had come to clear the forests (1) (2). His death was the first to attract international attention, although shootings are not uncommon. According to Amnesty International, 1,000 peasants, rubber tappers, indians, union officials, and their supporters have been killed in the Amazon since 1980 (3).

In March 1987, in Sarawak, Malaysia, members of the Penan, Kayan, and other native communities began a series of actions to keep loggers from entering tracts over which they assert ancestral domain. They set up some 25 blockades across roads into their home forests and temporarily halted cutting and log shipments from within. In October 1987, the government dismantled the blockades and arrested 42 Kayans, but the protests resumed in May 1988. Since then, the blockades are repeatedly taken down and reerected (4) (5). A Malaysian court exonerated the Kayans in 1989, but 128 Penans have been arrested at the barricades and charged under a new Sarawak law that makes interfering with logging a crime (6) (7). The Malaysian government defends its actions on the basis of the best interests of the forest and the native people (8).

Like the rubber tappers, the blockaders are seeking an end to forest clearing as a means of securing government recognition of their customary rights and livelihoods. It is a hard struggle, for the tradition of local autonomy over tropical forests is largely past, at least in the view of central governments; over 80 percent of closed-canopy forests in developing countries are now officially part of the public domain (9) (10).

In Kenya, women banded together in search of a different kind of social justice. A disproportionately large share of the chores of African daily life falls to women: they fetch water, farm the household crops, and gather the family's fuelwood. As trees become more scarce, women have to walk farther and farther to collect fuel. To alleviate this situation, the Green Belt Movement was funded in 1977 by a biologist, Wangari Maathai. Under her leadership, Green Belt has planted nearly 5 million trees in small wood lots across Kenya and has established more than 500 community nurseries. It has spread throughout the country at the request of local women's groups, who apply for and then oversee a nursery. By paying female tree tenders a premium for each seedling that survives, Green Belt attains a transplant survival rate of 80 percent. Not only does this arrangement bring women money, it ensures that the trees live long enough for local people to experience the benefits they provide. Maathai sees Green Belt as more than just an agroforestry project. In the hands of women, it is a way for them to accomplish something that male political leaders recognized as nationally important (11).

References and Notes

1. Richard House, "Leading Brazilian Ecologist Murdered at Home in Amazon," *Washington Post* (December 24, 1988), p. A10.
2. Susanna B. Hecht and Alexander Cockburn, *The Fate of the Forest: Developers, Destroyers and Defenders of the Amazon* (Verso, New York), pp. 161-191.
3. Amnesty International, "Amnesty International Brazil Briefing," background paper, Washington, D.C., September 1989.
4. Teresa Apin, "Local Action to Stop Deforestation," *D&C*, No. 5 (May 1987), pp. 24-25.
5. World Rainforest Movement and Sahabat Alam Malaysia, *The Battle for Sarawak's Forests* (World Rainforest Movement and Sahabat Alam Malyasia, Penang, Malaysia, 1989), pp. 6-7.
6. GPW, "Malaysia: Blockaders Exonerated," *Not Man Apart*, Vol. 19, No. 1 (1989), p. 16.
7. Alex Hittle, David Malakoff, and Sue Lincoln, "Loggers Threaten Malaysian Forests, Tribes," background paper Friends of the Earth, Washington, D.C., 1989.
8. Embassy of Malaysia "Information Paper on the Penan of Malaysia," Washington, D.C., 1988, pp. 2-9.
9. Jean-Paul Lanly, *Tropical Forest Resources* (Food and Agriculture Organization of the United Nations), Rome, 1982, cited in Robert Repetto, "Overview," in *Public Policies and the Misuse of Forest Resources*, Robert Repetto and Malcolm Gillis, eds. (Cambridge University Press, New York, 1988), pp. 2-9.
10. An important exception is Papua New Guinea, where 97 percent of the land is owned by individuals or clans. With an owner's consent, logging and other utilization rights may be purchased for a limited period by the government, but second-cut privileges are seldom ceded. Peter Burgess, "Asia," in "Natural Forest Management for Sustainable Timber Production," International Institute for Environment and Development, London, 1988, p. 5.
11. Maryanne Vollers, "Healing the Ravaged Land," *International Wildlife*, Vol. 18, No. 1 (1988), pp. 5-11.

leased. By comparison, the United States, the world's largest greenhouse gas emitter, released 1.2 million metric tons of carbon in 1987 by burning fossil fuels and through cement plant emissions. (See Table 24.1.)

Forests as a Carbon Sink: Tree Planting and Other Options

Offsetting the carbon emitted from global use of fossil fuels would require more than 1 billion hectares over the next 40–50 years (51). A potentially more realistic approach is to focus on smaller-scale plantings that will divert pressure from remaining tropical forests. Economic and land reforms that provide an agroforestry alternative to shifting agriculture would reduce carbon releases from tropical deforestation while increasing the number of trees. (See Box 7.5.) Similar synergistic benefits could be effected outside tropical forests as well. For example, every tree planted in an urban area not only takes up carbon dioxide but shades buildings in hot weather and

Box 7.5 Agroforestry in Guatemala to Offset Carbon Releases in the United States

Although the idea of planting trees to offset carbon emissions from fossil fuel consumption was raised as long ago as 1977 (1), the first project funded specifically for this purpose just got under way in mid-1989. A U.S. independent power producer, Applied Energy Services (AES), had undertaken construction of a new 183-megawatt coal-fired powerplant in Connecticut, but the company was concerned about the global warming effects of the approximately 15 million metric tons of carbon in the form of carbon dioxide that would be released over the plant's life.

AES approached the World Resources Institute (WRI) for ideas on how to neutralize the global warming impacts of the plant's carbon emissions most cost-effectively. After considering several forestry-related proposals, WRI recommended that AES fund a sustainable agriculture and agroforestry project being proposed by the international aid organization CARE for Guatemala. WRI recommended the project because of its likely ability over 40 years to offset the 15 million tons of carbon and the significant benefits to accrue to local residents. Over 10 years, it will bring private and community wood lots; agroforestry for food, fuel, and fodder; alley cropping; live fencing; and soil conservation practices to about 40,000 Guatemalan farm families. Although more than 52 million fast-growing trees will be planted, most of the project, a major portion of the carbon-sequestration benefits are expected to result from substituting these farmers' lands' increased productivity for forests destroyed for fuelwood and agricultural purposes (2).

References and Notes

1. Freeman J. Dyson, "Can We Control the Carbon Dioxide in the Atmosphere?" *Energy*, Vol. 2, No. 3 (1977), pp. 287-291.
2. Mark C. Trexler, Paul E. Faeth, and John Michael Kramer, "Forestry as a Response to Global Warming: An Analysis of the Guatemala Agroforestry and Carbon Sequestration Project," World Resources Institute, Washington, D.C. 1980, pp. 13-14, 18-19, and 26.

---

shields them from winds in cold weather, reducing air conditioning and heating demands—and thus electricity generation, much of which comes from burning fossil fuels (52).

Possible Responses of Forests to Climate Change

It is doubtful whether tree species can disperse fast enough to pursue their optimum climate conditions. If no other climatic variables were altered, higher temperatures alone would certainly reduce the extent of the world's forests (53).

When the potential natural ranges of all forests are matched with temperature changes corresponding to a doubling of current atmospheric carbon dioxide concentrations, the results are striking. The boreal forest is greatly reduced, warm temperate forests form pockets within cool temperate forests, and subtropical forests extend poleward. In South America and Africa, subtropical moist forest is often replaced by tropical dry forest. Grasslands and deserts, by contrast, increase their global ranges; tundra virtually disappears. For forests, changes in the amount of precipitation and its seasonality could be just as important (54). Scientists conjecture that droughts could be more devastating, windstorms more frequent, wildfires hotter, and insect pests more prevalent at certain times, in certain forests (55) (56).

In well-established forests, mature trees should withstand the effects of climate change for some time. Seedlings are more vulnerable. Thus forests may change slowly simply by not regenerating in their present locations (57) (58). Once mature trees are removed or die, species displacement proceeds more quickly and the forest is on its way to being transformed (59).

Any extreme climate change would place stress on a forest and make it more susceptible to infestations of pests and disease. Pestilence could result either because the warmer climate favors the spread of a given pest or disease vector or because greenhouse-induced stress in host species makes them more vulnerable to outbreaks (60) (61). For example, in the Pacific Northwest of the United States, the balsam wooly aphid might be able to extend its range upward into stands of subalpine firs, which have a low resistance to it (62). Simply stated, forest pests and pathogens are expected to do well in an era of rapid global change because they are well adapted to unsettled conditions. The effects of a greenhouse-induced parasite spread should be felt mostly in temperate forests because parasites from the tropics would be able to colonize new hosts (63).

Economic and Forest Management Implications

The overall effect could be a general forest malaise, with large reductions in the number of healthy stands in the temperate zone (64). It is uncertain how this change would affect timber yields in the short run; conceivably, there could be temporary gluts as the remaining marketable wood is removed from declining forests. Over the long term, yields would probably be lower than today's until forests again reach an equilibrium with the climate.

Yet it is well known that, under controlled conditions, increased carbon dioxide concentrations enhance plant growth; estimates range from 0.5–2.0 percent growth for each 10-parts-per-million increase in carbon dioxide (65). Whether this growth would occur in a natural forest is unknown, but there is some evidence that it could (66). Absolute reductions in the size of the natural temperate and boreal forests would probably countervail any carbon-induced growth enhancement.

Effects of Climate Change on Rangelands

In a warmer climate, grasslands, savannas, and desert areas are expected to expand, but they will be vulnerable to increased degradation through erosion and fire, and rangelands may suffer a decrease in forage plants. This trend is expected to be particularly strong in arid, semiarid, and subhumid regions such as the U.S. West, the Sahel, the Middle East, the temperate steppes of Asia and the Soviet Union, and Australia (67). All rangelands could experience a shift in community composition to species that tolerate water stress and that compete strongly for nutrients. In addition, the species that respond better than others to increased atmospheric carbon levels will grow faster and use water more efficiently, thus creating more biomass (68). Unfortunately, they may not be the species most suited for forage; if not, the number of livestock that can be raised on a given amount of rangeland may lessen.

Direct Effects of Rising Carbon Dioxide Levels

Experiments have shown that plants raised in an enriched carbon dioxide environment assimilate greater amounts of carbon, have larger leaves and grow faster than those raised at current atmospheric carbon dioxide levels. Species do not respond equally, however. Broad-leaved species grow faster than grasses (69). In addition, plants with the 3-carbon pathway (C_3) of photosynthesis (most crops) tend to grow faster than plants with the 4-carbon pathway (C_4) typical of many grasses (70) (71) (72) (73). However, these axioms may not hold at higher temperatures (74).

Many forage plants are expected to grow faster and larger because of carbon dioxide fertilization. This growth may not benefit livestock. Carbon dioxide-fertilized plants have been shown to have a lower proportion of nutrient-rich tissue to nonpalatable tissue than plants raised in a normal carbon dioxide environment. The nonpalatable tissue is composed of lignen, which decomposers break down slowly (75). The undecomposed fiber remains on the ground as litter, and an increase in loosely packed grassy litter can increase both the frequency and intensity of fires. There are even disturbing indications that carbon dioxide enrichment favors some highly flammable species. For example, exposure to higher concentrations of carbon dioxide substantially increases the productivity of *Bromus tectorum*, an invasive fire-tolerant grass in the United States (76).

The predicted increase in plants' growth rates could throw off their natural reproduction cycles in unpredictable ways. Earlier flowering or fruiting would affect pollinators, herbivores, and dispersers, altering the intricate relationships among plants and animals in this ecological community (77) (78).

Indirect Effects of Rising Carbon Dioxide Levels

Rising temperatures and changing precipitation—the indirect effects of increased carbon dioxide—may have more influence on rangeland environments than do the direct effects. The global climate models of the National Oceanic and Atmospheric Administration (NOAA) Geophysical Fluid Dynamics Laboratory and the Goddard Institute of Space Studies (GISS) have been used to predict precipitation changes that would leave some major rangelands drier and others

wetter. (See Chapter 2, "Climate Change: A Global Concern," Box 2.2.) Most of the midlatitude rangelands would be drier during the summer. But even areas with more rainfall may experience a critical drop in soil moisture because the increased temperatures would cause moisture to be lost through evaporation and transpiration (79). For example, in Saskatchewan, Canada, precipitation is predicted to increase 15 percent during global warming, but the net result of higher temperatures and evapotranspiration may reduce grassland productivity 5–20 percent (80).

The NOAA and GISS models predict that drier vegetation communities will replace wetter ones in many areas. For example, desert savanna ecosystems are expected to replace tallgrass ecosystems in southern Africa, bunchgrass-shortgrass regimes in southeast Australia, and tallgrass and shortgrass ecosystems in Texas and New Mexico (81). Such a change would mean a loss for the people who depend on these rangelands for livestock forage.

Effects of Fire

An important effect of climate change may be an increase in the frequency and severity of fires in rangelands. The reasons are complex. First, as noted above, some range areas will be prone to fire because they have less rainfall. Second, carbon dioxide-enriched plants may produce more fibrous litter that will serve as fuel. Third, in many regions, relative humidity could fall below 65 percent—the threshold at which range grasses and shrubs can become flammable—more frequently or for longer periods. Essentially, it is the absence of water that determines flammability. The rate at which rangeland plants dry out depends on the amount of solar energy striking the surfaces, wind movements, packing of the litter (i.e., dead wood), and relative humidity or vapor-pressure deficit. Systems are flammable when the litter layer's moisture content falls below 15 percent, a result of relative humidity below 65 percent (82).

Although fire is a natural and human tool by which grasslands are maintained, repeated fires can cause a net loss of nutrients (83). Then weedy, less-palatable species often become dominant. For example, in the foothills of the eastern Andes, repeated invasions of fire in rangelands have resulted in domination by *Imperata brasiliensis*, a highly flammable grass of little economic value. Soil erosion has increased and soil water storage has decreased (84). In the Great Basin of the United States, *Bromus tectorum*, a dominant weed in overgrazed rangelands, similarly promotes recurring wildfire and subsequent degradation of the landscape (85). Although it is not possible to predict accurately what percentage or which ecological communities might be more flammable as temperatures rise, fires will be an increasing component of many rangeland systems.

A worldwide decline in range productivity from increased grazing pressure, erosion, and drought is already apparent (86) (87) (88). This trend will likely intensify. The composition of vegetation might change in many rangelands as tallgrass savanna and open woodland give way to short grass and desert steppe. The uncertainties are many, and researchers are beginning to conduct carbon dioxide enrichment studies to assess the effects on soils, nutrient cycling, water relationships, ultraviolet exposure, fire regimes, and plant, animal, and microbial relationships (89). More knowledge on the direct and indirect effects of global change is essential; the possible effects of rangeland changes on human communities have not been assessed.

RANGELAND CONDITIONS IN LATIN AMERICA

Latin America spans a vast geographic and ecological range, from the temperate deserts of northwestern Mexico, across equatorial forests of the Amazon basin, to the tundra steppe of southern Patagonia. Within this immense region, livestock graze on arid grasslands and shrublands, seasonally dry and flooded savannas, alpine grasslands and tundra, and pastures hewn from temperate and tropical forests.

Rangeland covers about one third, or 700 million hectares, of Latin America's land surface. Permanent pasture, the most productive rangeland, occupies nearly 569 million hectares (90).

Types of Rangeland

Latin America has three major types of rangeland:
- Natural grasslands, woodlands, and savannas, where tree cover is limited by drought, fire, flooding, and poor soil.
- High-elevation natural grasslands and shrublands, where tree cover is generally limited by low temperatures.
- Cultivated pastures, established in areas once occupied by forests.

Seventy percent, or about 500 million hectares, of Latin America's rangeland is in the first category. It includes the *cerrado* of central Brazil, the *llanos* of the Orinoco River basin, the *chaco* of Argentina and Paraguay, the *matorral* of Chile and Peru, and the dry regions of north central Mexico. These rangelands have low livestock-carrying capacities, requiring from 15–50 hectares to support each animal. Although they are typically stocked with cattle or goats at low densities, they are often overgrazed. The result is soil erosion and the invasion of nonforage weeds in vast areas (91) (92). Soil salinity, which decreases soil productivity, has also increased on many arid ranges because decreased plant cover promotes evaporation at the soil surface.

Approximately 20 percent of Latin America's rangelands are at high elevations in the Andes. These native grasslands and shrublands extend from Venezuela to Argentina, and they include rangelands with possibly the highest livestock-production potential in all Latin America. Much of the Andean rangelands, however, are overgrazed by cattle, sheep, and the more common high-elevation livestock, llamas and alpacas. With reduced grazing pressure, the quality of some Andean rangelands could be restored (93) (94).

Livestock pastures derived from forests, which make up less than 10 percent of Latin American rangelands, are rapidly expanding. In Central America, for example, pastures are displacing permanent agriculture on fertile soils and are expanding into the less fertile soils of forests, where they are usually planted following slash-and-burn cultivation. On less fertile soils, cattle pastures are generally abandoned after a few years of grazing (95) (96).

Compared with rangelands in the United States, Australia, and Western Europe, Latin American rangelands are undermanaged; only a small percentage is improved through introduction of high-quality forage grasses and legumes or is preserved through herd rotation or restricted grazing densities. The most common management practice is burning, which is an inexpensive way to improve short-term productivity, but at the risk of diminished long-term productivity where soil fertility is low.

Rangeland conditions vary by geographic region and type of vegetation. Rangeland conditions in four regions of Latin America—Mexico, Central America and the Caribbean, tropical South America, and temperate South America—are summarized below.

Mexico

Mexico's rangelands cover 90 million hectares, which includes 13 percent, or 74 million hectares, of Latin America's permanent pasture. (See Table 7.4.) Nearly 80 percent of these rangelands are dry grasslands, shrublands, and savannas, primarily in northern and central Mexico.

The capacity of these dry lands to support livestock is low, and much of the range is overgrazed. For cattle, for example, the range can support one adult animal per 11–22 hectares, but average livestock densities are one per 3–6 hectares. Further, most of Mexico's 8 million sheep, the fifth largest flock in the world, and 10 million goats are raised on dry rangelands by campesinos with virtually no technical support. This overgrazing by cattle, goats, and sheep has provoked serious weed invasion and soil erosion (97) (98).

About one fifth of Mexico's rangelands were native forests. Cattle pastures continue to displace rapidly both native forests and agricultural crops in the humid and subhumid tropics of Mexico's southern states, particularly in Chiapas, Tabasco, and Veracruz. They have also expanded into large areas of subhumid tropical forests in central Mexico. In all, about 5.5 million hectares have been converted to pastures (99). These forest-derived cattle pastures are the most productive rangelands in Mexico, with carrying capacities of one animal per 0.5–6 hectares, but they too tend to be overgrazed and short lived.

Table 7.2 Rangeland Conditions in Selected Latin American Countries

Region/Country Size of Rangeland	Range Condition	Range Trend	Causes	Remedial Activities	Source
MEXICO					
Permanent Pasture[a]: 74,499,000 ha Total Rangeland: 90,000,000 ha	•Most rangelands are severely overgrazed. •Soil erosion provoked by grazing is common in dry regions; erosion and hydrologic changes accompany forest conversion to pastures, particularly on steep slopes. •Goat and sheep grazing on dry range is reducing tree cover and exacerbating soil loss. •Weed invasion limits range productivity in dry regions. •Grazing in pine/oak highlands is reducing the regenerative capacity of the forests.	•Rangeland is expanding, particularly in regions of humid tropical forest. •Overgrazing continues to degrade rangeland through soil loss and weed invasion.	•Livestock production has generally not benefited from technological advances and development of management systems. Only a small percentage of rangeland is improved with cultivated forage and fertilization or is managed through herd rotation. •Most rangeland is consolidated in large private estates, making intensification of livestock management difficult.	•Integration of cattle and agricultural production systems could reduce pressure on rangelands. For example, organic refuse from sugar cane operations could be employed as cattle feed. Pork production (which yields more meat nationally than cattle production) is based on a wide range of feeds and agricultural byproducts. •Cultivation of forage species, herd rotation, and development of alternative (nongrass) forage could improve rangeland quality.	1,2,3
CENTRAL AMERICA AND THE CARIBBEAN					
Permanent Pasture[a]: 19,455,000 ha	•Many pastures are on previously forested land and are rapidly degrading. •Pastures planted on rich coastal soils have low forage quality.	•Short-lived pastures continue to be formed on deforested land. •Agricultural lands are still being converted to pasture.	•Most pastures are formed with volunteer grasses (e.g., *Hyparhenia rufa*) that are less productive than cultivated forage. •Management is inadequate. •Many pastures are planted following slash-and-burn agriculture on poor soils.	•Cultivated forage species are used. •Pastures on forested highlands with poor soils are probably not sustainable	
Costa Rica Permanent Pasture[a]: 2,300,000 ha	•Cultivated pastures occupy up to 76 percent of the land that is suitable for annual crops. •Pastures on infertile lands with sharp relief are short-lived and provoke severe soil erosion.	•Expansion on forested land continues.	•Pasture formation is a means of securing land following abandonment of slash-and-burn agriculture.	•Cattle production could be increased through intensified management, such as use of improved forage species and herd rotation. •Small farms achieve higher livestock production rates through integration with crop production.	1,4,5,6,
Dominican Republic Permanent Pasture[a]: 2,092,000 ha	•Overgrazing is common in cattle pastures that are part of large estates. •Dry forest lands are used as open range for goats, which reduce tree cover.	•Overgrazing by cattle and open-range goat production continues. •Sugarcane is replacing cattle pastures in some areas.	•Goats are an important component of peasant economy. Combined with charcoal production and slash-and-burn agriculture, goat raising causes widespread land degradation.		1,7
Haiti Permanent Pasture[a]: 494,000 ha	•Dry-forest rangeland is degraded but is the base of subsistence goat production. •Cattle pastures occupying lowlands are poorly managed and unproductive.	•Subsistence goat production continues.	•Goat browsing habits are destroying dry forests on which they depend and are increasing erosion. •Pasture management is encumbered by the lack of technical support.		1,8
TROPICAL SOUTH AMERICA					
Bolivia Permanent Pasture[a]: 26,750,000 ha•	•Overgrazing dates back to Spanish colonization; most rangelands are affected by soil erosion. •The natural grasslands of Altiplano highlands are in surprisingly good condition, considering the region's history of overgrazing.	•Overgrazing and erosion continue.	•Complex cultural and economic incentives perpetuate overgrazing on communal lands of the Altiplano.	•Productivity of natural Altiplano grasslands might be increased several-fold with proper management, including reduction of grazing pressure, especially by sheep.	1, 9

Table 7.2

Region/Country Size of Rangeland	Range Condition	Range Trend	Causes	Remedial Activities	Source
Brazil Permanent Pasture[a]: 168,000,000 ha	•One half of the Brazilian cattle herd is supported on savanna, where poor forage quality in dry season limits yield. •Thorn-scrub rangeland in the northeast is over-grazed by goats and cattle and has suffered soil erosion and salinization. •Forest-derived pasture in the Amazon basin is productive for fewer than 10 years because of soil infertility.	•Extensive cattle grazing continues in savanna. •Goat production in thorn scrub is expanding to steeper slopes and erodible soils. •Pasture expansion in Amazon basin continues.	•Range management in savanna regions is not widely practiced. •Poor farmers are being displaced from high-quality ranges as large-scale farmers consolidate their holdings. •Pastures are used to secure land against expropriation in the Amazon basin.	•Fertilization and drought-resistant forage species can improve yields in savannas but are expensive. •Integrated brush management can reduce the negative impact of goats in the northeast. •Productivity of forest-derived pastures in the Amazon basin can be extended with mechanization, fertilization, and improved forage strains, but the cost is prohibitive.	1, 10, 11, 12, 13, 14
Colombia Permanent Pasture[a]: 40,083,000 ha	•Savannas in Orinoco basin have low rates of cattle production. •Forest-derived pastures in Amazon basin are short-lived and are subject to weed invasion. •Pastures of voluntary grasses support cattle at low grazing densities.	•Extensive cattle production with little management continues. •Amazon cattle pastures continue to be formed.	•High aluminum saturation and low soil fertility limit forage quality in savannas. •Amazon basin pastures are fragile because of soil infertility.	•The potential for range improvement in savannas is low except through fertilization. •The recuperation of degraded Amazon basin pasture using nongrass forage is being studied.	1, 12, 14
TEMPERATE SOUTH AMERICA					
Argentina Permanent Pasture[a]: 142,500,000 ha	•Dry *chaco* rangeland is degraded by shrub invasion and salinization. •Patagonia grasslands are overgrazed by sheep.	•High grazing pressure continues to degrade *chaco*. •The range in Patagonia is stable.	•Cattle grazing initially inhibited fires that kept weedy shrubs at bay. Now, fires are frequent but grasslands are unable to reestablish. •The lack of technical and infrastructural support has impeded range management in Patagonia.	•In the province of Salta, a *chaco* forest and rangeland were restored through control of livestock grazing and through grass and tree planting. •Research and extension are beginning to improve range management in Patagonia.	1, 15, 16, 17
Paraguay Permanent Pasture[a]: 19,960,000 Natural Grassland: 21,425,000	•Weed infestation is common in natural grasslands, particularly in *Chaco* region.	•Weed problems are expanding. •Soils are losing their fertility.	•Overgrazing permits weed infestation. •Frequent burning gradually degrades arid soils. •The majority of rangeland is not managed.	•Grazing limited to livestock carrying capacity (up to 0.5 animals per hectare in parts of the *Chaco*) could permit rangeland recuperation. •Weed control. •Grass cutting could prevent grass hardening.	1, 18

Sources:

1. Food and Agriculture Organization of the United Nations (FAO), unpublished data (FAO, Rome, July 1989).
2. Victor M. Toledo *et al.*, *La produccion rural en Mexico: Alternativas ecologicas* (Fundacion Universo Veintiuno, Mexico, D.F., 1989).
3. Jimmy T. LaBaume and Bill E. Dahl, "Communal Grazing: The Case of the Mexican Ejido," *Journal of Soil and Water Conservation*, Vol. 41, No. 1 (1986), pp. 24-27.
4. Gary Hartshorn *et al.*, "Costa Rica: Country Environmental Profile" (U.S. Agency for International Development, Washington, D.C., 1982), pp. 5-69.
5. H. Jeffrey Leonard, *Natural Resources and Economic Development in Central America: A Regional Profile* (International Institute for Environment and Development, Washington, D.C., 1987), pp. 73-130.
6. James D. Nations and Daniel I. Komer, "Central America's Tropical Rainforests: Positive Steps for Survival," *Ambio*, Vol. 12, No. 5 (1983), pp. 232-238.
7. Gary Hartshorn *et al.*, *Dominican Republic: Country Environmental Profile* (U.S. Agency for International Development, Washington, D.C., 1981) pp. 4-90.
8. Marko Ehrlich, *et al.*, *Haiti: Country Environmental Profile* (United States Agency for International Development, Washington, D.C., 1985), pp. 3, 36-39, 49-50.
9. Peter H. Freeman *et al.*, *Bolivia: State of the Environment and Natural Resources* (U.S. Agency for International Development, Washington, D.C., 1980), pp. 6.1 to 6.10.
10. Linda Howell Hardesty, "Multiple-Use Management in the Brazilian Caatinga," *Journal of Forestry*, Vol. 8 (August 1988), pp. 35-37.
11. Euclides Kornelius, "Pasture Establishment and Management in the Cerrado of Brazil," in *Pasture Production in Acid Soils of the Tropics: Proceedings of a Seminar Held at CIAT, Cali, Colombia, April 17-21, 1978*, P.A. Sanchez and L.E. Tergas, eds. (Centro Internacional de Agricultura Tropical, Cali, Colombia, 1979), pp. 147-166.
12. E.A.S. Serrao and J.M. Toledo, "Sustaining Pasture-Based Production Systems for the Humid Tropics," in T. Dowing, S. Hecht, and H. Pearson, *Development or Destruction of the Livestock Sector in Latin America*, Westview, Boulder, Colorado, 1990.
13. E.A.S. Serrao *et al.*, "Productivity of Cultivated Pastures on Low Fertility Soils in Amazon of Brazil," P.A. Sanchez and L.E. Tergas, eds. (Centro Internacional de Agricultura Tropical, Cali, Colombia, 1979) pp. 195-225.
14. Raul R. Vera and Carlos Sere R., eds., *Sistemas de produccion pecuaria extensiva: Brasil, Colombia, Venezuela* (Centro Internacional de Agricultura Tropical, Cali, Colombia, 1985), pp. 31-335 and 431-50.
15. Enrique H. Bucher and C.J. Schofield, "Economic Assault on Chagas Disease, *New Scientist*, Vol. 92 (October 29, 1981), pp. 321-324
16. Enrique H. Bucher, "Herbivory in Arid and Semi-Arid Regions of Argentina," *Revista Chilena de Historia Natural*, Vol. 60 (1987), pp. 265-273.
17. Guillermo E. Defossé and Ronald Robberecht, "Patagonia: Range Management at the End of the World, *Rangelands*, Vol. 9, No. 3 (1987), pp. 106-109.
18. International Institute for Environment and Development, *Paraguay: Environmental Profile of Paraguay* (U.S. Agency for International Development, 1985), pp. 101-128 and 143.

Subhumid tropical and temperate forests, often areas of rugged terrain and steep hillsides, are used as open range for goats that browse native trees and shrubs. Goat browsing and the widespread use of fire to promote grass cover degrade these forests by preventing the growth of new trees and killing mature trees (100).

Central America and the Caribbean

Permanent pastures, which occupy one fourth, or 19.5 million hectares, of the land surface in Central America and the Caribbean, account for 3.4 percent of Latin America's total permanent pastures. Six countries account for over 80 percent of the region's pastures: Nicaragua (27 percent), Cuba (14 percent), Honduras (13 percent), Costa Rica (12 percent), Dominican Republic (11 percent), and Guatemala (7 percent) (101).

Several characteristics of rangeland use and expansion in this region are leading to overgrazing and other management problems. First, livestock production is extensive rather than intensive; rarely practiced is pasture management through fertilization, introduction of high-quality forage species, or herd rotation. Second, the principal type of rangelands is beef-cattle pastures derived from tropical forests, which have erodible, marginally fertile soils. Third, pastures that have been planted on prime agricultural lands represent an underutilization of fertile soils. And fourth, farmers with little or no land of their own use dry forests and scrublands as open range for goats (102).

The impact of this pattern is exemplified in Costa Rica, where three fourths of the best agricultural land is devoted to cattle pastures (103). But pasture formation continues, despite the fact that logging and slash-and-burn deforestation are reducing tropical forests by roughly 60,000 hectares annually, and these practices are provoking erosion losses as high as 725.9 metric tons of soil per hectare per year in some areas (104) (105). By contrast, U.S. cropland was eroding at a rate of 16.3 metric tons per hectare in 1982 (106).

Most forest-derived pastures, which are productive for only a few years, are managed by occasional burning. The best managers of forest-derived rangelands appear to be small farmers, with less than 100 hectares, who produce up to twice the forage, meat, and milk per hectare produced by large-scale ranchers (107).

Tropical South America

This vast region encompasses the savannas, the Amazon basin, and the northern Andes. It comprises 1,348 million hectares, or two thirds of Latin America's land surface and one half its permanent pasture, about 287 million hectares (108).

The Savannas

The savannas of tropical South America extend over about 250 million hectares and include the Brazilian *cerrado*—the largest savanna in the Western Hemisphere (150 million hectares)—and the *llanos* of Colombia and Venezuela (more than 30 million hectares). Roughly one half the cattle in these three countries range on the native savanna vegetation, usually at low densities of about one animal per 5 hectares. Savanna soils, however, are acidic and not very fertile. They do not yield high-quality forage, and livestock production is impaired by nutritional deficiencies, particularly during the dry season.

Forage quality may be improved temporarily by burning, which releases nutrients in the vegetation to the soil as ash. The long-term effects of repeated burning are not yet known because most burning has taken place in the past decade. Significant improvements in the range quality of the *cerrado* and the *llanos* might be achieved through introducing cultivated grass and legume species (109) (110). (See Chapter 6, "Food and Agriculture," Sustainable Use of South America's Tropical Savannas.)

The *caatinga* of northeastern Brazil is also an important, though poorly studied, rangeland. A drought-stricken area covering 80 million hectares of thorn-scrub vegetation, it supports a large portion of Brazil's cattle and a large goat population.

Centuries of overgrazing, fire, and agricultural activity combined with episodic torrential rains and shallow soils have severely eroded the soil on most of the *caatinga*. Integrated brush management that promotes coppice growth for goat forage might improve livestock yields while conserving soil and vegetation (111).

The Amazon Basin

Since the 1960s, large tracts of forest in the Amazon basin have been converted to pastures. About 10 million hectares of Amazon forests in Brazil, 1.5 million hectares in Colombia, and 0.5 million hectares in Peru have given way (112). These pastures are created by cutting and burning the forests and then planting commercial grasses. Converting Amazon forests to pastures is encouraged by government subsidies; for farmers and ranchers, it establishes land ownership (113) (114). (See Box 7.3.)

Pastures carved from the Amazon forests are generally productive for less than 10 years, when weed invasion and phosphorus deficiencies limit forage growth. In parts of the western basin, however, volunteer grasses established in cultivated pastures can apparently support cattle for many years at low densities of one animal per 2 hectares (115). One negative impact of pasture expansion in the basin is large-scale burning of Amazon forests when fires escape during pasture burning.

Andean Rangelands

High-elevation grasslands and alpine vegetation of the Andes, the foundation of the Incan Empire, are still vital to the Andean economy. Almost the entire population of alpacas and llamas is found in these highlands; in Peru, goat, horse, and cattle raising is also concentrated in the mountains. Under proper management, alpine grasslands, such as the 17-million-hectare Altiplano of Bolivia and the *puna* of Peru, are among the richest, most productive ranges in the world (116) (117) (118). In general, the centuries of overgrazing since Spanish colonization have hidden the productive potential of Andean grasslands, but recent studies demonstrate the resilience of these ecosystems. In the Altiplano, for example, high-quality native forage is restored quickly with reduced grazing pressure (119).

Temperate South America

Latin American countries south of the Tropic of Capricorn comprise 20 percent of the region's land surface and one third, or 188 million hectares, of its permanent pastures (120). Three fourths of these pastures are in Argentina, where savannas, thorn scrubs, and natural grasslands dominate the landscape. Like other Latin American rangelands, it is marred with a legacy of degradation.

When Spanish colonists first penetrated the interior of Argentina, Paraguay, and Bolivia, the Gran Chaco—a region of about 100 million hectares—was predominately a park-like mosaic of dry woodlands interspersed with grasslands. Once cattle were introduced, wildfires had little fuel, unpalatable and thorny shrubs replaced the original vegetation, and populations of native mammals, such as the llama-like guanaco, declined (121)(122). Typical vegetation of Gran Chaco today is thorn scrub, with a cattle-carrying capacity of one adult animal per 2–20 hectares. Overgrazing by cattle and, in the semiarid region by goats, continues to degrade these rangelands, and forage production is limited in many areas by increased salinity (123).

Other rangelands of temperate South America have suffered similar fates. For decades, sheep have overgrazed the Patagonian rangelands of lower Argentina, which cover 90 million hectares. The 16 million sheep are managed much as they were at the turn of the century. Since the 1970s, studies to develop rangeland management practices for Patagonia have been done, but implementation of their recommendations is slow (124).

RECENT DEVELOPMENTS

THAILAND BANS LOGGING

In November 1988, the mountainous southern Thailand region was inundated with 40 inches of rain in five days. Thousands of cut logs, left on hillsides to dry, slid down to engulf entire villages. More than 350 people were killed.

In response to a public outcry that logging and the deforestation left in its wake, which greatly exacerbates soil erosion and water runoff, were to blame for the disaster, the Thai government in January 1989 banned all commercial logging, revoking all 301 logging concessions (125).

It reportedly approved the decrees after seeing satellite photographs indicating that the country's forest cover had declined to 19 percent in 1988 from 29 percent in 1985 (126). In the 1980s, Thailand lost about 2.5 percent of its forest area annually. (See Chapter 19, "Forest and Rangelands," Table 19.1.)

Some reports characterized the government action as a "well-calculated political move" taken in response to the public outcry over the flooding victims rather than as a policy decision to preserve the country's forest resources. Nevertheless, public sentiment seems strong enough that the decision will not be reversed easily (127).

Thai logging companies pressured the government for compensation, and they scrambled to find other sources of supply, an effort that may increase logging activity in other countries and at least partially offset the impact of the ban in Thailand. Myanmar and Laos have granted limited concessions to Thai logging firms (128). Illegal tree harvesting within Thailand is likely to increase as the ban pushes up the price of wood.

Other factors could undermine the ban. For example, the government, trying to reduce rural landlessness, settled some 1.2 million villagers on logged forest reserves in the past decade. Government policy also encouraged illegal squatters to settle along timber roads; an estimated 5 million people now inhabit forest reserves. These farmers may be required to plant trees on up to 30 percent of their land, but their cultivation practices could seriously threaten some of Thailand's forested watersheds (129).

Many other factors—harvesting forest products by rural dwellers for personal use, illegal commercial logging, and land clearing for cultivation—are cited as important contributors to Thailand's forest loss (130). Clearing and burning vegetative cover for agricultural plots is widespread in the north, west, and northeast regions of the country; the result is that some 500,000 hectares are cleared annually (131).

Forest losses have prompted action by other governments in the region. Indonesia, partly to develop its plywood processing capacity, stopped all log exports in 1985. According to one report, the ban stimulated development of pulp and plywood processing plants but is less effective as a conservation measure (132). There are some signs that the Indonesian government is getting tougher in enforcing its environmental laws; for example, the government reportedly revoked 70 forestry concessions in 1988–89 (133).

The Philippines, after heavily logging easily accessible areas in the 1960s, imposed a ban on logging in six provinces and for the nation on log exports in 1976; however, the log export ban was not fully applied until 1986. Log exports are under a near-total ban in peninsular Malaysia, but Sabah and Sarawak, which have their own forest departments and forestry codes, continue to be heavy exporters of logs (134).

Japan, which imports a large percentage of its lumber from Southeast Asia, has been criticized heavily. An April 1989 World Wildlife Fund analysis of Japan's tropical timber trade concluded that Japanese forest resource policies have affected the region negatively. It suggests that Japanese lumber companies, with heavy investments in the region, have promoted forest management systems that provide a fast return on investment at the expense of sustainable forest management (135).

BURNING OF AFRICAN SAVANNAS LINKED TO ACID RAIN

The frequent burning of savannas by farmers and herdsmen appears to contribute to unusually high levels of ozone and acid rain over the rainforests of Central Africa.

Scientists at the Max Planck Institute for Chemistry in Mainz, West Germany, found that fires in savanna regions appear to produce three to four times more emissions of carbon dioxide and other trace gases than deforestation in the tropics. Their study used FAO data on deforestation and agricultural land use changes for 1975–80 (136).

About one half the total emissions from biomass burning in the tropics are caused by fires in Africa's savannas; the fires set by farmers and herdsmen to kill insects and pests, remove dead grass, clear shrubs, and stimulate new growth. An estimated 75 percent of the entire 591 million hectares of Africa's savannas are burned annually. These fires consume about 2.4 billion metric tons of biomass each year, or

about 85 percent of the total biomass burned in tropical Africa (137). The frequency of savanna burning appears to be increasing; once burned every three years, many are now burned every one to two years (138).

Most of the carbon emissions from the savanna fires are ultimately reabsorbed when the savanna grasses come up the following year (139). The net release of carbon is caused mainly by deforestation, shifting cultivation, and burning fuelwood (140).

As the northeast winds blow over Africa's tropical forests, they carry ozone and hydrocarbons plus formic, acetic, and nitric acids and other pollutants released by fires. Surface ozone concentrations in excess of 40 parts per billion(ppb) (by comparison, the Canadian standard is 15 ppb) are frequently found in tropical Africa during the dry season (141). These emissions appear to contribute to high acid rain and ozone levels in the forests of Côte d'Ivoire, the Congo, Gabon, and Zaire. French scientists measured acid rain of pH 4.4–4.6 in the humid air over the forests of the north Congo and Côte d'Ivoire (142) (143). German researchers also found ozone at 60–70 parts per billion. The pollutants also affect the atmosphere thousands of kilometers from the fires (144).

Though they do not yet know how sensitive tropical forests may be to acid rain and ozone pollution, scientists are concerned that the high levels of pollution found by the recent studies may endanger the forests (145).

The section on forests was written by David Harmon, a conservation consultant in Houghton, Michigan, Mary Paden, managing editor of World Resources 1990-91, and Norbert Henninger, research associate for World Resources 1990-91. Judith Moore, a Washington, D.C., rangelands consultant authored the effects of climate change on rangelands. Daniel Nepstad, an Amazon expert with the National Wildlife Federation, authored the section on rangelands and the box on pastures in the Amazon, with assistance from Alberto Vargas, a conservation consultant in Madison, Wisconsin.

References and Notes

1. Jean-Paul Lanly, *Tropical Forest Resources* (Food and Agriculture Organization of the United Nations, Rome, 1982).
2. Food and Agriculture Organization of the United Nations (FAO), *An Interim Report on the State of Forest Resources in the Developing Countries* (FAO, Rome, 1988), pp. 1 and 8-9.
3. Douglas C. Daly and Ghilean T. Prance, "Brazilian Amazon," in *Floristic Inventory of Tropical Countries: The Status of Plant Systematics, Collection, and Vegetation, plus Recommendations for the Future.* David G. Campbell and H. David Hammond, eds. (New York Botanical Garden, New York, 1989), p. 402.
4. Alberto Waingort Setzer *et al.*, "Relatório de atividades do projeto IBDF-INPE 'SEQE'—ano 1987," abstract, National Space Research Institute of Brazil (INPE),— São José dos Campos, São Paulo, Brazil, 1988.

5. Santiago Mourao, Head of Environmental Affairs, Embassy of Brazil, Washington, D.C., 1990 (personal communication).
6. James Brooke, "Rain and Fines, but Mostly Rain, Slow Burning of Amazon Forest in Brazil," *New York Times* (September 17, 1989), international edition, p. 22.
7. Ricardo Bonalume Neto, "Burning Continues, Slightly Abated," *Nature*, Vol. 339, No. 6226 (June 22, 1989), p.569.
8. *Op. cit.* 5.
9. Roberto Pereira da Cunha, "Deforestation Estimates through Remote Sensing: The State of the Art in the Legal Amazonia." National Space Research Institute of Brazil (INPE), São José dos Campos, São Paulo, Brazil, 1989, n.p.
10. Philip Fearnside, "Deforestation in Brazilian Amazonia," in *The Earth in Transition: Patterns and Process of Impoverishment,* George Woodwell, ed. (Cambridge University Press, New York, in press).

11. *Op. cit.* 9.
12. Jose Goldemberg, President, University of São Paulo, São Paulo, Brazil, 1989 (personal communication).
13. B.B. Vohra, "Confusion on the Forestry Front," based on reconciled forestry figures in National Remote Sensing Agency, "Mapping of Froest Cover in India from Satellite Imagery" Government of India, Hyderbad, 1983).
14. Ibid., p. 4.
15. Jean-Paul Lanly, Director, Forest Resources Division, Forestry Department, Food and Agriculture Organiztion of the United Nations, Rome, 1989 (personal communication).
16. International Institute for Environment and Development and World Resources Institute, *World Resources 1988-89* (Basic Books, New York, 1987), pp. 195, 229.
17. "Thai Logging Ban Win Wide Acclaim," *New Straits Times* (January 12, 1989) reprinted

in *The Battle for Sarawak's Forests* (World Rainforest Movement and Sahabat Alam Malaysia, Penang, Malaysia, 1989), p. 191.

18. *Op. cit.* 14.

19. International Institute for Environment and Development and World Resources Institute. *World Resources 1987* (Basic Books, New York, 1987), pp. 175-176.

20. Jean-Paul Malingreau and Compton J. Tucker, "Large-Scale Deforestation in the Southeastern Amazon Basin of Brazil," *Ambio*; Vol. 17, No. 1 (1988), p. 49.

21. Incomparability was precisely the case when the FAO survey seemed to be at odds with a 1980 report to the U.S. National Academy of Sciences; later, the key conclusions of the two were generally reconciled. See J.M. Melillo *et al.*, "A Comparison of Two Recent Estimates of Disturbance in Tropical Forests," *Environmental Conservation*, Vol. 12, No. 1 (1985), pp. 37-40.

22. The census, whose reference year is 1990, will expand its use of satellite pictures and will combine computerized maps and various data bases into a geographical information system covering the world's tropical forests. Food and Agriculture Organization of the United Nations (FAO), *An Interim Report on the State of Forest Resources in the Developing Countries* (FAO, Rome, 1988), pp. 1-2 and Annex 1, pp. 12-13.

23. The Indonesian Forestry Community, "Indonesia: Tropical Forests Forever," *U.S. News and World Report* (December 18, 1989), pp. 80-81.

24. Robert Winterbottom, Forestry Program Director, World Resources Institute, Washington, D.C., 1990 (personal communication).

25. Norman Myers, *The Primary Source: Tropical Forests and Our Future* (W.W. Norton and Co., New York, 1984), pp. 144-146.

26. Nicholas Guppy, "Tropical Deforestation: A Global View," *Foreign Affairs*, Vol. 62, No. 4 (1984), pp. 928-965.

27. Robert Repetto, "Overview," in *Public Policies and the Misuse of Forest Resources,* Robert Repetto and Malcolm Gillis eds. (Cambridge University Press, New York, 1988), pp. 15-17.

28. Elaine Matthews, "Global Vegetation and Land Use," *Journal of Climate and Applied Meteorology*, Vol. 22, March 1983, pp. 474-487.

29. *Op. cit.* 27, pp. 10-12.

30. *Op. cit.* 27, p. 14.

31. *Op. cit.* 27, p. 14.

32. *Op. cit.* 27.

33. Robert Repetto, *World Enough and Time: Successful Strategies for Resource Management* (Yale University Press, New Haven, Connecticut, 1986).

34. Robert Repetto, *The Forest for the Trees? Government Policies and the Misuse of Forest Resources* (World Resources Institute, Washington, D.C., 1988).

35. International Conservation Financing Project, *Natural Endowments: Financing Resource Conservation for Development* (World Resources Institute, Washington, D.C., 1989).

36. Robert Repetto *et al.*, *Wasting Assets: Natural Resources in the National Income Accounts* (World Resources Institute, Washington, D.C. 1987).

37. *Op. cit.* 27, pp. 12.

38. Daniel Janzen, Honorary professor at the Universidad de Costa Rica, San Pedro, Costa Rica, and at the Universidad Nacional, Heredia, Costa Rica, 1989 (personal communication).

39. Food and Agriculture Organization of the United Nations (FAO) in cooperation with the World Resources Institute, the World Bank, and the United Nations Development Programme, *The Tropical Forestry Action Plan* (FAO, Rome, 1987).

40. John Spears, "A Reappraisal of Past and Future TFAP Objectives," Consultative Group on International Agricultural Research (CGIAR), Washington, D.C., 1989.

41. World Resources Institute (WRI), "Preliminary Analysis of 9 National TFAP Reports," presented at WRI workshop on country level TFAP exercises, Santo Domingo, October, 1989, p.7.

42. Louis Harris and Associates, Inc. (LHA), *Public and Leadership Attitudes to the Environment in Four Continents: A Report of a Survey in 14 Countries* (LAH, New York, 1988), Table 2-1A, p. 75; Table 2-1B, p. 77; and Table 2-1C, p. 79.

43. Robert Winterbottom, Forestry Program Director, World Resources Institute, Washington D.C., 1989 (personal communication).

44. Daniel B. Botkin and Robert A. Nisbet, "Projecting the Effects of Climate Change on Biological Diversity in Forests," in *Consequences of the Greenhouse Effect for Biological Diversity*, Robert L. Peters II, ed. (Yale University Press, New Haven, Connecticut, in press).

45. Mark C. Trexler, Paul E. Faeth, and John Michael Kramer, "Forestry as a Response to Global Warming: An Analysis of the Guatemala Agroforestry and Carbon Sequestration Project," World Resources Institute, Washington, D.C., 1989, p. 33.

46. U.S. Environmental Protection Agency, *Policy Options for Stabilizing Global Climate*, Daniel A. Lashof and Dennis A. Tirpak, eds. (U.S. Government Printing Office, Washington, D.C., 1989), Vol. 2, p. VII-195.

47. Richard A. Houghton, "Estimating Changes in the Carbon Content of Terrestrial Ecosystems from Historical Data," in *The Changing Carbon Cycle: A Global Analysis*, John R. Trabalka and David E. Reichle, eds. (Springer-Verlag, New York, 1986), p. 175.

48. *Op. cit.* 46, Vol. 2, p. VII-196.

49. G.M. Woodwell *et al.*, "Global Deforestation: Contribution to Atmospheric Carbon Dioxide," *Science*, Vol. 222 (12983), pp. 1081-1086; R.P. Detwiler, Charles A.S. Hall, and P. Bodgonoff, "Land Use Change and Carbon Exchange in the Tropics: Estimates for the Entire Region," *Environmental Management*, Vol. 9 (1985), pp. 335-344; R.P. Detwiler and Charles A.S. Hall, "Tropical Forests and the Global Carbon Cycle," *Science*, Vol. 239, No. 4835 (Jan. 1, 1988), pp. 42-47; R.A. Houghton *et al.*, "Net Flux of Carbon Dioxide from Tropical Forests in 1980," *Nature*, Vol. 316 (1985), pp. 617-620; Mark C. Trexler, Paul E. Faeth, and John Michael Kramer, "Forestry as a Response to Global Warming: An Analysis of the Guatemala Agroforestry and Carbon Sequestration Project," World Resources Institute, Washington, D.C., 1989, p. 6.

50. R.A. Houghton *et al.*, "The Flux of Carbon from Terrestrial Ecosystems to the Atmosphere in 1980 due to Changes in Land Use: Geographic Distribution of the Global Flux," *Tellus*, Vol. 39B, Nos. 1-2 (1987), Table 4, pp. 128-129.

51. Mark C. Trexler, Paul E. Faeth, "Can We Control the Carbon Dioxide in the Atmosphere?" *Energy*, Vol. 2, No. 3 (1977), pp. 287, 291.

52. *Op. cit.* 46, Vol. 2, pp. 206-217, and 234-235.

53. William R. Emanuel, Herman H. Shugart, and Mary P. Stevenson, "Climatic Change and the Broad-Scale Distribution of Terrestrial Ecosystem Complexes," *Climatic Change*, Vol. 7, No. 1 (1985), pp. 33-41.

54. *Ibid.*

55. Jerry F. Franklin *et al.*, "Effects of Global Climate Change on Forests in Northwestern North America," in *Consequences of the Greenhouse Effect for Biological Diversity*, Robert L. Peters II, ed. (Yale University Press, New Haven, Connecticut, in press).

56. Margaret B. Davis and Catherine Zabinski, "Changes in Geographical Range Resulting from Greenhouse Warming: Effects on Biodiversity in Forests," in *Consequences of the Greenhouse Effect for Biological Diversity*, Robert L. Peters II, ed. (Yale University Press, New Haven, Connecticut, in press).

57. *Op. cit.* 55.

58. Margaret B. Davis, "Lags in the Response of Forest Vegetation to Climatic Change," *Climate-Vegetation Interactions*, Cynthia Rosenzweig and Robert Dickinson, eds. (University Corporation for Atmospheric Research, Boulder, Colorado, 1986), p. 70.

59. F.I. Woodward, "Review of the Effects of Climate on Vegetation: Ranges, Competition, and Composition," in *Consequences of the Greenhouse Effect for Biological Diversity*, Robert L. Peters II, ed. (Yale University Press, New Haven, Connecticut, in press).

60. *Op. cit.* 53.

61. *Op. cit.* 55.

62. *Op. cit.* 53.

63. Andrew Dobson and Robin Carper, "Global Warming and Potential Changes in Host-Parasite and Disease-Vector Relationships," in *Consequences of the Greenhouse Effect for Biological Diversity*, Robert L. Peters II, ed. (Yale University Press, New Haven, Connecticut, in press).

64. See, for example, U.S. Environmental Protection Agency, *The Potential Effects of Global Climate Change on the United States*, Joel B. Smith and Dennis A. Tirpak, eds. (U.S. Government Printing Office, Washington, D.C., 1988), pp. 11-1, and 11-26.

65. "Executive Summary," in *Direct Effects of Increasing Carbon Dioxide on Vegetation*, Boyd R. Strain and Jennifer D. Cure, eds. (U.S. Department of Energy, Washington, D.C., 1985), pp. xvii and xxii.

66. *Op. cit.* 53.

67. Peter R. Crosson and Norman J. Rosenberg, "Strategies for Agriculture," *Scientific American*, Vol. 261, No. 3 (1989), p. 128.

68. *Op. cit.* 65, p. xxii.

69. Jennifer D. Cure, "Carbon Dioxide Doubling Responses: A Crop Survey," *Direct Effects of Increasing Carbon Dioxide on Vegetation*, Boyd R. Strain and Jennifer D. Cure, eds. (U.S. Department of Energy, Washington, D.C., 1985), p. 114.

70. S.D. Smith, B.R. Strain, and T.D. Sharkey, "Effects of CO_2 Enrichment on Four Great Basin Grasses," *Functional Ecology 1987*, Vol. 1, No. 2 (1987), p. 139.

71. S.M. Wray and B.R. Strain, "Interaction of Age and Competition under CO_2 Enrichment," *Functional Ecology 1987*, Vol. 1, No. 2 (1987), p. 149.

72. George H. Reichers and Boyd R. Strain, "Growth of Blue Grama (*Bouteloua gracilis*) in Response to Atmospheric CO_2 Enrichment," *Canadian Journal of Botany*, Vol. 66, No. 8 (1988), p. 1572.

73. Susan Marks and Boyd R. Strain, "The Effects of Drought and CO_2 Enrichment on Competition between Two Old-Field Perennials," *New Phytologist* Vol. 111, No. 2 (1989), p. 181.

74. James Coleman, Post-Doctoral Fellow, Department of Organismic and Evolutionary Biology, Harvard University, Cambridge, Massachusetts, 1989 (personal communication).

75. Walter Oechel and Boyd R. Strain, "Native Species Responses to Increased Atmospheric Carbon Dioxide Concentration," *Di-*

rect Effects of Increasing Carbon Dioxide on Vegetation (U.S. Department of Energy, Washington, D.C., 1985), p. 135.

76. Op. cit. 70.

77. Op. cit. 75, pp. 130-132.

78. Eric D. Fajer, M. Deane Bowers, and Fakhri A. Bazzaz, "The Effects of Enriched Carbon Dioxide Atmospheres on Plant-Insect Herbivore Interactions," Science, Vol. 243, March 3, 1989, p. 1200.

79. John R. Mather and Johannes Feddema, "Hydrologic Consequences of Increases in Trace Gasses and CO_2 in the Atmosphere," in Climate Change, Vol. 3 of Effects of Changes in Stratospheric Ozone and Global Climate, U.S. Environment Programme and United States Environmental Protection Agency, Washington, D.C., 1986), p. 267.

80. R.B. Stewart, "Climatic Change—Implications for the Prairies," in Climate Change, Vol. 3 of Effects of Changes in Stratospheric Ozone and Global Climate (United Nations Environment Programme and U.S. Environmental Protection Agency, 1986), pp. 123 and 132.

81. Op. cit. 79, Figure 3, p. 268, and Figure 4, p. 269.

82. Christopher Uhl, J. Boone Kauffman and Dian L. Cummings, "Fire in the Venezuelan Amazon 2: Environmental Conditions Necessary for Forest Fires in the Evergreen Rainforest of Venezuela," Oikos, Vol. 53, No. 2 (1988), p. 178.

83. Carl F. Jordan, Nutrient Cycling in Tropical Forest Ecosystems (John Wiley & Sons, New York, 1985), p. 124-128.

84. Ibid., p. 127.

85. J.A. Young and R.A. Evans, "Population Dynamics after Wildfires in Sagebrush Grasslands," Journal of Range Management, Vol. 31 (1978), pp. 283-289, cited in S.D. Smith, B.R. Strain, and T.D. Sharkey, "Effects of CO_2 Enrichment on Four Great Basin Grasses," Functional Ecology 1987, Vol. 1, No. 2 (1987), p. 142.

86. World Resources Institute and International Institute for Environment and Development, World Resources 1986 (Basic Books, New York, 1986), Table 6.3, p. 278.

87. World Resources Institute and International Institute for Environment and Development in collaboration with the United Nations Environment Programme, World Resources 1988-89 (Basic Books, New York), Table 20.4, p. 289.

88. World Resources Institute in collaboration with the United Nations Environment Programme, World Resources 1990-91 (Oxford University Press, New York, 1990), Table 18.5, p. 291.

89. James Detling, Professor, Department of Range Science, Colorado State University, Fort Collins, 1989 (personal communication).

90. Food and Agriculture Organization of the United Nations (FAO), unpublished data (FAO, Rome, July 1989).

91. Victor M. Toledo et al., La producción rural en México: Alternativas ecológicas (Fundacion Universo Veintiuno, Mexico, D.F., 1989).

92. Enrique H. Bucher, "Herbivory in Arid and Semi-Arid Regions of Argentina," Revista Chilena de Historia Natural, Vol. 60 (1987), pp. 270-272.

93. Peter H. Freeman et al., Bolivia: State of the Environment and Natural Resources (U.S. Agency for International Development,

Washington, D.C., 1980), pp. 6-1, 6-2, and 6-8.

94. Guillermo E. Defossé and Ronald Robberecht, "Patagonia: Range Management at the End of the World," Rangelands, Vol. 9, No. 3 (1987), pp. 107-109.

95. James D. Nations and Daniel I. Komer, "Central America's Tropical Rainforests: Positive Steps for Survival," Ambio, Vol. 12, No. 5 (1983), pp. 233-234, and Box 1, pp. 236-237.

96. Robert J. Buschbacher, "Tropical Deforestation and Pasture Development," BioScience, Vol. 36, No. 1, (1986), pp. 23-25, and 27.

97. Jimmy T. LaBaume and Bill E. Dahl, "Communal Grazing: The Case of the Mexican Ejido," Journal of Soil and Water Conservation, Vol. 41, No. 1 (1986), pp. 25-27.

98. Op. cit. 91.

99. A.M. Peralta and A.S. Ramos, "Diagnóstico de los sistemas de producción bovina en el trópico de México" (Centro Internacional de Agricultura Tropical, Cali, Colombia, 1988), as cited in E.A. Serrao and J.M. Toledo, Sustaining Pasture-Based Production Systems for the Humid Tropics," in T. Downing, S. Hecht, and H. Pearson, Development or Destruction of the Livestock Sector in Latin America, Westview, Boulder, Colorado, 1990.

100. Op. cit. 91.

101. Op. cit. 90.

102. H. Jeffrey Leonard, Natural Resources and Economic Development in Central America: A Regional Environmental Profile (International Institute for Environment and Development, Washington, D.C., 1987), pp. 86-92, 98-109, and 113-129.

103. Gary Hartshorn et al., "Costa Rica: Country Environmental Profile" (U.S. Agency for International Development, Washington, D.C., 1982), pp. 5 and 6.

104. James D. Nations and Daniel I. Komer, "Central America's Tropical Rainforests: Positive Steps for Survival," Ambio, Vol. 12, No. 5 (1983), Table 1, p. 232 and Box 1, pp. 236-237.

105. Op. cit. 103, p. 58.

106. Op. cit. 87, Table 17.6, p. 282.

107. Op. cit. 103.

108. Op. cit. 90.

109. Euclides Kornelius et al., "Pasture Establishment and Management in the Cerrado of Brazil," in Pasture Production in Acid Soils of the Tropics, P.A. Sanchez and L.E. Tergas, eds. (Centro International de Agricultura Tropical, Cali, Colombia, 1979), pp. 147-166.

110. Raúl R. Vera and Carlos Seré R., eds., Sistemas de Produccion Pecuaria Extensiva: Brasil, Colombia, Venezuela (Centro Internacional de Agricultura Tropical, Cali, Colombia, 1985), pp. 31-450.

111. Linda Howell Hardesty, "Multiple-Use Management in the Brazilian Caatinga," Journal of Forestry, Vol. 8 (August 1988), pp. 35-37.

112. E.A. Serrao and J.M. Toledo, "Sustaining Pasture-Based Production Systems for the Humid Tropics," T. Downing, S. Hecht, and H. Pearson, Development or Destruction of the Livestock Sector in Latin America, Westview, Boulder, Colorado, 1990.

113. Daniel C. Nepstad and Christopher Uhl, "Alternatives to Regional Forest Degradation in Eastern Amazonia," Ambio (in press).

114. E.A. Serrao et al., "Productivity of Cultivated Pastures on Low Fertility Soils in Amazon of Brazil," in Pasture Production in

Acid Soils of the Tropics, P. A. Sanchez and L.E. Tergas, eds. (Centro International de Agricultura Tropical, Cali, Colombia, 1979), pp. 195-225.

115. Op. cit. 112.

116. U.S. Library of Congress, "Draft Environmental Report on Peru," U.S. Agency for International Development, Washington, D.C., October 1979.

117. Brad Wilcox, "The Puna—High Elevation Grassland of the Andes," Rangelands, Vol. 6, No. 3 (1984), pp. 99-101.

118. Op. cit. 93, pp. 6-1 to 6-2.

119. Op. cit. 93, pp. 6-2 to 6-8.

120. Op. cit. 90.

121. Op. cit. 92, pp. 270-271.

122. E.H. Bucher and C.J. Schofield, "Economic Assault on Chagas Disease," New Scientist, Vol. 92 (October 29, 1981), pp. 322-324.

123. Utz Baum, "Evaluacion del estado actual de algunas recursos pastoriles en Argentina, Bolivia, Brasil y Paraguay," unpublished, Buenos Aires, 1989.

124. Op. cit. 94, pp. 106 and 108-109.

125. Paisal Sricharatchanya, "Getting Lumbered," Far Eastern Economic Review (February 2, 1989), p. 26.

126. Ibid.

127. Ibid.

128. Meri McCoy-Thompson, "Sliding Slopes Break Thai Logjam," World Watch (September/October 1989), p. 9.

129. Ibid., p. 8.

130. David Feeny, "Agricultural Expansion and Forest Depletion in Thailand, 1900-1975," in World Deforestation in the Twentieth Century, John F. Richards and Richard P. Tucker, eds. (Duke University Press, Durham, North Carolina, 1988), p. 130.

131. Ibid., p. 126.

132. Michael Vatikiotis, "Tug-of-War over Trees," Far Eastern Economic Review (January 12, 1989), p. 41.

133. Steven Erlanger, "Indonesia Takes Steps to Protect Rain Forests," New York Times (September 26, 1989), p. C4.

134. François Nectoux and Yoichi Kuroda, Timber from the South Seas: An Analysis of Japan's Tropical Timber Trade and Its Environmental Impact (WWF International, Gland, Switzerland, 1988), pp. 34-36.

135. Ibid., pp. 34-35.

136. W.M. Hao, M.H. Liu, and P.J. Crutzen, "Estimates of Annual and Regional Releases of CO_2 and Other Trace Gases to the Atmosphere from Fires in the Tropics, Based on the FAO Statistics for the Period 1975-80," in Proceedings of the Third International Symposium on Fire Ecology (Springer-Verlag, New York, forthcoming).

137. Ibid.

138. Marlise Simons, "High Ozone and Acid-Rain Levels Found Over African Rain Forests," New York Times (June 19, 1989), p. A1.

139. Meinrat O. Andrae, "Biomass Burning in the Tropics: Impact on Environmental Quality and Global Climate," Special supplement for Population and Development Review (in press).

140. Op. cit. 136.

141. Op. cit. 139.

142. Op. cit. 139.

143. Op. cit. 139.

144. Op. cit. 138.

145. Op. cit. 138.

8. Wildlife and Habitat

Wild plants and animals face an increasingly uncertain future, squeezed out of their habitats by both human development and by possible radical changes caused by global warming.

Humans have altered wildlife habitat for so long and so thoroughly that some long-populated, now-barren areas are hard to imagine as covered with ancient forests or other forms of foliage. This alteration has increased swiftly in recent decades and is now cause for deep concern. For example, the world's moist tropical forests—one of the richest habitats in terms of the number of species—have been reduced by 44 percent. (See Moist Forests, below.) Chapter 7, "Forests and Rangelands," discusses increasing deforestation rates for several tropical countries. This chapter describes what is known about the presettlement and current extents of a broad range of habitats.

To the threat of human encroachment on wildlife habitat must now be added the potential threat of climate change caused by global warming. Responding to climate computer models developed over the past few years, wildlife biologists and botanists have projected how the increase of a few degrees in temperature and possible changes in rainfall patterns might affect various species. One effect would be the break up of communities of plants and animals as each community member migrates at a different rate to follow its optimal conditions. In their migrations, most wildlife will encounter human-made obstacles such as cities, roads, and dams that may block their progress. (See How Might Communities and Species Respond? below.)

Attempts to preserve biological diversity range from establishing protected habitats, to incentives for protecting wildlife on private lands, to *ex situ* preservation in zoos and seed banks. None of these efforts can stand alone, but rather all must be integrated. Previous volumes of *World Resources* have discussed the world's systems of protected areas, ranching wildlife for profit and preservation, and combining wildlife preservation with sustainable development. This volume describes methods of conserving biological diversity outside protected areas and *ex situ* preservation of animals and plants.

CONDITIONS AND TRENDS

BIODIVERSITY AND HABITAT LOSS

Most people live, unwittingly, in a transfigured landscape. We give little thought to how profoundly our species has reapportioned Earth's natural habitats. Our ideas of the normal distribution of wild plants and animals are shaped during a brief life span of only a few decades. Yet today's eroding hillsides and degraded maquis vegetation of the Mediterranean re-

Datapoints Europe and North America are Major Importers of Wild Animals and Skins

Trade in animals and their products is especially worrisome when it leads to overexploitation or affects animals already threatened by loss of habitat. This graph shows the demand for four threatened animals or animal products.

The World Conservation Union, formerly the International Union for the Conservation of Nature and Natural Resources (IUCN), monitors the legal international trade in threatened animals. Signatories to the Convention on the International Trade in Endangered Species of Wild Flora and Fauna (CITES) supply to the World Conservation Union data on the import and export of listed species or products made from listed species. These data do not include illegal trade or legal domestic trade.

North America is the most important market for primates, used primarily in medical research, followed by Europe. The fur and leather trade supplies Europe, and to a lesser extent Asia with cat, crocodile, lizard, and snake skins used for fashionable clothing and accessories. The demand for parrots as pets is strongest in the developed world, especially North America and Europe.

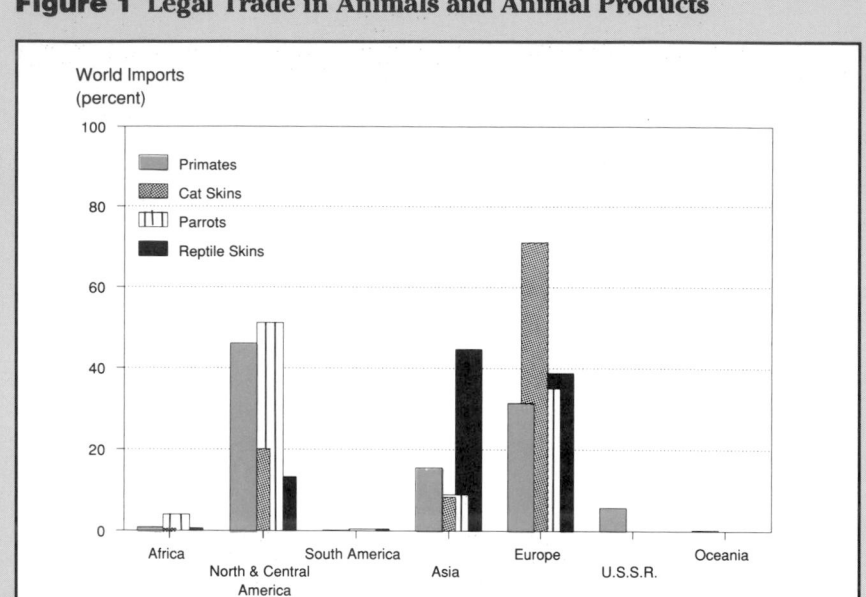

Figure 1 Legal Trade in Animals and Animal Products

Source: Chapter 20, "Wildlife and Habitat," Table 20.3.

gion give no hint of the forests that were known to antiquity. The Mediterranean may still be able to support wildlife, but in terms of its capacity for biological diversity, it is impoverished(1). So are the state of Iowa in the United States, the African Sahel, the Atlantic seaboard of Brazil, and countless other places where the needs of wild plants and animals are drastically subordinated to human demands for agriculture, transportation, energy, or housing (2) (3) (4).

Biological diversity, or *biodiversity*, is a global resource made up of the variety and variability of life forms on Earth, both wild and domesticated (5). The extent of this resource can be reckoned three ways. On the finest scale, there is *genetic diversity*, the differences in genetic makeup among individual organisms. On the broadest scale is *ecosystem diversity* comprising the distinctive assemblages of organisms that occur in different physical settings. In between is the most common measure, *species diversity* (6).

Each standard has its strengths. Genetic diversity is what allows organisms to adapt to environmental change—an important consideration as the world enters an unprecedented period of rapid global warming. Ecosystem diversity, because it is all-encompassing, is the best gauge of conditions and trends in global biodiversity. Species diversity has the advantage of being the most readily recognized by nonscientists and so can be used as a rallying point by advocates of conserving biodiversity.

By any of these measures, wildlife—both animals and plants—accounts for most of the world's biodiversity. Because the well-being of wildlife depends largely on the condition of its habitat (and because wildlife, in terms of wild plants, is an important component of habitat), habitat loss is a good yardstick of how much global biodiversity is declining.

No one expects civilization to be dismantled and the planet returned to a pristine state in the name of wildlife and its habitat. But diminishing biodiversity does carry a high cost (7). The cost is largely unnoticed precisely because most people live where the planet's biotic potential has long been obscured or because they could always take wildlife for granted. The concept of conserving biodiversity challenges this anthropocentrism because it implies (if not an equality among all living things) an imperative for humans to take the well-being of other species into account. There are cogent ethical and economic reasons for this thinking (see *World Resources 1988–89*, p. 90), but perhaps the most important is that science is constantly discovering ways biodiversity can alleviate human suffering and environmental destruction (8).

Table 8.1 Some Estimates of the Area of the World's Major Terrestrial Ecosystems, 1975-83

(million square kilometers)

	Whittaker & Likens (1975)(a)	Atlay, Ketner & Duvigneaud (1979)(b)	Olson, Watts & Allison (1983)(c)	Difference between high & low estimates
Forests	57.0	34.0	50.7	23.0
Tropical rainforest	17.0	10.0	12.0	7.0
Tropical seasonal forest	7.5	4.5	6.0	3.0
Temperate forest	12.0	6.0	8.2	6.0
Boreal forest	12.0	9.0	11.7	3.0
Woodland, shrubland, interrupted woods	8.5	4.5	12.8	8.3
Grasslands, Drylands	74.0	90.0	78.3	16.0
Savannah	15.0	22.5	24.6	9.6
Temperate grasslands	9.0	12.5	6.7	5.8
Tundra, alpine vegetation	8.0	9.5	13.6	5.6
Desert, semidesert	18.0	21.0	13.0	8.0
Extreme desert	24.0	24.5	20.4	4.1
Wetlands	4.0	5.5	6.1	2.1
Swamps, marshes, coastal land	2.0	2.0	2.5	0.5
Bogs, peatland	X	1.5	0.4	0.9
Lakes, streams	2.0	2.0	3.2	1.2
Developed Lands	14.0	18.0	15.9	4.0
Cultivated land	14.0	16.0	15.9	2.0
Built-up areas	X	2.0	X	NA
Total	149.0	147.5	151.0	3.5

Sources:
1. Robert H. Whittaker and Gene E. Likens, "The Biosphere and Man," in *Primary Productivity of the Biosphere*, Helmut Lieth and Robert H. Whittaker, eds. (Springer-Verlag, Berlin, 1975), Table 15-1, p. 306.
2. G.L. Atjay, P. Ketner, and P. Duvigneaud, "Terrestrial Primary Production and Phytomass," in *The Global Carbon Cycle*, B. Bolin et al., eds. (John Wiley & Sons, Chichester, U.K., 1979), pp. 129-182.
3. J.S. Olson, J.A. Watts, and L.J. Allison, *Carbon in Live Vegetation of Major World Ecosystems* (U.S. Department of Energy, U.S. Government Printing Office, Washington, D.C., 1983), cited in B. Bolin, How Much CO_2 Will Remain in the Atmosphere?" in *The Greenhouse Effect, Climatic Change, and Ecosystems*, Bert Bolin et al., eds. (John Wiley & Sons, Chichester, U.K., 1986), Table 3.2, p. 124.

Notes:
a. Data from Whittaker and Likens are unchanged.
b. "Mangrove forests" in Atjay, Ketner, and Duvigneaud are included here as tropical rainforest and their "forest plantation" divided between temperate and boreal forest.
c. "Interrupted woods" in Olson, Watts, and Allison are divided among tropical, temperate, and boreal forests here if the authors classified them as "second woods and field mosaic." "Tropical savanna and woodland" are included here as savanna.
X = not available.

Measuring Habitat Loss

Wildlife habitats are areas where nondomesticated species find the food, water, and other resources they need to survive (9). Humans have altered habitat unceasingly over the past 5,000 years. The broad habitat categories of forests, grasslands, and wetlands have all suffered significant losses. Most of it was converted to agriculture and pasture, more recently to urban development. The history of deforestation is generally known; 5,000 years ago, almost half the world's land (62-66 million square kilometers) may have been considerably wooded (10) (11). Even then, however, forests were being cleared in places such as Sumatra and Java (12). Over the next 4,500 years, the global magnitude of deforestation was small, though in certain countries, such as Togo, where forests were cut to obtain charcoal for smelting iron as early as the 1300s, the local impact was significant (13). Around 1500, European explorers

began laying the foundations of a global economy, and deforestation began to escalate. With the Industrial Revolution and the colonial expansion of the 1800s, forest clearing became pandemic (14). By 1988, only 34 percent of Earth's surface, or 45 million square kilometers, remained in forest and woodland–a loss of about 20 million square kilometers over 5,000 years, or an average of 4,000 square kilometers per year (15).

Such global accounts of long-term habitat loss, which proceed in orderly fashion from an "original" to a present-day extent, must be viewed with caution. Data from different regions, if they exist at all, are usually rough estimates and often are not directly comparable (16). Differences in terminology and measurement contribute to the large discrepancies in estimates of the area of the world's major ecosystems. (See Table 8.1.)

Likewise, the effect of habitat loss on wildlife is not easily summarized. Tropical forest, for example, is a broad category. It encompasses thousands of distinctive habitats, each with many species, all responding differently to deforestation (17). The notion of forest loss implies that once forests are cleared, the species living there cease to exist. But some adapt to another habitat or persist (though usually in smaller numbers and at the cost of some genetic depletion) in a seminatural forest setting.

Mapping Habitat Loss

The best attempts to map the world's habitat types do so by showing changes in the extent of vegetation formations. Maps of potential natural vegetation–that which would exist if it were not for human intervention–most closely approximate the extent of original habitats (18) (19) (20). Judging the potential area of each natural vegetation formation and then determining how much is left are as close as one can now get to a broad estimate of habitat loss. This work has been done on a regional basis only for sub-Saharan Africa and tropical Southeast Asia. (See Tables 8.2 and 8.3.) A similar study of Latin America would nearly complete calculations for tropical habitats (21). A detailed estimate for the United States shows changes to 1967 (22).

All the terrestrial habitat classification systems rely more or less on vegetation cover (23) (24). Two of the most widely used are Udvardy's division of the world into biogeographical realms and provinces and Holdridge's life-zone system (25) (26) (27). In the following discussion, some of Holdridge's terminology is used, but it is simplified (28).

LOSSES OF SELECTED HABITATS

Tropical Habitats

Moist Forests

Tropical moist forests–including evergreen rainforest, moist deciduous forest, and related forest types–

Table 8.2 Wildlife Habitats in the Afrotropical Realm, 1986

Vegetation Formation	Original Area (sq. km)	Area Remaining (sq. km)	Percent Remaining	Protected Area (sq. km)	Percent In Protected Areas
Dry Forests	8,216,808	3,415,988	41.6	512,965	15.0
upland montane	790,712	293,858	37.2	21,494	7.3
woodland	5,896,200	2,489,699	42.4	427,467	17.2
other	1,556,896	632,431	41.0	64,004	11.3
Moist Forests	4,699,704	1,867,629	39.7	132,457	7.1
Savanna/Grassland	6,954,875	2,835,196	40.8	296,957	10.5
Scrub/Desert	176,600	172,630	97.8	17,361	10.1
Wetland/Marsh	61,700	43,770	70.9	2,370	5.4
Mangroves	87,870	39,182	44.6	1,120	2.9

Source: Adapted by *World Resources* from John T. MacKinnon and Kathy MacKinnon, *Review of the Protected Areas System in the Afrotropical Realm* (International Union for Conservation of Nature and Natural Resources, Gland, Switzerland, 1986).

Note: The table is compiled from data on: Angola, Benin, Botswana, Burkina Faso, Burundi, Cameroon, Central African Republic, Chad (part), Côte d'Ivoire, Djibouti, Equatorial Guinea, Ethiopia, Gabon, The Gambia, Ghana, Guinea, Guinea Bissau, Kenya, Lesotho, Liberia, Madagascar, Malawi, Mali (part), Mauritania (part), Mauritius, Mozambique, Namibia, Niger (part), Nigeria, Rwanda, Senegal, Sierra Leone, Somalia, South Africa, Sudan (part), Swaziland, Tanzania, Togo, Uganda, Zaire, Zambia, and Zimbabwe. The source contained no data for a few small islands that are included in the Afrotropical Realm.

Totals in this table may not be the same as those in Chapter 20, "Wildlife and Habitat," Table 20.5, because of the use of different sources of information. For additional information, see Table 20.5 and the accompanying sources and technical notes.

Table 8.3 Wildlife Habitats in the Indomalayan Realm, 1986

Vegetation Formation	Original Area (sq. km)	Area Remaining (sq. km)	Percent Remaining	Protected Areas (sq. km)	Percent in Protected Areas
Dry Forests	3,414,064	940,145	27.5	99,471	10.6
mixed deciduous	456,727	94,563	20.7	5,955	6.3
tropical montane evergreen	279,769	295,542	70.0	26,100	5.5
subalpine	121,518	51,160	42.1	2,791	5.5
dry dipterocarp	284,581	57,554	20.2	7,557	13.1
subtropical/montane	144,350	125,741	87.1	1,100	0.1
other	2,127,119	315,580	18.6	55,968	14.2
Moist Forests	3,361,827	1,226,698	36.5	94,514	7.7
lowland rain	1,290,220	594,958	46.1	36,613	6.2
tropical semi-evergreen	940,298	253,073	26.9	34,511	13.6
tropical moist deciduous	24,374	56,406	23.4	1,790	3.2
subtropical pine	199,158	96,104	48	2,243	2.3
other	907,777	226,157	29	19,357	8.4
Savanna/Grassland	46,250	12,025	36.0	0	0.0
Scrub/Dessert	816,102	118,610	14.5	25,186	21.2
Wetland/Marsh	413,596	160,474	38.8	16,620	10.3
Mangroves	94,512	40,065	42.4	3,295	8.2

Source: Adapted by *World Resources* from John T. and Kathy MacKinnon, *Review of the Protected Areas System in the Indomalayan Realm* (International Union for Conservation of Nature and Natural Resources, Gland, Switzerland, 1986).

Data from the following countries or other entities are included in this summary: Bangladesh, Bhutan, Brunei, China (southern), Hong Kong, India, Indonesia, Japan (southern Ryukyu Archipelago), Cambodia, Laos, Malaysia, Myanmar, Nepal, Pakistan, Philippines, Sri Lanka, Taiwan, Thailand, and Viet Nam.

Note: Because this table and Chapter 20, "Wildlife and Habitat," Table 20.5 contain data from different sources, the figures in the two tables are not identical. See Table 20.5 and the related sources and technical notes for additional information.

are the most important habitat for protecting global biodiversity because they contain more than half of all species (29). These forests are found mainly in the Indomalayan realm of Southeast Asia, central and west-central Africa, and tropical Latin America. Clearing has taken a significant toll on all three areas. Only about 36 percent of the original 3.36 million square kilometers remains in Indomalaysia. (See Table 8.3.) Africa retains 1.87 million square kilometers, or 40 percent. (See Table 8.2.) Latin America still has 5.69 million square kilometers, or 81 percent (30). Together, the three major regions have only 8.79 million square kilometers, or 58 percent, of their original area of tropical moist forest.

This near-global loss of 6.30 million square kilometers can be attributed to a host of interconnected social, economic, and demographic causes. The demand for tropical timber and wood products in wealthy countries encourages deforestation in poor ones (31). In tropical Asia, deforestation is often caused by increased exports of hardwoods, and in Central America, forests have been cleared to plant export crops or raise cattle (32) (33). In the Amazon, resettlement projects have caused large-scale deforestation (34). (See Box 8.1.)

A minimum of 61,000 square kilometers of tropical moist forest is cleared each year around the world, and the total may be higher (35). (See Chapter 7, "Forests and Rangelands," New Estimates of Tropical Deforestation.) As the moist forests of the tropics are destroyed or fragmented into small patches, the world will experience a massive extinction of species (36).

Once tropical moist forest is cleared, little is done to restore lost habitat. Less than 10 percent of the area cleared is reforested, including plantation forests, which are usually monocultures and are of limited value as wildlife habitat (37) (38). Even if native moist-forest tree species were routinely regenerated on cleared sites, they would take up to a century to regrow and many centuries to mature fully. An extreme example is a patch of forest at Angkor, Cambodia, that still has not reached its climax stage 550 years after it was logged (39).

The situation is not hopeless, however. Studies suggest that many tropical animals survive partial clearing of their habitats, at least for a short time, as long as relatively large undisturbed areas remain nearby. Sizable numbers of birds of prey have been able to live at high densities in the few remaining primary forest patches of Java as well as to make use of cutover secondary forest (40). Tropical rainforest that has been selectively logged can support viable populations of a large number of species (41). But it is hard to predict how wildlife will respond to forest clearing and fragmentation. Some marsupials living in a fragmented tropical rainforest in northern Australia have persisted well in the forest remnants. The density of other species has lessened, however; presumably, their habitat requirements are more specialized (42). Biologists agree that large representative tracts of tropical moist forest must be kept intact if the range of its species diversity is to be maintained.

Dry Forests

There are nearly 8 million square kilometers of tropical dry forest around the world (43). Though they have not received nearly the same attention as rainforests, dry forests are also significant for biodiversity. Wildlife in dry forests exhibits a variety of behavior whose significance belies the relatively few species found there. For example, some plants bear fruit in seasons when most others do not, thus providing food to frugivores during a lean time of the year. Many dry-forest species are singularly resilient; they survive prolonged abnormal weather (either wet or dry) by becoming dormant (44) (45). As the world experiences global warming, this quality alone makes dry forests of primary biological interest.

Worldwide, about 38,000 square kilometers are cleared annually (46). In southeast Asia, a significantly higher percentage of dry forest than moist forest has been lost (73 percent versus 64 percent). (See Table 8.3.) Though data are incomplete for Latin America, it is almost certain that a higher percentage of dry than of moist forest has been destroyed so far. Because they are so accessible and easy to clear for farming, the dry forests of Central America are already gravely endangered. Less than 2 percent remain intact along the Pacific coast (47). In Africa the situation is different: A higher percentage of moist than of dry forest has been lost. (See Table 8.2.)

Other Habitats

Natural grassland and land with similar forms of vegetation have been under long-standing pressure throughout the tropics. In Latin America, large areas of grassland and savanna are being converted to other uses. Mining takes a heavy toll on the savanna-like *campo rupestre* in Brazil, and agriculture and grazing have lowered the extent of undisturbed *cerrado*, a grassland-woodland mosaic, to what is protected in national parks and other reserves (48). In Africa and Asia together, farming and livestock grazing are largely responsible for reducing natural grasslands to 41 percent of their original extent. (See Tables 8.2 and 8.3.) Desertification has reduced or destroyed the biological productivity of 83 percent, or 650 million hectares, of the rangelands in southern Asia, the Sahel, Mexico, and South America (49).

Selected Mediterranean-Climate Habitats

The Mediterranean climate is characterized by dry, hot summers and cool, moist winters (50). The climate is found around the sea for which it is named, stretching from Portugal to parts of Iraq, as well as in southern and southwestern Australia, the Cape province of South Africa, the state of California in the western United States, and central Chile (51).

Because their climate is so equable, these regions have long been settled and their natural vegetation altered (52). Nowhere is the change more apparent than around the Mediterranean Sea, where the forests, which totaled about 182,000 square kilometers in 1981, may once have covered 10 times that area. The rest of the plant communities of the Mediterranean basin have been completely transformed from their native state (53) (54). Desertification also affects 830,000 square kilometers of rangeland in the region— 75 percent of the total (55).

In South Africa, the Mediterranean-climate scrubland formation is called the *fynbos*. The flora of the *fynbos* is exceptionally rich; there are 8,550 vascular plants, 75 percent of them endemic (56) (57). Human activity (especially the introduction of exotics) has already destroyed one third of the original 75,000 square kilometers of *fynbos*, jeopardizing 1,585 plant species (58). All told, the *fynbos* accounts for 65 percent of the threatened plant species in the southern part of the continent, though it occupies just 1 percent of the area.

California also has a great diversity of plants, nearly one third of them endemic. But invading exotic species and rampant development have wiped out entire ecosystems. By 1967, 69 percent of the Central Valley's grasslands had been replaced by farms and other development (59).

Temperate and Boreal Habitats

Forests

Temperate and boreal forests cover more than 20.9 million square kilometers (60); at one time, they were much larger. In what is now the contiguous United States, forest (most of it temperate) once covered as much as 3.84 million square kilometers. Between 1630 and 1930, about 1.35 million of it was cleared— an average of 4,500 square kilometers per year. By 1977, tree planting and reversion of agricultural land had increased forest cover in the contiguous United States to 2.99 million square kilometers, but these processes have not replaced the native habitat lost when the virgin forest was destroyed (61) (62). Conifer forests, by and large, have remained intact, but the hardwood forests of the East and Midwest— especially the elm-ash, maple-basswood, beech-maple, and oak-hickory formations— have been greatly reduced, as have several open woodland types (63).

The history of the temperate Caucasian forests of the southern Soviet Union is similar; between 1700 and 1980, they were reduced from 147,000 square kilometers to 95,000 (64). Deforestation continues to affect the temperate forests of northern China. Especially hard-hit by clearing are the once-vast tracts of Heilongjiang, which also suffered a catastrophic fire (of human origin) in 1987 that burned more than 12,000 square kilometers of timber (65) (66).

Box 8.1. Biodiversity in Nine Key Latin American Countries

Latin America has perhaps the richest store of biodiversity in the world. The cause is chiefly the presence of Amazonia, but also the remarkable topographical variety, marked by dramatic transitions between mountains and lowlands. They are found in almost all the larger tropical countries. Nine Latin American countries are distinguished by great species richness or a high incidence of endemism, or both. (See Table 1.) All have considerable areas of contiguous tropical forest. (See Table 2.) And in all, the human populations are proliferating. From 1985 to 1990, their average annual growth rate was 2.24 percent—29 percent above the world average (1).

The Richness of Amazonia

In any ranking of biodiversity, Brazil stands out. It leads the world in the number of species of many kinds of organisms, including vascular plants, insects, terrestrial vertebrates, freshwater fish, amphibians, primates, and parrots. Largely because of habitat loss, Brazil also has the most endangered and vulnerable vertebrates—310 species in all (2).

Much of this richness is found in Amazonia. The basin of the Amazon River contains the world's largest expanse of tropical forest, most of it in Brazil. Estimates of the extent and rate of deforestation in the Brazilian Amazon vary widely. (See Chapter 7, "Forests and Rangelands.") It is clear, however, that although the forest is still comparatively intact, the rate of deforestation is increasing and its effects are severe in many places (3). In the early 1970s, for example, disturbed and cleared forest in the frontier state of Rondônia was only minor and scattered (4). Since then, government-induced immigration has caused massive changes in Rondônia's forest: by 1985, over 11 percent of the state, or 28,000 square kilometers, had been cleared outright, with another 87,000 disturbed (5).

Though international attention focuses on the Amazon, Brazil's most besieged habitats are elsewhere. As much as 93 percent of its moist forests along the Atlantic coast have been destroyed, and the rest is severely degraded (6). The temperate Araucaria (Paraná pine) forests of southern Brazil are even more decimated—with only 2 percent of their original extent (7). The dry caatinga woodland formation of northeastern Brazil is being drastically modified by fuelwood gathering and livestock grazing (8).

The Only True Rain Forest

Colombia, Peru, and Ecuador share some of the Amazon tropical forest with Brazil, but they have other biologically important habitats as well. Colombia's Chocó region along the Pacific coast is the wettest in the world and has the only true rainforest (in the strictest sense of the term) in Latin America. As yet little known to botanists, the Chocó may have 8,000-9,000 plant species, with perhaps 25 percent of them endemic. Both the southern Chocó and its northern reaches along the Panama border are suffering rapid deforestation (9) (10).

The Andes: Epicenter of Diversity

Peru's eastern uplands, where the Andes grade down into the Amazon basin, are also extremely rich in species; they have been described as a "global epicentre of diversity" (11). Only 35,000 square kilometers are intact throughout Colombia, Ecu-ador, and Peru, so effective conservation in existing protected areas is seen as crucial to preserving biodiversity in this area (12). In fact, Manu National Park has been called "the most important conservational unit in the world from the standpoint of preserving species diversity" (13). Yet the integrity of Manu is endangered by cattle grazing, cultivation, and a proposed highway and canal that would bisect the park and open the district to further settlement and oil exploration (14).

Forests Threatened by Logging, Land Clearing

The lowland forests on Ecuador's Pacific coast are characterized by high levels of vegetation endemism; in fact, three forest study areas that are only 100 kilometers apart share only 27 percent of the same species. The forests are under immense pressure from the country's fast-growing population and are threatened by poaching, timber extraction, and land clearing for agriculture (15).

Parts of Venezuela have also been extensively transformed in recent decades. In the western Llanos (at the foot of the Andes, southwest of Caracas), 33 percent of the forests disappeared between 1950 and 1975 (16).

Unlike the rest of tropical Latin America, the forests of Guyana and Suriname (as well as those of French Guiana) have not been subjected to intensive clearing. Timber and fuelwood cutting is limited to the coast, so over 85 percent of the Guianas remain forested, though in Guyana shifting cultivation by indigenous peoples may have altered the primeval vegetation heavily (17). Nonetheless, the high percentage of forest provides habitat for

---

Nowhere has the conversion of natural habitat been more ubiquitous than in the temperate portions of Europe, where essentially no natural habitat remains. (See Chapter 17, "Land Cover and Settlements," Table 17.1.) Wildlife still exists on the continent, of course, but the number of extinctions is significant. For example, over the past 2,000 years, the Netherlands lost one quarter of its native mammal species (67).

Grasslands

Natural grasslands and bushlands in temperate regions have suffered even more than those in the tropics. Of the 26 types of grasslands in the United States, at least 18 had shrunk 25 percent by 1967 (68). The tallgrass prairies of central North America once extended over 1 million square kilometers; today more than 99 percent are gone (69). Since the 16th Century, the native biota of the great pampa plains have given way to exotic species that followed Europeans and their livestock to South America (70). Thousands of square kilometers of steppe in the Soviet Union have been converted to agriculture, displacing grassland wildlife (71). Of the rangelands in Southern Africa, temperate and boreal Asia, Australia, and North America, 53 percent, or 8.8 million square kilometers, are now desertified (72).

Boreal habitats have not suffered as drastically, probably because their cold climate discouraged settlement, though these fragile ecosystems are threatened by increasing development of oil and other industries (73) (74). The sheer size and remoteness of the boreal forests can make them difficult to inventory. The Soviet Union, for example, has just over half the world's coniferous forest; its full size may not yet be known (75). The increases in national forest area reported for the Soviet Union between 1961 and 1978 are probably due to the inclusion of new areas of boreal forest in the inventory (76).

Box 8.1

a large number of mammal, bird, reptile, and amphibian species relative to the size of these countries. (See Table 1.)

Mexico's Topographic Diversity Breeds Biodiversity

Mexico's flora and fauna illustrate how a variety of topography contributes to biodiversity. The subtropical Sonoran and Chihuahuan deserts of the far north stand in sharp contrast to the tropical moist forests of the southern states of Chiapas and Veracruz. Unfortunately, much of the latter is steadily being converted to cattle pasture, coffee plantations, and croplands (18). These forests, which originally covered 110,000 square kilometers, were reduced to 16,000 square kilometers by 1983 (19).

Conserving Latin America's dwindling biodiversity will require, among other things, a great deal of money. There is a growing interest in this challenge among foreign donor organizations.

References and Notes

1. See Chapter 16, "Population and Health," Table 16.1.
2. Russell A. Mittermeier, "Biological Diversity in Brazil," World Wildlife Fund, Washington, D.C., 1988, p. 1.
3. Douglas C. Daly and Ghillean T. Prance, "Brazilian Amazon," in *Floristic Inventory of Tropical Countries: The Status of Plant Systematics, Collections, and Vegetation, plus Recommendations for the Future*, David G. Campbell and H. David Hammond, eds. (New York Botanical Garden, Bronx, New York, 1989), pp. 402 and 420-421.
4. Radam Brazil, *Levantamento de Recursos Naturais* (Ministry of Mines and Energy,

Rio de Janeiro, n.d.), pp. 1-23, cited in Jean-Paul Malingreau and Compton J. Tucker, "Large-Scale Deforestation in the Southeastern Amazon Basin of Brazil," *Ambio*, Vol. 17, No. 1 (1988), p. 53.
5. Jean-Paul Malingreau and Compton J. Tucker, "Large-Scale Deforestation in the Southeastern Amazon Basin of Brazil," *Ambio*, Vol. 17, No. 1 (1988), p. 53.
6. Scott A. Mori, "Eastern, Extra-Amazonian Brazil," in *Floristic Inventory of Tropical Countries: The Status of Plant Systematics, Collections, and Vegetation, plus Recommendations for the Future*, David G. Campbell and H. David Hammond, eds. (New York Botanical Garden, Bronx, New York, 1989), Table 2, p. 447.
7. John R. McNeill, "Deforestation in the Araucaria Zone of Southern Brazil, 1900-1983," in *World Deforestation in the Twentieth Century*, John F. Richards and Richard P. Tucker, eds. (Duke University Press, Durham, North Carolina, 1988), p. 17.
8. *Op. cit.* 6, pp. 428 and 430-432.
9. Alwyn Gentry, "Northwest South America (Colombia, Ecuador, and Peru)," in *Floristic Inventory of Tropical Countries: The Status of Plant Systematics, Collections, and Vegetation, plus Recommendations for the Future*, David G. Campbell and H. David Hammond, eds. (New York Botanical Garden, Bronx, New York, 1989), pp. 393-394.
10. International Union for Conservation of Nature and Natural Resources Tropical Forest Programme, *Colombian Chocó Conservation of Biological Diversity* (World Conservation Monitoring Centre, Cambridge, U.K., 1988), pp. 2 and 8.
11. C. Munn, Associate Research Zoologist, Wildlife Conservation International Unit, New York Zoological Society, 1987 (personal communication), cited in International Union for Conservation of Nature and Natural Resources Tropical Forest Programme, *Peru: Conservation of Biological Diversity* (World Conservation Monitoring Center, Cambridge, U.K., 1988), p. 3.
12. *Ibid.*

13. *Op. cit.* 9, p. 395.
14. International Union for Conservation of Nature and Natural Resources Tropical Forest Programme, *Peru: Conservation of Biological Diversity* (World Conservation Monitoring Centre, Cambridge, U.K., 1988), p. 4.
15. International Union for Conservation of Nature and Natural Resources Tropical Forest Programme, *Ecuador: Conservation of Biological Diversity* (World Conservation Monitoring Centre, Cambridge, U.K., 1988), pp. 1-2 and 7.
16. Jean-Pierre Veillon, "Las Deforestaciones en los Llanos Occidentales de Venezuela des de 1950 a 1975," in *Conservación de los Bosques Húmedos de Venezuela*, L.S. Hamilton, ed. (Sierra Club and Consejo de Bienstar Rural, Caracas, 1976), pp. 97-110, cited in Otto Huber and Dawn Frame, "Venezuela," in *Floristic Inventory of Tropical Countries: The Status of Plant Systematics, Collections, and Vegetation, plus Recommendations for the Future*, David G. Campbell and H. David Hammond, eds. (New York Botanical Garden, Bronx, New York, 1989), p. 368.
17. J.C. Lindeman and S.A. Mori, "The Guianas," in *Floristic Inventory of Tropical Countries: The Status of Plant Systematics, Collections, and Vegetation, plus Recommendations for the Future*, David G. Campbell and H. David Hammond, eds. (New York Botanical Garden, Bronx, New York, 1989), pp. 377 and 387.
18. International Union for Conservation of Nature and Natural Resources, *Mexico: Conservation of Biological Diversity* (World Conservation Monitoring Centre, Cambridge, U.K., 1988), p. 11.
19. Alejandro Estrada and Rosamond Coates-Estrada, "Rain Forest in Mexico: Research and Conservation at Los Tuxtlas," *Oryx*, Vol. 17 (October 1983), pp. 201-202.

Wetland and Aquatic Habitats

In terms of classification and mapping, aquatic habitats have received far less attention than terrestrial ones. Lakes are difficult to map at a small scale and are hard to fit into ecological land classifications; so too is riparian (streamside) vegetation (77). Estuarine and marine habitats have been neglected largely because they differ so radically from inland ecosystems; for this reason, relatively few specialists study them. It is only in the past 10 years that widely accepted classifications of aquatic habitats have been advanced (78) (79) (80).

Despite this lag, the innate biological richness of aquatic habitats cannot be doubted. In terms of biodiversity, the fauna of the deep sea, as yet largely unknown, may rival tropical forests (81). An indirect proof of abundance is the fact that 50 percent of the world's population lives in coastal areas and harvests the resources found there (82). Inland settle-

ment, too, has always followed the transportation lanes provided by rivers and other waterways.

The toll this development has taken on the original aquatic habitat is surely enormous, though global losses have not been estimated. But estimates of individual habitats indicate the general magnitude. For example, 70-90 percent of the U.S. natural riparian vegetation is thought to have been destroyed over the past 200 years; in certain areas of the West, the losses approach 99 percent (83) (84).

Tropical wetlands, which cover 2.64 million square kilometers worldwide, are highly productive habitats (85). In Africa, the small amount of tropical wetland is still fairly intact, but more than 60 percent has been lost in Asia. (See Tables 8.2 and 8.3.) The world's largest tropical wetland, the Pantanal floodplain in south-central Brazil, is nearly as large as all the remaining wetlands in Asia. This expanse of 140,000 square kilometers, one of the largest breeding grounds for waterfowl and a refuge for several endangered species, is being degraded by expanding agriculture, poaching, and chemical pollution (86). The most-decimated habitat in the United States is a wetland, the Mediterranean-climate Tule marshes of California. They once were extensive in the Central Valley, now an intensively farmed area. At least 89 percent of the Tule marshes have been degraded by polluted runoff from farms or have been destroyed outright (87).

Wetlands in temperate and boreal regions occupy about 5.72 million square kilometers (88). In mid-latitude areas, they have been routinely drained for cropland or other development (89). Of the nearly 7,800 square kilometers once found around southern Canada's urban areas, at most only 1,824 are left (90). In the contiguous United States, the original wetlands covered 870,000 square kilometers; now only 385,000 remain—a 56 percent loss (91).

Three important coastal and nearshore marine habitats are mangroves, seagrasses, and coral reefs. Worldwide, approximately 165,000 square kilometers of mangrove remain—58,000 in Asia, 39,000 in Africa, and 68,000 in the Americas (92). Fifty-eight percent of the mangroves in Indomalaysia have been lost (see Table 8.3), often by their conversion into brackish saltwater ponds for raising prawns and milkfish (93). In Africa, the loss is 55 percent. (See Table 8.2.) At least 26 percent of these habitats in the United States (excluding Puerto Rico) have been destroyed (94).

Seagrasses, salt-tolerant plants that grow underwater, often mingle with mangroves and coral reefs. Two major threats to seagrass beds are dredge-and-fill projects and pollutant discharges (95). Together, they caused declines of 41-79 percent in seagrass cover in Cockburn Sound, Australia, between 1954 and 1978 (96).

Coral reefs are the oceans' closest known counterparts to tropical rainforests in terms of species richness and biological productivity. Although there are no worldwide estimates of the extent of coral reefs, a comprehensive country-by- country directory was published in 1988 by the World Conservation Union (formally the International Union for Conservation of Nature and Natural Resources-IUCN) and the United Nations Environment Programme. Among the threats to reefs it lists are deforestation, which allows more eroded soil to flow down rivers and choke coastal coral reefs (97). Other threats include coastal development, chemical pollution, anchor damage, and dynamiting reefs for fishing (98). Of the three marine habitats discussed here, coral reefs appear to be the most threatened (99). It is not known what percentage of coral reefs have been lost worldwide. Concern about the Indian Ocean reefs was expressed as long ago as 1936; it is now estimated that 20 percent have been lost or seriously damaged (100).

CONSERVING BIODIVERSITY OUTSIDE PROTECTED AREAS

With so much of the world's wildlife habitat appropriated for human use, conservationists have spear-

headed efforts for the legal protection of representative samples of what is left (101). Delegates to the most recent World Congress on National Parks, held in Indonesia in 1982, voiced a fundamental objective of international conservation when they called for a global system of national parks and other protected areas. They recognized that establishing and managing such a network effectively is a major way to preserve species over the long term. The delegates hoped that this network, covering 10 percent of all terrestrial ecosystems, would be established within 10 years. But as 1992 draws near, the world is nowhere near the goal (102). In tropical Africa, for example, roughly 90,000 additional square kilometers would have to be placed under protection—an area almost the size of Malawi. In Indomalaysia, at least 44,000 square kilometers would be needed to reach the 10 percent target.

Because more than 96 percent of the world's land surface lies outside national protected areas, it is clear that, if we are to preserve biodiversity, we must protect wildlife habitat outside reserves and within human habitat. (See Chapter 20, "Wildlife and Habitat," Table 20.1.) Promoting sustainable development outside strict nature reserves is the best way to accomplish this end. (See *World Resources 1988-89*, pp. 99-102.)

In regions with a lengthy history of settlement, wildlife has relied for centuries on seminatural landscapes to survive. These include long-standing agroecosystems whose natural qualities derive from human interaction with the land as well as scenic areas managed intensively for recreation and tourism (103).

Protected Landscapes

One way to conserve wildlife is to designate these seminatural areas "protected landscapes." Unlike national parks and similar reserves, protected landscapes are dominated by private holdings interspersed with small publicly owned sites. Protected landscapes do not exclude local residents, for the concept affirms the belief that seminatural land must be tended by people—through husbandry, farming, and tourism development—if it is to retain its characteristic value as wildlife habitat (104). The moors of Great Britain are good examples of seminatural lands with an agriculture base. They are a mixture of grasslands, upland bogs, and wooded valleys maintained by low-intensity livestock grazing. The two main moorlands, Dartmoor and Exmoor, are managed as protected landscapes. Together they support 40,000 residents and a wide variety of wildlife (105). An example of a protected landscape encompassing a seminatural scenic area is the Parc Naturel Regional on Martinique in the Caribbean. There, seven vegetation types (including such important habitats as cloud forest, mangroves, and coastal wetlands) are managed for conservation in an area whose 80,000 residents rely heavily on tourism for income. At least 26 countries now use protected landscapes (106)

Multiple-Use Areas

In developing countries, the fate of wildlife is inextricably linked with the well-being of rural people. In general, plans that combine some form of multiple-use protected area with sustainable development projects on surrounding lands seem the most promising. Yet circumstances vary so much that conservation efforts need to be customized. The World Wildlife Fund has taken just such an approach since 1985 through its Wildlands and Human Needs Program. It gives money, advice, and other assistance to private groups that run sustainable development projects with wildlife components.

One such organization, Asociación de los Nuevos Alquimistas (ANAI), was formed in the 1970s to promote agroforestry and habitat protection in the Gandoca district on Costa Rica's Caribbean coast. Gandoca (which at one time was mostly a cacao plantation) is now controlled by small farmers who purchased their land or established property rights long ago. Because they value Gandoca's seminatural landscape, they have traditionally left part of their land uncultivated to benefit a diverse flora and fauna. With ANAI's help, the local people fended off a 1984 road-building proposal that would have opened Gandoca to squatters and changed its character forever.

From this incident, ANAI and the farmers realized that they needed to secure legal title to their land. Occupancy and boundaries are stable and well-recognized locally, but few people have the documents to substantiate their rights before the national government. In Costa Rica, land titling is expensive and complicated, beyond the reach of small farmers. The Wildlands and Human Needs Program has funded ANAI to hire professionals to coordinate the land titling, take aerial photographs, and map the district; to pay for fieldwork by Costa Rica's platting office; and to help claimants through the legal system (107).

A Last Resort: Zoos and Botanical Gardens

A third way to maintain biodiversity outside protected areas is for zoos and botanical gardens to save species whose survival in the wild is in immediate peril. Captive-breeding animals in zoos, growing plant specimens in botanical gardens, and storing seeds or other plant components in gene banks are referred to as *ex situ*, or offsite, conservation measures, as opposed to conserving wild plants and animals *in situ*, in their native habitats.

Zoos house about 540,000 individual mammals, birds, reptiles, and amphibians, but they have the capacity to sustain viable populations of, at most, only 900 species (108) (109). This is a small fraction of the number whose continued existence is threatened or endangered and is certainly not nearly enough to provide for the number of species expected to be in danger of extinction by 2050 (110). Currently, far fewer than 900 rare species are being bred and raised in zoos, though the number has risen steadily since the 1960s.

Wildlife in zoos is often bred using advanced technology. Artificial incubation and insemination are fairly common (111). Using frozen sperm or embryos is not yet widespread, though its potential is great (112). Embryo transfer, in which ova or embryos are taken from a donor and transplanted into the reproductive tract of a foster mother, is also relatively new. In 1987, a healthy calf of the endangered wild cattle species *Bos gaurus* was successfully delivered of a domestic Holstein cow, and in 1984 an embryo from the rare Przewalski's horse was transferred interspecifically. Overall, though, success with transfers involving wild animals is limited (113).

To help prevent inbreeding within zoo populations, a series of international studbooks (mostly for mammals) was developed. Since 1974, they have been supplemented by the International Species Inventory System (ISIS). ISIS is a centralized, computerized record of census, demographic, genealogical, and laboratory data on zoo animals (114). In addition to ISIS, many zoos participate in species survival plans that try to coordinate *ex situ* conservation efforts for individual reptiles, birds, and mammals (115) (116).

Zoo professionals point out that no nature reserve anywhere is big enough to ensure the long-term survival of large animals (117). Conceding that some animals are more popular than others, zoo administrators also contend that their efforts on behalf of charismatic species buy them time until these animals can be returned to their natural habitats. Despite a history of limited success, it is argued that reintroducing high-profile species to nature can galvanize support for habitat protection and restoration that will benefit less-favored wildlife. Zoos also instill visitors with an interest in and respect for wildlife, attitudes that may lead to support for conserving habitat. Finally, some zookeepers maintain that the likely sweeping habitat loss and mass extinctions over the next century mean that captive species will survive in a world of dwindling biodiversity (118).

The practical potential for *ex situ* conservation of plants is greater than it is for animals. Because they are immobile, plants are relatively easy to collect from the wild without disturbing the surrounding population. Once gathered, plant specimens can be transported to botanical gardens cheaply, and they require no expensive caging, elaborate habitat re-creations, or constant care. Propagation is relatively simple—through cuttings, root stock division, or tissue culture. Further, plants are generally viewed as benign, unlike animal pests and predators (119).

About 1,500 botanical gardens and arboreta hold an estimated 90,000 species (called accessions) of higher plants. But most collections are too small to be effective conservation tools. Of the 90,000 accessions, perhaps 1-2 percent are wild relatives of crops, and 35 percent have some other current or potential marketable value, mainly for medicine or horticulture. The remainder is dominated by rare or endangered species whose value is unknown or unquantified (120) (121).

There are three ways to conserve plants *ex situ*: keep the whole plant, its tissues, or its seeds. The advantage of whole plants is their usefulness in public exhibits, but the cost is high; maintaining the large numbers of specimens needed for conservation is expensive. *In vitro* tissue cultures can sustain genetic material for a long time. Seed banking under refrigeration is a third method. The sample size per species is much larger than for whole plants, and the amount of storage space required is about the same as that for tissue cultures (122). Only by reducing their moisture content can seeds be stored cold for a long period and remain viable. Unfortunately, not all seeds can withstand this treatment. Many valuable tropical species have "recalcitrant" seeds whose life spans are shortened by desiccation. This type of species must be stored as *in vitro* cultures (123).

A second problem with *ex situ* plant conservation is that basic botanical data and practical techniques of propagation and care have traditionally been passed along by word of mouth. Entire botanical collections sometimes fell into disuse upon the retirement or departure of the staff who built them up (124).

The Botanic Gardens Conservation Secretariat was created by IUCN in 1987 to address this problem and to promote and coordinate *ex situ* conservation of endangered wild plants (125). It concentrates on plants other than major crop species, which are already looked after by the Food and Agriculture Organization of the United Nations and other international boards (126). The Secretariat has established a data base of *ex situ* collections in botanical gardens and is functioning as a sounding board for them and other conservation organizations (127) (128). It is developing guidelines for conserving the germ plasm of wild plant species, including sampling and storing methods.

The Secretariat's challenge is formidable. An estimated 67,000 plant species are rare or under threat of extinction in the world's temperate, subtropical, and tropical regions (129). In the United States alone, 680 native plants are in danger of becoming extinct by 2000 (130). Some 2,300 more, or about 10 percent of all U.S. native plants, face a less-imminent extinction threat. Since the first Europeans arrived in North America, about 200 species have become extinct (131).

In response to the threat, botanical gardens and arboreta are increasing efforts to conserve rare native plants, in addition to their traditional research and service to horticulturists (132). U.S. efforts are being coordinated by the Center for Plant Conservation, which acts on behalf of a consortium of 19 botanical gardens and arboreta (133). As part of this initiative, botanical gardens are beginning to combine *ex situ* and *in situ* conservation. Some plant their own small

reserves with natives; others locate in national parks (134) (135). Ethnobotanical gardens, which specialize in cultivating plants traditionally used by indigenous peoples, are exploring coordination with *in situ* conservation projects, as several in the Asian tropics have already done (136). This shift complements efforts in a growing number of countries to conserve plant (especially crop) germ plasm using low-technology *in situ* strategies carried out by local communities (137).

FOCUS ON GREENHOUSE WARMING AND BIODIVERSITY

Until this century, wildlife habitat changed mostly because of direct human activity. Monumental as these changes have been, they would be dwarfed by the greenhouse effect. If the magnitude of change suggested by current climate models is accurate, we will have begun to alter the environment of all habitats significantly. As the relatively stable climate of the past 5,000 years comes to an end, global warming would begin changing the location, size, and character of wildlife habitat at an unprecedented rate. This fact, plus the fact that the changes would take place in a fragmented landscape, are a grave concern (138). Simply stated, if the world climate changes as fast as some scientists predict, many species would not be able to adapt quickly enough, and extinctions would be widespread because habitats would have shifted, shrunk, or disappeared (139).

It is probably too late to stop at least some atmospheric warming. Greenhouse gases (carbon dioxide, methane, chlorofluorocarbons, etc.) emitted since the Industrial Revolution may have already contributed to a global atmospheric warming of 0.5˚C (140). This warming alone could usher in a new world for wildlife because warming of less than 1˚C has been observed to cause substantial changes in the ranges of animals and plants. In England, for example, the range of the white admiral butterfly expanded northward between 1920 and 1961—when mean temperatures rose only 0.5˚C (141). If the global average rises 2˚C, plants and animals will encounter living conditions beyond any experienced for 10,000 years. Model projections indicate that, at some point during the 21st Century, the average global temperature may well rise 1.5-4.5˚C if current emission trends continue (142) (143). In evolutionary terms, this change would take place overnight. If the projections are correct, almost all children born in 1990 can expect to live long enough to witness profound changes on the planet (144).

Though scientists generally concur on the likely magnitude of global warming, they do not know how it would affect several facets of climate crucial to plant growth (and, therefore, to biodiversity). Included are cloud cover, evapotranspiration, storm intensity and frequency, and, most important, regional precipitation patterns (145) (146). Over much of the world, the consequences of the greenhouse effect for

Box 8.2 Biodiversity's Storehouse in a Warmer World: Climate Change in the Tropics

Scientists studying global warming's effects on biodiversity have focused on the temperate zone and high latitudes of the Northern Hemisphere, where climate change is expected to be the most abrupt. Temperatures near the Equator are projected to remain relatively stable, so less attention has been paid to the greenhouse effect in tropical forests.

In the absence of other changes, higher temperatures and changes in the total amount of rainfall could have little effect on biodiversity in the heart of the tropics. Tropical moist forests already have regular cloud cover, which moderates temperatures, and they commonly experience annual extremes of high or low rainfall. Rising temperatures could be negated if warming also caused more clouds over the forests. The normal swings in rainfall would seem to make moist forests resistant to greenhouse-induced precipitation changes. These observations do not necessarily hold, though, for tropical moist forests on the margin of their distribution—that is, at higher latitudes and altitudes. They are subject to cold air masses and so may be more sensitive to rising temperatures (1).

Changes in Wet and Dry Seasons

An important way that greenhouse warming could affect tropical biodiversity is by changing the length and intensity of wet and dry seasons. Most areas have one or two true dry seasons each year, and even the rainiest have periodic drier intervals. In moist forests near the Equator, two dry and two rainy seasons alternate annually. Moving out toward the subtropics, one of the dry seasons lengthens and the other shortens or appears only sporadically. Rainy seasons usually coincide with the passage of the sun overhead and peak about a month later. This order can be upset by unpredictable disturbances, such as an El Niño oceanic upwelling, outbreaks of polar air masses, and violent windstorms (2). Such tropical disturbances can be changed dramatically by slight alterations in temperature and precipitation. Under global warming, the number and intensity of hurricanes or wildfires might increase or decrease in a given location. These sorts of changes could deeply affect biodiversity because patterns of disturbance seem to be a primary contributor to high species diversity in the tropics (3).

Even under normal conditions, broad patterns vary considerably in tropical moist forests. Dry spells during the rainy season (and vice versa) are not uncommon. Such anomalies have a powerful influence on plants and animals. A dry spell of just a few weeks can kill tender seedlings and cause mature trees in a moist deciduous forest to drop their leaves. Full sunlight penetrating to the forest floor can dry out the soil, stunting or even killing shade-loving plants (4).

The Danger of a Wet Dry Season

On the other hand, if the dry season is abnormally wet, plants that need aridity to stimulate flowering may not bear fruit at all. This is what happened in 1970 at the Barro Colorado Island research station in Panama. Researchers could not walk 300 meters along the trails without finding the carcass of a fruit-eating animal. Agoutis, armadillos, howler monkeys, opossums, collared peccaries, porcupines—all starved to death. Frugivorous parrots and toucans flew from the island to escape the famine. Many coatis starved when their primary food source —arthropods living in decaying leaf litter on the forest floor—died out because the leaves did not fall during the wet "dry" season (5).

If dry seasons were lengthened by the greenhouse effect, the results for biodiversity could be equally catastrophic. A recent tropical drought offers clues to what might happen on a wide scale. Eastern Borneo was afflicted by a harsh drought in 1982 and 1983 that killed many trees. Wildfires burned hotter and longer than usual, destroying whole forests. As a result, animals that depend on plant reproductive parts for food had little to eat, primates and other canopy-dwellers found their arboreal pathways destroyed, and prey species lost the ground cover they needed to hide from predators (6). These examples show that any disruption of the normal cycle of wet and dry seasons hurts wildlife.

Breaking Up Co-Evolved Partners

Some species in the tropics are sensitive to climate change in a different way. Tropical forests provide the world's best examples of plants and animals that have coevolved so that key elements of their survival—such as pollination, seed dispersal, and herbivory—are interdependent. The lives of these mutualistic partners have become bound together, so to speak, under the stable climate of the past 5,000 years. If the climate changes beyond the tolerances of any one of the partners, all the others would be at risk. When a changing climate triggered just such a coevolutionary disequilibrium in the geological past, it may have contributed to the extinction of the mastodons, giant beavers, ground sloths, and other large land animals that once inhabited North America (7). It is possible that only slight changes in tropical climates could produce a disproportionately large number of local extinctions.

Tropical rainforests, which shrank dramatically during the last glacial period (10,000-70,000 years ago), were able to expand rapidly during the postglacial warm-up (8). In contrast, today's rainforests are under enormous human pressures. By the middle of the next century, when an equivalent doubling of atmospheric carbon dioxide is expected, the size of the rainforests will be far smaller still, and the landscape into which they would need to expand will be dominated by people as never before.

References and Notes

1. Gary S. Hartshorn, "Possible Effects of Global Warming on the Biological Diversity in Tropical Forests," in *Consequences of the Greenhouse Effect for Biological Diversity*, Robert L. Peters II, ed. (Yale University Press, New Haven, Connecticut, in press).
2. *Ibid.*
3. *Ibid.*
4. *Ibid.*
5. R.B. Foster, "Famine on Barro Colorado Island," in *The Ecology of a Tropical Forest: Seasonal Rhythms and Long Term Changes*, E.G. Leigh, Jr., A.S. Rand, and D.M. Windsor, eds. (Smithsonian Institution Press, Washington, D.C., 1982), pp. 201-212, cited in Gary S. Hartshorn, "Possible Effects of Global Warming on the Biological Diversity in Tropical Forests," in *Consequences of the Greenhouse Effect for Biological Diversity*, Robert L. Peters II, ed. (Yale University Press, New Haven, Connecticut, in press).
6. M. Leighton and N. N. Wirawan, "Catastrophic Drought and Fire in Borneo Tropical Rain Forest Associated with the 1982-1983 El Niño Southern Oscillation Event," in *Tropical Rain Forests and the World Atmosphere*, Ghillean T. Prance, ed. (Westview Press, Boulder, Colorado, 1986), p. 87.
7. Edward C. Wolf, "The Conversation between Earth, Sea, and Sun: Ecological Implications of a Warming Earth," *Orion Nature Quarterly*, Vol. 8, No. 1 (1989), p. 50.
8. R.A. Warrick *et al.*, "The Effects of Increased CO2 and Climatic Change on Terrestrial Ecosystems: Global Perspectives, Arms and Issues," in *The Greenhouse Effect, Climatic Change, and Ecosystems*, Bert Bolin *et al.*, eds. (John Wiley & Sons, Chichester, U.K., 1986), p. 365.

wildlife hinge on whether a given habitat will become warmer and drier or warmer and wetter (147)

Biodiversity would surely suffer more if the world became warmer and drier rather than warmer and wetter because many more species are adapted to an abundance of water than to a lack of it; tropical moist forests, not deserts, are the repository of most of the world's species. Yet predicting how green-

Figure 8.1 How Global Warming Could Force
Species to Shift Their Ranges to Higher
Altitudes

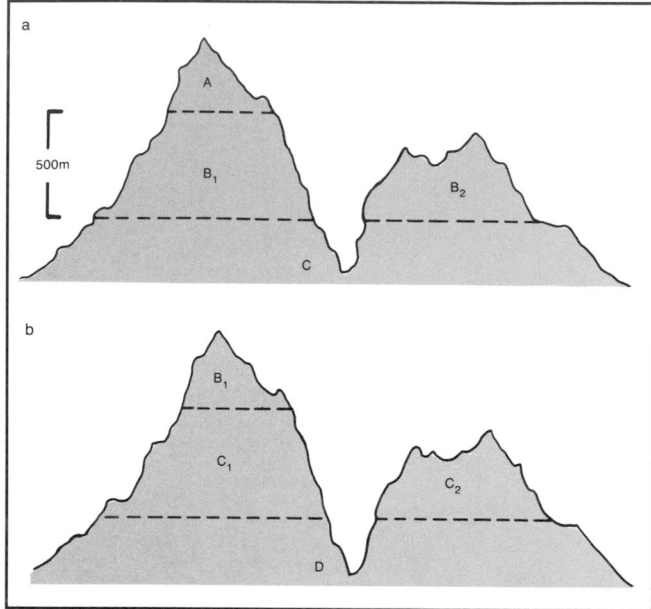

Source: Adapted from Robert L. Peters and Joan D.S. Darling, "The Greenhouse Effect and Nature Reserves," *BioScience*, Vol. 35, No. 11 (1985),
Figure 3, p. 714.
Notes:
a. Current altitudinal distribution of three species, A, B, and C. Species B is divided into two populations, B_1 and B_2.
b. Distribution after a 500-meter shift upward in response to an increase of 3˚C.
Species A cannot tolerate the higher temperature or is outcompeted by B_1
and becomes locally extinct. B_1 ascends to a new, smaller range; B_2 disappears. Species C fragments into two populations, C_1 and C_2, and becomes
restricted to a smaller area. A new species, D, colonizes the lower elevations.

house warming would affect the tropics is far more
complicated than simply determining whether rainfall will increase or decrease. (See Box 8.2.)

How Might Communities and Species Respond?

As global warming accelerates, plant and animal communities would gradually disintegrate (148) (149).
They would be broken up by the differing responses
of wildlife populations to a new climate. Associated
populations could not move in concert to a new suitable habitat because their capacities to migrate differ. Moreover, each may have unique genetic
adaptations as well as tolerances to changes in temperature and water availability that would govern its
responses (150) (151).

Species might counteract rapid global warming by
shifting their ranges to a higher altitude or a higher
latitude. Assuming an average global warming of 3˚C,
a 500-meter ascent or a poleward shift of 300 kilometers would be needed to compensate (152) (153). The
option to shift upward on the order of 500 meters is
open only to wildlife now occupying the foothills and
lower elevations of mountains. Supposing that alpine

species could tolerate higher temperatures, they
might still be squeezed out by migrants from below.
Even successful ascenders would not escape a fact of
topography: the peaks of mountains are smaller than
their bases. As they ascend, subalpine species with
formerly large ranges might not have enough habitat
to survive (154). (See Figure 8.1.)

The ability of most species to shift toward better
habitat would depend largely on how easily they can
migrate toward the poles (155). To do so, they must
overcome extensive artificial barriers to migration–
farmland, cities, roads, cleared areas in forests, and
so on–as well as natural east-west-lying landforms
such as the Alps, Pyrenees, Himalayas, Mediterranean Sea, and Black Sea. Moving between islands of
suitable habitat amidst a fragmented landscape
would be a severe test for survival. Some isolated
populations reduced to living in suboptimal habitats
may already occupy those parts of their historical
ranges most susceptible to climate change. And because deforestation adds materially to global warming, conventional habitat destruction and the
greenhouse effect would tend to intensify each other
(156).

In addition, wildlife would have to move quickly if
it were to find optimum climate. Plants especially
would be in trouble. The dispersal rate of tree species in eastern North America and in Europe, for example, is far too slow to keep pace with projected
range changes. The migration rates of some ostensibly mobile animals may not in fact be much faster
than plants if their distribution depends on the presence of slow- dispersing plants (157).

Wildlife in the mid-latitudes of the Southern Hemisphere faces an additional barrier to poleward shift.
The Southern Hemisphere has only a small amount
of land that can be classified as temperate, and the
vast boreal ecosystem has no austral counterpart
(158). Excluding Antarctica and Greenland, the area of
land north of 30˚N is at least 10 times larger than that
south of 30˚S (159). Poleward-shifting wildlife in South
America, for example, faces what could be termed
the "Patagonian funnel"–a neck of land that narrows
as it extends southward. Like the funneling effect of
mountain peaks, a lack of land would force migrating
South American temperate species into smaller
ranges; typical results would be smaller populations
that are more vulnerable to harmful chance events
and outbreaks of disease. African wildlife would encounter a similar, though less pronounced, latitudinal funnel.

Which Species Are Most Vulnerable to Global Warming?

Fragmentation and funneling effects certainly put alpine and temperate species at risk. Nevertheless, Arctic and boreal species would probably be the most
affected if current notions of greenhouse-induced
habitat change are correct because warming in the
northern high latitudes could be twice the global av-

Figure 8.2 Present-Day Ecoclimatic Provinces of Canada

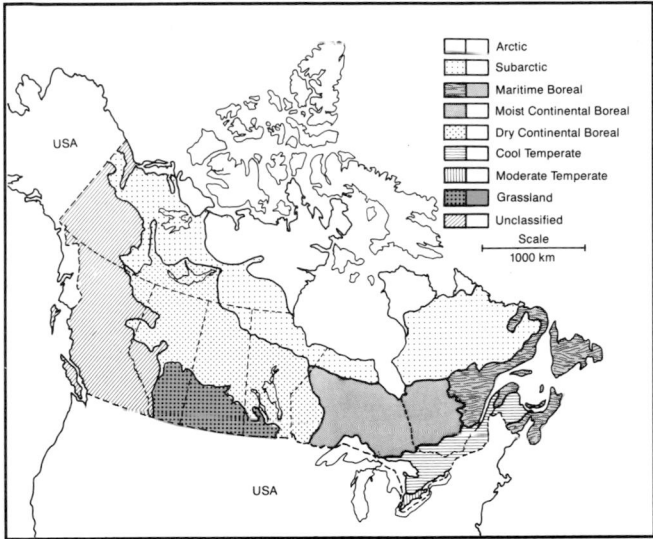

Legend:
- Arctic
- Subarctic
- Maritime Boreal
- Moist Continental Boreal
- Dry Continental Boreal
- Cool Temperate
- Moderate Temperate
- Grassland
- Unclassified

Scale
1000 km

Source: B. Rizzo, "The Sensitivity of Canadás Ecosystems to Climate Change," *Canada Committee on Ecological Land Classification Newsletter*, No.17 (Canadian Wildlife Service, Environment Canada, Ottawa, Ontario, 1988), Figure 1, p. 10.

Figure 8.3 Ecoclimatic Provinces of Canada under a Doubled Carbon Dioxide Climate

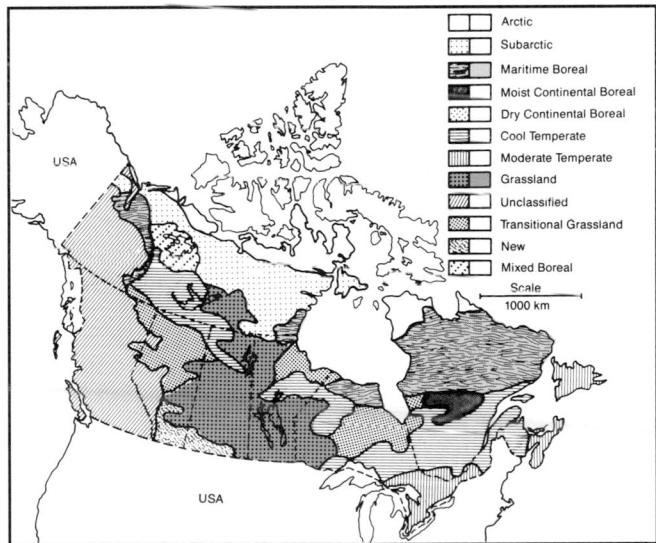

Legend:
- Arctic
- Subarctic
- Maritime Boreal
- Moist Continental Boreal
- Dry Continental Boreal
- Cool Temperate
- Moderate Temperate
- Grassland
- Unclassified
- Transitional Grassland
- New
- Mixed Boreal

Scale
1000 km

Source: B. Rizzo, "The Sensitivity of Canada's Ecosystems to Climatic Change," *Canada Committee on Ecological Land Classification Newsletter*, No.17 (Canadian Wildlife Service, Environment Canada, Ottawa, Ontario, 1988), Figure 2, p. 11.

erage (160). Canada's Arctic, subarctic, and continental boreal regions would shrink sharply, with big gains in the extent of maritime boreal, temperate, and grasslands regions. (See Figures 8.2 and 8.3.) Similar changes involving much larger land areas could be expected for the Soviet Union. On the other hand, the projected temperature increase in the southern polar region may not cause immediate changes in

Antarctic habitat. It has been estimated that a rise of 10-15°C would have to occur before the inland ice of east Antarctica would shrink significantly (161) (162).

Heat-intolerant species near the warmer margins of their range would be greatly affected. Organisms would be expected to retreat from areas whose temperatures are near their maximum tolerance. For widespread, abundant species, this range corrosion might pose little threat of extinction. Localized species living in a fragmented landscape could be vulnerable, however (163). In fact, any uncommon species with specialized habitat requirements is more vulnerable to unpredictable change than is a generalist (164). (See Box 8.3.) Even common species would face extinction if they have little genetic diversity or are poor at dispersing (165).

Changes in ocean currents would affect not only marine wildlife but terrestrial animals as well. If the Gulf Stream shifts, wildlife in northern Europe would have to contend with colder temperatures in an otherwise warmer world. Altered currents might also upset terrestrial migration in the Arctic; polunias (the natural open-water areas of the region) could grow larger, preventing animals from moving between ice floes (166).

If the seas rise, coastal wildlife habitats would be endangered. Low-lying barrier islands and atolls could be wholly inundated. Freshwater estuaries and littoral wetlands would be open to damage from impinging saltwater (167). In the Americas, vital bottleneck feeding sites along the coasts—places where huge numbers of migrating shorebirds gather simultaneously at certain times of the year—could be destroyed (168). Coral reefs and mangroves, two extremely productive habitats, might also be harmed (169). (See Chapter 11, "Oceans and Coasts," The Effects of Climate Change on Oceans and Coasts.)

The secondary effects of global climate change could become primary threats to biodiversity. If, as some models predict, soil moisture declines in the great grain belts of the world, thus leading to more droughts and lower crop yields, hunters might put pressure on wildlife to make up the shortfall (170). Another example is trypansomiasis, or sleeping sickness, which has long prevented people from colonizing parts of Africa where wild animals act as a reservoir of the disease. Heretofore, infected areas have served as *de facto* wildlife refuges and thus are important to conservation. The parasite that is the pathogen for sleeping sickness is carried by several species of tsetse fly. Studies of how the tsetses' ranges would be affected by global warming indicate that they would become less common in West and Central Africa—in other words, in the continent's chief remaining tropical moist forests. If the threat of sleeping sickness lessens, pressure to exploit these species-rich forests might increase (171).

In controlled experiments, increased atmospheric carbon dioxide causes certain plants to grow faster and more luxuriantly (172). Increased carbon dioxide may also actually help some plants compensate for

Box 8.3 How a Greenhouse Extinction Might Occur: The Case of the Kirtland's Warbler

Wildlife species with narrow habitat preferences would be at a serious disadvantage during a period of global warming. For some, such as the Kirtland's warbler, the greenhouse effect might be fatal. Kirtland's warblers nest only in dead ground-level branches of jack pine stands whose trees are 1.5-6 meters tall. Further, these stands must extend for at least 32 hectares, the average minimum territory of a male Kirtland's warbler. The bird nests only in jack pine growing on a coarse sandy soil that is found exclusively in a small area of the state of Michigan in the midwestern United States.

The Kirtland's warbler is protected by the federal Endangered Species Act, so the U.S. Forest Service periodically burns part of Huron National Forest to maintain young stands of jack pine, which depends on fire to open its cones and release seeds. However, if greenhouse warming proceeds according to a scenario put forward by the Goddard Institute for Space Studies (GISS) general circulation model, then the stands favored by Kirtland's warblers will no longer grow back after being burnt over because they are near the warm limit of the jack pine's range. It is thought that white pine and red maple will invade the bird's habitat in the years following the next burning. As the climate continues to warm, the habitat would be reduced to a treeless plain by 2070. It seems quite unlikely that the Kirtland's warbler could evolve a new nesting strategy over the next 80 years. So if the GISS model is correct, the chances are that the Kirtland's warbler would become extinct unless suitable habitat can be found (or created) farther north and the birds induced to use it (1).

References and Notes

1. Daniel B. Botkin and Robert A. Nisbet, "Projecting the Effects of Climate Change on Biological Diversity in Forests," in *Consequences of the Greenhouse Effect for Biological Diversity*, Robert L. Peters II, ed. (Yale University Press, New Haven, Connecticut, in press).

high temperature and water stress (173). (See Chapter 6, "Food and Agriculture," The Effects of Climate Change on Agriculture, and Chapter 7, "Forests and Rangelands," Possible Effects of Global Warming on Forests and Rangelands.)

Two important aspects of climate will not be affected by global warming: the length of the day and the angle that sunlight strikes Earth. These constants might help ameliorate the effects of climate change on plant life, particularly in the Arctic, where the length of exposure to sunlight is a more important determinant of the growing season than is ambient temperature (174).

Managing Biodiversity in a New Climate

Most wildlife conservation techniques now in use will have to be modified for a greenhouse climate. Basic concepts of nature-reserve management—wilderness, natural areas, and exotic species—might become obsolete unless they are radically rethought (175) (176) (177).

Protected areas were often set up explicitly to conserve certain species and communities. Many of these reserves have since become isolated by surrounding development. As climate changes, species' ranges may shift outside the reserve's boundaries, causing wildlife and human needs to conflict (178). The wildlife loses its protection and the reserve its purpose.

Guessing where particular species would thrive in years to come and then setting those areas aside as reserves is futile. As long as greenhouse gases are released into the atmosphere, global warming will continue, thus the problem is akin to hitting a moving target (179). Instead, biologists are suggesting that future protected areas include a variety of topography and soil types—two components of habitat that would remain relatively stable during rapid climate change. If the landscape consisted of an assortment of elevations, so much the better, for it is easier for species to migrate upward within a reserve than across unprotected lands between reserves. Larger reserves would be better than smaller ones, giving wildlife more room to shift ranges and find suitable microclimates, such as north-facing slopes (180).

But if biodiversity is to be preserved, national parks and other protected areas, long the backbone of wildlife conservation, may have to be supplemented by a system of mobile protections that follow species or species associations as they seek their optimal climate. For example, private lands might be assessed for their conservation value and owners paid a stipend to keep key habitats undeveloped. Along these lines, a system of development credits might be used to regulate land use over broad areas deemed important to wildlife (181). Alternatively, managed migration corridors might be created between remaining areas of natural habitat, though whether enough of them could be created is open to doubt (182).

Some scientists argue that the presence of genetic variation—and the inherent resiliency it affords—will ultimately decide whether any given population survives drastic climate change. In a greenhouse world, broad measures of the heat and moisture tolerance of an entire species would be less meaningful than exacting measures of how much a population has become genetically adapted to a specific site. Without human help, genes cannot diffuse through wildlife populations fast enough to allow adaptations prior to climate change (183). Therefore, we will have to become first-class ecological engineers (184). The suggestion is not entirely far-fetched. Techniques from the emerging field of restoration ecology indicate it may be possible to skip whole stages of plant succession and so produce mature vegetation formations more quickly (185). Such ideas, radical though they now seem, could become commonplace within our lifetimes. In fact, if the global climate changes, bold action would be absolutely necessary if our children—living in a new, warmer world—are to enjoy the same diversity of life our parents took for granted in theirs.

Figure 8.4 Change in African Elephant Populations, 1981-89

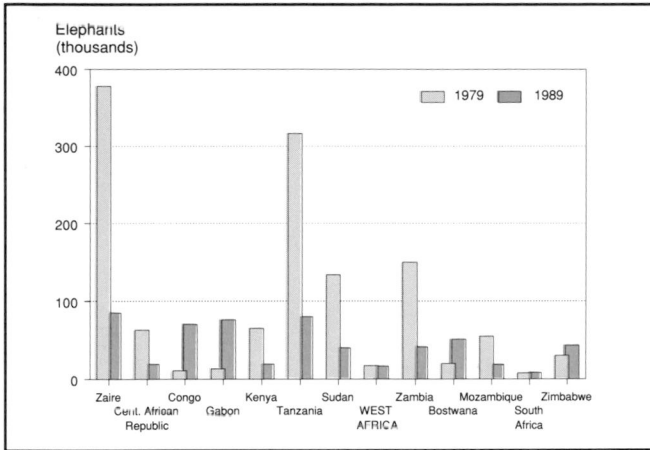

Source: *TRAFFIC USA*, World Wildlife Fund, Washington, D.C., 1989 (personal communiation).

Figure 8.5 Importers of Raw Ivory, 1989

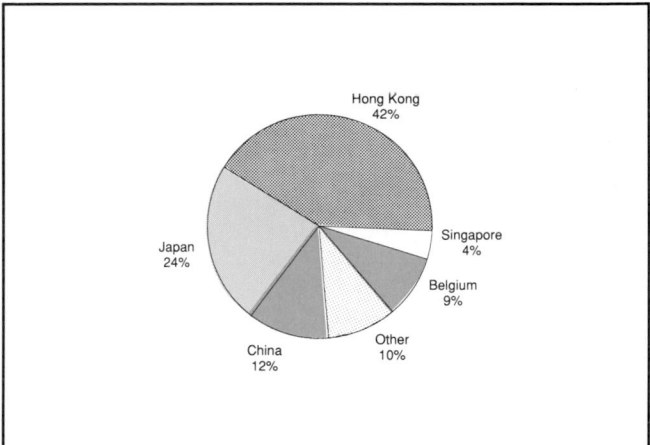

Source: Chapter 20, "Wildlife and Habitat," Table 20.3.

RECENT DEVELOPMENTS

CITES BANS IVORY TRADE IN ATTEMPT TO SAVE AFRICAN ELEPHANT

The extraordinary crash in elephant numbers–from 1.3 million in 1979 to 625,000 in 1989 (186)–has prompted a majority of nations to abandon a limited quota system in favor of a ban on all ivory trade.

The Convention on International Trade in Endangered Species (CITES) voted 76-11 in October 1989 to move the elephant from Appendix II of CITES, which permits limited trade, to Appendix I, which bans all trade. Five Southern African nations, including Zimbabwe and Botswana, which manage flourishing elephant herds partly supported by the sale of ivory, said they intended to file a "reservation" to the convention that would allow them to continue trading (187). Figure 8.4 shows the decline in elephant numbers in many African countries over the past decade.

Support for the trade ban gathered considerable momentum in the months prior to the CITES meeting. In June 1989, Britain banned all imports of raw and worked ivory; the rest of the European Community and the United States announced similar bans soon thereafter (188). Dubai, an important center of the ivory trade, also announced a ban. Japan, which imports 24 percent of the world's ivory, announced that it was banning all imports of worked raw ivory and all raw ivory that did not come directly from African producer countries authorized by CITES (189). Figure 8.5 shows the major importers of ivory.

Supporters of the ban hope it will increase pressures to control poaching and create similar pressures for importing nations to shut off imports. A ban also could help persuade some consumers not to buy ivory or to look for substitutes. Corian, a synthetic material, has been mentioned as a suitable substitute, and research is under way on ceramic substitutes (190) (191).

Consumer awareness helped reduce demand for leopard-skin coats, which was an important factor in the recovery of the leopard after it was listed on Appendix I. An Appendix I listing, however, has not helped the black rhino–down from 60,000 in 1970 to about 3,500 today (192)–in part because consumers in North Yemen and East Asia have not faced much domestic pressure to change their buying habits (193).

Failure of Legal Ivory Trade

The ban reflects the failure of the limited legal trade policy. Gaping loopholes in CITES' controlled trade system made the system largely ineffective. For example, the system lacked controls on worked ivory and established an export quota system that allowed the exporting country to set its own quota without regard to the number of elephants in that country (194). CITES in 1986 approved exports of some 108,000 tusks, representing more than 50,000 elephants, which some conservationists say is about 10 times more than the sustainable yield (195).

Even under the quota system, ivory smuggling has prospered. Poachers have killed about 22,000 elephants annually in Tanzania since 1981 (196). The elephant population in Zaire dropped from 376,000 in 1981 to 103,000 in 1989, with similar drastic declines over the same period in Kenya, Tanzania, and Zambia (197). Market prices for ivory are estimated to range between $63 and $260 per kilogram, so high that poachers are willing to risk their lives for it (198). Figure 8.5 shows the major importers of raw ivory.

Ban Critics Cite Good Management

Critics of the ban say the real problem is that the countries supporting the ban have ineffective wildlife management programs. Park rangers are poorly paid, often leading to collusion with poachers, and wildlife programs are poorly equipped (199). For example, Richard E. Leakey, Kenya's new director of wildlife conservation, said he had inherited a department that did not have a single working vehicle (200).

Botswana, South Africa and Zimbabwe all have stable or growing herd sizes. (See Figure 8.4.) Zimbabwe has found that big-game hunting, which may take 100-200 animals a year, is more lucrative than poaching (201). Dr. Anthony Hall-Martin, chief research officer of South Africa's national parks, attributes the success of the conservation program in South Africa to good pay and training for park rangers and to a policy of shooting poachers on sight (202).

Selective culling also brings in significant revenue for conservation. South Africa's Kruger National Park culls about 600 animals annually, which returns about 3-4 million from the sale of ivory and other elephant products. Zimbabwe earns somewhat more money through culling (203).

Critics also argue that a ban will make elephant products valueless, which will reduce the incentive for people to conserve them. Zambia, plagued by poor management for years, is having success with a sustainable management program in the Luangwa Valley, which gives much of the decision-making authority and returns much of the revenue from wildlife to local residents. Appendix I listing could threaten such programs (204).

Critics also fear that the effect of a ban will be to drive up the price of ivory, which would raise the profitability of poaching and increase the risk that poachers would venture into Southern African nations. A ban also will force the ivory trade further underground, making police detection more difficult.

Another concern is that a ban on elephant ivory could increase the demand for walrus ivory. Illegal shooting of walrus in Alaska is considered a serious problem (205).

SONGBIRDS DECLINE IN NORTH AMERICA

Loss of forested habitat appears to be decimating many populations of songbirds that breed in North America and winter in Central and South America.

Both continents are to blame. In North America, the breakup of contiguous forests and the resulting increase in forest-edge habitat have seriously harmed many songbird populations. In Central and South America, tropical deforestation was thought to be equally disastrous for many songbirds, but there was little evidence to prove it. But a 1989 study by scientists at the U.S. Fish and Wildlife Service and the National Zoological Park appears to confirm the idea that many songbird populations are declining and that tropical deforestation is apparently an important cause of that decline (206).

The research uses data from the North American Breeding Bird Survey, which since 1966 has annually deployed hundreds of veteran bird watchers to sam-

Table 8.4 Neotropical Migrant Songbird Population Trends, 1966-78 and 1978-87

(percent change/year)

Species	1966-78	1978-87
Olive-sided flycatcher	3.6	-5.7
White-eyed vireo	0.3	-1.2
Blue-winged warbler	1.0	-1.0
Northern parula	1.2	-2.1
Chestnut-sided warbler	2.2	-3.8
Cerulean warbler	-3.9	-0.9
Canada warbler	-2.7	-2.7
Scarlet tanager	2.6	-1.2
Rose-breasted grosbeak	6.1	-4.1

Source: Chandler S. Robbins *et al.*, "Population Declines in North American Birds That Migrate to the Tropics," Proceedings of the National Academy of Sciences, Vol. 86, No. 19 (1989)," Table 1, p. 14.

ple bird populations in 2,000 areas that cover some 80,000 kilometers of back roads. Each observer uses uniform procedures to sample bird populations at 50 stops at 0.8- kilometer intervals along secondary roads, counting all birds detected within a 0.4-kilometer radius during a three-minute period (207).

The new study examined data for 62 migrant species in the United States east of the Mississippi (and corresponding parts of eastern and central Canada) and calculated population trends for 1966-78 and 1978-87. During 1966-78, 15 species had population decreases, of which 6 were significant, and 47 species had population increases, of which 23 were significant. The picture changed dramatically in 1978-87: 44 species (71 percent) showed decreases, 20 of them significant, and only 18 species had increases, of which 4 were significant (208). (See Table 8.4.)

To examine the hypothesis that tropical deforestation is affecting populations, the researchers used recent censuses of wintering birds at the Sian Káan Biosphere Reserve in Quintana Roo, Mexico. The data distinguished between birds wintering in forest habitat and those using forest edge or scrub areas. The 16 species wintering at forest reserves showed a significant population decline; the group as a whole showed an average annual downward trend of 2.3 percent between the two periods. The 12 scrub species generally held their own (209).

The study also found that three species that breed in scrub but winter in forest (white-eyed vireo, blue-winged warbler, and chestnut-sided warbler) showed declines in the 1980s; three of five species that breed in forested habitat but winter in scrub habitat (eastern wood-pewee, least flycatcher, and blue-gray gnatcatcher but not yellow-throated warbler or rose-breasted grosbeak) showed increases (210).

The analysis "represents the strongest evidence to date that tropical deforestation is contributing to declines in migratory bird populations" (211). Many other factors may also contribute; for example, Chandler Robbins suggests that heavy use of pesticides and fungicides on South American plantations may be severely affecting migrant species. Many products used in Latin America, he notes, are manufactured in the United States but are banned for use there because of their toxic effects (212).

Tropical deforestation may be particularly critical to the songbird, other experts suggest, because songbirds spend much more time on their wintering grounds than on their breeding grounds and are much more densely packed in a relatively small area. Many North American songbirds migrate only to the near-tropics–Mexico, Cuba, Haiti, the Dominican Republic, and the Bahamas–so deforestation may be particularly harmful (213).

Previous studies have shown that forest fragmentation in North America provides access for predators and parasites and is an important contributor to the declining trends. For example, the parasitic brown-headed cowbird lives at the forest edge and lays its eggs in the nests of other species. The cowbird chicks grow faster than the young of the host species, which usually starve. These studies generally conclude that the preservation of remaining large forest areas in both the United States and the tropics is necessary to preserve songbird populations (214) (215).

"Wildlife and Habitat" was authored by David Harmon, a conservation consultant in Houghton, Michigan.

References and Notes

1. Ruggero Tomaselli, "The Degradation of the Mediterranean Maquis," *Ambio*, Vol. 6, No. 6 (1977), pp. 356-362.
2. Jeffrey M. Klopatek *et al.*, "Land-Use Conflicts with Natural Vegetation in the United States," *Environmental Conservation*, Vol. 6, No. 3 (1979), pp. 191-199.
3. James T. Thomson, "Deforestation and Desertification in Twentieth-Century Arid Sahelien Africa," in *World Deforestation in the Twentieth Century*, John F. Richards and Richard P. Tucker, eds. (Duke University Press, Durham, North Carolina, 1988), pp. 79-83.
4. John R. McNeill, "Deforestation in the Araucaria Zone of Southern Brazil, 1900-1983," in *World Deforestation in the Twentieth Century*, John F. Richards and Richard P. Tucker, eds. (Duke University Press, Durham, North Carolina, 1988), p. 15.
5. For biodiversity as a resource, see E.O. Wilson, "The Current State of Biological Diversity," in *Biodiversity*, E.O. Wilson, ed. (National Academy Press, Washington, D.C., 1988), p. 3.
6. David S. Wilcove, *Protecting Biological Diversity*, Vol. 2 of *National Forests, Policies for the Future* (The Wilderness Society, Washington, D.C., 1988), p. 3.
7. Jeffrey A. McNeely, *Economics and Biological Diversity: Developing and Using Economic Incentives to Conserve Biological Resources* (International Union for Conservation of Nature and Natural Resources, Gland, Switzerland, 1988), pp. 9-36.
8. E.O. Wilson, "The Current State of Biological Diversity," in *Biodiversity*, E.O. Wilson, ed. (National Academy Press, Washington, D.C., 1988), p. 3.
9. See U.S. Congress, Office of Technology Assessment, *Technologies to Maintain Biological Diversity* (U.S. Government Printing Office, 1987), Box 2-A, p. 38.
10. Elaine Matthews, "Global Vegetation and Land Use: New High Resolution Data Bases for Climate Studies," *Journal of Climate and Applied Meteorology*, Vol. 22, No. 3 (1983), pp. 474-487.
11. Jerry S. Olson, "Productivity of Forest Ecosystems," in *Productivity of World Ecosystems* (National Academy of Science Press, Washington, D.C., 1975), p. 37.
12. B.K. Maloney, "Man's Impact on the Rainforests of West Malesia: The Palynological Record," *Journal of Biogeography*, Vol. 12, No. 6 (1985), p. 537.
13. Candice L. Goucher, "The Impact of German Colonial Rule on the Forests of Togo," in *World Deforestation in the Twentieth Century*, John F. Richards and Richard P. Tucker, eds. (Duke University Press, Durham, North Carolina, 1988), pp. 58-60.
14. John F. Richards and Richard P. Tucker, "Introduction," in *World Deforestation in the Twentieth Century*, John F. Richards and Richard P. Tucker, eds. (Duke University Press, Durham, North Carolina, 1988), p. 3.
15. World Resources Institute and International Institute for Environment and Development in collaboration with the United Nations Environment Programme, *World Resources 1988-89* (Basic Books, New York, 1988), Table 5.1, p. 70.
16. *Op. cit.* 14, p. 11.
17. Gary S. Hartshorn, "Possible Effects of Global Warming on the Biological Diversity in Tropical Forests," in *Consequences of the Greenhouse Effect for Biological Diversity*, Robert L. Peters II, ed. (Yale University Press, New Haven, Connecticut, in press).
18. Definition from Reinhold Tuxen, "Die Heutige Potentielle Natürliche Vegetation als Gegenstand der Vegetationskartierung," *Angewandte Pflanzensoziologie*, Vol. 13 (1956), pp. 5-42, cited in A. W. Kuüchler, *Potential Natural Vegetation of the Conterminous United States: Manual to Accompany the Map* (American Geographical Society, New York, 1964), pp. 1-2.
19. David W. Crumpacker *et al.*, "A Preliminary Assessment of the Status of Major Terrestrial and Wetland Ecosystems on Federal and Indian Lands in the United States," *Conservation Biology*, Vol. 2, No. 1 (1988), p. 105.
20. One reason such maps are only approximations is that at no time would all of a given vegetation formation have been at its climax stage, even under wholly natural conditions, with no human interference. For example, the original extent of closed-canopy broad-leaved tropical moist forests was recently estimated at 90 percent of the climatic climax area. See Walter V. Reid and Kenton R. Miller, *Keeping Options Alive: The Scientific Basis for the Conservation of Biodiversity* (World Resources Institute, Washington, D.C., 1989), Table 18.
21. Areas of tropical habitat also occur in Irian Jaya (Indonesia), Papua New Guinea, Australia, and elsewhere in the Pacific.
22. *Op. cit.*, 19, Table 1, pp. 107-111.
23. It should be noted that classifications of vegetation are, inevitably, simplifications. See D. Muller-Dombois, "Classification and Mapping of Plant Communities: A Review with Emphasis on Tropical Vegetation," in *The Role of Terrestrial Vegetation in the Global Carbon Cycle: Measurement by Remote Sensing*, George M. Woodwell, ed. (John Wiley & Sons, Chichester, U.K., 1984), p. 27.
24. David W. Crumpacker, "Status and Trends of Natural Ecosystems in the United States," report to the U.S. Congress, Office of Technology Assessment, Washington, D.C., 1985, p. 2.
25. Miklos D. Udvardy, *A Classification of the Biogeographical Provinces of the World: Prepared as a Contribution to UNESCO's Man and the Biosphere Programme Project No. 8* (International Union for Conservation of Nature and Natural Resources, Morges, Switzerland, 1975).
26. For a world map of Udvardy's classifications, see World Resources Institute and International Institute for Environment and Development, *World Resources 1986* (Basic Books, New York), pp. 96-97.
27. L.R. Holdridge, "Determination of World Plant Formation from Simple Climatic Data," *Science*, Vol. 105, No. 2727 (1947), pp. 367-368.
28. When the Holdridge system is mapped worldwide, 20 classifications appear: tropical wet, moist, and dry forest; subtropical wet and moist forest; warm, dry-warm, and cool temperate forest; moist and wet boreal forest; tropical thorn woodland; temperate thorn steppe; cool temperate steppe; temperate desert bush; tropical desert bush; desert; boreal desert; tundra; ice; and high-altitude areas. For this map, see L.R. Holdridge, "Determination of World Plant Formation from Simple Climatic Data," *Science*, Vol. 105, No. 2727 (1947), pp. 367-368; it also appears as the frontispiece of Bert Bolin *et al.*, eds. *The Greenhouse Effect, Climatic Change, and Ecosystems* (John Wiley & Sons, Chichester, U.K., 1986).
29. *Op. cit.* 8, p. 8.
30. Walter V.C. Reid and Kenton R. Miller, *Keeping Options Alive: The Scientific Basis for the Conservation of Biodiversity* (World Resources Institute, Washington, D.C., 1989), Table 7.
31. *Op. cit.* 14, p. 4.

32. Jan G. Laarman, "Exports of Tropical Hardswoods in the 20th Century," in *World Deforestation in the Twentieth Century*, John F. Richards and Richard P. Tucker, eds. (Duke University Press, Durham, North Carolina, 1988), pp. 157-161.

33. Robert Buschbacher, "Tropical Deforestation and Pasture Development," *Bio Science*, Vol. 36, No. 1 (1986), p. 22.

34. Philip M. Fearnside, "Deforestation and International Development Projects in Brazilian Amazonia," *Conservation Biology*, Vol. 1, No. 3 (1987), pp. 214-221.

35. Alan Grainger, "Quantifying Changes in Forest Cover in the Humid Tropics: Overcoming Current Limitations," *Journal of World Forest Resource Management*, Vol. 1, No. 1 (1984), Table 8, p. 21, cited in World Resources Institute and International Institute for Environment and Development in collaboration with the United Nations Environment Programme, *World Resources 1988-89* (Basic Books, New York, 1988), p. 71.

36. *Op. cit.* 8, p. 11.

37. U.S. Congress, Office of Technology Assessment, *Technologies to Sustain Tropical Forest Resources* (U.S. Government Printing Office, Washington, 1984).

38. *Op. cit.* 14, p. 9.

39. *Op. cit.* 8, p. 9.

40. Jean-Marc Thiollay and Bernd U. Meyburg, "Forest Fragmentation and the Conservation of Raptors: A Survey on the Island of Java," *Biological Conservation*, Vol. 44, No. 4 (1988), p. 246.

41. Andrew D. Johns, "Selective Logging and Wildlife Conservation in Tropical Rain-Forest: Problems and Recommendations," *Biological Conservation*, Vol. 31, No. 4 (1985), pp. 355-375.

42. L.I. Pahl, J.W. Winter, and G. Heinsohn, "Variation in Responses of Arboreal Marsupials to Fragmentation of Tropical Rainforest in North Eastern Australia," *Biological Conservation*, Vol. 46, No. 1 (1988), pp. 78-80.

43. S. Brown and A.E. Lugo, "The Storage and Production of Organic Matter in Tropical Matter in Tropical Forests and their Role in the Global Carbon Cycle," *Biotropica*, Vol. 14 (1982), cited in Ariel E. Lugo, "Estimating the Reduction in Tropical Forest Species", in *Biodiversity*, E.O. Wilson, ed. (National Academy Press, Washington, D.C., 1988), p. 61.

44. Daniel H. Janzen, "Tropical Dry Forests: The Most Endangered Major Tropical Ecosystem," in *Biodiversity*, E.O. Wilson, ed. (National Academy Press, Washington, D.C., 1988), pp. 131-132.

45. *Op. cit.* 17.

46. Jean Paul Lanly, *Tropical Forest Resources* (Food and Agriculture Organization of the United Nations, Rome, 1982), pp. 84-85, cited in World Resources Institute and International Institute for Environment and Development in collaboration with the United Nations Environment Programme, *World Resources 1988-89* (Basic Books, New York, 1988), p. 71.

47. *Op. cit.* 44, p. 130.

48. Scott A. Mori, "Eastern, Extra-Amazonian Brazil," in *Floristic Inventory of Tropical Countries: The Status of Plant Systematics, Collections, and Vegetation, plus Recommendations for the Future*, David G. Campbell and H. David Hammond, eds. (New York Botanical Garden, Bronx, New York, 1989), pp. 432 and 434.

49. J.A. Mabbutt, "A New Global Assessment of the Status and Trends of Desertification," *Environmental Conservation*, Vol. 11, No. 2 (1984), p. 106.

50. Peter H. Raven, "The Evolution of Mediterranean Floras," in *Mediterranean Type Ecosystems: Origin and Structure*, Francesco di Castri and Harold A. Mooney, eds. (Springer-Verlag, New York, 1973), p. 214.

51. Norman J.W. Thrower and David E. Bradbury, eds., *Chile-California Mediterranean Scrub Atlas: A Comparative Analysis* (Dowden, Hutchinson & Ross, Stroudsburg, Pennsylvania, 1977), p. 7.

52. Homer Aschmann, "Man's Impact on the Several Regions with Mediterranean Climates," in *Mediterranean Type Ecosystems: Origin and Structure*, Francesco di Castri and Harold A. Mooney, eds. (Springer-Verlag, New York, 1973), pp. 363-371.

53. J.V. Thirgood, *Man and the Mediterranean Forest: A History of Resource Depletion* (Academic Press, London, 1981), p. 11.

54. Harold A. Mooney, "Lessons from Mediterranean-Climate Regions," in *Biodiversity*, E.O. Wilson, ed. (National Academy Press, Washington, D.C., 1988), p. 161.

55. *Op. cit.* 49.

56. I.A.W. Macdonald and M.L. Jarman, *Invasive Alien Organisms in the Terrestrial Ecosystems of the Fynbos Biome, South Africa* (Council for Scientific and Industrial Research, Pretoria, 1984), cited in Harold A. Mooney, "Lessons from Mediterranean-Climate Regions," in *Biodiversity*, E.O. Wilson, ed. (National Academy Press, Washington, D.C., 1988), p. 160.

57. M.L. Jarman, *Conservation Priorities in the Lowland Regions of the Fynbos Biome* (Council for Scientific and Industrial Research, Pretoria, 1986), cited in Harold A. Mooney, "Lessons from Mediterranean-Climate Regions," in *Biodiveristy*, E.O. Wilson, ed. (National Academy Press, Washington, D.C., 1988), p. 160.

58. *Op. cit.* 54, pp. 159-161.

59. *Op. cit.* 19, pp. 107-111.

60. *Op. cit.* 54, p. 157.

61. Marion Clawson, "Forests in the Long Sweep of American History," *Science*, Vol. 204 (June 15, 1979), Table 1, p. 1169.

62. Michael Williams, "The Death and Rebirth of the American Forest: Clearing and Reversion in the United States, 1900-1980," in *World Deforestation in the Twentieth Century*, John F. Richards and Richard P. Tucker, eds. (Duke University Press, Durham, North Carolina, 1988), pp. 213-214 and 223-225.

63. *Op. cit.* 19, Table 1, pp. 107-111.

64. Y.P. Badenkov *et al.*, "Caucasia: Regional Review," paper presented at the Earth as Transformed by Human Action Conference, Worcester, Massachusetts, October 1987, cited in World Resources Institute and International Institute for Environment and Development in collaboration with the United Nations Environment Programme, *World Resources 1988-89* (Basic Books, New York, 1988), Table 16.3 note, p. 269.

65. Vaclav Smil, "Deforestation in China," *Ambio*, Vol. 12, No. 5 (1983), p. 227.

66. Hal Bruno, "Inferno in China: Finally the Rains Came," review of *The Great Black Dragon Fire: A Chinese Inferno*, by Harrison E. Salisbury, *New York Times Book Review* (May 14, 1989), p. 7.

67. Netherlands Central Bureau of Statistics, *Environmental Statistics of the Netherlands 1987* (Staatsuitgeverij, The Hague, 1987), p. 124.

68. *Op. cit.* 19, Table 1, pp. 107-111.

69. The Conservation Foundation (CF), *State of the Environment: A View toward the Nineties* (CF, Washington, D.C., 1987), p. 546.

70. Alfred W. Crosby, *Ecological Imperialism: The Biological Expansion of Europe, 900-1900* (Cambridge University Press, Cambridge, U.K., 1986), pp. 159-161.

71. Philip R. Pryde, "Strategies and Problems of Wildlife Preservation in the U.S.S.R.," *Biological Conservation*, Vol. 36, No. 4 (1986), p. 352.

72. *Op. cit.* 49. The categories included here are Southern Africa; Western Asia; the U.S.S.R. in Asia, China, and Mongolia; Australia; and North America.

73. Zbigniew Karpowicz and Jeremy Harrison, "Circumpolar Protected Areas: An Overview, The North," International Union for Conservation of Nature and Natural Resources Conservation Monitoring Centre, Cambridge, U.K., n.d., p. 18.

74. J.G. Nelson, "Living with Exploitation in the Subarctic and Arctic of Canada," in *National Parks, Conservation, and Development: The Role of Protected Areas in Sustaining Society*, Jeffrey A. McNeely and Kenton R. Miller, eds. (Smithsonian Institution Press, Washington, D.C., 1984), pp. 527-533.

75. B.I. Yunov, "O lesopolzovanii v evropeyskoy chasti SSSR," *Lesnoye Khozyaystvo*, Vol. 10 (1978), p. 46, cited in Brenton M. Barr, "Perspectives on Deforestation in the U.S.S.R.," in *World Deforestation in the Twentieth Century*, John F. Richards and Richard P. Tucker, eds. (Duke University Press, Durham, North Carolina, 1988), p. 232.

76. J.H. Holowacz, "U.S.S.R.," *World Wood*, Vol. 8 (1975), p. 18, cited in Brenton M. Barr, "Perspectives on Deforestation in the U.S.S.R.," in *World Deforestation in the Twentieth Century*, John F. Richards and Richard P. Tucker, eds. (Duke University Press, Durham, North Carolina, 1988), pp. 232 and 248-249.

77. *Op. cit.* 24.

78. *Ibid.*, pp. 10-11 and 14.

79. Lewis M. Cowardin *et al.*, *Classification of Wetlands and Deepwater Habitats of the United States* (U.S. Government Printing Office, Washington, D.C., 1979).

80. Bruce P. Hayden, G. Carleton Ray, and Robert Dolan, "Classification of Coastal and Marine Environments," *Environmental Conservation*, Vol. 11, No. 3 (1984), pp. 199-207.

81. J.E. Grassle (Woods Hole Oceanographic Institution, Woods Hole, Massachusetts, 1987), cited in G. Carleton Ray, "Ecological Diversity in Coastal Zones and Oceans," in *Biodiversity*, E.O. Wilson, ed. (National Academy Press, Washington, D.C., 1988), p. 38.

82. G. Carleton Ray, "Ecological Diversity in Coastal Zones and Oceans," in *Biodiversity*, E.O. Wilson, ed. (National Academy Press, Washington, D.C., 1988), p. 37.

83. B.L. Swift and J.S. Barclay, "Status of Riparian Ecosystems in the United States," paper presented at the American Water Resources Association National Conference, Minneapolis, Minnesota, cited in David W. Crumpacker, "Status and Trends of Natural Ecosystems in the United States," report to the U.S. Congress, Office of Technology Assessment, Washington, D.C., 1985, p. 34.

84. F.E. Smith, "A Short Review of the Status of Riparian Forests in California," in *Proceedings: Riparian Forests in California–Their Ecology and Conservation Symposium*, A. Sands, ed. (University of California, Division of Agricultural Sciences, Berkeley, 1980), pp. 1-2, cited in David W. Crumpacker, "Status and Trend of Natural Ecosystems in the United States, report to the U.S. Congress, Office of Technology Assessment, Washington, D.C., 1985, p. 34.

85. International Institute for Environment and Development and World Resources Institute, *World Resources 1987* (Basic Books, New York, 1987), Table 6.6, pp. 86 and 87.

86. Cleber J.R. Alho, Thomas E. Lacher, Jr., and Humberto C. Gonsalves, "Environmental Degradation in the Pantanal Ecosystem," *BioScience*, Vol. 38, No. 3 (1988), p. 164.

87. Pollution of the Kesterson National Wildlife Refuge is a well-known example. Maura Dolan, "Pollution Endangers U.S. Refuges," *Los Angeles Times* (July 6, 1986), pp. 1 and 17-18.

88. *Op. cit.* 85, Table 6.6, p. 86.

89. *Op. cit.* 85, pp. 87-88.

90. National Wetlands Working Group, *Wetlands of Canada* (Environment Canada, Sustainable Development Branch, Ottawa, Ontario, 1988), Tables 10-8, 10-10, and 10-12 to 10-14, pp. 400-405.

91. *Op. cit.* 69, pp. 291-292.

92. The figures for Africa are from Table 8.2; for Asia (Indomalaysia, Japan, Taiwan, China, Australia and Oceania), from Table 8.3 and P. Saenger, E.J. Hegerl and J.D.S. Davie, eds. *First Report on the Global Status of Mangrove Ecosystems* (International Union for Conservation of Nature and Natural Resources Commission on Ecology, Working Group on Mangrove Ecosystems, Toowong, Australia, 1981), Figures 2-3, pp. 9-10; for the Americas, from Saenger, Hegerl, and Davie, Figures 4-5, pp. 11-12.

93. L.S. Hamilton and S.C. Snedaker, eds., *Handbook for Mangrove Area Management* (Environmental Policy Institute *et al.*, Washington, D.C., 1984), p. 25, cited in World Resources Institute and International Institute for Environment and Development, *World Resources 1986* (Basic Books, New York, 1986), p. 147.

94. *Op. cit.* 19, Table 1, pp. 107-111.

95. J.C. Zieman, "Tropical Sea Grass Ecosystems and Pollution," in *Tropical Marine Pollution*, E.J. Ferguson Wood and R.E. Johannes, eds. (Elsevier, Amsterdam, 1975), pp. 63-74.

96. M.L. Cambridge and A.J. McComb, "The Loss of Seagrasses in Cockburn Sound, Western Australia, Part 1: The Time Course and Magnitude of Seagrass Decline in Relation to Industrial Development," *Aquatic Botany*, Vol. 20, No. 3 (1984), p. 229.

97. International Union for Conservation of Nature and Natural Resources (IUCN) and United Nations Environment Programme (UNEP), *Atlantic and Eastern Pacific*, Vol. 1 of *Coral Reefs of the World*, Susan M. Wells, ed. (IUCN, Gland, Switzerland, and Cambridge, U.K./UNEP, Nairobi, 1988), p. xx.

98. International Union for Conservation of Nature and Natural Resources (IUCN) and United Nations Environment Programme (UNEP), *Central and Western Pacific*, Vol. 3 of *Coral Reefs of the World*, Susan M. Wells and Martin D. Jenkins, eds. (IUCN, Gland, Switzerland, and Cambridge, U.K./UNEP, Nairobi, 1988), Table 2, p. xxxv.

99. World Resources Institute and International Institute for Environment and Development, *World Resources 1986* (Basic Books, New York, 1986), p. 152.

100. International Union for Conservation of Nature and Natural Resources (IUCN) and United Nations Environment Programme (UNEP), *Management and Conservation of Renewable Marine Resources in the Indian Ocean Region: Overview* (IUCN and UNEP, 1985), cited in IUCN and UNEP, *Indian Ocean, Red Sea and Gulf*, Vol. 2 of *Coral Reefs of the World*, Susan M. Wells, ed. (IUCN, Gland, Switzerland, and Cambridge, U.K./UNEP, Nairobi, 1988), p. xx.

101. *Op. cit.* 99, Figure 6.2, p. 93, and pp. 94-99.

102. P.H.C. Lucas, "After Systems Reviews and Action Strategies: What Next for CNPPA?" in *New Challenges for the World's Protected Area System*, Jim Thorsell, comp. (International Union for Conservation of Nature and Natural Resources, Gland, Switzerland, 1988), p. 4.

103. See International Union for Conservation of Nature and Natural Resources (IUCN), *1985 United Nations list of National Parks and Protected Areas* (IUCN, Gland, Switzerland, 1985), p. 8.

104. John Foster, *Protected Landscapes* (International Union for Conservation of Nature and Natural Resources, Countryside Commission, Countryside Commission for Scotland, and Council of Europe, 1988), pp. 11-17.

105. Duncan Poore and Judy Poore, *Protected Landscapes: The United Kingdom Experience* (Countryside Commission, Countryside Commission for Northern Ireland, and International Union for Conservation of Nature and Natural Resources, Manchester, U.K., 1987), pp. 13 and 19-24.

106. International Union for Conservation of Nature and Natural Resources (IUCN) Conservation Monitoring Centre, *Protected Landscapes: Experience around the World* (IUCN, Gland, Switzerland, 1987), pp. iii, 1, and 346-349.

107. Dennis McCaffrey and Helena Landazuri, *The Matching Grant for Wildlands and Human Needs: A Program Evaluation* (World Wildlife Fund, Washington, D.C., 1987), pp. 1-2, 8, and 27-43.

108. "Viable population" refers to the minimum number needed to retain all the population's genetic diversity in captivity. For vertebrates, the figure is 250. Ulysses S. Seal, "The Realities of Preserving Species in Captivity," in *Animal Extinction: What Everyone Should Know*, R.J. Hoage, ed. (Smithsonian Institution Press, Washington, D.C., 1985), p. 88.

109. William Conway, "Can Technology Aid Species Preservation?" in *Biodiversity*, E.O. Wilson, ed. (National Academy Press, Washington, D.C., 1988), p. 266.

110. *Op. cit.* 9, p. 140.

111. *Op. cit.* 109, Table 30.1, p. 264.

112. *Op. cit.* 9, pp. 140 and 148-149.

113. Betsy L. Dresser, "Cryobiology, Embryo Transfer, and Artificial Insemination in Ex Situ Animal Conservation Programs," in *Biodiversity*, E.O. Wilson, ed. (National Academy Press, Washington, D.C., 1988), pp. 298-302.

114. Ulysses S. Seal, "Intensive Technology in the Care of Ex Situ Populations of Vanishing Species," in *Biodiversity*, E.O. Wilson, ed. (National Academy Press, Washington, D.C., 1988), p. 293.

115. *Op. cit.* 113, pp. 297-298.

116. Ulysses S. Seal, "The Realities of Preserving Species in Captivity," in *Animal Extinctions: What Everyone Should Know*, R.J. Hoage, ed. (Smithsonian Institution Press, Washington, D.C., 1985), pp. 90-92.

117. *Ibid.*, p. 71.

118. *Op. cit.* 109, pp. 265-266.

119. Peter S. Ashton, "Biological Considerations in *In Situ* vs *Ex Situ* Plant Conservation," in *Botanic Gardens and the World Conservation Strategy*, D. Bramwell *et al.*, eds. (Academic Press, London, 1987), p. 120.

120. Peter S.W. Jackson, Programme Director, Botanic Gardens Conservation Secretariat of the World Conservation Union (formerly International Union for Conservation of Nature and Natural Resources) Surrey, U.K., 1989 (personal communication).

121. Vernon H. Heywood, "Botanic Gardens and Germplasm Conservation," *Botanic Gardens Conservation News*, Vol. 1, No. 3 (Botanic Gardens Conservation Secretariat of the International Union for Conservation of Nature and Natural Resources, Surrey, U.K., 1988), p. 15.

122. *Op. cit.* 119, pp. 121-122.

123. *Op. cit.* 9, pp. 171-172.

124. J. Cullen *et al.*, *The Cultivation and Propagation of Threatened Plants: A Proposal for the Documentation of Botanic Garden Methods* (Botanic Gardens Conservation Secretariat of the International Union for Conservation of Nature and Natural Resources, Surrey, U.K., 1988), p. 4.

125. Peter S.W. Jackson, "Botanic Gardens Unite for Action," *Species*, No. 11 (International Union for Conservation of Nature and Natural Resources Species Survival Commission, 1988), p. 10.

126. International Union for Conservation of Nature and Natural Resources (IUCN), *Botanic Gardens Conservation Secretariat: A Prospectus* (Botanic Gardens Conservation Secretariat of the IUCN, Surrey, U.K., 1987), p. 1.

127. *Op. cit.* 124, pp. 5-9.

128. Vernon H. Heywood, "Implementing the Botanic Gardens Conservation Strategy: From Las Palmas to Reunion," paper presented at the Second International Botanic Gardens Conservation Congress, Reunion, April 1989.

129. Peter H. Raven, "The Scope of the Plant Conservation Problem World-Wide," in *Botanic Gardens and the World Conservation Strategy*, D. Bramwell *et al.*, eds. (Academic Press, London, 1987), pp. 23 and 27.

130. "CPC Survey Reveals 680 Native U.S. Plants May Become Extinct within 10 Years", *The Center for Plant Conservation Newsletter*, Vol. 3, No. 4 (Jamaica Plain, Massachusetts, 1988), p. 1.

131. Donald A. Falk and Linda R. McMahan, "Endangered Plant Conservation: Managing for Diversity," *Natural Areas Journal*, Vol. 8, No. 2 (1988), p. 91.

132. Donald A. Falk and Francis R. Thibodeau, "Saving the Rarest," *Arnoldia*, Vol. 46, No. 3 (Arnold Arboretum of Harvard University, Jamaica Plain, Massachusetts, 1986), pp. 3 and 6.

133. Center for Plant Conservation (CPC), *1987 Annual Report* (CPC, Jamaica Plain, Massachusetts, 1988), pp. 3-5 and 6-7.

134. Walter G. Berendsohn, "The Project 'Laderas de la Laguná: An Approach to Conservation at the Botanical Garden in El Salvador," *Botanic Gardens Conservation News*, Vol. 1, No. 2 (Botanic Gardens Conservation Secretariat of the International Union for Conservation of Nature and Natural Resources, Surrey, U.K., 1988), pp. 23-24.

135. N.E.G. Cruttwell, "The Lipizauga Botanical Sanctuary: A Conservation Project in Papua New Guinea," paper presented at the Second International Botanic Gardens Conservation Congress, Réunion, April 1989.

136. Brian A. Meilleur, "The Role of the Ethnobotanical Garden in Tropical Plant Conservation," paper presented at the Second International Botanic Gardens Conservation Congress, Réunion, April 1989.

137. See, for example, Miguel A. Altieri, M. Kat Anderson, and Laura C. Merrick, "Peasant Agriculture and the Conservation of Crop and Wild Plant Resources," *Conservation Biology*, Vol. 1, No. 1 (1987), pp. 49-58.

138. Daniel B. Botkin and Robert A. Nisbet, "Projecting the Effects of Climate Change on Biological Diversity in Forests," in *Consequences of the Greenhouse Effect for Biological Diversity*, Robert L. Peters II, ed. (Yale University Press, New Haven, Connecticut, in press).

8 Wildlife and Habitat

139. Robert L. Peters II, "Effects of Global Warming on Biological Diversity: An Overview," in *Preparing for Climate Change: Proceedings of the First North American Conference on Preparing for Climate Change: A Cooperative Approach*, October 27-29, 1987, Washington, D.C. (Government Institutes, Rockville, Maryland, 1988), pp. 171 and 182.
140. Tom M.L. Wigley, Director, Climatic Research Unit, University of East Anglia, Norwich, U.K., statement before the U.S. Senate Committee on Environment and Public Works, *Ozone Depletion, the Greenhouse Effect, and Climate Change*, Hearings, January 28, 1987 (U.S. Government Printing Office, Washington, D.C., 1987), p. 23.
141. Christopher Joyce, "Global Warming Could Wipe Out Wildlife," *New Scientist*, Vol. 117, No. 1598 (1988), p. 29.
142. Irving Mintzer, "Global Climate Change and Its Effects on Wild Lands," in *For the Conservation of Earth*, Vance Martin, ed. (Fulcrum, Golden, Colorado, 1988), pp. 61-62.
143. *Op. cit.* 139, pp. 171 and 173.
144. Only three countries–The Gambia, Sierra Leone, and Afghanistan–have life expectancies under 40 years.Chapter 16, "Population and Health," Table 16.2.
145. Stephen H. Schneider, Linda Mearns, and Peter H. Gleick, "Climate-Change Scenarios for Impact Assessment," in *Consequences of the Greenhouse Effect for Biological Diversity*, Robert L. Peters II, ed. (Yale University Press, New Haven, Connecticut, in press).
146. F.I. Woodward, "Review of the Effects of Climate on Vegetation: Ranges, Competition and Composition," in *Consequences of the Greenhouse Effect for Biological Diversity*, Robert L. Peters II, ed. (Yale University Press, New Haven, Connecticut, in press).
147. Carl H. Winget, "Forest Management Strategies to Address Climate Change," in *Preparing for Climate Change: Proceedings of the First North American Conference on Preparing for Climate Change: A Cooperative Approach*, October 27-29, 1987, Washington, D.C. (Government Institutes, Rockville, Maryland, 1988), p. 329.
148. Russell W. Graham, "The Role of Climate Change in the Design of Biological Reserves: The Paleoecological Perspective for Conservation Biology," *Conservation Biology*, Vol. 2, No. 4 (1988), p. 392.
149. Jerry F. Franklin *et al.*, "Effects of Global Climate Change on Forests in Northwestern North America," in *Consequences of the Greenhouse Effect for Biological Diversity*, Robert L. Peters II, ed. (Yale University Press, New Haven, Connecticut, in press).
150. Walter E. Westman and George P. Malanson, "Effects of Climate Change on Mediterranean-Type Ecosystems in California and Baja California," in *Consequences of the Greenhouse Effect for Biological Diversity*, Robert L. Peters II, ed. (Yale University Press, New Haven, Connecticut, in press).
151. *Op. cit.* 146.
152. *Op. cit.* 139, pp. 173 and 175.
153. In addition, some species might be able to negate warming by changing their aspect, that is, by moving from a south- to a north-facing slope in the Northern Hemisphere, and vice versa in the Southern.
154. *Op. cit.* 139, p. 175.
155. *Op. cit.* 139, p. 175.
156. *Op. cit.* 139, pp. 178 and 180.
157. *Op. cit.* 139, p. 177-178.
158. *Op. cit.* 50, p. 218.
159. Calculated using land areas from Chapter 17, "Land Cover and Settlements," Table 17.1, and the world map (van der Grinten's projection) in *Concise Earth Book Atlas*

(Graphic Learning, Boulder, Colorado, 1987), frontispiece.
160. Jill Jager, "Anticipating Climatic Change," *Environment*, Vol. 30, No. 7 (1988), p. 14.
161. M.Y. Verbitsky, *Influence of Antarctic and Greenland Icesheets on the World Oceanic Level in Sea Level and Oceanic Fluctuations for 15,000 Years*, (USSR Academy of Sciences, Institute of Geography, Moscow, *Nauka*, 1982), pp. 120-124, cited in G. deQ. Robin, "Changing the Sea Level," in *The Greenhouse Effect, Climatic Change, and Ecosystems*, Bert Bolin *et al.*, eds. (John Wiley & Sons, Chichester, U.K., 1986), p. 347.
162. G. deQ. Robin, "Formation, Flow, and Disintegration of Ice Shelves," *Journal of Glaciology*, Vol. 24 (), pp. 259-271 as cited in G. deQ. Robin "Changing the Sea Level," in *The Greenhouse Effect, Climatic Change, and Ecosystems*, Bert Bolin et al., eds. (John Wiley & Sons, Chichester, U.K., 1986), p. 347.
163. *Op. cit.* 139, pp. 169 and 177.
164. *Op. cit.* 138.
165. Robert L. Peters and Joan D.S. Darling, "The Greenhouse Effect and Nature Reserves," *BioScience*, Vol. 35, No. 11 (1985), pp. 713-714.
166. *Op. cit.* 142, pp. 63-64.
167. William R. Moomaw, "In Search of the Greenhouse Fingerprint," *Orion Nature Quarterly*, Vol. 8, No. 1 (1989), p. 10.
168. J.P. Myers *et al.*, "Conservation Strategy for Migratory Species," *American Scientist*, Vol. 75, No. 1 (1987), pp. 19-26.
169. Martin W. Holdgate, "Changing Habitats of the World," *Oryx*, Vol. 21, No. 3 (1987), p. 155.
170. *Op. cit.* 142, p. 64.
171. Andrew Dobson and Robin Carper, "Global Warming and Potential Changes in Host-Parasite and Disease-Vector Relationships," in *Consequences of the Greenhouse Effect for Biological Diversity*, Robert L. Peters II, ed. (Yale University Press, New Haven, Connecticut, 1989).
172. "Executive Summary," in *Direct Effects of Increasing Carbon Dioxide on Vegetation*, Boyd R. Strain and Jennifer D. Cure, eds. (U.S. Department of Energy, Washington, D.C., 1985), pp. xvii and xxii.
173. D. M. Gates, "An Overview," in *CO₂ and Plants: The Response of Plants to Rising Levels of Atmospheric Carbon Dioxide*, E.R. Lemon, ed. (Westview, Boulder, Colorado, 1983), pp. 7-20, cited in Margaret B. Davis and Catherine Zabinski, "Changes in Geographical Range Resulting from Greenhouse Warming: Effects on Biodiversity in Forests," in *Consequences of the Greenhouse Effect for Biological Diversity*, Robert L. Peters II, ed. (Yale University Press, New Haven, Connecticut, in press).
174. S.C. Zoltai, "Ecoclimatic Provinces of Canada and Man- Induced Climatic Change," *Canada Committee on Ecological Land Classification Newsletter*, No. 17 (Canadian Wildlife Service, Environment Canada, Ottawa, Ontario, 1988), p. 13.
175. *Op. cit.* 150.
176. *Op. cit.* 149.
177. *Op. cit.* 138.
178. Robert L. Peters II, "The Effect of Global Climatic Change on Natural Communities," in *Biodiversity*, E.O. Wilson, ed. (National Academy Press, Washington, D.C., 1988), pp. 454-455.
179. Margaret B. Davis and Catherine Zabinski, "Changes in Geographical Range Resulting from Greenhouse Warming: Effects on Biodiversity in Forests," in *Consequences of the Greenhouse Effect for Biological Diver-*

sity, Robert L. Peters II, ed. (Yale University Press, New Haven, Connecticut, in press).
180. *Op. cit.* 165, p. 715.
181. *Op. cit.* 150.
182. Russell W. Graham, "The Role of Climatic Change in the Design of Biological Reserves: The Paleoecological Perspective for Conservation Biology," *Conservation Biology*, Vol. 2, No. 4 (1988), p. 392.
183. *Op. cit.* 179.
184. *Op. cit.* 149.
185. W. Clark Ashby, "Forests," in *Restoration Ecology: A Synthetic Approach to Ecological Research*, William R. Jordan III, Michael E. Gilpin, and John D. Aber, eds. (Cambridge University Press, Cambridge, U.K., 1987), pp. 91-95.
186. Sue Armstrong and Fred Bridgland, "Elephants and the Ivory Tower," *New Scientist* (August 26, 1989), pp. 37-38.
187. Jane Perlez, "Global Trade in Ivory is Banned to Protect the African Elephant," *New York Times* (Oct. 17, 1989, New York), p. 1.
188. "Saving the Elephant: Nature's Great Masterpiece," *The Economist* (July 1, 1989), p. 16.
189. *Ibid.*
190. "Ceramics take plastic to tusk," *New Scientist* (August 26, 1989), p. 35.
191. Laurel Kornheiser, "The Evolution of Scrimshaw: The Art and the Ivory," *Cape Cod Life* (August/September 1989), p. 73.
192. Jeffrey P. Cohn, "Halting the Rhino's demise," *BioScience*, Vol. 38, No. 11, (1988), p. 740.
193. *Op. cit.* 188.
194. *Op. cit.* 186, p. 40.
195. *Op. cit.* 188.
196. *Op. cit.* 186, p. 40.
197. World Conservation Union, African Elephant and Rhino Specialist Group, cited in "Saving the Elephant: Nature's Great Masterpiece," *The Economist* (July 1, 1989), Table, p. 16.
198. *Op. cit.* 186, p. 40.
199. Jane Perlez, "Pretoria Rejects Ban on Ivory Trade," *New York Times* (June 22, 1989), p. A6.
200. *Op. cit.* 186, p. 39.
201. *Op. cit.* 188, p. 15.
202. *Op. cit.* 199.
203. *Op. cit.* 186, p. 38.
204. *Op. cit.* 186, pp. 40-41.
205. "Hunt for Walrus Ivory Worries Conservationists," *New York Times*, (August 15, 1989), p. C4.
206. Chandler S. Robbins *et al.*, "Population Declines in North American Birds that Migrate to the Tropics," *Proceedings of the National Academy of Sciences*, Vol. 86, No. 19 (1989), p. __.
207. *Ibid.*, pp. 2 and 3.
208. *Ibid.*, pp. 3 and 6.
209. *Ibid.*, pp. 5 and 7.
210. *Ibid.*, p. 8.
211. *Ibid.*, p. 10.
212. Chandler S. Robbins, Wildlife Biologist, U.S. Fish and Wildlife Service, Patuxent Wildlife Research Center, Laurel, Maryland, 1989 (personal communication).
213. Jack Connor, "Empty Skies: Where Have All the Songbirds Gone?" *Harrowsmith* (July/August 1988), pp. 40-41.
214. David S. Wilcove, "Nest Predation in Forest Tracts and the Decline of Migratory Songbirds," *Ecology*, Vol. 66, No. 4 (1985), pp. 1211-1214.
215. David S. Wilcove, "Changes in the Avifauna of the Great Smoky Mountains," *Wilson Bulletin*, Vol. 100, No. 2 (1988), pp. 256-271

9. Energy

In every part of the world, energy is used to cook food and warm dwellings. As a substitute for human labor, energy also plays a fundamental role in industrial development. In addition to these beneficial roles, however, energy is also a major cause of environmental degradation. Production and use of coal, oil, and natural gas—fossil energy sources—are the major sources of local air pollution and of regional problems such as acid rain. Fossil fuels also constitute the largest source of greenhouse gases that are changing the composition of the atmosphere and may lead to global climate change. Nonfossil energy sources such as nuclear power and large-scale hydroelectric dams also raise environmental concerns. Thus energy occupies a special place in any discussion of global environmental and resource problems and is central to strategies aimed at reducing air pollution and preventing global warming. If such strategies are adopted, energy production and use may undergo fundamental changes in coming decades.

The connection between energy and climate and strategies to reduce fossil fuel use are discussed in Chapter 2, "Climate Change: A Global Concern." Strategies and national policies designed to reduce air pollution (including that from energy production and use) are considered in Chapter 12, "Atmosphere." This chapter, "Energy," discusses current trends in energy production and use—including the reemer-

gence of rapid energy growth—and the implication of those trends, absent fundamental changes. Second, the chapter explores the potential of energy conservation and of more efficient use for slowing energy growth—and hence for reducing environmental impacts on both local and global scales. Energy use in cars, trucks, and buses is explored in more detail because, if current trends continue, light vehicles could become the world's dominant consumer of energy and the largest source of both global and local air pollution within 35 years. Finally, nonfossil sources of energy may assume particular importance in coming decades if international agreements curb emissions of greenhouse gases. This chapter reports on two such sources, nuclear and solar, and in particular on two specific families of technologies that might find widespread application if their economic and environmental problems can be overcome and their potential realized: advanced and potentially safer nuclear designs and photovoltaic solar energy cells.

CONDITIONS AND TRENDS

ENERGY PRODUCTION AND CONSUMPTION

World primary energy consumption rose 3.7 percent in 1988 (1). The rate of increase was greater than that experienced during the past several years—consumption of commercially traded energy grew 3.1 percent in 1987 and 2.4 percent in 1986. Energy use thus continues on an upward spiral, with annual consumption during the past two decades falling only in 1980, 1981, and 1982 (2).

Consumption and production of commercial energy by source and region in 1988 are shown in Table 9.1. The industrialized countries of the Organisation for Economic Co-operation and Development (OECD) (3) used roughly 50 percent of global primary energy, developing countries 16 percent, and countries with planned economies 34 percent. Energy consumption increased fastest in the industrializing countries of Southeast Asia—an average of 11.4 percent. South Korea's energy use, for example, shot up 13.1 percent. Consumption rose 3 percent in Latin America, 4.1 percent in Africa, and an average of 3 percent in countries with planned economies (4).

Among OECD countries, the United States experienced an increase of approximately 4 percent, boosting consumption to its highest level ever (5). In Western Europe, energy use grew an average of 1 percent. In Japan, energy use rose 6.2 percent (6).

Oil

Production and consumption increased for every type of primary energy. Oil consumption rose 3.1 percent globally, and oil production averaged 8.8 metric tons per day (mtd) in 1988. U.S. consumption accounted for roughly half of the oil used by all OECD countries, which in turn represented more than half of the world's demand. Southeast Asia recorded the greatest gain in oil use, 14 percent, with the rate jumping nearly 21 percent in Taiwan. If the world's use of oil rises as expected, the Organization of Petroleum Exporting Countries (OPEC), which in 1988 supplied 34 percent, is likely because of its huge reserves to provide an increasing share of the supply (7). While oil retained its position as the world's major fuel by a wide margin, supplying about 40 percent of world energy, consumption of other fuels increased at a faster pace (8).

Coal

Coal is the second-largest fuel source, supplying some 30 percent of global energy, and its use increased 3.7 percent between 1987 and 1988. China, which relies on coal for more than three quarters of its energy, consumed the most (24 percent), followed by the United States (19.8 percent) and the U.S.S.R. (12.8 percent). Western Europe consumed nearly 11 percent of the total (9). Because of widespread availability, abundant supplies, and generally low prices, coal consumption is likely to continue to increase significantly worldwide, absent major changes in the world energy picture. Developing countries are expected to experience the fastest growth and may soon consume roughly as much coal as Western Europe (10).

Reflecting the growing demand, world coal trade, which has expanded almost 40 percent since 1980, is projected to grow another 40 percent by the year 2000 (11). The traditional exporters—Australia is the largest, followed by the United States, South Africa, the U.S.S.R., Poland, and Canada—will face increas-

Table 9.1 Commercial Energy Production and Consumption by Region and Fuel, 1988

(petajoules) [a]

Region	Oil Production	Oil Consumption	Natural Gas Production	Natural Gas Consumption	Coal Production	Coal Consumption	Nuclear Power Production and Consumption	Hydropower Production and Consumption	Total Production	Total Consumption
North America	22,857	36,171	21,203	21,211	23,866	21,542	6,883	5,987	80,797	91,796
Latin America	14,278	9,551	3,601	3,308	888	959	96	3,902	22,765	17,816
Western Europe	8,290	24,875	6,297	8,332	7,821	11,033	6,155	4,509	33,073	54,904
Middle East	30,954	5,669	2,734	2,278	29	105	X	109	33,827	8,160
Africa	10,991	3,609	2,227	1,264	4,187	3,048	80	791	18,276	8,793
Asia and Australasia	6,816	19,507	4,539	4,091	9,877	12,779	2,759	2,491	26,483	41,627
Centrally Planned Economies										
U.S.S.R.	26,127	18,385	29,045	22,982	16,409	12,984	1,779	2,345	75,705	58,476
China	5,699	4,216	528	561	24,251	24,331	X	1,319	31,796	30,427
Other [b]	888	5,238	2,617	4,262	14,994	14,881	620	1,038	20,156	26,039
World Total [c]	126,900	127,222	72,791	68,290	102,322	101,660	18,373	22,493	342,878	338,037

Source: British Petroleum (BP), *BP Statistical Review of World Energy* (BP, London, 1989), pp. 4, 22, 26, 34.
a. Conversion factor: 1 million metric tons of oil equivalent = 41.87 petajoules.
b. Albania, Bulgaria, Czechoslovakia, German Democratic Republic, Hungary, Kampuchea, Laos, Mongolia, Democratic People's Republic of Korea, Poland, Romania, Viet Nam, and Yugoslavia.
c. Figures may not total because of rounding.
X = not available.

ing competition from new suppliers. These are countries such as Colombia, Venezuela, and Indonesia, which built dedicated export mines in response to the rise in world oil prices during the 1970s. China may also become a major exporter, but its domestic needs are huge and so far the country has never exported more than 1 percent of its total production (12).

Environmental concerns about sulfur dioxide and carbon dioxide emissions, however, could constrain the growth of coal consumption. Several European countries, as well as the United States and Japan, are carrying out extensive research and development on new technologies designed to remove sulfur dioxide and other pollutants while boosting energy efficiency. Two of the leading "clean coal" candidates for electrical generation are fluidized-bed combustion and coal gasification (13). (See Chapter 12 "Atmosphere.")

These and other clean-coal technologies are moving toward commercialization internationally and are likely to be adopted gradually over the next several decades as utilities refit current generating plants and add new capacity. A special need for such technologies exists in the developing countries trying to rapidly expand their electricity networks. Too often they have had to import older, inefficient power-generation equipment. Installing new plants based on these emerging technologies would not only reduce troublesome sulfur-dioxide and nitrogen-oxide pollution, but would help raise the overall efficiency of fossil-fuel electrical generation in the developing world from percentages in the high 20s to at least the high 30s. This would both save energy and reduce the growth of carbon dioxide emissions (14).

Natural Gas

Natural gas is generally expected to be the world's fastest growing energy source through the year 2000, as many countries are planning to use it to reduce dependence on oil and to reduce the environmental problems associated with other fossil fuels. Combustion of natural gas, for example, emits only about half the carbon dioxide of an equivalent amount of coal. Use of natural gas, which provides about 20 percent of global energy, rose 4.7 percent in 1988. The largest consumers by far were the U.S.S.R. (33.7 percent) and the United States (28.2 percent), while Western Europe used 12.2 percent. Consumption rates rose significantly in many developing countries—13.1 percent in Southeast Asia, 6.6 percent in Latin America, and 5.9 percent in Africa—and are expected to continue to grow fastest in the developing world (15).

Among the major gas producers, U.S. natural gas production—already 25 percent of the global total—is projected to increase (16). The U.S.S.R. accounted for nearly 40 percent of the world's gas production and plans to increase production for export, to replace existing domestic oil consumption where possible, and to meet much of the country's new energy demand (17).

Nuclear Power

Nuclear power, which amounted to over 5 percent of global energy production, posted the fastest growth rate at 8 percent. Much of this growth is an artifact, reflecting completion of reactors ordered many years ago rather than robust prospects for nuclear energy. Nonetheless, significant amounts of power are involved. The United States increased production nearly 16 percent and accounted for 33 percent of the global output of nuclear power, far outpacing France (12.4 percent), Japan (9.9 percent), and the U.S.S.R. (9.7 percent) (18). Nuclear energy supplied nearly 1,800 billion kilowatt-hours of electricity during 1988, approximately 17 percent of the global total. The world now produces as much electricity from nuclear energy as it did energy from all sources in 1957, the year the Shippingport station in Pennsylvania became the first reactor to feed electricity into a utility grid (19). The United States, with more reactors operating than any other country, produced about 20 percent of its electricity from nuclear energy, while the U.S.S.R., the second-leading producer, generated nearly 13 percent.

Overall, however, the outlook for nuclear energy has become considerably more clouded since the Chernobyl reactor accident in the U.S.S.R. in April 1986. Few new reactors are being ordered. As a result, nuclear generating capacity is now expected to grow slowly if at all in most countries. New limits on fossil energy sources and improved nuclear power technology, however, could conceivably change this outlook. (See Nonfossil Energy Sources: A Look Ahead, below.)

Hydropower

Hydroelectricity provides almost 7 percent of global energy, and its use increased slightly in 1988. Canada and the United States combined to use the most hydropower (26.6 percent), although U.S. consumption fell nearly 15 percent in 1988 as severe drought curtailed water supplies. Other major consumers were Western Europe (20.1 percent), Latin America (17.4 percent), and the U.S.S.R. (10.4 percent) (20).

Projections Forecast Increasing Consumption

Lower energy prices, led by a decline in the price of crude oil and robust economic growth in many countries, were key factors behind rising energy use. Projections of future energy consumption, given in Figures 9.1a and 9.1b by fuel and by major economic sector, forecast increasing consumption by market-economy countries of all primary energy resources, particularly during the early 1990s, although at a slower rate than experienced during the past several years (21). Total global energy use through the year 2000 is projected to grow at an annual rate of be-

Figure 9.1 Projected Energy Consumption in OECD Countries, 1985–2000

A. Projected Primary Energy Consumption

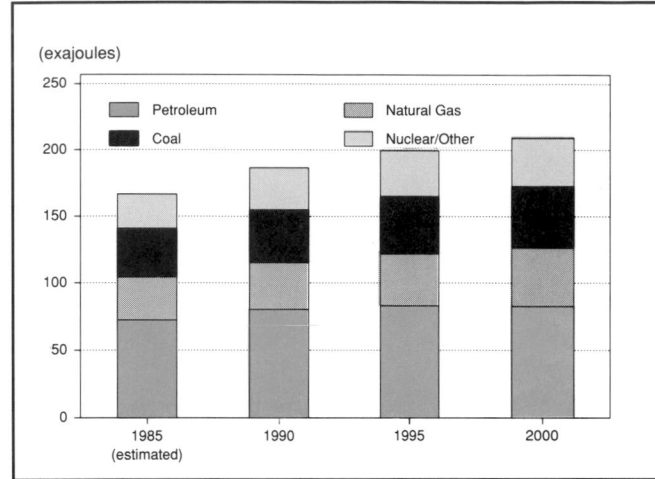

B. Projected Sectoral Energy Consumption

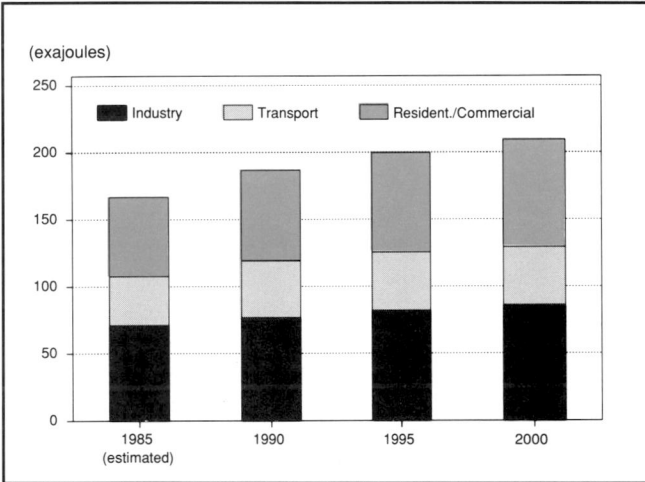

Source: U.S. Department of Energy (DOE), Energy Information Administration, Office of Energy Markets and End Use, *International Energy Outlook 1989: Projections to 2000* (U.S. DOE, Washington, D.C., 1989), p. 50.

tween 1.5 and 2.1 percent (22). Consumption is expected to increase the fastest in the developing world, as many countries build their industrial sectors and expand electricity and transportation networks. Many developing countries could grow at twice the rate of the OECD countries. These projections, however, do not take into account growing concern over global warming and the possibility, now being actively pursued, of international agreements that might stabilize or even reduce carbon dioxide emissions—and hence fossil energy consumption.

ENERGY RESOURCES

Proved reserves are those quantities that, under existing conditions, can be extracted in the future with reasonable certainty, according to geological and engineering information.

Oil Reserves

At the end of 1988, the world's proved reserves of oil were estimated at approximately 128 billion metric tons, of which OPEC members accounted for more than 70 percent. (See Table 9.2.) And early in 1989, Saudi Arabia adjusted its reserves upward by a dramatic 55 percent, or 11.4 metric tons, equal to about twice the existing reserves in all of North America (23).

Proved reserves have also been increasing in the market-economy countries outside OPEC, moving up to more than 156 billion barrels in 1988, a rise of some 1.4 billion metric tons since 1984. Still, this accounts for only 17 percent of the global supply (24).

Coal Reserves

China has 16 percent of proved global coal reserves, while the United States has 25 percent and the U.S.S.R. 25 percent. Australia, West Germany, South Africa, Poland, East Germany, and India also have large reserves, and many other countries have sizable deposits (25). As a measure of coal's immense abundance, the proved reserves in the United States alone are 43 percent greater than the world's oil and natural gas reserves combined (26).

Natural Gas Reserves

Natural gas is relatively abundant globally. (See Table 9.2.) The largest deposits are in the U.S.S.R., which has about 38 percent of global reserves, and in the OPEC countries in the Middle East, which have nearly 30 percent. But other areas are also richly endowed: the United States and Canada jointly have 7.1 percent of proved reserves, Latin America 6.1 percent, Western Europe 5.0 percent, Africa 6.5 percent, and Asia and Australia combine for 6.0 percent (27).

Nuclear Capacity

Installed nuclear capacity is now found in 25 countries and amounted to about 420 reactors in all operating commercially at the end of 1988. Roughly half of the countries derived at least one quarter of their electricity from nuclear power. France topped the list with nearly 70 percent, followed by Belgium (66 percent), Hungary (49 percent), Sweden (47 percent), South Korea (47 percent), Taiwan (41 percent), Switzerland (37 percent), Finland (36 percent), Spain (36 percent), Bulgaria (36 percent), West Germany (34 percent), Japan (28 percent), and Czechoslovakia (27 percent) (28). Argentina, Brazil, India, and Pakistan were the only developing countries that operated commercial reactors, as technical and financial problems have forced many countries to reduce

Table 9.2 Proved Commercial Energy Resources, Late 1988

(petajoules) [a]

Region	Oil Reserves	Oil R/P [b] (years)	Natural Gas Reserves	Natural Gas R/P [b] (years)	Hard Coal Reserves	Soft Coal Reserves	Coal R/P [b] (years)	Total Reserves
North America	230,285	10	310,720	14	3,746,080	1,876,699	286	6,163,784
Latin America	715,977	51	260,228	70	226,043	61,061	371	1,263,309
Western Europe	100,488	12	221,388	34	957,871	840,322	219	2,120,069
Middle East	3,236,551	100+	1,297,256	100+	X	X	X	4,533,807
Africa	314,025	29	275,764	100+	1,822,997	3,127	357	2,415,914
Asia and Australasia	113,049	18	264,112	57	1,217,685	700,918	228	2,295,764
Centrally Planned Economies	473,131	15	1,716,728	52	8,200,879	2,705,811	181	13,096,549
U.S.S.R.	334,960	13	1,650,700	55	3,003,144	1,892,404	X	6,881,208
China	129,797	23	34,956	64	4,316,561	187,832	X	4,669,145
Others [c]	8,374	11	31,072	12	881,175	625,576	X	1,546,196
World	5,183,506	41	4,346,196	58	16,171,556	6,187,938	218	31,889,196

Source: British Petroleum (BP), *BP Statistical Review of World Energy* (BP, London 1989), pp. 2, 20, 24.
Notes:
a. Conversion factors: 1 mtoe=41.87 PJ, 1 bcm natural gas=38.84 PJ, 1 mtce (hard coal)=27.91 PJ, 1 mtce (soft coal)=13.96 PJ.
b. R/P is the ratio of proved reserves to 1988 production rate.

or abandon their programs (29). Large uranium reserves are found in South Africa, Niger, Canada, the United States, Brazil, Australia, and the U.S.S.R.

Hydropower Reserves

Hydroelectric energy is a potentially vast resource in developing countries. While North America and Europe had by 1980 developed 59 and 36 percent of their large-scale hydropower potential respectively, Asia had harnessed just 9 percent, Latin America 8 percent, and Africa 5 percent (30). Developing large-scale hydroelectric production, however, could be slowed by financial problems as well as environmental and social concerns. Many countries, particularly in the developing world, are also installing small-scale hydroelectric units, which often are not connected to a central distribution grid but are used to run agricultural operations or provide power to isolated villages. A study for the U.S. Agency for International Development has projected that the capacity of stand-alone small hydro systems in developing countries could reach 29,000 megawatts in 1991, nearly triple the installed capacity in 1983. The worldwide potential is estimated to be well over 100,000 megawatts (31).

Other Renewables

Apart from hydroelectricity, other renewable energy sources—including geothermal, windpower, biomass, and solar power—already make sizable contributions in some countries. Among industrialized countries, for example, they now provide up to 5 per-

cent of total primary energy requirements in Australia, Austria, Canada, Denmark, Sweden, and Switzerland. In Ireland, peat resources provide about 8 percent of the country's commercial energy demand, and in Portugal biomass resources cover about 7 percent of those requirements (32). A number of developing countries have also expanded renewable energy production during recent years in order to increase their indigenous energy supplies. Still, the International Energy Agency (IEA) concludes that current economic and technical constraints will likely postpone a major global buildup of renewable energy production until sometime in the next century (33). But given their enormous potential and the public's heightened concern about the global environmental effects of fossil fuels, renewables seem likely to eventually become a significant part of the world's energy picture—sooner rather than later, if policies favoring use of these fuels, instead of fossil fuels, are widely adopted.

ENERGY EFFICIENCY AT THE CROSSROADS

Although developing new energy-producing technologies that will help protect the environment as well as propel economic growth is an important goal, this strategy addresses only one side of the equation. The other side, of course, is energy demand. Increased conservation in both developed and developing countries stands as perhaps the most promising option for mitigating the various risks associated with the current global energy habits.

Energy conservation was once unfairly linked to the need for drastic cutbacks in living standards. Although some changes in human behavior are clearly appropriate, conservation efforts are now strongly focused on introducing new technologies for producing and using energy more efficiently and on improving energy management. By increasing energy efficiency, demand can be reduced without adversely affecting personal lifestyles or a country's economic growth. In fact, increasing energy efficiency can even enhance them.

Conservation in Buildings, Industry, and Transportation

There are literally thousands of different types of technological and operational measures that can improve energy efficiency. They cover residential and commercial buildings, industry, transportation, and the electric utility sector, as well as the distribution of energy to end-users. Many of the efficiency improvements can be implemented much more economically than new energy supplies can be developed.

The world's buildings can be equipped with more efficient lighting, appliances, and heating and cooling systems. Advanced construction materials can also sharply reduce thermal losses through windows, doors, and walls. In "superinsulated" homes, for example, where normal insulation is doubled and walls have a liner that forms an airtight seal, the heat given off by people, lights, and appliances can provide all the warmth needed even on the coldest day. Scandinavian countries lead the way in designing energy-efficient buildings, with some Swedish homes requiring nearly 90 percent less energy than the average U.S. residence (34).

In industry, significant savings can be realized by using more efficient electric motors, new sensors and control devices, and advanced heat-recovery systems. Especially promising is cogeneration, in which steam produced to generate electricity is then used as a heat source for other industrial processes. By capitalizing on the otherwise wasted heat, cogeneration systems can deliver 50 percent or more of the heat in the original fuel to useful purposes, compared with about 33 percent for conventional steam-electric technologies (35). Japan is by far the world's leader in industrial energy efficiency (36).

Transportation constitutes the largest and most rapidly growing drain on the world's oil supplies. While the fuel efficiency of cars and trucks has improved considerably during the past 15 years, a variety of emerging technologies offer even more substantial gains. (See "Focus on The Automobile," below.)

Voluntary or Mandatory Measures

Conservation can also be fostered by voluntary or mandatory institutional changes. In the United States, for example, electric utilities are increasingly adopting "least-cost planning"—a process for examining all electricity-producing and electricity-saving options and selecting the mix that minimizes total consumer costs. Since conservation investments often cost considerably less than what a utility would otherwise have to spend in order to generate electricity, least-cost planning can benefit energy producers as well as consumers. Many utilities even find that investing in conservation measures on behalf of their customers—for example, supplying them with high-efficiency lightbulbs at little or no cost—proves financially beneficial. The utilities save in the short run through reduced operating expenses, and in the long run by not having to build new multimillion-dollar powerplants to satisfy expanded demand.

Energy Intensity

The success of energy conservation has already been amply demonstrated. From 1973 to 1985, OECD countries, who also belong to the International Energy Agency (IEA), collectively reduced their energy intensity—that is, the amount of energy used to produce a unit of gross domestic product (GDP)—by a remarkable 20 percent. Energy consumption within IEA countries grew by only 5 percent during this period, while their GDP grew by almost 32 percent (37). Clearly the countries were getting the same, or better, services from less energy. Higher energy prices may have played a crucial role in these changes.

With significantly lower energy prices during the past several years, however, many countries have retreated from earlier vigorous efforts to improve energy efficiency. In the United States, for example, energy intensity declined by about 25 percent between 1973 and 1986, saving consumers at least $160 billion per year on energy (38). But after world oil prices collapsed in 1986, U.S. energy intensity began moving slightly upward—by 0.1 percent in 1987 and 0.2 percent in 1988 (39).

The European Community has also experienced a recent slowing in national and local efforts to improve energy efficiency (40). From 1973 to 1982, the Community's 12 member nations achieved a combined 20-percent improvement—and Europe's energy intensity was already about half that of the United States. From 1982 to 1986, however, the members posted an improvement of only 2.4 percent (41). The process of decoupling economic growth from energy consumption has thus been slowed within the community, and has even been reversed in some countries such as France, Belgium, the Netherlands, Ireland, and Portugal (42). If this trend continues—as some analysts expect, given the termination of numerous conservation programs in many member countries—the Community will be unlikely to reach its target, set in 1986, of improving energy efficiency at least 20 percent by 1995 (43).

With earlier progress pointing the way, there remains impressive potential for further gains in energy efficiency. For example, studies published by World Resources Institute (WRI) suggest that per capita energy use in industrialized countries can be cut just about in half by the year 2020, while still maintaining strong economic growth and high standards of living (44).

Several other studies focused on IEA countries provide support for a conservative estimate that if all economically viable energy conservation measures were fully implemented by the year 2000, energy efficiency for the region would be 30 percent higher than current levels. Energy consumption would thus be more than 25 percent below what would have resulted if today's efficiency level remained unchanged (45). While major energy savings are possible in all IEA countries, the potential appears greatest in those regions that have been traditionally the most energy intensive, such as North America (46). The potential for improving efficiency is especially large in residential and commercial buildings, but major opportunities also exist in industry and transportation (47).

These studies provide useful approximations of conservation's potential. Most of them, however, were based on conditions in the 1970s—before the emergence of a number of promising new technologies but also during a period of higher energy prices. Energy markets and patterns of human behavior are also more thoroughly understood today, and the recent volatility in energy prices and global economics provides valuable information for assessing how consumers respond to market fluctuations and public policies. New studies are therefore urgently needed to assess more accurately how much energy can be saved through gains in energy efficiency and to help determine how the savings can best be achieved from country to country.

Energy Efficiency in Centrally Planned and Developing Countries

Improving energy efficiency presents a special challenge in centrally planned countries and in the developing world. The U.S.S.R. remains the least energy-efficient industrial economy. Its energy intensity has barely declined since 1980 and is estimated to be more than twice the average of OECD countries (48). In China, energy intensity has declined by over one quarter since 1980, but the country remains slightly less energy efficient than the U.S.S.R. China has announced plans to cut its current energy-intensity level at least 33 percent by the year 2000, but it remains to be seen whether the relatively limited measures instituted so far can achieve this ambitious goal (49). Unless major efficiency gains are made in these energy-intensive countries, the global energy outlook will remain clouded.

In developing countries, energy planning has focused primarily on expanding conventional energy supplies rather than on improving energy efficiency. Energy intensity in most countries has therefore risen steadily as economic activity has increased and populations have grown. For example, consumption of electricity in the major energy-using countries of

Latin America and Asia grew considerably faster than the countries' GDPs from 1980 to 1986. Consumption grew nearly twice as fast in India, Brazil, Pakistan, and the Philippines, and well over three times faster in Argentina and Venezuela (50). If historical trends in energy consumption continue, the developing world's per capita commercial energy demand in 2020 will be more than four times the level of 1980 (51).

With some help from the industrialized world, however, developing countries could apply technical energy-efficiency solutions to promote economic growth while keeping energy demand relatively low. This will mean adopting a host of energy-efficient technologies that are either already available commercially or could be commercialized in the near future.

According to WRI studies, such actions would allow a prototypical developing country to boost its standard of living to a level approaching modern Europe's by 2020, while increasing per capita energy use by only about 20 percent above the developing world's average rate in 1980 (52). This hopeful scenario should not obscure the challenge involved, however. Large amounts of capital and skilled management will be needed to shift to efficient end-use technologies and to build and maintain the required infrastructure. But strong evidence suggests that it would cost less to take this approach than it would to continue with conventional end-use technologies and increased energy supplies (53).

FOCUS ON THE AUTOMOBILE

A particular case where higher efficiency could have major impact, not only on global energy use but also on world environmental conditions, is the automobile. Cars, together with trucks and buses, consume a significant portion of global energy resources, contribute a major share of the greenhouse gases linked to global warming, and are a major source of air pollution. And their numbers are growing rapidly, especially in developing countries. Thus, their role in exacerbating energy and pollution problems can only be expected to increase, unless concerted actions are taken to improve their fuel efficiency.

Growth in the Vehicle Fleet

Almost half a billion vehicles were on the road throughout the world in 1986—more than three fourths of them were cars (54). The United States had the most, although, its share dropped from 75 percent in 1930 to 58 percent in 1960 and to 35 percent in 1986 (55). Europe accounted for just over one third of all light vehicles, with the rest divided among Africa, South America, Asia, and Oceania (56). (See Figure 9.2.)

The number of light vehicles is increasing at a rate faster than the rate of population growth—since

Figure 9.2 Number of Vehicles Worldwide by Region, 1986

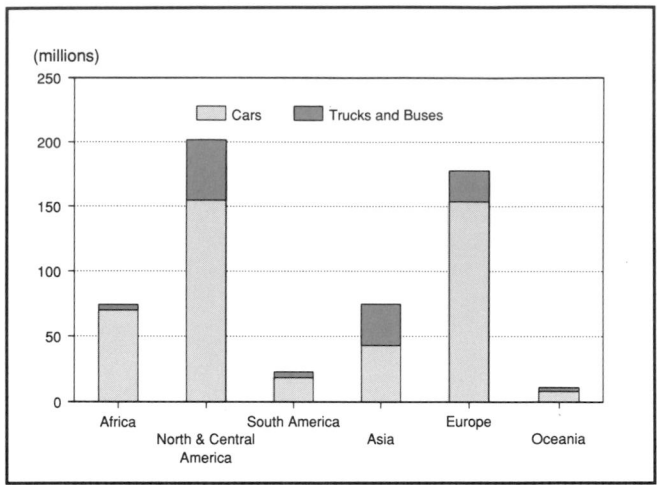

(millions)

Cars / Trucks and Buses

Africa / North & Central America / South America / Asia / Europe / Oceania

Source: Motor Vehicle Manufacturers Association of the United States (MVMA), *World Motor Vehicle Data*, 1989 Edition (MVMA, Detroit, Michigan, 1989), pp. 36-38.

1950, the average annual growth rate for cars has been 5.9 percent, with trucks and buses averaging a slightly lower 5.6 percent. Growth rates have abated somewhat in recent years, however, and the rate for trucks and buses has actually overtaken that for cars. Since 1970, trucks and buses have experienced an average annual growth of 5.1 percent, compared with 4.7 percent for cars (57).

The rate at which vehicle populations are growing varies greatly by region. Growth rates in Asia for both cars and commercial trucks and buses have exceeded the world average in recent years, as have the rates for commercial vehicles in South America. Growth rates in North and Central America, on the other hand, have been lower than the world average. If historic rates of vehicle growth are maintained, the global vehicle population will approach 650 million by the year 2000 and 1 billion by 2030 (58).

The Impact of Growth

The global vehicle fleet has an enormous impact on energy consumption. Light vehicles are by far the single largest component of the transportation sector, which accounts for roughly two thirds of the oil used in the United States, 40 percent in Western Europe, 25 percent in Japan, 50 percent in developing countries, and an estimated 51 percent in the U.S.S.R. (59). Indeed, future increases in global oil demand will likely be driven heavily by the rise in demand for road vehicles from developing countries (60). Projections from the U.S. Department of Energy estimate that nearly all of the increase in global oil demand (excluding the centrally planned economies) between 1986 and 2010 is expected to come from these countries (61). In the 15 largest developing countries, for example, approximately half of the increase in oil

use between 1970 and 1984 was for transportation, and this trend is expected to continue (62).

The growing vehicle fleet will also contribute substantially to global warming and to local air pollution problems. Burning a single tank of gasoline, for example, produces 300–400 pounds of carbon dioxide (CO_2), the most important greenhouse gas. Motor vehicles (excluding farm equipment) now account for almost 15 percent of the world's CO_2 output. The number of vehicles tends to increase as industrial and economic development occurs; thus, vehicles are expected to increase their relative contribution to global CO_2 emissions (63). In the United States, for example, vehicles account for almost 25 percent of the country's CO_2 emissions.

Vehicles also emit substantial amounts of carbon monoxide. These emissions, through chemical reactions in the atmosphere, are suspected of indirectly increasing the global warming phenomenon 20–40 percent (64). In addition, vehicles equipped with air conditioners contribute another important greenhouse gas, chlorofluorocarbons (CFCs). In the United States, which has the greatest percentage of air conditioned vehicles in its fleet, those vehicles provide 13 percent of the country's CFC emissions (65).

As a major source of local air pollutants, motor vehicles account for 47 percent of nitrogen oxide emissions and almost two thirds of carbon monoxide emissions within the OECD countries (66). In the United States, vehicles were responsible for 70 percent of the carbon monoxide, 45 percent of the nitrogen oxide, and 34 percent of the hydrocarbons emitted in 1985. In other regions where there are no major emission-control requirements, particularly developing and Eastern Bloc countries, motor vehicles figure even more dominantly in pollution problems (67). (See Chapter 12, "Atmosphere," Focus on the Automobile.)

Improving Fuel Economy

To mitigate the adverse effects of an expanding vehicle population, improving the fuel economy of those vehicles is critical. Not only will improved fuel economy reduce oil consumption, it will also directly reduce carbon dioxide emissions. The emission of other troublesome combustion products will generally be reduced as well, since less fuel will be burned per kilometer driven. This effort to improve fuel economy should focus primarily on automobiles, given their considerable predominance. But most of the technologies applicable to improving the fuel economy of cars can also be applied to light trucks (trucks weighing less than 3,900 kilograms). In the United States, light trucks consume more than 40 percent of the fuel used by all trucks and buses, while in other parts of the world—where vehicles generally are smaller—light trucks are estimated to account for even larger fractions (68).

The sales-weighted fuel economy of new automobiles in the United States, both domestic and imported, is 8.4 liters per 100 kilometers (l/100km); the figure for new light trucks is 11.2 l/100km (69). New automobile fuel economy in other OECD countries is estimated to be comparable to that of the United States or slightly higher; the rating for light trucks is not well-characterized (70). Data on new vehicle fuel economy in developing countries are poor, but it probably falls below that of the OECD countries. Western Europe and Japan, which produce more efficient cars than the United States, are the primary suppliers of vehicles to developing countries, but the cars tend to be less technically advanced. Poor road conditions, traffic congestion, and lack of vehicle maintenance also help hold down fuel economy in developing countries.

The potential exists to reach fuel economies that are much higher than today's, and a host of technologies are being pursued by engineers in many countries. Much of the work is concentrated on improving the efficiency of engines, since they lose 70–80 percent of the fuel's energy content, primarily through losses to exhaust gases and coolants (71).

Technology Options

In a gasoline engine, the highest losses occur when it is operating at "part load"—when the power it provides is significantly less than the maximum it could produce at a given engine speed. This is often the case during typical city driving. The problem is that at low speeds, the combustible air-fuel mixture is too "rich," with too much fuel for the power demanded. Using less fuel has generally been impossible, because the mixture burned too slowly and unevenly and had a proclivity to "knock" or ignite spontaneously in an uncontrolled manner. But better oxygen sensors and improved electronics have enabled automakers to develop what are called "ultra-lean-burn" engines. Toyota has already introduced such an engine, which can achieve 20-percent greater fuel efficiency, in some of the cars it sells in Europe (72).

The "stratified-charge" engine is also being reconsidered. Extensive research was conducted on this engine during the 1970s, because it combines the best features of gasoline engines with those of diesel engines. As in a diesel engine, the air and fuel are stratified in distinct layers in the combustion cylinder. Under part-load conditions, this permits the engine to run more efficiently. Moreover, because the stratified-charge engine is ignited by sparks like a gasoline engine, it does not have the cold-weather start-up problems typical of compression-ignited diesels (73).

While attractive, however, the stratified-charge engine was plagued by unacceptably high emissions. But recent advances in ignition and combustion techniques are enabling more complete combustion and hence lower emissions. Of particular interest today for application of the stratified charge approach is a "two-stroke" engine, in which fuel intake, compression, combustion, and exhaust occur in only two pistons strokes, rather than four strokes as in conventional engines. A factory to produce such an engine is presently being built in Michigan, and both Ford and General Motors have signed licensing agreements with the designer of the engine.

Ceramic Diesels

Another long-standing goal has been to develop a diesel engine that does not need cooling, which could significantly reduce energy losses. New generations of ceramic materials, which can withstand high temperatures and insulate the engine block from heat damage, may make such engines possible. Isuzu, for example, is said to be developing a ceramic diesel engine that works in concert with a system designed to recover energy from the exhaust. The engine is expected to produce a 30-percent improvement in fuel efficiency over today's diesel engines. Even more advanced ceramic engines may provide a 60-percent increase over today's gasoline engines (74).

Technological gains are also expanding the frontier for transmissions, which send the engine's power to the wheels. Conventional automatic transmissions are only 80–85 percent efficient, in part because of their limited gearing and tendency to lose energy when the fluid used in them moves, causing friction. The addition of more gears enables the engine to operate at its most efficient speed more of the time, and automatics are gradually evolving from three gears to four and, in the future, probably five. They can also be equipped with special clutches that lock into place in a given gear, which reduces friction caused by the fluid and improves efficiency. Some automakers are trying to develop these "lock-up clutches," now made almost exclusively for the top gear, for the lower gears as well (75).

Even greater efficiency improvements will be possible with continuously variable transmissions (CVTs). As the name suggests, these can continuously change gearing in order to keep the engine operating at full load, where it is most efficient. CVTs are not new, but they have been dogged by control and endurance problems. Solutions appear in sight, however, in the form of stronger construction materials and improved electronics (76).

Another opportunity for improving fuel economy lies in vehicle designs that reduce aerodynamic drag. Indeed, at highway speeds, more than 60 percent of the energy supplied to a car's wheels is used to overcome aerodynamic drag. Reducing drag by 10 percent can reduce highway fuel consumption by as much as 6 percent (77).

More Streamlined Profiles

In recent years, many cars and small trucks have taken on a more streamlined profile to reduce aerodynamic drag. The average drag coefficient (a measure of aerodynamic drag) for the U.S. car fleet is currently 0.37, while the world's best production vehicles have drag coefficients as low as 0.28 (78). Dropping the drag coefficient much below this level will involve significant vehicle redesign, such as putting "skirts" around wheel wells. The lead in aerodynamic research now belongs to Ford, which has developed an experimental vehicle, the Probe V, with a drag coefficient of 0.137 better than the coefficient of the U.S. military's sophisticated F-15 jet fighter (79).

Reducing vehicle weight by using lightweight materials can cut fuel use substantially—a 10 percent weight reduction yields on average a 6 percent improvement in fuel economy (80). Considerable progress has been made in developing plastics, aluminum, magnesium, ceramics, and various composite materials that are light yet capable of replacing steel. According to the U.S. Office of Technology Assessment, the substitution in vehicle parts of 20.4 kilograms of aluminum, 16 kilograms of composite plastic, or 11.34 kilograms of magnesium could potentially fulfill requirements that have historically been met with 45.3 kilograms of steel (81). Moreover, the potential exists for many of these materials to achieve similar or better collision resistance than the heavier steel they replace.

The world's automakers are applying these materials in numerous ways. For example, General Motors, Honda, and Renault have already produced vehicles with body panels composed almost entirely of composite plastics, and more companies are expected to follow. Audi is working cooperatively with Alcoa Aluminum to produce an aluminum-body vehicle by 1995 that is both cost-efficient and 10 percent lighter. And work is proceeding worldwide on the development of composite load-bearing parts. Volkswagen,

Ford, General Motors, and Chrysler have taken the lead in developing a composite chassis or chassis parts (82).

Fuel economy can also be improved by reducing the energy loss associated with braking. When a vehicle is braking or idling, the engine's power is needed only to run the vehicle's accessories, so much of its power is simply lost to the coolant and exhaust gases. Volkswagen, however, has developed a prototype system called "Glider Automatic," which uses the excess engine energy during these periods to charge a flywheel (83). When sufficient energy has been stored, the engine turns off and the flywheel runs the accessories. The flywheel's energy then restarts the engine when more power is needed. Volkswagen originally intended to introduce this system in the early 1980s, but the company has postponed plans indefinitely. Nissan is believed to be developing a device similar to Volkswagen's (84).

An even more advanced energy-storage system was under development for many years at the University of Wisconsin, with support from a Toyota affiliate. With this system, the vehicle's engine operates in its most efficient full-load mode all the time. When more power is produced than required, the excess energy is stored in a flywheel much like Volkswagen's. In addition, the system incorporates a continuously variable transmission that essentially can be run backward during braking to recapture a large fraction of the wheels' kinetic energy and store it in the flywheel. This stored energy is then used not only during deceleration and idling to power the vehicle's accessories but also to supplement the

Table 9.3 Fuel-Efficient Prototype Vehicles

Company	General Motors	British Leyland	Volkswagen	Volkswagen	Volvo	Renault	Renault	Peugeot	Peugeot	Ford	Toyota
Model	TPC (gasoline)	ECV-3 (gasoline)	Auto 2000 (diesel)	VW-E80 (diesel)	LCP 2000 (diesel)	EVE+ (diesel)	VESTA2 (gasoline)	VERA+ (diesel)	ECO 2000 (gasoline)	— (diesel)	AXV (diesel)
Number of Passengers	2	4-5	4-5	4	2-4	4-5	2-4	4-5	4	4-5	4-5
Aerodynamic Drag Coefficient	0.31	0.24-0.25	0.25	0.35	0.25-0.28	0.225	0.186	0.22	0.21	0.40	0.26
Curb Weight (kilograms)	472	662	778	699	705	853	475	789	449	851	649 target
Maximum Power (kilowatts)	28	54	40	38	39	37	20	37	21	30	42
Fuel Economy (liters per 100km)	3.9 city 3.2 highway	5.7 city 4.5 highway	3.7 city 3.3 highway	3.2 city 2.4 highway	3.7 city 2.9 highway	3.7 city 2.9 highway	3.0 city 2.2 highway	4.3 city 2.7 highway	3.4 city 3.1 highway	4.1 city 2.6 highway	2.6 city 2.1 highway
Innovative Feature	Aluminum body and engine.	High use of aluminum and plastics.	Advanced diesel engine; high use of plastic and aluminum; improved aerodynamics; flywheel energy storage.	Advanced diesel engine; high use of plastic and aluminum; flywheel energy storage.	Advanced diesel engine; high use of magnesium; improved aerodynamics.	Advanced diesel engine; improved aerodynamics.	High use of light material; highly improved aerodynamics.	Advanced diesel engine; high use of light material; improved aerodynamics.	High use of light material; improved aerodynamics.	Advanced diesel engine.	Advanced diesel engine; continuously variable transmission; high use of plastic and aluminum; improved aerodynamics.
Development Status	Prototype complete, no production plans.	Prototype complete.	Prototype complete.	Ongoing research possibility of production.	Prototype complete, adaptable to production.	Prototype complete.	Prototype complete.	Ongoing development.	Ongoing development.	Research.	Ongoing development.

Source: Deborah Lynn Bleviss, *The New Oil Crisis and Fuel Economy Technologies: Preparing the Light Transportation Industry for the 1990s* (Quorum Books, Westport, Connecticut, United States, 1988), p. 102.

engine's output when an additional boost is required. Preliminary laboratory tests of this system indicate a 50 percent improvement in fuel economy during city driving, and in theory the improvement could reach 100 percent (85).

Prototype Vehicles

While a number of fuel-savings advances have been commercially introduced in piecemeal fashion in recent years, no vehicles have been produced with the specific aim of saving as much fuel as possible. However, over the past decade numerous automobile manufacturers, primarily in Europe, have developed experimental prototype vehicles that use considerably less fuel. (See Table 9.3.) Most of these achieve a city fuel economy of at least 3.9 l/100km and a high-

way rating of at least 3.1 l/100km. Some do even better—Toyota's AXV has a combined fuel economy of 2.4 l/100km, and Renault's VESTA2 has a combined rating of 2.6 l/100km (86).

While most of these prototypes were designed as one-of-a-kind vehicles, at least two were designed with production in mind: Volkswagen's E-80 and Volvo's LCP 2000. Both companies clearly believed the vehicles could be brought to market at an affordable price. In fact, the LCP 2000, designed to carry two to four persons, was estimated to cost no more than today's average subcompacts. Yet, its ability to accelerate is the same as most conventional small cars, and it was designed to meet the world's most stringent crashworthiness requirements, those of the United States. The car also was designed to come close to the toughest emissions standards, again

Box 9.1 Policies to Encourage Vehicle Efficiency

At the core of most national programs to encourage efficiency has been the establishment of fuel economy targets for new vehicles. The United States set up the first target program in 1975. It was mandatory, covered both domestic and imported vehicles, and included light trucks as well as cars (1). The European, Japanese, and Canadian programs that followed were voluntary, covered only automobiles, and generally applied only to domestic vehicles, although many of these countries already produced more efficient cars than did the United States.

Several pricing mechanisms have also been used. Most of the countries belonging to the Organisation for Economic Co-operation and Development (OECD), with the notable exception of the United States, have substantial taxes on fuel. The taxes have had only limited success, however, in prompting consumers to purchase highly efficient vehicles—the fuel economies achieved in Europe and Japan are not significantly higher than in the United States, despite much higher fuel prices (2). Recent analysis indicates that in the United States, the mandatory fuel-economy targets played a much greater role in spurring efficiency than did the dramatic oil price hikes of the late 1970s and early 1980s (3). Generally, higher fuel prices are better at discouraging the purchase of very inefficient cars, for which fuel costs make up a large fraction of the operating costs (4).

A potentially more effective pricing mechanism is taxing vehicles directly. Most industrialized countries have some form of progressive tax on vehicle ownership, the most common being an annual registration fee. It is generally based on a vehicle's weight, engine size, horsepower, or some combination of these factors. The larger any of these factors is, the greater the fee imposed. Since these factors indirectly affect fuel-economy, if the fees are progressive enough, consum-

ers can be prompted to purchase more fuel-efficient cars.

Fees directly discouraging fuel inefficiency can also be added to the purchase price of vehicles. In the United States, a "gas guzzler" tax is assessed on the purchase of all new vehicles that fall below a threshold of fuel economy, and the tax increases as efficiency drops. Presently, cars with a fuel economy of less than 22.5 miles per gallon (mpg), or 10.4 liters per 100 kilometers, are assessed the tax, which starts at $500 and ranges up to $3,850 (5). Evidence suggests that many manufacturers have sought to maintain the fuel economy levels of their cars so that they will not be subject to this tax.

Many countries encourage the purchase of fuel-efficient cars through consumer information programs, providing prospective new car buyers with lists showing the tested fuel economy of all cars offered for sale within a given country. In the United States, this program is supplemented with a labeling requirement; each new vehicle must carry a sticker indicating its fuel economy and how its rating compares with other similarly sized vehicles.

Most countries also support some type of research program aimed at developing fuel-efficient automotive technologies. The most successful programs feature partnerships between government and private industry. Indeed, of the fuel-efficient prototype cars developed in the early 1980s, the French and West German vehicles were the product of jointly funded research programs (6).

Since the onset of the oil glut, national governments throughout the world have lost much of their interest in continuing to improve the efficiency of their light-vehicle fleets. None of the original fuel economy goals set for the mid-1980s has been updated by any of the participating countries. But this may well change, led by recent developments in the United States.

Spurred by concerns of global warming, a number of bills were introduced to the U.S. Congress in 1989 that would establish new fuel economy standards for the year 2000. Most would cover both new cars and new light trucks. One bill proposes increasing auto fuel economy standards by 20 percent between 1995 and 2000 and by 40 percent after 2000, with targets reaching as high as 45 mpg (5.2 l/100 km) (7). Some of the bills would also strengthen consumer information programs, increase the current gas guzzler tax, and establish significant rebates for purchasers of highly efficient vehicles.

Whatever the immediate fate of these bills, there is increasing belief that within the next several years the United States will adopt new fuel economy standards for the turn of the century. Other countries are likely to take similar steps.

References and Notes

1. Deborah L. Bleviss, *The New Oil Crisis and Fuel Economy Technologies: Preparing the Light Transportation Industry for the 1990s* (Quorum Books, Westport, Connecticut, 1988), p. 158.
2. *Transportation Energy Data Book: Edition 9*, Oak Ridge National Laboratory, Spring 1987, cited in Deborah L. Bleviss, *The New Oil Crisis and Fuel Economy Technologies: Preparing the Light Transportation Industry for the 1990s* (Quorum Books, Westport, Connecticut, 1988), p. 95.
3. David L. Greene, "CAFE or Price: An Analysis of the Effects of Federal Fuel Economy Regulations and Gasoline Price on New Car MPG," draft report, U.S. Department of Energy, May 10, 1989.
4. *Op. cit.* 1, p. 162.
5. *Op. cit.* 1, p. 167.
6. *Op. cit.* 1, pp. 219-220.
7. Senate Bill S.1224, "To Amend the Motor Vehicle Information and Cost Savings Act to require new standards for corporate average fuel economy, and for other purposes," June 22, 1989, 101st U.S. Congress, 1st session, pp. 7-8.

those of the United States, and promising new emission-control techniques should be able to lower levels even more (87).

Yet even the most innovative of these prototypes do not incorporate a number of promising technologies, such as ceramic engines and advanced energy-storage systems. Hence, by the turn of the century, vehicles may well be under development that will run on even less fuel than today's prototypes suggest.

While the technological potential for greater fuel economy is clear, it is unlikely that this potential will be reached in the near future without government intervention. (See Box 9.1.) With oil prices relatively low in recent years, consumers and manufacturers alike have reduced their interest in fuel economy and are turning instead to larger, more powerful vehicles.

Governmental action to encourage or require better fuel economy is badly needed, and soon. If no further improvements in fuel economy are achieved, the rate at which the global fleet consumes oil will accelerate rapidly. (See Figure 9.3.) By the turn of the century, demand for fuel from light vehicles could rise by as much as 25 percent. However, if fuel economy is improved such that by 2000 car fleets average 5.2 1/100km and light truck fleets average 6.7 1/100km, oil demand could drop below 1980 levels.

NONFOSSIL ENERGY SOURCES: A LOOK AHEAD

Although fossil fuels are likely to play a major role in meeting the world's commercial energy needs for years to come, nonfossil energy sources may become increasingly important as concern over global warming grows. Many nonfossil energy sources are already in use in various parts of the world, including solar-thermal heating, solar-thermal power, wind energy, geothermal heat, and especially wood and other renewable fuels derived from biomass—the primary energy source for a majority of the world's people.

For commercial energy, however, two families of energy technologies may have the potential to make especially significant contributions in the long term, if existing problems can be overcome. Both were once touted by enthusiasts as ready "solutions" to global energy problems, only to fall far short of such claims. But technological advances, some already achieved and some in the wings, may attract new attention in coming decades.

The two energy sources are nuclear energy, in the form of advanced reactors that advocates claim will be safer, simpler, and possibly less expensive than today's versions, and solar energy, in the form of improved photovoltaic systems that convert sunlight into electricity.

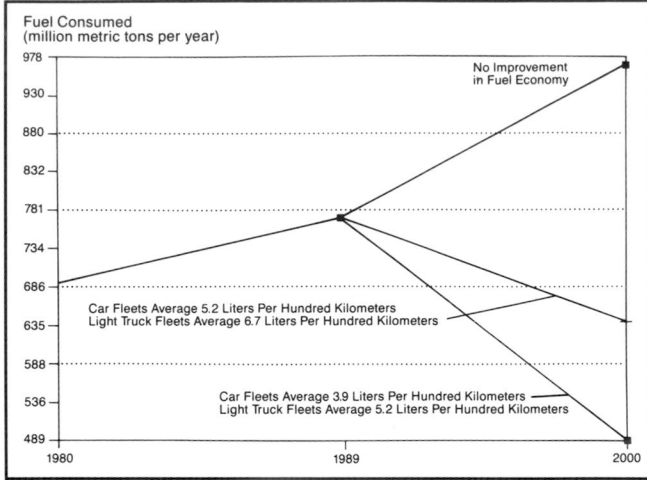

Figure 9.3 Projected Worldwide Light Vehicle Fuel Use, 1980—2000

Fuel Consumed (million metric tons per year)

No Improvement in Fuel Economy

Car Fleets Average 5.2 Liters Per Hundred Kilometers
Light Truck Fleets Average 6.7 Liters Per Hundred Kilometers

Car Fleets Average 3.9 Liters Per Hundred Kilometers
Light Truck Fleets Average 5.2 Liters Per Hundred Kilometers

Source: Deborah Lynn Bleviss, *The New Oil Crisis and Fuel Economy Technologies: Preparing the Light Transportation Industry for the 1990s* (Quorum Books, Westport, Connecticut, 1988).

Advanced Nuclear Power

Nuclear energy, once supposed to make electricity "too cheap to meter," has virtually stalled in many countries, including the United States. The causes include unfavorable economics and public concerns about safety. One result has been to focus new attention on improvements in nuclear technology that, advocates say, might make reactors easier to build and safer to operate.

Conventional light-water reactors are very complex devices. A typical U.S. reactor, for example, may have as many as 40,000 valves, while a coal-fired plant of similar size has only 4,000 (88). Such complexity means the reactors are relatively unforgiving of errors made during construction, maintenance, and operation, which renders them susceptible to possible catastrophic failure. Their complexity also makes them expensive, which forces utilities to build very large plants capable of generating lots of electricity in order to recover their investments. Many conventional nuclear plants have a generating capacity of 1,000 megawatts (MW) or more.

Passive Safety Systems

A newer generation of reactors, still under development, rely primarily upon "passive" safety systems, which substitute reliable natural forces like gravity and convection for the vast network of pumps, pipes, chillers, control systems and other components that protect today's nuclear plants. The new reactors may also prove more reliable, easier to operate and maintain, and less expensive to build, although this remains to be proven. Many of the new designs are "modular," meaning that they are composed of a number of identical but smaller subunits, for which major components could in many cases be fabricated at a factory. This could greatly simplify construction work at the site, and hence reduce costs, and it means that power plants could be smaller, and hence more easily matched to increases in demand for electricity or to the power grids of all but the largest countries (89).

Several new reactor concepts are being pursued actively in a number of countries. One approach is to develop advanced versions of light-water reactors that incorporate passive emergency cooling features. In the United States, General Electric Company and Westinghouse Electric Corporation are heading industrial teams working on two slightly different designs for such reactors (90). Both run at lower temperatures than current reactors and are 600 MW in size. The major new design feature, however, is emergency cooling systems that are intended to prevent damage to the core, in the event of an accident, by quickly and automatically flooding it with enough water to dissipate its heat for at least three days (91). In the Westinghouse design, for example, roughly 113,000 liters of water are contained in tanks located directly above the reactor vessel. During a severe accident, valves would open merely from the change in temperature or pressure alone, and gravity would deliver a deluge of coolant to the core. Today's light-water reactors, in contrast, depend on massive pumps powered by diesel generators to keep the fuel-bearing core covered with water from an external tank. Westinghouse hopes to have standardized designs certified by the Nuclear Regulatory Commission (NRC) by 1993–94, and General Electric expects certification by 1996–97.

Companies in several other countries are also developing small light-water reactors that use passive emergency cooling systems. In Great Britain, for example, Combustion Engineering, Stone and Webster, Rolls-Royce, and the U.K. Atomic Energy Agency recently began development of a 320-MW reactor that incorporates not only the core but also the steam generators, reactor coolant pumps, and other critical components within a single large reactor vessel. The vessel would be located in a cavity below ground, to offer protection from accidents such as airplane crashes. The core would be near the bottom, with 30 feet of water above it ready to provide instant cooling (92).

PIUS Advanced Light-Water Reactors

In Sweden, Asea Brown Boveri, the world's largest electrical equipment manufacturer and a major vendor of conventional reactors, has pioneered development of an even more revolutionary light-water reactor design (93). It is called PIUS, for Process Inherent Ultimate Safety, and two sizes of small reactors have been proposed (600 MW and 300–400 MW). Research on PIUS-type reactors has also spread to Japan, Italy, South Korea, and the Oak Ridge National Laboratory in the United States.

In a PIUS reactor, the core is actually submerged in a large pool of water containing the element boron, which can halt nuclear reactions by absorbing neutrons. Under normal conditions, the borated water is kept out of the fissioning core by the sheer pressure of the regular coolant. But if the regular coolant is suddenly lost through some sort of accident, the borated water would rush in without human intervention, keeping the core safely cooled for at least a week.

To ensure an adequate supply of borated water, the PIUS design calls for the pool to be contained in a prestressed concrete vessel lined with two leaktight layers of stainless steel. With walls some 10 meters thick, the vessel is so massive that powerful explosives will not be able to inflict enough damage to deplete the pool. In turn, the concrete vessel is to be housed in a containment structure built strong enough to withstand even an airplane crash (94).

PRISM Reactors: Liquid Metal Coolant

A fundamentally different strategy being explored is to use liquid metal rather than water as the reactor coolant. In the United States, for example, General Electric is designing a reactor known as PRISM—for power reactor inherently safe module—that would have up to nine identical reactor modules, each of 155-MW size and each submerged in a pool of liquid sodium, which also serves as the reactor coolant. The reactors are connected to one or more common turbine generators.

The design incorporates a number of passive safety systems. For one thing, the rise in temperature causes the fuel, sodium coolant, and other materials in and around the core to expand, which slows the heat-producing fission process. In addition, heat from the metallic uranium fuel passes readily into the surrounding coolant, and since sodium's boiling point is roughly 900 ° C, the coolant pool can absorb the excess heat. Additional passive systems are designed to carry away residual heat following an accident. Experiments with a sodium-cooled, 20-MW test reactor conducted by Argonne National Laboratory in the United States have established that the liquid metal can, in fact, contain the heat that results when normal cooling systems are suddenly shut down (95). Because of this safety feature, the design does not include a massive concrete containment building such as those required with light-water reactors.

Advanced Gas-Cooled Reactors

Yet another innovative design is the modular high-temperature gas-cooled reactor (MHTGR), which is under development in the United States, West Germany, and Japan. In this design, small modular reactors are cooled by helium. In the design for the U.S. effort headed by General Atomics, each reactor would be 139-MW in size and housed in an underground concrete silo (96).

The basic safety innovation of the MHTGR lies in the tiny fuel particles it uses. Smaller than a grain of sand, each particle consists of a kernel of uranium or thorium encapsulated in several layers of ceramic material that can withstand temperatures up to 2,000° C. This coating allows heat from the fissioning fuel to pass through it but traps the hazardous radioactive fission products tightly inside. In General Atomics' design, thousands of these particles are packed in small rods, which in turn are sealed in graphite fuel elements that go into the core. In the German design, the particles are packed in "pebbles," graphite spheres about the size of billiard balls that are used directly in the core. Because the graphite can absorb large amounts of heat, core temperatures cannot reach levels that might melt the fuel particles. These safety features have been demonstrated successfully on several occasions in West Germany, by subjecting a small test reactor to a total loss of coolant while operating at maximum power (97). As a result, the design has no emergency cooling system and does not include a massive concrete containment building.

These new reactor designs are still unproven in many respects—operation of full-scale prototypes, licensing questions, and costs need to be addressed. More fundamentally, opinion remains sharply divided on whether they represent so significant an advance to make nuclear power an environmentally attractive alternative to fossil fuels. The spectrum of opinion is wide. Some people maintain that modern light-water reactors do not need replacing since they have proven to be safe and efficient in many countries. Advocates see in these designs the means for nuclear energy to fulfill its original promise. Others are skeptical of the safety claims of these new designs, and some say that nuclear energy of any type is not acceptable, because of unresolved problems with nuclear waste disposal and the potential for nuclear proliferation.

Advocates state that the simplicity of the MHTGR minimizes the need for highly trained operating personnel and exotic equipment—thus making it more appropriate for developing countries than existing reactors—and the modularity of several designs will allow their components to be manufactured in factories and quickly assembled in the field, with savings in costs and improvements in quality.

Critics say that even reactor safety authorities have not yet concluded that these new designs are, in fact, as safe as proposed. They also point out that nuclear accidents even in light-water reactors have come largely from human error—in construction and operation—that no design can eliminate completely.

The debate seems likely to continue. But the new designs may at least help to broaden the world's energy options and, if they prove successful, may eventually provide an important alternative to burning the world's massive supplies of coal.

Photovoltaic Power

Photovoltaics, the process of converting sunlight directly into electrical energy using semiconductor devices, was supposed to bring virtually limitless streams of affordable electricity to much of the world. While photovoltaic (PV) systems have proved themselves in a variety of applications, their use has not spread as fast as some advocates predicted. As a source of electric power, they are still three of four times as expensive as other renewable energy sources that are in limited commercial use, such as wind energy and solar-thermal power. But research in recent years has greatly improved the efficiency, lowered the cost, and extended the reliable lifetime of PV systems, and most analysts agree that they are now at, or nearing, the point of being ready to rapidly expand their use in a number of important energy markets. There are prospects for continued technological advances that may eventually make photovoltaics a major source of electric power (98).

In addition, some photovoltaic energy systems lend themselves to continuous and even automated factory production to an extent unmatched by any other energy technology. Hence, once the technology is well-developed, production of photovoltaic power systems could expand rapidly and factories could be located in developing countries, just as many semiconductor factories are now. In the field, photovoltaics require only modest amounts of skilled labor to install or maintain, making them well-suited to village power systems.

Production and Use

In 1988, manufacturers worldwide shipped some 35 MW of PV modules, a gain of more than 20 percent over the previous year, according to one estimate (99). Japan produced nearly 37 percent of the world total, followed by the United States (32 percent) and Europe (20 percent). But Japanese sales have been relatively flat for several years, and in 1989 the United States was expected to regain its initial lead in the world market, with West Germany and several other European countries posting major gains as well. Australia, India, Brazil, Canada, and China also shipped significant quantities of PV modules in 1988, and their production is expected to increase steadily (100).

The majority of today's PV systems are used in signaling devices and telecommunications systems in remote areas, and in various "stand-alone" power applications, often in the developing world. Stand-alone systems have proved cost-effective for pumping water for drinking and irrigation, providing lighting, powering refrigerators for storing vaccines in health clinics, and even providing electricity to entire villages. In fact, developing countries use at least half of all PV-generated power—and they represent a promising market for the future (101) (102). For example, the United Nations estimates that more than 2 million villages worldwide lack electricity, and photovoltaics or PV-diesel hybrids are expected to prove more economical and reliable than other small power sources in many instances (103).

Approximately 36 percent of Photovoltaics used now go into consumer products—portable lights, hand-held calculators, battery chargers, and a host of other goods (104). While this market does not contribute greatly to meeting the world's key energy needs, it is expected to remain lucrative and yield revenues to help fuel the industry's advancement (105).

PV systems also provide electricity to about 15,000 homes worldwide, as well as many industries and institutions around the world (106). Many of these buildings are in remote areas and depend entirely upon solar power. In Italy, for example, the government has sponsored Project Ginostra on the island of Sicily, part of which consists of dozens of small PV systems providing power for several hundred homes to help define the optimum PV system for remote residential groups throughout Europe. An alternative approach is one in which houses with PV systems are connected to an electrical distribution grid, from which the consumers can draw backup power, and through which they can sell excess PV-generated electricity to the central utility. Small grid-connected systems like these are expected to multiply steadily over the coming decades (107). Ultimately, PV systems of perhaps 50–1,000 MW are expected to be built and operated as part of utility networks, producing power for sale (108). Several modest-sized demonstration PV plants are already providing commercial electricity, primarily in the United States and West Germany.

The key to these expectations is more efficient PV cells and, especially, continuing reductions in manufacturing costs. Toward this end, governments and industries in numerous countries are conducting research and development (R&D) programs, including the United States, West Germany, Japan, and Italy. The best commercial PV systems today produce electricity for roughly 35 cents per kilowatt hour. In the United States, this figure must come down to about 12–15 cents for Photovoltaics to be competitive with other energy sources for producing peaking power, and to 6–10 cents before Photovoltaics can enter widespread use in generating bulk electricity. The U.S. Department of Energy's National Photovoltaics Program has set a goal of reaching the first level by 1995 and the second by 2000 (109). Many industry officials agree that this schedule is possible, but they say it will only be met if the government increases its R&D funding substantially (110).

Technology

Efforts to improve PV technology are focusing on increasing solar cell efficiency, cutting the costs of the semiconductor materials used in solar cells, improving manufacturing processes, and improving the PV

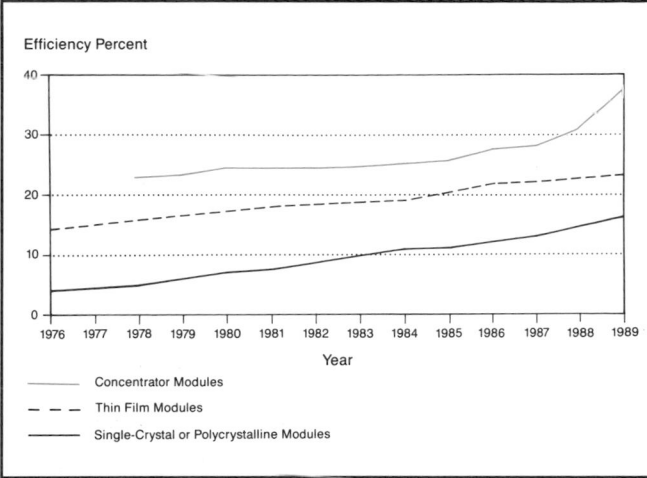

Figure 9.4 Advances in Photovoltaic Cell Efficiencies, 1976–89

Source: U.S. Department of Energy (DOE), Division of Photovoltaic Energy Technology, *Photovoltaic Energy Program Summary, Vol. 1: Overview Fiscal Year 1988* (U.S. DOE, Washington, D.C., 1989), p. 3.

modules made from those cells. Figure 9.4 illustrates the progress to date in improving the efficiencies of the three primary types of PV modules being developed: single-crystal or polycrystalline modules, concentrator modules and thin-film modules.

Silicon remains the industry's workhorse, with single-crystal and polycrystalline silicon modules accounting for about two thirds of global PV production. These technologies are relatively mature—commercial silicon modules have an efficiency of 12–14 percent (up from 8 percent in the 1970s) and a lifetime of more than 20 years. Still, there is room for improvement. Experimental silicon cells have recently reached nearly 23 percent efficiency, and opportunities exist for significantly reducing module costs by scaling up to larger production facilities, increasing automation, and transferring new methods for making large "ribbons" of polycrystalline silicon from R&D in manufacturing (111). Advanced silicon PVs will find increasing use in various remote-power applications, and the ribbon technology might yield PV systems suitable for the peak-power utility market.

An alternative approach is to use "concentrator" modules rather than the conventional type in which solar cells are simply arranged in a flat plate. Concentrators use lenses or mirrors to focus sunlight on small but highly efficient cells, and the units move in order to track the sun's progress across the sky, thereby keeping light focused on the cells at several hundred times normal intensity. These systems are far more efficient than flat-plate collectors. In 1989, a concentrator focusing sunlight on a stack of advanced cells reached a record efficiency of 38 percent (112). The tradeoff, however, is that they are significantly more expensive to build. Also, concentrators do not work as well when sunlight is diffuse, which may restrict them to geographic locations with abundant direct solar radiation.

Much of the attention in recent years, however, has shifted to "thin-film" photovoltaics, in which a semiconductor material is deposited in a layer many times finer than a human hair on an inexpensive substrate such as glass or stainless steel. These modules are less efficient than their conventional silicon cousins, but they hold great potential for low-cost mass production. In fact, thin-film modules are likely to be first in reaching the U.S. cost goals for utility peaking and baseload power generation.

Thin-film modules using a material called amorphous silicon are the furthest advanced and have already captured about 40 percent of the world PV market, mostly in consumer products (113). Commercial amorphous silicon modules have an efficiency of only 4–6 percent, however, and even though they cost much less to produce, their efficiency still must be more than doubled to be competitive for utility applications. In laboratories in the United States and Japan, experimental cells have already reached about 12 percent efficiency. And by stacking several cells containing amorphous silicon alloys on top of each other, efficiencies of 13.7 percent have been achieved (114).

Several companies are also studying prototype amorphous silicon production lines in order to identify ways to cut manufacturing costs in the United States, and Chronar began work on the first large-scale automated production plant in early 1989. Called the Eureka project, the California-based plant is projected to turn out 10 MW of modules annually (115). The experience gained with production plants such as this is expected to lead to even more efficient plants that will meet increasing module demand in the future.

Newer thin-film modules using other semiconductor materials may hold even more promise. For example, test cells made with an exotic polycrystalline compound called copper indium diselenide (CIS) have already recorded efficiencies of more than 14 percent, and are expected to top 15 percent in the near future (116). Experimental cells made with another material, cadmium telluride, have achieved efficiencies of more than 11 percent (117). Several international companies are planning on commercializing the first generation of PV products using these materials during the next few years.

While the potential efficiency of these new materials is certainly attractive, their real promise lies in low manufacturing costs. Modules can be produced using several common industrial processes, such as electroplating or vapor deposition, that are fast, efficient, and require only modest capital equipment. This feature may also make it easier for developing countries to install small-scale PV factories, which would make PV products even more affordable for widespread use.

Prototype Systems and Policy

Photovoltaic systems are being installed and tested in utility operations in order to monitor and demonstrate their performance. Among numerous projects in the United States, one of the most ambitious is called Photovoltaics for Utility-Scale Applications (PVUSA), directed by Pacific Gas & Electric (PG&E) and involving utilities across the country. A variety of 20-kilowatt (kW) PV systems representing a variety of emerging technologies, along with several 200–400 kW systems using more conventional PV technologies, will be evaluated over the next several years at PG&E's test site in Davis, California. The most promising technologies will also be tested by other utilities in different geographic areas.

An increasing number of PV systems are also being installed or planned for utilities in Europe and Japan. In Italy, for example, over the next several years one PV plant now producing 300 kW will be scaled up to 1 MW, and three new 100-kW plants, one 1-MW plant, and three 3-MW plants will be built. Combined with current efforts to promote the use of remote PV systems, Italy has a national goal of having 25 MW of PV systems installed by 1995—exceeding the PV capacities of the United States, Japan, and Germany (118).

One of the attractions of PV systems is that, in principle, the power they produce could be used to split water into its components and thus produce hydrogen, an extremely clean-burning, if difficult to store fuel. PV cells are in fact very well matched to the electrolysis cells used to split water. A recent World Resources Institute study suggests that PV-hydrogen could become attractive as a fuel for automotive fleets in some regions by the early 21st Century (119).

Whether the ambitious national goals for improving PV technology and reducing PV costs will be met is not yet clear, although there is considerable technological momentum. If they are, then growth could well explode. The director of the U.S. Solar Energy Research Institute maintains that photovoltaics have the potential to become the primary means of generating electricity worldwide by the end of the 21st Century (120). It is clear that this energy option deserves to be pursued vigorously, particularly in the face of growing concern over global warming and other environmental consequences of burning fossil fuels.

Both governmental and private-sector research budgets for Photovoltaics as for most other renewable energy sources are small compared with those for conventional energy sources. Economic policies could play a major role in the speed with which PV systems advance. Energy prices, for example, do not include such external costs as government subsidies or tax benefits favoring certain energy sources, the costs of pollution, and government support of industrial infrastructure and of research and development. Some estimates suggest that these external costs might nearly double the price of conventional energy sources if they were internalized (121) (122). Deliberate policies to accomplish this, such as by taxing carbon dioxide emissions or other economic or tax policies designed to put photovoltaics and other renewable energy sources on a more equal economic footing with conventional energy sources, deserve renewed consideration.

RECENT DEVELOPMENTS

CHERNOBYL FALLOUT WIDER THAN INITIAL ESTIMATES

Fallout caused by the April 1986 Chernobyl nuclear plant disaster near Kiev in the Soviet Ukraine has been far more extensive than the Soviet government's initial estimates.

In July 1986, the Soviet leadership announced that the accident had affected 1,000 square kilometers in the Ukraine. The Soviets said 28 people had died and 203 people had been diagnosed with radiation sickness (123). About 116,000 people living within 30 kilometers of the site were evacuated during the 10 days following the explosion (124).

In March 1989, the Soviet Communist Party newspaper *Pravda* published a report that the contamination was much more extensive than previously acknowledged. Contamination at unsafe levels—over 15 curies per square kilometer—was present over a 10,000 square-kilometer area, the report said, including about 7,000 square kilometers in Byelorussia, about 2,000 square kilometers in Russia, and about 1,500 square kilometers in the Ukraine. Pockets of contamination were found as far north as Orel, 482 kilometers from the plant and 322 kilometers south of Moscow (125). Milder contamination, between 5 and 15 curies per square kilometer, covered an additional 21,000 square kilometers (126).

Yuri Izrael, chairman of the state committee on hydrometeorology, said that some 230,000 people were living within the contaminated area (127). *Pravda* said that the health of the people living within the zone was not endangered so long as uncontaminated food supplies were brought in from outside the area and safety checks were carried out continually (128). Other reports by Soviet scientists and intellectuals and by American journalists suggest that there is growing frustration and anxiety among people living in the contaminated area and bitterness about the government's initial reluctance to disclose the true extent of the problem.

In the Ukrainian region of Narodichi, which is southwest of Chernobyl and slightly outside the immediate evacuation zone, radiation readings reportedly are as much as 30 times higher than normal and farm animals are being fed fodder from contaminated fields. The directors of the Petrovsky Collective Farm in Narodichi reported that 64 farm animals were born with serious deformities in 1987 and that the rate continued to rise in 1988. The farm had only

three or four similar cases before the accident (129). The region's farmers reportedly are still under pressure to meet the area's production schedule. A local party official has said that cows from the area are taken to graze in uncontaminated districts before being slaughtered (130).

The director of a hospital in the Nogilev region of Byelorussia said that "we cannot give any guarantees of a healthy life to the population of the contaminated areas." A Ukrainian legislator touring the republic's Poleski district reported seeing many residents, especially children, with swollen thyroid glands, sluggishness, cataracts, and cancers (131). Some 1.25 million people have undergone medical checkups, of which 627,000 are being placed under permanent observation (132).

There are strong disagreements about the number of additional people who should be evacuated. Izrael said in March 1989 that 3,000 additional people should be evacuated from Byelorussia, but the People's Deputies of Byelorussia favor immediate evacuation of 120,000 people (133).

Soviet government officials have spent vast sums to reduce the chances of contamination to food products and protect the local population from eating contaminated food, but their efforts are frequently thwarted. Local residents receive 30 rubles to buy clean food (134), but they often continue to eat home-grown fruits and vegetables. Large quantities of contaminated beef, pork, and potatoes from the affected region reportedly have been shipped to other parts of the country for processing (135).

SOLAR THERMAL TECHNOLOGY ADVANCES

A California-based firm is building commercial solar-thermal electric power plants that are increasingly cost-competitive with conventional power generation and could soon be constructed in several other locations around the world.

In December 1989, Luz International Limited completed construction of its first 80-MW solar thermal utility in the Mojave Desert northeast of Los Angeles. Without much fanfare, the company's systems now account for more than 90 percent of the world's solar electricity (136).

The company previously has built a 13.8-MW facility and six 30-MW plants in the same location. With the planned completion of four additional 80-MW plants by 1994, Luz will be generating nearly 600 MW of power, which is enough electricity to serve the residential needs of over 800,000 people.

The power helps Southern California Edison meet peak period demand, for instance, on hot summer days when many air conditioners are in use. Luz has signed a long-term power purchase agreement with Southern California Edison to sell electricity produced by these plants; by 1994 it will be providing about 3 percent of the utility's peak demand load of 20,000 MW (137).

Another 80-MW facility will be built for San Diego Gas and Electric, and the company could soon be building plants in Brazil, Spain, and India (138).

The Luz system uses rows of trough-like mirrors that track the sun with the help of light-sensing instruments and microprocessors. The mirrors focus the sun's light onto coated steel pipes, which are mounted at the base of the troughs inside vacuum-insulated glass tubes. A synthetic oil inside the pipes is heated to 411.6°C. The fluid passes through a heat exchanger, which generates superheated steam for electric turbine generators. A small natural-gas boiler serves as a backup or to sustain power generation at night when necessary (139).

The company has made substantial progress in bringing down costs. The first plant built in 1984 produced electricity at 24 cents per kilowatt-hour; the new 80-MW facility can generate power at just under 8 cents per kilowatt-hour, and the company expects that by 1994 costs will drop to about 5 cents per kilowatt-hour. That should make the technology fully competitive with new conventional coal- or oil-fired plants (about 6–9 cents per kilowatt-hour) and with nuclear plants (about 10 cents per kilowatt-hour) (140).

The company was founded in 1979 by Arnold J. Goldman, an Israeli electrical engineer born in the United States, and Patrick François, a French-born Israeli textile manufacturer. With some early help from a U.S.-Israeli research agency, the two men raised $40 million for research and development and eventually set up manufacturing facilities in Israel. The company's headquarters are now in Los Angeles. Luz operates the California plants, but it sees its future role largely as a supplier of solar energy systems and hardware (141).

Though a very promising and environmentally benign technology, solar thermal power plants appear to have their potential niche in supplementing fossil and nuclear plants rather than as base-load plants. Nevertheless, the cost-competitiveness, environmental advantages, and quick turnaround time—Luz plants are modular and require only 18 months to build—seem to assure the company a place in providing future energy supplies in sunny areas around the world (142).

The section "Focus on the Automobile" was written by Deborah Lynn Bleviss, executive director of the International Institute for Energy Conservation, Washington, D.C. The other sections of the chapter were written by Tom Burroughs, a science writer and consultant based in Durham, North Carolina.

References and Notes

1. Primary energy excludes fuels not traded commercially, such as wood and animal wastes, which are important in many countries, especially in the developing world.
2. British Petroleum Company, *BP Statistical Review of World Energy* (British Petroleum Company, London, July 1989), pp. 31-32.
3. OECD countries are Australia, Austria, Belgium, Canada, Denmark, the Federal Republic of Germany, Greece, Ireland, Italy, Japan, Luxembourg, the Netherlands, New Zealand, Norway, Portugal, Spain, Sweden, Switzerland, Turkey, the United Kingdom, and the United States.
4. *Op. cit.* 2.
5. U.S. Department of Energy (DOE), Energy Information Administration, *Annual Energy Review 1988* (U.S. DOE, Washington, D.C., May 1989), pp. 1-11.
6. *Op. cit.* 2.
7. The members of OPEC include Algeria, Ecuador, Gabon, Indonesia, Iran, Iraq, Kuwait, Libya, Nigeria, Qatar, Saudi Arabia, United Arab Emirates, and Venezuela.
8. *Op. cit.* 2, 7-8.
9. *Op. cit.* 2, pp. 27, 34.
10. U.S. Department of Energy (DOE), Energy Information Administration, *International Energy Outlook 1989, Projections to 2000* (U.S. DOE, Washington, D.C., 1989), p. 26 and Figure 10, p. 27. Note: The world oil price is defined as the average cost of imported crude oil to U.S. refiners.
11. U.S. Department of Energy (DOE), Energy Information Administration, *Annual Prospects for World Coal Trade* (U.S. DOE, Washington, D.C., 1989), p, vii.
12. *Ibid*, pp. 17, 20, 46.
13. David G. Streets, "Fulfilling the Promise of Clean-coal Technology," *Forum for Applied Research and Public Policy*, Vol. 4, No. 1 (Spring 1989), p. 30.
14. John Douglas, "Quickening the Pace in Clean Coal Technology," *EPRI Journal*, January/February 1989 (Electric Power Research Institute (EPRI), Palo Alto, California, 1989) pp. 13-15.
15. *Op. cit.* 2, pp. 23, 24.
16. *Op. cit.* 2, p. 22.
17. Randolf Gränzer, "Perestroika in Energy, The Soviet Union and Eastern Europe," *The OECD Observer*, No. 136, December 1988/January 1989, p. 25.
18. *Op. cit.* 2, p. 28, 34.
19. U.S. Council for Energy Awareness (USCEA), "USCEA Survey, World's Nuclear Capacity Grew Five Percent in Year," *INFO*, INFO 244 (June 1989), p. 2.
20. *Op. cit.* 2, pp. 30, 34.
21. The market-economy countries are defined as all countries other than the centrally planned economies of Eastern Europe, the Soviet Union, China, Cuba, Kampuchea, North Korea, Laos, Mongolia, Viet Nam, and Yugoslavia.
22. *Op. cit.* 10, p. 23.
23. "World Status Report: Oil Production and Reserves," *Energy Economist*, August 1989, p. 11.
24. *Ibid*, p. 15.
25. *Op. cit.* 2, p. 24.
26. National Coal Association, *Facts About Coal 1989* (National Coal Association, Washington, D.C., 1989), p. 5.
27. *Op. cit.* 2, p. 20.
28. U.S. Council for Energy Awareness (USCEA), "USCEA 1988 International Reactor Survey, World's Nuclear Power Plant Total Climbs to 417; China, France, India and West Germany Show Gains (news release, May 26, 1989).
29. Judith Perera, "Stunted growth of nuclear plants," South (April 1989), p. 65.30. World Energy Conference, Survey of Energy Resources 1980, cited in Cynthia Pollock Shea, *Renewable Energy: Today's Contribution, Tomorrow's Promise* (Worldwatch Institute, Washington, D.C., 1988) p. 9.
31. U.S. Agency for International Development (AID), *Decentralized Hydropower in AID's Development Assistance Program* (U.S. AID, Washington, D.C., 1986) p. 2.a, 2.3., cited in Christopher Flavin, *Electricity for a Developing World: New Directions*, Worldwatch Paper 70 (Worldwatch Institute, Washington, D.C., June 1986), p. 47.
32. International Energy Agency, *Renewable Sources of Energy* (Organisation for Economic Co-operation and Development, Paris, March 1987), p. 24.
33. *Ibid.*, p. 18.
34. John H. Gibbons, Peter D. Blair, and Holly L. Gwin, "Strategies for Energy Use," *Scientific American* (September 1989), p. 140.
35. Richard C. Dorf, *The Energy Fact Book* (McGraw-Hill Book Company, New York, 1981), p. 163.
36. U.S. Department of Energy (DOE), Office of Conservation and Renewable Energy, *Energy Conservation Trends* (U.S. DOE, Washington, D.C., August 1989), p. 15.

37. International Energy Agency (IEA), *Energy Conservation in IEA Countries* (Organisation for Economic Co-operation and Development, Paris, 1987), p. 29.
38. Howard S. Geller, "U.S. Energy Demand: Back to Robust Growth?" Energy Efficiency Issues Paper No. 1, American Council for an Energy-Efficient Economy, Washington, D.C., February 1989, p. 1.
39. *Op. cit.* 5, p.1.
40. Members of the European Community are Belgium, Denmark, the Federal Republic of Germany, France, Greece, Ireland, Italy, Luxembourg, Netherlands, Portugal, Spain, and the United Kingdom.
41. Commission of the European Communities, "The Main Findings of the Commission's Review of Member States' Energy Policies, the 1995 Community Energy Objectives," Brussels, May 3, 1988, p. 7.
42. *Ibid.*, p. 5.
43. *Ibid.*
44. Jose Goldemberg, Thomas B. Johansson, Amulya K.N. Reddy, *et al.*, *Energy for a Sustainable World* (World Resources Institute, Washington, D.C., September 1987), p. v.
45. *Op. cit.* 37.
46. *Op. cit.* 37, p. 82.
47. *Op. cit.* 37, p. 9.
48. *Op. cit.* 17, p. 26.
49. Randolf Gränzer, "The Energy Impediment to China's Growth," *The OECD Observer*, No. 157, April/May 1989, pp. 13-14.
50. S. Meyers and J. Sathaye, "Electricity in the Developing Countries: Trends in Supply and Use Since 1970," Lawrence Berkeley Laboratory, University of California, Berkeley, California, December 1988, p. 4-2.
51. *Op. cit.* 44, p. 2.
52. *Op. cit.* 44, p. 46.
53. *Op. cit.* 44, pp. 55-56.
54. Motor Vehicle Manufacturers Association of the United States (MVMA), *World Motor Vehicle Data, 1989 Edition* (MVMA, Detroit, Michigan, 1989) p. 38.
55. *Ibid.*, pp. 37-38.
56. *Ibid.*, pp. 36-38.
57. Michael P. Walsh, "Global Trends in Motor Vehicles and Their Use: Implications for Climate Modification," prepared for the World Resources Institute, December 19, 1988, p. 13.
58. Michael P. Walsh, "Global Trends in Motor Vehicles and Their Use: Implications for Climate Modification," prepared for the World Resources Institute, December 19, 1988.
59. U.S. Department of Energy (DOE), Office of Policy, Planning, and Analysis, *Patterns of U.S. Energy Demand* (U.S. DOE, Washington, D.C., August 1987), pp. 5, 31. World Resources Institute and International Institute for Environment and Development in collaboration with the United Nations Environment Programme, *World Resources 1988-89* (Basic Books, New York), p. 116.
60. Stephen Myers, "Transportation in the LDCs: A Major Area of Growth in World Oil Demand," International Energy Studies Group, Applied Science Division, Lawrence Berkeley Laboratory, University of California, Berkeley, California, March 1988. p. 1.
61. U.S. Department of Energy (DOE), Office of Policy, Planning, and Analysis, *Energy Projections to the Year 2010, A Technical Report in Support of the National Energy Policy Plan* (U.S. DOE, Washington, D.C., October, 1983), p. 5.
62. *Op. cit.* 60.
63. M.A. DeLuchi, R.A. Johnston, and D. Sperling, "Transportation Fuels and the Greenhouse Effect," Universitywide Energy Research Group, University of California,

UER-182, December 1987, cited in Michael P. Walsh, "Global Trends in Motor Vehicles and their Use: Implications for Climate Modification," prepared for World Resources Institute, December 1988, p. 6.
64. Gordon MacDonald, "The Greenhouse Effect and Climate Change," testimony presented to the Senate Committee on Environment and Public Works, January 28, 1987, cited in Michael P. Walsh, "Global Trends in Motor Vehicles and their Use: Implications for Climate Modification," prepared for World Resources Institute, December 1988, p. 5.
65. U.S. Environmental Protection Agency (EPA), *Regulatory Impact Analysis: Protection of Stratospheric Ozone* (U.S. EPA, Washington, D.C., December 1987), cited in Michael P. Walsh, "Global Trends in Motor Vehicles and their Use: Implications for Climate Modification," prepared for World Resources Institute, December 1988, p. 5.
66. *Op. cit.* 64, pp. 6-7.
67. *Op. cit.* 64, p. 6.
68. Oak Ridge National Laboratory, *1988 Automated Transportation Energy Data Book*, in press, 1988.
69. Oak Ridge National Laboratory, Patricia S. Hu, Linda S. Williams, and Dennis J. Beal, "Light Duty Vehicle MPG and Market Shares Report: Model Year 1988, April 1989, prepared for the U.S. Congress Office of Technology Assessment, p. xxxix.
70. Oak Ridge National Laboratory, *Transportation Energy Data Book: Edition 9* (Oak Ridge National Laboratory, April 1987), cited in Deborah L. Bleviss, *The New Oil Crisis and Fuel Economy Technologies: Preparing the Light Transportation Industry for the 1990s* (Quorum Books, Westport, Connecticut, 1988), p. 95.
71. D.E. Cole and J. H. Johnson, "Spark Ignition and Diesel Engines as Applied to Passenger Cars and Light Trucks," working paper prepared for the U.S. Congress Office of Technology Assessment, December 4, 1979, cited in Deborah L. Bleviss, *The New Oil Crisis and Fuel Economy Technologies: Preparing the Light Transportation Industry for the 1990s* (Quorum Books, Westport, Connecticut, 1988), p. 19.
72. Deborah L. Bleviss, *The New Oil Crisis and Fuel Economy Technologies: Preparing the Light Transportation Industry for the 1990s* (Quorum Books, Westport, Connecticut, 1988), pp. 23-24, 32.
73. John B. Heywood, "Alternative Automotive Engine and Fuels: A Status Review and Discussion of R&D Issues," unpublished report prepared for the U.S. Congress Office of Technology Assessment, November 1979, cited in Deborah L. Bleviss, *The New Oil Crisis and Fuel Economy Technologies: Preparing the Light Transportation Industry for the 1990s* (Quorum Books, Westport, Connecticut, 1988), p. 40.
74. *Op. cit.* 72, pp. 37-38.
75. *Op. cit.* 72, pp. 20, 43-44.
76. Energy and Environmental Analysis, Inc., "Developments in the Fuel Economy of Light-Duty Highway Vehicles," prepared for the U.S. Congress Office of Technology Assessment, August 1988, pp. 3-9.
77. *Op. cit.* 72, p. 63.
78. *Op. cit.* 72, p. 62.
79. *Op. cit.* 72, p. 63.
80. *Op. cit.* 72, p. 52.
81. Merton C. Flemings *et al.*, *Materials Substitution and Development for the Light Weight, Energy Efficient Automobile*, report prepared for the U.S. Congress Office of Technology Assessment, February 8, 1980, cited in Deborah L. Bleviss, *The New Oil Crisis and Fuel Economy Technologies: Preparing

the Light Transportation Industry for the 1990s* (Quorum Books, Westport, Connecticut, 1988), p. 52.
82. *Op. cit.* 72, pp. 53, 59-60.
83. Peter Hofbauer *et al.*, "GA 1-Glider Automatic," SAE Paper No. 830529, cited in Deborah L. Bleviss, *The New Oil Crisis and Fuel Economy Technologies: Preparing the Light Transportation Industry for the 1990s* (Quorum Books, Westport, Connecticut, 1988), p. 76.
84. *Op. cit.* 72, pp. 75-76.
85. R. Qian and A.A. Frank, "Theoretical Considerations for Kinetic and Potential Energy Storage Systems for Motor Vehicles," paper presented at the 20th IICEC '85, cited in Deborah L. Bleviss, *The New Oil Crisis and Fuel Economy Technologies: Preparing the Light Transportation Industry for the 1990s* (Quorum Books, Westport, Connecticut, 1988), pp. 77-78.
86. *Op. cit.* 72, p. 102. Note that while a high miles per gallon figure indicates energy efficiency, in liters per hundred kilometers, a lower number indicates greater efficiency.
87. Volvo LP2000 Light Component Project, Volvo literature, 1985, cited in Deborah L. Bleviss, *The New Oil Crisis and Fuel Economy Technologies: Preparing the Light Transportation Industry for the 1990s* (Quorum Books, Westport, Connecticut, 1988), p. 102, 103, 118.
88. Robert Livingston, "The Next Generation," *Nuclear Industry*, July/August 1988 (U.S. Council for Energy Awareness, Washington, D.C., 1988), p. 3.
89. "AP 600" A New Generation of Nuclear Power...Naturally Safe," *Energy Digest* (Westinghouse Electric Corporation, Pittsburgh, Pennsylvania, March 1989), p. 6.
90. *Op. cit.* 88, pp. 2-15.
91. Jack Catron, "New Interest in Passive Reactor Designs," *EPRI Journal*, April/May 1989 (Electric Power Research Institute (EPRI), Palo Alto, California, 1989), p. 9.
92. Margaret L. Ryan, "Outlook on Advanced Reactors," *Nucleonics Week*, March 30, 1989 (McGraw-Hill, Inc., New York, 1989) p. 17.
93. *Op. cit.* 91, p. 13.
94. Ulf Bredolt, Jan Fredall, Käre Hannerz, *et al.*, "PIUS—The Next Generation Water Reactor" (Asea Brown Boveri company (formerly ABB Atom AB), unpublished paper, Vasteras, Sweden).
95. H.P. Planchon Jr., "EBR-11 Tests Prove Reactors Can Be Made Inherently Safe," *logos*, Progress through Science, Vol. 4, No. 1 (Argonne National Laboratory, Argonne, Illinois, 1986), p. 11.
96. *Op. cit.* 91, p. 13.
97. William J. Broad, "Experts Call Reactor Design 'Immune' to Disaster, *New York Times* (November 15, 1988).
98. H.M. Hubbard, "Photovoltaics Today and Tomorrow," *Science*, Vol. 244, April 21, 1989 (American Association for the Advancement of Science, Washington, D.C., 1989) pp. 297, 300.
99. Paul Maycock, "World PV-Module Production Up," *Solar Today*, March/April 1989 (American Solar Energy Society, Boulder, Colorado, 1989), p. 12.
100. Paul Maycock, president of Photovoltaic Energy Systems, October 1989 (personal communication).
101. "U.S. Third World PV Efforts 'Making It,'" *Solar Today*, September/October 1988, pp. 11-12.
102. Catherine Healy, "Sunshine Lights Dark Corners," *Americas*, January/February 1988 (Organisation of American States, Washington, D.C., 1988), p. 54.
103. *Op. cit.* 32, p. 166.

104. *Op. cit.* 100.
105. *Op. cit.* 32, p. 164-165.
106. *Op. cit.* 32, p. 157.
107. *Op. cit.* 98, p. 300.
108. *Op. cit.* 98, p. 300.
109. U.S. Department of Energy, Photovoltaic Energy Technology Division, *Investing in Success* (Washington, D.C., January 1989).
110. John Corsi, Richard Blieden, and Scott Sklar, *Photovoltaics: Solar Electricity in the 1990s*, Solar Energy Industries Association (Arlington, Virginia, November 1988).
111. *Op. cit.* 100.
112. *Op. cit.* 100.
113. *Op. cit.* 99.
114. Taylor Moore, "Thin Films: Expanding the Solar Marketplace," *EPRI Journal*, March 1989 (Electric Power Research Institute (EPRI), Palo Alto, California, 1989), p. 7.
115. R. H. Annan, *et al.*, *Department of Energy Review of the U.S. Photovoltaics Industry, November and December 1988* (Solar Energy Research Institute, Golden, Colorado, February 1989), p. 7.
116. *Op. cit.* 98, 299.
117. *Op. cit.* 115.
118. "Italian PV Program Explodes!" *PV News*, Vol. 8, No. 10 (Energy Systems, Inc., Casa

nova, Virginia, October 1989), p. 3.
119. Joan M. Ogden and Robert H. Williams, *Solar Hydrogen: Moving Beyond Fossil Fuels* (World Resources Institute, Washington, D.C., 1989).
120. *Op. cit.* 98. p. 299.
121. *Op. cit.* 98, p. 300.
122. *Op. cit.* 109.
123. Paul Quinn-Judge, "New Soviet report tallies costs of Chernobyl nuclear accident," *The Christian Science Monitor* (July 21, 1986, Boston), p. 12.
124. Vladimir Kolosov, Senior Researcher, Department of Global Problems, Institute of Geography of the Soviet National Academy of Sciences, Moscow, 1989 (personal communication).
125. Michael Dobbs, "Pravda Says Chernobyl Fallout Wider Than Acknowledged," *The Washington Post* (March 21, 1989, Washington, D.C.), p. A12.
126. *Op. cit.* 124.
127. *Op. cit.* 125.
128. *Op. cit.* 125.
129. David Remnick, "Chernobyl's 'Coffin Bonus,'" *The Washington Post* (November 24, 1989, Washington, D.C.), p. A1.
130. Peter Gumbel, "Villagers Suffering

Chernobyl's Fallout Face Soviet Silence," *The Wall Street Journal* (March 6, 1989, New York), p. 1.
131. *Op. cit.* 129.
132. *Op. cit.* 124.
133. *Op. cit.* 124.
134. *Op. cit.* 124.
135. *Op. cit.* 129.
136. Luz International Limited, "Luz in Brief" (Luz International, Los Angeles, May 1989), p. 1.
137. Louis Sahagun, "Farming the Power of Light," *The Los Angeles Times* (January 9, 1989, Los Angeles).
138. John Wilson, Vice President, Conway and Company, Washington, D.C., 1989 (personal communication).
139. Daniel B. Wood, "State-of-the-Art Solar Collectors for Clean Energy," *The Christian Science Monitor* (June 13, 1989, Boston), p. 13.
140. Nancy Rader, "Power Forecast: Sunny, Breezy, Wet," *The Washington Post* (July 23, 1989, Washington, D.C.), p. D3.
141. James Cook, "Warming Trend," *Forbes* (February 20, 1989), p. 68.
142. *Ibid.*

10. Freshwater

Water is an essential resource for life. Although the availability of freshwater varies widely with geographic location, Earth's water cycle is an abundant provider. Both natural conditions and human activities affect the quantity and quality of available water.

Water pollution sometimes renders the water supply unfit for various human uses, including drinking. It can also profoundly affect natural biological systems, leading to the overfertilization or eutrophication of lakes and coastal seas or to the accumulation of unsafe levels of organic residues and metals in fish and other marine life, for example. The first global assessment of freshwater quality, recently carried out under the auspices of the Global Environmental Monitoring System (GEMS), found that contamination of water resources continues to increase in much of the world. There is some evidence of improvement in industrial countries (1). Overall, monitoring of water quality is inadequate and control measures are weak.

The quantity of the resource is fast becoming an issue in some areas. Although essentially a renewable resource on a global scale, freshwater is being extracted from some river basins at rates approaching those at which the supply is renewed and from some underground aquifers at rates exceeding natural replacement. Many human activities have high water use rates. As the human population has grown, so have withdrawals of water for agriculture, industry, and municipal use. A new element of uncertainty is potential changes in precipitation and hence in freshwater resources due to changes in climate caused by human activities.

Managing water quality will be increasingly important. Traditionally, effluents from human activities have been diluted by the volume of river flow and often eliminated by the self-cleansing action of streams. But as populations and economic activity increase, these effluents threaten the life of some river basins. Attention to wastewater recovery and treatment methods and to the potential for changing industrial processes to produce less wastewater and fewer effluents is needed.

CONDITIONS AND TRENDS

FRESHWATER QUALITY: A GLOBAL PERSPECTIVE

Historical Perspective

Freshwater pollution arising from human activities began with the first settlements and increased in severity as populations grew. A schematic representation of the expansion of pollutant effects from local to regional to global is given in Figure 10.1. Initially, pollution from organic wastes and the salinization of irrigation systems were the major problems. To

Figure 10.1 The Evolution of Water Pollution Problems

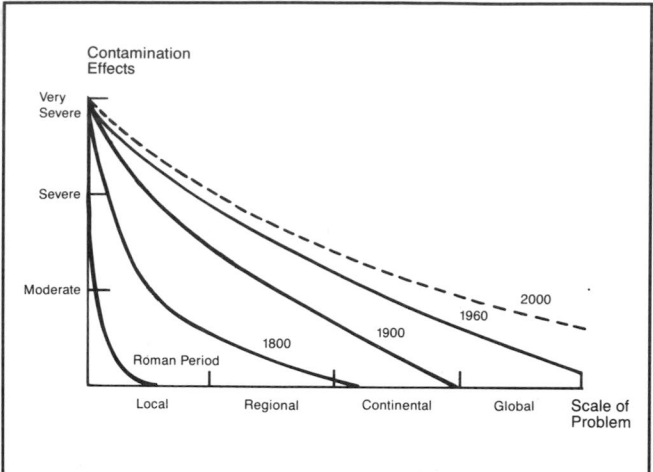

Source: Michel Meybeck, Deborah Chapman, and Richard Helmer, eds., *Global Freshwater Quality: A First Assessment* (Blackwell Reference, Oxford, U.K., 1990), Figure 19-22, p. 288.

them must now be added concern about suspended solids, heavy metals, radioactive wastes, nitrates, and organic micropollutants and about the acidification of lakes and streams and the eutrophication of lakes and coastal waters.

To address these problems, many detailed assessments of water pollution have been undertaken at the local level and to some degree at the national or regional level, as within the European Community (2). The only systematic attempt to gather water quality data worldwide is being made by the Global Environmental Monitoring System (GEMS). The GEMS water monitoring programme was launched in 1977 by the United Nations Environment Programme (UNEP) and World Health Organization (WHO). The monitoring data are based on voluntary contributions from 59 countries.

The financial resources available are not comparable with the immensity of the task of monitoring water quality globally. The result is that comparable data are usually collected from only a few sites in each country. Moreover, because natural water quality depends on rainfall, geographic location, vegetation type, flooding frequency, etc., a global assessment can look only for trends and compare levels of selected contaminants where measurements have been taken. Data from regional assessments would strengthen knowledge of water quality.

There are three major sources of water pollution: domestic wastewater, industrial effluents, and land use runoff, although leaching from mine tailings and solid waste dumps and the atmospheric deposition of pollutants into water bodies are of growing importance.

Figure 10.2 The Long-term Development of Pathogen Pollution in the River Seine

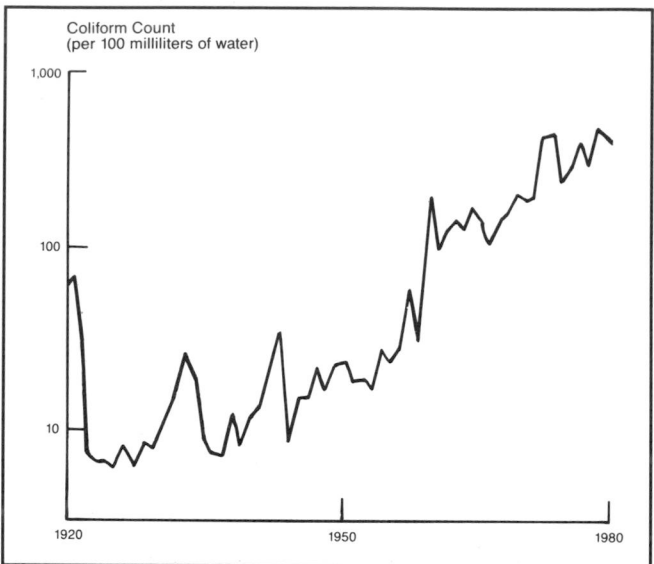

Source: Michel Meybeck, Deborah Chapman, and Richard Helmer, eds., *Global Freshwater Quality: A First Assessment* (Blackwell Reference, Oxford, U.K. 1990), Figure 19–4, p. 276.

Domestic Wastewater

Urbanization results in the concentration of human wastes and other domestic wastewaters that, when disposed of in waterborne sewage systems, are usually discharged into nearby bodies of water. As it decays, this organic waste depletes the water of oxygen essential to aquatic life and upsets the natural balance of the aquatic ecosystem.

Sewage contains pathogenic bacteria and viruses derived from human feces. Untreated wastewater is often the carrier of viruses and bacteria (e.g., *Salmonella typhosa*) and, along with household sanitation practices, has been linked to high infant mortality rates in developing countries. Even where most sewage is treated, as in the developed world, recent measurements of fecal coliforms in some countries indicate increasing pollution. In the River Seine just upstream from Paris, for example, concentrations of bacteria increased from fewer than 10 coliforms per 100 milliliter (ml) in the mid-1920s to approximately 500 per 100 ml in 1980. (See Figure 10.2.) Overloading sewage treatment systems, in part from growing urban population densities, appears to be the cause. In other river systems, however—such as the Thames—coliform counts have decreased.

Domestic waste problems are especially critical in Latin America, where it is thought that little if any of the urban sewage is treated (3), and twice as many rivers as in other regions have fecal coliform counts of more than 100,000 per 100 ml (8 percent versus 4 percent) (4). The World Health Organization (WHO) recommends a coliform count of 0 per 100 ml for drinking water (5).

In addition to coliform counts, such pollution can usually be measured in terms of the organic matter content, suspended solids in the water, or the biochemical oxygen demand (BOD), the amount of oxygen removed from the water as the organic material in it degrades and decays. BOD is one of the most commonly measured water quality variables in the GEMS network; 10 percent of all river stations report BOD levels above 6.5 milligrams (mg) per liter, indicating organic pollution (6).

Domestic wastewater usually contains high levels of nitrogen and phosphorus compounds, both essential to aquatic life. In the absence of human influences, low levels of such nutrients often limit primary productivity within water bodies. However, when levels are artificially raised, algae grow in dense blooms, increasing the suspended organic matter content and decreasing oxygen levels as it decays. This enrichment process, known as eutrophication, is further enhanced by runoff of fertilizers from agricultural and urban areas.

In the developed world, water quality legislation has led to extensive development of wastewater treatment facilities that remove much of the organic matter before the wastewater is discharged. Nevertheless, some organic pollution still occurs, much of it from nonpoint sources such as surface runoff from the watershed. As a result, stretches of many rivers (e.g., in the United Kingdom) have unacceptable water quality according to national criteria (7). In the developing world, human settlements and industry are growing fast, often faster than wastewater treatment facilities can be provided (8). Thus, much untreated domestic wastewater is discharged into rivers, making the water unsuitable for drinking or for water contact uses, at least locally.

Industrial Effluents

Industrial processes frequently produce waste, some of it toxic even in small quantities. Major contributors are the pulp and paper, chemicals, petrochemicals and refining, metalworking, food processing, and textile industries. The wastes, broadly categorized as heavy metals or synthetic organic compounds, reach bodies of water either through direct discharge from the atmosphere, or by leaching from waste dumps.

In the developed regions of the world, many industrial discharges are strictly controlled. Yet, pollution of water bodies continues from accumulations of wastes discharged over the past 100 years. In the developing regions, industrial discharges are largely uncontrolled, and water quality is directly affected. In Chile's Maipo River basin, for example, a very small percentage of the effluents produced by the chemical, plastic, and rubber industries are treated, and

none of the pulp and paper industry's waste discharges are processed in any way (9).

The effects of heavy metal pollution on water quality vary considerably, depending on local geochemical factors, but can cause very serious health problems; natural levels of trace metals vary substantially. It may therefore be difficult to pinpoint contributions of local industrial sources or to obtain reliable and comparable data on a global scale.

Nevertheless, it is possible to obtain an indication of the trends in the pollution of water by heavy metals at different stages of industrial development. Analyses focus on sediments, where metals may ultimately accumulate, and biota that take in metals through the food chain, the water, or suspended particulates. The Rhine, for example, runs through a highly industrialized region. Measurements over an 80-year period beginning in 1900 show that concen-

Figure 10.3 Changes in Metal Contamination Associated with Suspended Matter in the Rhine River

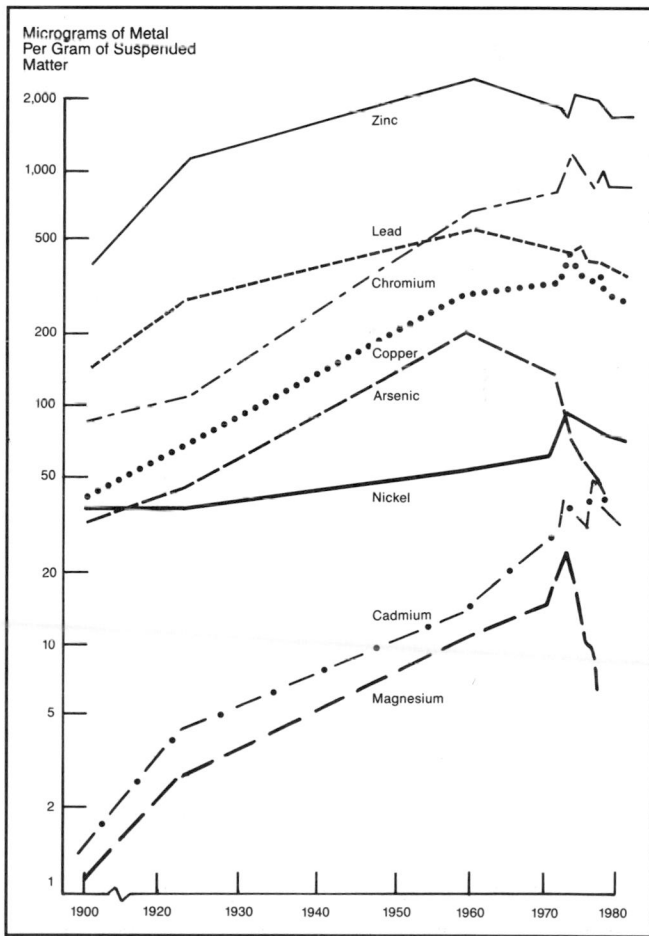

Source: W. Salomons *et al.*, "Help! Holland Is Plated by the Rhine (Environmental Problems Associated with Contaminated Sediments)," in *Effects of Waste Disposal on Groundwater.* Proceedings of Exeter Symposium (International Association of Hydrological Sciences), Figure 3, reproduced in Michel Meybeck, Deborah Chapman, and Richard Helmer, eds., *Global Freshwater Quality: A First Assessment* (Blackwell Reference, Oxford, U.K. 1990), Figure 19-12, p. 280.

Figure 10.4 Increases in Chloride in the Rhine River

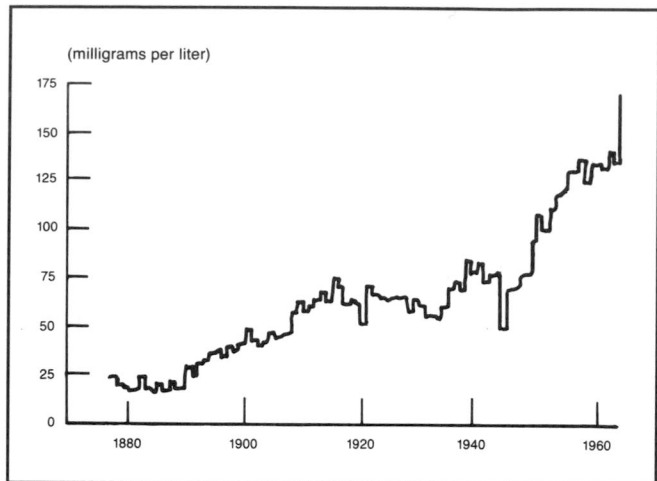

(milligrams per liter)

Source: Commission Internationale pour la Protection du Rhin contre la Pollution, *Annual Report 1988* (Koblenz, 1988), Figure 3, reproduced in Michel Meybeck, Deborah Chapman, and Richard Helmer, eds., *Global Freshwater Quality: A First Assessment* (Blackwell Reference, Oxford, U.K. 1990), Figure 19-15, p. 282.
Note: Chloride concentrations are estimated for an average discharge of 2,000 cubic meters per second.
The Alsace (potash mines) and Lorraine (salt) are one of the major point sources of Cl- in the Rhine basin.

trations of heavy metals increased during most of this century but have decreased in the past 10–15 years, as heavy metal wastes were reduced and treatment improved (10). (See Figure 10.3.)

Mining operations can result in metals' leaching into the acidic effluents or runoff water, thus adding to the metal load in rivers, lakes, and groundwater. Discharge of mercury from gold mining activities is contaminating some streams in Brazil and Ecuador, for example, creating serious health risks for villages that depend on those streams for drinking water. Earlier episodes of cadmium discharges in Japan contaminated fish that, when eaten by local inhabitants, caused severe health problems that became known as Itai-Itai disease (11). Mining can also increase salinity, as it did in the Rhine. (See Figure 10.4.) Because the demand for metals is leading to increased production in some areas, this source of pollution may remain a problem for some time (12).

Combustion of fossil fuels in automobiles and in industry releases sulfur and nitrogen compounds and, in some regions, lead, all of which enter bodies of water via the atmosphere. The most widespread effect has been the acidification of rivers and lakes, resulting in disruption of aquatic ecosystems and the death of many aquatic species. Although acidification is primarily a European and North American problem, southern China's extremely high sulfate deposition rates indicate that similar acidification problems may become severe in developing countries, depending on soil pH, future emissions, and population density (13).

Land Use Runoff

The clearing of land for agriculture and agricultural practices such as irrigation and the use of fertilizers and pesticides have had a major impact on water quality in many parts of the world. Although deforestation of woodlands has a long history, the current demand for wood-, range-, and farmland has led to deforestation on an unprecedented scale. Soil erosion in deforested watersheds due to accelerated water runoff can load rivers with sediment; leaching of soil nutrients often accelerates immediately after cutting. These effects are particularly severe in tropical regions especially during the rainy season. Deforestation can then temporarily cause an estimated increase of more than 100 times the normal sediment load in affected rivers (14). Adequate evidence of widespread degradation of surface waters is difficult to obtain because monitoring of suspended matter in rivers is often inadequate.

In some areas where irrigation for agriculture is common, increased salinity is an acute problem. Salinity levels in the Kent River, western Australia, have more than doubled over 40 years, from an annual weighted mean of approximately 0.5 grams per liter in 1940 to 1.2 in 1980 (15). In the United States, the average monthly salinity level in the San Joaquin River in California increased from approximately 0.28 grams per liter in the late 1930s to 0.45 in the mid-1970s (16). These increases may have harmful effects on downstream agriculture or on the areas abutting the rivers. In addition, the withdrawal of too much water for irrigation in coastal areas can lead to saltwater intrusion, creating salinity problems in coastal aquifers.

New technology and the demand for greater agricultural productivity have led to an exponential increase in the use of fertilizers and pesticides. The runoff of these chemicals, particularly nitrogenous fertilizers, creates one of the most widespread and serious of all water quality problems, particularly in industrialized countries. Evidence from long-term monitoring in the United Kingdom suggests that nitrate levels have been building up for many years. (See Figure 10.5.) Data from the GEMS water program suggest that over 90 percent of the rivers monitored in Europe show some evidence of nitrate pollution; 5 percent have nitrate concentrations over 200 times the unpolluted level. Worldwide, 10 percent of rivers monitored at one or more stations exceed nitrate levels established by WHO and, therefore, the water is unfit for human consumption unless it is treated (17).

Use of chemical fertilizers is still relatively low in the developing world; the per-hectare consumption of fertilizers in Latin America during 1984–87 amounted to only 44.7 kilograms (kg) per year versus 228 kg in Europe and 93 kg in the United States. (See Chapter 18, "Food and Agriculture," Table 18.2.) Water pollution from fertilizer runoff is therefore expected to be relatively less severe in developing countries. Nevertheless, fertilizer consumption in

Latin America and the Caribbean increased 97 percent from 1973 to 1985 and in some countries has reached developed-country levels (18).

Organic micropollutant data obtained by the GEMS water program suggest that pesticides, PCBs, and other synthetic organic chemicals are found in many rivers around the world (19). In the United States, for example, a five-year study (completed in 1980) of more than 150 rivers showed that 42–82 percent of all water and sediment samples were contaminated by organochlorine insecticides and 2–7 percent contaminated by organophosphate insecticides (20). Developing countries are thought to have lower levels of such contamination because they use fewer synthetics. But there are important exceptions. Brazil, for example, ranks among the top five countries in the world in terms of pesticide use; it consumes 150,000 metric tons annually (21). GEMS data suggest that some Asian bodies of water, most noticeably in Indonesia and Malaysia, have extremely high levels of PCBs and some pesticides. (See Table 10.1.) A more complete picture of freshwater quality problems is available from recent reviews (22) (23) (24) (25).

Success Stories

For specific types of pollution, many countries are making substantial improvements in water quality. Most industrialized countries now treat domestic wastes. The Federal Republic of Germany, Switzerland, Denmark, and Sweden have made great strides; over 80 percent of their populations are connected to sewage treatment plants.

Table 10.1 Levels Of Chlorinated Hydrocarbons (Insecticides and PCBs) Detected by the Global Environmental Monitoring System Water Program, 1979–84

| Continent | Level of Contamination | | | |
	<10 nanograms per liter	10–50 nanograms per liter	100–1,000 nanograms per liter	1,000 nanograms per liter
North America	United States [12] Canada [5]			Colombia [1] (dieldrin, DDT)
Africa				Tanzania [1] (dieldrin)
Europe	Netherlands [6] UK [7] Finland [5]	UK [1] (DDT, aldrin, dieldrin, HCH) Finland [1] (DDT) Belgium [1] (DDE) Spain [6] (DDT)	UK [1]: PCB	
Asia	Thailand [3] Japan [5] Malaysia [5]	Thailand [1] (DDE) China [4] (HCH) Japan [3] (PCB)	Malaysia [1] (dieldrin) Indonesia [11] (PCBs)	
Oceania	Australia [1]			

Source: Global Environmental Monitoring System, *Assessment of Freshwater Quality* (United Nations Environment Programme and World Health Organization, Nairobi and Geneva, 1988), Table 25, p. 48.
Note: Numbers in brackets indicate the number of monitoring stations.

Figure 10.5 Nitrate Concentration Trends in Four United Kingdom Rivers

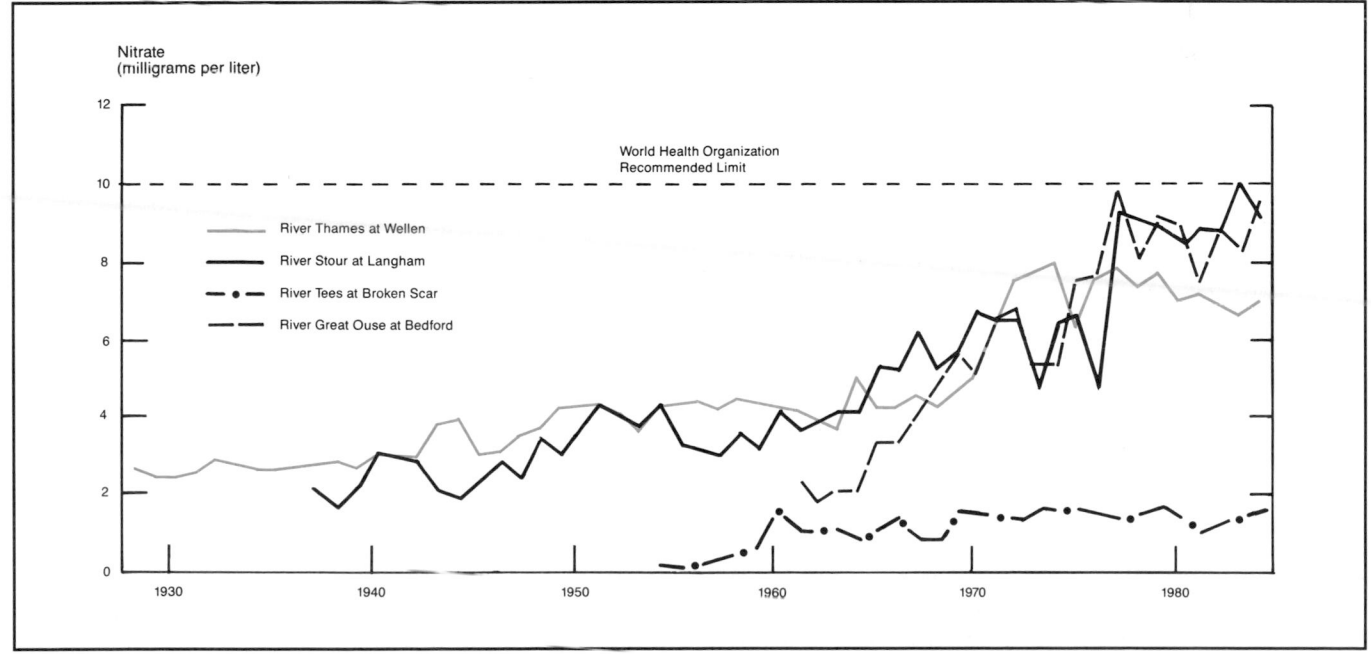

Source: G. Roberts and T. Marsh, "The Effects of Agricultural Practices on the Nitrate Concentrations in the Surface Water Domestic Supply Sources of Western Europe," in *Water for the Future: Hydrology in Perspective* (International Association of Hydrological Sciences 1987), Figure 5, reproduced in Michel Meybeck, Deborah Chapman, and Richard Helmer, eds., *Global Freshwater Quality: A First Assessment* (Blackwell Reference, Oxford, U.K. 1990), Figure 9-7, p. 127.

Sweden is a leader in the treatment of industrial effluents; the BOD released by pulp and paper mills into Swedish rivers dropped 50 percent over 10 years. Drastic treatment efforts and the introduction of industrial inplant water recycling caused significant decreases in heavy metal levels in the Rhine (26). The introduction of lead-free gasoline in the United States, following passage of the Clean Air Act in 1979, significantly lowered lead levels in U.S. rivers (27). Banning DDT in the early 1970s in the developed countries has led to documented decreases in pesticide concentrations in the Great Lakes and Lake Geneva (28).

The Future

Despite the progress in some areas, there are clearly grounds for growing concern about water quality. The GEMS attempt to assess global water quality has made evident the paucity of data in much of the world, particularly in the developing world (29). Additional monitoring efforts and more attention to water quality in general will be required to obtain an accurate picture of the state of the world's freshwater.

Nonetheless, some overall trends in water quality issues can be identified. In Organisation for Economic Co-operation and Development (OECD) countries, where regulations are strict, domestic and industrial effluents are now generally controlled. Yet gaps and accidents, such as the 1986 chemical spill in the Rhine and the oil spill in the Ohio River in 1988, continue to occur. Runoff from urban and agricultural areas is a major continuing problem, as is pollution from atmospheric sources (acid rain).

Industrialized, non-OECD countries, particularly Eastern Europe and the U.S.S.R., are experiencing extremely high levels of pollution from all sources— domestic, industrial, and agricultural—with industrial pollutants the most obvious and serious problem.

In developing countries, domestic sewage is a major problem, particularly in urban areas. Industrial pollution is a significant and growing problem where industry is concentrated; although industrial effluents are beginning to be targeted by regulatory agencies, most are not treated before they are dumped into water bodies. Runoff from the increasing and generally uncontrolled use of fertilizers and pesticides is likely to generate major water quality problems. Because large segments of the population in developing countries do not have access to treated water for drinking, bathing, and preparing food, polluted water poses a significant public health problem.

FRESHWATER QUANTITY

Water is one of the most pervasive substances on Earth—its total volume is 1.41 billion cubic kilometers (km^3) (30) (31). If this amount of water were

Table 10.2 Freshwater in Rivers, Volume by Continent
(cubic kilometers)

Continent	Freshwater in Rivers
Europe	76
Asia	533
Africa	184
North America	236
South America	946
Australia	24
Total	1,999

Source: Adapted from *Mirovoi vodnyi balans i vodnye resursy Zemli (The world water balance and the water resources of the Earth)* (Leningrad, Gidromoteoizdat, 1974).

spread evenly over Earth's surface, the water layer would be nearly 3,000 meters deep.

But 98 percent of this water can only excite the eye and the imagination, because it is the saltwater of the oceans, inland seas, and deep underground basins. The remaining 2 percent is freshwater. Nature guards jealously the greater part of this freshwater; 87 percent is locked in ice caps and glaciers and most of the rest is underground (32), in the soil, in the atmosphere, and in living things. Aside from lakes, only about 2,000 km^3 of freshwater, mostly in rivers, is available to satisfy human needs (33). These freshwater resources are unequally distributed among the continents (see Table 10.2) and within them.

Human demands for water are growing rapidly, in part because of population growth, but also because people are using more water per capita every year. At present, almost 10 km^3 of freshwater is withdrawn daily; the annual volume used is almost twice the store of freshwater in the rivers at any given instant. What makes this storage possible is a grand yet simple process—the water cycle.

With the Sun's warming of Earth, water evaporates from the surface of land and sea into the atmosphere. The moisture is then transported great distances before it falls back to earth as precipitation. Once back in the terrestrial part of the cycle, it either flows off the surface or permeates the ground. The runoff helps to replenish rivers and lakes. The water in the ground is taken up and transpired by plants, evaporated from the soil surface or percolates downward to the water table where it is stored in groundwater aquifers. Rivers carry water to lakes and seas, and the process renews itself.

The replacement times for water at various stages in the cycle differ greatly. The global average for water in rivers is 18–20 days, but the atmospheric moisture is replaced even more rapidly—every 12 days. Deep groundwater requires several hundred years or more to renew (except for fossil water, which does not renew). With rivers renewing so rapidly, humans have access not only to 2,000 km^3 of river water each year but to more than 40,000 km^3.

The components of the water cycle are closely interrelated; they cannot exist without renewing others and in turn being renewed by them. Hence any change at one stage of the cycle inevitably has an impact on the others. In addition, through evaporation, water rids itself of numerous substances contained within it. Thus, those water resources with shorter periods of renewal contain less salt, giving rise to the description of the water cycle as the "great distill-ator."

FRESHWATER RESOURCES

Water resources are those that can be used by people. They include practically all the waters of Earth: river, lake, sea, and groundwaters, soil moisture, glacial ice, and atmospheric vapor. They may be fresh or mineralized (salinated). The latter is difficult to use directly without removing the salts, an expensive and highly energy-consumptive process. (See *World Resources 1987*, pp. 121–122.)

An important characteristic of water resources is their variability over time. Seasonal fluctuations of river runoff can be extreme, for example, a factor influencing utilization strategies. More regular distribution can sometimes be created with water storage basins.

Traditionally, freshwater resources are divided into nonrenewable fixed reserves and renewable resources. The division is somewhat artificial because all water is conserved and is thus renewable, although not all on a human time scale. Thus, if the amount of water locally withdrawn from rivers and underground sources does not exceed its renewal,

humans can use all sources of freshwater indefinitely. It is important to distinguish between withdrawals and consumption. Much of the water that is withdrawn is later returned to its source, say, a river and thus can be withdrawn several times during its journey to the sea. Consumptive uses of water include evaporation from industrial cooling towers or irrigation systems and are usually far smaller than water withdrawals.

Sources of River Runoff

River runoff is the most widely studied water resource. The water in rivers is derived from precipitation. Rainfall may run off directly to streams or recharge groundwaters, which eventually discharge to a stream. Precipitation may be temporarily stored as snow or ice before it is released to streams in the spring and summer (34) (35).

River basins fed mainly by rainfall occupy 60 percent of the land mass and support 90 percent of the world population. These rivers are found in ecosystems ranging from the most humid equatorial regions, where frequent flooding takes place on high-water rivers, to tropical deserts, where once in several years a local runoff creates small streams. Snow is the primary source for river water in 25–30 percent of the world, mainly in the U.S.S.R., Scandinavia, Canada, and Alaska. Rivers fed mainly by glaciers are found in mountainous areas such as the Alps, the Caucasus, high-mountain regions in Middle and Central Asia and Alaska, and the Andes, espe-

Table 10.3 River Runoff and Its Use

(cubic kilometers per year)

Elements	Europe	Asia	Africa	North America	South America	Australia Oceania	U.S.S.R.	World Total
Total River Runoff	2,321	10,485	3,808	6,945	10,377	2,011	4,350	40,673
Groundwater Discharge to Rivers	845	2,879	1,464	2,222	3,736	483	1,020	12,689
Surface Runoff	1,476	7,606	2,720	4,723	6,641	1,528	3,330	27,984
1980s								
Water Withdrawal	364	1,591	176	767	161	26	443	3,528
Consumptive Use	134	1,145	146	339	110	19	239	2,120
Waste and Returning Waters	230	446	42	428	51	7	204	1,408
Use of Resources (percentage of total river runoff)	16	15	4	11	2	1	10	9
Year 2000 Projection								
Water Withdrawal	404	2,160	289	946	293	35	533	4,660
Consumptive Use	158	1,433	201	434	165	22.5	286	2,699
Waste and Returning Waters	246	727	88	512	128	12.5	247	1,960
Use of Resources (percentage of total river runoff)	17	21	7	14	3	2	12	11

Sources:
1. A.V. Belyaev, U.S.S.R. Academy of Sciences, *Institute of Geography*, Moscow.
2. *Vodnyi balans S.S.S.R. i ego preobrazovaniya (Water budget of the U.S.S.R. and its transformations)* (Moscow, Nauka, 1969).
3. G. Ya. Karasik, *Vodnyi balans Afriki (Water budget of Africa)* (Moscow, 1970).
4. G.M. Chernogayeva, *Vodnyi balans Yevropy (Water budget of Europe)* (Moscow, 1971).
5. G. Ya. Karasik, *Vodnyi balans Yuzhnoi Ameriki (Water budget of South America)* (Moscow, 1974).
6. G.M. Nikolayeva and G.M. Chernogayova, *Vodnyi balans Azii (Water budget of Asia)* (Moscow, 1977).
7. N.N. Dreyer, *Vodnyi balans Severnoi Ameriki (Water budget of North America)* (Moscow, 1978).

Figure 10.6 Global River Runoff

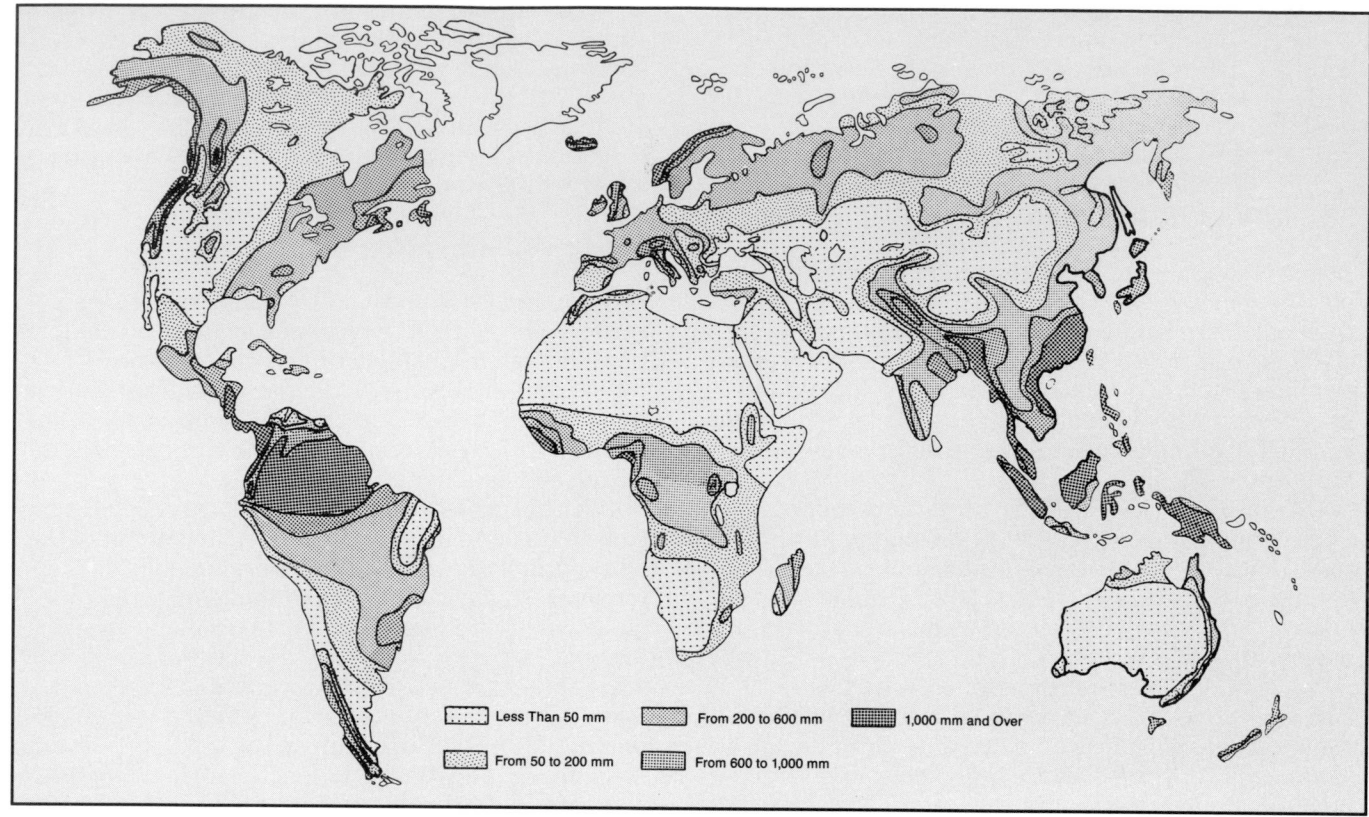

Less Than 50 mm   From 200 to 600 mm   1,000 mm and Over
From 50 to 200 mm   From 600 to 1,000 mm

Source: Institute of Geography, U.S.S.R. Academy of Sciences, Moscow.

cially in the lower part of the Patagonian Cordilleras. Groundwater discharge predominates in foothill regions around Middle and Central Asia. Here water percolates downward, forming more or less abundant groundwaters that feed rivers called karasus. Karasus are not water abundant, but they are relatively stable and, what is most important, they bring water to arid regions around the mountains.

The annual river runoff worldwide is estimated at 31,000–47,000 km^3 (36) (37) (38), with an annual variation of not more than 6 percent (39). Table 10.3 shows that South America has the most abundant runoff (26 percent of the world total) of any continent, Europe (6 percent) and Australia, New Zealand/Oceania (5 percent) the least.

South America enjoys the most abundant river flows. If the ratio of total river runoff to total land area is used as an index, South America has twice the runoff of all the other continents taken together. Africa, with a runoff index only half the global value, has the least abundant flows. Rivers in North America and Asia correspond to the mean; Europe's indexes are somewhat higher.

Within the continents are great contrasts as well. Distribution of world river runoff is shown in Figure 10.6. The map shows that runoff depends on geographic zones. Values are the highest in the humid equatorial belt, especially in the basin of the Amazon, the largest river in the world; in the near-equa-

tor regions of Africa, where the largest African rivers collect their water; and in Southeast Asia, where monsoons of the Indian and Pacific oceans are intense. River runoff in the temperate belt (Europe, Asia, and North America) is not as diverse as in the tropical regions, although exceptional climatic conditions such as those found on the northwestern coast of Scandinavia or the deserts of Central Asia may create more extreme runoff. In the subarctic belt, river runoff is fairly low.

Although overall river runoff in a country may be high, seasonal fluctuations may result in water scarcities during parts of the year. Thus, from an economic point of view, groundwater storage is particularly important because it remains stable over time (40).

The world distribution of groundwater discharge to streams (see Figure 10.7) is similar to overall river runoff (41). For example, in the taiga and mixed forests of the temperate belt, the groundwater discharge is between 40–47 percent of the total runoff. In the permanently humid evergreen forests of the equatorial belt (tropical forests), its share of the total river runoff reaches 50 percent.

Forest regions generally have a more stable pattern of river runoff, thanks to a higher proportion of groundwater discharge to surface runoff that is due to the catchment capabilities of the forest ecosystem. Deforestation may therefore lead to significant changes in patterns of river runoff, with

Figure 10.7 Global Groundwater Flow to Rivers

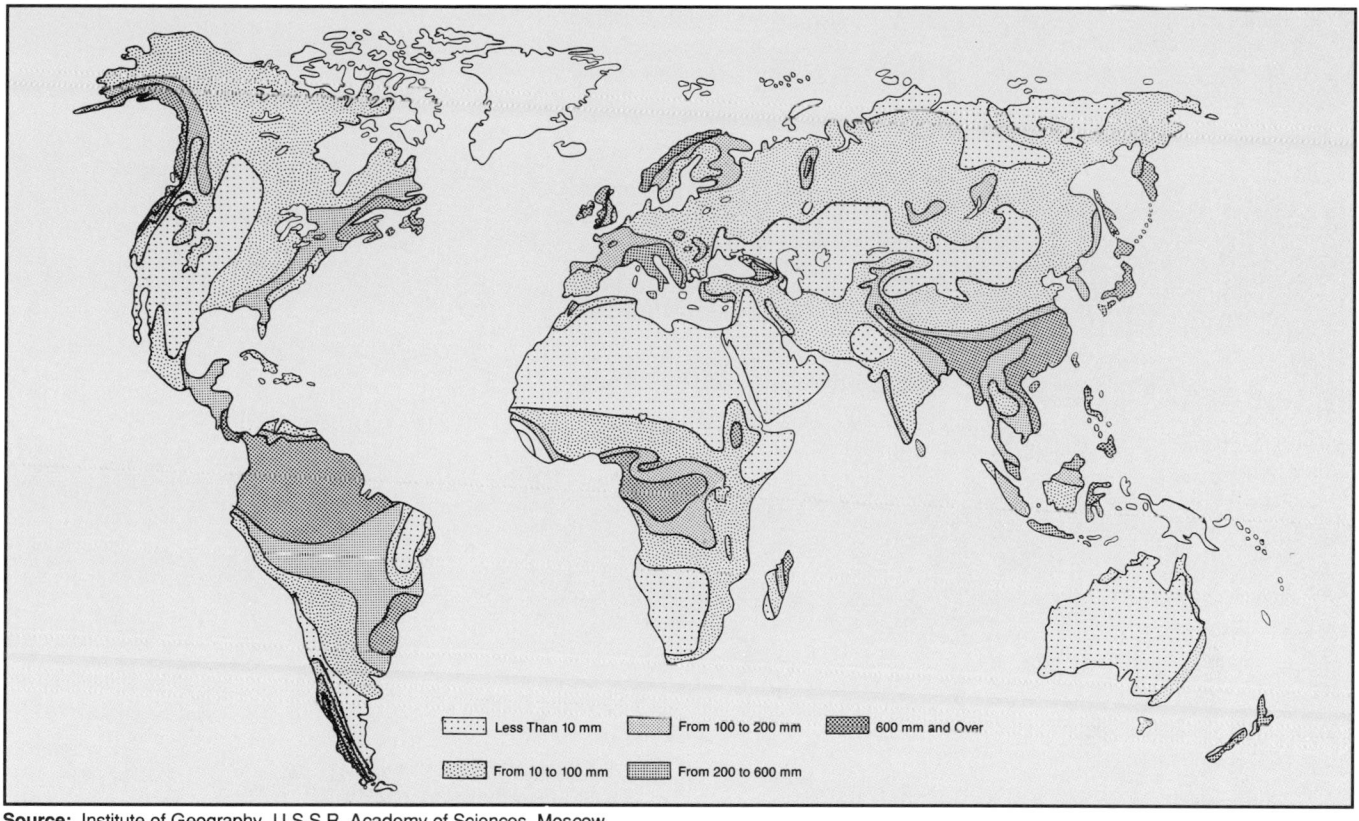

Less Than 10 mm From 100 to 200 mm 600 mm and Over

From 10 to 100 mm From 200 to 600 mm

Source: Institute of Geography, U.S.S.R. Academy of Sciences, Moscow.

Figure 10.8 Global Variations in Soil Moisture

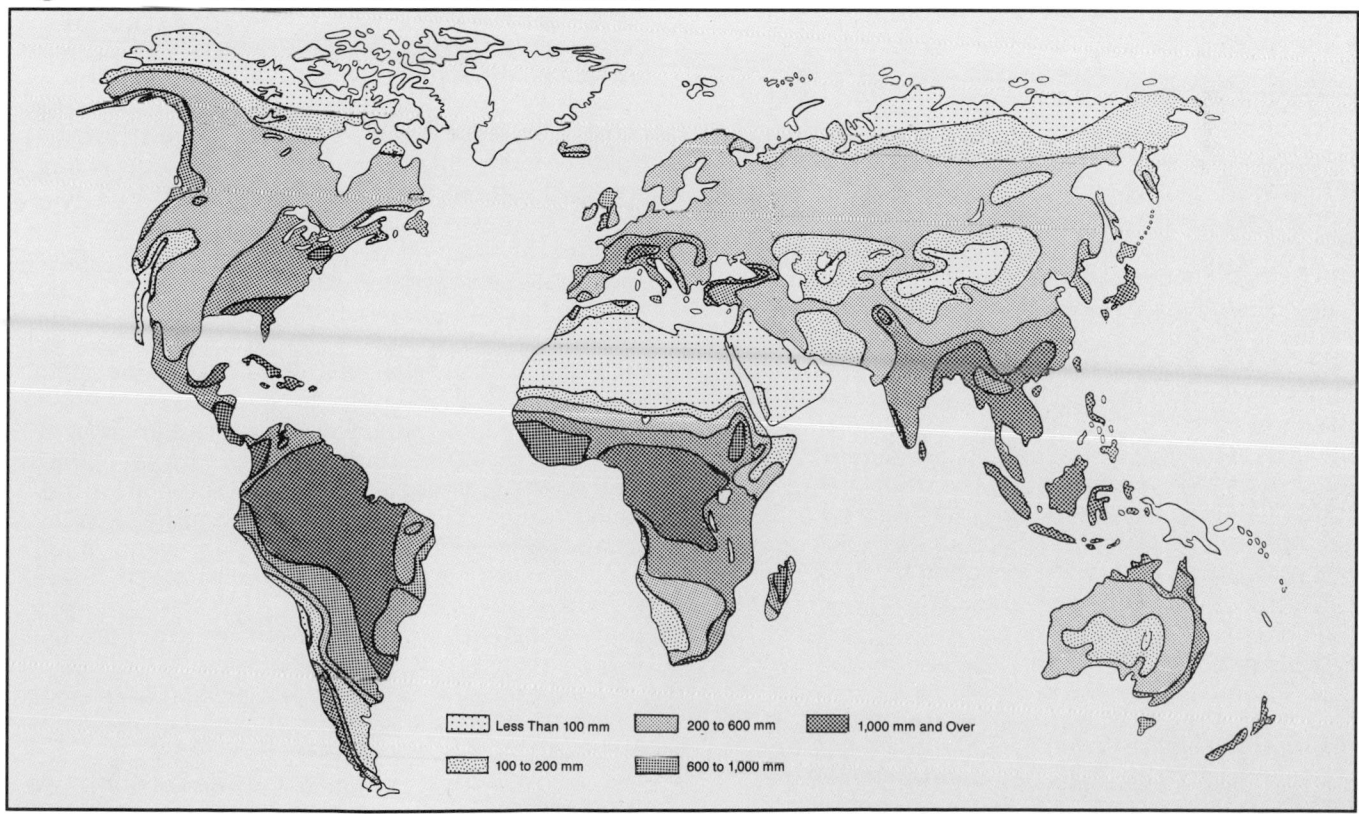

Less Than 100 mm 200 to 600 mm 1,000 mm and Over

100 to 200 mm 600 to 1,000 mm

Source: Institute of Geography, U.S.S.R. Academy of Sciences, Moscow.

Box 10.1 Water Storage Basins

The efficient use of river runoff is difficult because of seasonal and climatic irregularities in flow. Flooding and prolonged periods of low water levels can cause problems of catastrophic proportions. Water-storage basins help; they store water in times of surplus and disperse water in times of scarcity. They also prevent flooding.

But by definition large-scale dams flood selected areas, some ecologically sensitive lands; they displace human and wildlife populations and slow the water cycle, resulting in siltation, eutrophication, and changed microclimates (1).

More than 30,000 water storage basins have been built around the world, 75 percent in the past 35 years. These basins have a total filled capacity of about 6,000 cubic kilometers (km^3) (2) and their total area comprises 400,000 square kilometers (km^2), an area the size of Norway. The latter figure represents only the dam area proper; it doubles with inclusion of the flooded lake area behind the dam.

In Africa, the storage basin at Lake Victoria has a total volume of 200 km^3 and a total surface water area, including the lake, of 76,000 km^2. This dam is the largest water storage basin in the world. The six largest storage basins in the world are in Africa.

In the U.S.S.R., more than 1,250 storage basins have been built, and the Volga, Kama, Dnieper, Yenisei, and other large rivers have been harnessed into cascades of storage basins. The current rapid rate of water storage basin construction globally may well continue because of increasing demand for hydropower, flood regulation, and basic water supplies (3).

The negative side of large storage basins includes rates of high evaporation. According to estimates prepared by the Institute of Geography of the Soviet Academy of Sciences, the water evaporated from the surface of storage basins is about 240 km^3 per year. The U.S.S.R., the United States, and Canada account for

over 40 percent of this loss, and, combined with China, Brazil, India, Iraq, and Egypt, it accounts for over 75 percent. By country, the U.S.S.R. holds the record for water evaporation rates at 37.9 km^3 per year, with the United States a close second at 34.8 km^3.

References and Notes

1. *Vodokhranilishcha i ikh vozdeistviye na okruzhayushchuyu sredu (Water storage basins and their impact on the environment)* (Moscow, Nauka, 1986).
2. A.B. Avakyan, *et al., Vodokhranilishcha* (Water storage basins) (Moscow, Mysl, 1987).
3. For a review of forecasts, see I.A. Shilkomanov and O.L. Markova, *Problemy vodoobespecheniya i perebroski rechnogo stoka v more (The Problems of Water Availability and Transfer of River Runoff in the World)* (Leningrad, Gidrometeoizdat, 1987), p. 63.

higher surface runoff rates and flooding in wet seasons and a greater likelihood of dried-up rivers in dry seasons.

Moisture absorbed by soil and then released through evaporation and transpiration in plants or stored as groundwater is another source of freshwater. This process provides vegetation with water and is essential to agricultural production. Figure 10.8 shows how soil moisture varies with geographical zone.

The desert areas of the temperate and tropical belts and the polar deserts and tundra of the far north are the most arid, the former from a lack of moisture, the latter from a lack of heat. In virtually all regions with soil moisture of less than 300 millimeters (mm) per year, vegetation is scarce and the productivity of biomass is less than 3,000 kg per hectare (ha) (42). Under these conditions, only irrigated cultivation is possible.

In the temperate zone, soil moisture may reach 550 mm per year and biological productivity is high, between 10,000 and 12,000 kg per ha. Soil moisture is greatest in the equatorial belt, where the productivity of permanently wet forests and tropical forests exceeds 40,000 kg per ha. In South America, for example, soil moisture averages almost 1,500 mm per year, which is equal to irrigation levels of 15,000 cubic meters (m^3) per ha.

FRESHWATER USE

During the past three centuries, the amount of water withdrawn from freshwater resources by mankind has increased more than 35-fold, reaching the current rate of more than 3,500 cubic km^3 per year in

1970 (43). In the second half of the 20th Century, water withdrawals increased 4–8 percent annually. Water use is still growing in the developing world but is stabilizing in the industrial countries, with the result that the withdrawal rate increase is slowing. Withdrawals are expected to rise only 2–3 percent annually from now until year 2000.

Local river runoff is used most intensively in the Netherlands, Belgium, East Germany, Romania, southern and southwestern U.S.S.R., the United States, Cuba, the Eastern Mediterranean, Iran and the south of the Arabian peninsula, Pakistan, the northwestern states of India and those adjoining the Ganges, the northern and eastern provinces of China, and Japan. In parts of these regions, withdrawals are nearly equal to the locally generated river runoff. A similar situation is rapidly developing in Eastern Europe, western Germany, Spain, Italy, the Sudan, Mali, Morocco, South Africa, the Far and Near East, the southeastern and northwestern provinces of China, and the southern part of India.

The river runoff resources of the Scandinavian countries, the northern and eastern regions of the U.S.S.R., the greater part of Canada and Alaska, the Himalayan and Tibetan mountains, Southeast Asia, and the equatorial and subequatorial regions of Africa and South America substantially exceed current withdrawals.

Human activities affect levels of river runoff principally by the direct withdrawal of water but also by the regulation of river runoff and by land uses that change the surrounding environment and affect watershed dynamics (44) (45). (See Box 10.1.)

Of the 3,500 km^3 of water withdrawn for human use each year, some 2,100 km^3 of water are for consump-

Box 10.2 The Dying Aral Sea

The Aral Sea is an inner basin in the sands of the Turan lowland within the Soviet Union. Until 1960, the Aral was the fourth largest inland body of water in the world—smaller only than the Caspian Sea (between the U.S.S.R. and Iran), Lake Superior (in the United States), and Lake Victoria (in East Africa). The Aral had covered an area of about 67,000 square kilometers (km^2) and contained 1,050-1,100 cubic kilometers (km^3) of water, but over the past 30 years, the sea lost 40 percent of its area and 67 percent of its volume (1).

At equilibrium, the sea received 60–65 km^3 of water per year, of which 90 percent came from the runoff of the two largest rivers of Soviet Central Asia, the Amu Darya and the Syr Darya. Under these conditions, evaporation was equal to precipitation and river runoff. The mean level of the Aral Sea, averaged over many years, remained at about 53 meters above sea level, rising or falling only 1–1.5 meters, depending on regional rainfall.

After completion of the first section of the Karakum Canal in 1959, which withdrew Amu Darya water for irrigating the piedmont agricultural regions of Turkmenia, an extensive development of irrigated fields began in Uzbekistan, Turkmenia, and Kazakhstan. From 2.9 million hectares (ha) in the 1950s, the area of irrigated lands has grown to 7.5 million ha. The principal diversions of water are into the Karakum Canal (which withdraws at least 12 km^3 annually); the Amu-Bukhar and Karshinsky canals (5 km^3

each) on the Amu Darya River; and irrigation systems of the Cergana Valoley, the Golodnaya, and the Dzhizakskaya steppes on the Syr Darya. Along the whole length of both rivers, considerable amounts of water are withdrawn by individual consumers.

As a result, by 1990, the level of the Aral is expected to drop to about 39 meters above sea level, or 14 meters below its 1960 level. The sea will then cover less than 40,000 km^2, and its volume will be reduced to about 350 km^3. Depletion of the sea by human use was aided by a relatively dry regional climate in the first half of the 1980s. An analysis by the Institute of Geography of the U.S.S.R. Academy of Sciences indicates that 80 percent of the Aral depletion is due to economic activity and 20 percent by the natural climate variations.

With shrinkage to nearly one third of the Aral's original volume, its mineral (including salt) concentration nearly tripled. The result is almost a complete extinction of basin animal life and a profound disturbance of the marine ecosystem. The surrounding land area has also been greatly disturbed, with 30,000 km^2 of the former sea bottom turned into desert, intensive wind erosion, and large quantities of salt and dust blown into the atmosphere. The width of the dried-up zone in some places reaches 80 km, so that the former maritime towns of Muinak and Aralsk, with populations of several thousands, are now in a desert. The drop in sea level resulted in the disappearance of numerous lakes and

swamps in the Amu Darya and Syr Darya deltas, where 75–80 percent of the animal species became extinct.

In the Aral region, drinking water is a crucial problem. Human health, unemployment due to the closing of traditional marine trades, and other social problems are acute. In 1987, river runoff reaching the Aral totaled about 11 km^3; in 1988, it increased to more than 20 km^3. Perhaps dry years are at an end. Yet, even if the high level of rainfall persists in the coming years, stabilizing the sea level requires no less than 35 km^3 of freshwater per year. Under current conditions, that seems unrealistic; more likely is the continued retreat of the Aral for another seven or eight years before it stabilizes at 20 meters or more below the 1960 level.

A resolution adopted by the Central Committee of CPSU and the Soviet Government in September 1988 is directed at improving economic and ecological conditions for people in the Aral region and preserving the Aral Sea, even if in a form more contracted than at present.

The Aral Sea may never regain its position as the fourth largest closed basin of the world. But it could become a landmark of disturbing environmental effects of a deeply erroneous agricultural policy.

References and Notes

1. Official Report of Government Committee on Aral Sea Regions, *Meteorology and Hydrology*, 1988, Vol. 9.

---

tive use; the remaining 1,400 km^3 of wastewater are returned to rivers and other water sources, frequently in a polluted condition. (See Table 10.3.) This returned water is often reused farther downstream.

How much of the use is consumptive and how much is returned as wastewater? The answer depends on the particular end use of the water. (See Table 10.3.) In Europe, for example, industrial and energy production use dominates and the consumptive use is relatively small (37 percent of withdrawals). As a result, Europe produces relatively large flows of wastewaters; in fact, Europe, North America, and the U.S.S.R. together produce 55 percent of the world's wastewaters. In Africa and Asia, substantial withdrawals for irrigation result in greater consumptive use (71 and 76 percent, respectively, of withdrawals). By the year 2000, developing countries are projected to increase industrial and energy water use and to decrease the relative proportion of consumptive use to 56–70 percent. (See Table 10.3.)

Wastewater is sometimes treated before it is returned to rivers and other freshwater resources, but

all too often it is returned without treatment. Even when it has been treated, wastewater usually needs to be diluted to reduce concentrations of pollutants still further before the water is fit for reuse. The potential for dilution of wastewater is increasingly limited because as water withdrawals increase, the amount of clean water remaining decreases and the volume of wastewater grows. Globally, wastewater can now be diluted by clean water in a ratio of 1:25. In Europe, the ratio is 1:8; in Australia and New Zealand, it is 1:250. By the year 2000, these ratios are projected to worsen slightly in Europe and substantially in Asia.

The structure of the water economy differs among countries, depending on natural climatic conditions; the availability, accessibility, and quality of water resources; and the economic and social development of a country. The principal uses are agricultural, domestic, and industrial.

Agricultural Use

Irrigation was the basis of life as far back as the ancient civilizations of Egypt, Mesopotamia, India, and

Table 10.4 Water Withdrawals for Irrigation

	1980s				Year 2000 Projection			
Region	Areas of Irrigation (million hectares)	Water withdrawal (cubic kilometers)	Consumptive use (cubic kilometers)	Recycled water (cubic kilometers)	Areas of irrigation (million hectares)	Water withdrawal (cubic kilometers)	Consumptive use (cubic kilometers)	Recycled water (cubic kilometers)
Europe	17	110	95	15	19	125	105	20
Asia	140	1,300	980	320	165	1,500	1,150	350
Africa	11	120	85	35	15	160	110	50
North America	29	330	215	115	35	390	260	130
South America	8.5	70	55	15	11	90	70	20
Australia and Oceania	2.0	16	13	3	2.5	20	15	5
U.S.S.R.	20	260	180	80	23.5	300	210	90
World Total	227.5	2,206	1,623	583	271	2,585	1,920	665

Source: A.V. Belyaev, U.S.S.R. Academy of Sciences, Institute of Geography, Moscow in consultation with other international sources.
Note: Oceania is defined as Australia, Fiji, New Zealand, Papua New Guinea, and Solomon Islands.

China. Some form of it is practiced today in every country on the planet. Irrigation is extremely water intensive—to grow one-half metric ton of grain per person—enough grain to supply 50 percent of a person's diet for 1,000 days—can require as much as 1,700 km^3 of water per capita per year (46). In Asia, irrigation comprises 82 percent of total water withdrawals; in the United States, 41 percent; in Europe, 30 percent. (See Tables 10.3 and 10.4.) Water withdrawals for irrigation can have very destructive environmental effects. (See Box 10.2.)

As the tables show, water withdrawals for irrigation are declining compared with other uses; worldwide, today's 63 percent is projected to decline to 55 percent by the year 2000. This change in water withdrawal patterns will occur mainly in Asia, Africa, and Latin America because of increased industrial withdrawals. Asia has by far the largest area of land under irrigation, 140 million ha versus 29 million ha in North America, which is the second largest user. China, India, and Pakistan account for most of Asia's irrigation water withdrawals, with 1,300 km^3, more than half the world total.

The total area of irrigation outside the Asian continent is nearly 90 million ha and requires about 900 km^3 of water annually, or 42 percent of the total withdrawals for irrigation. (See Tables 10.3 and 10.4.) The largest areas of irrigation are concentrated in the United States, the U.S.S.R., and Mexico. Irrigation is not widely practiced in Europe, with the exception of Italy, Spain, Romania, and Bulgaria in the south and east.

In Africa, irrigation is most widespread in the Nile basin, primarily in Egypt and the Sudan. Madagascar, South Africa, and many countries in North Africa also irrigate extensively.

As mentioned above, the United States and Mexico dominate the use of irrigation water in North and Central America; Cuba and western Canada are also heavy users. But, despite their use of irrigation, these countries do not have an overall water deficit.

In South America, the largest areas of irrigation are concentrated in Argentina and Chile. Initially, irrigation on this continent developed in the temperate zones, but in the 1950s, its use spread to the tropical belt as well, where today the area irrigated is twice that in the temperate zone (47).

Australia has one of the highest ratios of irrigated land per capita in the world.

Until recently, irrigation contributed little to the pollution of water basins. But the wide application of fertilizers and pesticides has heavily polluted irrigation return flows, presenting a significant threat to the aquatic environment.

The importance of livestock to water quality and quantity problems would seem to be less than other sectors of the economy. Its total use is only 65 km^3 per year (2 percent of the total). Nevertheless, the industry produces large amounts of water polluted by organic wastes (48) and represents a substantial proportion of total water use in many developing countries—40 percent in Mongolia and over 60 percent in Lesotho, Mauritania, Namibia, Bhutan, Paraguay, and Brazil.

The use of irrigation is likely to continue to grow, but not anywhere near the rates experienced over the past century nor as fast as predicted (49). It may well increase in Africa and also in South America about 30 percent by the year 2000 and between 15–18 percent in Asia, North America, and the U.S.S.R. The total irrigated area is projected to increase 19 percent, or about 44 million ha. (See Table 10.4.)

Projections of future water use in this and subsequent tables are based on a detailed model developed at the Institute of Geography in Moscow. The model considers such factors as population growth

Table 10.5 Domestic and Municipal Water Consumption

Region	1980s				Year 2000 Projection			
	Population (millions)	Water withdrawal (cubic kilometers)	Consumptive use (cubic kilometers)	Waste water (cubic kilometers)	Population (millions)	Water withdrawal (cubic kilometers)	Consumptive use (cubic kilometers)	Waste water (cubic kilometers)
Europe	496	48	10	38	512	56	8	48
Asia	2,932	88	53	35	3,612	200	100	100
Africa	589	10	7	3	853	30	18	12
North America	411	66	20	46	489	90	22	68
South America	279	24	14	10	367	40	20	20
Australia and Oceania	26	4.1	1.2	2.9	30	5.5	1.5	4
U.S.S.R.	282	23	5	18	310	35	5	30
World Total	5,015	263.1	110.2	152.9	6,173	456.5	174.5	282

Source: A.V. Belyaev, U.S.S.R. Academy of Sciences, Institute of Geography, Moscow, in consultation with other international sources.
Note: Oceania is defined as Australia, Fiji, New Zealand, Papua New Guinea, and Solomon Islands.

Table 10.6 Water Use in Industry
(cubic kilometers)

Region	1980s			Year 2000 Projection		
	Water Withdrawal	Consumptive use	Waste Water	Water Withdrawal	Consumptive use	Waste Water
Europe	193	19	174	200-300	30-35	170-175
Asia	118	30	88	320-340	65-70	255-270
Africa	6.5	2	4.5	30-35	5-10	25
North America	294	29	265	360-370	50-60	310
South America	30	6	24	100-110	20-25	80-85
Australia and Oceania	1.4	0.1	1.3	3.0-3.5	0.5	2.5-3.0
U.S.S.R.	117	12	105	140-150	20-25	120-125
Total	759.9	98.1	661.8	1,153-1,308.5	190-225.5	962.5-993

Source: A.V. Belyaev, U.S.S.R. Academy of Sciences, Institute of Geography, Moscow.

and trends in economic development, water use levels, water resources, and irrigation and energy use factors by country.

Domestic and Municipal Use

Domestic water use includes water for drinking, food preparation, sanitation, washing, cleaning, watering gardens, and the service industry (e.g., laundries, pools, heating systems, restaurants, medical services).

Domestic and municipal water needs have always been modest in relation to others, accounting for only about 7 percent of total withdrawals. Domestic use comprises 13–16 percent of withdrawals in Europe, South America, and Australia/Oceania; by the year 2000, domestic use will increase to 9 or 10 percent of withdrawals in Asia, Africa, and North America and 14–16 percent in Europe, South America, and Australia/Oceania. (See Table 10.5.)

Although the quantity of water required for domestic needs is not large, the quality must be high. Domestic water use tends to increase with living standards; it may be as low as 20 liters or even more than 500 liters per person per day.

Present withdrawals of water for domestic use exceed 250 km[3] a year. Only about 4 percent of the population uses as much as 300–400 liters per day per person, and two thirds of the population, concentrated in Africa and Asia, use less than 50 liters per day. By the year 2000, a projected 17 percent of the population will be using more than 300 liters per day, but 30 percent, some 1.8 billion people, will still be using fewer than 50 liters per day.

Domestic and municipal water use is highest in the United States—53 km[3] a year. In Canada and Switzerland, domestic water use is also high. Perhaps the most serious problem occurs in developing countries—providing potable water for rapidly growing populations—a problem that will require vast amounts of capital. (See Chapter 5, "Human Settlements," Financing Water and Sanitation.)

An increase in domestic water use is accompanied by an increase in the amount of wastewater, which usually comprises not less than 70 or 80 percent of water withdrawals (50) (51) (52). With water use per capita expected to increase substantially in developing countries with high economic or population growth rates, the volume of wastewater requiring treatment is expected to present a growing problem. By the year 2000, for example, domestic water use in Asia is projected to reach 150 liters per person per day. The wastewater generated will exceed that generated today by the combined populations of Europe and North and South America. Accordingly, wastewater treatment techniques should have a high priority (53) (54).

Industrial Use

Industry uses substantial amounts of water for cooling, processing, cleaning, and removing industrial wastes. With industrial use, most of the water is returned to the water cycle, but it is often heavily polluted with chemicals and heavy metals.

Use in energy production and other industry, comprising 21 percent of total water withdrawal, is expected to increase to 24–28 percent by the year 2000. (See Tables 10.3 and 10.6.) Industrial water use is most marked in Europe and North America. Water use differs widely by industry (55). For example, producing 1 metric ton of linen requires about 250 m^3 of water (not counting the water required to grow the flax); producing 1 metric ton of synthetic fibers requires 2,500–5,000 m^3. The chemical and metal industries require substantial water inputs, for example, 1,000 m^3 to produce 1 metric ton of ammonia and 2,000 m^3 per metric ton of synthetic rubber; smelting 1 metric ton of nickel requires 4,000 m^3 of water. The amount of water needed depends on the technology used as well as climatic conditions; use is generally higher in warm climates.

Worldwide, the water withdrawn for industry and energy production now totals 760 km^3, which is second to irrigation. Only in Europe, where irrigation is relatively less common, does industrial water use equal the other uses taken together.

Water withdrawals for industrial use in North America have increased almost 20-fold during this century (56). Present industrial water use is greatest in the United States and the U.S.S.R. (260 and 117 km^3 per year, respectively), together comprising half the world's industrial water use. Next come Japan (37 km^3 per year), the Federal Republic of Germany (35 km^3), and China (35 km^3). Table 10.6 shows estimates for current and future industrial water use.

Wastewater from industrial use now comprises 660 km^3 per year, or 87 percent of total water withdrawals by this sector. By the end of the 20th Century, water withdrawals in Asia, Africa, and Latin America

Figure 10.9 River Runoff Changes in the Holocene Optimum (6,200-5,300 Years Ago)

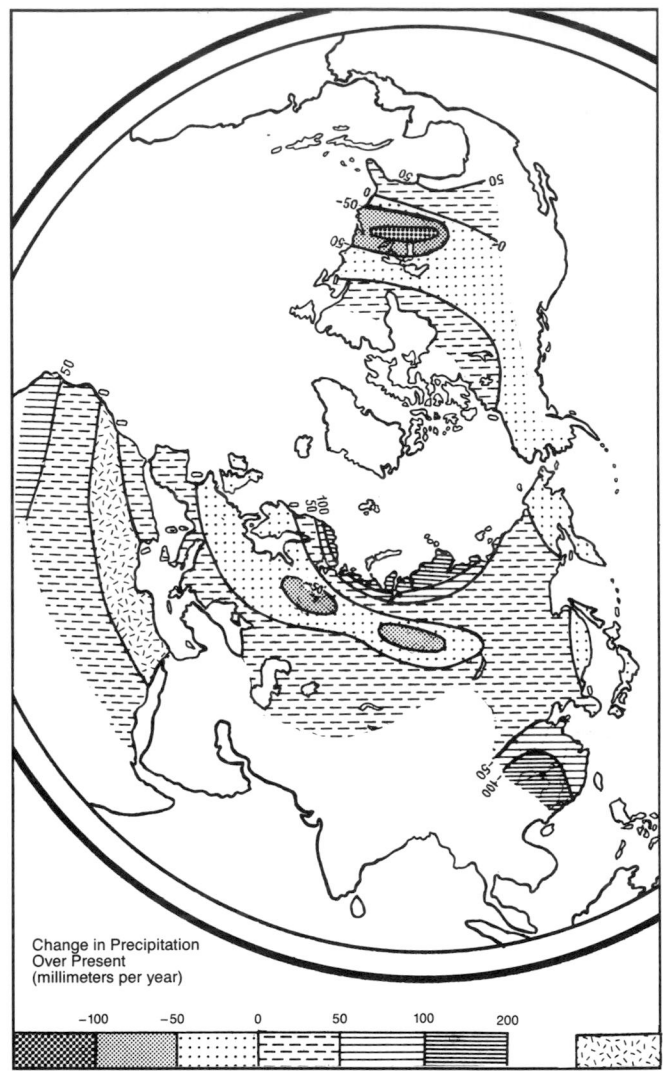

Change in Precipitation Over Present (millimeters per year)

| −100 | −50 | 0 | 50 | 100 | 200 |

Source: U.S.S.R. Academy of Sciences, Institute of Geography, Moscow.
Note: Runoff exceeded today's levels in both high and low altitudes, with changes similar to those for precipitation.

are projected to increase three- to five-fold, and in the developed world, 10–25 percent. Wastewater will increase by the same proportions—underscoring the need for regulation and control.

FOCUS ON GLOBAL CLIMATE CHANGE

Global warming is not yet a proven phenomena, but the world may be at the threshold of a period of rapid change in climate. (See Chapter 2, "Climate Change: A Global Concern.") If such a warming does occur, it is likely to affect greatly the flux of water through the hydrological cycle and hence the world's freshwater resources. A warmer climate would lead to increased evaporation from the sea and hence probably to increased precipitation and

Figure 10.10 River Runoff in the Mikuline Interglacial Optimum (125,000 Years Ago)

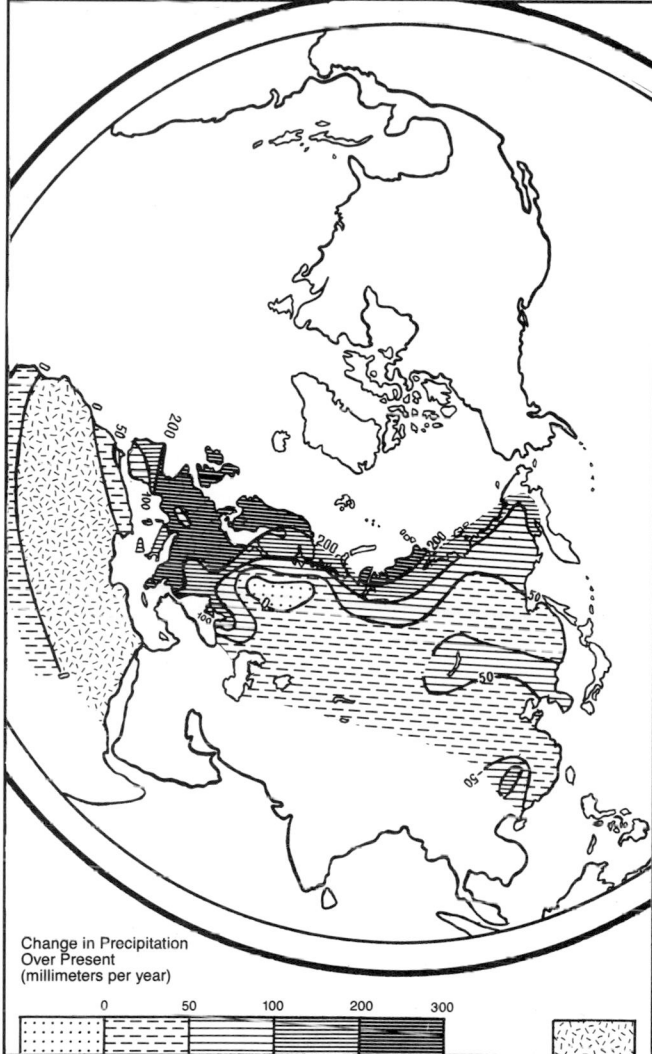

Change in Precipitation
Over Present
(millimeters per year)

0   50   100   200   300

Source: U.S.S.R. Academy of Sciences, Institute of Geography, Moscow.
Note: In this warmer climate, runoff exceeded present levels in Europe and much of Asia; levels for much of Africa have not changed.

river runoff, globally, but with quite varied and uncertain regional and local changes.

The changes could be quite significant. An increase in the mean global atmospheric surface temperature of 0.5°C, for example, could increase annual atmospheric precipitation as much as 10 percent. Moreover, according to some analyses, precipitation is likely to increase mainly at high and low latitudes in continental regions of the Northern Hemisphere and to decrease in middle latitudes—regions of intensive agricultural production (57). Thus, changes in temperature and precipitation could decrease soil moisture and annual river runoff in large areas of the Northern Hemisphere.

Estimating hydrological effects of changed climates is done in two ways. One method is to use general circulation models of the atmosphere (see

Chapter 13, "Global Systems and Cycles") to predict precipitation patterns and amounts. The second is to reconstruct precipitation patterns with the help of paleoclimatic data from past geologic eras in which comparable climatic conditions prevailed. Such reconstructions can give an independent indication of what selected climate changes may mean for the hydrological cycle.

In particular, the following geological periods provide useful analogs to stages of global temperature increases 1–5°C: the Holocene optimum (6,200–5,300 years ago), a 1–2°C increase; the Mikulino interglacial optimum (125,000 years ago), a 2–3°C increase; and the Pliocene optimum (4,300,000–3,300,000 years ago), a 3–5°C increase (58). Maps of annual and seasonal values of atmospheric precipitation and temperature in the Northern Hemisphere have been reconstructed for these periods with paleoclimatic data.

During the Holocene optimum, precipitation was greater than at present in high and low latitudes, by 50–100 mm per year in high latitudes and 200–300 mm farther south, for example, in the Sahara. In the middle latitudes of Eurasia (50–60° north latitude) and North America (30–60° north latitude), however, were large areas in which precipitation decreased (59). River runoff changed in comparable ways. (See Figure 10.9.) In the main agricultural zone of Europe, the river runoff was lower than at present by only about 50 mm. In the southern part of Europe, runoff was near present values. Runoff increased considerably, 50–100 mm, in the narrow northern belt of the modern forest-tundra and tundra.

In Asia, runoff was about 50 mm lower than at present in a wide east-west belt of the taiga zone of western and eastern Siberia and only slightly lower along the northeast margin of the Asian continent and in the regions adjoining the Sea of Japan. Along the coasts of the polar ocean as well as in zones of forest-steppe and forests of East Asia, runoff increased 50–100 mm over present conditions. In the rest of the temperate belt, the river runoff was near today's amounts, exceeding them only insignificantly.

North America's pattern is similar to Asia's. The region in which runoff decreased significantly, up to 50 mm and more, includes the southeast zone of forests as well as the eastern part of the subtropical belt. To the north and south of this region, the runoff exceeded today's but by no more than 25–50 mm.

In Africa, the area with no runoff was much smaller than at present. The Sahara runoff was higher, although still small. To the south of 15° north latitude, across Senegal and the southern parts of Mali, Niger, Chad, and Sudan, runoff reached 50 mm and more above the present. Considerable increase in runoff occurred in northern Africa and on the coasts of the Mediterranean Sea.

In the warmer climate of the Mikulino interglacial optimum, 1–2°C above present temperatures, the relationship between temperature and precipitation patterns—and hence river runoff—was much different from the Holocene optimum (60). (See Figure 10.10.) In Europe, river runoff considerably surpassed present values—it was 100–200 mm higher within Western and Eastern Europe and in the western part of the European U.S.S.R. At the same time, in the greater part of the Russian plain, runoff increased no more than 50 mm over present values. Large areas of reduced runoff were found in the same locations as during the Holocene optimum.

During the Mikulino interglacial, the runoff was higher than at present in almost all of Asia, especially north of the polar circle, in the northeastern part of the continent, and in a small area of humid mixed forests of East Asia. In a large latitudinal belt in the north of the continent and in the eastern regions of the temperate and subtropical belts, the runoff exceeded present levels by 50 mm or more. In most of Asia, the runoff was somewhat higher than that of today.

In Africa, the area of no runoff during the Mikulino optimum coincided closely with today's. Only in the north was runoff higher by 25–50 mm or more.

During the Pliocene optimum, a period of temperatures 4°C or more above the present, paleoclimatic reconstructions found even larger changes in precipitation (61). In this period, precipitation increased in nearly all of Eurasia. The greatest increase, up to 300 mm, occurred in the northeastern coast of the Asiatic U.S.S.R. In the basin of the Baltic Sea, the north of west Siberia, Yakutia, and the Yana and Indigirka river basins, precipitation increased 200 mm. In the Caspian region, the south of west Siberia, north Kazakhstan, and Soviet Central Asia, the increase was also 200–250 mm. North Africa also showed increases of 200–300 mm during this period.

Global circulation models have predicted some important possible changes in the hydrological cycle under conditions of global warming. Although such models disagree on the details of precipitation changes, they generally predict that global atmospheric precipitation and evaporation will increase 4–12 percent (62). Experiments with one model suggest that in high latitudes—north of 40–50° north latitude and south of 40° south latitude—the increase in precipitation averaged by latitudinal belts may reach 40 percent (63). Such experiments also suggest that with increased temperatures, snow melting would come earlier in the season; soil moisture would decrease considerably in middle and high altitudes during the summer but would increase during the coolest months.

The preceding discussion was based partly on analogies with climate change in past geological epochs and partly on experiments with global circulation models of the atmosphere coupled to models of the hydrological cycle. It presents some idea of the scale and geological distribution of possible changes in river runoff. In many cases, the estimates often agreed quite closely, but in others, the differences were considerable. Clearly, more needs to be learned before such estimates can be relied on.

RECENT DEVELOPMENTS

AGRICULTURE MAY THREATEN THE SAUDI WATER SUPPLY

Saudi Arabia's remarkable success in increasing its agricultural production threatens to drain the country's underground water resources completely in the next 10–20 years.

Expanding agricultural production has been a key component of the Saudi government's effort to redistribute oil revenues and diversify the country's economy. Results of the effort are impressive; nearly 3 million hectares were under cultivation by 1988, a dramatic increase from the 150,000 hectares farmed in 1975 (64). Pushed by government subsidies of $400–530 per metric ton, Saudi farmers were expected to produce more than 3 million metric tons of wheat in 1989, an extraordinary increase from the 3,000 metric tons produced in 1976. Domestic consumption takes about 800,000 metric tons a year; the rest is sold or given away. The Saudis also produce surpluses of eggs and dairy products. Overall, Saudi farmers now produce 35 percent of the country's food, up from 15 percent in 1984 (65).

The government's agricultural policy has significant costs, however. For example, farm subsidies for wheat alone cost the government about $1 billion annually, more than eight times the cost of importing the wheat (66).

Perhaps more significant, the farm sector is a voracious user of water. It is estimated that Saudi Arabia used about 20,520 million m^3 of water in 1988. Agriculture used about 90 percent of the total, or 18,633 million m^3, with wheat accounting for about 35 percent of agricultural water use (67).

Roughly 90 percent of water use is from nonrenewable groundwater sources; renewable groundwater, surface water, and desalinated water make up the balance (68).

Virtually all the groundwater used comes from aquifer storage that accumulated thousands of years ago; average annual recharge is thought to be negligible (69). In 1980, the country's estimated reserves totaled 497,500 million m^3: 337,500 million m^3 in primary aquifers and 160,000 million m^3 in deeper secondary reserves; since 1980, total reserves have declined to an estimated 385,000 million m^3. According to a confidential U.S. government agency report obtained by the *Middle East Economic Digest*, at the current rate of depletion, nonrenewable fossil groundwater will be exhausted by 2007 (70).

The country has the most ambitious desalinated water program in the world, with 29 plants producing about 500 million m^3 a year, but the cost of production (about $2.70 per m^3) is prohibitive for growing crops such as wheat with moderate water requirements (71).

Saudi officials stoutly deny that extractions over the past decade have affected groundwater resources. In defense, they point to the country's efforts to install water-saving irrigation techniques and the rapid growth of its desalination program (72).

The government of Libya also has ambitious plans to exploit its groundwater resources. Libya reportedly is planning a massive $25 billion project to pipe water from under the Sahara to the more fertile coastal region. The project is intended to double the country's water supply, with most of the water planned for irrigation for cereal production and sheep and cattle grazing. Like Saudi Arabia, the Saharan aquifers are nonrenewable, but there is little agreement about how long they may last (73).

UNDERGROUND ECOSYSTEMS

Rivers have long been known to be full of life, but that fact is only part of the story—river ecologists are now finding an abundance of life forms in the ground beneath river channels. These underground ecosystems are important to the health of rivers.

Traditional research has focused mainly on river channels and shoreline vegetation. The new studies suggest that river ecosystems are much larger than the river channel area and that there are crucial interactions between the aboveground and underground systems. The findings could have significant implications for efforts to protect river ecosystems; for example, it may be necessary to expand protection several miles from the river bank to include the underground system.

The studies suggest active life in the gravel, sand, and rock (the hyporheic zone) beneath gravel-bed rivers, which are common worldwide. Worms, shrimp, bacteria, algae, and immature insects live in a maze of underground spaces.

A study of the Flathead River in Montana in the western United States suggested a zone about 10 meters deep and 3 km wide, far wider than the roughly 50-meter width of the river. Researchers found stoneflies and other creatures within a grid of shallow (10-meter) wells located on the floodplain up to 2 km from the river channel (74). Immature stoneflies are voracious predators in the underground zone, but when they emerge as adults, they develop wings and mate along the river banks, then become prey for fish.

In the Flathead study, Drs. Jack Stanford of the University of Montana and James Ward of Colorado State University found that underground aquifers are important in recharging the river and providing nutrients during periods of low river flows. The underground zone provides significant amounts of phosphorus and nitrogen to the river system during periods of low river flows, and the underground aquifer helps recharge the river during low-flow periods (75).

Dr. Ward is conducting a similar study of the South Platte River in Colorado. The results are generally similar so far, but the underground zone of the South Platte is apparently not as wide as the Flathead's (76).

In a study of the Sycamore River in southeast Arizona, where the climate is dry and streams often experience severe flash floods and periods of extensive drying, the hyporheic zone was found to be an important refuge for creatures during times of drought or stress. Researchers also found that the underground zone plays an important role in the recovery of a river after a flood (77) and may also help clean streams of pollutants such as hydrocarbons, which are broken by microbial activity in the hyporheic zone (78).

The Freshwater Quality section was authored by Debbie Chapman of the Global Environmental Monitoring System's Monitoring and Assessment Research Centre, London, with the assistance of Michel Meybeck of the Laboratoire de Geologie Appliquée, Universite de Paris, as principal consultant. The sections on Freshwater Quantity, Freshwater Use, and Global Climate Change were authored by Alexander V. Belyaev and his colleagues of the Institute of Geography, Soviet Academy of Sciences, Moscow, where Dr. Belyaev is Deputy Director for Scientific Affairs.

References and Notes

1. Michel Meybeck, Deborah Chapman, and Richard Helmer, eds., *Global Freshwater Quality: A First Assessment* (Blackwell Reference, Oxford, U.K., 1989), pp. 279-80.
2. Docter Institute for Environmental Studies, *European Environmental Yearbook 1987* (DocTer International, London, 1987), pp. 487-509.
3. World Health Organization (WHO) and United Nations Environment Programme (UNEP), "Global Pollution and Health," in *Global Environmental Monitoring System (GEMS)* (UNEP and WHO, London, 1988), pp. 9-11.
4. United Nations Economic Commission for Latin America and the Caribbean, *The Water Resources of Latin America and the Caribbean: Water Pollution* (United Nations, New York, 1989), p. 9.
5. World Health Organization (WHO), *Guidelines for Drinking Water Quality*, Vol. 1: Recommendations. (WHO, Geneva, 1984), Table 1, p. 5.
6. Global Environmental Monitoring System, *Assessment of Freshwater Quality* (United Nations Environment Programme and World Health Organization, Nairobi and Geneva, 1988), p. 33.
7. Department of the Environment, *Digest of Environmental Protection and Water Statistics* (Her Majesty's Stationery Office, London, 1988), pp. 22-24.
8. World Health Organization (WHO), *Environmental Pollution Control in Relation to Development* (WHO, Geneva, 1985), cited in R. Helmer, "Socio-Economic Development Levels and Adequate Regulatory Policy for Water Quality Management," *Water Science and Technology*, Vol. 19, No. 9 (1987), pp. 258-59.
9. *Op. cit.* 4, Figure 3, p. 11.
10. *Op. cit.* 1, p. 166.
11. J. Kobayashi, "Relations between the 'Itai-Itai' Disease and the Pollution of River Water by Cadmium from a Mine," cited in *Advances in Water Pollution Research*, Proceedings of the 5th International Conference, San Francisco and Hawaii (1971), pp. 1-7.
12. United Nations Environment Programme, *Environmental Data Report*, 2nd ed. (Blackwell Reference, Oxford, U.K., 1989), Table 3.24, pp. 315-19.
13. Henning Rodhe *et al.*, "Acidification and Regional Air Pollution in the Tropics," in *Acidification in Tropical Countries*, Henning Rodhe and Rafael Herrera, eds. (John Wiley and Sons Ltd., Chichester, U.K., 1988), pp. 26-27.
14. *Op. cit.* 6, p. 48.

15. P.D.K. Collins and W.G. Fowlie, *Denmark and Kent River Basins Water Resources Survey* (Public Works Department, 1981), Figure 10, p. 30.

16. G.T. Orlob and A. Ghorbanzadeh, "Impact of Water Resource Development on Salinization of Semi-Arid Lands," in *Land and Stream Salinity*, J.W. Holmes and T. Talsma, eds. (Elsevier, Amsterdam), Figure 4, p. 280.

17. Michel Meybeck, "The Water Quality of World Rivers through the GEMS Program," in *Transport of Carbon and Minerals in Major World Rivers*, Part 4 (Mitteilungen aus dem Geologisch-Palaeontologischen Institut der Universitat Hamburg, Hamburg, 1987), pp. 14-15.

18. *Op. cit.* 4, p. 25.

19. M. Marchand, "La contamination des eaux continentales par les micropollutants organiques," *Revue des sciences de l'Eau*, Vol. 2 (1989), pp. 229-264.

20. Robert J. Gilliom, Richard B. Alexander, and Richard A. Smith, *Pesticides in the Nation's Rivers, 1975-1980, and Implications for Future Monitoring* (Washington, D.C., U.S. Government Printing Office, 1985), Figures 3 and 6, pp. 12 and 19.

21. Maria Elena Hurtado, "Agrotoxics: Blight on the Next Generation," *Earth*, No. 77 (March 1987), p. 97.

22. Michel Meybeck and Richard Helmer, "The Quality of Rivers: From Pristine Stage to Global Pollution," *Paleogeography, Paleoclimatology, Paleoecology* (in press).

23. "Inland Waters," *The World Environment, 1972-1982*, Martin W. Holdgate, Mohammed Kassas, and Gilbert F. White, eds. (Dublin, Tycooly International, 1982), pp. 121-170.

24. *Op. cit.* 6.

25. *Op. cit.* 1.

26. A.V. Belyaev, U.S.S.R. Academy of Sciences, Institute of Geography, Moscow.

27. *Op. cit.* 22.

28. R.L. Thomas, J.P. Vernet, and R. Frank, "DDT, PCBs and HCB in the Sediments of Lake Geneva and the Upper Rhone, *Environmental Geology*, Vol. 5 (1984), 103-113, cited in Michel Meybeck and Richard Helmer, "The Quality of Rivers: From Pristine Stage to Global Pollution," *Paleogeography, Paleoclimatology, Paleoecology* (in press).

29. *Op. cit.* 6, pp. 73-74.

30. *Mirovoi vodnyi balans i vodnye resursy Zemli (The World water balance and the water resources of the Earth)* (Leningrad, Gidrometeoizdat, 1974).

31. Mark I. L'vovich, *Mirovye vodnye resursy i ikh budushchee* (Moscow, Mysl, 1974), Raymond L. Nace, translator, *World Water Resources and Their Future* (American Geophysical Union, Washington, D.C., 1979), p. 15.

32. *Ibid.*, pp. 21-22.

33. G.P. Kalinin, *Problemy globalnoi gidrologii (Problems of global hydrology)* (Leningrad, Gidrometizdat, 1968).

34. Mark I. L'vovich *Elementy rezhima rek zemnogo shara* (Elements of the regime of the World's rivers) (Sverdlovsk and Moscow, 1945).

35. *Fiziko-geograficheskiy atlas mira (Physico-Geographical atlas of the World)* (Moscow 1964).

36. *Op. cit.* 30, Table 188, p. 570.

37. *Op. cit.* 31, Table 8, p. 64.

38. Asit K. Biwas, ed., *United Nations Water Conference: Summary & Main Documents; Water Development, Supply & Management*, Vol. 2 (Pergamon Press, Oxford, U.K., 1978), p.39.

39. *Op. cit.* 30, Table Z62, p. 502.

40. *Metody issledovaniya vodnogo balansa territorii i kartirovarinya ego elementov (Methods for studying water balance of a territory and mapping of its elements)* (U.S.S.R. Academy of Sciences, Institute of Geography, Moscow, 1973).

41. A.V. Belyaev *Kompleksnye zavisimosti vodnogo balansa osnovnykh geograficheskikh zon zemnogo shara (Integrated dependence of the water budget of the principal geographical zones of the World)*, Izv. AN SSSR ser. geograf. No. 1 (1977).

42. A.M. Ryabchikov, *Struktura i dinamika geosfery (The structure and dynamics of geosphere)* (Moscow, Mysl, 1972), Table 6, p. 48.

43. Mark I. L'vovich and Gilbert F. White, Use and Transformation of Terrestrial Water Systems," in *The Earth Transformed by Human Action: Global and Regional Changes in the Biosphere over the Past 300 Years*, B.L. Turner, ed. (Cambridge University Press, Cambridge, U.K. 1990).

44. I.A. Shiklomanov, *Antropogennye izmeneniya vodnosti rek* (Anthropogenic impact on the abundance of water in rivers) (Leningrad, Gidrometeoizdat, 1979).

45. Mark I. L'vovich, *Voda i zhizn' (Water and life)* (Moscow, Mysl, 1986), Table 8, p. 65.

46. Peter P. Rogers, "Fresh Water," in *The Global Possible*, Robert Repetto, ed. (Yale University Press, New Haven, Connecticut, 1985), p. 256.

47. See "Water-Meliorative Transformations" in *The Earth Transformed by Human Action: Global and Regional Changes in the Biosphere over the Past 300 Years* (Cambridge University Press, Cambridge, U.K., 1990).

48. V.S. Kaminsky, "Okhrana poverkhnnostnykh vod v SSSR" (Conservation of surface waters in the U.S.S.R.), *Vodnye resursy*, No. 6 (1982).

49. For a review of forecasts, see I.A. Shiklomanov and O.L. Markova, *Problemy vodoobespecheniya i perebroski rechnogo stoka v mire (The Problems of Water Availability and Transfer of River Runoff in the World)* (Leningrad, Gidrometeoizdat, 1987).

50. *Op. cit.* 30.

51. I.A. Shiklomanov, "Dinamika vodopotrebleniya i vodoobespechennost' v mirĕ ("Dynamics of water consumption and availability of water in the world"), *Vodnye resursy*, No. 6, (1986) Figures 5-7, pp. 136-138.

52. *Op. cit.* 49.

53. *Op. cit.* 43.

54. A.I. L'vovich, *Zashchita vod ot zagryazneniya (The protection of water against pollution)* (Leningrad, 1977).

55. *Op. cit.* 45.

56. *Op. cit.* 43.

57. K.L. Vinnikov, *Chuvstvitel'nost' klimata* (Sensitivity of climate) (Leningrad, Gidrometeoizdat, 1986).

58. *Antropogennye izmeneniya klimata* (Anthropogenic changes of climate) (Leningrad, Gidrometeoizdat, 1987).

59. A.A. Velichko, V.P. Grichuk, E.E. Gurtovaja *et al.*, Paleogeographic Atlas of the Northern Hemisphere, charts on Paleoclimatic and Paleoenvironmental Reconstructions from the Pleistocene to Holocene (INQUA, Budapest, in press).

60. A.V. Belyaev and A.G. Georgiadi, "The River Runoff of the Northern Hemisphere during the Optimum of Holocene and Mikulino Interglacial," in *Paleogeographic Atlas of the Northern Hemisphere* (INQUA, Budapest, in press).

61. *Op. cit.* 58.

62. *Dinamika klimata* (Leningrad, Gidrometeoizdat, 1988).

63. S. Manabe and R.J. Stouffer, "A CO2-Climate Sensitivity Study with a Mathematical Model of the Global Climate," *Nature* (London), 1979, Vol. 282, No. 5738, pp. 491-93.

64. "Hopes Dry Up for Food Security," *Middle East Economic Digest*, Vol. 33, No. 40 (1989), p. 15.

65. "Just Add Water," *The Economist* (July 15, 1989), p. 51.

66. *Ibid.*

67. *Op. cit.* 64.

68. *Op. cit.* 64.

69. Saudi Arabia Ministry of Agriculture, *Water Atlas of Saudi Arabia* (U.S. Geological Survey, Reston, Virginia, 1984), p. 45.

70. *Op. cit.* 64.

71. *Op. cit.* 64.

72. *Op. cit.* 64.

73. *Op. cit.* 65.

74. J.A. Stanford and J.V. Ward, "The Hyporheic Habitat of River Ecosystems," *Nature*, Vol. 335, No. 6185 (1988), pp. 64-66.

75. *Ibid.*

76. J.V. Ward, Professor of Biology, Colorado State University, Fort Collins, Colorado, 1989 (personal communication).

77. Nancy B. Grimm *et al.*, "Contribution of the Hyporheic Zone to Stability of an Arid-Land Stream," Arizona State University, Department of Zoology, Tempe, Arizona, 1989, p. 8.

78. William Booth, "Beneath Rivers, Another Realm," *Washington Post* (October 26, 1989), p. A1.

11. Oceans and Coasts

In early 1990, a group of scientists—called the Group of Experts on the Scientific Aspects of Marine Pollution (GESAMP)—issued an assessment of the health of the world's oceans for the United Nations. They concluded that "chemical contamination and litter can be observed from the poles to the tropics and from beaches to abyssal depths," but that the distribution of pollutants was uneven. Many coastal areas are significantly polluted while the open ocean is still relatively clean. (See New Assessment of Ocean Pollution, below.)

Of greatest concern is that the trend is toward more pollution and if unchecked, it will lead to "global deterioration in the quality and productivity of the marine environment." The major marine contaminants in their order of importance are: nutrients from urban sewage and rural runoff; microbial contaminations from sewage; plastics from land and sea disposal; synthetic organic compounds such as pesticides and industrial chemicals; and oil from routine transport and spills, according to the GESAMP report.

The contamination of coastal areas is understandable since these are the waters nearest the sources of pollution. But it is especially unfortunate because these are also the waters that shelter the vast majority of the oceans' life. Although the global fish catch continues to climb, certain fisheries in heavily fished and polluted areas are declining. (See Datapoints, below.) And microbial contamination from sewage

has affected shellfish beds in some areas. Plastics and pesticides have affected marine mammals even on remote islands.

The GESAMP scientists concluded their assessment with a warning and a call for action: "We fear, especially in view of the continuing growth of human populations, that the marine environment could deteriorate significantly in the next decade unless strong, coordinated national and international action is taken now."

Each year *World Resources* takes a close look at one area of the sea, including its resources, pollution problems, and governance. The wider Caribbean, including the Gulf of Mexico, is an especially rich and diverse area that illustrates most of the problems highlighted in the GESAMP report. Ironically, this region's economic vitality depends on its ability to maintain its image as a tropical paradise for tourists. Pollution from oil and sewage and other sources threaten that image. Governments of the Caribbean's many tiny islands, Latin America's larger coastal nations, and the United States are beginning to learn how to work together on marine problems.

Far from the tropics, in frigid Antarctica, innovations in marine resource management may provide a guide for other regions. Under a treaty, the countries that share an interest in Antarctica manage the continent's marine resources as an *ecosystem*, rather than maximizing the production of one or two com-

Datapoints: World Fishery Resources

Figure 1 Global Marine and Freshwater Catch 1950—88

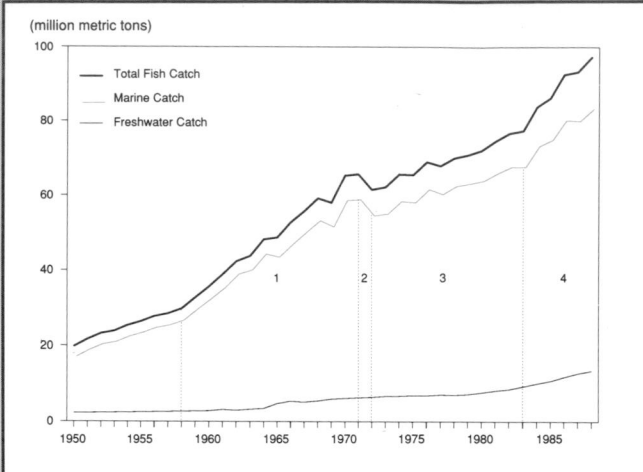

Source: Chapter 23, "Oceans and Coasts," Table 23.2

Figure 2 World Fish Catch by Use 1950—88

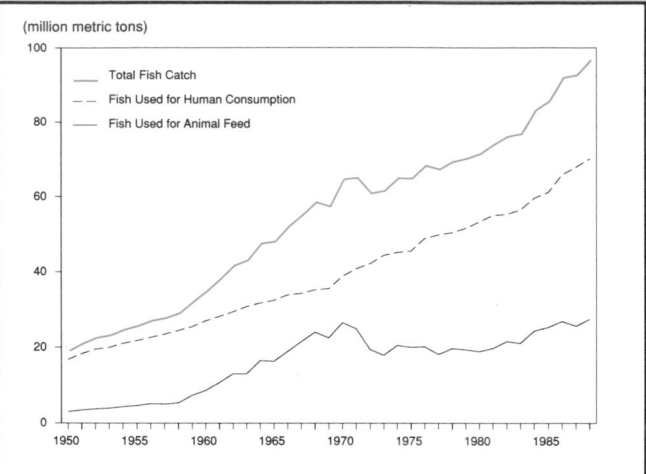

Source: Chapter 23, "Oceans and Coasts," Table 23.2.

Since 1950, the world's total recorded landings of marine and freshwater fish have almost quintupled, from 19.8 million metric tons to 97.4 million metric tons in 1988. (See Figure 1.) Both marine and freshwater landings increased, with a rise in landings of marine fish from 17.6 to 84 million metric tons, and of freshwater fish from 2.2 to 13.4 million metric tons. The vast majority of the world's fish is caught in the ocean. The marine catch grew faster than the inland catch in the 1950s and early 1960s. Since the mid 1970s, landings from inland waters have grown faster than those from the sea.

Trend data of marine catch show four phases. First, between 1958 and the early 1970s, the rate of increase of the early 1950s doubled because of heavier exploitation of existing stocks, discovery of new stocks, and improved fishing technology. Second, in the early 1970s, El Niño, a climatological event that mutes the cold-water upwelling off the coast of Peru, caused a collapse of the Peruvian anchovy stocks. Third, from the early 1970s until the early 1980s, the world's landings of marine fish increased moderately, with more or less constant landings of bottom dwelling fish (e.g., cods,

hake, haddocks) and slightly increased landings of pelagic stocks (e.g., anchovies, sardines). In the final phase, since 1982, trend data show a steep rise. This can be attributed primarily to increased landings of four species, Peruvian anchovy, South American sardine, and Japanese sardine and the Alaska pollock, which account for 70 percent of the rise in marine catch during the 1980s. Populations of anchovies, sardines, and other "shoaling pelagic" species are notoriously unstable. In the past, they have shown large fluctuations because of changes in currents, upwellings, or other environmental factors.

Since 1950, freshwater catch has increased by a factor of six. However, it is not known whether capture fisheries or aquaculture made up most of this increase, because the Food and Agriculture Organization of the United Nations (FAO) doesn't separate data by those methods. In 1984, FAO began to collect global statistics on aquaculture by questionnaire. (See Chapter 23, "Oceans and Coasts," Table 23.2.) Asia reports the largest share of the world's inland catch, about 67 percent, followed by Africa with 15 percent. A major omission from all fishery statistics are subsistence

catches, an important source of animal protein, usually not covered in FAO statistics.

Figure 2 shows the total world fish catch divided into fish used for food, mainly the bottom-dwelling species, and fish used for animal feed, mainly the shoaling pelagic species. In the mid-1950s improved fishing technology increased the catch of shoaling pelagics: their share of the total catch climbed from 15 percent in 1950 to 40 percent in 1970, and has since fluctuated around 27 percent.

The world's marine and freshwater fisheries are reaching the limit of sustainable yields, which has been estimated by FAO at 100 million metric tons annually. As the fishery is approaching this boundary, environmental pressures such as eutrophication, chemical pollution, and destruction of nursery grounds will have increasing impacts on the resource productivity. The combined effects of heavy fishing pressure and pollution can already be observed regionally. Most local fisheries are moderately to heavily exploited. (See Chapter 23, "Oceans and Coasts," Table 23.3.)

mercially valuable species as is done elsewhere. This system may avoid the sharp population declines of overexploited species that are common in most fisheries.

Climate change would have a marked effect on the world's oceans. The warming of sea water would

cause it to expand, thus raising sea levels 30–100 centimeters. Warmer sea water could also change weather patterns resulting in more severe tropical storms. Changing weather and wind patterns could change the locations of currents, and thus of fishery areas.

CONDITIONS AND TRENDS

NEW ASSESSMENT OF OCEAN POLLUTION

Most of the wastes and contaminants produced by human activities eventually reach the sea. Every year, billions of metric tons of silt, sewage, industrial waste, chemical residues, and oily runoff from urban streets pour into the world's oceans.

Rivers alone carry 9.3 billion metric tons of silt and waste to coastal waters annually. Storm-drain and outfall pipes, direct dumping in oceans, and the settling of airborne pollutants, such as particulate matter from smoke and engine exhaust, contribute additional contaminants.

Some pollutants are highly visible. Wastes from ships, oil from spills, and plastic debris not only contaminate the oceans and endanger wildlife, but they also spoil the aesthetic value of the oceans.

Other pollutants are highly dangerous. Human-disease pathogens accumulate in shellfish, persistent pesticides and other toxic chemicals and metals invade the food web, and medical wastes wash up on beaches.

Are the oceans becoming irretrievably polluted? Are they at serious risk? Or are they still able to absorb these insults and cleanse themselves?

No single entity tracks the world's oceans, but the United Nations (U.N.) periodically diagnoses their condition. In 1969, eight U.N. agencies assigned scientists to the Group of Experts on the Scientific Aspects of Marine Pollution (GESAMP) to advise other agencies. Following the 1972 United Nations Conference on the Human Environment in Stockholm, GESAMP was asked to review the state of marine pollution periodically. The group's first report, *The Health of the Oceans*, was published in 1982. It pointed out the fundamental role that oceans play in maintaining conditions for life on earth, interacting closely with the atmosphere, the sea floor, and the continental land masses. Although marine pollution had not yet been detected on a global scale, warning signs could be seen in some coastal areas (1).

In 1986, a second GESAMP group was convened. Its report, issued in early 1990, concludes that marine pollution is worsening.

■ Most of the world's coastal areas are polluted. The open ocean, however, is relatively clean except for floating tar and plastic debris, found mainly in shipping lanes and drift lines where currents converge.
■ The most widespread and serious sources of pollution are not large oil spills or toxic-waste dumping; they are sewage disposal and sedimentation from land clearing and erosion—both exacerbated by growing coastal populations coupled with the lack of sewage treatment and erosion control.
■ Alteration of the coastal habitat, through both pollution and mechanical destruction of wetlands, mangroves, coral reefs, and sand dunes for the sake of coastal development, is a serious threat; destruc-

tion of feeding and nursery areas has imperiled and reduced fish and wildlife populations in many areas.
■ Oil spills, together with oil leaks from ships and offshore drilling platforms, and dumping of synthetic organic compounds, such as pesticides, are locally damaging pollution problems but are not pervasive environmental threats.
■ Damage to marine birds and mammals and to beaches by plastic litter from garbage, discarded fishing nets and gear, and tar balls from oil spills is a growing concern worldwide.
■ Of less immediate concern, because they are less pervasive, are toxic metals and radioactive substances (2).

In sum, GESAMP found widespread pollution in most coastal waters, mainly from sewage and sediments, but that the open oceans are largely unpolluted.

Why the Open Ocean Stays Clean

Most pollution sources are on the continents and most contaminants are waterborne in rivers, outfall pipes, or storm runoff. Because most contaminants settle out as water movement slows, there is a gradient of reduced concentration away from the source. It is not expected, then, that the deep seas would be as contaminated as coastal waters.

For the most part, human fouling of the open ocean, occasional deep-water dumping notwithstanding, results either directly from shipping activities or indirectly from atmospheric pollution. Although the human presence can be detected in the deep waters—by measuring organochlorines, metals, radionuclides, and oily residues, particularly in the shipping lanes—the contaminants occur only in low concentrations.

Why the Coastal Areas Are Polluted

Coastal areas are polluted by runoff from land, sewage, and industrial discharge as well as by industrial and municipal dumping.

Historically, people have gathered near coastal areas to take advantage of water transportation, seafood, and aesthetics.

The effect of this pollution on life in the sea is disproportionate, because most of ocean life is found in these nearshore waters. Whereas most open ocean is a biological desert, the coastal areas, including upwelling areas and shallow continental shelves, bloom with life. Especially in upwelling areas, where currents bring bottom nutrients near the sunlight of the surface, algae and aquatic plants prosper and other organisms feed. (See *World Resources 1988-89*, Figure 9.1, p. 145.) More than 99 percent of the global catch of marine fish is taken within 320 kilometers of a coastline (3).

By polluting coastal waters, we are having a far greater effect on life in the sea than if the ocean were polluted evenly.

Table 11.1 Number of Red Tides and Their Adverse Effects in Tolo Harbour, Hong Kong, 1977—87

Year	Number of Red Tides	Number of Adverse Effects Recorded
1977	2	1
1978	1	x
1979	1	x
1980	4	2
1981	3	x
1982	3	x
1983	11	1
1984	16	2
1985	16	2
1986	26	4
1987	19	3

Source: United Nations Environment Programme (UNEP), "State of the Marine Environment in the East Asian Seas Region," UNEP, Nairobi, Kenya, draft 1987, p.36.
X = not available.
Note: a. Adverse effects include fish kills, algal blooms, and oxygen depletion.

Eutrophication

Eutrophication—the overenrichment of waters—is widespread globally and is getting worse. In the process of eutrophication, an overabundance of nutrients, mainly nitrogen and phosphorus, cause algal blooms and rapid growth of other aquatic plants. When these plants die, decomposing bacteria can deplete the water of oxygen, killing fish and other marine life. Under extreme conditions, eutrophication can severely deplete the water of dissolved oxygen. Mass kills of fish and other organisms can result (4).

The major sources of nutrients are urban wastes, particularly sewage; agricultural runoff from heavy fertilizer use; and wastes from intensive livestock husbandry and fish farming. GESAMP found an increasing rate of nutrients and organic matter entering the oceans.

Over the past decade, marine scientists have noted increasing numbers of unusual blooms of species not previously observed in a location—some of which are toxic. (See Table 11.1.) These novel blooms can harm other organisms in several ways; they can cause eutrophication, they can be directly toxic, or they can disrupt the food chain because they are not edible. These red, green, and brown tides have occurred more frequently in recent years, but their link to increasing levels of pollution has not been established (5). (See Recent Developments, below.)

Eutrophication is especially a problem in enclosed or semienclosed bays or estuaries where water is not flushed out frequently. In Northern Europe, eutrophication is found in the bays and inlets of the Baltic Sea and in the German Bight of the North Sea, which is influenced by large rivers carrying industrial wastes (6).

On Mediterranean coasts, the Gulf of Lyons, the Lake of Tunis, and the Bay of Ismir often show extreme effects. In the Northern Adriatic, heavy algal blooms appear along the Emilia-Romagna coastal waters in late summer and fall (7) (8).

Along the central and western Pacific coast of Japan—particularly in Tokyo Bay, Ise Bay, and parts of Seto Inland Sea—where 72 million people live, above-normal nutrient concentrations are common, and algal-cell counts are 10 times those of 20 years ago (9) (10).

Along U.S. coasts, the extremely oxygen-deficient waters are most extensive along the southern coast of Louisiana, in the Chesapeake Bay, and in the New York Bight. They are least extensive along the Pacific coast (11).

Sources of Pollution

During the 1970s and 1980s, water pollution control centered on industrial and municipal sources, including ocean dumping and discharges from pipes. A third source of pollutants, runoff from land, receives little regulatory attention even in developed countries. Runoff, also called nonpoint pollution, contributes 34 percent of the total pollution in U.S. coastal waters. (See Table 11.2.)

Nonpoint pollution is a major source of suspended solids that block sunlight to aquatic plants, clog filter-feeding organisms, and carry phosphorus, fecal-coliform bacteria, and other pollutants. Nonpoint source pollution from agricultural areas also contains fertilizers, which further increase the nutrient load, and pesticides, most of which are toxic to marine organisms in high concentrations. The worst offender in the United States study—municipal sources of pollution— contribute mainly nitrogen and phosphorus, which are contained in sewage. Industrial sources contribute most of the arsenic, mercury, chlorinated hydrocarbons, and other toxic chemicals (12).

Table 11.2 Major U.S. Sources of Ocean Pollutants, 1986
(percent)

Pollutant Source	Biochemical Oxygen Demand	Total Suspended Solids	Total Kjedahl Nitrogen	Total Phosphorous	Cadmium	Chromium	Copper	Lead	Arsenic	Iron	Mercury	Zinc	Oil and Grease	Chlorinated Hydro-carbons	Fecal Coliform Bacteria
Industrial	11	<1	9	5	89	15	18	20	100	13	91	25	11	94	<1
Municipal	56	1	46	36	1	13	11	8	<1	5	5	9	41	6	16
Nonpoint	34	99	45	59	10	72	71	73	<1	82	3	66	47	<1	84

Source: Office of Technology Assessment (OTA), *Wastes in Marine Environments* (U.S. Government Printing Office, Washington, D.C., 1987), p. 62, based on Resources for the Future, *Pollutant Discharges to Surface Waters in Coastal Regions* (OTA, Washington, D.C., 1986).

Types of Marine Pollutants

Sewage

Sewage is a major cause of ocean pollution. It contributes to eutrophication and carries disease-causing pathogens.

Raw sewage, consisting of human excreta and domestic wastes, contains an average 3.2 kilograms of nitrogen and 0.6 kilograms of phosphorus per person per year (13). Although it may be disposed of in landfills or by incineration, much is discharged into rivers or directly into the oceans. Many parts of the world have no sewerage systems, and beaches or open water courses are used as toilets. Even where sewerage systems are available, the untreated wastes are usually piped into rivers or the sea, often with industrial effluents and storm runoff added.

Usual treatment options range from screening, grit removal, and primary sedimentation (primary treatment) to biological oxidation (secondary treatment). Advanced tertiary treatment, which precipitates more solids and removes more nitrogen, is less common. Even after secondary treatment, the effluent still contains oxygen-demanding wastes, suspended solids, nitrates, phosphates, viruses, heavy metals, pesticides, and radioactive isotopes (14).

Wastewater Pathogens. A 1987 study by the United Nations Environment Programme (UNEP) found that most of its 10 Regional Seas Programmes, covering the majority of the world's marine regions, identified microbial contamination of the oceans as the subject of highest concern (15). But although the situation is widely documented, statistics on the occurrence of diseases are limited to those from occasional studies of specific locations.

Sewage from urban areas almost inevitably contains pathogenic organisms, such as enteric bacteria, viruses, protozoans, and helminth worms; they cause bacillary and amoebic dysentery, cholera, typhoid and paratyphoid fevers, salmonella gastroenteritis, infectious hepatitis, viral gastroenteritis, and other diseases. The number of pathogens in the raw sewage depends on the health of the population from which it originated, but it is likely to be high in countries where waterborne diseases are endemic. In the sewage treatment process, the bacterial content of the sewage is reduced up to 100-fold.

Until recently, it was believed that pathogens could not survive more than a few days in salt water. But human pathogenic viruses have survived for 17 months after sewage sludge was dumped and researchers have discovered viable pathogens in sea water even though they cannot be detected by conventional testing methods (16). (See Recent Developments, below.)

The vector by which the pathogens transfer to people is usually shellfish, although consumption of other seafood and exposure to contaminated sea water during recreational activities have also been implicated.

Lack of Sewerage Systems and Sewage Treatment. Although the number of pathogens is significantly reduced by treatment, most sewage is not treated. Indeed, only a small proportion of the world's sewage even enters sewerage systems where treatment is possible. In Europe, for example, where sewerage systems are more prevalent than in most of the world, it is estimated that only 72 percent of the population is served by such systems (17); the proportion served varies from 25 percent in Greece to 95 percent in the United Kingdom. In addition, even in areas with secondary treatment systems, problems arise when storm runoff overloads the system and when the population served increases greatly, such as by an influx of tourists during a holiday season.

In the Mediterranean, the problem of treatment and disposal of municipal liquid wastes is considered urgent. During the tourist season, the coastal regions have a population of 200 million people, 50 percent of whom are not served by sewerage systems. Eighty-five percent of the effluents from large cities are discharged into the sea untreated, causing extensive beach pollution and contaminating shellfish beds (18).

In West and Central Africa, few large cities have either proper drainage facilities or sewage treatment plants. For the majority of city dwellers, no toilet system of any kind exists; bucket collection and disposal of sewage are common in coastal areas (19). In Bangladesh, the rate of fecal-transmitted disease from urban wastes is high. Directly or indirectly disposing of raw sewage into the sea is the general practice in Eastern Asia, where it is the major source of ocean pollution (20).

Finding the Effects. In several areas, concentrations of fecal coliform bacteria exceed national standards, indicating a potential human risk, but records of diseases contracted by swimmers do not exist. In the coastal waters of Jakarta, for example, pathogens are reported in fish and shellfish; in the Straits of Malacca, high levels of coliform bacteria occur in shellfish beds; and in the Gulf of Thailand, oysters and mussels are contaminated by sewage. A 1988 epidemic of hepatitis associated with the consumption of shellfish occurred in Hong Kong, with nearly 1,400 cases reported (21).

In Australia, where almost all large cities are located on the coast and their sewage is disposed of in the ocean, extensive beach contamination is a major concern (22).

In Latin America, municipal discharges from coastal cities of Central and South America are the main sources of pollution. Their high coliform counts account for intestinal infections, conjunctivitis, and other public health problems common among beach and marina users (23). Sewage is also the greatest single pollutant of coastal waters in the wider Caribbean (24). (See Regional Seas: The Wider Caribbean, below.)

Only in regions with small populations such as East Africa and the South Pacific is sewage pollution

Box 11.1 Oil Spill Damage Depends on Time, Place, Conditions

U.S. oil spills grabbed the headlines in the first half of 1989 with four accidents, dumping 47,000 cubic meters of crude oil into the nation's coastal waters (1). The largest spill in U.S. history occurred when nearly 38,000 cubic meters poured from the *Exxon Valdez* after it grounded on a reef in Prince William Sound in Alaska in March; the oil covered more than 2,600 square kilometers of coast and nearby waters (2). By comparison, in the world's largest tanker spill, the 1978 grounding of the *Amoco Cadiz* off the coast of France, almost seven times as much oil was lost.

The *Exxon Valdez* accident drew widespread attention for several reasons. First, it occurred in a pristine environment and caused serious short-term environmental damage and unknown long-term impacts (3).

Second, the *Valdez* spill was the result of human error, compounded by inefficient cleanup by all parties involved (4)

Third, prior to construction of the Alaskan pipeline, environmentalists voiced major concerns about possible oil spills once tankers began transporting Prudhoe Bay oil from Prince William Sound. Steps were purportedly taken to ensure that the oil industry would be fully prepared to handle oil spills quickly and efficiently. But the *Exxon Valdez* accident clearly indicated that neither the industry nor the state and federal government agencies were prepared (5).

Although tanker accidents receive widespread public attention, in fact, they are responsible for a relatively small percentage of the oil dumped into the oceans each year. As shown in Table 11.2, tank cleaning, ballasting, and other routine tanker operations are responsible for more oil pollution than are accidents. Other sources of marine oil pollution are offshore oil production activities, land-based industrial effluents and municipal wastes, and natural seeps.

Because of local, often dramatic impacts of a tanker spill or offshore plat-form accident plus the fact that many spills occur nearshore, these accidents have been studied intensively. As shown in Figure 1, most oil spills occur along Northern Hemisphere coasts, which have the heaviest tanker traffic. The eastern coast of the United States and European coastal waters suffer the heaviest concentration of oil spills.

ASSESSING THE DAMAGE

The severity of environmental damage from a spill depends on a variety of factors. The type of petroleum is a major factor because the toxicity of crude and refined products differs widely.

The time of year also influences severity of the damage. A coastal spill that occurs during a spawning season, for example, could wipe out an entire year's production. If the spill coincided with the bird or marine mammal migration, it would be more serious than had it occurred at another time.

Further, wind direction and force, atmospheric and sea temperatures, and other weather conditions determine how the spill will spread and therefore will influence cleanup attempts.

And, although no locations are "good" for oil spills, some are assuredly desirable. Areas with high concentrations of marine life are particulary sensitive to pollution. A spill in coastal waters, for example, is more serious than an open-ocean spill. Most spills occur in coastal waters.

Oil spill calamities rivet public attention on dead birds, fish, seals, sea otters, and other animals covered with oil. The March 1989 *Exxon Valdez* accident killed 36,471 birds, 1,016 sea otters, and 144 raptors; these figures represent only corpses found by the following November, perhaps 6–10 percent of the animals killed (6). In the *Torrey Canyon*'s 1967 English Channel oil spill, more than 25,000 birds were suffocated by the oil or poi-soned by chemicals used to disperse the oil. And in 1974, when the tanker *Metula* sank near the Strait of Magellan, 3,000–4,000 birds were killed (7). All these counts underestimate the toll because those killed shortly after the spill either sank or were consumed by predators before they could wash ashore.

Damage below the surface is usually more significant. Oil spills destroy the eggs, larvae, and adults of plankton, krill, coral, shellfish, and a wide array of benthic and coastal species. Stock on 809 hectares of French oyster beds was wiped out by the *Amoco Cadiz* spill, and the beds produced few oysters for several years (8). The 1979 blowout of Ixtoc I, an offshore oil platform in the Gulf of Mexico, is the largest marine oil spill to date; it destroyed the crab and mollusc population along several hundred kilometers of gulf coast. The lack of data and unreliable fishery statistics, however, have made it difficult to assess the full impacts of the Ixtoc I spill on gulf fisheries (9).

Some problems persist for years. Oil pollution causes chronic problems among marine organisms, including changed feeding and reproductive patterns and abnormal behavior and growth. It also interrupts chemical communication among animals and the senses they use for migration (10).

REVISITING THE SCENES OF OIL SPILLS

The status of the environment years after oil spills gives insight into the long-term problems they can cause. In 1969, for example, the barge *Florida* ran aground near West Falmouth, Massachusetts, and spilled 663 cubic meters of No. 2 fuel oil (11). Seven years later, the population of marsh fiddler crabs was still depressed; surprisingly, 10 years after the spill, fresh oil was found in nearby

---

not a major problem. Even there, however, it is a problem around urban areas (25) (26).

Marine Litter

In recent years, growing amounts of litter have been found in the marine environment. Litter harms marine life, interferes with shipping and other activities, and impairs the aesthetic quality of beaches. Its increased presence results largely from the rapid replacement of natural materials with synthetic compounds in manufacturing processes. Natural materials disintegrate quickly, but plastics are relatively nonbiodegradable. They persist for up to 50 years and, because they are usually buoyant, they are widely distributed by ocean currents and wind (27) (28).

A major source of plastic debris is the land; plastics—everything from six-pack-container rings to plastic bags, sheeting, and pellets of the raw materials used in manufacturing—wash down rivers from landfills and land-based operations (29). Another major source is the fishing industry, whose nets, ship lines, buoys, and other equipment are now manufactured largely from synthetic materials. When these items are lost or discarded at sea, they may sink, but more often they float or drift beneath the surface. Worldwide, the annual loss of fishing gear alone is estimated at more than 150,000 metric tons (30). Sections of netting continue for several years to

Box 11.1

marshes, which acted like sponges, creating a continuous source of oozing oil.

The largest oil spill in the tropical Americas occurred in 1986 near the Caribbean entrance to the Panama Canal. Eight thousand cubic meters of medium-weight crude oil poured from a ruptured storage tank. The location of the spill, near the Smithsonian Tropical Research Institute, provided a unique opportunity for study. The area had been studied before and again after a smaller oil spill in 1968. It is a typical Caribbean coastal ecosystem of sea grass beds, mangroves, algal flats, and coral reefs (12).

The spill covered mangroves roots along 27 kilometers of coast. The trees died within a year, affecting all the marine organisms that depend on them. It will take years for new mangroves to reclaim the area. Some meadows of subtidal sea grasses, although damaged, survived but all intertidal beds were destroyed. The coral reefs hit by the oil were heavily damaged, particularly those in the subtidal zone. Subtidal and intertidal shrimp and snails were also affected (13). These findings contradict widely held beliefs that spills in tropical ecosystems are not as serious as those in colder climates and that coral reefs are not affected by oil spills.

Figure 1 Main Tanker Routes and Major Spills

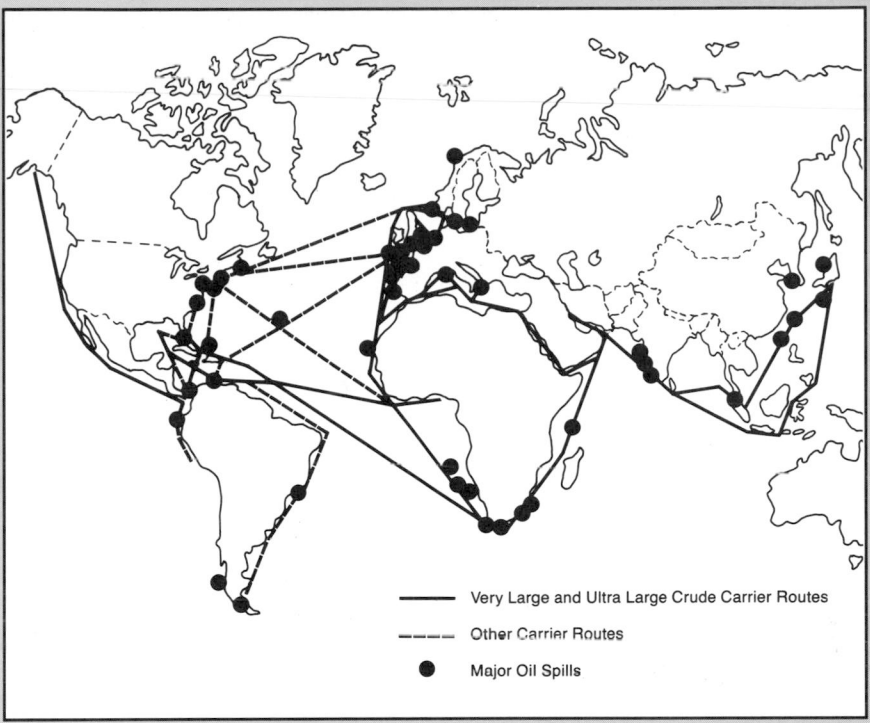

— Very Large and Ultra Large Crude Carrier Routes

---- Other Carrier Routes

● Major Oil Spills

References and Notes

1. Ken Wells, "Oil Industry's Inability to Contain Spills at Sea Poses Political Trouble," *Wall Street Journal* (June 26, 1989), p. A1.
2. "The One That Got Away," *Hazardous Cargo Bulletin* (Interpress Publishing Limited, London) (May 1989), p. 69.
3. Joyce W. Hawkes, "Morphological Effects of Petroleum and Chlorobiphenyls on Fish Tissues," in *Animals as Monitors of Environmental Pollutants* (National Academy of Sciences, Washington, D.C., 1979), p. 381.
4. Richard Townsend, "The *Exxon Valdez* Oil Spill: A Management Analysis," Center for Marine Conservation, Washington, D.C., 1989, pp. 1-20.
5. *Ibid.*, pp. 21-48.

6. Alaska Department of Environmental Conservation, *Oil Spill Chronicle*, Vol. 1, No. 17 (1989).
7. Bill Shaw, Brenda J. Winslett, and Frank B. Cross, "The Global Environment: A Proposal to Eliminate Marine Oil Pollution," *Natural Resources Journal*, Vol. 27, No. 1 (1987), p. 158.
8. *Ibid.*, pp. 158-159.
9. Arne Jernelöv and Olof Lindén, "Ixtoc I: A Case Study of the World's Largest Oil Spill," *Ambio*, Vol. 10, No. 6 (1981), pp. 304-306.

10. Group of Experts on the Marine Environment, *The Health of the Oceans* (United Nations Environment Programme, Nairobi, 1982), p. 45.
11. U.S. Environmental Protection Agency (EPA), *Decision Series: A Small Oil Spill at West Falmouth* (EPA, Washington, D.C., 1979), p. 9.
12. J.B.C. Jackson *et al.*, "Ecological Effects of a Major Oil Spill on Panamanian Coastal Marine Communities," *Science*, Vol. 243 (January 6, 1989), pp. 37-38.
13. *Ibid.*, pp. 40-42.

entangle fish, birds, and marine mammals in a kind of "ghost fishing." A third source of plastic debris is the material used for banding and strapping cargo and other packages.

Deaths of Marine Mammals. Reports come from all over the world of deaths of marine mammals that became entangled in discarded nets. In the North Pacific, for example, populations of Dall's porpoises are at risk from floating nets. Entanglement is the suspected cause in the deaths of fur seals in the eastern Aleutian Islands (31) and Hawaiian monk seals (32). Discarded plastic bands encircle mammals, fish, and birds and tighten as their bodies grow. Turtles, whales, and other marine mammals have died after

eating plastic sheeting. A recent study found plastic particles in the digestive tracts of 25 percent of the world's seabird species (33). Although information is not sufficient to quantify the impacts of such litter on these animal populations, experienced observers have expressed concern.

Beach Litter. Plastic litter also degrades the aesthetic qualities of beaches and nearshore waters. Plastic debris pervades every sector of the world's oceans, even the remote shores of the Antarctic. Up to 70 percent of the debris examined in the Mediterranean and more than 80 percent of it in the Pacific were plastic (34). A three-hour survey of 563 kilometers of Oregon beaches in 1984 yielded more than 26

Table 11.3 Sources of Petroleum Inputs to the Marine Environment

Source	Petroleum (million metric tons annually)	Percent
Municipal and industrial waste discharge and runoff	1.18	36.3
Municipal wastes	0.70	
Refineries	0.10	
Nonrefining industrial wastes	0.20	
Urban runoff	0.12	
River runoff	0.04	
Ocean dumping	0.02	
Offshore oil production	0.05	1.5
Transportation	1.47	45.2
Tanker operations	0.70	
Tanker accidents	0.40	
Bilge and fuel oils	0.30	
Other transportation activities	0.07	
Natural sources (seeps and erosion)	0.25	7.7
Atmosphere	0.30	9.2
Total	3.20	

Source: National Research Council, *Oil in the Sea: Inputs, Fates, and Effects* (National Academy Press, Washington, D.C., 1985), Table 2-22, p. 82

Table 11.4 Number of Oil Spills Worldwide 1974-86

Year	Number of Spills		
	All Sizes	6,800-68,000 Million Metric Tons	+68,000 Million Metric Tons
1974	1,450	91	26
1975	1,350	98	23
1976	1,099	66	25
1977	956	66	20
1978	746	57	24
1979	695	54	37
1980	554	48	13
1981	401	48	5
1982	247	44	3
1983	216	52	11
1984	146	25	7
1985	137	25	8
1986	118	24	6
Total	8,115	698	208

Source: C. Walder, *Marine Transportation of Oil and Other Hazardous Substances: Annex to the Report on the State of the Environment*, draft, United Nations Environment Programme, Nairobi, Kenya, 1989, Table X, p. 10.

metric tons of plastic materials, mostly pieces of polystyrene, food utensils, bags and sheet material, and bottles (35).

Petroleum

Oil is a widespread pollutant in the oceans—and a high-profile one. When it is concentrated, its presence is obvious. The most recent comprehensive study of oil in the oceans estimates that 3.2 million metric tons enter the marine environment annually. (See Table 11.3.) Of this amount, nearly 45 percent is from marine transportation, including spills, 36 percent is from municipal and industrial discharges.

Recent political and economic events have changed oil use and subsequently reduced the volume transported. World exports of crude and finished oil products declined 25 percent between 1977 and 1986 (36). Associated with the decline in oil transportation is the reduction in the number of oil spills; annual spills involving 6.8 metric tons or more dropped 74 percent between 1974 and 1986. (See Table 11.4.)

The most dramatic occurrences of oil pollution result from drilling-platform accidents or shipwrecks that spill large quantities of oil. (See Box 11.1.)

Synthetic Organic Compounds

Synthetic organic compounds include chlorinated hydrocarbon pesticides; industrial chemicals, such as chlorinated biphenyls (PCBs); and organometals, such as tributyl tin (TBT). Like many other contaminants, these substances reach the sea through direct or indirect industrial discharges, rivers, storm and irrigation runoff from land, ocean dumping, and atmospheric deposition. They are widely distributed in the oceans.

Chlorinated hydrocarbons are not biodegradable. Because they are carried by wind and water, they are found from the poles to the tropics. Chlorinated hydrocarbons have been detected, usually at low levels, in most species of organisms, including human beings. They are lipid-soluble and accumulate in fatty tissues. The amount of accumulation increases up the food chain so that high concentrations are found in the body fat of the top predators among birds, fish, and mammals.

Among the more widespread chlorinated hydrocarbons are the insecticides DDT and lindane (Y-hexachlorocyclohexane) and the fungicide HCB (hexachlorobenzene). Many others, however, including chlordane, toxaphene, and dieldrin, are also present in the oceans. Chlorinated hydrocarbons inhibit photosynthesis and movement in plankton and cause a wide range of damage in other organisms, such as wasting syndrome, tumors, liver damage, reproductive failure, and birth defects. It was demonstrated recently in the Netherlands' Wadden Sea that uterine occlusions, which prevent the birth of seal pups, result from PCB contamination of the mother seals' fish diet (37).

In many countries, particularly in Europe and North America, use of the more hazardous chlorinated hydrocarbons has been banned, greatly reducing their input to the sea and lowering concentrations found in sea life. For example, in the United States, where the use of DDT was banned in 1972, levels of the pesticide in fish and shellfish declined markedly from the mid-1970s through the mid-1980s. PCB levels, on the other hand, have declined

Box 11.2 International Control of Marine Pollution

Reducing marine pollution requires both national and local action. Action at both levels is most productive when it is supported by international agreement.

A network of international treaties and conventions exists. Some are regional; some are global. Others are under discussion.

In the 1950s and 1960s, when marine pollution was first recognized as a significant problem, major public concern focused on oil spills and routine discharges from ships. As a result, several multicountry conventions were adopted:

■ The International Convention for the Prevention of Pollution of the Sea by Oil, London, 1954, amended 1962, 1969, and 1971.

■ The International Convention Relating to Intervention on the High Seas in Cases of Oil Pollution Casualties, Brussels, 1969.

■ The Agreement for Co-Operation in Dealing with Pollution of the North Sea by Oil, Bonn, 1969.

■ The International Convention on Civil Liability for Oil Pollution Damage, Brussels, 1969.

■ The International Convention on the Establishment of an International Fund for Oil Pollution Damage, Brussels, 1971.

■ The Agreement between Denmark, Finland, Norway and Sweden concerning Co-Operation in Measures to Deal with Pollution of the Sea by Oil, Copenhagen, 1971.

By the early 1970s, public concern was growing. The United Nations Conference on the Human Environment held at Stockholm in June 1972 set forth two principles regarding the obligation of preserving the marine environment:

Principle 7: "States shall take all possible steps to prevent pollution of the seas by substances that are liable to create hazards to human health, to harm living resources and marine life, to damage amenities or to interfere with other legitimate uses of the sea."

Principle 21: "States have, in accordance with the Charter of the United Nations and the principles of international law, the sovereign right to exploit their own resources pursuant to their own environmental policies, and the responsibility to ensure that activities within their jurisdiction or control do not cause damage to the environment of other States or of areas beyond the limits of national jurisdiction."

After the conference and subsequent meetings that focused on ocean dumping of wastes, two conventions were adopted:

■ The Convention for the Prevention of Marine Pollution by Dumping from Ships and Aircraft, Oslo, 1972.

■ The Convention on the Prevention of Marine Pollution by Dumping of Wastes and Other Matter, London, 1972.

On the recommendation of the Stockholm conference, the United Nations General Assembly established the United Nations Environment Programme (UNEP). Reduction and control of marine pollution continued to receive attention, particularly pollution from onshore activities, exploration, and exploitation of the seabed, and more recently, the atmosphere. One result of this attention was the 1973 International Convention for the Prevention of Pollution from Ships (modified by the protocol of 1978). The convention lays groundwork for a comprehensive system to control ship discharges of oil, noxious substances, sewage, and garbage.

Among other actions, international conventions have established dumping rules for substances by category. One category, substances on a so-called grey list, may be dumped if they are in trace amounts or if they are rapidly rendered harmless in the sea; dumping those on a black list requires special authorization.

Except for shipping, worldwide agreement on rules and procedures to control marine pollution have not been developed. The need for them has long been recognized. The 1982 United Nations Convention on the Law of the Sea provided a broad agenda for such action. One initiative—to control marine pollution from land activities—was addressed by UNEP and resulted in the 1985 adoption of the "Montreal Guidelines" that propose methods of dealing with sewage, oily urban runoff, and other sources of pollution.

Since 1972, several regional conventions have been adopted:

■ The Convention for the Prevention of Marine Pollution from Land-based Sources, Paris, 1974.

■ The Convention on the Protection of the Marine Environment of the Baltic Sea Area, Helsinki, 1974.

■ The Convention on the Protection of the Environment between Denmark, Finland, Norway and Sweden, Helsinki 1974.

■ The Agreement for Co-Operation in Dealing with Pollution of the North Sea by Oil and Other Harmful Substances, 1983.

In addition, UNEP has established conventions for developing rules and standards to protect the special characteristics of regional areas from pollution.

Much has yet to be done on atmospheric pollution, and the area north of the East China Sea is not covered by any of these conventions. Nevertheless, the range of international conventions and agreements now in force lays a broad, if incomplete, foundation for controlling marine pollution.

References and Notes

1. M. Nauke, *International Conventions on the Prevention of Marine Pollution: Coastal Strategies*, annex to State of the Marine Environment (United Nations Environment Programme, Nairobi, Kenya, in press).

only slightly since the United States banned most uses in 1976 (38).

Globally, pesticides make a major contribution to human food production. Although accurate production and use data are difficult to obtain, pesticide use appears to be increasing, especially in developing countries. In West and Central Africa, for example, pesticide use is growing rapidly, and, despite increased reliance on more readily degradable organophosphorus and carbamate pesticides, organochlorines such as DDT and chlordane are still widely used.

In South Asia, pesticide use is considerable, with India alone using 55,000 metric tons a year (39).

About 25 percent of these pesticides end up in the sea (40).

Organometal compounds, which have been used for some time as fungicides and biocides, present a different problem. TBT was introduced as an antifoulant in marine paints in the mid-1960s. As a marine antifoulant, it is extremely effective, but some 10 years ago, it was shown to cause shell malformations and mortality in oysters (41). TBT is now known to affect a wide range of invertebrates. TBT concentrations can be high in marinas where many small boats are moored. It is used also on the nets and structures of mariculture facilities and on ships. France, the

United Kingdom, and several U.S. states have taken steps to restrict the use of TBT.

Metals

The mercury poisoning at Minamata, Japan, in the mid-1950s, spurred widespread concern over metal pollution in the oceans. Metals entering the marine environment do so directly through industrial discharge, dumping, and mining and indirectly through rivers. Some, particularly lead and mercury, are carried in significant amounts in the atmosphere.

Apart from isolated hot spots near industrial discharges, metal concentrations in the sea are considerably below those that produce toxic effects in organisms. At most, metal pollution is a minor hazard in the marine environment (42). To ensure protection of public health, however, monitoring should be continued where marine concentrations are elevated.

Radioactive Substances

Radioactive substances are present everywhere in the environment, including the oceans. They are derived naturally from Earth's crust and cosmic radiation. Since the 1940s, however, people have contributed radionuclides of their own making—through fallout from nuclear weapons testing. Because the fallout is dispersed globally in the atmosphere, concentrations in the oceans are negligible compared with concentrations of natural radionuclides.

Accidents at nuclear power plants are a potential source of radiation pollution. Of the cesium-137 released during the 1986 accident at Chernobyl, 6.7 percent reached the oceans. Although increased levels were detected in the Baltic Sea, the North Sea, and the Mediterranean, the effects on marine organisms and human exposure through consumption of seafood were negligible (43).

The two other sources are discharges from nuclear power and nuclear reprocessing plants and the disposal of low-level radioactive material such as that used in research and medicine. These discharges are under international control, which includes authorization and monitoring. Dumping radioactive material at sea was suspended under the London Dumping Convention in 1982 (44).

Pollution Control

Although pollution of many coastal areas is serious, corrective actions are possible. The decrease in DDT concentrations in fish and shellfish in the United States and Sweden demonstrates the possibilities of change (45) (46). Similarly, recent reports indicate that TBT-damaged oyster beds are recovering in France and the United Kingdom since the compound was banned there (47). These successes involved single substances or groups of substances in individual situations. Any strategy for protecting the marine en-

vironment must be broader, dealing with a range of pollution sources from sewage to runoff to industrial discharges. Box 11.2 lists international agreements on marine pollution.

Efforts to control marine pollution are presently thwarted by serious gaps and other deficiencies in scientific knowledge, interjurisdictional policymaking, and political will. GESAMP identified several actions needed to address the global problem of worsening coastal pollution:

■ The vulnerability of the coastal zone demands its protection by policies that call for both national and international action.

■ Coastal effects of various inland activities, including land-use operations and manipulation of water cycles, should be taken into account at the planning stage of these activities.

■ To control eutrophication, changes are needed in sewage disposal techniques and agricultural practices.

■ Public health standards regarding sewage pollution of beaches and shallow waters should be upgraded and their enforcement strengthened; consideration should be given to monitoring seafood sold for public consumption.

■ The problem of plastic litter should be assessed; the production and use of alternative materials less harmful to the environment should be encouraged.

In addition to these actions, GESAMP cited three critical gaps in current scientific knowledge of the ocean environment:

■ Research is urgently needed on the nature of unusual phytoplankton blooms—red tides—and the events that trigger them.

■ Further research should be encouraged on the relationships between chemical-contaminant concentrations in water and sediments and their effects on organisms.

■ Silt *per se* (mainly from runoff) should be recognized as a marine pollutant and its effects on oceans should be assessed (48).

REGIONAL SEAS: THE CARIBBEAN

The Caribbean is called the American Mediterranean—the sun-and-sand vacation spot convenient to both North and South Americans. The islanders of the Caribbean depend mainly on tourism for their livelihood; they brought in $4.4 billion in tourist revenue in 1988 (49). The area's major pollution problems might be seen as those threatening the beaches and crystal-clear waters that attract tourists, for without these natural assets, the economies of most of the tiny islands would be seriously depressed (50).

The wider Caribbean is also a major oil-producing area and, not surprisingly, oil is a major pollutant. Around urban areas, sewage threatens the tourist trade— most sewage in the Caribbean is discharged untreated into rivers or the sea (51). In fact, in all Latin America, an estimated 98 percent of the sewage is not treated (52).

The wider Caribbean region is composed of 40 states and territories, which encircle two connected basins: the Gulf of Mexico (U.S. gulf states and Mex-

Figure 11.1 The Wider Caribbean: Distribution of Marine Living Resources

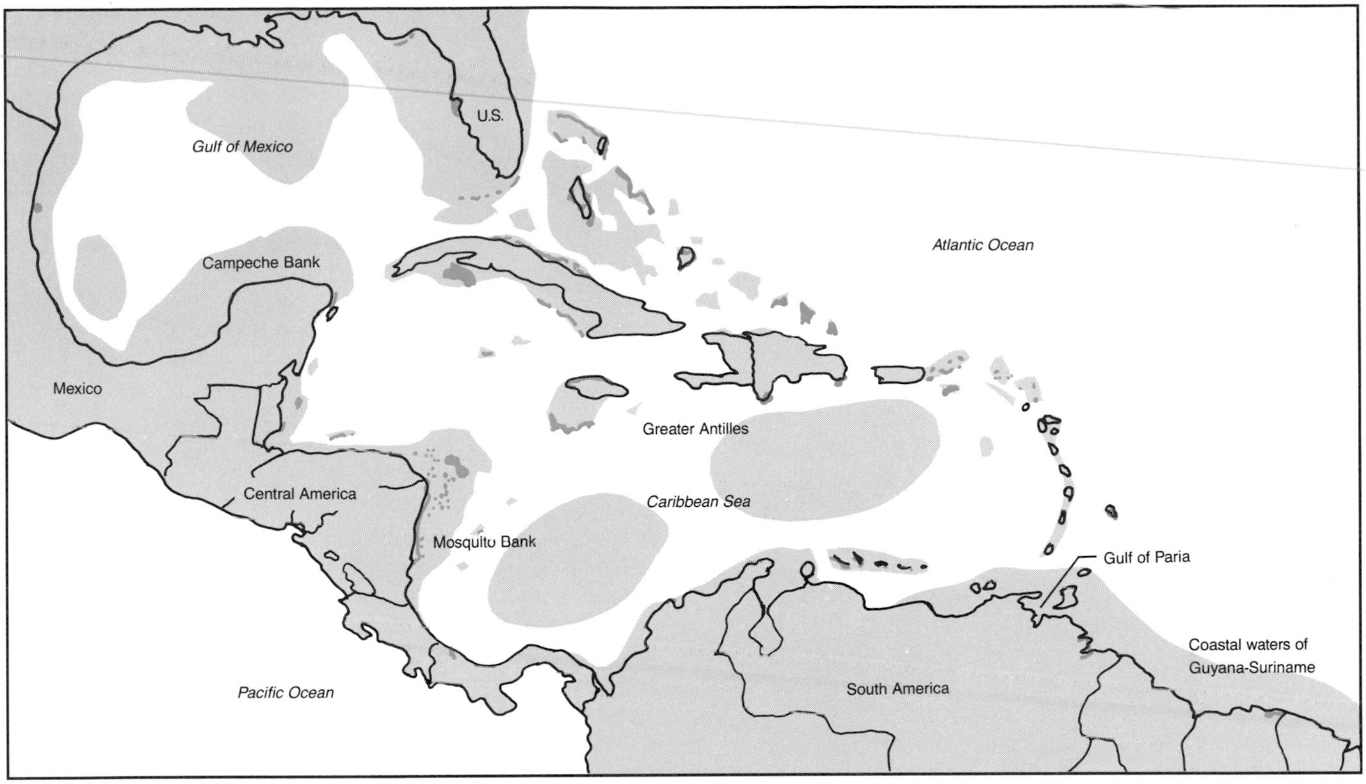

Source: World Conservation Union, Gland, Switzerland, and James Dobbin Associates, Alexandria, Virginia.
Note: Marine Living Resources include conch, shrimp, lobsters, finfish, crocodiles, sea turtles, seabirds and wading birds, manatees, and whales. Not all species are found in all shaded areas.

ico) and the Caribbean Sea (the Central American countries except El Salvador; the South American countries of Colombia, Venezuela, Guyana, Suriname, and French Guiana; and the Caribbean islands from the Bahamas through the Greater and Lesser Antilles. (See Figure 11.1.) These nations are culturally and economically diverse (53).

In 1989, the countries and the U.S. states bordering the Caribbean had a population of 236 million (54) (55) (56). All the islands in the Antilles chain have more than 100 inhabitants per square kilometer. Colombia and Venezuela, on the other hand, have vast unsettled interior areas.

The Physical Setting

The water flow in the wider Caribbean is dominated by two major currents, the Caribbean and the Antilles; both flow from east to west, then clockwise around the basin. These currents determine the general dispersal patterns of pollutants (57). The drainage area of the wider Caribbean region is about 7.5 million square kilometers and includes the basins of two of the world's largest rivers: the Mississippi and the Orinoco (58).

Both the Gulf of Mexico and the Caribbean Sea have deep basins in their centers that are fairly unproductive. Except in a few places, there are no sig-

nificant upwellings that bring nutrients from deep waters to support a fishery. Further, because the surface water temperature remains about 27 °C year round, there is no forced turnover in which the upper waters cool and sink, pushing nutrient-rich water to the surface (59). These central basins are not polluted, but conditions indicate that were they to become polluted, they would remain so for a long time. Most marine organisms in the Caribbean depend on coastal mangroves, estuaries, and coral reefs for feeding grounds (60).

Marine Resources

Figure 11.2 shows the distribution of living marine resources: fish, marine mammals, sea birds, and turtles. The darkest areas indicate the regions of highest biological diversity. The most significant fisheries in the area are on the Campeche Bank in the Gulf of Mexico, the Mosquito Bank in the Caribbean off the coasts of Honduras and Nicaragua, the Gulf of Paria between Trinidad and Tobago, and the coastal waters of the Guyana-Suriname area. The United States and Mexico maintain large mechanized shrimp fisheries in the gulf, and vessels from Japan, Korea, Taiwan, the United States and other countries fish for yellowfin tuna, swordfish and dolphin in deeper Caribbean waters (61). Most coastal fishing is artisanal. About 200 species are found near coral reefs including grouper, snapper, grunt, goatfish, and parrotfish. Lobster, conch, and other valuable shellfish are also found nearshore. Fish landings in the wider Caribbean in 1987 totaled 3 million metric tons (62) (63).

Many larger forms of marine life have been hunted to near extinction. The West Indian manatee is legally protected in the Dominican Republic, the United States, and Jamaica, but it is still hunted for food in some areas (64). Six species of sea turtle are found in the wider Caribbean, including the endangered Kemp's Ridley turtle. Turtles are threatened by shrimp boat nets, the development or disturbance of their nesting beaches, ingestion of tar balls and plastic debris, and, in some areas, direct hunting (65). In 1989, the United States required turtle excluder devices on all shrimp boats more than 25 feet long during certain months (66)(67).

Spectacular coral reefs throughout the region, especially off the coasts of Belize, Bonaire, the Bahamas, and the Lesser Antilles, attract tourists and divers from the Americas and Europe. Unfortunately, one survey found that reefs declined significantly between 1975 and 1985 in most countries (68). Reefs are degraded by silt and sewage from coastal areas and are destroyed by fishermen and tourists (69).

The wider Caribbean is a major oil-producing area (70). In 1988, onshore and offshore production in the Mexican gulf region, Venezuela, Colombia, Trinidad and Tobago, and the U.S. gulf states was 403 million metric tons, compared with 198 million metric tons from Saudi Arabia (71). The Caribbean is also a major shipping crossroads for oil tankers. The areas at most risk from accidental oil spills are narrow passages along the tanker routes where an error in judgment or navigation could be disastrous (72).

Pollution Problems

Spilled oil is the area's most widespread pollution problem. The International Maritime Organization (IMO) estimates that 6.7 percent of the total offshore production is lost through spills into the marine environment as a consequence of pipeline accidents, blowouts, platform fires, overflows, and malfunctions and other minor occurrences. In 1989, the 1978 blowout of a Mexican oil company's offshore oil platform that spilled 470,000 metric tons of oil, still held the record as the world's largest offshore oil spill (73). High levels of petroleum are generally found in Caribbean waters, and windward-exposed coasts are contaminated with tar. Many beaches have been severely damaged by recreational activities, and others in Curacao, Bonaire, and Grand Cayman are totally unusable (74).

Other pollution problems are localized, but they can be severe, especially around large urban or industrial areas (75). A 1974 Pan American Health Organization (PAHO) survey found that less than 10 percent of the area's sewage was treated, and a PAHO official believes that the situation had changed little in 1989 (76) (77). The bulk of the sewage is discharged untreated into rivers or harbors or is piped to cesspools, where it seeps out and pollutes the groundwater. Metropolitan centers with populations of more than 1 million, such as Havana and Caracas, have serious water pollution problems. Panama has polluted Panama Bay with inadequately treated sewage and industrial discharges; in Venezuela, beaches north of Caracas have been polluted; and in Jamaica, the pollution of Kingston Harbor is well documented (78). In Colombia, Cartegena and Barranquilla have heavily polluted bays and coasts (79).

In addition to urban sewage, direct discharge from resort hotels onto swimming beaches can be a health and pollution problem. Many of these hotels are equipped with prefabricated treatment plants, but the plants are often overloaded or are maintained inadequately and thus become ineffective (80).

Another major coastal pollution problem is sedimentation caused by land clearing. More than 2 million hectares of Caribbean tropical forests are cleared annually, but only 70,000 hectares are replanted (81). Erosion rates are high in some Caribbean and Central American countries, where soil from denuded steep slopes is washed to the sea by heavy tropical rains. In several Central American countries, erosion rates reached 500 metric tons per hectare (82). By comparison, U.S. cropland is eroding at a rate of 18 metric tons per hectare (83).

Sedimentation also stems from agricultural runoff and mining wastes. Sediments can smother sea grass beds and coral reefs, clog streams, and disrupt water flow. Fertilizers can cause algal blooms and resultant eutrophication of enclosed bays. Pesticides can be absorbed by fish, especially groupers and other large fish. One study found DDT and DDE in the grouper tissue in the Gulf of Mexico and the Grand Bahamas (84). (See New Assessment of Ocean Pollution, above.)

The mineral causing the most pollution in the region is bauxite. Used in aluminum production, bauxite is mined in three islands and two South American countries. "Red mud," a waste product from bauxite processing, is highly alkaline and toxic. It is traditionally disposed of in limestone pits, where it leaches into groundwater. Mining beach sand is the single most destructive human activity affecting beach erosion, according to UNEP (85). Sand is mined mainly for road building materials used on the islands. Beach mining operations have seriously disturbed coastal and marine ecosystems on several islands (86).

A recent threat is the transboundary dumping of hazardous wastes. UNEP reports Haiti's complaints that a foreign ship had illegally offloaded wastes onshore and that several Caribbean governments have been approached by private entities seeking disposal sites (87).

Regional Seas Programme

To help coastal area governments to assess their marine resources and pollution problems and then act on these problems, UNEP established 10 Regional Seas Programme involving 120 countries. Each *World Resources* volume examines one of these program areas.

The Caribbean Environment Programme, begun in 1976, is a joint project of UNEP, the Economic Commission for Latin America and the Caribbean, and the 28 concerned countries. In 1981, 22 Caribbean countries adopted an action plan that sets priorities for regional cooperation. Two years later, at Cartagena, Colombia, 15 countries signed the Convention for the Protection and Development of the Marine Environment of the Wider Caribbean Region. As of 1989, 16 had ratified the convention, which pledges signatories to control pollution and protect the environment.

Each regional seas action plan, outlines the region's environmental priorities and identifies areas in which countries can readily cooperate. The action plan of the Caribbean Environment Programme, like many others, stresses the need for monitoring and assessing pollution, cooperating on a plan to respond to oil spills, forming watershed management guidelines, improving environmental health services such as clean drinking water and sanitation, and promoting environmental education.

A central office, the Regional Coordinating Unit, was set up in Kingston, Jamaica, in 1987. Participating nations are committed to contribute to a trust fund; UNEP also provides some funding (88).

Most of the work of the program is in the area of education and training, and several specific areas have been addressed. A protocol incorporated into the convention pledges ratifiers to cooperate to clean up oil spills and notify other countries of spills near their borders. With IMO, the Caribbean Environment Programme held workshops to train 35 officials in major spill response (89). A cooperative effort with the Intergovernmental Oceanographic Commission involved training scientists how to monitor pollution in their areas. Participants were encouraged to form a network of institutions to communicate on local environmental problems. For policymakers, the program has put forth 39 recommendations for the management of threatened or vulnerable sea turtles (90). Working with the Caribbean Conservation Association, a nongovernmental organization, the program has tried to raise public environmental awareness levels.

In assessing the progress of the wider Caribbean toward environmentally sustainable development, the Caribbean Environment Programme recently reported that Caribbean countries face "an overwhelming array of environmental problems resulting from intensive exploitation of coastal and marine resources, coupled with ineffective and often inappropriate approaches to development planning."

Constraints include weak public support, inadequate funding, lack of trained personnel and equipment, uncertainty over proposed pollution standards, unsupportive judicial systems, and a reluctance in many countries and territories to confront powerful economic interests (91). The report saw raising public awareness of environmental issues, obtaining financial and technical assistance from richer nations, and enhancing local educational systems as the keys to spurring political action on environmental problems in the Caribbean (92).

FOCUS ON ANTARCTICA: SHARING RESOURCES IN THE LAST WILDERNESS

Antarctica is a vast silent continent, virtually barren inland but teeming with birds and marine life along its coasts and in the surrounding Southern Ocean.

In 1959, 12 countries agreed to preserve the region south of 60° south latitude—about 36 million square kilometers or 10 percent of Earth's surface—as a unique scientific laboratory and to designate the

Figure 11.2 Distribution of Krill in the Southern Ocean around Antarctica

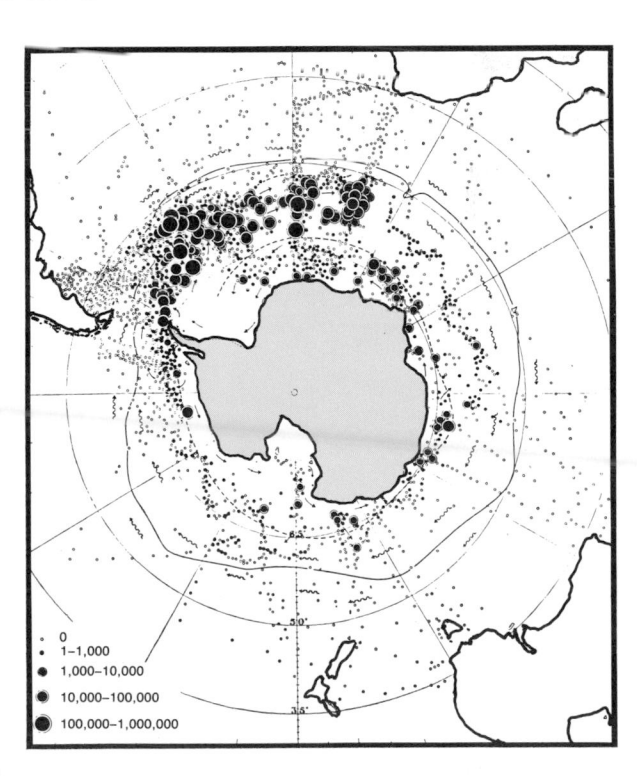

0
· 1–1,000
● 1,000–10,000
● 10,000–100,000
● 100,000–1,000,000

Source: *Antarctic Science,* D.W.H. Walton ed. (Cambridge University Press, Cambridge, U.K., 1987), p. 50.

Figure 11.3 Fishery Harvest in the Southern Ocean, 1970–87

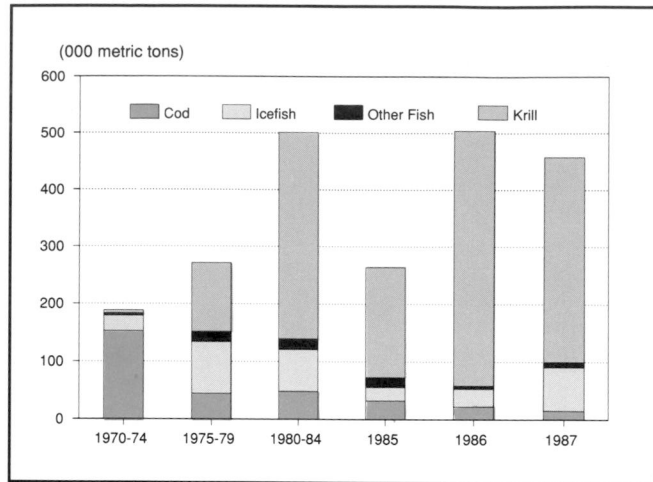

Source: Chapter 23, "Oceans and Coasts," Table 23.4.

area an international zone of peace (93). They concluded an agreement, the Antarctic Treaty, which gives equal standing to seven countries claiming parts of the region and five countries that do not recognize these claims. Today 39 countries are party to the treaty, and, since its signing, three more international agreements have been concluded.

The 1980 Antarctic Marine Living Resources Convention establishes one of the most advanced systems for management of living resources in the world. Whereas most countries manage their fisheries and wildlife for the health of a few target species, the 1980 convention calls for management of all species to maintain the ecosystem.

The 1980s presented new, often competing interests in Antarctica that challenge its governance. The three greatest challenges are the possibility of mineral-extraction activities, growth in the number and scale of scientific research programs, and increased tourism.

Because the original 1959 Antarctic Treaty may come up for review after 1991, it is time to assess the effectiveness of Antarctica's unique system of governance—and its success in promoting peaceful scientific research, conserving and managing Antarctic resources, and preventing changes that could adversely affect Antarctica's vital influence on global weather, climate, and ocean circulation.

Antarctic Marine Living Resources

The interior of Antarctica is almost barren, but the coasts and coastal waters support 35 species of penguins and other birds, 6 varieties of seals, 12 whale species, and nearly 200 types of fish (94). For more than 200 years, seal hunters, whalers, and fishermen have been drawn to the remote but productive South-

ern Ocean. Often the result was a serious depletion of species. But in the late 1970s, spurred by a growing commercial interest in harvesting krill, a small (4–7 centimeter) shrimp-like crustacean, Antarctica's governing countries negotiated the Convention on the Conservation of Antarctic Marine Living Resources (CCAMLR). Unlike any previous wildlife management agreement, it sets conservation of the entire ecosystem as its primary goal.

Seals

As early as 1784, large numbers of seals were hunted in the Southern Ocean (95). By the early 19th Century, the elephant seal population was substantially reduced and fur seals were near extinction (96). Because seal hunting there has been of minor importance during the 20th Century, the seal populations have largely recovered (97). To protect them from further exploitation, the parties to the Antarctic Treaty agreed to the Convention for the Conservation of Antarctic Seals (CCAS), which entered into force in 1972 (98). CCAS bans hunting Ross, elephant, and fur seals and sets quotas for other species; it establishes closed hunting seasons and areas and sets up three seal reserves. Each year, however, some seals are hunted for dog food and scientific research; in 1987–88, for example, Soviet researchers reported killing 4,000 seals for experimental purposes (99) (100).

Whales

Whaling began in Antarctica in the early 1900s, initially with shore-based processing facilities that by the mid-1920s were replaced by more efficient shipboard processing (101). Whaling has taken a severe toll on the marine mammals. In 50 years, the population of blue whales in the Southern Ocean was only 5 percent of its original size, humpbacks 3 percent, fin whales 21 percent, and Sei whales 54 percent (102). Recent assessments suggest that this bleak picture may overestimate whale populations (103). Antarctic whalers hunted the largest, most valuable species first, then turned to progressively smaller species as they depleted the stocks. Only the smallest of the Southern Ocean whales, the Minke, has not been substantially reduced.

Since 1946, Southern Ocean whaling has been regulated by the International Whaling Commission (IWC) (104). In 1982, IWC issued a moratorium on commercial whaling to begin in the 1985–86 Antarctic summer. Four countries objected to the moratorium, and whalers from two of them—Japan and the Soviet Union—continued to harvest Minke whales in the Southern Ocean through the 1986–87 season (105). Today, Japanese whalers hunt Minke whales for purportedly scientific purposes, taking some 300 in 1987–88 (106).

Whales have low reproductive rates, raising concern that some species may not recover from exploitation. For example, the sub-Antarctic right whales, which have been protected since the 1930s, showed no evidence of recovery until the 1970s (107).

Fish and Krill

More recently, bottom-dwelling fish and krill have become the target of the Southern Ocean harvest. Harvesting bottom fish began in the late 1960s with cod fishing near the island of South Georgia (108). Since then, fishing harvests increased to a peak of about 400,000 metric tons in the 1969–70 season; by the late 1970s, several fish populations were severely depleted (109). (See Figure 11.3.)

Krill fishing began in 1976 and rapidly reached a peak harvest of 528,201 metric tons in 1981–82 (110). The annual catch, mainly by the Soviet Union and Japan (77 percent and 20 percent, respectively, of the 1988–89 total), fluctuates because of marketing and processing problems (111). The Soviets process krill into fishmeal for animal feed and the Japanese use it as a shrimp substitute for human consumption. Well managed, Antarctic krill could become an important world source of protein. The krill biomass is estimated at 250–600 million metric tons, with an annual production of roughly 100–500 million metric tons (112) (113). Such a resource could likely sustain a harvest of tens of millions of metric tons annually (114). By comparison, the 1988 worldwide fresh- and saltwater fish harvest was 97 million metric tons (115). But because krill are a key in the Southern Ocean food web, any reduction in their abundance could profoundly affect all mammals, birds, and fish in the ecosystem. It is thought that before their depletion, whales consumed more than 150 million metric tons of krill each year; today they consume an estimated 40 million metric tons (116).

Environmental change threatens the krill population. Biologists fear that ozone depletion in the stratosphere could reduce krill populations indirectly by increasing ultraviolet (UV) radiation in the Antarctic. High UV radiation retards the photosynthetic rate of phytoplankton, reducing the amount of food available to krill and their other predators (117).

CCAMLR: A New Method of Marine Management

International fishery-management agreements usually try to maximize or sustain yields by managing the stock of a target species. These agreements largely ignore the fact that changes in the population of one species will likely affect other species, particularly the prey and predators of the target species. Recent advances in understanding marine ecosystems have made fishery managers increasingly aware of the need to consider interactions among species when setting management policies, but implementing this policy is still considered innovative.

Because krill are the primary food source for many Southern Ocean species, multispecies considerations are clearly important in managing their fishery. A heavy krill harvest, for example, could substantially affect the populations of some whales, seals, and marine birds in the Southern Ocean. Instead of focusing on krill management, CCAMLR's primary objective is conserving the Southern Ocean ecosystem; it defines the area covered by the agreement as the entire ecosystem. It requires managers to evaluate the impact of the krill harvest on other species in the ecosystem as well as on the krill fishery. This approach would maintain the balance among ecosystem species. Further, ecosystem changes that are not reversible within two or three decades will be minimized or prevented.

CCAMLR established a scientific committee to analyze fishery and ecosystem data and make management recommendations; its commission sets management policies (118).

The success of CCAMLR's ecosystem approach is hindered by two factors. First, obtaining the basic research needed to understand the structure and dynamics of the ecosystem is difficult to obtain and is costly; the Southern Ocean is remote, weather conditions are severe, and it is more than twice the size of South America. Second, CCAMLR management decisions must be made by consensus—all parties must agree to fishing restrictions before they can be imposed. Observers question the adequacy of this arrangement because of initial delays in imposing harvest restrictions on seriously overfished species (119).

Since the 1987 CCAMLR annual meeting, regulations closing certain zones to harvesting and restricting fishing seasons have been adopted (120) (121). Working groups have been established to evaluate technical information on the status of species, and the Commission has paid close attention to recommendations made by its scientific committee.

Pollution Threats in a Pristine Environment

Antarctica is a unique natural laboratory for scientific research. Its relatively pristine environment, free from most sources of pollution, provides a baseline for detecting the chronology and effects of both natural phenomena (e.g., volcanic eruptions) and human activities (e.g., nuclear explosions) in other parts of the world. Antarctic ice cores yield atmospheric records covering millions of years, offering clues to past and future climate changes. Monitoring pollution in this relatively untouched environment is an early warning system of increasing global hazards (122). For example, detection of DDT in penguin fat and eggs indicates the distances the chemical has been transported through the marine food web (123). Studying the movement of the cold Antarctic waters is fundamental to understanding ocean circulation and heat balances between oceans and atmosphere— a crucial ingredient to predicting global warming (124).

Pollution is by no means widespread in Antarctica, but the situation may be changing. Both the number of countries establishing research programs and the number of tourists seeking to visit are increasing. These developments increase the need for energy, which raise the risk of oil spills and exacerbate the problem of waste disposal; they also subject an ever greater part of the continent to human impacts, undermining its value as a scientific reserve and diminishing its natural beauty. That minerals may be discovered and mined in Antarctica is a major cause for concern.

The Bahia Paraiso Oil Spill

In January 1989, the Argentine ship *Bahia Paraiso*, carrying fuel to Argentine research stations in Antarctica, was grounded on an underwater reef near the Antarctic peninsula and spilled an estimated 693 cubic meters of diesel fuel (125) (126). The spill may have compromised some long-term studies of Antarctic species; it may also make it impossible to interpret research on the effects of increased UV radiation produced by the ozone hole over Antarctica (127).

Although a scientific assessment of the impacts of the *Bahia Paraiso* oil spill is incomplete, two conclusions are apparent; first, the spill adversely affected marine species and fouled the environment in the area. It killed all skua chicks and caused significant mortality among cormorant chicks, intertidal limpets, and seaweeds in some areas (128). Second, and perhaps more significant in the long term, effects of the oil on marine life may distort results from research and monitoring programs, some that have been ongoing for 20 years (129). On the other hand, the spill offers opportunities to document long-term effects of oil spills in the Antarctic environment. Natural recovery from spilled oil takes place more slowly in low polar temperatures than in temperate climates (130).

Research Station Pollution

In the past decade, the number of Antarctic research programs has nearly doubled, and the number of investigators who remain during winter months has risen from about 800 to more than 1,000 annually. The impact of this increase is concentrated along the coasts where most research stations are located. These same ice-free areas are the natural habitat of native species (131). But as the community grows, exhaust from vehicles, disposal of solid and toxic wastes, and other common forms of pollution from human settlements increase. When Greenpeace, an international environmental group, spotlighted practices at the U.S. McMurdo Sound Station of dumping untreated sewage into the sea and burning garbage in an open pit, the station installed a primary sewage treatment facility and began recycling solid waste (132).

Tourism

In 1987, 2,400 tourists visited Antarctica (133). Large tour ships carried upward of 100 persons each (134), and small sailing, mountaineering, and skiing expeditions brought others. Passengers on the tour ships usually land on relatively circumscribed areas or at research stations, often disturbing local breeding sites, trampling vegetation, or interfering with scientific research (135). In general, their impacts on the Antarctic environment are produced by the ships on which they travel. The effects of small expeditions depend on the area visited and the purpose of the visit.

A resurgence of commercial tourist flights over Antarctica began in 1987–88, following a moratorium in the wake of the tragic Air New Zealand DC-10 crash on Mount Erebus in November 1979 that killed all 279 passengers and crew. The real threat of air-traffic growth, however, is posed by flights to blue-ice airfields. Blue ice is highly compacted and is strong enough to permit wheeled aircraft to land at any time of the year. In 1988, an interim stop at a blue-ice airfield allowed the first commercial tourist trip (for a fee of up to $35,000 per person) to reach the South Pole (136). Large-scale hotels in Antarctica are also proposed to take advantage of blue-ice tourism possibilities.

Mineral Exploration

Without doubt, the most serious threat facing the Antarctic environment is the possibility of mining operations. Although no commercially attractive minerals have been discovered in Antarctica, evidence of offshore oil has sparked interest in exploration. Oil drilling raises the specter of massive crude-oil spills similar to the 1989 *Exxon Valdez* spill in Prince William Sound, Alaska. Mineral extraction could significantly affect Antarctica's air and water quality, ice flow patterns, native species, terrestrial and marine environments, and ongoing scientific research. Its potential impacts on regional and global marine environments and climate and weather patterns are more speculative. Antarctica is a massive heat sink that exerts a major influence on global weather patterns. Possible cumulative effects of particulate air pollutants—in part derived from mineral exploration—on this process should be investigated (137).

Antarctic Treaty System

Four international treaties govern activities in Antarctica and the Southern Ocean. They build directly on the 1959 Antarctic Treaty, which establishes the area south of 60° south latitude as a demilitarized zone and a place for cooperative scientific investigation. It establishes rules for countries conducting scientific investigations in Antarctica and lays a foundation for assessing and regulating other activities that affect the continent.

Disputed territorial status is still an obstacle to regulating activities in Antarctica. Seven countries claim 85 percent of the continent, but no others recognize these claims; worse, the claims of Argentina, Chile, and the United Kingdom overlap (138).

Additional Agreements

As the possibility of new Antarctic activities has arisen, additional agreements, some discussed above, have been concluded.
■ The 1972 Convention for the Conservation of Antarctic Seals.
■ The 1980 Convention on the Conservation of Antarctic Marine Living Resources.
■ The 1988 Convention on the Regulations of Antarctic Mineral Resource Activities.

The first two agreements are in force; CRAMRA will enter into force as soon as 16 nations ratify it. In fall 1989, Australia and France declined to sign the minerals agreements and proposed that the continent be declared a wilderness reserve. If the two countries won't sign, CRAMRA cannot take effect. Without CRAMRA, Antarctica could be left open to unregulated mining. Some treaty nations, such as the United Kingdom, oppose any permanent ban on mineral development, which presumably would be part of a wilderness reserve treaty (139).

Several global agreements on ship safety and marine-pollution control in the Southern Ocean are administered by IMO. The 1989 UNEP Convention on the Control of Transboundary Movements of Hazardous Wastes and their Disposal, not yet in force, includes banning the export of hazardous and other wastes for disposal in the area south of 60° south latitude. The 1946 IWC whaling convention regulates whaling by its signators in the Southern Ocean. Other broad international agreements that apply to Antarctica include the 1958 Geneva conventions on ocean law and the 1982 United Nations Convention on the Law of the Sea, not yet in force.

Conservation Strategies

In 1986, the World Conservation Union, formally the International Union for Conservation of Nature and Natural Resources, joined the Scientific Committee on Antarctic Research to produce a conservation plan for Antarctica and is completing a long-term strategy for managing and conserving terrestrial and marine resources in the entire region (140).

In 1964, parties to the 1959 Antarctic Treaty declared Antarctica a special conservation area; they have since adopted measures to preserve its unique environment and value for scientific research. In response to mounting criticism over the effectiveness of these measures, the countries expanded environmental protection, most notably by adopting environmental impact assessment procedures in 1987. In

light of increasing human activity, however, a comprehensive approach to Antarctic conservation is clearly needed. The possible renewal of the Antarctic Treaty after June 1991 provides an opportunity to do so.

On the agenda of the Antarctic Treaty Consultative Meetings are proposals to upgrade the code of conduct for waste disposal, reduce the adverse impacts of tourism, and develop further the protected-areas system for safeguarding scientific research, biologically important areas, historic sites and monuments, and outstanding geological, scenic, recreational, and wilderness areas. The *Bahia Paraiso* oil spill is expected to generate additional proposals for prevention, response, and liability for marine pollution incidents and for improvement of weather and charting services for ships and aircraft. The larger question is whether the parties will integrate their previous agreements into a comprehensive environmental protection strategy.

EFFECTS OF CLIMATE CHANGE ON OCEANS AND COASTS

The oceans face two major uncertainties in regard to climate change. The first is the extent to which the oceans would moderate global warming (see Chapter 2, "Climate Change: A Global Concern"); the second is the extent to which climate change would cause the sea level to rise, increase severity of storms, and change fish abundance and location.

Sea-Level Rise

Global climate and sea levels have varied considerably over geologic time. Oceans once covered vast areas of what is now land, only to recede as huge amounts of water were frozen in glaciers during ice ages. For the past 100 years, the sea level has been rising 1–1.5 millimeters per year (141) (142). This rise is attributed to thermal expansion of the upper waters of the oceans, which results from global warming (143).

A number of scientists believe global warming will accelerate a sea-level rise by further expanding sea water and melting alpine glaciers and polar ice sheets. Sea levels could rise 50–150 centimeters by the year 2087, according to a National Academy of Sciences study (144). In late 1989, after considering new measurements of snow and ice accumulation in Greenland and Antarctica, scientists scaled back their assessment of sea-level rise to 30 centimeters by 2050. In one study, a University of Colorado team used a pair of satellites to measure the Greenland ice sheet and determined that it was growing—not shrinking, as had been thought—in a warming climate. A University of Wisconsin team found that snow and ice were also accumulating in Antarctica (145).

Effects on Coastlines

A sea level rise of up to 100 centimeters over the next 100 years would inundate up to 20 kilometers inland from the current shoreline and severely affect the world's densely populated delta areas (146). The very existence of low-lying island nations, such as the Maldives, would be threatened.

If the sea rises more rapidly than the marshes can reestablish themselves inland, great losses of wetlands would occur, mostly in developed areas where protective structures prevent the inland migration of wetlands (147) (148).

Over the past several thousand years, coastal marshes have generally kept pace with sea-level rises through sedimentation and peat formation (149). But throughout the world, major rivers have been dammed, leveed, and channeled, curtailing the amount of sediment that reaches the deltas. Even at today's rate of sea-level rise, the sea is reclaiming substantial amounts of coastal land in Egypt and Mexico; at the same time, marshes in Louisiana are being starved from a lack of sediment (150).

About 70 percent of the world's sandy coasts have eroded in the past few decades (151). Any sea-level rise is certain to exacerbate this loss. Global warming can be expected to increase rainfall and flooding. Coastal flooding increases the vulnerability of the land to erosion; a higher sea level results in higher storm surges and decreased drainage.

Coastal barrier islands, continually rebuilt by the landward transport of sand during major storms, will likely survive a rising sea level. However, erosion may undermine developments on coastal and barrier islands (152). The potential for increased erosion would be particularly important to seaside resorts, which include some of the world's most economically valuable and intensively developed land (153).

Climate Change and Tropical Storms

Although the complexities of ocean dynamics make it difficult to project the effects of global warming on ocean-driven storms, it is generally agreed that increases in sea-surface temperatures will profoundly affect storm variability (154). Atmospheric temperature and the temperature of the top 100 meters of the ocean's water column are closely related (155). Higher air temperatures mean higher sea-surface temperatures (156), and higher sea-surface temperatures markedly alter ocean currents (157). Changes in the ocean currents, in turn, alter atmospheric conditions and storm patterns.

As the major ocean currents circulate, they distribute heat from the equatorial zone to the poles. Tropical storms also distribute heat. Atlantic and eastern Pacific hurricanes and western Pacific typhoons are important influences on atmospheric circulation, transporting precipitation and latent heat toward the poles.

One frequently cited study concludes that changes in the maximum intensity of tropical cyclones are strictly related to changes in sea-surface temperature (158). It concludes that a doubling of atmospheric carbon dioxide will increase sea-surface temperatures of 2.3°–4.8° C. If such an increase occurs, the destructive potential of tropical storms could increase as much as 60 percent (159). A recent atmospheric model predicts that a rise in sea-surface temperature may strongly influence the size as well as the intensity of tropical storms (160).

Because a warmer ocean surface would accelerate normal evaporation, it is also likely that these storms would increase precipitation. If an increase in the kinetic energy and sustained winds of a storm were coupled with an increase in sea level, the potential damage to coastal zones could be severe.

Effects of Global Change on Currents and Fisheries

Fisheries provide a significant source of human nutrition; they supply 14 percent of the animal protein in the world's diet and almost 60 percent in Japan's (161). Climate change can affect fisheries in several ways. First, by warming the surface waters, it could change the distribution of plankton, the beginning of the marine food web upon which all fish ultimately depend for food. Second, again by warming surface waters, it could change the global wind and weather patterns, which in turn could affect ocean currents. Many currents scour nutrients from the bottom or maintain a temperature suitable for fish or larvae, thus altering currents could destroy or change the location of major fisheries. Third, climate change could destroy fish breeding grounds in estuaries either by raising the sea level or by reducing the freshwater input that maintains a salinity level critical for certain organisms.

Effects of Ozone Depletion

Although not linked with global warming, another atmospheric change—ozone depletion—can affect marine life. Ozone depletion increases the amount of biologically harmful solar radiation, specifically, ultraviolet-B (UV-B) rays, that penetrate the ocean's surface. UV-B radiation damages fish larvae and juveniles as well as the phytoplankton base of the food web (162). Shallow-water plants and animals appear to be at greatest risk. Some studies suggest that increased UV-B radiation alters the genetic material of marine plants and animals, thus affecting their growth and reproduction (163). Mutations in phytoplankton could lower the rate and amount of primary production, leading to a base-level nutrition shortage that would reverberate throughout the food web. In addition, because phytoplankton take up large amounts of carbon dioxide during photosynthesis, any decrease in their population would exacerbate the rising levels of carbon dioxide in the atmosphere (164).

RECENT DEVELOPMENTS

VIRUSES ARE ABUNDANT IN OCEANS

New discoveries that viruses are far more abundant in freshwater and sea water than previously thought are causing scientists to revise their thinking about life cycles in these environments.

Four Norwegian scientists found counts of virus particles ranging from 5–15 million per milliliter in marine waters during warmer seasons. These counts are some 1,000 to 10 million times higher than other studies had reported. Marine samples taken during the winter months had low numbers of viruses, suggesting a seasonal variation in virus populations (165). The study found that most virus particles appeared to be free in the water; some were associated with bacteria. Whereas previous studies generally found relatively large viruses in marine environments, the Norwegian study, using new techniques and powerful electron microscopes found relatively small viruses dominant (166).

The results may be important for the study of how oceans affect global climate change. They serve as a repository for carbon, which builds up in the atmosphere because of fossil fuel burning. The cycling of nutrients and carbon by these small organisms may prove to be important to scientists trying to understand the oceans' ability to store carbon (167).

The abundance of viruses may have important implications for the evolution of bacteria. Viruses, bundles of genetic material that cannot reproduce on their own, invade bacteria and other living cells. These cells respond to the instructions of the viral genes by manufacturing new viruses. Bacteria, weakened by manufacturing viruses, eventually burst. The process may help keep bacterial populations in check (168).

Viruses can assimilate genes from one bacterium and pass them on to another, so they may be important in creating new combinations of genes and new species of bacteria. If this process *is* a common mechanism of gene transfer, it might mean that bacteria can adapt more rapidly to new situations than had been thought. For example, they might be able to develop resistance to the antibiotics used in aquaculture. Another possibility is that the traits of pathogenic bacteria in sewage released into lakes or coastal waters might be transferred to indigenous bacteria (169).

RED AND GREEN TIDES INCREASINGLY THREATEN MARINE LIFE

Coastal waters are increasingly plagued by toxic and nontoxic algal blooms that pose a serious threat to marine life.

The frequency of these blooms has been increasing over the past two decades in the coastal seas of both hemispheres. At the same time, the occurrence of paralytic shellfish poisoning (PSP) is increasing globally (170). For example, the dinoflagellate *Gyrodinium aureolum*, which first appeared in the northeast coast of the United States, in a massive bloom along the Norwegian coast in the fall of 1966, has since spread throughout northern European waters. Another PSP-producing dinoflagellate, *Gymnodinium catenatum*, previously found in southern California waters, has spread to the coasts of Spain and Tasmania and to eight new areas in Japanese waters. It is not yet clear whether a bloom is caused by the species moving to new areas or by the proliferation of tiny indigenous populations. Considerable evidence suggests that the increased incidence of blooms is related to the nutrient enrichment of coastal waters and inland seas on a global scale (171).

The red tide is a special plankton bloom in which the dominant species is toxic. Red tides are annual events in many parts of the world. They often kill other marine organisms, damage mariculture production, and cause illness and death among humans who eat shellfish that have accumulated these toxins.

The incidence of red tides seems to be increasing, as shown in the records for Hong Kong harbor. (See Table 11.1.) They are also spreading to new areas. In late 1987, for example, more than 150 people in Canada suffered acute shellfish poisoning from eating blue mussels harvested from eastern Prince Edward Island. The poisoning was caused by a neurotoxin, domoic acid, previously found only in red seaweeds of Japan and the Mediterranean; the neurotoxin was traced to the bloom of a pelagic diatom (172).

Massive Nontoxic Bloom in Adriatic

A recent massive algal bloom in the northern Adriatic was not toxic, but it damaged marine life and coastal activities. In summer 1989, for the second consecutive year, a massive algal bloom suffocated the canals and the 549-square-kilometer lagoon of Venice (173). Similar problems occurred over roughly a 1,000-kilometer stretch of Adriatic coastline from Trieste south to Pescara, threatening both the commercial fishing industry and the region's important tourism industry (174). Adriatic fishermen complained that the algae clogged their engines and tore their fishnets (175).

As algae decompose, they deplete the water of oxygen, thus threatening fish and shellfish. Eutrophication was reportedly so severe along the Adriatic coast near Rimini in mid-1989 that biologists found a large area virtually devoid of oxygen (176).

Two factors are largely to blame for the problem; the load of nutrients and waste carried by the Po and other tributaries into the Adriatic is heavy, and the enclosed Adriatic ecosystem keeps pollutants from being dispersed into the Mediterranean.

11 Oceans and Coasts

The section on the New Assessment of Ocean Population was authored by Professor Alasdair D. McIntyre of the University of Aberdeen, Scotland, who was chairman of the working group on the state of the marine environment, which prepared the report State of the Marine Environment. Box 11.1 on oil spills was written by Hope Robertson, a consultant in Stockton, New Jersey. The regional seas section on the Caribbean was written by Mary Paden, World Resources 1990-91 managing editor. Focus on Antarctica: Sharing Resources in the Last Wilderness was authored by Lee Kimball of the Center for Ocean Law in Washington, D.C., and Walter Reid, an associate with the World Resources Institute. The effects of Climate on Oceans and Coasts was written by Margaret Davidson, executive director of the South Carolina Sea Grant Program, Charleston, South Carolina.

References and Notes

1. Group of Experts on the Scientific Aspects of Marine Pollution (GESAMP), *The Health of the Ocean* (United Nations Environment Programme, Nairobi, Kenya, 1982).
2. Group of Experts on the Scientific Aspects of Marine Pollution (GESAMP), *The State of the Marine Environment* (United Nations Environment Programme, Nairobi, Kenya, 1990).
3. Colin W. Clark, "Bioeconomics of the Ocean," *BioScience*, Vol. 31, No. 3, p. 231.
4. U.S. Office of Technology Assessment, *Wastes in Marine Environments* (U.S. Government Printing Office, Washington D.C., 1987), p. 100.
5. Theodore J. Smayda, "Novel and Nuisance Phytoplankton Blooms in the Sea: Evidence for a Global Epidemic," in E. Granéli and L. Edler, eds., *Proceedings of the Fourth International Conference on Toxic Marine Phytoplankton* (Elsevier Science Publishing Company, New York, 1990) n.p.
6. Christine Lancelot *et al.*, "*Phaeocystis* Blooms and Nutrient Enrichment in the Continental Coastal Zones of the North Sea," *Ambio*, Vol. 16, No. 1 (1987), p. 41.
7. Marco Vighi, "Preliminary Draft Report on Eutrophication in Italian Lakes and Marine Waters," and J. Stern, "Eutrophication in the Mediterranean Sea," papers presented to The Group of Experts on the Scientific Aspects of Marine Pollution, Working Group on Nutrients and Eutrophication in the Marine Environment, Paris, September 1987.
8. I. Koike, "Experiences from Coastal Areas Round Japan," and J. Stern, "Eutrophication in the Mediterranean Sea," papers presented to the Group of Experts on the Scientific Aspects of Marine Pollution, Working Group on Nutrients and Eutrophication in the Marine Environment, Paris, September 1987.
9. *Ibid.*, p. 9.
10. J. Stern, "Eutrophication in the Mediterranean Sea," paper presented to the Group of Experts on the Scientific Aspects of Marine Pollution, Working Group on Nutrients and Eutrophication in the Marine Environment, Paris, September 1987.
11. U.S. Office of Technology Assessment, *Wastes in Marine Environments* (U.S. Government Printing Office, Washington, D.C., 1987), p. 100.
12. *Ibid.*, pp. 62-64.
13. H.J. Lidgate, "Nutrients in the North Sea—A Fertilizer Industry View," in *Environmental Protection of the North Sea*, P.J. Newman and A.R. Agg, eds. (Heinemann Professional Publishing, Ltd., Oxford, U.K., 1988), p. 190.
14. "Human Impact on the Biosphere," *Biology: The Unity and Diversity of Life*, 3rd ed., Cecie Sturr and Ralph Taggart, eds. (Wordsworth Publishing Co., Belmont, California, 1984), pp. 671-672.
15. Notes of the Rapporteur, "Meeting of the Rapporteurs of the Regional Task teams for the Review of the State of the Marine Environment," United Nations Environment Programme, Geneva, November 14-18, 1987, p. 15.
16. R. R. Colvell, "Microbiological Effects of Ocean Pollution," in *Environmental Protection of the North Sea*, P. J. Newman and A. R. Agg, eds. (Heinemann Professional Publishing, Ltd., Oxford, U.K., 1988), pp. 385-386.
17. P.C. Wood, "Sewage Sludge Disposal Options," in *The Role of The Oceans as A Waste Disposal Option*, G. Kullenberg, ed. (D. Reidel Publishing Company, Dordrecht, the Netherlands, 1986), p. 111.
18. United Nations Development Programme (UNEP), "Review of the State of the Mediterranean Marine Development," UNEP, Nairobi, Kenya, draft of November 1987, pp. 18, 22.
19. United Nations Environment Programme (UNEP), "State of The Marine Environment: West and Central African Region," UNEP, Nairobi, Kenya, draft of November 1988, p. 10.
20. United Nations Environment Programme (UNEP), "State of The Marine Environment in The East Asian Seas Region," UNEP, Nairobi, Kenya, draft of November 1988, p. 61.
21. *Ibid.*, p. 39.
22. M.W. Whyte, "Marine Disposal of Sewage and Sludge— Australian Practice," paper presented at a meeting of the Institute of Civil Engineers, April 1989, Brighton, U.K., p. 63.
23. United Nations Environment Programme (UNEP), "The State of The Marine Environment in the Eastern African Region," UNEP, Nairobi, Kenya, draft of November 1988, p. 40.
24. United Nations Environment Programme (UNEP), "State of the Marine Environment in the Caribbean Region," UNEP, Nairobi, Kenya, draft of December 1987, p. 17.
25. United Nations Environment Programme (UNEP), "The State of The Marine Environment in the Eastern African Region," UNEP, Nairobi, Kenya, draft of August 1988, p. 9.
26. United Nations Environment Programme (UNEP), "State of the Marine Environment in the South Pacific Region, UNEP, Nairobi, Kenya, 1987, p. 1.
27. D.H.S. Wehle and F.C. Coleman, "Plastics at Sea," *Natural History*, Vol. 92, No. 2 (1983), pp. 20- 26.
28. Nancy Wallace, "Debris Entanglement in the Marine Environment: A Review, " in *Proceedings of the Workshop on the Fate and Impact of Marine Debris*, 27-29 November 1984, Honolulu, Hawaii, Richard S. Shomura and Howard O. Yoshida, eds. (National Oceanic and Atmospheric Administration, Rockville, Maryland, 1985), pp. 263-264.
29. Michael J. Bean, "Legal Strategies for Reducing Persistent Plastics in the Marine Environment," *Marine Pollution Journal*, Vol. 18, No. 6B (1988), p. 359.
30. R. V. Arnaudo, "The Problem of Persistent Plastics and Marine Debris in The Oceans," Annex to Group of Experts on the Scientific Aspects of Marine Pollution (GESAMP) *The State of the Marine Environment* (United Nations Environment Programme, Nairobi, Kenya, 1990).
31. Joe Scordino, *Proceedings of the Workshop on the Fate and Impact of Marine Debris*, November 27-29, 1984, Honolulu, Hawaii, Richard S. Shomura and Howard O. Yoshida, eds. (National Oceanic and Atmospheric Administration, Rockville, Maryland, 1985) p. 279.
32. R.T. Tinney, testimony before U.S. Senate Committee on Commerce, Marine Mammal Protection Act Reauthorization, April 26, 1974, cited in Nancy Wallace, "Debris Entanglement in the Marine Environment: A Review," in *Proceedings of the Workshop on the Fate and Impact of Marine Debris*, 27-29 November 1984, Honolulu, Hawaii, Richard S. Shomura and Howard O. Yoshida, eds. (National Oceanic and Atmospheric Administration, Rockville, Maryland, 1985), p. 69.
33. *Op. cit.* 30.
34. *Op. cit.* 30.
35. Judie Neilson, The Oregon Experience," in *Proceedings of the Workshop on the Fate and Impact of Marine Debris*, 27-29 November 1984, Honolulu, Hawaii, Richard S. Shomura and Howard O. Yoshida, eds. (National Oceanic and Atmospheric Administration, Rockville, Maryland, 1985), pp. 154-158.
36. C. Walder, "Marine Transportation of Oil and other Hazardous Substances," Annex to *The State of the Marine Environment* (United Nations Environment Programme, Nairobi, Kenya, 1990).
37. J.H. Reijnders, "Reproductive Failure in Common Seals Feeding on Fish from Polluted Coastal Waters," *Nature*, Vol. 324, No. 6096 (1986), p. 456.
38. Alan J. Mearns *et al.*, *PCB and Chlorinated Pesticide Contamination in U.S. Fish and Shellfish: A Historical Assessment Report*, National Oceanic and Atmospheric Administration, Seattle, 1988, pp. 32, 47, 113 and 114.
39. United Nations Environment Programme (UNEP), "Problems of the South Asian Seas Region, UNEP, Nairobi, Kenya, draft 1988.
40. United Nations Environment Programme (UNEP) *Regional Report of the Marine and Coastal Environmental Problems of the South Asian Seas Region*, UNEP, Nairobi, Kenya, draft 1988, p. 10.
41. *Op. cit.* 2, p. 38.
42. *Op. cit.* 2.
43. *Op. cit.* 2, p. 42.

44. Beth Millemann, *And Two if by Sea: Fighting the Attack on America's Coasts* (Coast Alliance, Washington, D.C., 1986), pp. 62-71.

45. *Op. cit.* 38, p. 47.

46. *Monitor*, The National Swedish Environmental Programme (PMK), National Swedish Environmental Protection Board, 1985, pp. 133-134.

47. Dr. John Portmann, Ministry of Agriculture Fisheries and Food, Burnham, U.K., 1989 (personal communication).

48. *Op. cit.* 2.

49. United Nations Economic Commission for Latin America and the Caribbean (ECLAC), *Statistical Yearbook for Latin America and the Caribbean, 1988* (ECLAC, New York, 1987), Table IV, pp. 429-485.

50. Arsenio Rodriguez, Regional Advisor for Latin America and the Caribbean, United Nations Environment Programme, Mexico, D.F., 1989 (personal communication).

51. Raymond Reid, Regional Advisor, Water Supply and Sanitation, Pan American Health Organization, Washington, D.C., 1990 (personal communication).

52. Economic Commission for Latin America and the Caribbean (ECLAC), "*El Medio Ambiente en America Latina*," 1976, cited in ECLAC, "The Water Resources of Latin America and the Caribbean: Water Pollution," Washington, D.C., 1989.

53. Barry B. Levine, "Abundance and Scarcity in the Caribbean," *Ambio*, Vol. 10, No. 6 (1981), pp. 275-279.

54. *Op. cit.* 49, Table 106, p. 165.

55. United Nations (U.N.) *World Population Prospects, 1988* (U.N., New York, 1989), Table 2, p. 80.

56. Susan Weber, ed., *U.S.A. by Numbers: A Statistical Portrait of the United States* (Zero Population Growth, Washington, D.C., 1988), pp. 19-20.

57. Arsenio Rodriguez, "Marine and Coastal Environmental Stress in the Wider Caribbean Region," *Ambio*, Vol. 10, No. 6 (1981), pp. 283-289.

58. *Ibid.*, pp. 284, 286.

59. *Ibid.*, p. 284.

60. United Nations Environment Programme (UNEP), *Development and Environment in the Wider Caribbean: A Synthesis* (UNEP and Economic Commission for Latin America, Nairobi, 1982), p. 12.

61. Mel Goodwin, Coordinator of International Programs, South Carolina Sea Grant Consortium, Charleston, South Carolina, 1989 (personal communication).

62. Food and Agriculture Organization of the United Nations (FAO), *FAO Yearbook: Fisheries Statistics*, Vol. 64 (FAO, Rome, 1987), Table A-4, pp. 107-109.

63. National Oceanic and Atmospheric Administration, *Fisheries of the United States 1988* (U.S. Government Printing Office, Washington, D.C., 1989), p. 4.

64. United Nations Environment Programme (UNEP), *Regional Overview of Environmental Problems and Priorities Affecting the Coastal and Marine Resources of the Wider Caribbean* (UNEP, Nairobi, 1989), p. 8.

65. Greenpeace International, *Caribbean Seas: Environmental Assessment of the Wider Caribbean Region for the United Nations Environment Programme's Caribbean Action Plan* (Greenpeace, Washington, D.C., 1988), p. 24.

66. 50 C.F.R. (Code of Federal Regulations), Section 227- 272 (U.S. Government Printing Office, Washington, D.C.) 1989.

67. 50 C.F.R. (Code of Federal Regulations), Section 227- 272, United States Public Law 100-478, Section 1008 (U.S. Government Printing Office, Washington, D.C.) 1989.

68. C.S. Rogers, "Degradation of Caribbean and Western Atlantic Coral Reefs and Decline of Associated Fisheries," in *Proceedings of the Fifth International Coral Reef Congress*, Tahiti, Vol. 6.

69. International Union for Conservation of Nature and Natural Resources (IUCN), *Atlantic and Eastern Pacific*, Vol. 1 of *Coral Reefs of the World Atlantic and Eastern Pacific* (IUCN, Cambridge, U.K., 1988), pp. xx-xxi.

70. United Nations Environment Programme (UNEP), *The State of Marine Pollution in the Wider Caribbean Region* (UNEP, Nairobi, 1984), p. 24.

71. Oil and Gas Journal, *1989 Energy Statistics* (Pennwell Publishing Company, Tulsa, Oklahoma, 1989).

72. *Op. cit.* 70, pp. 27-32.

73. Arne Jernelöov and Olof Lindén, "Ixtoc I: A Case Study of the World's Largest Oil Spill, *Ambio*, Vol. 10, No. 6 (1981), p. 299.

74. United Nations Environment Programme (UNEP), *State of the Marine Environment in the Caribbean Region* (UNEP, Nairobi, Kenya, 1988).

75. *Op. cit.* 57, p. 289.

76. Raymond Reid, "Environment and Public Health in the Caribbean," *Ambio*, Vol. 10, No. 6 (1981), pp. 315-316.

77. Raymond Reid, Regional Advisor, Water Supply and Sanitation, Pan American Health Organization (PAHO) Washington, D.C., In late 1989, PAHO was conducting another survey that most of the region now has piped water and sewer service is beginning. Although sewerage has been extended, population growth has kept the percentage of the population served at about the same level as in 1974.

78. *Op. cit.* 76, p. 316.

79. Salvano Briceño, Coordinator, United Nations Environment Programme, Regional Coordinating Unit, Caribbean Action Plan, Kingston, Jamaica, 1989 (personal communication).

80. *Op. cit.* 70, p. 22.

81. *Op. cit.* 64, p. 4.

82. *Op. cit.* 64, p. 4.

83. World Resources Institute, International Institute for Environment and Development in collaboration with the United Nations Environment Programme, *World Resources Report 1988-89* (Basic Books, New York, 1988) Table 17.6, p. 282.

84. *Op. cit.* 70, p. 22.

85. United Nations Environment Programme (UNEP), *Development and Environment in the Wider Caribbean Region: A Synthesis* (UNEP, Nairobi, Kenya, 1982), pp. 16-17.

86. International Union for Conservation of Nature and Natural Resources, Susan Wells, ed., Conservation Monitoring Centre, *Atlantic and Eastern Pacific*, Vol. 1 of *Coral Reefs of the World* (IUCN, Cambridge, U.K., 1988), p. xxi.

87. *Op. cit.* 64, p. 34.

88. Salvano Briceño, Co-ordinator, United Nations Environment Programme, Regional Seas Programme, Caribbean Action Plan, Kingston, Jamaica, 1989 (personal communication).

89. United Nations Environment Programme (UNEP) *Summary of Projects Being Undertaken by the Regional Co-ordinating Unit* (UNEP, Nairobi, Kenya, 1989), n.p.

90. *Ibid.*, p. 1-2.

91. *Op. cit.* 64, pp. 34-35.

92. *Op. cit.* 64, p. 37.

93. Argentina, Australia, Belgium, Chile, France, Great Britain, Japan, New Zealand, Norway, South Africa, the Soviet Union, and the United States.

94. Michael D. Lemonick, "Antarctica," *Time* (January 15, 1990), p 58.

95. D.W.H. Walton, "Geography, Politics and Science," in *Antarctic Science*, D.W.H. Walton, ed. (Cambridge University Press, Cambridge, U.K., 1987), p. 6.

96. Deborah Shapley, *The Seventh Continent: Antarctica in a Resource Age* (Resources for the Future, Washington, D.C., 1985), p. 7.

97. James K. McElroy, "Antarctic Fisheries: History and Prospects," *Marine Policy*, Vol. 8, No. 3 (1984), p. 239.

98. *Op. cit.* 96, p. 107.

99. U.S. Department of State, unclassified cable, ref. Moscow 13225, "Soviet Sealing in Antarctica: Offical Report on 1986-87 Season," November 1987.

100. *Op. cit.* 97, p. 240.

101. *Op. cit.* 97, p. 239.

102. Martin W. Holdgate, "Regulated Development and Conservation of Antarctic Resources," in *The Antarctic Treaty Regime*, Gillian D. Triggs, ed. (Cambridge University Press, London, 1987), p. 130.

103. Martin W. Holdgate, Director General, The World Conservation Union (formally the International Union for Conservation of Nature and Natural Resources (IUCN)), Gland, Switzerland, 1989 (personal communication).

104. J.A. Gulland, *The Management of Marine Fisheries* (University of Washington Press, Seattle, 1974), p. 18.

105. Sara L. Ellis, "Japanese Whaling in the Antarctic: Science or Subterfuge?" *Oceanus*, Vol. 31, No. 2 (1988), p. 68.

106. *Ibid.*, p 69.

107. Douglas G. Chapman, "Living Resources: Whales," *Oceanus*, Vol. 31, No. 2 (1988), p. 67.

108. K.H. Kock, G. Duhamel, and J.C. Hureau, *Biology and Status of Exploited Antarctic Fish Stocks: A Review* (Scott Polar Research Institute, Cambridge, U.K. 1985), p. 9.

109. J.A. Gulland, "The Development of Fisheries and Stock Assessment of Resources in the Southern Ocean," *Memoirs of the National Institute of Polar Research*, Special Issue No. 27 (1983), p. 235.

110. Scientific Committee for the Conservation of Antarctic Marine Living Resources (SC-CCAMLR), *Report of the Fourth Meeting of the Scientific Committee* (CCAMLR, Hobart, Tasmania, Australia, 1985), Table 1, p. 205.

111. Scientific Committee for the Conservation of Antarctic Marine Living Resources (SC-CCAMLR), *Report of the Seventh Meeting of the Scientific Committee* (CCAMLR, Hobart, Tasmania, Australia, 1988), Table 2.1, p. 5.

112. Sayed Z. El-Sayed, "Living Resources: The BIOMASS Program," *Oceanus*, Vol. 31, No. 2 (1988), p. 76.

113. Robin M. Ross and Langdon B. Quentin, "Euphausia Superba: A Critical Review of Estimates of Annual Production," *Comparative Biochemical Physiology*, Vol. 90B, No. 3 (1988), pp. 502, 504.

114. Dietrich Sahrhage, "Fisheries Overview," in *Antarctic Politics and Marine Resources: Critical Choices for the 1980s*, Lewis M. Alexander and Lynn Carter Hanson, eds. (University of Rhode Island, Kingston, Rhode Island, 1984), p. 107.

115. *Op. cit.* 62, p. 93.

116. I. Everson, "Life in a Cold Environment," in *Antarctic Science*, D.W.H. Walton, ed. (Cambridge University Press, Cambridge, U.K., 1987), p. 120.

117. Sayed Z. El-Sayed, F.C. Stephens, R.R. Bidigare, *et al.*, "Proceedings of the Fifth SCAR Symposium on Antarctic Biology" [in press]. (Springer-Verlag, New York.)

118. Robert J. Hofman, "The Convention on the Conservation of Antarctic Marine Living Resources," in *Antarctic Politics and Marine Resources: Critical Choices for the 1980s*, Lewis

M. Alexander and Lynn Carter Hanson, eds. (University of Rhode Island, Kingston, Rhode Island, 1984), pp. 113-115.

119. J.A. Heap, "Current and Future Problems Arising from Activities in the Antarctic," in *The Antarctic Treaty Regime*, G.D. Triggs, ed. (Cambridge University Press, London, 1987), p. 205.

120. Commission for the Conservation of Antarctic Marine Living Resources (CCAMLR), *Report of the Sixth Meeting of the Commission* (CCAMLR, Hobart, Tasmania, Australia, 1987), pp. 21, 22.

121. Commission for the Conservation of Antarctic Marine Living Resources (CCAMLR), *Report of the Seventh Meeting of the Commission* (CCAMLR, Hobart, Tasmania, Australia, 1988), p. 28.

122. Richard Fifield, *International Research in the Antarctic* (Oxford University Press, Oxford, U.K., 1987), pp. *v*, 20, 81.

123. *Ibid.*, p. 130.

124. *Ibid.*, p. 68.

125. Boyce Rensberger, "Disaster Narrowly Averted in Antarctic Ship Sinking," *Washington Post* (February 4, 1989), p. A17.

126. *Ibid.*

127. "The Sinking of the Argentine Polar Ship *A.R.A. Bahia Paraiso* and Its Consequences on the Local Environment," Supplemental Information Provided by the U.S. and Argentine Governments, at the preparatory meeting for the 15th Antarctic Treaty Consultative Meeting, Paris, May 12, 1989.

128. "The Sinking of the *Bahia Paraiso*," Supplemental Information Provided by the U.S. Government, Paris, May 12, 1989.

129. *Op. cit.* 125.

130. Robert H. Rutford, ed., *Reports of the SCAR Group of Specialists on Antarctic Environmental Implications of Possible Mineral Exploration and Exploitation (AEIMEE)* (International Council of Scientific Unions, Scientific Committee on Antarctic Research, 1986), p. 20.

131. International Union for Conservation of Nature and Natural Resources (IUCN) and International Council of Scientific Unions, Scientific Committee on Antarctic Research, *Conservation in the Antarctic* (IUCN, Gland, Switzerland, May 1986), pp. 3-9.

132. *Op. cit.* 94, p. 61.

133. National Science Foundation, U.S. Antarctic Program Safety Review Panel, *Safety in Antarctica* (National Science Foundation, Washington, D.C., 1988), pp. 9-2 to 9-3.

134. *Op. cit.* 131, p. 28.

135. *Op. cit.* 131, pp. 29-30.

136. *Op. cit.* 133, pp. 9-5 to 9-6.

137. Final Environmental Impact Statement on the Negotiation of an International Regime for Antarctic Mineral Resources, U.S. Department of State, Washington, D.C., 1980, pp. 6-26 to 6-28.

138. Argentina, Australia, Chile, France, New Zealand, Norway, and the United Kingdom.

139. *Op. cit.* 94, p. 62.

140. *Op. cit.* 131, p. 2.

141. T.P. Barnett, "Global Sea Level: Estimating and Explaining Apparent Changes," in *Coastal Zone 83: Proceedings of the Third Symposium on Coastal and Ocean Management*, O.T. Magoon, ed. (American Society of Civil Engineers, New York, 1983), pp. 2777-2782.

142. V. Gornitz, S. Lebedeff, and J. Hansen, "Global Sea Level Trends in the Past Century," *Science*, Vol. 215, No. 4540 (March 26, 1982), pp. 1611-1614.

143. T.P. Barnett, "The Estimation of Global Sea Level Change: A Problem of Uniqueness," *Journal of Geophysical Research*, Vol. 89, No. C5 (1984), pp. 7980-7988.

144. National Academy of Sciences, Responding to Changes in Sea Level: Engineering Implications (National Academy Press, Washington D.C., 1987), p. 29.

145. William Booth, "Projected Sea-Level Rise Scaled Back by Scientists," *Washington Post* (December 22, 1989), p. A3.

146. P. Vellinga and Stephen P. Leatherman, "Sea Level Rise, Consequences and Policies," *Climatic Change*, Vol. 15, pp. 175-189.

147. W. Roland Gehrels and Stephen P. Leatherman, *Sea Level Rise: Animator and Terminator of Coastal Marshes* (Vance Bibliography, Monticello, Indiana, in press).

148. Timothy W. Kana, Bert J. Baca, and Mark L. Williams, *Potential Impacts of Sea Level Rise on Wetlands around Charleston, South Carolina* (U.S. Environmental Protection Agency, Washington, D.C.), 1986, pp. *iii* and 7-8.

149. Richard A. Davis, *Coastal Sedimentary Environments* (Springer-Verlag, New York, 1985).

150. John D. Milliman and Robert H. Meade, "World-Wide Delivery of River Sediment to the Oceans," *Journal of Geology*, Vol. 91, No. 1 (1983), pp. 1-2.

151. E.C.F. Bird, *Coastline Changes: A Global Review* (London, U.K., John Wiley and Sons, 1985).

152. Robert G. Dean and E.M. Maurmeyer, "Models for Beach Profile Response," in *CRC Handbook of Coastal Processes and Erosion Control* (CRC Press, Boca Raton, Florida, 1983), pp. 151-165, cited in National Research Council, *Responding to Changes in Sea Level: Engineering Implications* (National Academy Press, Washington, D.C., 1987), pp. 56-57.

153. Stephen P. Leatherman, "Shoreline Response to Sea Level Rise: Ocean City, Maryland," in *Proceedings of Icelandic Conference on Coasts and Rivers, Reykjavik, Iceland*, 1986, cited in Pier Vellinga and Stephen P. Leatherman, "Sea Level Rise, Consequences and Policies," *Climatic Change*, Vol. 15, pp. 175-189.

154. Eugene M. Rasmusson, "Potential Shifts of Monsoon Patterns Associated with Climate Warming," in *Coping with Climate Change: Proceedings of the Second North American Conference on Preparing for Climate Change*, John C. Topping, Jr., ed. (The Climate Institute, Washington, D.C., 1989), pp. 120, 122.

155. Reginald E. Newell, "Climate and the Ocean," *American Scientist*, Vol. 67, No. 4 (1979), p. 405.

156. Wayne M. Wendland, "Tropical Storm Frequencies Related to Sea Surface Temperatures," *Journal of Applied Meteorology*, Vol. 16, No. 5 (1977), p. 479.

157. Wallace S. Broecker, "The Biggest Chill," *Natural History*, Vol. 96, No. 10 (1987), p. 77.

158. J. Hansen *et al.*, *Monthly Weather Review*, Vol. 3 (1983), pp. 609-662, cited in Kerry A. Emanuel, "The Dependence of Hurricane Intensity on Climate," *Nature*, Vol. 326 (April 2, 1987), p. 485.

159. *Ibid.*

160. Jay S. Hobgood and Randall S. Cerveny, "Ice-Age Hurricanes and Tropical Storms," *Nature*, Vol. 333 (May 19, 1988), p. 244.

161. C.W. Clark, "Bioeconomics of the Ocean," *Bioscience*, Vol. 31, No. 34 (1981), p. 231.

162. R.C. Worrest, "The Effect of Solar UV-B Radiation on Aquatic Systems: An Overview," in *Overview*, Vol. 1 of Effects of Changes in Stratospheric Ozone and Global Climate, James G. Titus, ed. (U.S. Environmental Protection Agency, Washington, D.C., 1986), p. 175.

163. *Ibid.*, p. 178.

164. Ben Patrusky, "Dirtying the Infrared Window," *Mosaic*, Vol. 19, No. 3/4 (1988), pp. 27, 37.

165. Oivind Bergh, *et al.*, "High Abundance of Viruses Found in Aquatic Environments," *Nature*, Vol. 340 (August 10, 1989), p. 467.

166. *Ibid.*

167. William Booth, "Viruses Central to Ocean Life," *Washington Post* (August 13, 1989), p. A1.

168. *Op. cit.* 165.

169. Evelyn B. Sherr, "And Now, Small is Beautiful," *Nature*, Vol. 340, No. 6233 (August 10, 1989), p. 429.

170. Serge Gosselin, Louis Fortier, and Jacques A. Gagné, "Vulnerability of Marine Fish Larvae to the Toxic Dinoflagellate *Protogonyaulax tamarensis*," *Marine Ecology Progress Series*, Vol. 57, 1-10 (September 15, 1989), p. 1.

171. *Op. cit.* 5.

172. Atlantic Research Laboratory Shellfish Toxin Team, "Solving the Toxic Mussel Problem," *Canadian Chemical News* (October 1988), p. 15.

173. Marlise Simons, "Now, Venice Is under Attack by Giant Algae," *New York Times* (June 13, 1989), p. A13.

174. Marlise Simons, "Adriatic Fouled by Vast Sea of Algae," *New York Times* (August 16, 1989), p. A6.

175. Karen Wolman, "Italy Copes with Summer Slime," *Christian Science Monitor* (August 3, 1989).

176. *Op. cit.* 174.

12. Atmosphere

When the battle to curb air pollution began in earnest 20 years ago, the stakes were measured in terms of the soot density in the smoke-filled industrial valleys of the world, ranging from Yokkaiichi to the Ruhr. Today, concern has shifted to new pollutants. But even so, despite remarkable progress in cleaning up some forms of air pollution, much of the world's urban population breathes air that is unhealthy at least some of the time. In addition, some air pollutants are now recognized as having a regional and even a global impact. Increasingly at issue is not only the quality of human life in major urban areas but also environmental damage on a global scale.

This chapter surveys trends in urban air pollution, although the survey is incomplete. What little comprehensive data there is comes from industrialized countries; in developing countries, the data are generally unavailable. Since most of the world's population growth is occurring in the urban areas of the developing countries, this lack of data may conceal a significant public health problem.

The chapter also surveys approaches to air pollution control, focusing on existing and proposed regulatory strategies. Four case studies, all in developed countries, exemplify and analyze the evolution of air pollution control strategies. Again, the lack of information for developing countries points to an area that may need additional attention and support by international lending agencies.

Finally, the chapter focuses on the role of cars, trucks, and buses in urban air pollution—a role which is growing rapidly because the number of vehicles and the number of miles they travel are also growing rapidly. (See Chapter 9, "Energy," Focus on the Automobile.)

Progress over the past 20 years has produced some encouraging trends. Annual mean concentrations of common air pollutants such as sulfur dioxide, which causes respiratory problems and forms acid rain, have declined in 20 of the 33 cities which participate in the Global Environment Monitoring System (GEMS) of the United Nations Environment Programme; 26 of 37 cities reported declines in suspended particulates (1). In Toronto, for example, mean daily levels of sulfur dioxide were cut over 60 percent between 1976 and 1985 (2).

Some sources of toxic air pollutants are being eliminated in parts of the world. Lead pollution, for example, which at even low levels of exposure can cause cognitive deficits in children (3) and raise cardiovascular health risks in adults (4), should decline from the urban air of most industrialized nations as government bans on leaded petrol take effect. Lead petrol is not yet banned in many developing countries, however.

Spurred by increasing awareness that pollution recognizes no national boundaries, the 1979 convention on Long-Range Transboundary Air Pollution set in place a "thirty percent club" of countries pledged to reduce sulfur dioxide emissions (5), and a comparable though less stringent plan to control oxides of nitrogen (6)—a key ingredient in the formation of both smog and acid rain. The widely-publicized 1987 "Montreal Protocol" (7) restricted the production and use of chlorofluorocarbons, which are industrial chemicals that destroy stratospheric ozone.

Progress in some countries has been nothing short of spectacular. Sweden cut its sulfur dioxide emissions by over two-thirds between 1970 and 1985 (8) and plans more reductions. Switzerland and Austria have gone further. With 43 percent of Switzerland's trees and 29 percent of Austria's trees damaged by air pollution (9), the two countries have adopted the world's toughest regulations—even motorcycles must meet emissions standards (10).

Germany cut total annual emissions of sulfur dioxide by 64 percent between 1983 and 1988 by requiring power plants to install devices that cleanse stack gases (11). Now a movement is sweeping through Western Europe to demand state-of-the-art pollution controls for cars and trucks as well. And in the United States, in the state of California—where the regulation of automotive air pollution first began in the 1970s and where automobile emission standards for oxides of nitrogen are already 60 percent below what U.S. national laws require (12)—there are proposals to reduce emissions still further (13). Ozone levels in the Los Angeles basin reach levels that, in animal tests, have caused damage to cell walls within the respiratory system (14) (15). Oxides of nitrogen react in the atmosphere to form ozone, a major component of smog.

Yet, side by side with these success stories are other countries in which there has been no progress in cleaning up the air or in which cleanup proposals are encountering great opposition. Vehicle emissions remain essentially uncontrolled, for example, in much of the developing world, in Eastern Europe, and in the Soviet Union—which already is the world's fifth largest vehicle manufacturer and is planning to double its capacity (16).

The World Health Organization (WHO) estimates that 70 percent of the global urban population, primarily in developing countries, breathes air that, at least some of the time, has unhealthy levels of suspended particles, while another 10 percent breathes air that has "marginal" levels (17). Studies of sulfur dioxide and sulfate particulates have also linked these pollutants to bronchial diseases in children (18).

Between 1979 and 1985, 14 of the 20 GEMS monitoring stations in Chinese cities reported increases in mean sulfur dioxide pollution. At some of these sites, concentrations were three to five times those found at any GEMS monitoring station in North America (19). Beijing, Xian, Guangzhou and Shenyang in China all have sulfur dioxide pollution above WHO's maximum annual guidelines, as do Tehran, Manila, Seoul, and São Paulo. (See Table 24.6.) Many cities in developing countries lack monitoring stations. (See Box 12.1.)

In some industrialized nations, too, pollution levels are dangerously high. In Krakow, Poland, damage to stone monuments and buildings is so severe that the stone is described as dissolving (20). In Czechoslovakia, over 70 percent of the trees surveyed were damaged (21). Ozone and acid rain have been implicated in forest decline throughout parts of both Europe and North America (22) (23).

Even in areas remote from industrial facilities, air pollution can be damaging. In parts of Africa, for example, scientists report acid rain and smog levels as high as those of central Europe, probably from the regular burning of the vast grasslands to clear land. (See Chapter 7, "Forests and Rangelands," Recent Developments.)

CONDITIONS AND TRENDS

APPROACHES TO AIR POLLUTION CONTROL

In the absence of effective international cooperation and coordination, nations are acting individually to curb air pollution. The approaches can vary remarkably from one nation to another as governments seek to accommodate variations in their political and social institutions. Although differing in their details, the programs share certain basic features which allow them to be compared with one another.

Ambient Air Quality Standards

The most common fundamental approach to controlling air pollution is to set ambient air quality standards; that is, controls are imposed on specific sources of air pollution to reduce ambient levels below a maximum allowable concentration, measured at some distance away from the source. This system of working backwards from air quality in an area to pollution sources was developed by the U.S. Public Health Service, then incorporated by the U.S. Congress into the 1970 Clean Air Act. So many other nations have followed the pattern of this U.S. law that the ambient standards approach has become widespread.

Usually, government officials establish two levels of ambient air quality: a "primary" standard, which is designed to protect human health; and a "secondary" standard, which is to protect welfare and the environment. The sampling interval (e.g., one hour versus eight hours) can vary, as can other factors (e.g., the height and location of monitors, the type of measuring device, etc.). Generally, however, there are two standards for each of several different but common air pollutants. Almost invariably, the pollutants regulated include sulfur dioxide, oxides of nitrogen, carbon monoxide, and particulates of various sizes. Many nations regulate hydrocarbons and various heavy metals (e.g., lead and cadmium) as well.

Box 12.1 Gaps in Air Pollution Information

Despite a growing amount of evidence that developing nations face severe air pollution threats, this chapter is weighted toward developed nations. The reason is simple and important: although some developing nations monitor urban air quality, data remain limited and difficult to obtain, even by established international agencies such as the United Nations Environment Programme (UNEP). Comprehensive information on air pollution control policies is also lacking.

During the past five years the number and quality of monitors in developing nations have increased (1). However, the data are still difficult to obtain (2), worldwide estimates of emissions are of uneven quality, thus coverage cannot be considered reliable for measuring global trends (3). Data from developing nations—and, to a degree, from developed nations as well—are collected and analyzed almost on an *ad hoc* basis.

For example, most of the parties to the 1979 Convention on Long-range Transboundary Air Pollution participate in a program to monitor air pollution in Europe (of the signatories, only the United States and Canada are outside Europe) (4). Similarly, the Organisation for Economic Co-operation and Development collects data, but only of member nations, not developing countries (5).

THE GLOBAL ENVIRONMENT MONITORING SYSTEM

The most comprehensive data collection system is UNEP's Global Environment Monitoring System (GEMS). GEMS monitoring networks include the following:
■ GEMS/Air, which provides data on sulfur dioxide and particulates in urban air and is possibly the best known of the GEMS networks (6); and,

■ The Background Air Pollution Monitoring Network (BAPMoN), which provides data on carbon dioxide, atmospheric aerosol optical depth, precipitation chemistry and suspended particles in rural and remote areas (7).

GEMS/Air data are collected on emissions, urban concentrations, and source standards from 50 countries, of which only about 35 provide representative data sets for major urban areas (8).

BAPMoN data are collected on six continents at, as of December 1989, 196 stations in 57 countries. Although both BAPMoN and GEMS/Air sites are selected to be as geographically representative as possible, the numbers vary too widely to be characterized as comprehensive; there are, for example, over 20 BAPMoN precipitation chemistry monitoring stations in the United States and Canada, but only two in Africa (9).

MONITORING IMPROVEMENTS PLANNED

Even in nations that are the focus of concerted international efforts to curb air pollution, monitoring data may be scanty. In Mexico, for example, the World Bank is preparing a project to tackle air pollution, particularly in the capital, through public transportation improvements, emissions standards, and other actions. First, however, the World Bank is supporting improvements for air quality monitoring in the Mexico City urban area (10).

Data on national pollution control policies are even less available than that on emissions and ambient concentrations. Though some information exists, policies are expressed in such disparate ways that meaningful comparison is a complex, demanding and often fruitless task (11). As a result, comparative data are usually, of necessity, qualitative and somewhat subjective.

In his preface to a 1987 report, UNEP Executive Director Mostafa Tolba termed broadened data collection "of the utmost importance...particularly in the developing countries (12)." Until that broadening comes to pass, reports such as this will continue to suffer from the same inadequacies as the data on which they must ultimately rely.

References and Notes

1. Ann Willcocks, Global Environment Monitoring System (GEMS) Monitoring and Assessment Research Centre, London, January 1990 (personal communication).
2. *Ibid.*
3. United Nations Environment Programme, *Environmental Data Report* (Basil Blackwell, Oxford, 1987), p. 3.
4. Economic Commission for Europe, United Nations, *National Strategies and Policies for Air Pollution Abatement* (United Nations, New York, 1987), Table 6, p. 34 and p. 56.
5. Organisation for Economic Co-operation and Development (OECD), *OECD Environmental Data/Données OCDE sur l'Environnement: Compendium 1989* (OECD, Paris, 1989), pp. 7-8.
6. *Op. cit.* 3.
7. *Op. cit.* 3.
8. Global Environment Monitoring System, United Nations Environment Programme and World Health Organization, *Assessment of Urban Air Quality* (United Nations Environment Programme and World Health Organization, London, 1988), p. 3.
9. Rumen Bojkov, Chief, Environmental Division, World Meteorological Organization, Geneva, January 1990 (personal communication).
10. Carl Heinz Mumme, The World Bank, Washington, D.C., January 1990 (personal communication).
11. *Op. cit.* 8, Table 3, p. 47, Table 4, pp. 48-55.
12. *Op. cit.* 3, vii.

The ambient standards approach proved to be a reasonably effective mechanism for dealing with urban air pollution during the 1970s. But it encouraged the use of tall smokestacks to export pollution from one area to another (24) and did not usually require every pollution source to be controlled. With growth in the number of pollution sources, in the complexity of air pollution, and of the scientific data needed to characterize it, the ambient standards approach is increasingly giving way to a second major approach—technology-based pollution control regulation.

Technology-Based Regulation

Technology-based regulation essentially requires *every* source of an air pollutant either to install a specific control technology or to meet certain emission limits. This approach can even require the abandonment of a technology— as in the case of the Montreal Protocol for chlorofluorocarbons—or a switch to new industrial procedures or energy technologies. Often the regulatory requirement is phrased not in terms of a specific technology, but rather in terms of performance standards or emission limits that acceptable technologies must meet—leaving the choice of technology to accomplish the task up to individual manufacturers. By requiring adherence to a performance level, governments can use the technology-based approach to encourage innovation and to "force" the development of new and better technologies.

One of the earliest and most common technology-based requirements was the tailpipe standards for automobiles. They were adopted by the U.S. Congress at a time when U.S. car manufacturers contended that the means of complying with them did not exist (25). The Congress rejected these arguments, with Senator Edmund Muskie, the chief sponsor of the requirements, saying that "Predictions of technological impossibility or infeasibility are not sufficient as reasons to avoid tough standards and deadlines and thus to compromise the public health." (26) Echoes of this thinking can be found in other technology-based regulatory efforts, most recently in the Federal Republic of Germany in what is probably the world's most sweeping attempt to control air pollution by retrofitting existing power plants.

The exact technologies—or the emissions limits which are their surrogates—vary from country to country. The sources most commonly regulated in this manner, however, are motor vehicles and electric power plants, the two main sources of air pollution in most developed nations. In the United States, for example, power plants create 65 percent of sulfur dioxide emissions and 29 percent of nitrous oxide emissions; motor vehicles emit 41 percent of nitrous oxide pollution and 41 percent of organic compounds emissions (27).

COMPARING APPROACHES

Each of these two approaches—ambient standards and technology-based regulation—has its strengths and weaknesses. The overwhelming advantage of the technology approach is its certainty and ease of administration—any given source either is equipped with a device or is not, and the device either works properly or does not. Arguments can be resolved through the simple expedients of inspection and testing.

The ambient standards approach allows flexibility. "Over control"—that is, the removal of more pollution than is necessary to protect either health or the environment—is avoided and sources retain the flexibility of choosing the approach which is the cheapest and most cost-effective.

The advantages of the two differing approaches are also their flaws. The "certainty" of a technology-based strategy is to some merely mindless rigidity which wastes limited societal resources. To others, the "flexibility" of an ambient standards strategy is merely a system that always errs on the side of under-control; it is economically but not environmentally prudent, and can open the door to environmental surprises, such as the Antarctic ozone hole, linked to emissions of chloroflourocarbons. (See Recent Developments, below.)

Most nations have adopted regulatory programs which blend these two approaches, making quantitative comparison virtually impossible. The Federal Republic of Germany, for example, has relatively weak ambient standards. But technology requirements that are collectively the world's most stringent have been used to achieve reduction of sulfur dioxide emissions by over 50 percent within five years (28). Canada, in contrast, has no uniform technology re-

quirements for old power plants. But its ambient standards program—designed to prevent a level of acid precipitation that would harm sensitive resources—will also achieve significant results: a 40 percent reduction from 1980 to 1994 in sulfur dioxide emissions in Canada's eastern provinces (29).

Trend Toward Tighter Controls

In recent years there is an apparent trend in industrial countries toward not only increasingly stringent air pollution controls, but also toward the greater certainty that characterizes technology-based control systems. This can be seen most clearly in the spread of tighter controls over cars and trucks. A striking but quite different example is the Montreal Protocol, the international agreement negotiated in September 1987 for limiting the use of chlorofluorocarbons, which initially required only a 50 percent reduction in production and use of ozone-depleting chemicals by July 1998 (30). Within 18 months, however, Canada, the United States, and the European Community called for a total ban on ozone-depleting chlorofluorocarbons (31) (32).

For fixed sources of air pollution such as power plants, most nations chose to require technology-based limits solely on new power plants, allowing existing units to comply with only the more flexible rules based on ambient standards. But since 1980, The Federal Republic of Germany, Switzerland, the Netherlands, and Austria have adopted technology requirements for existing plants (33). Japan is implementing such requirements on a case-by-case basis (see below). The U.S. Environmental Protection Agency is authorized to require them when plants undergo major modifications (34). Uniform technologies are a key method employed by many nations to implement the recently adopted NOx protocols or other NOx reduction strategies. (See Box 12.2.) The European Community is discussing a similar approach.

Standards for the sulfur content of fuels are widely used in developed countries as means of curbing air pollution. According to the United Nations Economic Commission for Europe, "a gradual tightening of standards is apparent in a majority of countries." Those reported to be toughening fuel standards include Denmark, the Federal Republic of Germany, Luxembourg, the Netherlands, Sweden, and Switzerland, as well as the European Community as a whole (35).

A notable exception to this trend is the package of proposed amendments to the U.S. Clean Air Act (see below). These proposed amendments rely heavily on "market-based" controls that would result in requirements that would vary widely from source to source. The proposals would establish a system of pollution "allowances," under which permits to emit certain amounts of pollutants could be traded much like monetary instruments, although the government would retain the right to further control pollution. The proposals would also allow a company or a whole industry to average pollutant emissions, so that emissions of existing plants could be offset by gains from efficiency improvements or from switching to cleaner fuels somewhere else.

Whether the proposals are a temporary anomaly or the first evidence of a new direction in pollution control philosophy that relies on market incentives remains to be seen.

POLICY CASE STUDIES

A wide variety of policy approaches to air pollution control have been pursued in industrial countries. These experiences, reviewed below, may be useful for countries still formulating air pollution control strategies.

The Federal Republic of Germany

Since 1983 Germany has developed one of the world's most aggressive and resolute regulatory systems for air pollution. Although the principal responsibility for air pollution control lies with the 11 German states, called Länder (36), the central government has dictated what probably is the world's largest pollution retrofit program—an estimated 21 billion Deutschmarks in power plant pollution control equipment between 1983 and 1993 (37). It requires every large power plant to install state-of-the-art technologies, and steadily ratchets permissible levels downward as those technologies improve (38).

This is a sharp contrast to the prevailing attitude during the 1970s when German pollution control was inconsistent, and the use of tall chimney stacks to dispense pollutants over wide areas was the main mechanism of reducing local air pollution (39). The abrupt change can be explained in a single word: Waldsterben. Since it was coined in 1980-81, "Waldsterben"—which translated literally means "forest death"—has galvanized the German people and pushed politicians inexorably toward tighter and tighter controls. (See *World Resources 1986*, pp. 203-226.)

The earlier German regulatory effort could fairly be described as relaxed. Prior to 1983, the ambient standard for sulfur dioxide was 140 micrograms per cubic meter ($\mu g/m^3$), compared with the European Community (EC) standard of 80-120 $\mu g/m^3$ and the World Health Organization's recommendation of 40-60 $\mu g/m^3$ (40). An automobile controls program was virtually nonexistent.

But this generally sanguine attitude changed as reports of Waldsterben mounted. Initial reports about damage to Germany's forests were continually revised upward during the early 1980s, with researchers concluding in 1985 that 50 percent of the nation's trees were injured (41).

Mounting public concern caused the government to order the development of an air pollution reduction program. The final result was twofold. First, the

Table 12.1 Emission Standards for Stationary Sources in the Federal Republic of Germany

	Size of Source	Maximum Allowable Emissions[a]
Sulfur Dioxide		
Solid fuel	less than 300 MW	2.0
	more than 300 MW	0.4
Liquid fuel	less than 300 MW	1.7
	more than 300 MW	0.4
Oxides of Nitrogen		
Solid fuel	less than 300 MW	0.4
	more than 300 MW	0.2
Liquid fuel	less than 300 MW	0.3
	more than 300 MW	0.15

Source: Economic Commission for Europe, United Nations, *National Strategies and Policies for Air Pollution Abatement* (United Nations, New York, 1987), Table 4, p. 49.
Note: a. grams of pollutant per normal cubic meter of flue gas.

government's stationary source control program set emission limits so low that adoption of state-of-the-art controls—wet scrubbers or their equivalent for SO_2 control and selective catalytic reduction or its equivalent for elimination of oxides of nitrogen—was required by all medium- to large-sized power plants in Germany. (See Table 12.1.) Second, even though European Community rules preclude Germany from unilaterally legislating auto emissions standards, the government adopted tax incentives to encourage the purchase of low emission vehicles (42). Simultaneously, Germany began pressing relentlessly for European Community adoption of stringent tailpipe standards. In June 1989, the EC adopted standards for small cars (engines smaller than 1.4 liters), effective beginning in 1992 (43) (44). The EC countries also agreed in principle to enact similarly strict standards for medium- and large-sized cars, beginning in 1992 (45).

The battle within the EC has now shifted to the questions of assuring that the standards are complied with after cars roll off the assembly line. Germany, the Netherlands, and Denmark are pushing for adoption of other provisions of the long-established U.S. programs: manufacturer warranties that the pollution control systems will last for five years or 50,000 miles; recall programs for defective cars; and emissions testing (46).

Japan

Few nations have deployed pollution control technologies and requirements as aggressively as Japan. State-of-the-art controls are installed on most cars, power plants, steel mills, incinerators and other facilities. Yet there are surprising exceptions. For example, although Japan perfected selective catalytic reduction (SCR) as a technology for reducing power plant emissions of oxides of nitrogen, only about one-third of the nation's electric capacity is equipped with it: 32,100 megawatts of a total of 88,294 in March

Box 12.2 Efficient Combustion Methods for Fossil-Fuel Power Plants

Power plants burning fossil fuels are one of the main sources of air pollutants such as sulfur dioxide and nitrogen oxides; they also are a major contributor of greenhouse gases. Ways to minimize their pollution include: reducing energy demand through increased energy conservation; switching from heavily polluting fossil fuels such as coal to less polluting natural gas; removing pollutants from exhaust gases through mechanisms such as scrubbers; and converting to cleaner or more efficient combustion technologies.

Much of the focus of reducing air pollution from power plants has been on removing pollutants from exhaust gases. Scrubbers, devices that remove sulfur dioxide from the flue gas of coal power plants, can remove up to 90 percent of sulfur dioxide in power plant emissions (1). Catalytic reduction is used to remove nitrogen oxides from flue gas; selective catalytic reduction, a technology used with fossil fuels, typically reduces emissions of nitrogen oxides by 80 percent (2). Selective catalytic reduction has been deployed extensively in Japan and the Federal Republic of Germany.

An alternative approach, often overlooked in public debate, is to replace existing fossil fuel burners with new power plant systems that are both more efficient and release fewer pollutants. Several such systems that are now being tested and entering service in the United States, Japan, and Europe are described below. An integrated strategy to introduce these systems (see Box 12.4) can significantly reduce air pollution.

HIGH-EFFICIENCY GENERATORS

Most fossil-fuel power plants burn natural gas, oil, or coal to heat water into steam that drives a turbine; the turbine is attached to an electric generator. Their thermal efficiency—the fraction of energy released that is turned into electricity—averages 35 percent at best in most developed countries (3).

The hot exhaust gases from the combustion of oil and natural gas can also be used to drive a turbine directly. The newer gas turbines have thermal efficiencies of 37 percent and above. Their efficiency can be increased further by injecting steam into the combustion area, boosting effeciencies to an estimated 45 percent. Steam can also be injected to lower emissions of nitrogen oxides by 75 percent (4). Still newer gas turbines, derived from aircraft jet engines, are projected to have even higher efficiencies when they become available. (5). Adding an intercooler, a modification for gas turbines that exchanges pressure and heat between high- and low-pressure compressors, may be another method of increasing efficiency: from 44 to 55 percent in one application studied (6).

Table 1 The Effect of Fuel Type and Efficiency on Carbon Dioxide Emissions
(percent)

Conversion Efficiency of Fuel to Electric Power	Kilograms of Carbon Dioxide Released per Kilowatt Hour of Electricity[1]		
	Coal	Oil	Natural Gas
30	1.08	0.88	0.59
40	0.81	0.66	0.45
50	0.65	0.53	0.36
60	0.54	0.44	0.30

Source: Calculations based on Gordon J. MacDonald. et al., *The Long-Term Impacts of Increasing Atmospheric Carbon Dioxide Levels* (Ballinger, Cambridge, Massachusetts, 1982), Table 2-12, p. 30.

Note: 1 Values for oil and coal are based on averages of fuel types. Actual values depend on the type of oil or coal used.

Steam and gas-fired turbines can be used together in a combined cycle system: fuel is burned in a gas turbine, then the exhaust gases heat water to generate steam and drive a steam turbine. A 2,000 MW combined cycle power plant in Futsu, Japan, achieves 47 percent efficiency with low emissions (7), and combined cycle systems could reach 55 percent efficiencies in the 1990s (8).

These techniques burn oil and natural gas. Coal and Solid fuels, however, are used to generate about 57 percent of electricity in the United States and Canada, over 45 percent in Europe, and large fractions in countries such as China and Poland. Two efficient low-emissions means of generating electricity from coal are now entering use: fluidized bed combustion and coal gasification.

In fluidized-bed combustion, a strong upward current of air suspends crushed coal mixed with a "bed" of limestone particles in a furnace. The limestone reacts with and removes more than 90 percent of the sulfur in the coal, producing a benign dry waste that can be used as a construction material.

One type of fluidized-bed system, atmospheric fluidized-bed combustion (AFBC), burns coal at or near atmospheric pressure and produces electricity through a steam generator. AFBC systems can emit up to 95 percent less sulfur dioxide and 80 percent less nitrogen oxides than equivalent current coal-burning facilities. In the United States over 100 AFBC systems are in nonutility industrial use or on order, and the U.S. Department of Energy has sponsored two AFBC power plant pilot projects (9).

In another type of fluidized-bed system, pressurized fluidized-bed combustion (PFBC), the furnace is pressurized at 8 to 16 times atmospheric pressure. The coal combustion heats water that passes through tubes in the combustion chamber into steam that drives a steam turbine, while exhaust gases drive a gas turbine (10). Current PFBC pilot plants generate up to 700 MW of electricity and have efficiencies of 42.5 to 44.1 percent (11); PFBC systems should remove up to 95 percent of sulfur dioxide and 90 percent of the oxides of nitrogen from combustion (12).

In coal gasification, coal is not burned directly but is first converted into a clean-burning synthesis gas, which is then used in an efficient combined-cycle gas-turbine system. One such integrated gasification-combined cycle (IGCC) system has been tested in the "Cool Water" project in California (13). In an IGCC system, a sorbent removes over 95 percent of the sulfur from the synthesis gas before it is burned (14). General Electric Co. in the United States estimated that an IGCC system should be able to achieve a 43 percent thermal efficiency (15).

ALTERNATIVES: FUEL CELLS

Fuel cells use an electrochemical process to convert a fuel's energy directly into electricity, rather than burning it to drive a turbine and generator. In construction, a fuel cell is similar to a battery, with an electrode and a cathode separated by an electrolyte. Unlike a battery, a fuel cell consumes a fuel and an oxidant to produce electricity; so long as these are supplied to the two electrodes, the cell will provide direct current electricity. Most fuel cells use hydrogen-rich gases, such as natural gas and methane; possible fuels also include methanol and synthetic gas from coal.

Fuel cells produce only trace levels of pollutants, well below U.S. new-source emissions standards (16).

There are several fuel cell designs; efficiencies can range from 40 to 70 percent. Since some types of fuel cells generate waste heat that may be usable for cogeneration (see below), further increases in efficiency are possible (17).

Fuel cells' advantages include low emissions, high efficiency, nearly silent operation, and small unit size. Fuel cell power plants can thus be located in the center of cities, and small units could even be placed in individual buildings. Such local use would eliminate energy losses from long-distance transmission of electricity. Over 65 fuel cell demonstration power plants, ranging in output size from 10 kilowatts (KW) to 4.5 MW plants have been built in Tokyo and New York City by United Technologies Corporation. Based on experience from the Tokyo plant, United Technologies and Toshiba

Box 12.2

Corporation plan to jointly begin selling commercial fuel cell plants in the near future [18].

COGENERATION

One further method that could provide a significant improvement in efficiency, and thus a large reduction in air pollution, is not a new technology but a simple, long-known technique: cogeneration. Cogeneration refers to a variety of systems that both generate electric power and provide direct heat, usually in the form of steam. The heat can be used either for residential and commercial heating—called "district heating" because a group of buildings or even a whole neighborhood can be heated by one power plant—or for industrial applications. In a district heating system, for example, a power plant will both generate electricity and provide a nearby neighborhood with steam or hot water, heated at the power plant and distributed through underground pipes. Waste heat also can be used to operate absorption chillers to provide cooling with two benefits: enhanced efficiency and cooling without the use of the ozone-destroying chlorofluorocarbons used in conventional air conditioning units. Cogeneration captures some of the energy normally wasted as heat in power plants; it can yield overall efficiencies (counting both electricity and heat generation) up to 85 percent [19].

Cogeneration has been used for over 100 years, but the 1970s oil crises and renewed concern over air pollution have prompted a revival of cogeneration systems. In industrial uses, cogeneration can involve individual factories with power plants that supply their own heat and electricity, and that sell additional electricity to local utilities. District heating is most effective in densely populated areas with significant heating requirements. Sweden, in particular, has promoted district heating in its cities: in 1985, for example, 22.8 million square meters of residential and office space in Stockholm were heated through cogeneration systems, up from 2.6 million in 1965 [20].

POWER PLANT EFFICIENCY AND CARBON DIOXIDE EMISSIONS

The use of fossil fuels inevitably adds to the buildup of carbon dioxide and other greenhouse gases, increasing the likelihood of bringing global warming. Nonetheless, until the complete introduction of renewable energy sources that do not add to global warming, the use of more efficient fossil fuel plants can reduce this impact. (See Table 1.) The efficiencies of the electric generating systems described above are shown in Table 2.

Table 2 Energy Efficiencies of Fossil-Fuel Power Plant Systems
(percent)

Power Plan System	Current Efficiency[1]
Current U.S. powerplants (overall average)	33
Oil and Natural Gas Combustion	
Aircraft-derivative turbines	40
Combined cycle systems	47
Coal Combustion	
Atmospheric Fluidized Bed Combustion	38
Pressurized Fluidized Bed Combustion (using a combined cycle)	42-44
Integrated gasification combined cycle	43
Fuel Cells	40-60
Cogeneration	up to 85

Sources: Current U.S. powerplant efficiency: Department of Energy, *Monthly Energy Review*, July 1989, Table A9, p. 129. Other Figures: Sources in text

Note: 1. Efficiency of commercial or pilot plants currently in use. See text for sources.

References and Notes

1. Mike McQueen, "Clean Coal Technologies: A Key Clean Air Issue," Environmental and Energy Study Conference special report, October 31, 1989, p. 3.
2. Steven G. Nanos, Johnson Matthey Co., "Emission Controls for Stationary Engines," paper presented at the State and Territorial Air Pollution Program Administrators and the Association of Local Air Pollution Control Officials (STAPPA/ALAPCO) Workshop on Innovative Air Pollution Control Technologies, New Orleans, Louisiana, October 1989, pp. 2-3.
3. Masatoshi Furuichi, "Environmental Issues in the Global Community—Striving for Clean Air and Power: The Japanese Experience," paper presented at the Environmental Technology Seminar for U.S.-Japan Collaboration, The Japan External Trade Organization, Long Beach, California, November 30, 1989, Figure 2-2-3, p. 25.
4. Project Development Organization, Southern California Edison, *Steam Injected Gas Turbine (STIG) Application Study at the Redondo Beach Generating Station—Volume I, Final Report* (Southern California Edison, Rosemead, California, April 1986), pp. 1-2.
5. Tedd R. McCormack, General Electric Marine and Industrial Engines, "Aircraft Derivative Gas Turbines: Higher Efficiencies, Lower NOx," paper presented at the State and Territorial Air Pollution Program Administrators and the Association of Local Air Pollution Control Officials (STAPPA/ALAPCO) Workshop on Innovative Air Pollution Control Technologies, New Orleans, Louisiana, October 1989, p. 2.
6. Sam Rashkin (principal author), *Energy Technology Status Report* (California Energy Commission, Sacramento, California, October 1988), pp. 4-71.
7. Curtis Moore, *Trip report: Japanese Energy Technologies* (Center for Global Change, University of Maryland, College Park, Maryland, forthcoming).
8. C.E. Maslak and E.W. Zeltmann, General Electric Co., "Combined Cycles—Emissions and Economics," paper presented at the State and Territorial Air Pollution Program Administrators and the Association of Local Air Pollution Control Officials (STAPPA/ALAPCO) Workshop on Innovative Air Pollution Control Technologies, New Orleans, Louisiana, October 1989, p. 2.
9. *Op. cit.* 1, p. 7.
10. Office of Fossil Energy, U.S. Department of Energy, *The Role of Repowering in America's Power Generation Future* (U.S. Department of Energy, Washington, D.C., 1987) pp. 9-10.
11. K.K. Pillai, Asea Brown Boveri, "Repowering with PFBC for Efficiency, Emissions and Heat Rate Improvement," paper presented at the State and Territorial Air Pollution Program Administrators and the Association of Local Air Pollution Control Officials (STAPPA/ALAPCO) Workshop on Innovative Air Pollution Control Technologies, New Orleans, Louisiana, October 1989, Table 2, p. 4.
12. *Op. cit.* 1, p. 7.
13. The name "Cool Water" has nothing to do with the system itself, but refers to the test site, the former Cool Water Ranch on the edge of the Mojave Desert, the deed to which requires that all facilities bear the ranch's name.
14. *Op. cit.* 1, p. 7.
15. *Op. cit.* 8, p. 7.
16. A.J. Appleby, "Advanced Fuel Cells and Their Future Market," in *Annual Review of Energy*, Vol. 13 (1988), Jack M. Hollander, Harvey Brooks, and David Sternlight, eds. (Annual Reviews, Inc., Palo Alto, California, 1988), pp. 267-268.
17. Leo J.M.J. Blomen, "Fuel Cells: A Review of Fuel Cell Technology and its Applications," in *Electricity: Efficient End-Use and New Generation Technologies and Their Planning Implications*, Thomas B. Johansson, Birgit Bodlund, and Robert H. Williams, eds. (Lund University Press, Lund, Sweden, 1989), p. 631.
18. *Ibid.*, pp. 629-630.
19. The Swedish District Heating Association and Swedish Trade Council, *District Heating: Clean Heat for Urban Areas* (The Swedish District Heating Association and Swedish Trade Council, Stockholm, 1986), p. 6.
20. *Ibid.*, pp. 3, 5.

Figure 12.1 Average Sulfur Dioxide Emissions of Thermal Power Plants in Japan

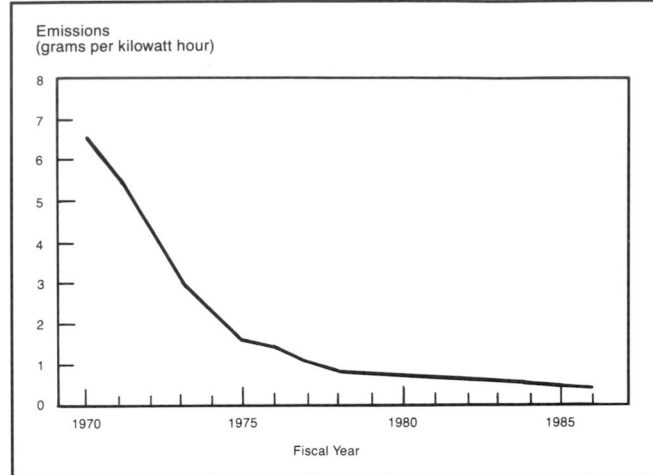

Source: Masatoshi Furuichi, "Environmental Issues in the Global Community—Striving for Clean Air and Power: The Japanese Experience," paper presented at The Environmental Technology Seminar for U.S.-Japan Collaboration, The Japan External Trade Organization, Long Beach, California, November 30, 1989, p. 11.

Figure 12.2 Average Nitrogen Oxides Emissions of Thermal Power Plants in Japan

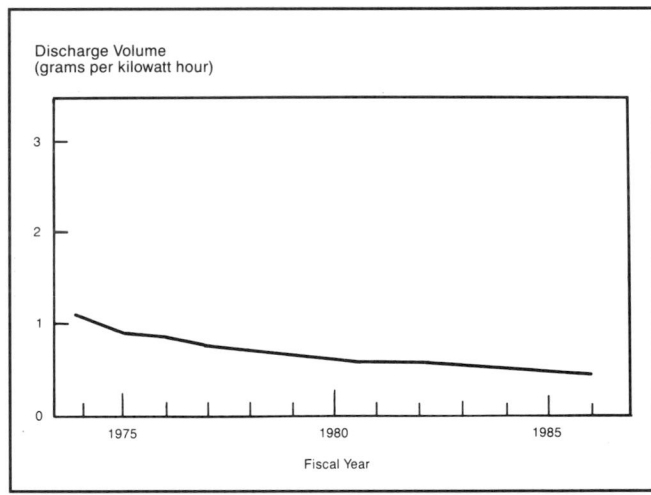

Source: Masatoshi Furuichi, "Environmental Issues in the Global Community—Striving for Clean Air and Power: The Japanese Experience," paper presented at The Environmental Technology Seminar for U.S.-Japan Collaboration, The Japan External Trade Organization, Long Beach, California, November 30, 1989, p. 14.

1987. Similarly, "scrubbers" could be found on only 14,760 megawatts of capacity (47). Nonetheless, this state of affairs compares favorably with most of the world.

There are several explanations for this partial application of advanced technology. First, some local communities exercise their considerable leverage to compel the installation of advanced pollution controls. Their power flows not so much from a formal delegation of authority, but from a moral suasion,

Table 12.2 Ambient Air Quality Standards in Japan

Sulfur dioxide: hourly values may not exceed 0.1 parts per million (ppm) and the daily average of hourly values may not exceed 0.04 ppm.

Oxides of nitrogen: the daily average of hourly values must be within or below the range of 0.04 ppm to 0.06 ppm.

Photochemical oxidants: hourly values may not exceed 0.06 ppm.

Carbon monoxide: the daily average of hourly values may not exceed 10 ppm, and the average of hourly values for eight consecutive hours may not exceed 20 ppm.

Particulate: the average daily concentration of airborne particles of 10 microns in diameter or less may not exceed 0.10 micrograms per cubic meter ($\mu g/m^3$) and hourly values may not exceed 0.20 $\mu g/m^3$.

Source: International Affairs Division, Environment Agency, Government of Japan, *Quality of the Environment in Japan 1988* (Environment Agency, Government of Japan, Tokyo, 1989) Appendix 2, p. 303.

which is part of the Japanese tradition. Second, Japan is the only major industrialized nation to have adopted a system for identifying and compensating the victims of air pollution. Victim payments are made from a trust fund that derives its revenue from a tax on sulfur dioxide emissions. Here again, the system creates both moral and economic pressure on polluting industries to adopt aggressive controls (48). Finally, Japan has a strict ambient air quality system: for example, the level of allowable sulfur dioxide pollution varies according to a region's "K-value." The more polluted the air, the lower the K-value; the lower the K-value, the more stringent the emissions limit (49). Through this approach, Japan uses ambient air quality standards to drive technology, yielding significant reductions in average emissions. (See Figures 12.1 and 12.2.)

Japan has established health-based ambient standards for most conventional pollutants from stationary sources. (See Table 12.2.) In addition, there are vehicle emission controls as well. On paper, the Japanese tailpipe emissions standards are the same as the United States "49-state requirements" (as the 1983 U.S. requirements are known). In reality, however, Japanese standards are somewhat more lenient because they use a different set of tests to simulate driving (50).

Japan controls air pollution through two other major mechanisms. Following the oil shocks of the 1970s, the government adopted a national policy of reducing dependence on oil. This policy caused industries and utilities (see Table 12.3) to increase their efficiency—thus lowering fuel consumption and the amount of pollution emitted—and to switch to cleaner fuels, such as nuclear-generated electricity and imported natural gas (51). Earlier, Japanese electric utilities had embarked on a program of fuel desulfurization. This significantly reduced the average sulfur content of all fossil fuels from 1.5 percent in 1970 to 0.28 percent in 1986 (52).

The United States

As the 1980s began, the United States was widely regarded as the world's leader in air pollution control.

Table 12.3 Efficiencies of Thermal Power Plants, Developed Countries.

(percent)

	Japan	United States	Canada	United Kingdom	France	Federal Republic of Germany	Italy
Thermal Power Plant Thermal Efficiency (based on 1985 data)	36.4	32.7	32.0	32.6	33.0	33.2	35.1

Source: Masatoshi Furuichi, "Environmental Issues in the Global Community—Striving for Clean Air and Power: The Japanese Experience," paper presented at the Environmental Technology Seminar for U.S.-Japan Collaboration, The Japan External Trade Organization, Long Beach, California, November 30, 1989, Figure 2-2-3, p. 25.

The 1970 Clean Air Act had imposed a system for achieving levels of ambient air quality that protected the health of sensitive groups such as children and the elderly. The law supplemented this approach with a technology-based program which forced the development of catalytic converters for cars and promised to achieve the same result for stationary sources of air pollution such as power plants.

Although admired outside the United States, the Clean Air Act was beset by controversy and opposition within the country. The tailpipe standards had been written directly into the law and ultimately went into force despite repeated attempts to weaken them. The stationary source program, especially for power plants, had an even more troubled history.

The power plant program was intended to phase in pollution reductions by requiring that new power plants—or those repaired so extensively that their lifetimes were significantly extended—meet relatively stringent "New Source Performance Standards." What actually happened was that for a variety of reasons ranging from the economic downturn and decrease in demand for power to the widespread adoption of conservation programs following the two oil shocks, relatively few new plants were built. Simultaneously, regulations were issued which allowed old ones to continue operation with only modest controls or, in many cases, none at all.

In 1981, the U.S. government adopted an agenda that included a widespread relaxation of environmental control programs. None was enacted into law by the Congress, but they were implemented to some extent by the Executive Branch. For example, implementation of new carbon monoxide tailpipe standards was waived twice. In fact, a number of indicators of air quality—which had been gradually improving in the United States after the 1970 law—began turning down. For example, in 1981 the National Commission on Air Quality, a nonpartisan body created by the Congress to review the Clean Air Act, estimated that by 1987 there would be eight urban areas where the air remained unhealthy (53). In fact, in 1988 there were 90 urban areas in violation of the ozone or "smog" standard and 40 areas violating the health standard for carbon monoxide (54).

Ozone levels during the summer of 1988 were the worst in a decade in most U.S. cities, partly as a result of unusual weather conditions (55), leading to a widespread public outcry for newer and tougher standards. During this same period, the demand for a new program to control acid rain also gained public support.

The amendments proposed in 1989 were not clearly based on either technology or ambient standards. Instead, they embodied a mix of approaches to reduce pollution through least-cost means, such as switching to cleaner fuels in heavily polluted areas.

In dealing with urban smog, the 1970 Clean Air Act established deadlines for the attainment of ambient air quality standards and attached penalties (e.g., bans on the construction of new pollution sources) for failure to meet them. The new plan proposed repealing the deadlines and replacing them with "attainment dates" that could be extended up to 20 years (56). Instead of penalties, failure to achieve healthful air by the specified date would result in the assignment of a new date for attainment (57).

The proposal called for increasingly stringent control measures based on the severity of an urban area's air pollution. The plan called for reducing motor vehicle pollutants—hydrocarbons, nitrogen oxides, and carbon monoxide—in nine major urban areas through the introduction by the year 2005 of over 9 million vehicles capable of running on so-called "clean fuels" such as methanol (58). The proposal also called for tightening permissible auto emissions of oxides of nitrogen from 0.62 grams per kilometer to 0.43 in 1995 (59); however, the plan called for repeal of the every-car-complies requirement of the law and replacement with a system of averaging emissions, which many critics contend would effectively relax the limits.

To reduce acid rain, the plan proposed reductions in emissions of both sulfur dioxide and oxides of nitrogen to be accomplished mainly by reducing the emissions of fossil-fuel power plants. For sulfur dioxide, the plan called for an overall reduction of 9.07 million metric tons per year from 1980 levels—some of which has already been achieved—to be reached by January 1, 2001 (60), but with possible individual two-year extensions to 2003 (61). For oxides of nitrogen, the proposal was a 1.82 million metric ton reduction from projected year 2000 levels (62). As the program allows utilities to use an averaging approach in achieving reductions (63), some analysts predict that compliance would be through switching to lower-sulfur fuels and, in the case of oxides of nitrogen, combustion or burner modifications.

World Resources 1990–91

Box 12.3 Air Pollution Control in California

Nowhere in the United States is air pollution worse than in the Los Angeles basin. Nowhere are more innovative—some might say extreme—measures being considered to cure that problem.

The Los Angeles Basin is technically known as the South Coast Air Basin. It consists of all of Orange County and the nondesert portions of Los Angeles, Riverside, and San Bernardino Counties. It is projected to have a population of 15.5 million by 2010.

Air pollution is so severe in the Los Angeles Basin that the Basin fails to meet the federal air quality standards for four of the six federal criteria pollutants (1). The Basin meets federal standards for sulfur dioxide and lead. But ozone concentrations sometimes reach three times the federal health standard and carbon monoxide and fine particulate matter concentrations reach approximately double the legal maximum (2). The Basin is the only area in the United States that still fails to meet the nitrogen dioxide standard (3).

Air quality is unhealthy two days out of three. Researchers have estimated that meeting federal air pollution standards would reduce the medical costs of area residents by at least $9.4 billion a year; in fact, these researchers calculate that residents of San Bernardino and Riverside counties, where levels of suspended particulates are highest, face a 1 in 5,000 risk of meeting a premature death due to particulates pollution, approximately the same as their risk of dying in an automobile accident (4).

In 1988, the South Coast Air Quality Management District, which regulates air pollution in the Los Angeles Basin, adopted a policy crafted to achieve all federal and state health standards at the earliest practicable date, but no later than December 31, 1996, for nitrogen dioxide, December 31, 1997, for carbon monoxide, and December 31, 2007, for ozone and fine particulates (5).

The plan includes interim goals for ozone and fine particulates to be met by the year 2000. For ozone, the interim goal is to reduce maximum average concentrations over a one-hour period to no higher than 0.20 parts per million (ppm) and to reduce by 70 percent—compared with 1985—the average per capita exposure to ozone levels above the federal standard. For fine particulates, the interim goal is to attain the federal standards (6).

Researchers believe air pollution in the Basin will decline until the year 2000, then begin rising again—inexorably, unless steps are taken now. Without such action, they say that by 2010 emissions of some pollutants will have returned to 1985 levels and will continue to worsen. This resurgence in emissions would be due entirely to the impact of population growth in the Basin (7).

THREE TIERS OF CONTROLS

In developing the plan, California officials identified the potential measures to reduce air pollution that could be available by the year 2007. They grouped the measures into three "tiers," based on the speed with which they could be implemented.

Tier I controls are those feasible within the next five years, using currently available technology (8). They include tighter tailpipe standards for new cars and trucks; the introduction of buses using "clean" fuels such as methanol; more rigorous inspection and maintenance of diesel buses and trucks (9); shifts away from oil-based paints and glues; restrictions or outright bans on consumer products, such as some cleaning solutions and deodorants that contain smog-forming gases (10); tighter regulations for oil refineries and other major industrial sources of air pollution; and phasing out the use of coal and fuel oil in power plants and industrial burners (11).

Tier II measures include ozone and particulates control technologies that can reasonably be expected to be available in the next 10 to 15 years, as well as government activities requiring new funding or legislative approval. These measures include conversion of 40 percent of cars and 70 percent of trucks and 100 percent of buses to "low-emitting vehicle technologies," such as motors powered by natural gas, methanol, electricity or other alternative fuels. They also contemplate a reduction in the remaining emissions from aircraft, ships, locomotives, solvents, coatings, and consumer products by 50 percent (12).

Tier III programs are designed to bring about major technological breakthroughs to further reduce emissions of reactive organic gases. Unlike the first two tiers, Tier III requires commitments to research, development, and widespread commercial application of technologies that may not exist yet, but may be expected given the rapid technological advances experienced over the past 20 years. For example, Tier III programs contemplate the widespread use of fuel cells in both mobile and stationary source applications; conversion of electricity production to non-fossil sources such as solar, geothermal and wind energy; and utilization of intrinsically clean fuels such as hydrogen (13).

Despite these intentions, there is evidence that the Los Angeles area governments may find implementing these plans to be more difficult than developing them. The only major contest so far has concerned how soon NOx controls should be imposed on existing power plants and how stringent they should be.

One proposal would have required approximately a 9 parts per million (ppm) emissions limit by 1993 (14). Although ambitious, power plants now exist that come close to achieving this level of control; the 2,000 megawatt Futtsu power station operated by the Tokyo Electric Power Company, for example, achieves 10 ppm NOx with technology that both the manufacturers and plant operators say could be improved further (15). However, Southern California Edison, the region's largest utility and the second largest in the United States, opposed the standard. Ultimately, the board adopted a requirement that power plants meet a 113 g/mw-hr NOx standard (approximately 25 ppm) by 1999 (16).

References and Notes

1. South Coast Air Quality Management District and Southern California Association of Governments, *Air Quality Management Plan: South Coast Air Basin* (South Coast Air Quality Management District and Southern California Association of Governments, El Monte and Los Angeles, California, 1989), pp. 1-6, 1-7, 2-1.
2. *Ibid.*, pp. 2-1, 2-7, 2-8, 2-9.
3. John Calcagni, Director, Air Quality Management Division, Environmental Protection Agency, Durham, North Carolina, January 1990 (personal communication).
4. Jane V. Hall *et al.*, *Economic Assessment of the Health Benefits from Improvements in the South Coast Air Basin* (South Coast Air Quality Management District, Fullerton, California, 1989), pp. E-11, E-12.
5. *Op. cit.* 1, p. ii.
6. *Op. cit.* 1, p. ii.
7. *Op. cit.* 1, pp. v, vi, 2.
8. *Op. cit.* 1, p. vii.
9. *Op. cit.* 1, pp. 4-17, 4-18.
10. South Coast Air Quality Management District, *Final Air Quality Management Plan, 1989 Revision—Final Appendix N-A: Tier I, Tier II, and Contingency Control Measures* (South Coast Air Quality Management District, El Monte, California, 1989), pp. A-1, A-55– A-61.
11. *Op. cit.* 1, pp. xv-xvi.
12. *Op. cit.* 1, pp. 4-24, 4-25.
13. *Op. cit.* 1, pp. 4-30, 4-39.
14. Gene Howard, Rules Development Division, South Coast Air Quality Management District, El Monte, California, January 1990 (personal communication).
15. Curtis A. Moore, *Trip Report: Japanese Energy Technologies* (Center for Global Change, University of Maryland, College Park, Maryland, forthcoming).
16. *Op. cit.* 14.

Other comprehensive sets of amendments to the Clean Air Act were introduced in both houses of the Congress by various legislative leaders. Judging from the common elements of these proposals, together with reactions to the government's proposals, it seems clear that the United States is likely to tighten the emissions control programs for cars and trucks through a combination of more stringent tailpipe standards and use requirements, and to require substantial reductions in emissions of sulfur dioxide and oxides of nitrogen from stationary sources. The United States is also likely to establish a new set of dates for achieving health-based ambient air quality standards, with the actual date varying from one location to another based on the severity of the air pollution problem. But it seems unlikely that the current amendments to the Clean Air Act will require widespread adoption of advanced pollution control or energy efficiency technologies. In California, however, regulatory authorities are pursuing a far more aggressive approach, one that may set the pattern for other individual states to follow. (See Box 12.3.)

Canada

Air pollution control in Canada during the 1980s has largely been an outgrowth of the nation's increasing concern over the damages caused by acid rain, called by the Canadian Environment Minister in 1980 "the most serious environmental threat ever to face the North American continent" (64). Scientific journals and the popular press contained not only reports of damages to Canadian lakes, but also forests and human health (65) (66).

In the United States (as in most industrialized nations), utility power plants account for about two thirds of sulfur dioxide emissions and almost one third of the total emissions of nitrogen oxides. In mineral-rich Canada, however, smelters account for 50 percent of the sulfur dioxide emissions (67). Until the public outcry over acid rain, these emissions were either largely uncontrolled or exported to other regions by tall smokestacks. (The chimney at the Inco smelter in Sudbury, Ontario is the world's tallest at 380 meters.)

Fossil-fuel-fired power plants are comparatively less important sources of air pollution in Canada than in the United States. They account for only 17 percent of Canada's sulfur dioxide emissions, compared with 66 percent in the United States (68). This is partially because of the greater Canadian reliance on nuclear power (which supplies almost 50 percent of Ontario's and 15 percent of all Canada's electricity) and hydropower (which accounts for over 60 percent of Canada's electricity). Nevertheless, these nonscrubbed units acquired a symbolic importance to U.S. politicians, some of whom repeatedly observed that Canada was asking the United States to install hundreds of "scrubbers" even though none was operating in Canada.

Responding to pressure on both sides of the border, Canada's government announced in March 1985 a two-part acid rain control program. The first part involved an agreement among the eastern provincial governments to reduce aggregate sulfur dioxide emission from stationary sources by 1.4 million metric tons by 1994, which is about 37 percent below 1980 actual levels. In addition, the federal government implemented controls of motor vehicle emissions standards, beginning in 1988 (69).

Under the first part of the program, for example, Canada's largest electric utility, Ontario Hydro, will have to reduce its sulfur dioxide emissions to 175,000 metric tons, and total sulfur dioxide plus nitrogen oxides emissions to 215,000 metric tons; in 1987 the utility emitted 340,000 metric tons of sulfur dioxide and 60,000 metric tons of nitrogen oxides (70). Since it faces a system-wide limit, Ontario Hydro plans to meet the requirements by taking coal plants out of continuous service as its Darlington nuclear plant comes on line, installing scrubbers on only 2,000 MW of coal plant capacity by 1994. Ontario Hydro plans to install scrubbers on 4,000 MW more coal capacity only as electricity demand increases and the older plants are brought back into full service (71).

The second prong of the Canadian acid rain control program was the adoption of the U.S. 49-state auto emission standards effective in 1988. Although motor vehicles account for 41 percent of U.S. emissions of oxides of nitrogen, cars and trucks contribute roughly 60 percent of Canada's NOx (72).

At a national level, Canada's regulations for controlling power plant emissions at existing power plants, set in 1980, are similar to those in the United States. And, as in the United States, very few new plants are planned for the foreseeable future, so existing facilities will emit most of the air pollution. For new power plants, NOx emission limits of 258, 129, and 86 nanograms per Joule (ng/J) are required of power plants that burn coal, oil, and gas respectively. These limits would not require the use of the technology known as selective catalytic reduction. Limits for sulfur dioxide are based on a sliding scale which requires plants with emissions in excess of 258 ng/J to remove 90 percent of the stack gas sulfur dioxide (73). As a practical matter, this limit requires either stack gas scrubbing to remove sulfur dioxide from emissions or fluidized-bed combustion or other similar advanced combustions systems.

Although the Canadian central government sets "national emissions guidelines," the provinces are not obliged to adopt them. Most have, however. The ambient air quality standards themselves are quite stringent in comparison to those of other nations. For example, the U.S. standard for sulfur dioxide allows up to 365 micrograms per cubic meter ($\mu g/m^3$) during a 24-hour period, while that of Ontario, Canada's largest province, limits concentrations to 275 $\mu g/m^3$ In addition, Ontario has adopted limits for one-hour periods for both oxides of nitrogen and sulfur dioxide, a step the U.S. Environmental Protection Agency has not taken (74).

FOCUS ON CARS AND TRUCKS

One of the most remarkable and unexpected chain of events in the 1980s was the widespread adoption of state-of-the-art emissions controls for cars and trucks. The decade began with the United States and Japan standing alone, as they had for 10 years, as the only nations with advanced pollution control programs for cars. Nor was there any indication that this might change. Yet, by 1990 there had been a 180-degree reversal; every major developed nation had adopted tougher tailpipe controls, except the Soviet Union and the Eastern Bloc nations. Some developing nations had also adopted controls.

This turn-about came none too soon, for in 1988, the global car population exceeded 400 million for the first time in history. The highest growth in cars took place in the rapidly industrializing areas of Asia, but there were also record sales of new cars in Western Europe and other highly developed areas. With commercial vehicles included, a total of over one half billion motor vehicles were on the world's roads in 1988—a 10-fold increase since 1950 (75).

There seems to be no end in sight for this phenomenal growth in the number of cars and trucks, caused in part by a rapid increase in human population, especially in the developing regions of Africa, Asia, and Latin America. (See Chapter 4, "Population and Health.")

Much of this increased population is centered in urban areas, thus increasing vehicle use even further. One result is that global automobile production and use are projected to continue to grow substantially over the next decade. Some analysts expect that by the year 2000, there will be over 600 million vehicles in the world (76).

This growth makes widespread adoption of emission controls imperative if vehicular pollution is to be controlled. But as the decade began there was considerable doubt whether even the United States would maintain stringent controls.

In Western Europe, however, the Germans, Swiss, Austrians and Swedes became increasingly alarmed at the rapidly accelerating environmental damages attributed to air pollution. Germany began lobbying within the EC for tighter standards, while the remaining three nations—all nonmembers of the European Community, and thus free to act on their own—made it clear that they would unilaterally require catalyst-based controls. In June 1989, the EC countries decided to require U.S. 49-state standards for new cars beginning with model year 1992. A comparable policy was adopted in Canada. Australia has also established controls.

Only a few developing nations, including Mexico, Brazil, South Korea, and Taiwan, have adopted the U.S. 49-state controls, despite air pollution probably being worse in many cities in developing countries than in cities of the developed world.

The large number of nations with uncontrolled vehicles include: China, India, the Soviet Union, and all other Eastern Bloc nations; Chile, Argentina, and the remaining Central and South American countries; and virtually all African countries. In effect, there was only one continent where every major nation had adopted catalyst-based controls: North America.

Thus, in the 1990s a majority of the world's new vehicles are likely to be manufactured for use in countries with emission controls. However, the greatest growth in vehicles is occurring in the regions of the world without controls.

Future standards might be even more stringent. In late 1989, the U.S. Congress discussed tightening the standard for emissions of oxides of nitrogen from 0.62 grams per kilometer to 0.25. Because the administration has advocated a 0.43 standard for oxides of nitrogen, some tightening of the tailpipe standards seemed inevitable in the United States.

In California, there is serious discussion of reducing the oxides of nitrogen standard—already 0.25 grams per kilometer—to 0.12 grams per kilometer (77). The California Air Resources Board has already agreed, in June 1989, by a vote of 8 to 0, to tighten the hydrocarbon standard for cars from 0.25 to 0.15 grams per kilometer for the 1993 model year (78). In addition, California is requiring that car makers build autos to last longer, increasing the legal "useful life" from 80,460 kilometers to 160,920 kilometers for certification and 120,700 kilometers for recalls.

Only these increasingly tighter standards have made it possible to hold automotive air pollution in check as the number of vehicles on the road—and most importantly, the number of kilometers they travel—has risen rapidly. In the United States., for example, oxides of nitrogen emissions from mobile sources have declined by only 13 percent over the period 1977 to 1986, despite the introduction of exhaust controls. Inevitably, either tailpipe emissions must begin to approach zero or growth has to be controlled, or both, if major cities are to have air that is healthy to breathe. (See also Box 12.4.)

RECENT DEVELOPMENTS

ARCO INTRODUCES CLEANER PETROL

Facing strong political pressures to reduce air pollutants from motor vehicles, the U.S. oil industry is moving to develop new lead-free low-emission petrols that could significantly cut pollution from cars and trucks (79).

On September 1, 1989, ARCO Corporation introduced one such new fuel into the southern California market. The lead-free petrol is designed to replace leaded regular petrol used by older cars and trucks that are not equipped with catalytic converters (80). Though relatively few in number, old cars neverthe-

Box 12.4 Strategies for Reducing Air Pollution

It is possible to design control strategies for industrial countries, based on currently available technologies, which would sharply reduce air pollution within the foreseeable future. One such strategy, limited to power plants, would be the following.

Fossil fuels differ in their contribution to air pollution. Natural gas is the cleanest, then oil, with coal the most polluting. Natural gas is also the most flexible fuel, easily used in a variety of combustion devices, from boilers to turbines. Most power plants now burning natural gas to generate electricity use conventional boilers that operate with efficiencies of around 35 percent, at best. (See Box 12.2.)

If these units were replaced with higher efficiency equipment, such as combined cycle turbines that operate at about 50 percent efficiency, the same amount of natural gas would yield more than 40 percent more power. This additional electricity, if used to replace coal-fired power plants, could thus reduce air pollution significantly. Oil-burning power plants could also be converted to combined cycle turbines, at somewhat greater cost, thus making it possible to retire still more coal-fired capacity. The economic costs of these conversions—and the speed of the cleanup—would de-

pend on whether they are done in the near future, requiring a forcible shutdown of existing plants, or gradually, when existing coal-fired plants reach the end of their productive life.

Conventional coal-burning power plants average about 34 percent efficiency. Newer systems for burning coal, such as pressurized fluidized-bed units or plants that gasify coal and then burn it in a combined cycle turbine, are projected to have efficiencies between 42 and 44 percent. These newer systems also greatly reduce emissions of sulfur dioxide and oxides of nitrogen, compared with existing coal-fired plants. (See Box 12.2.) If the remaining coal-burning plants were converted, pollutant emissions would drop further.

Some analysts have concluded that these steps, applied in the United States, could dramatically cut emissions from the country's power plants, reducing carbon dioxide by 40 percent, oxides of nitrogen by nearly 60 percent, and sulfur dioxide by about 90 percent (1). But since they would require rebuilding every fossil-fired utility plant in the country, the economic cost—to companies and to their customers in the form of higher electricity rates—could be substantial. Shifting to such highly efficient technologies, however, might in the long run

prove economically advantageous to the country as a whole—especially if fuel costs rise in coming years.

For developing countries and for international development agencies, the best strategy appears to be to adopt or support the use of these more efficient and less-polluting technologies as new investments are made to build up a country's industrial base. By leapfrogging to newer technology, developing countries would avoid both a significant increment of additional pollution and the possible expense of future retrofits, while building a more efficient economy. Industrial countries could also help reduce not only local air pollution in developing countries but also the emission of carbon dioxide and other greenhouse gases that contribute to global climate change by facilitating transfer of advanced energy technologies whenever possible.

References and Notes

1. Curtis Moore, "A Path Toward Zero Air Pollution: The Role of Today's Technologies," paper presented at the State and Territorial Air Pollution Program Administrators and the Association of Local Air Pollution Control Officials (STAPPA/ALAPCO) Workshop on Innovative Air Pollution Control Technologies, New Orleans, Louisiana, October 1989.

---

less are heavy polluters: a 20-year-old car puts out about 8 to 10 times as much pollution as a new car sold in California (81). The 1.2 million older cars and trucks are only 15 percent of the 8 million vehicles in the southern California region, but they cause more than 30 percent of the region's vehicular air pollution. If all current users of leaded fuel in the region switched to the new fuel, the effect would be to remove about 315 metric tons of pollutants from the air every day, according to ARCO (82).

Other companies are expected to follow suit. In the fall of 1988, auto makers and the oil industry started a research project to develop low-pollution cars and reformulated fuels (83). ARCO and other refiners in the next few years also may introduce reformulated petrol for newer vehicles that run on unleaded petrol (84). According to some preliminary estimates, reformulated fuels might be able to reduce vehicle pollutants by as much as 30 percent (85).

The potential benefits for global motor vehicle emissions could be substantial. There are now over 500 million cars and trucks registered worldwide (86), and it is estimated that the number of vehicles could approach 1 billion by the year 2030 (87). Though they have a relatively small fraction of all cars, Third World countries have a disproportionately high

share of older cars that use leaded petrol and could use the reformulated fuel developed by ARCO (88).

The new fuel eliminates lead and uses an additive, methyl tertiary butyl ether (MTBE), which includes an oxygen molecule that promotes better engine combustion. Olefins and aromatics, which react in the presence of sunlight to form smog, are reduced by one third. Benzene is cut by about 50 percent, and sulfur content is reduced by up to 80 percent. To reduce evaporative emissions, Reid Vapor Pressure is lowered by 70 grams per square centimeter below the current standard (89). ARCO said that tests of older cars indicated the new fuel would reduce carbon monoxide by 10 percent, nitrogen oxide by 6 percent, hydrocarbons by 5 percent, evaporative emissions by 22 percent (90), and sulfur dioxide by 80 percent (91).

The move by ARCO is an outgrowth of proposals under consideration in the state of California and in the U.S. Congress to phase in much lower standards for motor vehicle emissions in the 1990s and beyond. ARCO officials think that reformulated fuels have the potential to meet these tougher standards, though it remains to be seen whether the standards under consideration by California for the year 2000 and beyond can be met with reformulated fuels or whether refiners and auto makers will have to turn to alternative

Figure 12.3 Concentration of Ozone Above Antarctica, 1979-89

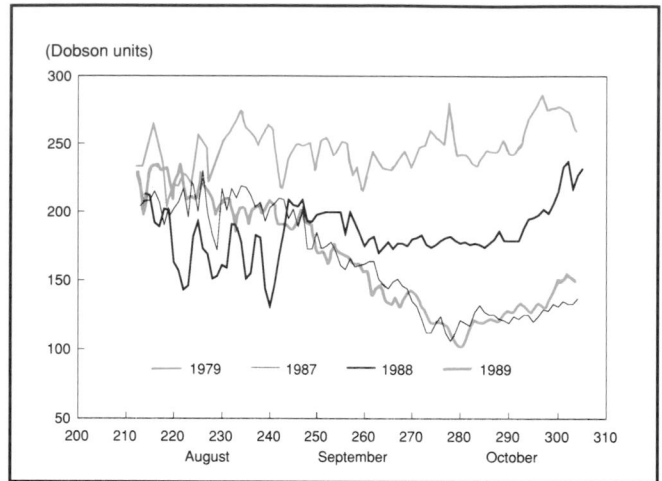

(Dobson units)

— 1979   — 1987   — 1988   — 1989

August   September   October

Source: Dr. Arlin J. Krueger, Laboratory for Atmosphere, U.S. National Aeronautics and Space Administration, Goddard Space Flight Center, Greenbelt, Maryland, 1989 (personal communication).
Note: a. A Dobson unit is a standard measure of ozone loss.

fuels such as methanol, ethanol or compressed natural gas (92).

SEVERE OZONE DEPLETION RETURNS IN 1989

The seasonal ozone hole over Antarctica returned with a vengeance in 1989 after a relatively moderate depletion in 1988.

Scientists at the National Aeronautics and Space Administration's (NASA) Goddard Space Flight Center near Washington, D.C., in Greenbelt, Maryland, have been monitoring ozone levels over the Southern Hemisphere with an instrument on board NASA's NIMBUS-7 satellite. By October 5, 1989, the minimum value within the ozone hole had decreased by about 45 percent from early August, dropping during September about 1.5 percent per day. The decline was nearly identical to the severe depletion that occurred in October 1987 (93). The 1989 hole covered some 26 million square kilometers (94). (See Figure 12.3.) In contrast, ozone depletion was relatively slight, about 15 percent in September 1988 (95).

Ozone depletion over the Southern Hemisphere was first observed in 1985. The phenomenon origi-

nates with the long, dark Antarctic winter, when polar air forms into an isolated vortex. Temperatures drop so severely that icy clouds form in the lower stratosphere at altitudes of 15 to 23 kilometers (96). The ice in the clouds reacts with chlorine nitrate and hydrogen chloride molecules that are present in the stratosphere from the breakdown of chlorofluorocarbons, or CFCs; the reaction creates chlorine molecules. When sunlight returns in the austral spring in August and September, it triggers photochemical reactions that break down the chlorine molecules into individual chlorine atoms, which react with and break up ozone molecules. Warm air eventually dissolves the vortex: the icy clouds disappear, and chemical reactions bind chlorine again into chlorine nitrate and hydrogen chloride molecules (97).

The severity of the depletion appears to be related to temperatures in the vortex. In 1987 and 1989, temperatures in the vortex were between -85°Centigrade and -90°Centigrade, whereas in 1988 temperatures were between -78°Centigrade and -80°Centigrade (98). The severity of the cold temperatures in 1989 actually took some scientists by surprise; they theorized that the two-year pattern of stratospheric tropical winds blowing from the east would warm the polar vortex and cause a repeat of 1988's shallow hole, but the theory was shaken by the severity of the 1989 temperatures (99).

Scientists currently do not believe the ozone hole will get much bigger than the 1989 occurrence because the destruction takes place within the polar vortex and the vortex is constrained by other weather systems (100). The ozone hole nevertheless poses a health threat. Reduced ozone increases the amount of ultraviolet light entering the atmosphere, which increases the risk of skin cancer especially among lighter-skinned people (101). Ozone-depleted air can migrate out of the ozone hole during December. Such a phenomenon occurred in December 1987, when ozone levels over New Zealand and southern Australia dropped by 10 percent for a period of about three weeks (102).

Scientists are still investigating the effects and implications of these seasonal and yearly variations in Antarctic ozone depletion (103).

Curtis Moore is an independent consultant on high-efficiency energy technologies and on air pollution control technologies and writes frequently on these subjects. He is based in McLean, Virginia.

References and Notes

1. Global Environment Monitoring System, United Nations Environment Programme and World Health Organization, in cooperation with the Monitoring and Assessment Research Center, *Assessment of Urban Air Quality* (United Nations, London, 1988), p. 15, Figure 3.3, p. 27, Figure 4.2.
2. World Resources Institute and International Institute for Environment and Devel-

opment in collaboration with the United Nations Environment Programme, *World Resources 1988-89*, World Resources Institute (Basic Books, New York), Table 23.4, p. 338.
3. D. Bellinger, J. Sloman, A. Leviton *et al.*, "Low-level lead exposure and child development: assessment at age 5 of a cohort followed from birth," in *International Confer-*

ence: *Heavy Metals in the Environment* (CEP Consultants, Edinburgh, 1987), pp. 49-53, cited in Agency for Toxic Substances and Disease Registry, U.S. Department of Health and Human Services, *The Nature and Extent of Lead Poisoning in Children in the United States: A Report to Congress*, draft copy (U.S. Department of Health and Human Services, Atlanta, 1987), p. IV-11.

4. Environmental Protection Agency (EPA), *Air Quality Criteria for Lead*, Vol. I (draft final copy) (EPA, Research Triangle Park, North Carolina, 1986), Table 1-16, p. 143.
5. Protocol on the Reduction of Sulphur Emissions or their Transboundary Fluxes by at least 30 percent, 1985, to the Convention on Long-Range Transboundary Air Pollution, 1979, reproduced in *International Environment Reporter: Reference File* (Bureau of National Affairs, Inc., Washington, D.C., 1989), pp. 21:3021-21:3022.
6. 1988 Protocol to the 1979 Convention on Long-Range Transboundary Air Pollution Concerning the Control of Emissions of Nitrogen Oxides or their Transboundary Fluxes, reproduced in *International Environment Reporter: Reference File* (Bureau of National Affairs, Inc., Washington, D.C., 1989), pp. 21:3041- 21:3050.
7. Montreal Protocol on Substances that Deplete the Ozone Layer (1987) to the Vienna Convention for the Protection of the Ozone Layer (1985), Article 2, Paragraph 4, reproduced in *International Environment Reporter: Reference File* (Bureau of National Affairs, Inc., Washington, D.C., 1989), pp. 21:3153- 21:3158.
8. United Nations Environment Programme, *Environmental Data Report* (Basil Blackwell, Oxford, 1989), Table 1.15, p. 30.
9. U.N. Economic Commission for Europe, "The 1988 Forest Damage Survey in Europe: Executive Summary." (U.N. Economic Commission for Europe, Geneva, September 1989), cited by Global Environment Monitoring System, United Nations, New York, January 1990, (personal communication).
10. Michael P. Walsh, international motor vehicle emissions consultant, Washington, D.C., 1989 (personal communication).
11. Jürgen Schmölling, "Air Quality Control Policy with Respect to Retrofitting in the Federal Republic of Germany," paper presented at the State and Territorial Air Pollution Program Administrators and the Association of Local Air Pollution Control Officials (STAPPA/ALAPCO) Workshop on Innovative Air Pollution Control Technologies, New Orleans, Louisiana, October 1989.
12. Michael P. Walsh, "Global Trends in Motor Vehicle Pollution Control—A 1988 Perspective," Technical Paper Series, Society of Automotive Engineers, Warrendale, Pennsylvania, 1989, pp. 8-9.
13. Air Resources Board, State of California, "Public Consultation Meeting to Discuss Regulations for Gasoline and Clean Fuels/Low Emission Vehicles," November 13, 1989 (Notice of Public Hearing on November 29, 1989), p. 3.
14. South Coast Air Quality Management District, Southern California Association of Governments, *Air Quality Management Plan: South Coast Air Basin* (South Coast Air Quality Management District and Southern California Association of Governments, El Monte and Los Angeles, California, March 1989), pp. 2-7, 2-8.
15. Morton Lippman, "Ozone Health Effects and Emerging Issues in Relation to Standard Setting," *Atmospheric Ozone Research and Its Policy Implications: Proceedings of the 3rd U.S.-Dutch International Symposium, Nijmegen, the Netherlands, May 9-13, 1988*, T. Schneider, S.D. Lee, G.J.R. Woleers, L.D. Grant, eds. (Elsevier Science Publishers, Amsterdam, 1989), pp. 27-29.
16. *Op. cit.* 10.
17. Global Environment Monitoring System, United Nations Environment Programme and World Health Organization, *Global Pollution and Health* (Yale Press Ltd., London, 1987), p. 7.
18. F.E. Speizer, "Studies of acid aerosols in six cities and in a new multi-city investigation," *Environmental Health Perspectives*, Vol. 79 (February 1989), pp. 61-67, cited in Morton Lippmann, "Health Benefits from Controlling Exposures to Criteria Air Pollutants," in *Health Benefits of Air Pollution Control: A Discussion* (Congressional Research Service, Library of Congress, Washington, D.C., 1989), p. 112.
19. *Op. cit.* 2.
20. Don Hinrichsen, "In Krakow Even the Buildings Dissolve," *International Wildlife*, Vol. 17 (National Wildlife Federation, Washington, D.C., March- April 1987), pp. 12-15.
21. *Op. cit.* 9.
22. Sten Nilsson and Peter Duinker, "The Extent of Forest Decline in Europe: A Synthesis of Survey Results," *Environment*, Vol. 29, No. 9 (November 1987), pp. 4-9, 30-31.
23. James J. MacKenzie and Mohammed T. El-Ashry, *Ill Winds: Airborne Pollution's Toll on Trees and Crops* (World Resources Institute, Washington, D.C., 1988), pp. 13-24.
24. James R. Vestigo, "Acid Rain and Tall Stack Regulation Under the Clean Air Act," *Environmental Law*, Vol. 15, No. 4 (1985), p. 730.
25. Former U.S. Senator Robert P. Griffin, quoted in U.S. Senate, Committee on Environment and Public Works, *A Legislative History of the Clean Air Act Amendments of 1970* (U.S. Government Printing Office, Washington, D.C., 1974), Vol. 1, p. 237.
26. Former U.S. Senator Edmund Muskie, statement on Senate floor, September 21, 1970, quoted in U.S. Senate, Committee on Environment and Public Works, *A Legislative History of the Clean Air Act Amendments of 1970* (U.S. Government Printing Office, Washington, D.C., 1974), Vol. 1, p. 229.
27. James T. MacKenzie, *Breathing Easier: Taking Action on Climate Change, Air Pollution, and Energy Insecurity* (World Resources Institute, Washington, D.C., 1989), p. 13, Figure 10.
28. *Op. cit.* 11, p. 2.
29. Mira Courpas and Larry B. Parker, *Canada's Progress on Acid Rain Control: Shifting Gears or Stalled in Neutral?* (Congressional Research Service, Library of Congress, Washington, D.C., April 1988), p. 10, Table 4.
30. *Op. cit.* 7, p. 21:3154.
31. "EC Council Agrees on Total Ban of CFCs by 2000; Bush Says US Goal Dovetails," *International Environment Reporter*, Vol. 12, No. 3 (March 8, 1989), pp. 105-106.
32. "Canada Aiming for Total CFC Ban by 1999, Environment Minister Says," *International Environment Reporter*, Vol. 12, No. 3 (March 8, 1989), pp. 107-108.
33. Economic Commission for Europe, United Nations, *National Strategies and Policies for Air Pollution Abatement* (United Nations, New York, 1987), p. 4.
34. Don R. Clay, Acting Assistant Administrator for Air and Radiation, "Applicability of Prevention of Significant Deterioration (PSD) and New Source Performance Standards (NSPS) Requirements to the Wisconsin Electric Power Company (WEPCO) Port Washington Life Extension Project," EPA, Washington, D.C., September 1988 (internal memorandum).
35. *Op. cit.* 33, pp. 6-7.
36. Helmut Weidner, *Air Pollution Control Strategies and Policies in the Federal Republic of Germany: Laws, Regulations, Implementation, and Principal Shortcomings* (Edition Sigma Bohn, Berlin, 1986), p. 65.
37. *Op. cit.* 11, p. 6.
38. *Op. cit.* 33, pp. 17, 19.
39. E. Rehbinder, "Implementation of Air Pollution Control Programs Under the Law of the Federal Republic of Germany," in *Air Pollution Control: National and International Perspectives* (American Bar Association, Washington, D.C., 1980), p. 31.
40. Gregory S. Wetstone and Armin Rosencranz, *Acid Rain in Europe and North America: National Response to an International Problem* (Environmental Law Institute, Washington, D.C., 1983), p. 84.
41. Peter Schütt and Ellis B. Cowling, "Waldsterben a General Decline of Forests in Central Europe: Symptoms, Development and Possible Causes," *Plant Disease*, Vol. 69, No. 7 (July 1985), p. 548.
42. *Op. cit.* 36, pp. 90, 95.
43. Michael P. Walsh, "Europe on the Verge of State of the Art Auto Standards," *Car Lines*, Vol. 1, No. 2 (April 1989), p. 2.
44. "European Community Environment Ministers Agree on New Emission Levels for Small Cars," *International Environment Reporter*, Vol. 12, No. 6 (June 14, 1989), p. 283.
45. *Op. cit.* 10.
46. *Op. cit.* 10.
47. Japan Electric Power Information Center, *Environmental Pollution Control for Thermal Power Stations in Japan* (Japan Electric Power Information Center, Tokyo, 1987) pp. 17-24.
48. Helmut Weidner, "Environmental Protection in Japan: Development and Prospects," in *Trends of Economic Development in East Asia*, W. Klenner, ed. (Springer Verlag, Berlin, 1989), pp. 291-293.
49. International Affairs Division, Environment Agency, Government of Japan, *Quality of the Environment in Japan 1988* (Environment Agency, Government of Japan, Tokyo, 1989), pp. 129-131.
50. *Op. cit.* 10.
51. International Affairs Division, Environment Agency, Government of Japan, *Quality of the Environment in Japan 1980*, (Environment Agency, Government of Japan, Tokyo, 1980), p. 44.
52. Masatoshi Furuichi, "Environmental Issues in the Global Community —Striving for Clean Air and Power: The Japanese Experience," paper presented at the Environmental Technology Seminar for US-Japan Collaboration, The Japan External Trade Organization, Long Beach, California, November 30, 1989, p. 9.
53. National Council on Air Quality, *To Breathe Clean Air: Report of the National Council on Air Quality* (U.S. Government Printing Office, Washington, D.C., 1981), p. 121.
54. Environmental Protection Agency, "EPA Lists Places Failing to Meet Ozone or Carbon Monoxide Standards," *Environmental News* (press release), Environmental Protection Agency, July 27, 1989, attached Tables 1, 2.
55. John Calcagni, Director, Air Quality Management Division, Environmental Protection Agency, Durham, North Carolina, January 1990 (personal communication).
56. U.S. Environmental Protection Agency, U.S. Office of Management and Budget, The White House Office, "A Bill to amend the Clean Air Act to provide for the attainment and maintenance of the national ambient air quality standards, the control of toxic air pollutants, the prevention of acid deposition, and other improvements in the quality of the nation's air" (introduced as H.R. 3030 on July 27, 1989, and as S. 3030 on August 3, 1989), Section 102(b), p. 33.
57. *Ibid.*, p. 36.
58. *Ibid.*, Section 201(b), p. 124.
59. *Ibid.*, Section 202(a), Table 1, p. 135.
60. *Ibid.*, Section 501, p. 217.

61. *Ibid.*, Section 501, p. 237.
62. *Ibid.*, Section 501, p. 217.
63. *Ibid.*, Section 503.
64. John R. Roberts, Canadian Environment Minister, quoted in Gregory S. Wetstone and Armin Rosencranz, *Acid Rain in Europe and North America: National Response to an International Problem* (Environmental Law Institute, Washington, D.C., 1983), p. 114.
65. Michael Keating, "Withering Death of Forests Spreading Across the East," *The Globe and Mail* (August 6, 1987, Toronto, Canada), p. A1.
66. David Bates and Ronnie Sizto, "Air Pollution Levels and Hospital Admissions in Southern Ontario," *Environmental Research*, Vol. 43, No. 2 (August 1987), pp. 317-331.
67. United States-Canada memorandum of Intent on Transboundary Air Pollution, Work Group 3B, June 1982, cited in Larry B. Parker, *Canada's Acid Rain Control Program: Catching Up or Pulling Away?* (Congressional Research Service, Library of Congress, Washington, D.C., 1985), Table 1, p. 3.
68. *Ibid.*
69. Environment Canada, *Acid Rain: The Canadian Approach*, March 1985, cited in Mira Courpas and Larry B. Parker, *Canada's Progress on Acid Rain Control: Shifting Gears or Stalled in Neutral?* (Congressional Research Service, The Library of Congress, Washington, D.C., April 1988), p. 3.
70. *Op. cit.* 29, pp. 21-22.
71. Alec Mansen, Director of Issues Management, Environment Canada, Hull, Quebec, January 1990 (personal communication).
72. *Op. cit.* 67.
73. Larry B. Parker, *Canada's Acid Rain Control Program: Catching Up or Pulling Away?* (Congressional Research Service, Library of Congress, Washington, D.C., 1985), pp. 9-10.
74. *Ibid.*, Table 8, p. 18.
75. *Op. cit.* 12, p. 1.
76. *Op. cit.* 12, p. 7.
77. Air Resources Board, State of California, "Public Consultation Meeting to Discuss Regulations for Gasoline and Clean Fuels/Low Emissions Vehicles," November 13, 1989 (Notice of Public Hearing on November 29, 1989).
78. Jerry Martin, Information Officer, Air Resources Board, State of California, January 1990 (personal communication).
79. Matthew L. Wald, "That 'Cleaner Fuel' May Be Gasoline," *The New York Times* (August 23, 1989), p. D1.
80. Linda K. Cohu, Larry A. Rapp, and Jack S. Segal, "EC-1–Emission Control Gasoline," paper presented at the State and Territorial Air Pollution Program Administrators and the Association of Local Air Pollution Control Officials (STAPPA/ALAPCO) Workshop on Innovative Air Pollution Control Technologies, New Orleans, Louisiana, October 1989, pp. 3-4.
81. Jannane Sharpless, Chairwoman, California Air Resources Board, quoted in Matthew L. Wald, "ARCO Offers New Gasoline to Cut Up to 15% of Old Cars' Pollution," *The New York Times* (August 16, 1989), p. A1.
82. ARCO Corporation, *ARCO's New Low-Emission Gasoline is First Phase in Helping Clean Up Southern California Air Pollution* (ARCO Corporation, Los Angeles, August 15, 1989), pp. 1-2.
83. Doron P. Levin, "Gasoline Changes Discussed as Way to Cut Pollution," *The New York Times* (September 11, 1989), p. D1.
84. Matthew L. Wald, "ARCO Offers New Gasoline to Cut Up to 15% of Old Cars' Pollution," *The New York Times* (August 16, 1989), p. A1.
85. *Op. cit.* 79.
86. *Op. cit.* 12, p. 1.
87. Michael P. Walsh, "Global Trends in Motor Vehicles and Their Use: Implications for Climate Modification," paper prepared for World Resources Institute, Washington, D.C., December 1988, pp. 13-15.
88. *Ibid.*, p. 13.
89. *Op. cit.* 82, p. 3.
90. *Op. cit.* 80.
91. ARCO Corporation, "EC-1 Regular—ARCO's Emission-Control Gasoline," attachment to *ARCO's New Low-Emission Gasoline is First Phase in Helping Clean Up Southern California Air Pollution* (ARCO Corporation, Los Angeles, August 15, 1989).
92. Charles McCoy, "California Air-Quality Regulators Mull Plan to Radically Lower Auto Emissions," *The Wall Street Journal* (December 18, 1989, New York), p. 18.
93. National Aeronautics and Space Administration (NASA), *NASA Confirms '89 Ozone Hole Matches '87 Record* (NASA, Washington, D.C., October 12, 1989), p. 1.
94. William Booth, "Severe Ozone Depletion Likely Again This Year," *The Washington Post* (October 6, 1989, Washington, D.C.), p. A18.
95. *Op. cit.* 93.
96. Richard A. Kerr, "Ozone Hits Bottom Again," *Science*, Vol. 246, No. 4928 (October 20, 1989), p. 324.
97. Mark Schoeberl, Atmospheric Scientist, Goddard Space Flight Center, Greenbelt, Maryland, January 1990 (personal communication).
98. Surendra Verma, "...as Antarctica's ozone hole grows," *New Scientist*, Vol. 124, No. 1685 (October 7, 1989), p. 27.
99. *Op. cit.* 96.
100. *Op. cit.* 94.
101. Janice Longstreth, "Global Climate Change: Potential Impacts on Public Health," paper presented at Conference on Global Climate Change and Life on Earth, Albany, New York, April 1989, p. 2.
102. *Op. cit.* 98.
103. *Op. cit.* 97.

13. Global Systems and Cycles

Climate change caused by human activity may become a central environmental problem in the 21st century. But as discussed in Chapter 2, "Climate Change: A Global Concern," major scientific uncertainties about climate change still exist. Reliable predictions of the impact of climate change—particularly at regional levels—are not yet possible. Enough is simply not yet known about Earth's climatic cycles.

Within the past decade, a consensus has developed in the world scientific community that to fully understand the intricate workings of Earth's systems and cycles, such as the climate cycle, a global and international approach to their study must be taken (1).

Key to this study is continuous monitoring of these systems on a worldwide basis. Without basic measurements of how each global variable changes over timescales ranging from days to decades, no comprehensive vision of whole-Earth dynamics can develop. Lacking such a global view, we can neither assess the impact of human activities on critical systems such as the climate, nor plan wisely to moderate these activities or to mitigate their effects (2).

Monitoring climate change—and the elaborate and still evolving system designed to accomplish that task—is the focus of this year's Global Systems and Cycles chapter. Although some global monitoring systems exist, most notably the Global Environment Monitoring Systems (GEMS) effort of the United Nations Environment Programme, these efforts have been handicapped by a shortage of resources. Fortunately, however, a working prototype of a global climate monitoring system has existed for years: the World Weather Watch (WWW). This international network consists of ground-, sea-, air-, and space-based observing systems—from surface weather stations to sophisticated satellites—as well as computers to analyze the data obtained and communication links to make both data and weather analyses widely available.

Climate is usually defined as the average weather over a period of years. Thus weather data, monitored year after year, is also climate data. The atmospheric circulation that powers storm systems, the ocean currents that transport heat, and the solar radiation that drives the atmospheric heat engine all influence the balance of winds, humidity, air pressure, and temperature that determine daily weather (3) (4) (5) (6). Thus, via the world meteorological network, many of the factors affecting climate are probed directly or indirectly. Still others, of course, need specialized monitoring systems. Monitoring, as described here, deals with documenting long-term changes and understanding the process that causes them.

There is another sense in which the world's weather system is a prototype for monitoring and, eventually, predicting climate. The flow of data from the World Weather Watch is so vast that high-speed computers are needed to interpret it in a timely manner. These same computers also produce weather forecasts, using atmospheric models to predict trends in temperature, pressure, and so on. The weather forecasting models are closely related to the global circulation models used to simulate changes in climate—and both are subject to the same kinds of limitations or inherent uncertainties for long-range forecasts. Thus, in understanding these models and their limitations, we can understand the difficulties involved in moving from monitoring to climate prediction.

Examining the World Weather Watch allows us to glimpse the dimensions of the global climate monitoring task. The proven ground- and satellite-sensing techniques, computer models, and communication systems—not to mention the spirit of cooperation—of today's weather watch together form the solid core of the more comprehensive monitoring network planned for the 1990s and beyond.

THE WORLD WEATHER WATCH

Of all global phenomena, weather has perhaps the most immediate impact on living things. No aspect of human activity, social or economic, escapes its influence, and every country, developed or developing, shares a concern for what the weather may bring. It is this common concern, and the knowledge that weather originates from global-scale interactions and follows a worldwide pattern, that bases the international effort known as the World Weather Watch.

WWW consists of three basic elements: a worldwide network of facilities and instruments for tracking the weather, called the Global Observing System; a network of computer centers—the Global Data-Processing System—to analyze the weather data and produce short- and medium-range forecasts on a global and regional basis; and a Global Telecommunication System to assure the rapid exchange of all data and information among the 160 member nations contributing to the WWW (7). (See Table 13.1.)

The WWW has been a remarkably successful demonstration of international cooperation. While it acts under the auspices of the World Meteorological Organization (a specialized agency of the United Nations) it does not exist as an independent system of sensors, computers, and data links. Rather, it is a carefully integrated union of national weather services acting at their own expense and using their own facilities to accomplish a specified portion of the global weather task.

The guiding philosophy of the WWW is that each member nation contributes services according to its means, while all nations have equal access to the data and forecasts produced. The level of assigned activities thus varies from country to country. Those with the means to launch weather satellites or provide computer power and analytical expertise do so, but every member nation contributes to the network of local ground and air observations that produce much of the raw data used to forecast weather at the local, regional, and global level (8).

Table 13.1 The World Weather Watch at a Glance

Membership	160 nations
Surface-Based Observing Stations	
Land Stations (all types)	9,525
Ocean Vessels	7,424
Upper-Air Observing Stations (Land and Sea)	2,306
Ground Weather Radar Stations	644
Meteorological Satellites	
Polar Orbiters	4
Geostationary Orbiters	5
Satellite Ground Stations	325
Meteorological Centers	
World Centers	3
Regional Centers	25
Total Forecasts and Analyses Produced Daily	2,895
Daily Data Movement Throughout the Global Telecommunication Network	15 million characters

Sources:
1. World Meteorological Organization (WMO), *World Weather Watch: Fourteenth Status Report on Implementation*, WMO Publication No. 714 (WMO, Geneva, 1989), pp. II-6 Table 1, II-7, II-10–II-11, II-23–II-24, III-3, Annex I p. III-10, and Annex VI p. III-15.
2. World Meteorological Organization (WMO), *The World Weather Watch*, WMO Publication No. 709 (WMO, Geneva, 1988), pp. 15-21.

The Global Observing System

The Global Observing System (GOS) acts as the eyes of the WWW. It consists of several complementary observing networks that combine a variety of methods and facilities to give a timely view of weather patterns at all scales. The GOS is organized at three levels: national, regional, and global. National networks of weather stations provide local coverage, with some of these stations selected to feed into regional networks. Some of the regional stations, in turn, feed into the global network under the auspices of the WWW (9). Special efforts are made to obtain weather data from the ocean regions, where regularly spaced observing stations do not exist: buoys, volunteer ships, and satellites all play important roles in ocean weather observations. The data they collect are particularly useful in climate studies.

The GOS yields both qualitative and quantitative information. By measuring air temperature, relative humidity, atmospheric pressure, wind speed, and other parameters at varying altitudes, the GOS allows a quantitative assessment of the atmosphere's physical state. But GOS observations of cloud forms and movements and precipitation types, mostly from satellites, make possible a more qualitative assessment of the state of the sky as well (10)—one that is equally indispensable in following the weather, especially in tracking severe weather such as tropical cyclones.

Surface-Based Observations

The GOS observing networks fall neatly into two categories: surface-based and space-based. Surface-based networks cover a wide spectrum and include:

■ Weather stations on land and sea for surface and upper-air observations,
■ Commercial and other aircraft observing weather variables during flight,
■ Ground weather radar stations, and
■ Special stations such as climatological and agricultural stations, tidal-gauge stations, background air pollution monitoring stations, ozone monitoring stations, lightning detection stations, meteorological rocket stations, and radiation monitoring stations (11).

The number of land-based stations making observations at the surface or in the upper air using weather balloons totaled more than 9,500 in 1989. Nearly 4,000 of these stations make up the surface portion of the basic synoptic network that provides much of the WWW's core data (12).

Stations in this basic network are spaced to guarantee minimum coverage over all the continental regions. The WWW relies on such stations to make at least five basic measurements every three hours at the surface: humidity, temperature, wind direction and speed, pressure, and precipitation. Weather stations not in the basic network are members of various national weather networks that freely contribute their observations to improve the coverage of the WWW system even though they are not under obligation to do so (13).

Also essential to the basic synoptic network are the nearly 2,200 (14) (15) ground stations equipped to make upper-air observations. Many of these stations release balloon-borne packages of instruments called radiosondes that transmit their observations of temperature, pressure, humidity, and wind speed by radio as they ascend. Radiosonde data remains the most accurate method of probing the atmosphere at altitudes of up to 30 kilometers, though it is relatively expensive to obtain and its coverage is very limited compared to, for example, satellite observations.

Surface and upper-air observations over the ocean are problematic and coverage tends to be sparse. These observations rely on a combination of several weather ships at fixed ocean stations, about 7,000 merchant vessels plying commercial routes, and some 350 automated weather facilities, including fixed platforms, moored buoys, and drifting buoys (16) (17).

Unfortunately, only about 40 percent of the merchant ships volunteering to file weather reports are at sea at any one time. In addition, they may experience difficulty filing their reports through coastal radio stations or may spend much of their time in coastal waters. Further, only about 60 ships are currently equipped to make upper-air observations (18).

The number of fixed automated marine stations is increasing rapidly and many can measure a host of environmental factors, such as sea temperatures, wave height and direction, surface and underwater currents, and air and water pollution, in addition to the standard weather program. The number of drifting buoys, sending observations about 10 times a day via satellite, is also on the rise (19).

Ground weather radar stations provide one of the best means to follow small- or mid-scale (mesoscale) weather systems, especially severe thunderstorms. Some 600 of these stations are active worldwide and

Figure 13.1 Weather Watching Satellites: Polar Orbiters and Geostationary Satellites

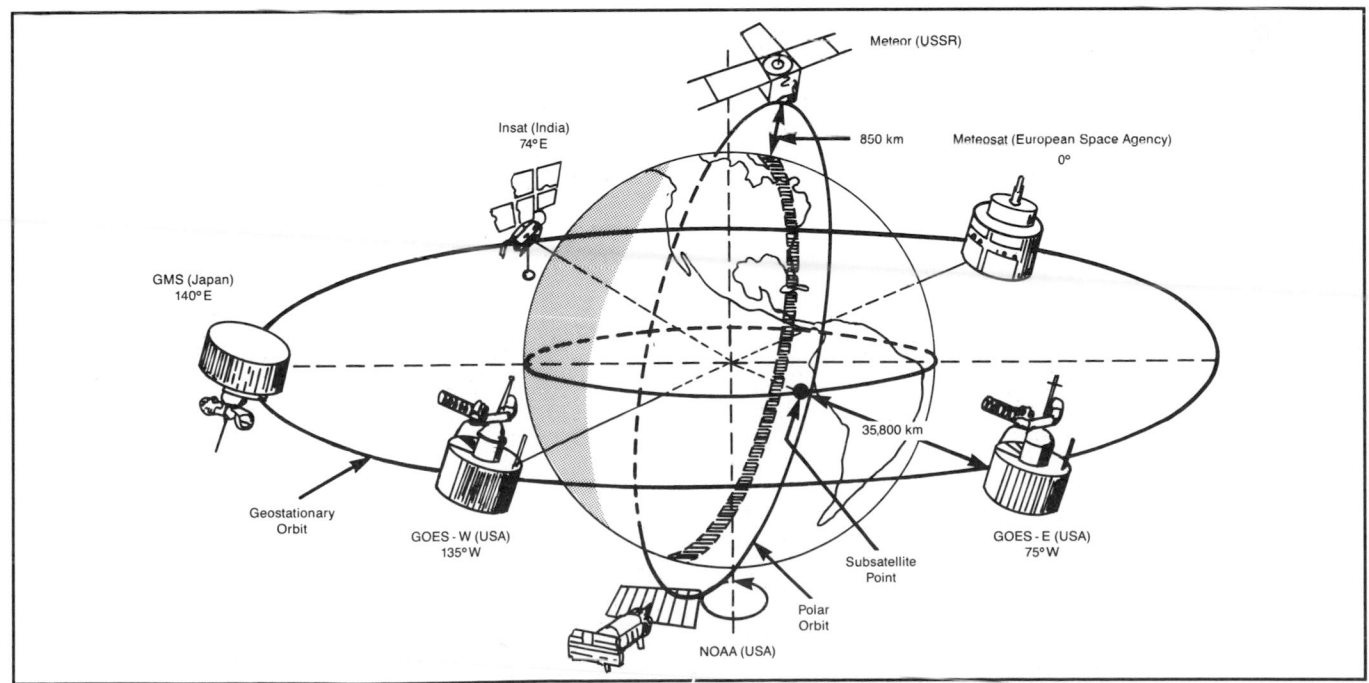

Source: World Meteorological Organization (WMO) *The World Weather Watch* (WMO, Geneva, 1988), p. 16.

are particularly useful in detecting, tracking, and forecasting tornadoes, severe ice or hailstorms, or tropical cyclones (20).

The prevalence of commercial aviation makes observations from aircraft yet another practical and valuable source of weather data. A new program to place automated weather equipment in some wide-body jets (21) is already augmenting the manual reports provided by thousands of willing pilots.

Space-Based Observations

While improvements in ground observations have been steady over the years, it is the revolution in satellite observations that has contributed most to the tremendous progress in weather forecasting since the mid-1960s. The launch of the first weather satellite–TIROS-1–in April 1960 led quickly to a network of satellites that began to fill the existing holes in weather coverage, particularly over the two thirds of the planet covered by oceans (22) (23).

These satellites now comprise an international network that provides temperature and moisture profiles of the atmosphere, cloud patterns and wind profiles, ocean and land surface temperatures, and several other terrestrial measurements of interest, such as vegetation mapping and snow cover analyses. Satellite data are used directly in numerical weather models and to track severe storms, among other applications (24).

Two basic satellite systems cover the world's weather: polar orbiters and geostationary satellites. (See Figure 13.1.) Polar orbiters perform the bulk of the weather monitoring. These satellites follow an 850 kilometers-high orbit that nearly crosses the poles on each orbital pass. Such an orbit allows each spacecraft to view every point on Earth twice a day as the world rotates beneath it. Both the United States and the Soviet Union operate a pair of these satellites. The U.S. satellites orbits are spaced such that measurements over any region are no more than six hours old. In addition to viewing cloud cover and taking temperature and humidity readings from the Earth's surface up through the atmosphere, these satellites also relay data broadcast by many automatic weather stations on land and sea. They also provide a crucial link in an international search and rescue tracking system (25) (26) (27).

The second set of satellites–geostationary satellites–observes the Earth from a much greater distance–about 36,000 kilometers. They maintain an orbit about the equator and their orbital speed matches the Earth's spin, so that these satellites are geostationary, or fixed relative to the Earth. Because of their fixed location above a single position on the Earth, each geostationary satellite can monitor one large area constantly, allowing 24-hour coverage of weather developing in that region and making them especially useful for tracking storm systems such as tropical cyclones (28).

The WWW geostationary network consists of five satellites at different positions about the globe and originating from four different parties: the United States, the European Community, India, and Japan (29). All are equipped to observe cloud motions both day (via visual and infrared observations) and night (via infrared observations). Some give quantitative data on vertical temperature and moisture profiles and act as data relay stations from Earth. The nearly continuous monitoring (every 30 minutes or so) provided by geostationary satellites helps fill the time gaps in the observations of the polar satellites and allows an intensive data analysis critical to modern forecasting (30).

The Global Data Processing System

The prodigious stream of weather data provided continuously by the Global Observing System must be gathered and analyzed in a variety of ways to produce useful forecasts. This is the task of the Global Data Processing System.

As with the other components of the WWW, the Global Data Processing System is organized at the global, regional, and national levels. Three world meteorological centers, located in Melbourne, Moscow, and Washington, D.C., provide global-scale analyses

Table 13.2 Daily Output of World and Regional Meteorological Centers

	Type of Product		
World Centers	**Analyses**	**Forecasts**	**Total**
Melbourne	18	40	58
Moscow	32	32	64
Washington, D.C.	109	303	412
Total World Center Products			534
Regional Centers	**Analyses**	**Forecasts**	**Total**
Algiers	76	94	170
Antananarivo	36	4	40
Beijing	33	27	60
Bracknell	168	556	724
Brasilia	17	2	19
Buenos Aires	14	6	20
Cairo	44	50	94
Dakar	16	12	28
Darwin	10	X	10
Jeddah	59	48	107
Khabarovsk	41	49	90
Lagos	X	X	X
Melbourne	26	42	68
Miami	10	X	10
Montreal	26	99	125
Moscow	57	74	131
Nairobi	17	18	35
New Delhi	16	16	32
Novosibirsk	36	57	93
Offenbach	38	87	125
Rome	32	32	64
Tashkent	46	48	94
Tokyo	35	64	99
Tunis/Casablanca	44	24	68
Wellington	19	36	55
Total Regional Center Products			2,361

Source: Adapted from World Meteorological Organization (WMO), *World Weather Watch: Fourteenth Status Report on Implementation* (WMO, Geneva, 1989), Annex I and VI, pp.III-10 and III-15.
X = not available

Figure 13.2 Configuration of the Main Telecommunication Network

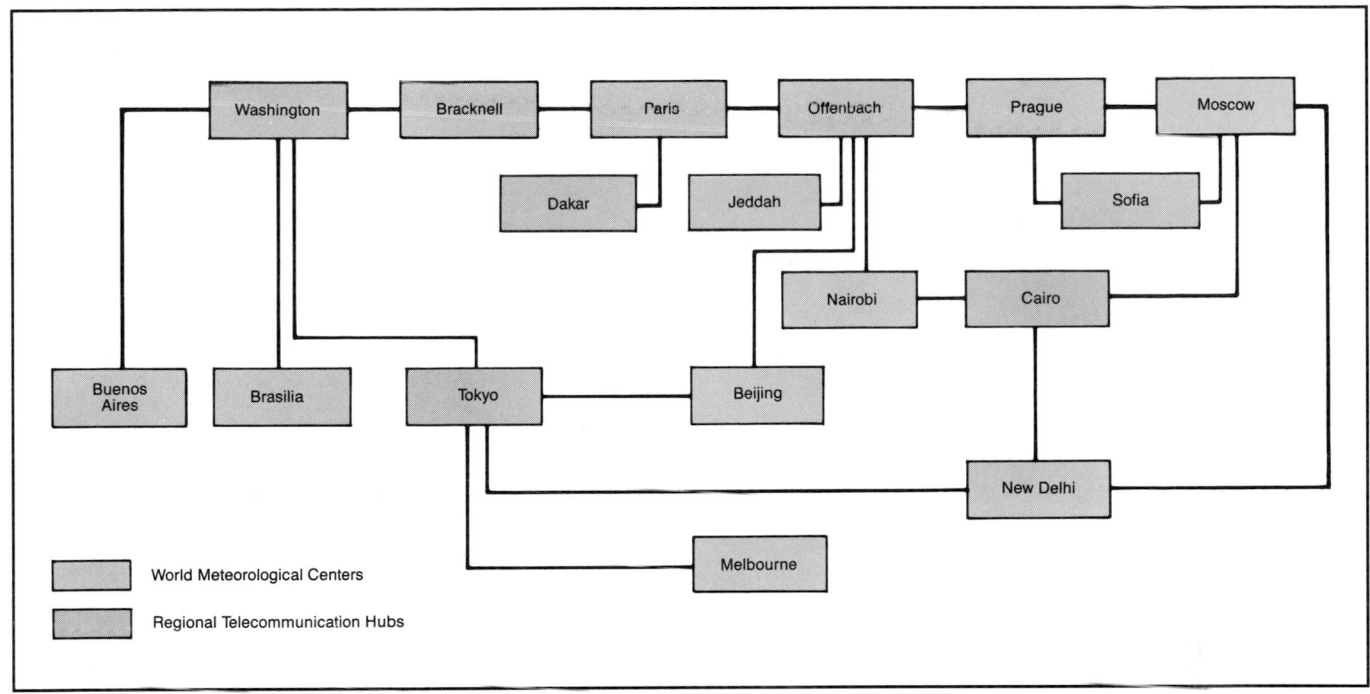

Source: World Meteorological Organization

to be used for short-, medium-, and long-range forecasting of large-scale weather systems (31).

The products of the world weather centers are available to all WWW members, but are particularly directed to the 25 regional meteorological centers. Regional centers are responsible for area-wide analyses, which in turn are directed to the various national meteorological centers in that area for use in forecasting small-, medium-, and large-scale weather patterns for local information (32).

The three world centers maintain the latest in computer technology to allow them to cope with the roughly 8 million characters of alphanumeric data provided by the Global Observing System every day (33). Between them, they produce almost 500 analyses and forecasts daily, valid from one to seven days (34). To do this, they make use of sophisticated global weather models that simulate future weather patterns from current atmospheric conditions as measured by satellites and ground stations.

Many of the regional centers use computer models with a finer resolution than the global models to forecast weather over their areas. These regional models, however, depend on the output of the coarser global models to set their initial or "boundary" conditions, again demonstrating the intimate tie between the world and regional centers. Together, the regional centers produce over 2,300 analyses and forecasts each day (35). (See Table 13.2.)

In addition, several specialized meteorological centers help track tropical cyclones and produce medium-range forecasts. For example, the European Center for Medium-Range Weather Forecasts, with the greatest computing capacity of any weather facility in the world, daily provides over 150 six-day forecasts for distribution worldwide (36).

The Global Telecommunication System

A reliable communication network is essential to the functioning of the World Weather Watch. The Global Telecommunication System provides for the collection, exchange, and distribution of both raw weather data and the analyses and forecasts produced by the various meteorological centers. (See Figure 13.2.) Over this, the WWW's nerve system moves 15 million characters of alphanumeric data and more than 2,000 weather products daily (37).

The heart of the Global Telecommunication System is a series of 22 point-to-point circuits that comprises the main telecommunication network. This network directly links the three world meteorological centers with 15 regional telecommunication hubs. Additional communication circuits connect these and 16 other regional hubs with 149 national centers (38). The data capacity of these circuits varies widely, from 50 bits per second to 19.2 kilobits per second, with most lines on the main telecommunication network at 9.6 kilobits per second (39).

The increasingly rapid speeds at which the vast quantity of weather data is exchanged has necessitated the development of internationally accepted data gathering and transmission procedures to keep the data from all segments of the system compatible. This process of standardizing the collection, sorting, and flow of data is a continuing challenge to a system

Box 13.1 Developing Nations and the Weather Watch

The World Weather Watch (WWW) is a global and cooperative effort involving most of the world's nations. But a significant disparity exists in the ability of its members to contribute to the system and to make use of its output. Nearly every country has a national meteorological service, but many developing countries lack the facilities to efficiently collect and distribute data or to use the regional analyses and forecasts to derive locally useful products (1).

A case in point is the variability within the global telecommunication system. While the data handling capacity of the main telecommunication line between the world meteorological center in Washington, D.C., and the regional center in Bracknell, England, is 19,200 bits per second, the capacity of the line from Washington, D.C., to the regional center in Brasilia, Brazil, is only 75 bits per second. The bulk of the local and regional circuits throughout Africa, Central and South America, and much of Asia have a capacity of only 50 bits per second (2). Indeed, the upgrading of the communication system is one of the most pressing needs faced by WWW members, but the expense involved is prohibitive (3).

The quality of the weather analyses and forecasts available to many developing countries, especially those in the tropics, is also markedly lower than for those analyses targeted for the northern hemisphere. This is both because weather data–especially upper-air data–are sparser over these regions, and because the computer weather models used by world and regional centers are optimized for mid-latitudes, and do not accurately model tropical weather patterns. Consequently, forecasts produced for the tropics are typically of shorter range than those for mid-latitudes. Yet, the quality of these forecasts has improved somewhat since 1980, and the number directed toward the region has increased in the last few years (4) (5). (See Table 1.)

In an attempt to address some of these concerns, the World Meteorological Organization–the coordinating force behind the WWW–maintains a variety of programs to assist developing nations to gather and

Table 1 Daily Output of Forecasts and Analyses from World Meteorological Centers, 1986 and 1988a

Center	Areas Covered	1986	1988
Melbourne	Northern Hemisphere	0	0
	Tropical Belt	0	0
	Southern Hemisphere	45	45
Moscow	Northern Hemisphere	29	29
	Tropical Belt	4	4
	Southern Hemisphere	2	2
Washington, D.C.	Northern Hemisphere	158	201
	Tropical Belt	14	30
	Southern Hemisphere	86	185

Source: World Meteorological Organization (WMO), World Weather Watch: Fourteenth Status Report on Implementation, WMO Publication No. 714 (WMO, Geneva, 1989), p.III-14.
Note: a. Forecasts and analyses produced at 12 AM Greenwich mean time only. Figures do not reflect all products issued by world centers.

use weather data. Under the Voluntary Cooperation Program, WWW members commit funds and donate services and equipment for upgrading technology and for training. In conjunction with the United Nations Development Programme, projects involving 133 countries were supported through this program in 1988 (6).

In addition, the planned establishment of the African Center for Meteorological Applications for Development (ACMAD) should greatly advance the usefulness of WWW products in Africa. ACMAD is expected to become a focal point for data collection and region-wide analysis–including the development of a regional computerized weather model–to support critical activities such as agriculture in the region. Recent advances in modeling the tropical atmosphere will also figure in improving the area's forecasting benefits (7).

In South America, Brazil has taken the lead in developing a regional computer model and is currently seeking funds for the required computer facilities (8). Moreover, the next generation of weather satellites planned for the 1990s will lend more accurate atmospheric soundings of temperature and moisture, improving the synoptic coverage of the southern hemisphere particularly (9). Thus, the prospect of improved products and more useful applications for many WWW members

in the tropics or south of the equator exists, though budgetary restrictions loom large as a continuing obstacle (10).

References and Notes

1. World Meteorological Organization (WMO), World Weather Watch:Fourteenth Status Report on Implementation, WMO Publication No.714 (WMO, Geneva, 1989), p. III-8.
2. *Ibid.*, Annex 1, p. IV-11.
3. James Rassmussen, Director, World Weather Watch, World Meteorological Organization, Geneva, July 28,1989 (personal communication).
4. *Ibid.* 13. Global Systems & Cycles.
5. *Op. cit.* 1, p. III-7.
6. World Meteorological Organization (WMO), Annual Report of the World Meteorological Organization, 1988, WMO Publication No.713 (WMO, Geneva,1989), pp. 72-77 and 99-182.
7. *Op. cit.* 3.
8. *Op. cit.* 3.
9. National Oceanic and Atmospheric Administration (NOAA) and National Aeronautics and Space Administration (NASA), Space-BasedRemote Sensing of the Earth: A report to the Congress, 1987 (U.S. Department of Commerce, Washington, D.C., 1987), p. 85.
10. *Op. cit.* 3.

of over 220 circuits and a vast array of differing equipment. This effort represents, perhaps, one of the WWW's most valuable contributions to the field of global monitoring.

Though the process of modernizing telecommunication facilities is slow, particularly among developing nations without the funds to adopt more efficient technology, progress in data management is steady. For example, the recent adoption of a binary universal code for computer data will triple communication capacities for this type of information (40). Nonethe-

less, upgrading the Global Telecommunication System is among the most pressing needs faced by the majority of WWW member nations.

MODELING THE GLOBAL ATMOSPHERE

Comprehensive global monitoring means not just more and better physical observations, but improved methods for using these observations to understand global cycles and to predict how these cycles will behave in the future. This is the province of modeling.

An accurate, quantitative model of a natural system is a powerful tool that allows observations of the system's present state to be used to forecast the future state of the system. For example, a numerical model of the global circulation of the atmosphere is essential both for modern weather predictions and for studies of how the global climate may change in response to increasing levels of greenhouse gases in the atmosphere.

A numerical model begins with a theory of how a given system works–its fundamental processes and how they interact. These processes are then expressed as simply as possible in mathematical terms; that is, as mathematical equations with as few variables as possible. In weather models, these variables correspond to the pressure, temperature, wind speed, humidity, and other factors measured every day via the Global Observing System. When the current weather values are entered in numerical form and the model equations allowed to interact over a simulated time period, the result is a new set of values representing the predicted weather for the specified future date –usually one to 10 days ahead (41).

Because of the size and complexity of natural systems, computers are required for even the simplest of numerical models. In fact, computer capacity represents one of the real limiting factors in the current ability to model global processes. In a sense, every numerical model is a compromise between accurately describing reality–all the details of how the system works–and respecting the practical limits of computation (42).

Numerical Weather Models

The atmospheric processes that determine weather were first expressed mathematically using the basic laws of physics around the turn of the century, and then were refined in the 1920s. These equations were not mathematically well-behaved; that is, they did not lend themselves to easy analytical solutions and could only be approximated by arduous calculations. Thus, while the possibility of predicting weather using these equations was recognized, the mathematical calculations involved were so intractable and the critical weather data so sparse that accurate prediction was impossible at the time (43).

With the improvement of upper-air observations in the 1940s and the birth of the digital computer, the use of these equations became practical, leading to the first numerical weather prediction in 1950. These early weather models attempted to forecast only a single variable–the atmospheric pressure–but still gave reasonable approximations of the weather (44). Modern weather models forecast all the common weather variables: temperature, pressure, winds, and humidity.

To model weather on a global scale, current numerical models divide the global atmosphere into con-centric layers, stacked like the layers of an onion from the surface into the stratosphere (45). The layers nearer to the surface–those that most affect short-term weather processes–are made thinner to better simulate the important processes that take place there, while the upper stratosphere and stratospheric layers are generally thicker (46).

Since no computer could hope to calculate weather variables at every point over the globe at all times, the model confines its computations to evenly but widely spaced locations, describing a three-dimensional gridwork over the entire globe. In the most sophisticated current operational global weather models, there may be as many as 19 layers, with an equivalent gridwork of points spaced about 140 kilometers apart throughout the atmosphere (47) (48).

The spacing between these gridpoints is crucial, since this determines the resolution of the model: how well it can discern small-scale weather phenomena and, ultimately, how accurately it can forecast. The global model presently in use at the National Meteorological Center in Washington, D.C., has an equivalent horizontal spacing between gridpoints of about 160 kilometers and a vertical spacing ranging from about 200 meters in the bottom layers to over 9 kilometers in the stratospheric layers (49).

The volume of atmosphere surrounding each gridpoint is called a cell, and the weather values computed for the gridpoint represent a blending of all the weather processes at work within the cell. To begin the calculation, an initial value of the temperature, pressure, winds, and humidity must be given for each gridpoint, gleaned from the combined measurements from all observing systems–ground stations to satellites–within the cell area. The computer then uses its mathematical expressions of the basic weather processes to let each weather cell interact with its neighbors on all sides, creating weather patterns as computer time moves forward (50) (51).

The response of the atmosphere to the sun, the transfer of heat, the movement of air masses, the friction between the atmosphere and the Earth's surface, and the movement of moisture through evaporation, condensation, and precipitation are all accounted for in the model. By recalculating values for each gridpoint again and again as the cells interact, the forecast slowly builds (52) (53). At the Washington, D.C., Center, each forecast day takes about 22 minutes of supercomputer time with three-day forecasts produced twice daily, and a complete 10-day global forecast is run once a day (54).

Many important weather phenomena occur on a scale smaller than the horizontal spacing of the gridpoints. For instance, individual clouds, which are important in regulating surface temperatures and in generating severe weather, are invisible to current models. Since these phenomena cannot be ignored completely in an accurate model, they must be represented indirectly. This is done by describing them in relation to one of the variables that can be measured, such as temperature or pressure (55).

For example, cloud formation is related to the average temperature and humidity in a given area. If the relationship that links them is known, then a special modification can be made to the equations that include temperature and humidity in the model to represent an average of the area's cloudiness. This process is known as parameterization and it is essentially a means to represent the complete climatic picture within the limits of the grid. In doing so, however, it is also a significant source of error in the forecast, since it is at best an approximation (56) (57).

The forecasts produced by a global model are synoptic-scale forecasts; that is, they describe weather patterns on the largest scale, tracking the movements of major storm systems and large masses of air. For regional forecasts describing medium-scale weather patterns, a model with a finer meshed grid is required, typically with a horizontal resolution of 20-60 kilometers. Such a model "sees" local weather events better and can account for such factors as mountains and smaller weather fronts that the global model might miss. Regional models also contain more realistic parameterizations and treat precipitation more accurately (58).

To obtain the best forecast, a model requires that the initial values for the weather variables at each gridpoint be as accurate as possible. This is a difficult task, since weather observations are of variable quality and may be sparse in some areas, especially over the oceans.

Fortunately, today's models contain a kind of "memory function." They are refined enough that, based on a previous forecast, the model itself can predict, with a high degree of confidence, an approximation of the initial values that should be used for the current run. The raw data from the field are then used to correct these approximations. In this manner, the model can provide information in areas where observations are either sparse or absent altogether and provide continuity from one analysis time to the next. Thus, the model acts to enhance the observing system and the observing system acts as a check on the performance of the model (59).

Over the years, the forecasts that models produce have improved steadily in quality. Today's five-day forecasts are as accurate as the three-day forecasts of the mid-1970s (60). This has come about through the continued refinement of the models, the computers they run on, and the weather observing systems. Current models have better horizontal and vertical resolution and represent weather processes more realistically—something that would be impossible without faster computers. Meanwhile, satellites and other technology advances have led to more accurate weather data (61).

In the near future, model sophistication will continue to improve dramatically. Plans for the early 1990s in European and North American modeling centers call for acquisition of the next generation of supercomputers. This will allow at least a doubling of current resolutions, both horizontally and vertically. Global models will become as accurate as regional models are today, and regional models will become much more capable of capturing important small-scale phenomena. For instance, the future North American regional model will have 30 layers and a horizontal resolution of 30 kilometers (62). At the same time, the capabilities of ground stations and satellites will continue to increase, making weather data more available and reliable. In fact, many experts believe that progress in forecasting depends more on this gradual increase in the quality of weather data than on increasingly sophisticated models.

In any case, continued progress in weather modeling beyond that planned for the next decade may be difficult, since the dynamics of the atmosphere at small scales vary markedly from those at larger scales. As the grid size approaches these small scales, modelers will have to reevaluate how adequately the equations used in current models are representing the atmospheric physics taking place within each cell. Reformulating these equations, if necessary, may prove an overwhelming task and will likely result in drastically increased computer requirements as well (63).

Global Climate Models

The models used to study climate changes induced by the release of greenhouse gases into the atmosphere are similar in principle to the numerical models used to forecast daily weather. Both are so-called general circulation models (GCMs) that use the same mathematical equations to simulate the dynamics of the atmosphere (64). Climate models, however, are generally more complex and involve a wider range of processes than weather forecasting models. In addition, the applications of the two models are quite different and this difference leads to a divergence of end products.

The fundamental difference in how climate models and weather models behave stems from the timescales over which they are required to act. Weather models simulate atmospheric dynamics for short periods of time—usually one to 10 days. But a typical run of a climate model exploring the greenhouse effect is 10 to 100 years and simulations of ancient Earth climates may treat time periods of thousands of years (65) (66).

Over such lengthy periods of time, climatic variables with little influence on short-term weather can become major factors. For instance, the transfer of heat from the ocean surface to its deep waters, the waxing and waning of the polar ice sheets, and the growth or die-off of forests are all factors with definite effects on climate. Weather models routinely ignore these processes by substituting a single unchanging value for these variables in their equations—for instance, by using a single estimate of sea ice instead of letting it vary as the simulation proceeds (67).

But climate models must account for all these phenomena and many more, such as the minute changes in the output of solar radiation over cycles of thousands of years, or the action of volcanic aerosols. Because of the length of the simulation, climate models are very sensitive to how these processes are parameterized. If the phenomena are not represented accurately in the model equations, then the complexity of the climate system–its many feedbacks and interactions–will not be reproduced and the model results will be spurious (68).

Perhaps the most immediate challenge faced by climate modelers today is to adequately account for the effects of ocean dynamics, cloud behavior, and terrestrial ecosystems on the global climate (69) (70) (71). In each case, submodels treating each of these areas must be constructed and coupled to the larger model to reduce the uncertainty of the model results. These refinements will eventually be incorporated into weather models as well, but their development is much more crucial to the success of climate models.

Chaos and the Limits to Prediction

Despite the many advances in weather observing and atmospheric modeling over the years, our ability to predict weather remains very limited. Local weather forecasts maintain useful accuracy for only a few days, and, even at their most reliable, they are hardly flawless. In predicting rain, for example, two-day local forecasts in the United States are only about 30 percent better by virtue of modern observations, theory, and computers than an informed guess made by simply extrapolating from past weather records (72).

Unfortunately, simply sharpening computer models and improving weather observations will not ultimately lead to foolproof weather forecasting. Since the 1960s, meteorologists have realized that atmospheric processes are inherently unpredictable for more than two to three weeks in advance (73). This is not simply because the physical climate is extremely complex and requires dozens of equations to represent mathematically. It is more in the nature of these equations: they are inherently nonlinear. In other words, they do not yield solutions with any periodic pattern–nothing that could be guaranteed to repeat or regularly recur (74).

This apparently random behavior, which is typical of many natural phenomena, is termed "chaotic," and the discipline that treats it is called chaos theory, or simply chaos (75). Chaos has had profound repercussions on the science of meteorology and the effort to predict weather. In the 1950s, as the computer age dawned, most meteorologists believed that long-range weather prediction was primarily a problem in physical calculation of the complex math needed to describe weather processes (76). It was assumed that this problem would ultimately yield to faster computers and the better resolution they would allow in computer models. However, a simple weather model in the early 1960s proved otherwise (77).

The model contained only three equations representing temperature, wind speed, and pressure. Nonetheless, the simplistic atmosphere portrayed in the model exhibited chaotic behavior. If the initial values plugged in for temperature, wind, and pressure were varied ever so slightly from run to run, a completely different forecast would appear. Two forecasts with almost negligible differences in initial values would start out similarly, as expected, but would quickly diverge and rapidly lose all similarity (78).

The large changes induced in the forecast were completely out of proportion with the size of the difference in initial values, and the divergence was magnified with each day of continued forecast. This ability of small-scale changes in initial values to ultimately affect the largest scale events was dubbed the "butterfly effect," since it implied that something as insignificant as the turbulence created by a butterfly beating its wings could theoretically change world-scale weather events in just a few days (79).

When the results of the simple weather model and the existence of the butterfly effect became widely known to meteorologists in the early 1970s, the implications for forecasting were clear. Such a sensitivity to initial conditions meant that, no matter how good the weather model, if the values plugged into the model varied even slightly from those that actually existed in nature, a completely accurate weather prediction was impossible (80). In essence, this meant that perfect measurements of weather variables–not to mention a truly representative weather model– were required in order to predict weather for indefinite lengths of time. Conversely, the greater the error in weather observations, and thus in initial values, the more quickly the forecast accuracy degraded (81).

The process of parameterizing weather processes that are smaller than the grid size of the model may have the same disastrous effect on forecast accuracy as measurement errors do. Initially, parameterization induces errors on the smallest scale of the model, but these errors quickly cascade up through the model until even global-scale events are contaminated. The doubling time for small errors in today's atmospheric models is estimated at two to two and a half days (82).

The sensitivity of weather forecasts to initial values in the computer model is a powerful argument for making weather observations as accurate as possible by upgrading the observing system (83). Existing weather data suffer from a variety of ills, including incomplete surface and upper-air coverage and marginal accuracy for some measurements. For example, current temperature soundings from geostationary and satellites are only accurate within a range of about 2°C (84), and satellite wind measurements, which rely on tracking cloud movements, are unable to pin down wind altitudes with precision (85).

Further, the regional synoptic network does not produce uniform observations at all stations at all the recommended observing times. Some stations can only make a partial set of observations at any one time and others cannot observe at all the recommended hours (86). Surface observations over ocean areas are still sporadic and widely spaced and tend to occur on those shipping lanes most traveled, leaving many areas with poor coverage (87).

Also, as described above, computer models suffer from several inadequacies, both in terms of spatial resolution and in how well they describe weather-making processes at every scale. All in all, meteorologists acknowledge that they are still far from the theoretical prediction limit of approximately two weeks and that better observations and more refined models could yield improved forecasts for some time to come (88) (89).

While chaos sets rather severe limits on weather prediction, it also provides some hope for forecasters. For beneath the apparent disorder of weather, chaos theory argues that there is an underlying order, implied by the very fact that the atmosphere can be described by mathematical equations, no matter how complex (90). Chaos theory states that the seemingly random behavior of chaotic systems actually occurs within definite bounds and that these systems tend toward a limited set of states. For instance, snowflakes generally have six arms, though no two snowflakes are alike (91).

The set of states toward which a system tends is called an "attractor," since the behavior of the system is drawn or attracted to it (92). Attractors are mathematical abstractions expressing something about the complexity of and the relationships among the equations that determine the weather. The existence of one or more weather attractors is by no means certain (93). But if they do exist and can be adequately described, weather attractors could be enormously useful in long-range forecasting by defining the underlying patterns within which the weather varies (94). To be useful, however, they must be relatively simple, and many meteorologists feel that the inherent complexity of the physical climate rules out such simple attractors (95).

If chaos theory confines useful forecasting to a narrow temporal range of no more than a few weeks, how can general climate models hope to describe the climate 100 years or more in the future? The answer lies in the fact that climate models do not produce a forecast, per se.

While weather models seek to predict the exact values of weather variables for some limited time period in the future, climate models are only interested in estimates of what the climate will be in the future. Will the mean global temperature be warmer or cooler? Will there be more or less rain, on average? Will the climate show greater variability than today or will it be more regular? Climate models are concerned only with statistical averages of weather variables like surface temperature, not with the precise

details of storm systems on a given day in the future. Only in this way can they escape the chaos dilemma and the long-term unpredictability of the weather (96).

THE FUTURE OF GLOBAL MONITORING

Even as the global monitoring activities of the world weather network continue to improve, there are plans to go well beyond these meteorology-based observations in the years ahead. If current plans are carried through, the 1990s will see the development of the first comprehensive global monitoring effort in history.

The impetus for such an effort comes from the convergence of three factors. First, there is a growing awareness of the coupling between Earth systems: there is a complex system of forcing factors and feedback mechanisms that define an integrated and ever-changing whole (97). Realizing this, scientists have begun to reach beyond the barriers of their individual disciplines to embrace the concept of an integrated "Earth system science"–a science that can only proceed if all aspects of the global system are studied in tandem and in depth over long timescales. Such a study, in turn, requires a well-developed monitoring system (98).

A second factor is that the technology to conduct such a monitoring program is now at hand, or will be shortly. Prospects for global monitoring have risen most with the tremendous increase in satellite capabilities in the past 20 years, but surface-based technologies have also evolved (99).

Perhaps the most crucial factor in establishing the global monitoring effort is the recent international concern over the possibility of rapid changes in the global climate. Whether brought on by natural or human causes, such as the release of greenhouse gases into the atmosphere, most nations have begun to realize that the social and ecological consequences of such changes could be devastating. Unfortunately, current knowledge is insufficient to judge the magnitude or timing of either the climate change itself or its global repercussions (100).

To bridge this gulf of ignorance, Earth system science assumes a fourfold strategy. It describes Earth processes through global observations, uses these descriptions to understand the dynamics behind Earth processes, constructs conceptual and computer models with this understanding, and, finally, uses these models to predict future Earth changes (101).

This strategy translates into an ambitious global monitoring scheme. Not only must observations of the biosphere, geosphere, cryosphere, and hydrosphere be made on a global basis, they must be made simultaneously, since the physical, chemical, and biological processes that drive these Earth systems are all interactive and cannot be understood in isolation. Further, observations must continue over an extended period–at least 20 years. During such a 20-year period we could reasonably expect to see the first manifestations of "greenhouse"-induced global warming and the effects of widespread deforestation, and we could follow such natural cycles as sunspots, El Niño events, and the circulation of deep ocean currents (102).

Examples of the measurements to be made by a global monitoring system include (103): high-resolution vertical profiles of surface and atmospheric temperatures, pressures, and winds; the extent of global cloud cover; the extent of global ice and snow cover; sea surface temperatures and topography (to help follow ocean currents); ocean chlorophyll concentrations; wind stress at the sea surface and ocean wave heights; an index of vegetation cover; the extent and productivity of terrestrial biomes; global precipitation and the moisture content of the atmosphere; motions and deformations of the Earth's crust; the planetary gravitational and magnetic field; and the solar constant (the energy output of the sun in the direction of the Earth).

Already, a somewhat bewildering array of national and international programs have begun to address the newfound interest in global dynamics, especially the prospects for climate change. Oldest of the international efforts is the World Meteorological Organization's World Climate Programme, initiated nearly a decade ago to explore climate processes with implications for long-range forecasting and other meteorological tasks (104). Some of the World Climate Programme's major observing programs now under way or planned for the near future include:

■ The World Ozone Program, which seeks to monitor and define the atmospheric chemistry associated with stratospheric ozone and other associated chemical species (105);
■ The International Satellite Cloud Climatology Project, which monitors global cloud fields to measure the interaction of clouds and radiation from either the sun or reflected from the Earth (106). Its main application is to refine the way clouds are represented in weather and climate models;
■ The International Satellite Land Surface Climatology Project, which measures the interaction of land surface processes with climate in specific biomes (107);
■ The Tropical Ocean Global Atmospheric Program, which focuses on the dynamics of the upper tropical ocean and overlying atmosphere in order to determine their coupling (108). This has special relevance to predicting El Niño events;
■ The World Ocean Circulation Experiment, which aims to study ocean circulation for the purpose of modeling this crucial climate element on timescales of decades and longer (109); and
■ The Global Energy and Water Cycle Experiment, which will attempt to determine the hydrological cycle and energy fluxes through direct observations, with an eye toward the modeling of these cycles (110).

Two other related international measurement programs are:
■ The Joint Global Ocean Flux Study, which will trace the production and fate of biogeochemical materials in the global oceans (111); and
■ The International Global Atmospheric Chemistry Program, which is designed to study the large-scale chemistry of the lower atmosphere (112).

Although these projects are wide-ranging, most are primarily meteorological in orientation and barely begin to address the broad mandate of Earth system science. Thus, in 1986, the International Council of Scientific Unions—an umbrella organization for professional societies in all scientific disciplines—adopted the concept for an International Geosphere-Biosphere Program (IGBP) to study, in cooperation with the World Climate Program, global change at all levels and across all timescales. The effort has received broad support from the international scientific community and has been formally endorsed by the World Meteorological Organization and the United Nations Environment Programme and adopted by many nations. It will involve a series of simultaneous research missions funded and implemented by single nations or through bilateral or multilateral agreements, but coordinated through IGBP, and with access to data assured to all (113).

The effort to marshal international involvement and to establish a coordinating mechanism for the many individual investigations required is far from trivial, and the pace has been slow so far in defining the exact scientific program the IGBP will follow. Nonetheless, much progress has been reported in stimulating discussion within and between disciplines and in reorienting scientists to look upon their work from a global perspective (114).

That the topics of global change and global monitoring are being taken seriously in government and scientific circles is clear by the funding commitments being made by many nations. For instance, in 1989, the United States adopted a landmark budget that united the many separate government-supported programs relating to global studies into one cohesive program—the U.S. Global Change Research Program—that will contribute to IGBP and World Climate Program activities (115). At the same time, the Soviet Union has committed itself to convert the data format from many of its scientific instruments to a more universally acceptable form, and to undertake several crucial ecosystem studies as part of the international effort (116). Similar commitments have been made by many other nations.

Future Observing Systems

An impressive array of satellites and surface monitors will support the many global system studies already under way and planned under the IGBP. Space officials from 23 nations are now coordinating their plans for more than 20 satellites with varying capabilities for Earth viewing over the next six years (117). By making simultaneous and continuous global monitoring a reality, the satellite era has played a fundamental role in the genesis of Earth system science. Satellites will be the workhorses of global monitoring, probing land, sea, and atmospheric processes, with calibration and corroboration provided from ground measurements (118). Table 13.3 lists a selection of the Earth-observing satellites planned for the next 15 years.

In the near term, several important satellite missions will add significantly to global measurements from existing satellites already under way. In 1990, the European Earth Remote Sensing satellite (ERS-1) will join U.S.(Landsat), French (SPOT), and Indian (IRS-1) satellites in producing detailed images of continental regions for resource planning, vegetation mapping, and a host of other research applications. The Japanese (JERS-1) will follow suit in 1991. ERS-1 will also significantly enhance detailed ocean and coastal zone monitoring, as will the Canadian Radarsat satellite (119) (120).

The United States' Upper Atmosphere Research Satellite (UARS), planned for 1991, will explore the coupled chemistry and dynamics of the stratosphere and the mesosphere and the role of solar radiation in these processes. In 1991, the joint French-U.S. TOPEX/Poseidon satellite will measure the height of the world's oceans with great accuracy to reveal the structure of ocean currents (121), directly feeding data to the World Ocean Circulation Experiment.

Beyond 1995, global monitoring will make a quantum jump with the birth of the international Earth Observing System (EOS). EOS will consist of a series of four or five large space platforms in a 705 kilometers-high polar orbit (122) (123). The first platform, due to be launched by the United States in 1996, will contain 19 precision Earth-viewing sensors designed to operate synergistically by viewing precisely the same area of the Earth and atmosphere with complementary technologies (124) (125). European and Japanese platforms will follow in 1997 and 1998 respectively, carrying slightly different complements of instruments (126). Separate international teams are currently developing the first platform's sensors, and nearly 30 additional teams are designing interdisciplinary studies to be conducted using the data from these instruments (127).

Earth Observing System sensors can be divided into three classes: a group of instruments that images the Earth's surface and conducts temperature and moisture soundings of the lower atmosphere; a group of radar instruments that will probe the character and structure of the surface; and a group of instruments to study atmospheric composition and dynamics (128). As an example, one of the sensors will allow the detailed identification of minerals and soils, the estimation of the grain size of snow, the monitoring of phytoplankton productivity in coastal waters, and the study of biochemical processes in vegetation canopies (129).

The number and rate of data produced by EOS platforms will be prodigious—about a thousand times greater than present satellite data—and planners know that data management will be one of EOS' greatest challenges. To accommodate this embarrassment of riches, EOS planners are creating the EOS Data

Table 13.3 Launch Schedule for Selected Earth-Observing Satellites

October 1990	ERS-1 Earth Remote Sensing Satellite (Europe)
September 1991	UARS Upper Atmosphere Research Satellite (United States)
February 1992	JERS-1 Japanese Earth Remote-Sensing Satellite (Japan)
June 1992	TOPEX/Poseidon (France-United States)
Undecided	GREM Geopotential Research Explorer Mission (United States)
Undecided	Radarsat (Canada)
December 1997	EOS-A Earth Observing System, First Polar-Orbiting Platform (United States)
1996	EPOP-1A European Polar-Orbiting Platform; Second EOS Platform (Europe)
December 1999	EOS-B Earth Observing System, Third Polar Orbiting Platform (United States)
1998	JPOP Japanese Polar-Orbiting Platform; Fourth EOS Platform (Japan)
2002	EOS-C Earth Observing System, Replacement Platform for EOS-A (United States)
2004	EOS-D Earth Observing System, Replacement Platform for EOS-B (United States)

Source: Condensed from National Aeronautics and Space Administration (NASA),"Earth Observer Launch Schedule," The Earth Observer, Vol. 1, No. 3 (June 1989). p. 6, and Gerard Soffen, Project Scientist, Earth Observing System, NASA, 1990 (personal communication).

and Information System to gather, archive, and distribute data to the scientific community (130).

Both before and during the EOS era, weather satellites and ground stations will continue to play an important role in monitoring not only short-term weather processes, but also other atmospheric and surface processes of interest to researchers. All along, weather satellites have served this double role as operational tools for weather watching and observation stations for unrelated research (131). In many cases, the same data can be used for both purposes, and in other cases special instruments have been placed aboard weather satellites, such as the Earth radiation budget monitor aboard current U.S. polar orbiters, or the total ozone mapping spectrometer to be placed aboard a Soviet polar orbiter in 1991 (132) (133).

Significant upgrades in weather satellites began in 1989 when the first of a new generation of geostationary satellites was orbited. Both the sensors that produce weather pictures and the sounders that collect temperature profiles throughout the atmosphere were significantly enhanced. In 1992, U.S.polar orbiters will begin carrying an advanced microwave sounding unit that allows higher resolution temperature and moisture soundings than at present, even in cloud-covered regions. This will give forecasters the ability to look within storm systems for the first time, which should greatly improve the forecasting of severe weather such as tornadoes and cyclones (134).

As EOS platforms are launched, many new capabilities will be available to forecasters as part of the total monitoring effort. For instance, a laser atmosphere wind (LAWS) sounder aboard the second EOS platform will supplant presently scanty wind data that are critical to the accuracy of numerical weather models (135).

Along with the upgrading of existing ground weather stations with increasing automation and better radars, a series of new computerized ground stations has been proposed to include measurements of seismic activity, soil properties, and local hydrology,

in addition to weather variables (136). In combination with the planned increases in satellite capabilities, these improvements in global monitoring should begin to manifest in superior weather forecasting and improved understanding of global systems and cycles within the first two decades of the next century (137) (138).

This chapter was authored by Gregory Mock, a science writer and editor in Ben Lomond, California.

References and Notes

1. Edward Edelson, "Laying the Foundation, " *Mosaic*, Vol. 19, No.3/4 (National Science Foundation, Washington, D.C., Fall/Winter 1988), pp. 4-11.
2. Earth System Sciences Committee, NASA Advisory Council, *Earth System Science: A Closer View*, (National Aeronautics and Space Administration, Washington, D.C., 1988), pp. 11-21.
3. Earth System Sciences Committee, NASA Advisory Council, *Earth System Science: Overview* (National Aeronautics and Space Administration, Washington, D.C., 1988), pp. 17-18.
4. Bette Hileman, "Global Warming, " *Chemical and Engineering News*, Vol. 67, No.11 (March 13, 1989), p. 40.
5. Wallace S. Broecker, "The Biggest Chill, " *Natural History*, Vol. 96, No.10 (1987), pp. 74-82.
6. William K. Stevens, "Sunspot Activity May Influence Our Weather and Improve Forecasting, " *Santa Cruz Sentinel* (July 14, 1989, Santa Cruz, California), p. E-1.
7. World Meteorological Organization (WMO), *The World Weather Watch*, WMO Publication No.709 WMO, Geneva, (1988), p. 7.
8. *Ibid.*
9. World Meteorological Organization (WMO), *World Weather Watch: Fourteenth Status Report on Implementation*, WMO Publication No.714 (WMO, Geneva, 1989), pp. II-3 and II-4.
10. World Meteorological Organization (WMO), *Annual Report of the World Meteorological Organization 1988*, WMO Publication No.713 (WMO, Geneva, 1989), p. 14.
11. *Op. cit.* 9, pp. II-3, II-9, and II-10.
12. *Op. cit.* 9, pp. II-4 and II-6, Table 1.
13. *Op. cit.* 10, p. 18.
14. *Op. cit.* 9, pp. II-5, II-6, and II-8.
15. *Op. cit.* 10, p. 18.
16. *Op. cit.* 10, pp. 18-20.
17. *Op. cit.* 9, pp. II-7 to II-9.
18. *Op. cit.* 9, pp. II-7 to II-8.
19. *Op. cit.* 9, p. II-8.
20. *Op. cit.* 10, p. 19.
21. *Op. cit.* 9, p. II-8.
22. National Aeronautics and Space Administration and National Oceanic and Atmospheric Administration, *Sentinels in the Sky: Weather Satellites*, NASA Facts Publication No.NF-152(S), pp. 3-4, n.d.
23. *Op. cit.* 7, p. 15.
24. Edward S. Epstein, William M. Callicott, Daniel J. Cotter, *et al.*, "NOAA Satellite Programs, " *IEEE Transactions on Aerospace and Electronic Systems*, Vol. AES-20, No.4 (1984), pp. 325-328.
25. *Op. cit.* 9, pp. II-10 to II-11.
26. *Op. cit.* 24, p. 327.
27. *Op. cit.* 22, pp. 11-13.
28. *Op. cit.* 22, p. 14.
29. *Op. cit.* 9, pp. II-11 to II-13.
30. *Op. cit.* 22, p. 14.
31. *Op. cit.* 9, pp. III-3 to III-4.
32. *Op. cit.* 9, pp. III-3 to III-4.
33. *Op. cit.* 7, pp. 17-18.
34. *Op. cit.* 9, Annex IV, p. iii-13.
35. *Op. cit.* 9, Annex VI, p. III-15.
36. *Op. cit.* 7, p. 20.
37. *Op. cit.* 7, p. 21.
38. *Op. cit.* 7, p. 21.
39. James R. Neilon, Chief, International Activities Division, National Weather Service (National Oceanic and Atmospheric Administration), Silver Spring, Maryland, July 25, 1989 (personal communication).
40. *Op. cit.* 7, pp. 22-24.
41. Joseph J. Tribbia and Richard A. Anthes, "Scientific Basis of Modern Weather Prediction, " *Science*, Vol. 237, No.4814 (1987), p. 496.
42. *Op. cit.* 2, p. 95.
43. *Op. cit.* 41, pp. 493-494.
44. *Op. cit.* 41, p. 494.
45. *Op. cit.* 41, p. 496.
46. Ralph Petersen, Chief, Short-Range Modeling Branch, National Meteorological Center, National Weather Service, Washington, D.C., August 16, 1989 (personal communication).
47. *Ibid.*
48. Joseph Tribbia, Deputy Head, Global Dynamics Section, National Center for Atmospheric Research, Boulder, Colorado, August 16, 1989 (personal communication).
49. *Op. cit.* 46.
50. *Op. cit.* 41, p. 496.
51. *Op. cit.* 48.
52. William Booth, "Computers and the Greenhouse Effect:The Genesis of Understanding" (June12, 1989, Washington, D.C.), p. A-3.
53. *Op. cit.* 41, p. 496.
54. *Op. cit.* 48.
55. Stephen H. Schneider, "Climate Modeling, " *Scientific American*, Vol. 256, No.5 (May, 1987), p. 73.
56. *Ibid.* , pp. 73-74.
57. Wayman Baker, Deputy Chief, Development Division, National Meteorological Center, National Weather Service, Washington, D.C., August 16, 1989 (personal communication).
58. *Op. cit.* 41, p. 496.
59. *Op. cit.* 46.
60. *Op. cit.* 46.
61. *Op. cit.* 41, p. 496.
62. *Op. cit.* 46.
63. *Op. cit.* 48.
64. *Op. cit.* 48.
65. *Op. cit.* 55, pp. 75-76.
66. James Hansen, *et al.* , "Global Climate Changes as Forecast by Goddard Institute for Space Studies Three-Dimensional Model, " *J.Geophys.Res.*, Vol. 93, No.D-8(1988), p. 9342.
67. *Op. cit.* 55, p. 72.
68. *Op. cit.* 55, p. 72.
69. *Op. cit.* 48.
70. Stephen H. Schneider, "The Greenhouse Effect: Science and Policy," *Science*, Vol. 243 (1989), pp. 771-777.
71. Richard A. Houghton and George M. Woodwell, "Global Climate Change, " *Scientific American*, Vol. 260 (April, 1989), pp. 36-44.
72. H.R.Glahn, Trends in Skill and Accuracy of National Weather Service PoP Forecasts (NOAA Technical Memorandum NWS TDL73, Department of Commerce, Washington, D.C., 1984); Bull. Am. Meteorological Society, in Richard A. Kerr, "Pity the Poor Weatherman," *Science*, Vol. 228 (1985), p. 704.
73. Robert Pool, "Is Something Strange About the Weather?" *Science*, Vol. 243, No.4896 (1989), p. 1290.
74. *Op. cit.* 41, p. 497.
75. Ann Gibbons, "Chaos and the Real World, " *Technology Review*, Vol. 191, No.5 (July, 1988), p. 11.
76. *Op. cit.* 41, p. 497.
77. *Op. cit.* 75, p. 10.
78. *Op. cit.* 75, p. 10.
79. *Op. cit.* 77, pp. 10-11.
80. *Op. cit.* 41, p. 497.
81. A.A.Tsonis, "Chaos and the Unpredictability of Weather, " *Weather*, Vol. 44, No.6 (June, 1989), p. 261.
82. *Op. cit.* 41, p. 497.
83. *Op. cit.* 81, p. 263.
84. W.L.Smith, *et al.*, "The Meteorological Satellite:Overview of 25 Years of Operation," *Science*, Vol. 231(1986), p. 461.
85. National Aeronautics and Space Administration (NASA), *LAWS: Laser Atmospheric Wind Sounder, Earth Observing System*, Vol. IIg, Instrument Panel Report (NASA, Washington, D.C., 1987), p. 3.
86. *Op. cit.* 9, p. II-19.
87. U.S.Committee on Earth Sciences (CES), the U.S. Global Change Research Program, *Our Changing Planet:The FY1990 Research Plan* (CES, Washington, D.C., July 1989), p. 31.
88. *Op. cit.* 81, p. 263.
89. *Op. cit.* 85.
90. *Op. cit.* 73.
91. *Op. cit.* 81, p. 260.
92. *Op. cit.* 81, p. 260.
93. *Op. cit.* 73, pp. 1291-1293.
94. *Op. cit.* 73, p. 1291.
95. *Op. cit.* 73, pp. 1292-1293.
96. *Op. cit.* 48.
97. *Op. cit.* 2, pp. 10-19.
98. *Op. cit.* 2, pp. 10-19.
99. Committee on Space Research, "Potential of Remote Sensing for the Study of Global Change, " S.I. Rasool, ed., *Advances in Space Research*, Vol. 7, No.1 (1987), pp. 4-87.
100. *Op. cit.* 4, pp. 25-44.
101. *Op. cit.* 3, p. 21-23.

102. Committee on Science, Engineering, and Public Policy, National Academy of Sciences, *Research Briefings, 1985* (National Academy Press, Washington, D.C., 1985), quoted in Earth System Sciences Committee, NASA Advisory Council, *Earth System Science: Overview* (National Aeronautics and Space Administration, Washington, D.C., 1988), p. 29.

103. *Op. cit.* 3, p. 28.

104. Francis Bretherton, Director, Space Science and Engineering Center, University of Wisconsin-Madison (former Director of the National Center for Atmospheric Research), August 23, 1989 (personal communication).

105. *Op. cit.* 3, p. 34.

106. *Op. cit.* 10, p. 58.

107. *Op. cit.* 3, p. 43.

108. *Op. cit.* 10, p. 59.

109. *Op. cit.* 10, p. 61.

110. *Op. cit.* 10, p. 53.

111. *Op. cit.* 3, p. 36.

112. *Op. cit.* 104.

113. *Op. cit.* 1, pp. 4-10.

114. *Op. cit.* 104.

115. *Op. cit.* 87 (entire).

116. Charles Redmond, Space Sciences Public Information Officer, National Aeronautics and Space Administration, (NASA), Washington, D.C., August 2, 1989 (personal communication).

117. Craig Covault, "Major Space Effort Mobilized to Blunt Environmental Threat," *Aviation Week and Space Technology* (March 13, 1989), p. 39.

118. *Op. cit.* 99, p. 11.

119. National Oceanic and Atmospheric Administration (NOAA) and National Aeronautics and Space Administration (NASA), *Space-Based Remote Sensing of the Earth: A Report to the Congress* (U.S. Department of Commerce, Washington, D.C., 1987), p. 75.

120. *Op. cit.* 99, p. 68.

121. *Op. cit.* 2, 162.

122. *Op. cit.* 117, p. 38.

123. National Aeronautics and Space Administration (NASA),*Earth Observing System: 1989 Reference Handbook* (NASA,Washington, D.C., 1989), p. 10.

124. *Op. cit.* 117, p. 38.

125. *Op. cit.* 119, p. 81.

126. *Op. cit.* 117, pp. 38-39.

127. *Op. cit.* 123, pp. 12-67.

128. *Op. cit.* 2, p. 165.

129. Philip H. Abelson, "Earth Observation From Space," *Science,* Vol. 244, No.4907 (1989), p. 901.

130. *Op. cit.* 117, p. 40.

131. *Op. cit.* 119, p. xxi.

132. *Op. cit.* 117.

133. *Op. cit.* 2, p. 161.

134. *Op. cit.* 22, p. 15.

135. *Op. cit.* 85, p. vi.

136. Space Science Board, National Research Council (NRC), *Space Science in the Twenty-First Century: Imperatives for Two Decades; Mission to Planet Earth* (NRC, Washington, D.C., 1988), pp. 93-95.

137. National Aeronautics and Space Administration (NASA), *Earth Observing System, Volume 1: Science and Mission Requirements, Working Group Report,* Technical Memorandum 86129 (NASA, Washington, D.C., 1984), pp. v-vi.

138. Space Science Board, National Research Council (NRC), *Space Science in the Twenty-First Century: Imperatives for Two Decades: Overview* (NRC, Washington, D.C.,1988), pp. 5-14.

14. Policies and Institutions

Natural Resources Accounting

Whatever their shortcomings and however little their construction and use are understood by the general public, the gross national product (GNP) and related aggregate economic accounts are undoubtedly among the most significant social inventions of the 20th Century. Their political and economic impact can scarcely be overestimated. No matter how inappropriately, they serve to divide the world into "developed" and "less developed" countries. In the "developed countries," whenever the quarterly GNP figures emerge, policymakers stir. If the figures are lower, even marginally, than those of the preceding three months, a recession is declared, the strategies and competence of the administration are impugned, and public political debate ensues. In the "developing" countries, the rate of growth of GNP is commonly perceived as the principal measure of economic progress and transformation. National income accounts have, in short, become an institution.

As the most visible and widely used measure of economic progress, GNP also plays an important role in national policies and in the policies of international lending agencies. Thus, it is a matter of concern if those accounts produce seriously distorted or inaccurate measures of economic performance and lead to misguided and ultimately unsustainable policies. And yet, that appears to be the case.

The present national income accounts, and hence the gross national product and such related measures as the net national product, do not reflect the depletion and degradation of natural resources—even where, as is the case in many developing countries, those resources are a primary source of national income. If a country whose primary wealth is in its forests, for example, were to cut down those forests and sell off the lumber, under the current system its GNP would increase. Yet such a country, having consumed its source of wealth, would be poorer and would have a less promising economic future.

This chapter considers changes in natural resource accounting that could remedy the problems inherent in the present system. It illustrates how revised accounts might be constructed and the differences such revisions might make in the specific test case of Indonesia. The chapter also reports on the progress of many countries toward modifying their national accounts to integrate either the costs of environmental pollution or the depletion or degradation of natural resources into the decisionmaking process.

THE NEED FOR MORE ACCURATE ECONOMIC INDICATORS

The aim of national income accounting is to provide an empirical framework for analyzing the performance of the macroeconomic system. The current system reflects the Keynesian theories that were dominant in the 1930s and 1940s when the system was developed. The great aggregates of Keynesian analysis—consumption, savings, investment, and government expenditures—are defined and measured carefully. Earlier classical economists had regarded income as the return on natural resources, human resources, and invested capital (land, labor, and capital, as they put it). But Keynes and his contemporaries were preoccupied with the depression of the early 1930s; specifically with explaining how an economy could remain at less than full employment for long periods of time. During the years of the depression, prices for many commodity products based on natural resources were at an all-time low. Natural resource scarcities were the least of Keynes' concerns. As Keynesian analysis largely ignored the role of natural resources in generating income, so does the current system of national accounts. After World War II, when these theories were applied to problems of economic development in the Third World, human resources also were left out on the grounds that labor was always "surplus"; development came to be seen almost entirely as a matter of savings and investment in physical capital. Low-income countries—which are typically most dependent on natural resources for employment, revenues, and foreign exchange earnings—were instructed to use a system for national accounting and macroeconomic analysis that almost completely ignores their principal assets.

As a result, there is a dangerous asymmetry today in the way we measure and, hence, in the way we think about the value of natural resources. Manufactured assets—buildings and equipment, for example—are valued as productive capital. As they wear out, a depreciation charge is taken against the value of the production that these assets generate. This practice recognizes that a consumption level maintained by drawing down the stock of capital exceeds the sustainable level of income. Natural resource assets are not so valued, and their loss entails no debit charge against current income that would account for the decrease in potential future production. A country could exhaust its mineral resources, cut down its forests, erode its soils, pollute its aquifers, and hunt its wildlife and fisheries to extinction, but measured income would not be affected as these assets disappeared.

Codified in the United Nations System of National Accounts (SNA), which is followed more or less closely by many countries, this difference in the treatment of natural resources and other tangible assets gives the wrong signals to policymakers. It reinforces the false dichotomy between the economy and the "environment" that leads policymakers to ignore or destroy the latter in the name of economic development. It confuses the depletion of valuable assets with the generation of income. Thus, it promotes and seems to validate the idea that rapid rates of economic growth can be achieved and sustained by exploiting the resource base. The result can be il-

lusory gains in income and permanent losses in wealth (1).

A considerable and growing body of expert opinion has recognized the need to remove this anomaly from the accounting framework by accounting for the depreciation of natural resource assets like the depreciation of other physical capital (2). In June 1985, the member governments of the Organisation for Economic Co-operation and Development (OECD) adopted a "Declaration on Environment: Resources for the Future." They declared that they will "[e]nsure that environmental considerations are taken fully into account at an early stage in the development and implementation of economic and other policies...by...[inter alia]...improv[ing] the management of natural resources, using an integrated approach, with a view to ensuring long-term environmental and economic sustainability. For this purpose, they will develop appropriate mechanisms and techniques including more accurate resource accounts" (3).

Our Common Future, the 1987 report of the World Commission on Environment and Development, stated, "Thus, figuring profits from logging rarely takes full account of the losses in future revenue incurred through degradation of the forest. Similar incomplete accounting occurs in the exploitation of other resources, especially in the case of resources that are not capitalized in enterprise or national accounts: air, water, and soil. *In all countries, rich or poor, economic development must take full account in its measurements of growth of the improvement or deterioration in the stock of natural resources*" (4).

Similarly, academic experts and such international agencies as the OECD have recommended that capital consumption allowances be extended to natural resource assets, such as mineral deposits (5). The World Bank and the United Nations Environment Programme have emphasized the deficiencies in the current accounting system and have sponsored work on improvements. According to a recent World Bank publication, "Gross Domestic Product...is essentially a short-term measure of total economic activity for which exchange occurs in monetary terms....it is less useful for gauging long-term sustainable growth partly because natural resource depletion and degradation are being ignored" (6).

NATURAL RESOURCE ACCOUNTING

Although the need for natural resource accounting is widely recognized, there remain questions regarding the most conceptually sound and the most practical method for representing in national income accounts the value of natural resources consumed or degraded and the economic costs of pollution. Economists associated with the United Nations Statistical Office have tentatively proposed a system of satellite environmental accounts supplementing the United Nations System of National Accounts, rather than changing the basic accounts themselves. (See Box

Box 14.1 The United Nations Role

The United Nations Statistical Office has an important role to play. The United Nations System of National Accounts (SNA) provides a standard and a model that, at least in its core flow accounts, is followed more or less closely by many countries. The United Nations Statistical Office is also a worldwide source of expertise and guidance in the development of national income and other statistical systems.

The system of national accounts published by the United Nations Statistical Office (1) is more complete with respect to natural resource accounting than are the accounting systems implemented by most national governments. The SNA provides for balance sheets that record opening and closing stocks, and sources of increase and decrease. Such accounts are included for reproducible tangible assets, such as tree plantations, and nonreproducible tangible assets, such as agricultural land and subsoil minerals. The criterion for inclusion in the SNA is whether the assets are privately owned and used in the commercial production of goods and services so that economic values can be established. Natural resources in the public domain, such as surface waters, atmosphere, and wilderness, are excluded on the grounds that the SNA deals with the market economy and that the economic values of natural resources outside the market system cannot be established readily.

For natural resource assets included in the SNA, the accounting framework provides for "reconciliation accounts" that link balance sheet and flow accounts. These revaluation accounts encompass changes in opening stocks resulting from changes in prices during the period, and from physical changes such as growth, discoveries, depletion, extraction, and natural losses. The valuation principle endorsed by the United Nations for use in

these accounts is market asset value, when possible. When direct asset value cannot be established, the United Nations guidelines endorse the economic asset-valuing principle discussed above: the present value of the expected future income stream obtainable from the resource is the measure of the resource's asset value.

The United Nations Statistical Commission, advised by a number of expert working groups, is currently considering changes in the SNA, as it does periodically. Dissatisfaction stems from many inconsistencies and omissions in the current system. For example, production of goods and services outside the enterprise sector, notably by households, is largely omitted. Also, along with natural resources, other kinds of capital assets, such as knowledge and the stock of skills possessed by the work force, are ignored. Furthermore, in the government sector, the goods and services produced are not measured directly, but are valued at their factor cost. These and many other deficiencies have led to a long agenda of suggested improvements.

Although deliberations will continue until 1991, the United Nations Statistical Commission has evidently reached the decision already that there should be no fundamental changes in the existing SNA. The existing accounting methodology is protected, in a sense, by its very inadequacy: wholesale reform is a large task, and improvement limited to just one aspect is hard to justify when so many other problems would remain. Moreover, at both the national and international levels, decisions regarding the accounting system are in the hands of the *producers* of statistics, not the users. The national income accounts are like sausages: there are many consumers, but few who want to know how they are put together. Partly for this reason, decisions are domi-

nated by the concerns of national income statisticians who are typically handicapped by shortages of staff, budgets, and raw data. These statisticians resist recommending changes when so much work remains to be done before the *existing* SNA can be implemented fully. The rationale for this position is pragmatic: until more national statistical offices are capable of estimating depreciation accounts for natural resource assets, the core national income accounts should not be modified.

With respect to depreciation accounts for natural resources, the expert committees of the United Nations Statistical Office have taken the position that countries should be encouraged to implement balance sheet accounts for reproducible and nonreproducible tangible assets and link them to conventional national income measures through "satellite accounts," as indicated in the present system. In other words, their position is that depletion accounts for natural resources should be calculated but kept apart from the main tables. This approach means that depreciation in the national income accounts would not be extended to include depreciation of natural resources and that the present misleading indicators of economic performance would be maintained (2). As argued elsewhere in this chapter, this attitude has the potential to do great harm to the developing countries of the world and their endangered natural resources.

References and Notes

1. United Nations, Department of Economic and Social Affairs, *A System of National Accounts*. Statistical Papers, Series F, No. 2, Rev. 3 (United Nations, New York, 1968).
2. Robert Repetto *et al.*, *Wasting Assets: Natural Resources in the National Income Accounts* (World Resources Institute, Washington, D.C., 1989), pp. 10-11.

14.1.) Various methods for adjusting the fundamental concepts and measurements that constitute the national accounts also have been proposed. Before reviewing these approaches, however, it is useful to consider the current concepts and functions of national income accounting and to consider how they might be applied in a specific example.

Income Statements and Balance Sheets

A complete system of financial accounts consists of two parts: one (e.g., the *income statement*) dealing with transaction flows in a period of time, and the other (e.g., the *balance sheet*) with stocks of tangible and financial assets at points in time. The concepts of production, consumption, revenues, and costs relate to transaction flows within accounting periods. The national economic accounts in which they ap-

pear are comparable to income statements in business accounting. In contrast, balance sheets record stocks or levels of assets, liabilities, and net worth at the end of accounting periods. Flows and stocks are linked, in that flows are equal to differences between stocks, and that stocks are equal to accumulated past flows.

National balance sheets record a country's tangible and financial wealth at a point in time, facilitating economic comparisons among different years or among different countries. Changes in a country's stock of manufactured tangible capital goods over time are reflected in the flow accounts.

The value of capital goods, such as structures and equipment, declines over time with use because of physical wear and obsolescence. This gradual decrease in the future productive potential of capital goods is reflected in the national accounts by a depreciation allowance that amortizes the asset's value over its useful lifetime. Depreciation of tangible reproducible capital is subtracted from gross national product in calculating the net national product (NNP) and national income. Thus, NNP should provide a more useful measure of economic performance than GNP, but NNP generally receives less attention in economic policy planning. Ignoring or underestimating the deterioration or depletion of the capital stock can lead to economic policy errors with serious, long-term consequences (7).

The fundamental definition of income encompasses the notion of sustainability. In accounting and economics textbooks, income is defined as the maximum amount that the recipient could consume in a given period without reducing the amount of possible consumption in a future period. Business income is defined as the maximum amount a firm could pay in current dividends without reducing net worth. This income concept encompasses not only current earnings, but also changes in asset positions: real capital gains are a source of income, and capital losses are a reduction in income. Depreciation accounts reflect the fact that, unless the capital stock is maintained and replaced, future consumption possibilities inevitably will decline (8).

In resource-dependent countries especially, natural resource assets are legitimately drawn upon to finance economic growth. The revenues derived from resource extraction finance investments in industrial capacity, infrastructure, and education. A reasonable accounting system would recognize that one kind of asset has been exchanged for another, which is expected to yield a higher return. Should a farmer cut and sell the timber in his woods to raise money for a new barn, his private accounts would reflect the acquisition of a new asset, the barn, and the loss of an old asset, the timber. He thinks himself better off because the barn is worth more to him than the timber. In the present national accounts, however, income and investment would rise as the barn is built, but income also would rise as the wood is cut; nowhere is the loss of a valuable asset reflected. This can lead to serious miscalculation of the development potential of resource-dependent economies by confusing gross and net capital formation. Even worse, if the proceeds of resource depletion were used to finance current consumption, the economic path is ultimately unsustainable, whatever the national accounts say. If the same farmer used the proceeds from his timber sale to finance a winter vacation, he would be poorer on his return and no longer able to afford the barn. But the present national income accounts would register only a gain, not a loss, in wealth.

Thus, in resource-dependent countries, failure to extend the concept of depreciation to the capital stock embodied in natural resources, which are such a significant source of income and consumption, overstates the level and growth of income. (See Indonesia: A Case Study, below.)

INDONESIA: A CASE STUDY

The potential for natural resource accounting is illustrated using Indonesia as an example and following the natural resource accounting method developed in a recent report by the World Resources Institute (9). Over the past 20 years, Indonesia has drawn heavily on its considerable natural resource endowment to finance development expenditures. Revenues from the production of oil, gas, hard minerals, timber, and forest products have offset a large share of governmental development and routine expenditures. Primary production contributes 44 percent of gross domestic product (GDP) (10), 84 percent of exports, and 55 percent of total employment. (See Table 14.1.) Generally, Indonesia's economic performance over this period is judged to have been successful: per capita GDP growth averaging 4.6 percent per year from 1965 to 1986 has been exceeded by only a handful of low- and middle-income countries, and is far above the average for those groups. Gross domestic investment rose from 8 percent of GDP in 1965, at the end of the Sukarno era, to 26 percent of GDP (also well above average) in 1986, despite low oil prices and a difficult debt situation (11).

Estimates derived from the Indonesian case study illustrate how much this evaluation is affected by "keeping score" more correctly. Table 14.2 and Figure 14.1 compare the gross domestic product at constant prices with the net domestic product, derived by subtracting estimates of net natural resource depreciation for only three sectors: petroleum, timber, and soils. It is clear that, after accounting for consumption of natural resource capital, conventionally measured gross domestic product overstates net in-

Table 14.1 Direct Contribution of Primary Production to Gross Domestic Product (GDP)
(percent)

	Share of GDP 1983-87	Growth Rate 1983-87	Share of Merchandise Export 1987-88	Share of Employment 1985
Renewable Resources	24.2	3.2	30.4	54.6
Agriculture	21.3	3.5	13.7	
Food crops	-14.8	-2.2	-0.6	
Other crops	-4.0	-6.4	-12.3	
Livestock	-2.5	-6.6	-0.8	
Fishing	1.7	0.6	2.3	
Forestry[a]	1.2	2.5	14.4	
Exhaustible Resources	19.7	3.0	53.3	0.8
Oil and natural gas[b]	18.5	2.9	47.7	
Other mining	0.8	5.6	5.6	
Total Primary Sectors	43.9	3.1	83.7	55.4

Source: Indonesian Central Bureau of Statistics and Bank Indonesia. Jakarta, as cited in Robert Repetto *et al.*, *Wasting Assets: Natural Resources in the National Income Accounts* (World Resources Institute, Washington, D.C., 1989), p. 5.
Notes:
a. Includes logs, sawn timber, and plywood.
b. Includes crude oil and condensates, natural gas, liquefied-natural gas, and liquefied petroleum gas, but excludes other oil products.

Table 14.2 Comparison of Gross Domestic Product (GDP) and Net Domestic Product (NDP)
(billions of constant 1973 Rupiah)

| Year | GDP[a] | Net Change in Natural Resources | | | Net Change | NNP |
		Petroleum	Forestry[b]	Soils		
1971	5,545	1,527	-312	-09	1,126	6,671
1972	6,067	337	-354	-83	-100	5,967
1973	6,753	407	-591	-95	-279	6,474
1974	7,296	3,228	-533	-90	2,605	9,901
1975	7,631	-787	-249	-85	-1,121	6,510
1976	8,156	-187	-423	-74	-684	7,472
1977	8,882	-1,225	-405	-81	-1,711	7,171
1978	9,567	-1,117	-401	-89	-1,607	7,960
1979	10,165	-1,200	-946	-73	-2,219	7,946
1980	11,169	-1,633	-965	-65	-2,663	8,506
1981	12,055	-1,552	-596	-68	-2,215	9,840
1982	12,325	-1,158	-551	-55	-1,764	10,561
1983	12,842	-1,825	-974	-71	-2,870	9,972
1984	13,520	-1,765	-493	-76	-2,334	11,186
Average Annual Growth	7.1%					4%

Source: Robert Repetto *et al.*, *Wasting Assets: Natural Resources in the National Income Accounts* (World Resources Institute, Washington, D.C., 1989), p. 6.
Notes:
a. From the Indonesian Central Bureau of Statistics.
b. Includes logs, sawn timber, and plywood.

come and its growth substantially. In fact, while GDP increased at an average annual rate of 7.1 percent from 1971 to 1984, the period covered by this case study, the adjusted estimate of NDP rose by only 4 percent per year. If 1971, a year of significant additions to petroleum reserves, is excluded, the respective growth rates from 1972 to 1984 are 6.9 percent and 5.4 percent per year for gross and net national products.

In actuality, the overstatement of income and its growth may differ considerably from these estimates since only three natural resources are covered: petroleum, timber, and soils on Java. Other important exhaustible resources that have been exploited over the period, such as natural gas, coal, copper, tin, and nickel, have not been included in the accounts yet. The depreciation of other renewable resources, such as nontimber forest products and fisheries, also has not been taken into account. When complete depreciation accounts are available, they probably will show a greater divergence, on balance, between gross output and net income.

Other important macroeconomic estimates are even more badly distorted. Table 14.3 and Figure 14.2 compare estimates of gross and net domestic investment, the latter reflecting depreciation of natural resource capital. This statistic is central to economic planning in resource-based economies. Countries such as Indonesia that are heavily dependent on exhaustible natural resources *must* diversify their asset base to preserve a sustainable, long-term growth path. Extraction and sale of natural resources must finance investments in other productive capital. It is relevant, therefore, to compare gross domestic investment with the value of natural resource depletion. If gross domestic investment is less than resource depletion, then, everything considered, the

country is drawing down, rather than building up, its asset base, and using its natural resource endowment to finance current consumption. If domestic net investment is positive, but less than required to equip new labor force entrants with at least the capital per worker of the existing labor force, then increases in output per worker and income per capita are unlikely. In fact, the results from the Indonesian case study show that the adjustment for natural resource asset changes is large in many years relative to gross domestic investment. In 1971 and 1973, the

Figure 14.1 Comparison of Gross Domestic Product (GDP) and Net Domestic Product (NDP), 1971–84

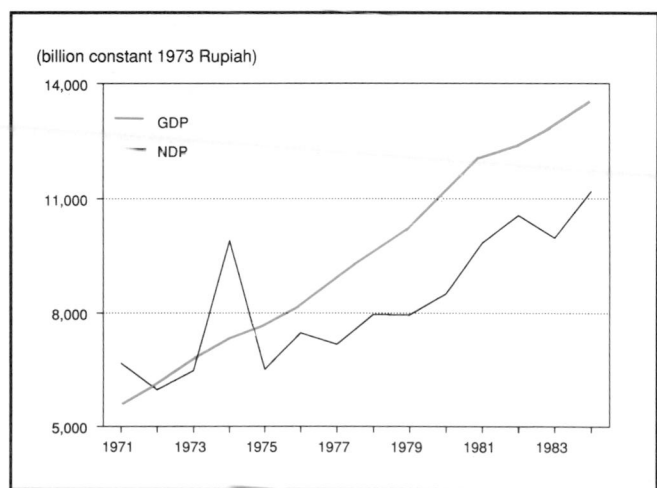

Source: Robert Repetto *et al.*, *Wasting Assets: Natural Resources in the National Income Accounts* (World Resources Institute, Washington, D.C., 1989), p. 7.

Table 14.3 Comparison of Gross Domestic Investment (GDI) and Net Domestic Investment (NDI), 1971–84

(billions of constant 1973 Rupiah)

Year	GDI[a]	Resource Depletion[b]	NDI
1971	876	1,126	2,002
1972	1,139	-100	1,039
1973	1,208	-279	929
1974	1,224	2,605	3,829
1975	1,552	-1,121	431
1976	1,690	-684	1,006
1977	1,785	-1,711	74
1978	1,965	-1,607	358
1979	2,128	-2,219	-91
1980	2,331	-2,663	-332
1981	2,704	-2,215	489
1982	2,783	-1,764	1,019
1983	3,776	-2,870	906
1984	3,551	-2,334	1,217

Source: Robert Repetto *et al., Wasting Assets: Natural Resources in the National Income Accounts* (World Resources Institute, Washington, D.C., 1989), *p. 8.*
Notes:
a. From the Indonesian Central Bureau of Statistics, Jakarta.
b. Includes depletion of forests, petroleum, and the cost of erosion on the island of Java.

Figure 14.2 Comparison of Gross Domestic Investment (GDI) and Net Domestic Investment (NDI), 1971–84

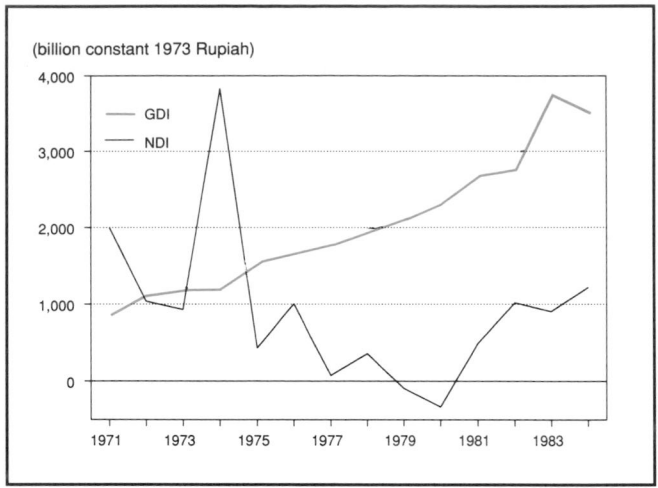

Source: Robert Repetto *et al., Wasting Assets: Natural Resources in the National Income Accounts* (World Resources Institute, Washington, D.C., 1989), p. 9.

adjustment is positive, owing to additions to petroleum reserves (12). In most years during the period, however, the depletion adjustment offsets a good part of the gross capital formation. In some years, net investment was negative, implying that natural resources were being depleted to finance current consumption expenditures.

Such an evaluation should flash an unmistakable warning signal to economic policymakers that they are on an unsustainable course. An economic accounting system that does not generate and highlight such evaluations is deficient as a tool for analysis and policy in resource-based economies and should be amended.

The same holds true with respect to the evaluation of performance in particular economic sectors, such as agriculture. Almost three quarters of the Indonesian population live on the fertile, but overcrowded, "inner" islands of Java, Bali, and Madura, where lowland, irrigated, rice paddies are farmed intensively. In the highlands, population pressures have brought steep hillsides into use for the cultivation of maize, cassava, and other annual crops. As hillsides have been cleared of trees, erosion has increased, now averaging over 54 metric tons per hectare per year, according to estimates by the World Resources Institute.

Erosion's economic consequences include loss of nutrients and soil fertility and increased sedimentation in reservoirs, harbors, and irrigation systems. Increased silt concentrations affect fisheries and downstream water users. Although crop yields have improved in the hills because farmers have used better seed and more fertilizer, the estimates presented in a recent report by the World Resources Institute imply that the annual depreciation of soil fertility, cal-

culated as the value of lost farm income, is about 4 percent of the value of crop production, which is as large as the annual production increase. In other words, these estimates suggest that *current* increases in farm output in Indonesia's uplands are being achieved almost wholly at the expense of potential *future* output. Since the upland population is unlikely to be smaller in the future than it is now, the process of soil erosion represents a transfer of wealth from the future to the present. By ignoring the future costs of soil erosion, the sectoral income accounts overstate significantly the growth of agricultural income in Indonesia's highlands (13).

APPROACHES TO NATURAL RESOURCE ACCOUNTING

The United Nations System of National Accounts

The United Nations System of National Accounts (SNA) already provides for national balance sheets that record opening and closing stocks and sources of change. At least in concept, the United Nations has endorsed accounting for certain natural resources. The SNA specifically includes forests and subsoil assets (e.g., oil and gas reserves) in model, national balance sheets.

However, neither the United Nations SNA nor the national income accounts of any country integrates the treatment of natural resources on both income and balance sheet accounts by including capital formation and depreciation in the income accounts. Instead, these important statistics are relegated to "reconciliation accounts," leaving GNP and NNP unchanged.

In arguing for keeping such asset revaluations in satellite accounts, the United Nations' guidelines point out that large and sudden changes in subsoil assets as a result of extensive new discoveries, changes in technology increasing the range of exploitable reserves, or changes in market conditions could affect estimates of current income markedly. This argument ignores the fact that changes in technology or market conditions do, in fact, affect other kinds of capital stock, such as buildings and equipment, in just that way. Energy price increases in the 1970s, for example, made firms write off most older, heavy, industrial equipment as economically worthless because, with high energy prices, those plants could not produce at a profit. The same fluctuations in energy prices led to drastic inflation, then deflation, in real estate values in oil-producing regions. Changes in technology and market conditions, along with discoveries and depletion, influence the value of natural resources as well as other kinds of economic assets.

In essence, reconciliation or satellite accounts provide a means of recording changes in the value of net assets between successive measurement dates *without having to show any effect on income in the intervening period*. Recording these adjustments in reconciliation accounts is likely to minimize their consideration in national policy analysis. Therefore, while it is significant that the United Nations has specifically endorsed the principle of valuing natural resource assets and changes in the value of those assets in the system of national accounts, its system still leaves the income account seriously biased as an estimate of economic performance (14). (See Box 14.1.)

The importance of bringing some measure of natural resource depletion and degradation into the main national income accounts, rather than relegating them to "satellite" or "reconciliation" accounts, is demonstrated by events of the past decade. While virtually all countries maintain national income accounts and calculate GNP or a comparable index, few have implemented the United Nations recommendations with respect to satellite tables in the system of national accounts because, with limited resources, they have had to "stick to the basics." It is perhaps indicative that the most work to date on satellite resource accounts has been carried out by European countries with ample statistical and financial resources, and that work has had relatively little impact on policy. (See Box 14.2.) Furthermore, while neither business firms nor households would ignore significant changes in their balance sheets, few national governments even calculate theirs. Similarly, despite their recognized deficiencies, politicians, journalists, and even sophisticated economists in official agencies continue to use GNP growth as the prime measure of economic performance. Only if the basic measures of economic performance, as codified by the official national accounting framework, are brought into conformity with a valid definition of

income will economic policies be influenced toward sustainability.

New Proposals

Economists associated with the United Nations Statistical Office, building on the results of a number of expert workshops sponsored by the World Bank and the United Nations Environment Programme, have tentatively proposed a system of additional satellite accounts for integrating environmental accounts with the United Nations SNA. It would construct an "environmentally adjusted gross national product" by subtracting from conventional GNP the identifiable direct costs, sometimes called "defensive expenditures," of protecting the environment or preventing environmental damage. It would construct a measure of "sustainable gross national product" by further subtracting from conventional GNP estimates of the depletion of nonrenewable and some renewable natural resources and the qualitative degradation of natural resources because of air, water, and soil pollution. Depletion estimates would be derived from detailed resource accounts for land, minerals, and nonrenewable resources.

This proposal raises methodological difficulties. Rather than treating natural resources in a manner consistent with economic assets, it introduces two new and somewhat self-contradictory forms of natural resource capital, both of which are handled quite differently from other tangible capital. By considering natural resource depletion as an intermediate cost to be subtracted from gross revenues in calculating GNP, the proposal treats both renewable and nonrenewable natural resources in the private sector as working capital, with no asset value. Rather, such assets are dealt with as if their destruction were just another business expense.

An alternative to satellite accounts would be to construct a framework that is consistent with the treatment of other types of assets and is adopted more readily into core national accounts by redefining net national product. That is the approach proposed in a recent report from the World Resources Institute (WRI) (15).

The basic approach in the WRI report is to unify the treatment of assets by putting natural resource accounts into the same balance sheets used for other tangible capital and to treat depletion and degradation of natural resources like depreciation of other forms of physical capital. This has the advantage of consistency and, moreover, would highlight the value of natural resource assets. In applying this approach, the gross national product is unchanged, but net national product is corrected to take into account the depletion of natural resources in a manner consistent with other depreciation charges. For this technique to be effective, it would require that policymakers focus their attention on NNP, not only on GNP.

Another method proposed in a recent symposium sponsored by the United Nations Environment Programme and the World Bank would estimate the portion of the income from natural resource extraction that would have to be set aside and reinvested in order to produce an equivalent income stream to replace the depleted resource (16). This portion would be subtracted from GNP and excluded from income. Under certain circumstances, this method would give the same results as that used in the WRI study, but it requires estimates of future extraction rates and the remaining years of resource availability—estimates that depend on future market conditions and cannot be known exactly. This approach also treats natural resources in a manner inconsistent with other tangible assets.

A recent Harvard study proposes still another tack, which would adjust GNP itself by an amount representing the loss (or gain) in future income from extracting the resource more rapidly (or slowly) than optimal (17). This course, while also recognizing the basic point that natural resources are productive assets, would require statistical offices to estimate "optimal" rates of resource depletion, which depend both on future market conditions and on social choices between current and future consumption.

The discussion of natural resource accounting is just beginning and it is too early to say which approach will prevail. What is clear is that an improvement is badly needed over the current approach, which is to treat natural resources as a gift of nature and to ignore completely—in our most widely used

Box 14.2 The Experience of Other Countries

Several developed countries have proposed or set up systems of environmental accounts, among them Norway, France, Canada, Japan, and the Netherlands. These systems have been reviewed in detail and evaluated for the United Nations Environment Programme (1). In addition, resource accounts are being estimated in several developing countries, including Costa Rica, the Philippines, and the People's Republic of China.

While natural resources take priority in Norway and France, the focus in Japan has been on pollution and environmental quality. Canada and the Netherlands combine elements of both approaches, while United States efforts have focused only on the costs of pollution control. A natural resource approach is being followed in the developing country studies.

In Norway and France, extensive systems of resource accounting have been established to supplement their economic accounts. In the Norwegian system of natural resource accounting, accounts have been compiled for "material" resources, such as fossil fuels and other minerals, such "biotic" resources as forests and fisheries, and such "environmental" resources as land, water, and air. The accounts are compiled in physical units of measurement, and are not integrated with the national income accounts. However, resource accounts, especially those for petroleum and gas, have been expressed in value terms for use in macroeconomic planning and projection models maintained by the Central Bureau of Statistics (2).

The French natural patrimony accounts are intended as a comprehensive statistical framework to provide the authorities with the facts and data they need to monitor the state and changes in "that subsystem of the terrestrial ecosphere that can be quantitatively and qualitatively altered by human activity"

(3). They are conceptually broader than the national income accounts: material and energy flows to and from economic activities form only a subset of the accounts. Methodology and empirical estimates have been under development since the early 1970s, and they now cover the same range of resources as Norway's: nonrenewables, the physical environment, and living organisms. The basic accounting units are physical, with provision for monetary valuation of stocks and flows that are marketed or contribute directly to market production (4).

In industrialized countries experiencing increasingly acute problems of pollution and congestion while becoming less dependent on agriculture, mining, and other forms of primary production, attention has focused on "environmental," rather than natural resource, accounting. Two issues demand attention. First, expenditures by governments and households to prevent, avoid, or mitigate environmental damages are currently considered final expenditures that increase income, rather than intermediate expenditures that prevent a reduction in income or welfare. Second, damages to transient resources, such as rivers, coastal waters, and the atmosphere, are not reflected in national accounts unless they reduce household earnings or business profits.

Attempts have been made already in several countries to identify environmental protection costs separately. In the United States, for example, expenditures for pollution abatement exceeded $9.1 billion in 1987 (5).

The challenge is to segregate the costs of environmental protection services, so-called "defensive expenditures," from other spending. To the extent that this can be done, these defensive expenditures should be treated as intermediate costs, not final expenditures. Subtraction of environmental damages to transient re-

sources, such as air and water from national income, is more problematic.

There has been little consensus on the principles or quantification of proposals for broader environmental quality accounting so far, although the discussion has helped to highlight the importance of incorporating environmental protection and effective natural resource management in national economic planning. However, for most developing countries and other resource-based economies, it is more relevant to think of natural resources as productive assets than as consumer goods. The first priority is to account for these disappearing assets in a way that gives due emphasis to the costs.

References and Notes

1. Edward Weiller, *The Use of Environmental Accounting for Development Planning. Report to the United Nations Environment Programme* (United Nations, New York, 1983). Anthony Friend, "Natural Resource Accounting: International Experience," paper presented at the Consultative Meeting of the United Nations Environment Programme, Geneva, February 23-25, 1983. Quoted in Robert Repetto *et al.*, *Wasting Assets: Natural Resources in the National Income Accounts* (World Resources Institute, Washington, D.C., 1989), p. 14.
2. "The Norwegian Systems of Resource Accounts," *Statistical Journal of the United Nations*, ECE Vol. 1 (North-Holland, 1983).
3. P. Corniere, "Natural Resource Accounts in France, An Example: Inland Waters," in *Information and Natural Resources* (Organisation for Economic Co-operation and Development, Paris, 1986), p. 45.
4. J.L. Weber, "The French Natural Patrimony Accounts," Statistical Journal of the United Nations, ECE Vol. 1 (1983).
5. Gary L. Rutledge and Nikolaos A. Stergioulas, "Plant and Equipment Expenditures by Business for Pollution Abatement, 1987 and planned 1988," *Survey of Current Business* (U.S. Department of Commerce, Bureau of Economic Analysis, November 1988), p. 26.

indicators of economic progress—the consequences of their overexploitation.

There is ample time to explore fully the implications of extending the concept of depreciation to natural resource assets and to investigate the theoretical consistency and practical applicability of other proposed approaches to natural resource accounting before the United Nations Statistical Office announces revisions to the SNA. The Statistical Office would use this period to prepare for that change.

At the same time, key international economic institutions, such as the World Bank, other multilateral development banks, the International Monetary Fund, and the OECD, should begin to compile, use, and publish estimates of net national product and national income. All these institutions should ready themselves to provide technical assistance to the growing number of national statistical offices in the developing world that wish to adopt these changes and to make such estimates for themselves.

References and Notes

1. Robert Repetto et al., Wasting Assets: Natural Resources in the National Income Accounts (World Resources Institute, Washington, D.C., 1989), pp. 1-3.
2. F. Thomas Juster, "The Framework for the Measurement of Economic and Social Performance," in The Measurement of Economic and Social Performance. Studies in Income and Wealth, M. Moss, ed., Vol. 38 (National Bureau of Economic Research, New York, 1973), as cited in Robert Repetto et al., Wasting Assets: Natural Resources in the Natural Accounts (World Resources Institute, Washington, D.C., 1989) p. 17.
3. Organisation for Economic Co-operation and Development (OECD), "Declaration on Environment: Resources for the Future," in OECD and the Environment (OECD, Paris, 1986), pp. 19-20.
4. World Commission on Environment and Development, Our Common Future (Oxford University Press, Oxford, U.K., 1987), p. 52.
5. Op. cit. 3.
6. Ernest Lutz and Salah El-Serafy, "Environmental and Resource Accounting: An Overview," in Environmental Accounting for Sustainable Development, Yusuf J. Ahmad, Salah El-Serafy, and Ernest Lutz, eds. (The World Bank, Washington, D.C., 1989), p. 1.
7. Op. cit. 1, pp. 12-13.
8. Op. cit. 1, p. 4.
9. Op. cit. 1.
10. Gross domestic product (and net domestic product) differ from gross national product (and net national product) in that the domestic product refers to the production of goods and services within a country's border. National product refers to production owned by that country's citizens–and excludes, for example, production by foreign-owned enterprises and interest paid abroad. For the purposes of this discussion, however, domestic and international products can be used interchangeably.
11. The World Bank, World Development Report (Oxford University Press, Oxford, U.K., 1988) pp. 222, 230.
12. It may seem anomalous that in 1971 and 1973 depreciation was a negative number; that is, net capital consumption was added to gross domestic product and investment. The reason for this is that the value of additions to petroleum reserves in these years was considerably larger than all categories of depletion combined, leading to "negative" depreciation. One way of resolving this apparent anomaly would be to account separately for additions and subtractions from natural resource assets. Real capital gains (as distinct from those resulting from price changes) can be accounted for as gross income and gross capital formation. This is consistent with the earlier definition of income in the chapter because additions to resources during the current year augment the amount that could be consumed currently without reducing potential consumption in future years. This is obvious in the case of forest growth, but less obvious for mineral discoveries, since current discoveries may leave less to be discovered later. However, insofar as additions to mineral reserves reflect advances in the technology of exploration or extraction, the total potential resource base will have expanded.
13. Op. cit. 1, pp. 4-7.
14. Op. cit. 1, pp. 13-14.
15. Op. cit. 1.
16. Yusuf J. Ahmad, Salah El-Serafy, and Ernest Lutz, eds. Environmental Accounting for Sustainable Development (The World Bank, Washington, D.C., 1989), pp. 10-18.
17. Shantayanan Devarajan and Robert J. Weiner, "Natural Resource Depletion and National Income Accounts," unpublished (Harvard University, John F. Kennedy School of Government, Cambridge, Massachusetts, 1988) as cited in Robert Repetto, et al., Wasting Assets: Natural Resources in the National Income Accounts (World Resources Institute, Washington, D.C., 1989).

Data on the World's Resources

There remain gaping holes in our knowledge about the world's resources and how those resources change over time. These holes must be filled if governments and institutions are to act wisely in shaping the economic and environmental policies that will guide us into and through the 21st Century. The data that do exist—and meet minimum criteria of quality, relevance, and coverage—are often nearly inaccessible to citizens, analysts, and policymakers and so are outside the common knowledge of humankind.

In Part IV of *World Resources 1990-91*, we present data—selected from the World Resources Database—that describe what is known about the current state of the world's human and natural resources, and environmental conditions. We believe these data are essential to the analysis and understanding of global environmental issues. By presenting what is known, we hope to illuminate what is not known and encourage greater effort in the collection of globally relevant environmental data.

We chose these data on the basis of coverage, historical base, and quality. The ideal, but unattainable, data set would be well-referenced, current country-level statistics on a relevant global environmental topic for which data have been collected annually. We have assembled from among the available data sets (from international organizations, scientists, and others) those that come closest to this ideal.

This report would not be possible without the good will and active cooperation of the many people and organizations that assemble and report data on a global basis. The Food and Agriculture Organization of the United Nations (FAO) collaborated closely with us, provided timely data, and helped to shape the reporting of the world's agricultural, marine, rangeland, and forestry resources. Underfunded and dependent on independent national statistical offices for much of its data, FAO maintains the world's primary data bases on these topics. The World Bank provided us access to and advice on its global economic and financial data sets. The United Nations Statistical Office of the Department of International Economic and Social Affairs provided data on population, trade, and energy use and consumption. The World Health Organization assisted with data it assembled on health and factors that might affect health and mortality. The World Conservation Union and the World Conservation Monitoring Center provided data on wildlife and wildlife

conservation efforts. We thank them and others for their help.

We did not, however, elect to report all their, or others', data. Much of the available data are not relevant to global environmental questions. Other data have deficiencies of method or source. Because of space considerations, we often can show only a part of more extensive data sets. The Sources and Technical Notes at the end of each chapter contain citations of those data sets and comments on the methods used to generate the data.

Many of the questions one can ask about the world's resources cannot be answered because the necessary data are not collected globally or in a way that they can be compared meaningfully. For example, no international organization reports annually on the extent of tropical deforestation, primarily because most countries do not make annual assessments of that particular resource topic. FAO assesses deforestation only once a decade, and that effort is especially difficult because of national differences in method, definition, expertise, and material resources. Similarly, global urban air quality and global water quality are difficult to assess because the stations reporting to the United Nations Environment Programme's Global Environment Monitoring System are relatively few and, in any case, are heavily weighted toward northern industrialized countries.

Like a musical note, a datum of information is not a pure tone. It has harmonics of purpose, inherent noise, inaccuracies of tuning, and idiosyncrasies of performance. All data have a provenance. They are collected or estimated by someone with particular skills, certain questions in mind, a notion of an acceptable level of accuracy, a limited budget, and, sometimes, social, cultural, or political constraints. To use and improve global data, we need to understand these origins, limitations, and constraints.

Many international organizations publish data reported to them by their member nations. These countries vary widely in knowledge of their resources and the desire to make that knowledge public. Nations also vary in their interpretation of definitions, the methods they use for estimation, the training they provide their personnel, and in the resources they can dedicate to data collection. When countries fail to respond to data requests, global data compilers commonly use earlier years or the experience of similar countries to estimate the resource. International organizations rarely fund basic ongoing data collection efforts at the national level. Data collection is both costly and mundane. Yet complete data sets are vital to our understanding world trends and hence to formulating appropriate policies. We selected data sets for *World Resources* that are complete and contain the minimal number of third-party estimates.

Intensive users of data will soon discover inconsistencies between different sources because of differing practices and assumptions. But these inconsistencies can reflect different views of the same reality. For example, one can compare the data collected by the private sector (e.g., up-to-date energy statistics from British Petroleum in Chapter 9, "Energy") with those assembled from the reports of governments (e.g., the older but more complete energy statistics from the United Nations Energy Statistics Yearbook shown in Chapter 21, "Energy, Materials, and Waste"). The comparison shows that there are two ways to estimate the relative importance of primary electricity production to the world's energy supply. One can use actual production or one can use the oil equivalents to produce the same energy in thermal electric plants. Both measures are valid and together provide a fuller picture of a resource question.

There are also problems in definition and in scale. In Chapter 23, "Oceans and Coasts" for example, we list the length of each country's coastline. The length depends entirely on the scale arbitrarily chosen by the investigator. A coastline has a fractal geometry; the closer one looks, the greater the number of indentations and extensions one can see and therefore measure. These coastlines were measured at various fractal scales. Some, from especially crude baselines that cut across embayments, were originally used to define zones of national influence. Nonetheless, given their limitations, these measurements are still useful and relevant.

All data are only approximate measures of the phenomena they purport to describe. Data should be as accurate as possible, but data can never be perfectly accurate. In a well-designed sample survey based on well-known statistical criteria, a measure of accuracy can be calculated.

Even in countries with highly developed national statistical institutions, the uncertainty attached to numerical estimates can be large in absolute terms. For example, the 1987 United States cereal production estimate of 279,122,000 metric tons is designed to have 95 percent confidence levels of ±3 percent (i.e., there are 19 chances in 20 that the *actual* production figure is within this range). For perspective, this range of uncertainty is greater than estimates of the total cereal production in such countries as Egypt, Nigeria, Japan, or Yugoslavia. Global environmental data sets rarely carry such estimates of their inherent accuracy.

Nonetheless, a trend of increase or decrease over time—assuming similar measurement, data collection, and data reporting schemes—is relatively trustworthy, and it is extremely valuable in analyzing

change. When possible, we selected data with a historical series for this reason. The change from year to year of emissions of carbon dioxide from fossil fuels, for example, is known with more accuracy than is the absolute amount of those emissions. Of course, changes in data might not represent changes in the underlying phenomenon. For example, many more protected areas are shown in Chapter 20, "Wildlife and Habitat," than were shown in a similar chapter in *World Resources 1988–89* (pages 294-295). Although a few protected areas are new, most of the additional areas this year are simply newly compiled. Such changes are not reported as time-series.

Data series are generally repeated from volume to volume of *World Resources*. In Chapter 15, "Basic Economic Indicators," we show the total external debt of many developing countries as estimated by the World Bank. We chose these data over alternative measures because of timeliness and precedent. Other measures of external debt are available from the Organisation for Economic Co-operation and Development, the International Monetary Fund, and the Bank of International Settlements.

With few exceptions, data in Part IV are reported in the most appropriate units, just as they were received or calculated, and are always in the metric system. Thus a resource might be reported to the nearest metric ton, thousand hectares, or petajoule. But readers should not take the precision of data in most of these tables as statements of accuracy. In Chapter 17, "Food and Agriculture," for example, we show average cereal yields to the nearest kilogram. A yield of 867 kilograms per hectare, however, should not be thought to be very different from a yield of 848 kilograms per hectare. Each is an estimate based (often loosely) on survey results, and each has inherent uncertainties attached. We selected this data set because it is useful for comparisons and the calculation of change over time. Unfortunately, because of the precision of these estimates, many readers impute unwarranted accuracy to them. Rounding data, while removing precision, can distort the data by overly emphasizing or minimizing real differences. The data in these tables are usually left as precise as they were when calculated to facilitate further analysis.

Seemingly simple questions about the global environment are often really complex. But, regardless of the complexity of the question, or the quality and extent of available data with which to forge an answer, global environmental issues are too important to ignore. These data provide readers with the common background necessary to help build our understanding of the global environment and to help shape global environmental policy.

Please use these data and the detailed Sources and Technical Notes when formulating environmental questions, comparisons, and analyses. We invite your comments, additions, suggestions and corrections, as well as your identification of data series that might be included in future volumes or made available through the World Resources Database.

15. Basic Economic Indicators

Economic considerations drive many of the decisions that affect the world's resources. This chapter contains a set of basic economic data that provide the context for understanding some of those decisions.

Table 15.1 provides several indicators of the health of economies. One, gross national product (GNP), is a measure of the total economic activity in a country. GNP per capita gives a relative measure of the population's wealth, and the rate of change of GNP gives a relative measure of an economy's health. Unfortunately, none of these GNP-based indicators accounts for the depletion or deterioration of natural resources. This type of measurement, however, is still useful in showing the great differences between the wealthiest and poorest countries. In the Americas, for example, the economies of Canada and the United States overshadow those of countries farther to the south. While the growth of GNP slowed throughout the hemisphere, the slowdown was most dramatic in Latin America.

Another indicator of economic status is the provision or receipt of official development assistance (ODA), which is given to promote economic development and general welfare. In 1987, Saudi Arabia gave the most assistance per capita ($230), although the United States gave the most overall ($9.3 billion). Israel received the highest ODA per capita ($286), but was second to Egypt as the largest recipient ($1.7 billion)–although Egypt received only $35 per capita. Donations appear to bear little relation to the receiving country's GNP or its GNP per capita. ODA given as a percentage of GNP is often used as a measure of a developed country's commitment to the poorest on the planet, but such a measure is somewhat removed from everyday experience. ODA per capita is probably a more intuitive indicator of that commitment, as well as of the relative success of countries in attracting ODA. ODA is important to some developing countries (ODA represents over 20 percent of the GNP in 11 countries in Africa), but it is dwarfed by international borrowing.

Table 15.2 shows some trends in international debt for 111 developing countries. Unlike developed countries, developing countries are required by their foreign creditors to repay their debts in an acceptable foreign currency. Total external debt for these countries totaled over $1 trillion in 1987 (the most recent year with complete data) and is growing. In fact, many of these debt indicators show a decline in economic health over the past 10 years. Many Latin American countries pay out more each year to service their existing debt than they borrow. Indeed, news reports of preliminary figures for 1988 show that developing countries borrowed $92 billion and paid out, including prepayments, $142 billion in interest and principal on old loans. Much of this negative capital flow is accounted for by highly indebted, middle-income countries. The poorest countries still borrow more than they pay. For many developing countries, external debt is a large proportion of GNP and debt service takes a significant portion of the total foreign exchange earned from the export of goods and services. Mexico's external long-term public debt was 59 percent of its GNP in 1987, for example, and its debt service represented 30 percent of its total export earnings. This amount leaves less foreign exchange to fund development, import essential goods, and import goods to improve people's material lives.

Developed countries, as well, have external obligations. In 1987, for example, the United States had $387 billion more in foreign liabilities than it had in foreign assets. Though huge compared with the debts of developing countries, this represented only 8.4 percent of its GNP.

Table 15.3 provides a picture of the relative importance of international trade to a country's economy, as well as the contribution of trade in various goods and services to the current account balance. The current account balance is the sum of net exports of goods and services, and capital transfers. Capital importing countries always have a negative current account balance. The decline in commodity prices (shown in Table 15.4) contributes to the deteriorating current account balance for those countries that depend on exports of food, raw materials, and fuels to fund their foreign-exchange requirements. Nowhere, however, was the decline in the current account balance more dramatic than in the United States, where imports far outweighed exports.

Developing countries depend on the international trade in commodities to earn foreign exchange. Table 15.4 shows that price indexes for all categories of commodities (except timber) have declined over the past 14 years. As a result of declining prices, developing countries must export more just to maintain static national export incomes. Though the demand for foreign exchange is growing because of increased population and debt service, all commodities listed in Table 15.4, whether extracted (copper, tin), grown (wool, cotton), or manufactured (urea), have shown net declines on world markets through the end of 1988.

Table 15.1 Gross National Product and Official Development

	Gross National Product 1987		Average Annual Change in Real GNP (percent)		Distribution of Gross Domestic Product, 1987 (percent)			Average Annual Official Development Assistance (ODA) (million $US) {a}		ODA as a Percentage of GNP {a}		1987 ODA Per Capita ($US) {a}
	Total (million $US)	Per Capita ($US)	1967-77	1977-87	Agriculture	Industry	Services	1980-82	1985-87	1980-82	1985-87	
WORLD												
AFRICA												
Algeria	60,728	2,629	0.0	10.2	12.0	42.0	45.0	72	124	0.2	0.2	9
Angola	X	X	(4.5)	2.8 b	X	X	X	58	119	X	X	15
Benin	1,315	305	2.0	3.9	46.0	14.0	39.0	84	123	7.9	9.1	31
Botswana	1,202	1,059	15.0	11.9	3.0	57.0	40.0	102	118	12.1	11.0	136
Burkina Faso	1,589	191	3.0	5.1	38.0	25.0	38.0	214	255	15.9	16.4	34
Burundi	1,205	241	4.4	2.7	59.0	14.0	27.0	122	174	12.9	15.3	39
Cameroon	10,494	966	4.1	9.5	24.0	31.0	45.0	225	199	3.4	1.8	20
Cape Verde	170	494	X	5.3	X	X	X	57	88	43.8	51.5	251
Central African Rep	909	334	3.2	0.6	41.0	13.0	46.0	101	139	13.5	15.5	64
Chad	733	139	X	X	43.0	18.0	39.0	53	182	X	23.8	38
Comoros	159	372	1.5	5.6	X	X	X	42	49	37.1	31.1	125
Congo	1,761	873	6.2	7.6	12.0	33.0	55.0	89	111	4.9	6.1	75
Cote d'Ivoire	8,262	743	7.0	1.7	36.0	25.0	39.0	157	188	1.9	2.3	23
Djibouti	372	1,005	X	X	X	X	X	65	96	X	25.9	247
Egypt	33,986	678	4.8	5.9	21.0	25.0	54.0	1,365	1,758	5.6	3.5	35
Equatorial Guinea	114	293	X	X	X	X	X	11	27	X	25.1	110
Ethiopia	5,621	126	3.2	1.8	42.0	18.0	40.0	219	662	5.1	12.9	14
Gabon	2,869	2,733	14.9	(0.8)	11.0	41.0	48.0	54	74	1.5	2.4	78
Gambia, The	174	219	6.2	3.2	X	X	X	57	85	26.6	47.8	130
Ghana	5,329	393	1.7	0.2	51.0	16.0	33.0	161	316	0.7	5.7	28
Guinea	2,050	316	3.9	1.6	X	X	X	95	169	6.0	8.7	33
Guinea-Bissau	145	157	2.2 b	2.1	X	X	X	64	78	36.4	53.3	113
Kenya	7,300	330	7.3	3.8	31.0	19.0	50.0	444	486	6.9	7.1	26
Lesotho	579	355	11.3	2.3	21.0	28.0	51.0	96	97	14.4	17.0	67
Liberia	1,048	451	3.2	(2.2)	37.0	28.0	35.0	105	89	9.8	8.6	34
Libya	22,326	5,453	3.0	(4.2)	X	X	X	(213)	(47)	(0.7)	(0.2)	(8)
Madagascar	2,258	207	1.7	(0.7)	43.0	16.0	42.0	238	277	8.1	12.5	30
Malawi	1,292	164	6.1	2.3	37.0	18.0	45.0	134	197	11.8	17.5	35
Mali	1,551	200	3.6	2.5	54.0	12.0	35.0	236	372	16.7	25.4	47
Mauritania	816	439	2.3	1.9	37.0	22.0	41.0	201	202	28.8	26.2	96
Mauritius	1,557	1,500	5.8	3.2	15.0	32.0	53.0	46	50	4.3	3.6	63
Morocco	14,324	615	5.7	2.9	19.0	31.0	50.0	900	535	5.9	3.9	17
Mozambique, People's Rep	2,123	146	X	(5.7)	50.0	12.0	38.0	174	457	7.4	16.0	45
Niger	1,752	258	(1.7)	(0.1)	34.0	24.0	42.0	208	320	9.5	18.2	51
Nigeria	39,216	368	8.5	(0.9)	30.0	43.0	27.0	(41)	11	(0.0)	0.0	0
Rwanda	1,939	301	3.6	4.7	37.0	23.0	40.0	153	212	11.8	11.1	38
Senegal	3,543	510	2.5	2.4	22.0	27.0	52.0	315	501	12.3	14.6	92
Sierra Leone	956	249	3.5	1.5	45.0	19.0	36.0	78	74	6.8	12.3	18
Somalia	1,560 c	290	3.1	0.3	65.0	9.0	26.0	423	482	14.3	27.9	102
South Africa	59,910 c	1,870	3.7	1.8	6.0	44.0	50.0	X	X	X	X	X
Sudan	7,647	331	4.6	(1.4)	37.0	15.0	48.0	652	992	8.4	13.9	39
Swaziland	497	699	8.2	4.2	X	X	X	38	35	7.0	7.4	64
Tanzania	5,370 c	180	4.5	1.3	61.0	8.0	31.0	688	683	11.9	15.0	37
Togo	930	286	4.2	0.2	29.0	18.0	54.0	77	137	8.3	14.9	38
Tunisia	9,007	1,182	7.6	4.2	18.0	32.0	50.0	227	223	2.8	2.6	37
Uganda	X	260	1.5	(0.1)	76.0	5.0	19.0	127	219	0.0	0.1	18
Zaire	4,977	153	2.2	0.2	32.0	33.0	35.0	390	465	4.2	9.8	19
Zambia	1,785	248	2.3	(0.8)	12.0	36.0	52.0	286	407	7.7	23.6	59
Zimbabwe	5,288	585	6.4	3.8	11.0	43.0	46.0	197	252	3.3	4.9	33
NORTH & CENTRAL AMERICA												
Barbados	1,358	5,346	2.7	1.3	X	X	X	15	6	1.6	0.5	26
Canada	391,928	15,160	5.0	2.7	3.3	34.8	61.8	(1,154)	(1,737)	(0.4)	(0.5)	(73)
Costa Rica	4,193	1,608	6.3	0.9	18.0	29.0	53.0	67	235	2.2	6.0	88
Cuba	X	X	X	X	X	X	X	21	22	X	X	3
Dominican Rep	4,930	734	7.6	2.0	17.0	30.0	53.0	122	143	1.9	3.1	19
El Salvador	4,176	842	4.8	(2.3)	14.0	22.0	64.0	162	371	4.7	7.9	86
Guatemala	7,989	947	6.2	0.1	X	X	X	71	153	0.9	1.8	29
Haiti	2,221	362	2.9	1.1	X	X	X	113	182	7.8	8.5	35
Honduras	3,788	808	4.5	1.9	22.0	24.0	55.0	123	272	4.8	7.6	55
Jamaica	2,256	940	3.0	(1.2)	6.0	41.0	53.0	157	172	5.6	8.0	70
Mexico	149,396	1,825	6.9	2.5	9.0	34.0	57.0	99	184	0.0	0.1	2
Nicaragua	2,903	829	4.6	(2.3)	21.0	34.0	46.0	172	131	0.0	0.0	40
Panama	5,084	2,239	5.2	4.0	9.0	18.0	73.0	42	54	1.2	1.1	18
Trinidad and Tobago	5,130	4,199	3.5	(0.3)	4.0	39.0	57.0	3	20	0.0	0.4	28
United States	4,516,739	18,529	2.5	2.4	2.1	30.1	67.8	(7,041)	(9,304)	(0.2)	(0.2)	(37)
SOUTH AMERICA												
Argentina	74,491	2,394	3.6	(0.9)	13.0	43.0	44.0	31	75	0.0	0.1	3
Bolivia	3,341	496	4.9	(2.0)	24.0	24.0	53.0	162	281	3.0	7.1	47
Brazil	285,877	2,021	10.0	3.2	11.0	38.0	51.0	176	196	0.1	0.1	2
Chile	17,029	1,358	0.2	1.5	X	X	X	(8)	19	(0.0)	0.1	2
Colombia	36,491	1,238	6.0	2.9	19.0	35.0	46.0	96	68	0.3	0.2	3
Ecuador	10,333	1,044	9.4	1.7	16.0	31.0	53.0	53	162	0.4	1.4	21
Guyana	310	389	4.4	(4.8)	X	X	X	48	29	9.7	8.6	36
Paraguay	3,903	995	6.6	4.0	27.0	26.0	47.0	57	66	1.0	1.4	21
Peru	29,682	1,467	4.4	1.8	11.0	33.0	56.0	208	293	0.9	1.0	14
Suriname	952	2,268	10.1	(3.2)	X	X	X	94	16	9.5	1.5	53
Uruguay	6,556	2,198	2.6	(0.5)	13.0	32.0	55.0	7	16	0.1	0.3	6
Venezuela	58,941	3,226	4.8	(0.9)	6.0	38.0	56.0	(103)	(32)	(0.1)	(0.1)	(0)

Assistance, 1967–87

Table 15.1

	Gross National Product 1987		Average Annual Change in Real GNP (percent)		Distribution of Gross Domestic Product, 1987 (percent)			Average Annual Official Development Assistance (ODA) (million $US) {a}		ODA as a Percentage of GNP {a}		1987 ODA Per Capita ($US) {a}
	Total (million $US)	Per Capita ($US)	1967-77	1977-87	Agriculture	Industry	Services	1980-82	1985-87	1980-82	1985-87	
ASIA												
Afghanistan	3,080 d	220 d	3.3	X	X	X	X	22	21	X	X	2
Bahrain	3,670 c	X	X	X	X	X	X	131	57	X	X	1
Bangladesh	17,387	164	1.5	3.7	47.0	13.0	39.0	1,243	1,415	9.9	9.0	15
Bhutan	202	150	X	X	51.0	16.0	32.0	10	35	X	20.0	31
China	313,672	294	6.9	9.3	31.0	49.0	20.0	119	933	0.0	0.3	1
Cyprus	2,920 c	5,200	X	5.8	X	X	X	43	38	2.0	1.2	61
India	248,073	311	3.6	4.8	30.0	30.0	40.0	1,722	1,718	0.9	0.7	2
Indonesia	76,038	444	8.0	5.1	26.0	33.0	41.0	944	853	1.1	1.2	7
Iran, Islamic Rep	86,400 e	1,756 e	11.2	0.1 b	X	X	X	14	38	0.0	X	1
Iraq	40,000 e	2,400 e	X	X	X	X	X	(349)	50	X	X	5
Israel	29,804	6,812	7.5	3.0	X	X	X	841	1,722	3.6	6.1	286
Japan	1,924,663	15,764	6.1	4.0	2.8	40.6	56.7	(3,182)	(5,628)	(0.3)	(0.3)	(61)
Jordan	5,500 d	1,560	6.3 b	5.4	9.0	28.0	64.0	1,046	567	29.6	13.1	157
Kampuchea, Dem	X	X	X	X	X	X	X	152	13	X	X	2
Korea, Dem People's Rep	19,000 d	910 d	X	X	X	X	X	X	X	X	X	X
Korea, Rep	113,153	2,689	9.5	6.9	11.0	43.0	46.0	168	(5)	0.3	(0.0)	0
Kuwait	27,324	14,589	3.6	(1.6)	1.0	51.0	48.0	(1,146)	(596)	(3.7)	(2.4)	(167)
Lao People's Dem Rep	627	166	X	X	X	X	X	38	48	X	6.3	16
Lebanon	X	X	X	X	X	X	X	293	82	X	X	38
Malaysia	30,075	1,820	7.2	5.2	X	X	X	138	262	0.6	0.9	22
Mongolia	X	X	X	X	X	X	X	X	X	X	X	X
Myanmar	8,319	212	3.9	4.8	X	X	X	304	379	5.2	4.8	9
Nepal	2,836	161	2.5	3.7	57.0	14.0	29.0	181	294	8.1	11.0	20
Oman	7,847	5,826	11.2	11.2	3.0	43.0	54.0	177	59	2.8	0.8	12
Pakistan	36,149	353	4.5	6.8	23.0	28.0	49.0	915	876	3.5	2.7	8
Philippines	34,427	589	6.2	1.3	24.0	33.0	43.0	336	739	0.9	2.3	13
Qatar	4,130	12,430	2.2 b	(5.4)	X	X	X	(220)	(8)	(3.3)	(0.2)	(3)
Saudi Arabia	83,270 c	X	14.0	(0.1) b	4.0	50.0	46.0	(4,903)	(2,990)	(3.4)	X	(230)
Singapore	20,884	7,992	9.9	7.5	1.0	38.0	62.0	19	26	0.1	0.1	9
Sri Lanka	6,645	406	4.1	5.0	27.0	27.0	46.0	394	519	9.0	8.2	31
Syrian Arab Rep	18,501	1,645	9.9	2.6	27.0	19.0	54.0	1,383	678	8.8	2.6	62
Thailand	45,542	850	6.8	5.7	16.0	35.0	49.0	405	494	1.2	1.2	9
Turkey	63,812	1,213	6.9	3.8	17.0	36.0	46.0	778	314	1.4	0.5	8
United Arab Emirates	22,955	15,787	X	1.4	2.0	57.0	41.0	(773)	(30)	(2.5)	(0.1)	67
Viet Nam	12,400 d	200 d	X	X	X	X	X	202	126	X	X	2
Yemen Arab Rep	4,955	585	10.7 b	6.3	28.0	17.0	55.0	431	298	10.5	6.6	41
Yemen, People's Dem Rep	956	421	X	(1.0)	16.0	23.0	61.0	110	88	12.1	9.1	35
EUROPE												
Albania	2,800 d	930 d	X	X	X	X	X	X	X	X	X	X
Austria	90,726	11,980	4.6	1.8	3.3	36.8	59.9	(211)	(214)	(0.3)	(0.2)	(26)
Belgium	113,201	11,476	4.2	1.4	2.1	30.9	67.1	(556)	(559)	(0.5)	(0.5)	(70)
Bulgaria	61,200 d	6,800 d	X	X	X	X	X	X	X	X	X	X
Czechoslovakia	143,900 d	9,280 d	X	X	X	X	X	X	X	X	X	X
Denmark	76,590	14,939	2.7	1.8	4.9	28.9	66.3	(433)	(665)	(0.7)	(0.9)	(168)
Finland	71,330	14,463	4.3	3.5	6.5	35.4	58.1	(130)	(319)	(0.3)	(0.5)	(88)
France	711,503	12,789	4.6	1.7	3.7	30.5	65.8	(4,124)	(5,208)	(0.7)	(0.7)	(117)
German Dem Rep	187,500 d	11,300 d	X	X	X	X	X	X	X	X	X	X
Germany, Fed Rep	880,813	14,399	3.4	1.6	1.8	38.2	60.0	(3,300)	(3,722)	(0.5)	(0.4)	(72)
Greece	40,163	4,015	5.9	1.2	16.0	29.0	56.0	22	23	0.1	0.1	3
Hungary	23,759	2,237	8.4	1.5	15.0	40.0	44.0	X	X	X	X	X
Iceland	4,083	16,596	5.5	3.0	X	X	X	X	(2)	X	(0.1)	(8)
Ireland	21,795	6,129	4.3	0.8	9.9	37.4	52.7	X	(51)	X	(0.2)	(14)
Italy	593,885	10,355	3.5	2.3	4.3	34.4	61.3	(720)	(2,039)	(0.2)	(0.3)	(46)
Luxembourg	6,866	18,506	5.5	4.7	X	X	X	X	X	X	X	(16)
Malta	1,446	4,191	10.0	3.6	X	X	X	31	8	2.5	0.6	(3)
Netherlands	173,813	11,856	4.4	1.3	4.0	29.9	66.2	(1,537)	(1,657)	(1.0)	(1.0)	(143)
Norway	72,028	17,203	4.3	3.8	3.5	34.7	61.8	(504)	(754)	(0.9)	(1.1)	(213)
Poland	72,541	1,926	X	2.6 b	X	X	X	X	X	X	X	X
Portugal	28,725	2,827	5.3	2.1	9.0	40.0	51.0	82	102	0.4	0.4	6
Romania	138,000 d	6,030 d	0.0	3.9 b	X	X	X	X	X	X	X	X
Spain	231,897	5,972	5.3	1.5	6.2	36.8	57.0	(78)	(181)	(0.0)	(0.1)	(4)
Sweden	130,642	15,636	3.0	1.7	3.3	34.8	61.9	(956)	(1,102)	(0.9)	(0.9)	(165)
Switzerland	139,468	21,332	2.1	2.0	X	X	X	(247)	(424)	(0.2)	(0.3)	(84)
United Kingdom	592,764	10,419	2.1	1.8	1.8	37.9	60.3	(1,949)	(1,711)	(0.4)	(0.3)	(33)
Yugoslavia	57,985	2,477	6.3	1.4	11.0	43.0	45.0	(13)	21	(0.0)	0.0	1
U.S.S.R.	2,356,700 d	8,375 d	X	X	X	X	X	(2,458)	(3,804)	X	X	(15)
OCEANIA												
Australia	180,413	11,103	4.4	2.6	4.1	32.5	63.4	(733)	(709)	(0.5)	(0.4)	(39)
Fiji	1,137	1,575	6.7	0.7	X	X	X	37	37	3.1	3.2	49
New Zealand	25,901	7,764	3.6	1.9	8.1	30.5	61.4	(68)	(72)	(0.3)	(0.3)	(26)
Papua New Guinea	2,706	730	4.0	1.7	34.0	26.0	40.0	324	282	13.6	11.1	87
Solomon Islands	123	421	X	5.9	X	X	X	35	36	27.9	30.2	194

Source: The World Bank; Organisation for Economic Co-operation and Development; and U.S. Central Intelligence Agency.
Note: a. For ODA, flows to recipients are shown as positive numbers; flows from donors as negative numbers (in parentheses). b. Based on less than 10 years of data.
c. 1986 World Bank estimate. d. 1986 Central Intelligence Agency estimate. e. 1987 Central Intelligence Agency estimate.
0 = zero or less than half of the unit of measure; X = not available; negative numbers are shown in parentheses.
For additional information, see Sources and Technical Notes.

Table 15.2 External Debt Indicators, 1977–87

	Total External Debt (million $US)			Disbursed Long-Term Public Debt (million $US)			Long-Term Public Debt as a Percentage of GNP			Debt Service as a Percentage of Exports of Goods and Services			Debt Service as a Percentage of Current Borrowing			Current Borrowing Per Capita ($US)
	1977	1982	1987	1977	1982	1987	1977	1982	1987	1977	1982	1987	1977	1982	1987	1987
111 COUNTRIES	263,945	665,037	1,079,072	197,015	475,253	905,851										21.04
AFRICA	54,283	124,040	212,406	44,118	101,517	182,284										28.70
Algeria	10,318	16,683	22,881	8,632	13,932	19,240	42	32	31	16	30	51	34	194	117	181.67
Benin	170	661	1,133	143	583	929	23	57	56	2	7	16	12	7	50	15.65
Botswana	184	214	518	181	211	514	52	28	35	2	2	4	37	24	68	89.55
Burkina Faso	164	378	861	134	346	794	17	27	42	5	8	11	12	19	28	13.47
Burundi	46	227	755	41	201	718	8	20	65	3	5	38	15	11	30	28.12
Cameroon	1,014	2,356	3,508	861	1,959	2,785	27	30	22	6	13	16	17	134	111	27.85
Cape Verde	14	58	131	14	58	121	16	44	58	X	X	X	4	6	93	21.53
Central African Rep	113	251	586	103	208	520	21	28	49	3	3	12	18	25	29	27.95
Chad	148	168	318	130	153	269	20	X	38	3	2	4	9	26	13	9.68
Comoros	25	69	203	23	68	188	42	62	95	X	4	4	6	5	9	30.03
Congo	732	1,941	4,636	590	1,756	3,679	81	88	195	11	19	19	26	46	37	263.63
Cote d'Ivoire	2,391	6,656	10,291	1,897	5,071	8,450	33	72	90	10	34	22	33	68	118	54.12
Djibouti	27	32	181	19	24	152	X	X	39	X	X	6	88	33	45	75.09
Egypt	8,777	25,713	39,166	8,467	21,214	34,515	41	77	61	29	24	19	37	66	116	25.75
Equatorial Guinea	49	116	193	32	89	175	31	X	130	X	20	28	21	9	38	62.03
Ethiopia	500	1,237	2,590	447	1,039	2,434	14	23	46	6	10	28	48	48	45	9.01
Gabon	1,477	1,009	2,071	1,256	825	1,605	48	25	54	9	12	5	56	326	27	252.48
Gambia, The	40	208	319	27	155	273	18	81	158	1	13	13	6	37	40	46.70
Ghana	1,065	1,381	3,094	781	1,153	2,207	8	4	47	3	10	19	35	70	47	26.92
Guinea	915	1,356	1,784	827	1,245	1,616	66	79	78	X	X	X	132	89	76	22.55
Guinea-Bissau	27	161	424	26	143	391	20	72	320	X	14	38	3	11	18	56.35
Kenya	1,475	3,129	5,453	1,114	2,453	4,482	26	40	58	4	21	29	20	85	112	20.34
Lesotho	24	121	241	24	118	237	7	17	35	0	2	4	4	19	36	24.99
Liberia	298	903	1,619	267	658	1,152	31	63	108	6	7	3	36	52	35	13.64
Madagascar	296	1,920	3,377	225	1,679	3,113	12	61	162	4	20	35	31	32	64	21.03
Malawi	453	870	1,363	369	703	1,155	47	64	98	10	23	24	22	86	54	16.66
Mali	460	877	2,016	429	823	1,847	43	67	99	5	5	9	12	8	27	15.09
Mauritania	526	1,142	2,035	462	1,001	1,868	92	143	215	22	13	18	47	18	61	75.12
Mauritius	125	560	729	71	363	545	9	35	31	2	12	6	30	79	106	67.84
Morocco	4,839	12,151	20,334	4,100	10,251	18,468	38	72	116	15	46	30	15	64	100	54.24
Niger	170	728	1,425	118	604	1,258	9	32	60	4	25	30	32	95	69	22.93
Nigeria	3,016	11,595	28,364	855	9,076	25,707	2	10	112	1	11	10	118	39	76	9.58
Rwanda	87	218	583	78	197	544	10	14	26	1	3	11	5	15	22	14.11
Sao Tome & Principe	2	41	87	2	40	84	9	113	338	0	19	42	0	19	42	79.93
Senegal	583	1,621	3,653	436	1,235	3,067	23	50	70	7	5	23	54	15	76	51.84
Seychelles	109.2	48.4	119.1	1.2	36.4	84.1	2	25	35	0	1	6	0	7	94	156.48
Sierra Leone	253	512	659	189	401	513	30	32	90	11	7	X	52	13	218	0.57
Somalia	400	1,248	2,534	388	1,125	2,288	36	42	149	3	6	8	4	10	12	12.38
Sudan	2,534	6,641	10,754	1,930	5,165	7,876	29	76	97	11	17	7	28	12	29	7.29
Swaziland	61	184	293	52	170	273	17	33	46	1	4	6	20	60	105	41.46
Tanzania	1,723	2,913	4,324	1,446	2,405	4,068	42	38	144	6	12	19	16	17	78	4.47
Togo	369	953	1,223	321	805	1,042	42	104	91	24	10	14	33	84	126	15.39
Tunisia	2,109	3,652	6,683	1,879	3,516	6,189	38	45	67	11	16	27	20	74	113	105.78
Uganda	360	922	1,405	278	598	1,116	0	0	0	4	16	22	48	53	38	11.92
Zaire	3,430	4,750	8,630	2,900	4,104	7,334	34	47	143	9	8	12	21	78	50	15.10
Zambia	2,188	3,662	6,400	1,403	2,379	4,354	58	65	244	20	16	14	107	51	99	18.07
Zimbabwe	197	1,804	2,461	151	1,181	2,044	4	18	36	1	9	23	X	27	138	30.68
NORTH & CENTRAL AMERICA	34,854	100,420	130,034	26,683	68,642	112,544										81.12
Bahamas	88	303	233	88	230	175	14	16	7	5	4	3	113	51	1,181	17.35
Barbados	58	336	621	50	226	501	10	23	37	3	4	X	34	38	235	118.50
Belize	21	69	139	14	62	113	12	38	51	X	X	7	9	38	93	68.67
Costa Rica	991	3,249	4,437	733	2,378	3,629	24	109	89	9	12	12	37	79	211	33.05
Dominican Rep	818	2,266	3,563	610	1,666	2,938	15	23	63	7	22	11	46	62	113	21.42
El Salvador	482	1,286	1,692	266	972	1,597	9	28	35	6	8	20	121	25	150	24.16
Grenada	10	40	72	9	33	67	22	37	49	1	3	8	20	9	50	109.33
Guatemala	472	1,369	2,709	217	1,144	2,345	4	13	34	1	8	25	28	31	234	14.79
Haiti	157	536	804	138	416	674	14	28	30	11	5	7	29	23	24	15.30
Honduras	612	1,686	3,188	458	1,431	2,681	29	53	71	7	19	23	31	64	124	39.34
Jamaica	1,179	2,793	4,389	969	2,112	3,511	31	67	139	16	18	27	133	46	140	129.96
Mexico	26,665	78,011	93,734	20,703	51,642	82,771	25	32	59	43	34	30	53	79	108	101.43
Nicaragua	1,274	3,331	7,291	845	2,488	6,150	0	0	0	13	36	X	41	55	7	141.31
Panama	1,684	3,923	5,323	1,333	2,917	3,722	66	74	73	12	7	6	49	80	276	61.21
St. Vincent	5	21	39	5	19	36	13	24	30	X	3	4	0	37	39	61.72
Trinidad and Tobago	339	1,203	1,801	247	907	1,635	8	11	39	1	4	23	8	71	298	105.49
SOUTH AMERICA	68,616	170,520	266,896	46,976	107,437	225,962										30.31
Argentina	8,178	32,407	53,955	5,036	15,886	47,451	10	31	62	15	24	46	88	49	134	93.69
Bolivia	1,713	3,184	5,348	1,428	2,861	4,599	46	49	115	23	31	22	35	127	65	31.00
Brazil	28,392	68,799	109,497	22,401	50,798	91,653	13	20	29	21	43	27	51	96	493	10.99
Chile	4,904	8,588	18,773	3,675	5,243	15,536	28	23	89	34	20	21	132	79	235	46.38
Colombia	4,664	9,114	15,482	2,700	5,990	13,828	14	16	41	9	18	34	84	71	195	41.30
Ecuador	1,967	6,233	10,407	1,111	4,042	9,026	17	32	93	7	42	20	20	410	76	65.87
Guyana	481	922	1,285	416	678	874	100	158	353	12	18	X	54	65	122	26.10
Paraguay	390	1,166	2,419	336	940	2,218	16	16	49	6	10	21	24	28	104	54.58
Peru	7,054	10,621	16,625	4,711	6,956	12,485	37	28	29	30	36	12	50	72	91	24.29
Uruguay	1,022	2,441	4,091	736	1,700	3,048	18	19	42	30	13	24	108	52	171	79.25
Venezuela	9,852	27,045	29,015	4,426	12,342	25,245	10	16	52	8	16	23	41	166	910	17.26

Table 15.2

	Total External Debt (million $US)			Disbursed Long-Term Public Debt (million $US)			Long-Term Public Debt as a Percentage of GNP			Debt Service as a Percentage of Exports of Goods and Services			Debt Service as a Percentage of Current Borrowing			Current Borrowing Per Capita ($US)
	1977	1982	1987	1977	1982	1987	1977	1982	1987	1977	1982	1987	1977	1982	1987	1987
ASIA																
Bangladesh	2,442	5,056	9,506	2,181	4,434	8,851	32	37	51	15	14	24	22	18	35	8.71
Bhutan	0	1	41	0	1	41	X	1	20	X	X	X	X	0	3	11.99
China	0	8,359	30,227	0	5,562	23,659	0	2	8	0	7	7	X	100	50	5.34
Cyprus	275	730	2,016	162	609	1,419	15	28	38	3	8	13	27	53	135	284.56
India	15,234	24,388	42,928	14,647	19,685	37,325	14	11	15	10	10	19	61	46	61	6.76
Indonesia	13,595	23,300	48,477	11,670	18,513	41,284	25	21	63	12	11	29	64	54	103	30.77
Israel	10,071	18,998	20,604	8,086	14,614	16,767	55	59	50	11	23	18	52	93	233	240.39
Jordan	735	2,413	4,564	623	1,841	3,518	39	49	75	5	9	22	18	41	148	92.14
Korea, Rep	13,500	33,877	34,357	8,679	20,191	24,541	24	29	21	9	14	22	55	94	555	52.70
Lao People's Dem Rep	173	365	736	173	365	736	X	131	105	X	X	X	14	9	11	31.16
Lebanon	53	643	496	39	213	236	X	X	X	X	X	X	135	376	238	4.69
Malaysia	2,706	10,141	21,091	2,007	8,144	19,065	15	32	64	7	6	14	70	28	216	83.14
Maldives	2	65	73	2	42	62	17	107	98	0	2	6	0	22	173	18.88
Myanmar	594	2,043	4,348	505	1,961	4,257	12	33	47	11	26	59	20	30	54	8.56
Nepal	101	353	947	72	298	902	5	12	33	2	2	10	9	9	22	8.66
Oman	654	958	2,879	473	724	2,474	20	10	34	5	2	14	54	41	179	253.82
Pakistan	7,513	11,594	16,234	6,001	9,708	13,150	45	36	38	23	18	26	42	41	125	9.18
Philippines	6,056	21,070	28,446	3,032	8,912	22,321	15	23	65	7	13	23	35	53	211	17.41
Singapore	1,053	1,729	2,848	998	1,460	2,543	15	10	12	1	1	1	32	77	114	169.50
Sri Lanka	1,131	2,623	4,616	779	1,971	4,109	19	41	62	14	11	19	81	30	87	23.66
Syrian Arab Rep	1,826	2,996	4,678	1,495	2,229	3,648	22	13	11	7	13	19	20	105	68	48.04
Thailand	2,465	9,881	17,603	1,119	6,138	14,023	6	17	30	3	8	12	39	56	148	24.46
Turkey	10,940	19,283	39,952	4,438	16,064	30,490	9	31	46	16	29	32	43	112	109	79.47
Yemen Arab Rep	432	1,414	2,389	346	1,316	2,155	21	27	47	4	7	25	18	21	126	13.58
Yemen, People's Dem Rep	329	808	1,724	239	749	1,669	44	75	178	0	18	38	0	15	31	100.59
EUROPE	13,670	65,569	125,828	10,202	50,794	103,842										85.99
Greece	4,603	9,522	21,691	2,669	6,663	17,437	10	17	37	11	14	34	105	70	128	267.52
Hungary	43	8,989	18,957	43	6,739	15,931	0	30	63	0	17	27	0	148	102	298.22
Malta	74	167	295	53	121	112	9	10	7	0	0	1	85	73	4,267	0.87
Poland	1,740	15,051	42,135	1,740	15,051	35,569	X	24	56	3	4	13	79	36	390	13.09
Portugal	2,619	12,812	17,615	2,322	8,963	14,922	14	41	43	7	31	38	24	80	174	272.87
Romania	288	10,003	6,662	288	7,797	5,425	1	16	X	0	21	X	12	93	341	20.87
Yugoslavia	4,303	9,025	18,473	3,087	5,460	14,446	7	9	24	5	6	13	91	130	676	13.38
OCEANIA	642	1,402	2,132	471	1,120	1,977										43.77
Fiji	121	302	362	87	265	334	12	23	30	3	6	11	40	60	351	24.65
Papua New Guinea	477	999	1,576	341	768	1,471	21	34	50	4	10	13	58	52	100	47.51
Solomon Islands	9	28	89	9	23	85	14	18	77	0	0	9	0	2	53	52.20
Vanuatu	5	4	25	5	4	15	X	5	X	X	1	X	60	16	29	32.02
Western Samoa	31	69	80	29	60	72	X	54	72	12	12	14	24	37	68	39.09

Source: The World Bank.
Note: 0 = zero or less than half of the unit of measure; X = not available.
For additional information, see Sources and Technical Notes.

Table 15.3 World Trade

		Trade (millions $US)											
		Food, Live Animals, etc.		Raw Materials		Fuels		Manufactures		Services (1987)		Current Account Balance	
	Year {a}	Imports	Exports	Imports	Exports	Imports	Exports	Imports	Exports	Imports	Exports	1977	1987
WORLD													
AFRICA													
Algeria	1987	1,754.3	29.8	629.7	41.3	160.4	7,973.5	4,484.2	141.3	3,302.5	1,012.4	(2,323.0)	(405.6)
Angola	1981	435.2	98.1	119.2	7.3	12.9	1,539.4	1,038.0	229.5	X	X	X	X
Benin	1982	117.9	10.4	6.4	10.3	22.5	1.7	328.2	20.3	81.0	43.0	(56.7)	(208.0)
Botswana {b}	X	X	X	X	X	X	X	X	X	553.9	284.1	25.2	597.2
Burkina Faso	1984	79.2	11.9	15.6	59.8	37.3	0.0	120.3	6.9	X	X	(83.0)	X
Burundi	1987	19.2	64.3	7.9	5.5	20.5	0.0	153.8	14.5	109.8	13.7	16.2	(132.0)
Cameroon	1987	250.1	322.6	37.4	209.7	24.0	145.2	1,437.5	151.9	1,385.0	403.0	(85.6)	(1,112.0)
Cape Verde	1980/84	27.6	2.2	2.7	0.3	6.2	42.8	31.2	4.3	X	X	X	X
Central African Rep	1980	16.4	34.1	2.5	48.0	1.4	0.0	60.1	33.1	174.9	53.8	(18.9)	(96.0)
Chad	X	X	X	X	X	X	X	X	X	219.2	57.0	(28.5)	(83.3)
Comoros	1980	10.0	7.1	0.0	0.4	6.4	0.0	0.0	4.1	39.3	11.6	X	(22.6)
Congo	1985	106.8	15.8	13.2	19.5	18.0	1,013.9	442.3	38.1	801.4	133.1	(193.3)	(244.6)
Cote d'Ivoire	1985	293.9	1,730.5	49.2	405.8	381.3	259.1	3,333.0	273.5	2,040.5	663.2	(176.7)	(623.8)
Djibouti	1979	47.5	0.7	1.1	0.0	7.4	0.0	112.1	10.7	X	X	X	X
Egypt	1987	3,997.4	492.0	1,555.5	809.3	449.8	1,555.0	10,223.0	1,495.3	6,231.0	4,021.0	(1,074.0)	(2,705.0)
Equatorial Guinea	X	X	X	X	X	X	X	X	X	X	X	X	X
Ethiopia	1985	245.9	235.2	82.7	65.5	146.2	33.1	513.8	4.0	359.8	248.9	(47.3)	(263.6)
Gabon	1983	118.2	6.2	16.5	208.8	12.1	1,172.4	538.7	87.9	X	X	98.9	(209.7)
Gambia, The	X	X	X	X	X	X	X	X	X	5.1	X	(7.8)	6.3
Ghana	1983/82	73.4	434.2	19.5	198.6	54.9	69.3	394.5	14.3	X	X	(79.7)	(275.0)
Guinea	X	X	X	X	X	X	X	X	X	X	X	X	X
Guinea-Bissau	1980	10.7	4.7	1.7	5.2	3.4	0.0	38.7	0.8	33.5	X	X	(25.8)
Kenya	1986	92.1	839.0	114.8	90.6	300.2	134.9	1,138.5	151.4	828.3	830.9	27.5	(496.7)
Lesotho {b}	X	X	X	X	X	X	X	X	X	55.0	305.0	(9.2)	(12.0)
Liberia	1984	86.1	30.9	11.1	412.6	71.7	0.0	194.2	4.5	X	X	(18.6)	X
Libya	1982	1,053.0	0.0	258.5	0.0	103.7	15,513.3	5,760.3	57.8	X	X	2,159.1	X
Madagascar	1986	52.6	256.4	13.5	28.1	83.2	7.0	224.3	25.1	397.0	106.0	(16.0)	(135.0)
Malawi	1983	20.5	225.1	13.3	4.4	56.0	0.0	220.6	14.2	216.5	28.2	(61.8)	(23.7)
Mali	1982	81.8	84.6	6.9	142.4	110.6	0.0	202.6	6.5	X	X	5.4	(111.0)
Mauritania	1975/84	48.3	148.2	5.4	146.9	1.3	0.0	96.5	0.0	302.0	68.0	(122.2)	(73.0)
Mauritius	1985	89.8	216.3	45.0	2.2	75.1	0.0	313.4	215.8	315.0	328.0	(78.9)	72.0
Morocco	1987	476.9	773.2	708.2	588.6	739.6	77.0	2,304.0	1,368.1	1,994.6	1,468.6	(1,855.1)	164.1
Mozambique, People's Rep	1984	121.6	57.7	0.0	12.1	90.9	5.4	124.3	11.2	310.0	X	X	(372.0)
Niger	1982	125.8	16.6	24.4	279.9	54.3	5.1	261.3	29.5	176.9	42.0	(95.6)	(66.9)
Nigeria	1984	1,103.4	307.0	431.3	20.2	68.1	11,371.5	4,251.5	228.2	2,343.0	307.0	(1,009.1)	(380.0)
Rwanda	1985	44.9	110.6	32.8	18.3	44.9	0.0	172.8	2.0	167.3	55.0	20.8	(130.7)
Senegal	1984	206.2	148.7	92.8	189.8	282.3	98.6	405.4	97.3	803.0	560.0	(66.9)	(316.0)
Sierra Leone	1986/84	80.0	40.9	11.0	58.6	30.8	2.3	154.7	45.0	X	X	(49.4)	X
Somalia	1981	96.1	148.2	40.6	2.9	11.2	0.3	365.0	0.0	78.1	X	(32.8)	247.7
South Africa {b}	1985	496.9	765.2	627.6	2,048.5	59.5	1,457.9	9,126.8	12,070.9	7,539.4	3,168.8	533.9	3,026.8
Sudan	1982	227.9	198.6	13.1	277.1	348.1	14.0	663.9	7.5	832.0	X	(291.9)	(422.0)
Swaziland {b}	X	X	X	X	X	X	X	X	X	143.1	143.9	11.4	39.7
Tanzania	1981	48.5	343.3	26.1	148.3	267.0	1.0	524.7	71.2	249.0	101.6	(70.2)	(128.4)
Togo	1985	67.4	40.4	2.2	131.5	18.1	2.9	176.1	12.2	243.6	142.4	(87.3)	(73.2)
Tunisia	1987	310.3	193.2	379.8	143.6	326.8	508.4	2,004.9	1,307.2	1,120.4	1,260.9	(520.4)	(62.4)
Uganda	X	X	X	X	X	X	X	X	X	146.1	24.7	68.1	(107.0)
Zaire	1978	164.1	251.8	30.4	572.5	60.2	12.6	541.5	61.8	1,372.0	203.6	(1,451.2)	(705.0)
Zambia	1982	53.1	1.7	30.2	951.5	207.7	0.0	707.0	27.8	345.0	52.0	(217.2)	21.0
Zimbabwe	1985	21.2	367.9	32.0	236.9	201.9	9.6	301.4	214.0	572.0	192.0	(14.0)	50.0
NORTH & CENTRAL AMERICA													
Barbados	1986	87.1	43.0	25.1	1.3	60.7	45.0	420.3	188.1	X	X	(51.4)	(82.8)
Canada	1987	5,039.3	8,024.9	4,102.3	18,825.9	4,248.1	9,382.3	73,420.1	56,652.6	34,488.5	16,187.7	(4,112.5)	(7,963.1)
Costa Rica	1984	96.9	682.9	56.4	25.5	166.7	18.1	766.3	223.7	541.4	389.8	(225.6)	(225.4)
Cuba	1986	853.6	5,415.2	539.2	407.7	3,075.3	315.5	4,694.5	159.0	X	X	X	X
Dominican Rep	1985	83.4	334.5	101.8	0.0	439.7	0.0	623.0	137.2	X	X	(128.6)	X
El Salvador	1984	135.8	415.7	72.7	29.5	497.1	16.6	608.4	152.9	372.0	355.0	30.8	127.0
Guatemala	1984	81.3	619.1	67.1	191.4	485.7	26.2	837.7	256.9	467.9	189.0	(31.9)	(464.1)
Haiti	1984	95.8	63.9	46.8	2.7	60.8	0.0	268.7	100.6	222.4	116.0	(37.5)	(31.2)
Honduras	1985	73.5	579.7	20.9	86.7	226.3	5.9	552.9	27.1	429.2	131.0	(128.7)	(183.3)
Jamaica	1984	197.3	141.7	73.5	489.8	352.0	18.3	521.4	92.7	849.7	932.3	(42.1)	(96.2)
Mexico	1985	1,365.2	1,940.8	1,985.0	1,176.9	705.5	14,639.8	12,096.2	6,607.0	14,320.0	9,104.0	(1,849.4)	3,884.0
Nicaragua	1984	84.2	208.9	39.2	145.6	146.0	0.1	555.7	31.9	X	X	(181.9)	X
Panama	1985	148.2	228.9	37.5	11.6	293.0	21.8	904.7	38.8	2,612.9	3,401.6	(155.4)	341.5
Trinidad and Tobago	1987	239.5	65.8	91.3	10.3	52.4	1,041.4	835.5	344.9	X	X	174.1	(184.0)
United States	1987	26,675.6	23,191.3	20,717.6	23,976.4	46,742.8	7,803.5	328,271.1	197,596.1	155,530.0	175,270.0	(14,490.0)	(153,950.0)
SOUTH AMERICA													
Argentina	1987	260.3	3,012.9	552.3	1,249.0	664.3	85.4	4,340.9	2,012.8	7,076.0	2,243.0	1,126.0	(4,285.0)
Bolivia	1984/85	66.7	25.2	17.4	270.3	2.2	374.5	330.2	2.5	577.0	145.5	(117.9)	(484.6)
Brazil	1987/85	1,195.1	7,812.9	1,390.1	4,692.7	5,395.4	1,624.5	8,598.0	11,464.0	14,996.0	2,467.0	(5,112.0)	(1,275.0)
Chile	1986	122.8	1,165.3	180.3	2,604.6	440.5	2.5	2,220.5	385.2	3,421.0	1,264.0	(551.0)	(811.0)
Colombia	1986	259.3	3,433.5	368.7	216.8	153.2	665.0	3,070.9	792.7	3,976.0	1,404.0	375.4	255.0
Ecuador	1984	142.3	735.2	167.0	24.1	27.2	1,797.4	1,379.2	20.9	1,421.0	362.0	(343.4)	(1,176.0)
Guyana	1979	43.8	137.3	12.1	135.4	63.1	0.0	171.1	17.1	X	X	(97.5)	(170.0)
Paraguay	1986	52.1	59.7	11.3	151.8	124.0	0.0	390.6	21.1	531.7	250.8	(58.7)	(411.1)
Peru	1986/84	511.7	400.9	169.2	1,098.7	68.1	651.6	1,616.8	373.3	X	1,029.0	(922.4)	(1,914.0)
Suriname	X	X	X	X	X	X	X	X	X	73.8	81.7	(3.5)	75.7
Uruguay	1987	86.8	373.4	86.0	290.6	178.7	2.1	790.4	525.0	709.2	467.6	(167.1)	(124.4)
Venezuela	1985/83	730.4	86.2	833.3	330.8	176.2	13,838.8	5,678.3	187.5	4,833.0	2,123.0	(3,179.0)	(1,125.0)

Table 15.3

		Trade (millions $US)											
		Food, Live Animals, etc.		Raw Materials		Fuels		Manufactures		Services (1987)		Current Account Balance	
	Year {a}	Imports	Exports	Imports	Exports	Imports	Exports	Imports	Exports	Imports	Exports	1977	1987
ASIA													
Afghanistan	1981	100.3	225.5	26.3	94.6	112.1	272.6	383.7	100.9	X	X	X	X
Bahrain	1985	252.1	0.0	54.5	178.3	1,449.0	2,553.5	1,336.7	157.6	X	X	(324.0)	X
Bangladesh	1987	572.1	203.5	346.4	97.1	352.4	15.2	1,301.7	878.7	413.1	261.8	(174.4)	(309.4)
Bhutan	X	X	X	X	X	X	X	X	X	X	X	X	(56.2)
China	1987	2,707.5	4,969.7	4,412.1	4,332.8	540.3	4,550.5	35,733.4	25,688.5	3,676.4	5,413.1	911.0	299.8
Cyprus	1987	185.0	187.8	65.8	31.5	182.5	32.3	1,030.1	369.5	568.2	1,376.1	(89.9)	94.0
India	1985	727.6	2,211.1	2,331.0	991.5	4,297.8	540.4	8,867.1	5,245.8	4,516.0	4,911.0	1,967.0	(3,750.0)
Indonesia	1986	637.5	1,841.1	996.6	1,984.3	1,105.1	8,097.0	7,979.2	2,864.3	7,489.0	1,551.0	(46.7)	(1,837.0)
Iran, Islamic Rep	X	X	X	X	X	X	X	X	X	X	X	X	X
Iraq	X	X	X	X	X	X	X	X	X	X	X	X	X
Israel	1987	787.4	794.0	698.3	443.2	997.1	1.2	9,269.2	7,236.4	6,711.0	4,693.0	(287.0)	(999.0)
Japan	1987	22,395.3	1,548.5	27,677.5	3,387.5	39,136.8	782.2	56,838.4	223,336.3	85,490.0	79,730.0	10,910.0	87,660.0
Jordan	1986	488.1	128.3	132.4	287.3	346.0	0.7	1,446.2	298.0	1,464.4	1,444.4	(10.2)	(349.5)
Kampuchea, Dem	X	X	X	X	X	X	X	X	X	X	X	X	X
Korea, Dem People's Rep	X	X	X	X	X	X	X	X	X	X	X	X	X
Korea, Rep	1987	1,644.7	2,167.7	7,098.9	704.8	5,993.0	719.8	26,188.8	43,579.5	9,034.0	10,011.0	12.0	9,854.0
Kuwait	1984	1,170.7	122.2	166.0	39.1	39.1	10,160.7	5,520.6	1,948.3	4,776.4	6,904.4	4,560.9	4,413.9
Lao People's Dem Rep	X	X	X	X	X	X	X	X	X	X	X	X	X
Lebanon	X	X	X	X	X	X	X	X	X	X	X	X	X
Malaysia	1987	1,253.1	997.9	800.5	6,273.8	899.9	3,572.5	9,635.3	7,066.3	6,757.3	3,126.6	436.4	2,336.4
Mongolia	X	X	X	X	X	X	X	X	X	X	X	X	X
Myanmar	X	X	X	X	X	X	X	X	X	X	X	X	X
Nepal	1986	62.2	39.4	22.4	14.4	59.5	0.0	350.0	99.3	103.4	213.4	20.5	(132.8)
Oman	1986	381.8	56.5	43.2	29.1	66.9	3.4	1,892.3	201.4	X	X	276.8	X
Pakistan	1987	466.4	558.9	829.9	561.0	1,026.6	27.8	3,506.3	3,030.0	2,013.0	1,031.0	(902.5)	(336.0)
Philippines	1986	533.8	904.7	331.3	1,067.6	918.5	61.5	3,608.9	2,808.0	3,573.0	3,497.0	(753.0)	(539.0)
Qatar	1982/81	198.8	0.0	73.7	0.0	12.8	5,050.2	1,656.8	329.1	X	X	X	X
Saudi Arabia	1986	3,029.5	X	505.3	X	47.4	X	15,496.3	X	X	X	11,991.2	X
Singapore	1987	2,163.7	1,477.6	1,944.3	2,113.9	5,950.2	4,523.5	22,416.0	20,477.4	6,691.9	9,970.6	(295.2)	538.9
Sri Lanka	1986	315.5	428.7	64.3	169.9	230.5	68.5	1,221.3	527.1	523.8	346.0	142.0	(378.0)
Syrian Arab Rep	1985	682.0	44.9	233.3	179.9	1,162.3	1,212.5	1,869.6	199.0	1,451.0	941.0	(232.6)	(465.0)
Thailand	1987	665.1	4,283.6	1,157.6	1,165.8	1,729.2	85.0	9,403.2	6,124.7	3,926.4	3,420.0	(1,094.1)	(585.6)
Turkey	1987	560.8	2,487.4	1,859.9	714.6	3,167.2	233.4	8,574.7	6,754.3	4,282.0	4,111.0	(3,138.0)	(984.0)
United Arab Emirates	1982/78	800.1	109.5	208.2	0.0	518.9	8,617.5	7,073.3	320.3	X	X	2,434.0	X
Viet Nam	X	X	X	X	X	X	X	X	X	X	X	X	X
Yemen Arab Rep	1985	398.1	7.3	49.1	2.3	86.9	0.0	756.2	3.4	368.0	517.0	292.0	(607.0)
Yemen, People's Dem Rep	X	X	X	X	X	X	X	X	X	216.0	115.5	(67.9)	(122.5)
EUROPE													
Albania	X	X	X	X	X	X	X	X	X	X	X	X	X
Austria	1987	1,879.3	920.4	2,461.0	2,180.0	2,358.2	479.1	25,939.5	23,583.3	14,547.5	18,849.6	(2,192.6)	(226.0)
Belgium	1987	8,361.7	8,146.0	7,635.6	5,028.5	7,688.3	3,058.0	58,912.8	66,718.5	X	X	(558.1)	2,920.0
Bulgaria	X	X	X	X	X	X	X	X	X	X	X	X	X
Czechoslovakia	1986	1,362.7	557.4	2,200.9	696.9	6,413.9	724.0	11,111.3	18,478.7	X	X	X	X
Denmark	1987	2,994.6	7,059.0	1,647.6	1,865.0	1,996.0	734.3	18,696.2	15,038.5	14,135.6	10,570.0	(1,723.0)	(2,950.8)
Finland	1987	1,101.3	441.5	1,437.7	3,058.9	2,671.6	444.4	14,649.5	16,094.6	7,507.0	4,707.8	(107.3)	(1,937.5)
France	1987	16,707.2	21,517.4	11,008.7	8,689.1	16,933.5	3,165.3	112,874.3	109,704.7	69,903.0	79,873.0	(428.6)	(4,087.9)
German Dem Rep	X	X	X	X	X	X	X	X	X	X	X	X	X
Germany, Fed Rep	1987	24,897.7	13,494.1	20,257.3	11,187.7	21,964.8	3,889.9	160,214.5	265,217.8	90,000.9	80,890.6	3,928.7	44,955.9
Greece	1987	2,295.8	1,722.2	1,011.7	856.9	1,785.9	436.3	7,814.8	3,474.0	3,364.0	4,528.0	(1,078.0)	(1,298.0)
Hungary	1987	689.6	1,680.0	945.1	706.0	1,675.0	401.7	6,544.9	6,783.1	3,093.4	2,240.9	(864.0)	(676.2)
Iceland	1987	124.3	1,070.9	55.5	176.4	117.3	0.0	1,284.3	127.8	X	X	(48.3)	X
Ireland	1987	1,621.5	4,338.7	650.6	561.7	1,005.0	115.0	10,336.4	10,955.1	6,553.3	3,027.5	(570.8)	391.0
Italy	1987	16,512.8	7,243.2	14,186.6	3,206.8	16,664.0	2,842.5	74,847.3	103,289.8	39,019.7	39,471.5	2,449.0	(1,058.6)
Luxembourg	X	X	X	X	X	X	X	X	X	X	X	X	X
Malta	1986	124.0	25.6	32.8	5.0	53.4	10.8	669.5	455.7	375.1	692.6	46.4	11.0
Netherlands	1987	11,870.5	19,067.0	6,589.0	7,493.2	9,935.9	10,089.3	62,921.0	56,232.3	36,390.7	36,639.8	1,224.7	3,372.2
Norway	1987	1,222.6	1,657.7	1,468.4	2,898.3	1,192.6	8,689.9	18,664.0	8,171.3	13,855.3	11,560.0	(5,034.7)	(4,111.3)
Poland	1987	1,128.8	1,276.2	1,195.7	1,326.0	1,872.5	1,373.7	6,646.7	8,228.9	3,836.0	1,479.0	(2,392.0)	(578.0)
Portugal	1987	1,416.6	674.5	1,402.9	948.4	1,563.4	149.2	9,054.6	7,394.7	3,342.1	3,634.6	(960.3)	641.3
Romania	X	X	X	X	X	X	X	X	X	X	X	X	X
Spain	1987	4,791.8	5,522.5	4,856.0	2,304.1	7,992.6	2,000.9	31,368.4	24,271.3	13,200.9	23,439.5	(2,144.2)	(51.1)
Sweden	1987	2,601.3	791.8	2,507.4	4,933.8	3,626.9	1,267.6	31,885.2	37,319.8	16,137.5	12,216.2	(2,189.4)	(852.9)
Switzerland	1987	3,611.4	1,380.9	2,975.0	1,664.3	2,245.1	61.2	41,725.7	42,250.5	19,257.7	32,167.7	3,394.8	5,879.3
United Kingdom	1987	16,681.3	9,187.9	12,394.6	6,207.6	9,984.6	14,364.9	115,327.4	101,367.4	106,120.3	124,872.4	(164.5)	(2,621.2)
Yugoslavia	1987	931.2	1,265.3	1,771.5	1,586.9	2,803.5	280.3	10,588.2	11,410.2	3,317.0	3,896.0	(1,346.2)	819.0
U.S.S.R.	1984	16,213.5	0.0	2,189.1	6,219.4	0.0	49,998.4	46,877.4	20,405.2	X	X	X	X
OCEANIA													
Australia	1987	1,386.2	5,749.1	1,063.4	8,170.5	1,306.4	5,064.2	23,077.8	6,299.7	16,447.6	7,238.4	(3,140.2)	(8,687.9)
Fiji	1985	73.3	126.8	15.8	11.1	100.2	47.2	251.3	32.4	222.3	242.4	(26.1)	(0.9)
New Zealand	1987	457.0	3,248.4	305.1	2,087.2	484.0	64.4	6,008.4	1,779.4	4,656.8	2,504.0	(692.3)	(1,367.7)
Papua New Guinea	1984	182.5	241.1	14.5	600.4	171.8	0.0	608.5	55.3	569.4	155.4	99.1	(325.8)
Solomon Islands	1984	13.2	25.5	1.8	64.3	14.9	0.0	35.6	2.4	X	X	5.8	(84.0)

Sources: United Nations Statistical Office; and The World Bank.
Notes:
a. When two years are given for trade statistics, the first refers to imports, the second to exports.
b. South African trade (except services) includes: Botswana, Lesotho, Namibia, South Africa, and Swaziland, all members of a customs union.
0 = zero or less than half of the unit measure; X = not available; negative numbers are shown in parentheses.
For additional information, see Sources and Technical Notes.

Table 15.4 World Commodity Indexes and Prices, 1975–88

Commodity Indexes (based on constant prices with 1979-81 = 100)

	1975	1976	1977	1978	1979	1980	1981	1982	1983	1984	1985	1986	1987	1988
33 NONFUEL COMMODITIES	101	112	123	102	105	105	91	82	89	92	81	69	63	71
Total Agriculture	100	116	134	108	106	104	91	81	89	94	81	71	59	63
Total Food	103	117	143	111	106	104	90	81	88	95	83	75	58	63
-Beverages	71	134	206	132	121	99	82	85	88	104	95	99	59	61
-Cereals	142	115	96	100	92	101	107	79	87	85	74	55	47	59
-Fats and Oils	104	110	112	112	114	96	92	76	92	110	76	50	53	65
-Other Foods	129	95	78	79	86	121	92	79	85	77	74	67	66	68
Nonfood Agriculturals	89	111	103	97	102	106	92	82	95	90	75	58	65	62
Timber	53	70	74	68	104	110	87	88	84	99	80	75	100	98
Metals and Minerals	113	110	104	92	103	105	92	84	90	86	81	63	66	84

Commodity Prices (in constant 1980 $US per unit measure) {a}

	1975	1976	1977	1978	1979	1980	1981	1982	1983	1984	1985	1986	1987	1988
Cocoa (kg), New York & London	1.98	3.21	5.41	4.23	3.61	2.60	2.07	1.75	2.20	2.53	2.35	1.83	1.62	1.19
Coffee (kg), Brazil	2.94	5.17	9.70	3.97	4.26	4.58	3.85	3.20	3.26	3.48	3.49	4.50	1.90	2.02
Tea (kg), World Average	2.21	2.41	3.84	2.72	2.36	2.23	2.01	1.95	2.41	3.64	2.07	1.70	1.38	1.34
Rice (t), Thailand	578.2	399.5	388.9	456.5	363.3	433.9	480.4	295.6	286.6	265.6	225.10	185.60	186.61	226.23
Grain Sorghum (t), U.S.	178.2	165.2	126.3	116.5	118.8	128.9	125.8	109.5	133.3	124.6	107.40	72.80	58.96	73.90
Maize (t), U.S.	190.5	176.5	136.1	125.1	126.6	125.3	130.2	110.3	140.8	143.2	117.00	77.20	61.34	80.23
Wheat (t), Canada	288.7	234.1	165.4	167.5	189.0	190.8	195.4	168.0	175.5	174.3	180.70	141.60	108.19	134.74
Sugar (kg), World	0.72	0.40	0.26	0.21	0.23	0.63	0.37	0.19	0.19	0.12	0.09	0.12	0.12	0.17
Beef (kg), U.S.	2.11	2.48	2.15	2.66	3.16	2.76	2.46	2.41	2.53	2.40	2.25	1.85	1.93	1.89
Lamb (kg), New Zealand	2.27	2.42	2.36	2.70	2.60	2.89	2.73	2.40	2.00	2.02	1.92	1.90	1.75	1.81
Bananas (kg), Any Origin	0.39	0.40	0.39	0.36	0.36	0.38	0.40	0.38	0.44	0.39	0.40	0.34	0.29	0.27
Black Pepper (kg), Any Origin	3.19	3.08	3.59	2.91	2.32	1.99	1.58	1.57	1.75	2.40	3.98	4.26	4.31	X
Copra (t), Philippines	408.1	431.7	574.7	584.4	737.6	453.8	377.0	317.1	513.5	748.2	402.5	174.6	250.4	298.7
Coconut Oil (t), Philippines & Indonesia	626.6	656.2	826.0	848.7	1,079.5	673.8	567.1	468.6	755.6	1,216.7	615.2	261.9	358.2	424.1
Groundnut Meal (t), Any Origin	222.9	276.3	311.4	254.7	231.4	240.3	236.7	190.8	207.0	187.6	153.3	145.5	131.3	157.6
Groundnut Oil (t), Nigeria	1,364.7	1,163.3	1,217.6	1,340.6	974.5	858.8	1,037.8	590.2	735.9	1,071.3	943.7	501.8	405.2	442.8
Linseed (t), Canada	538.2	478.8	388.4	310.1	366.0	350.9	352.5	300.7	287.1	314.0	285.7	183.4	136.9	220.7
Linseed Oil (t) Any Origin	1,116.2	857.1	659.9	539.4	700.7	697.1	656.7	523.3	501.7	602.7	654.8	369.5	254.4	391.8
Palm Kernels (t), Nigeria	329.3	361.1	466.1	451.9	548.3	345.1	315.7	267.4	378.2	556.4	303.4	125.2	146.7	200.4
Palm Oil (t), Malaysia	691.4	638.2	757.1	745.6	716.9	583.5	567.9	449.1	519.0	768.2	522.4	226.6	277.9	328.0
Soybeans (t), U.S.	350	363	400	333	327	296	287	247	292	297	234	183	175	228
Soybean Oil (t), Any Origin	986	688	823	754	726	597	504	451	545	763	596	302	271	348
Soybean Meal (t), U.S.	247	311	329	265	267	262	251	220	246	208	164	163	164	201
Fish Meal (t), Peru	390	590	649	509	433	504	466	356	469	393	292	283	310	408
Cotton (kg), Index	1.85	2.66	2.22	1.95	1.85	2.05	1.84	1.61	1.92	1.88	1.37	0.93	1.34	1.05
Burlap (meter), U.S.	0.35	0.30	0.31	0.27	0.33	0.40	0.27	0.27	0.30	0.39	0.36	0.22	0.22	0.23
Jute (t), Bangladesh	590.8	464.1	458.4	540.0	424.8	308.0	274.6	288.4	312.8	559.3	607.8	238.2	319.2	329.9
Sisal (t), East Africa	924	736	733	609	775	765	642	598	583	615	548	453	415	413
Wool (kg), New Zealand	4.37	5.35	5.11	4.66	4.86	4.60	4.25	3.94	3.77	3.83	3.71	2.92	3.66	4.35
Natural Rubber (kg), New York	10.49	13.71	13.10	13.75	15.61	16.24	12.46	10.11	12.82	11.55	9.64	8.33	9.05	9.66
Logs (cubic meter), Malaysia	X	X	132.3	121.0	186.4	195.5	154.8	157.3	150.3	176.2	142.0	133.4	179.4	176.3
Plywood (sheet), Philippines	1.94	2.32	2.36	2.35	2.88	2.74	2.44	2.35	2.38	2.39	2.20	2.41	3.23	2.69
Sawnwood (cubic meter), Malaysia	265.0	263.9	220.1	255.2	371.8	365.1	312.6	304.7	315.0	323.3	288.1	234.7	223.7	230.1
Tobacco (t), India	2,416	2,215	2,386	2,099	2,336	2,300	2,338	2,432	2,324	2,097	2,034	1,688	1,499	1,555
Coal (t), U.S.	86.4	84.1	77.5	69.2	62.0	55.7	57.8	56.9	59.5	60.9	58.0	46.6	X	X
Petroleum (barrel), OPEC	16.7	18.1	17.7	15.8	19.0	29.4	33.0	34.3	30.5	30.6	29.3	12.1	X	X
Gasoline (t), Europe	191.6	216.5	188.0	198.8	367.3	358.0	352.3	326.5	293.2	271.2	266.1	128.7	X	X
Jet Fuel (t), Europe	180.3	187.8	185.0	181.6	384.1	349.3	333.7	325.3	287.4	275.7	276.1	142.3	X	X
Gas Oil (t), Europe	159.2	168.6	168.3	159.8	340.9	307.1	297.2	292.3	256.9	251.1	250.1	125.5	X	X
Fuel Oil (t), Europe	98.9	107.5	108.9	94.0	146.5	170.2	182.6	165.5	168.7	187.8	158.2	64.7	X	X
Aluminum (t), Europe	1,099	1,353	1,416	1,298	1,667	1,730	1,331	1,071	1,548	1,445	1,160	1,112	1,303	1,910
Bauxite (t), Jamaica	40.29	42.70	44.00	42.61	40.13	41.20	39.80	36.33	35.92	34.77	31.28	24.69	X	X
Copper (t), London	1,970	2,199	1,870	1,696	2,177	2,183	1,733	1,493	1,648	1,453	1,478	1,212	1,420	1,953
Lead (t), London	664	699	883	822	1,325	906	723	551	440	468	408	358	484	493
Tin (t), Malaysia	10,656	11,730	15,304	15,547	16,252	16,437	13,992	13,066	13,493	13,124	12,032	5,433	5,421	5,327
Zinc (t), New York	1,368	1,292	1,083	849	901	825	977	856	944	1,130	928	739	749	996
Iron Ore (t), Brazil	36.0	34.4	30.9	24.1	25.6	26.7	24.2	26.1	24.8	24.3	23.7	19.4	18.0	17.4
Manganese Ore (10 kg), India	2.19	2.28	2.11	1.78	1.51	1.57	1.67	1.66	1.57	1.51	1.47	1.22	1.07	X
Nickel (t), Canada	7,277	7,808	7,433	5,729	6,563	6,519	5,924	4,881	4,837	5,008	5,108	3,422	800	1,034
Steel (t), U.S.	461.9	464.2	460.9	452.1	444.4	452.8	505.4	539.5	565.3	569.1	563.2	354.8	X	X
Phosphate Rock (t), Morocco	106.7	56.5	43.6	36.0	36.2	46.7	49.3	42.8	38.2	40.4	35.3	30.2	25.1	27.0
Diammonium Phosphate (t), U.S.	386.9	188.4	190.0	173.7	212.0	222.2	194.0	184.5	190.0	199.3	176.2	136.0	140.3	147.5
Potassium Chloride (t), Canada	129.5	86.3	72.9	70.1	84.1	115.7	111.8	82.3	78.0	88.2	87.6	60.7	55.9	65.7
Triple Superphosphate (t), U.S.	322	143	139	122	161	180	160	139	140	138	126	107	112	119
Urea (t), Any Origin	315.3	175.8	182.0	179.9	189.6	222.1	214.9	160.2	140.2	180.5	142.1	94.4	94.5	99.0

Source: The World Bank.
Notes: a. Log, petroleum, and tin price series replace discontinued series for similar commodities found in previous editions.
t = metric ton; X = not available.
For additional information, see Sources and Technical Notes.

Sources and Technical Notes

Table 15.1 Gross National Product and Official Development Assistance, 1967–87

Sources: Gross national product (GNP), GNP per capita, change in GNP: The World Bank, unpublished data (The World Bank, Washington, D.C., May 1989). GNP for certain centrally managed economies: U.S. Central Intelligence Agency (CIA), *The World Factbook 1988* (CIA, Washington, D.C., 1988). Share of gross domestic product (GDP): The World Bank, *1989 World Development Report* (The World Bank, Washington, D.C., 1989). Official development assistance (ODA): Organisation for Economic Co-operation and Development (OECD), *Development Co-operation* (OECD, Paris, 1984, 1986, 1987, 1988); OECD, *Geographical Distribution of Financial Flows to Developing Countries 1981/84* and *1984/87* (OECD, Paris, 1986, 1989).

Gross national product (GNP) is the sum of two components: the *Gross domestic product* (GDP, the final output of goods and services produced by the domestic economy, including net exports of goods and nonfactor services) and *net factor income* from abroad. Net factor income from abroad is income in the form of overseas workers' remittances, interest on loans, profits, and other factor payments that residents receive from abroad less payments made for factor services (labor and capital). Most countries estimate GDP by the production method. This method sums the final outputs of the various sectors of the economy (agriculture, manufacturing, government services, etc.) from which the value of the inputs to production has been subtracted.

GNP is calculated as in the World Bank's *World Bank Atlas.* GNP in domestic currency was converted to U.S. dollars using a three-year average exchange rate, adjusted for domestic and U.S. inflation. However, the strong appreciation of the U.S. dollar through 1985 affected the conversion factor and may mask real growth in GNP and GNP per capita in some countries. Alternative measures of GNP exist. Of special interest are the United Nations' International Comparison Program (ICP) estimates of GDP (in U.S. dollars) using purchasing power parity for a basket of common goods and services instead of exchange rates. These ICP GDP estimates tend to raise GNP estimates for the poorest countries and lower them slightly for most developed countries. Traditional measures of the health of an economy, while counting the income generated by the exploitation of natural resources, do not take into account their depletion or destruction. Many believe that this oversight leads to an overestimation of economic growth and its sustainability and that these factors should be included in any calculation of GDP.

In many centrally managed economies, such as the U.S.S.R., the net material product (NMP) is a primary economic statistic instead of GNP. The estimation of GNP in U.S. dollars for such economies is difficult and is an ongoing research project of the World Bank. Estimates of GNP for centrally managed economies, especially members of the Council for Mutual Economic Assistance (CMEA), are included as an attempt to show the relative magnitudes of those important economies. These GNP estimates, however, are not directly comparable to estimates made by the World Bank for other economies. Few organizations are willing to publish estimates of GNP for CMEA countries.

The *average annual change in real GNP* was calculated by fitting a least-square regression line to the logarithmic values for GNP in constant prices.

Share of GDP does not always add to 100 percent because of rounding errors.

Net *average annual official development assistance* (ODA) is the net amount of disbursed grants and concessional loans given or received by a country. Grants include gifts, in money, goods, or services, for which no repayment is required. A concessional loan has a grant element of 25 percent or more. The grant element is the amount by which the face value of the loan exceeds its present market value because of below-market interest rates, favorable maturity schedules, and repayment grace periods. Nonconcessional loans are not a component of ODA.

ODA contributions are shown as negative numbers (in parentheses); receipts are shown as positive numbers. Data for donor countries include contributions directly to developing countries and through multilateral institutions. The GNP data used to calculate *ODA as a percentage of GNP* were derived by using single-year exchange rates, not the three-year exchange rates discussed previously.

Sources of ODA include the development assistance agencies of OECD, CMEA, and OPEC members. Grants and concessional loans to and from multilateral development agencies are also included in contributions and receipts.

OECD gathers ODA data through questionnaires and reports from countries and multilateral agencies. Only limited data are available on ODA flows among developing countries.

Data for OPEC countries–and some other countries that both contribute to and receive ODA–were determined by subtracting ODA contributions from ODA receipts for each country, where possible. Sufficient data also exist to subtract ODA contributions from ODA receipts for several other countries, including China and India, the two largest donors among developing countries. Limited data are available on flows of ODA from CMEA countries; thus, data on ODA donations in this table are restricted to those of the U.S.S.R., which typically comprise 85–90 percent of the CMEA total. Data on net ODA from the U.S.S.R. are available for countries receiving the largest amounts.

These data have been added to their net ODA receipts.

Table 15.2 External Debt Indicators, 1977–87

Sources: The World Bank, unpublished data (The World Bank, Washington, D.C., May 1989).

The World Bank operates the Debtor Reporting System (DRS), which compiles reports supplied by 111 of the Bank's member countries. Countries submit detailed reports on the annual status, transactions, and terms of the long-term external debt of public agencies and of publicly guaranteed private debt. Additional data are drawn from the World Bank, the International Monetary Fund (IMF), regional development banks, government lending agencies, and the Creditor Reporting System (CRS). The CRS is operated by the Organisation for Economic Co-operation and Development (OECD) to compile reports from the members of its Development Assistance Committee.

Total external debt includes long-term debt, use of IMF credit, and short-term debt. Long-term debt is an obligation with a maturity of at least one year that is owed to nonresidents and is repayable in foreign currency, goods, or services. Long-term debt is divided into long-term public debt and long-term publicly guaranteed private debt. Short-term debt is public or publicly guaranteed private debt that has a maturity of one year or less. This class of debt is especially difficult for countries to monitor. Only a few countries supply these data through the DRS; the World Bank supplements these data with creditor-country reports, information from international clearinghouse banks, and other sources to derive rough estimates of short-term debt. Use of IMF credit refers to all drawings on the Fund's General Resources Account. Use of IMF credit is converted to dollars by using the average Special Drawing Right exchange rate in effect for the year being calculated.

Private debt is an external obligation of a private debtor that is not guaranteed by a public entity. Data for this class of debt are less extensive than those for public debt; many countries do not report these data through the DRS. Complete or partial data are currently available from 24 countries, and World Bank estimates are available for 25 others. These 49 countries account for the majority of the private nonguaranteed debt of developing countries. In 1987, the private nonguaranteed debt of the 111 members of the DRS was estimated at about 10 percent of all long-term debt.

Disbursed long-term public debt is outstanding public and publicly guaranteed long-term debt. Public debt is an obligation of a national or subnational government or its agencies and autonomous bodies. Publicly guaranteed debt is an external obligation of a private debtor that is guaranteed for repayment by a public entity.

Long-term public debt as a percentage of GNP is calculated using the disbursed long-term public debt described previously. GNP is defined in the Technical Note for Table 15.1. The GNP data used to derive debt-to-GNP ratios were converted from local currencies at a single-year exchange rate, rather than the three-year average exchange rate used to determine total GNP in Table 15.1. Total debt service comprises actual interest payments and repayments of principal made on the disbursed long-term public debt in foreign currencies, goods, and services in the year specified. Exports of goods and services are the total value of goods and all services sold to the rest of the world. Current borrowing is the total long-term debt disbursed during the specified year. *Current borrowing per capita* was calculated using the 1987 mid-year population.

Debt data are reported to the World Bank in the units of currency in which they are payable. The World Bank converts these data to U.S. dollar figures, using the IMF par values, central rates, or the current market rates, where appropriate. Debt service data are converted to U.S. dollar figures at the average exchange rate for the given year. Comparability of data among countries and years is limited by variations in methods, definitions, and comprehensiveness of data collection and reporting. Refer to the World Bank's *World Debt Tables* for details.

Table 15.3 World Trade

Sources: Food/Live Animals/Etc., Raw Materials, Fuels, and Manufactures: United Nations, personal communication and the *International Trade Statistics Yearbook* (U.N., New York, 1989). Services and Current Account Balances: The World Bank, *World Tables*, 1988–89 Edition (The World Bank, Washington, D.C., 1989).

The United Nations Statistical Office, Transport and Trade Division, compiles detailed data on international trade from data (usually customs) submitted by reporting countries. Descriptive categories are used to convert the data to the lowest possible level of the Standard International Trade Classification (SITC). For example, cut flowers would be classified 29271 (cut flowers), which is under 2927 (cut flowers and foliage), part of 292 (crude vegetable materials, not elsewhere stated), within Division 29 (crude animal and vegetable materials, not elsewhere stated), and within Section 2 (crude materials, excluding fuels).

Trade figures have been aggregated to show imports and exports of larger categories that should help to characterize each country's economy (e.g., as agricultural, extractive, or industrial). The categories shown in this table follow the scheme used by the World Bank in its *World Tables*, which aggregates SITC data by main section divisions (except that the World Bank's category "Nonfuel Primary Products" is divided here into *food/live animals/etc.* and *raw materials*). *Food/live animals/etc.* corresponds to SITC Sections 0 (food and live animals) and 1 (beverages and tobacco); *raw materials* corresponds to SITC Sections 2 (crude materials, excluding fuels) and 4 (animal and vegetable oils, fats, and waxes), and Division 68 (nonferrous metals); *fuel* corresponds to SITC Section 3; and *manufactures* corresponds to SITC Sections 5 (chemicals), 6 (basic manufactures, excluding nonferrous metals), 7 (machines and transport equipment), 8 (miscellaneous manufactures), and 9 (goods not classified).

Trade in *services* includes nonfactor and factor services. Nonfactor services (i.e., transport services, travel, and other transactions not otherwise accounted for, including transactions with nonresidents by the government or its personnel abroad, and transactions with foreign governments by private citizens) are a growing component of international trade. Factor services are the services of labor and capital (e.g., income from direct foreign investment, dividends, interest, rents, and labor payments). These data on services stand apart from the other data on trade because a transport component (which is a nonfactor service) is often included by countries in their statistics on the importation of goods.

The *current account balance* is the sum of net exports of goods and nonfactor services, net factor service income, and net transfers. In principle, world financial flows would net to zero if all countries' current account balances were known accurately.

Table 15.4 World Commodity Indexes and Prices, 1975–88

Sources: The World Bank, *Commodity Trade and Price Trends* (The World Bank, Washington, D.C., 1988); The World Bank, unpublished data (The World Bank, Washington, D.C., June 1989).

Price data are compiled from major international market places for standard grades of each commodity. For example, the gasoline series refers to 91/92 octane regular gasoline, in barges, f.o.b. (free on board) Rotterdam.

The 1980 U.S. constant dollar figures were derived by converting current average monthly prices in local currencies to U.S. dollars using the monthly average exchange rate. These monthly average U.S. dollar figures were then averaged to produce an annual average dollar figure, which was adjusted to 1980 constant dollars using the Manufacturing Unit Value (MUV) index. The MUV index is a composite price index of all manufactured goods traded internationally.

The aggregate price indexes have the following components:
1. 33 Nonfuel commodities: individual items listed under items 4–10.
2. Total agriculture: total food and nonfood agricultural products.
3. Total food: beverages, cereals, fats and oils, other foods.
4. Beverages: coffee, cocoa, tea.
5. Cereals: maize, rice, wheat, grain sorghum.
6. Fats and oils: palm oil, coconut oil, groundnut oil, soybeans, copra, groundnut meal, soybean meal.
7. Other food: sugar, beef, bananas, oranges.
8. Nonfood agricultural products: cotton, jute, rubber, tobacco.
9. Timber: logs.
10. Metals and minerals: copper, tin, nickel, bauxite, aluminum, iron ore, manganese ore, lead, zinc, phosphate rock.

The commodity prices reported here are specific to the markets named. The commodities themselves are often defined more specifically than suggested by the table (e.g., Meranti logs [Sabah SQ best quality]; Imported-frozen-boneless, 90-percent-visible-lean beef; Santos 4 coffee).

16. Population and Health

By mid-1990, the world's human population will total 5,292,200,000, an increase of 75 percent since 1960. At projected growth rates, the population will reach 8,488,600,000 by the year 2025–a 60-percent increase over 1990 and nearly triple the 1960 population. Demands on resources also grow, causing pressure on, for example, wildlife and wildlife habitat, forests, and coastal areas. This chapter presents data on the demographic and health considerations that underlie decisions in these areas and others like food and agriculture, education, health care, and urban policy.

While the world's population is still growing (nearly 88 million people annually), at least the annual rate of growth (percentage increase) has slowed, from 2.06 percent in 1965–70 to 1.73 percent in 1985–90. This slowdown in the rate of population growth is the result of lower fertility rates that accompany economic development (the demographic transition), and of increasing use of contraceptive practices in poorer countries. However, a large portion of the population of most developing countries is in its peak reproductive years; thus, the number of children born each year will continue to increase, albeit at a lower percentage rate than in previous years.

Table 16.1, which presents data on world population, as well as its labor force, shows that Africa has the world's highest annual rates of growth (3 percent overall, with two thirds–21–of its countries above 3 percent), followed by South America (over 2 percent) and Central America (above 2 percent in most countries). Asia's rate of increase—1.85 percent—is slightly higher than the world average. Even there, the lower birth rates in China, Japan, and 7 other countries are offset by 14 countries with growth rates above 3 percent, as well as 14 countries with growth rates between 1.85 and 3 percent.

The relatively high growth rates in developing countries result partly from declines in infant and child mortality rates ascribable to improving public health practices. Table 16.2 shows increases in life expectancy in all areas of the world. Children born today can expect to live longer than those born 20 years ago: 7.8 years longer in Africa, 5.3 years in North and Central America, 7.1 years in South America, 7.8 years in Asia, 3.4 years in Europe, 2.8 years in the U.S.S.R., and 4.9 years in Oceania.

Increases in life expectancy and continuing high fertility rates in much of the developing world have changed the age structure of the population. World-wide, people 15 years of age and under will form 32.4 percent of the population in 1990, a decrease of over 5 percent from 1970. In much of the developing world outside the Americas, however, the percentage of the population in this age group continues to rise. This has significant implications for the economic, health, and welfare policies of those countries and of the world.

Life expectancy, especially in developing countries, increased between 1965–70 and 1985–90 because of declining infant and child mortality. Table 16.3 shows, however, that child death rates remain high in Africa and parts of Asia. Over 30 percent of children in Afghanistan will die before age 5; 19 other countries (all in Africa) will lose over 20 percent of their children by age 5. Globally, over 14 million children under the age of 5 die each year, according to United Nations estimates. Maternal death rates (for which data are incomplete) remain high in many developing countries.

Much illness and many deaths could be prevented by ensuring access to safe drinking water, adequate sanitation, and health care. Table 16.4 shows the availability of these basic requirements in many countries.

Table 16.5 shows that over half the world's children were immunized by 1988, a dramatic improvement over the 5 percent of a decade before. It also presents the extent of use of oral rehydration salts (ORS), a simple and inexpensive procedure to restore bodily fluids and thereby treat diarrheal illnesses. (Use of ORS is growing; according to UNICEF, the percentage use per 1,000 episodes of diarrhea in children under 5 years of age in 1986 was nearly double that in 1984.) The table also contains information on the regional variations in access to health care providers.

Only about 50 countries, half of them in Europe, report death registry data to the World Health Organization. The available information, presented in Table 16.6, shows that death rates from infection are higher in the developing countries than in Europe, while death rates for cancer (especially for men) are higher in developed countries than in the developing world. Over the past 35 years, death rates from malignant neoplasms of the trachea, bronchus, and lung have increased significantly (especially among women). While death rates from suicide and self-inflicted injury have also increased, death rates have decreased from ischemic heart disease, cerebrovascular disease, and motor vehicle accidents.

Table 16.1 Size and Growth of Population and Labor Force,

	Population (millions)			Average Annual Population Change (percent)			Average Annual Increment to the Population (thousands)			Average Annual Growth of the Labor Force (percent)		
	1960	1990	2025	1965-70	1975-80	1985-90	1965-70	1975-80	1985-90	1960-70	1970-80	1980-90
WORLD	3,019.4	5,292.2	8,466.5	2.06	1.74	1.73	72,398	74,081	87,666	1.7	2.1	1.9
AFRICA	281.1	647.5	1,581.0	2.63	2.95	3.00	8,928	13,185	18,015	2.2	2.5	2.5
Algeria	10.8	25.4	50.6	2.85	3.06	3.12	365	530	733	0.3	3.2	3.7
Angola	4.8	10.0	24.7	1.52	3.39	2.70	82	241	253	1.2	2.8	1.8
Benin	2.3	4.7	13.0	2.06	2.77	3.15	53	90	138	1.5	2.0	2.1
Botswana	0.5	1.3	3.4	2.54	3.54	3.51	15	29	41	1.2	3.0	3.3
Burkina Faso	4.5	9.0	22.7	2.14	2.30	2.67	114	151	225	1.4	1.8	2.0
Burundi	2.9	5.5	13.1	1.45	1.80	2.88	48	71	146	1.3	1.3	2.2
Cameroon	5.5	11.2	26.2	2.11	2.57	2.60	135	208	274	1.8	1.6	1.9
Cape Verde	0.2	0.4	0.9	3.04	0.87	2.81	8	3	10	3.2	1.0	3.3
Central African Rep	1.6	2.9	6.8	1.63	2.22	2.46	29	48	67	0.9	1.2	1.4
Chad	3.1	5.7	13.2	1.82	2.10	2.47	64	89	132	1.4	1.7	1.9
Comoros	0.2	0.5	1.3	2.43	3.40	3.11	6	12	15	2.1	3.1	2.5
Congo	1.0	2.0	5.0	2.18	2.46	2.73	25	35	51	1.9	2.1	1.9
Cote d'Ivoire	3.8	12.6	39.8	4.05	4.19	4.12	202	315	469	2.9	2.6	2.6
Djibouti	0.1	0.4	1.1	7.68	4.48	2.96	11	12	11	X	X	X
Egypt	25.9	54.1	94.0	2.35	2.69	2.55	733	1,046	1,296	2.0	2.1	2.6
Equatorial Guinea	0.3	0.4	1.0	1.55	1.99	2.34	4	7	10	0.9	1.1	1.4
Ethiopia	24.2	46.7	112.3	2.41	2.44	2.01	695	888	894	2.3	2.0	1.9
Gabon	0.5	1.2	2.9	0.36	4.70	3.45	2	34	37	0.5	0.8	0.7
Gambia, The	0.4	0.9	1.9	2.77	3.15	2.83	12	19	23	1.9	2.0	1.3
Ghana	6.8	15.0	37.0	1.91	1.76	3.14	157	181	436	1.5	2.4	2.7
Guinea	3.7	6.9	15.7	1.95	2.17	2.48	81	111	160	1.5	1.8	1.7
Guinea-Bissau	0.5	1.0	2.2	0.06	5.04	2.08	0	36	20	(0.6)	3.9	1.3
Kenya	6.3	25.1	77.6	3.30	3.82	4.22	350	578	955	3.3	3.6	3.5
Lesotho	0.9	1.8	4.3	2.01	2.41	2.85	20	30	47	1.4	2.0	2.0
Liberia	1.0	2.6	7.2	2.85	3.09	3.18	36	53	75	2.4	2.6	2.3
Libya	1.3	4.5	12.8	4.04	4.37	3.65	73	119	152	3.3	3.8	3.6
Madagascar	5.3	12.0	33.0	2.28	2.90	3.18	145	237	352	1.9	2.2	2.0
Malawi	3.5	8.4	22.8	2.56	3.00	3.31	109	169	257	2.1	2.3	2.6
Mali	4.6	9.4	24.1	2.15	2.19	2.94	116	146	256	1.6	1.7	2.6
Mauritania	1.0	2.0	5.0	2.17	2.46	2.73	25	36	52	1.9	1.8	2.8
Mauritius	0.7	1.1	1.5	1.83	1.91	1.25	15	17	13	2.8	2.5	2.9
Morocco	11.6	25.1	44.4	2.78	2.27	2.56	397	415	604	1.9	3.5	3.2
Mozambique, People's Rep	7.5	15.7	34.4	2.39	2.83	2.65	212	320	389	1.8	3.8	2.0
Niger	3.2	7.1	18.9	2.08	2.59	3.01	82	129	199	2.2	1.9	2.4
Nigeria	42.3	113.0	301.3	3.24	3.49	3.43	1,709	2,577	3,564	2.7	3.1	2.7
Rwanda	2.7	7.2	18.1	3.16	3.27	3.40	109	156	226	2.5	3.1	2.8
Senegal	3.0	7.4	16.4	2.89	3.46	2.69	108	180	185	2.6	3.3	1.9
Sierra Leone	2.2	4.2	9.6	1.79	2.15	2.49	45	66	97	0.8	1.0	1.2
Somalia	2.9	7.6	18.9	2.31	5.06	3.32	80	239	231	1.7	3.7	1.7
South Africa	17.4	35.2	63.2	2.49	2.23	2.19	526	595	731	2.8	1.3	2.8
Sudan	11.2	25.2	59.6	2.29	3.08	2.88	300	534	675	1.8	2.7	2.9
Swaziland	0.3	0.8	2.2	2.52	3.12	3.43	10	16	25	1.8	2.1	2.3
Tanzania	10.0	27.3	84.8	3.08	3.42	3.67	385	593	915	2.7	2.9	2.9
Togo	1.5	3.5	9.5	4.33	2.52	3.09	79	60	99	2.6	2.1	2.3
Tunisia	4.2	8.2	13.3	2.04	2.58	2.35	99	155	181	1.2	3.7	3.1
Uganda	6.6	18.4	55.2	3.96	3.20	3.49	352	387	590	3.9	2.6	2.8
Zaire	15.9	36.0	99.5	2.11	3.27	3.17	390	796	1,056	1.3	1.9	2.3
Zambia	3.1	8.5	25.5	2.96	3.40	3.76	115	179	290	2.6	2.7	3.3
Zimbabwe	3.8	9.7	22.6	3.28	2.96	3.15	159	197	283	3.5	2.9	2.8
NORTH & CENTRAL AMERICA	269.5	427.2	594.9	1.64	1.47	1.28	5,007	5,253	5,276	2.0	2.7	1.6
Barbados	0.2	0.3	0.3	0.30	0.28	0.62	1	1	2	(0.1)	2.7	1.5
Canada	17.9	26.5	32.1	1.61	1.04	0.88	329	243	229	2.7	3.1	1.2
Costa Rica	1.2	3.0	5.3	3.11	2.98	2.64	50	63	75	3.4	3.9	2.8
Cuba	7.0	10.3	12.0	1.87	0.84	0.75	153	80	76	1.0	3.1	2.3
Dominican Rep	3.2	7.2	11.4	3.00	2.42	2.22	123	130	151	2.2	3.1	3.4
El Salvador	2.6	5.3	11.3	3.54	2.05	1.93	117	88	97	3.5	3.0	3.1
Guatemala	4.0	9.2	21.7	2.77	2.77	2.88	136	179	247	2.5	2.2	2.9
Haiti	3.7	6.5	11.5	2.12	1.76	1.88	90	91	117	1.3	0.9	2.0
Honduras	1.9	5.1	11.5	2.71	3.46	3.18	67	116	151	2.5	3.2	3.9
Jamaica	1.6	2.5	3.8	1.20	1.24	1.52	22	26	37	0.7	2.9	2.8
Mexico	38.0	88.6	150.1	3.30	2.57	2.20	1,604	1,699	1,844	2.7	4.4	3.2
Nicaragua	1.5	3.9	9.2	3.19	2.81	3.36	60	73	120	2.8	2.9	3.8
Panama	1.1	2.4	3.9	2.88	2.26	2.07	41	42	47	3.0	2.5	2.9
Trinidad and Tobago	0.8	1.3	1.9	1.27	1.64	1.59	12	17	20	1.2	2.3	2.4
United States	180.7	249.2	300.8	1.08	1.06	0.82	2,150	2,357	1,990	1.8	2.3	1.1
SOUTH AMERICA	146.8	296.8	498.4	2.47	2.27	2.07	4,421	5,148	5,835	2.5	2.9	2.3
Argentina	20.6	32.3	45.5	1.45	1.61	1.27	336	437	398	1.4	1.0	1.1
Bolivia	3.4	7.3	18.3	2.37	2.59	2.76	97	135	189	1.8	2.1	2.8
Brazil	72.6	150.4	245.8	2.57	2.31	2.07	2,311	2,651	2,961	3.1	3.4	2.2
Chile	7.6	13.2	19.8	2.05	1.48	1.66	185	159	210	1.7	2.5	2.4
Colombia	15.5	31.8	51.7	2.77	2.14	2.05	538	523	621	2.7	2.5	2.7
Ecuador	4.4	10.8	22.9	3.18	2.88	2.79	178	218	281	2.6	2.7	3.0
Guyana	0.6	1.0	1.6	1.90	2.07	1.74	13	17	17	2.1	3.8	2.8
Paraguay	1.8	4.3	9.2	2.74	3.20	2.93	60	93	117	2.4	3.5	3.0
Peru	9.9	22.3	41.0	2.80	2.63	2.51	345	427	527	2.0	3.4	2.9
Suriname	0.3	0.4	0.6	2.28	(0.54)	1.46	8	(1)	6	2.1	0.5	2.6
Uruguay	2.5	3.1	3.9	0.84	0.56	0.76	23	16	23	0.8	0.2	0.7
Venezuela	7.5	19.7	38.0	3.35	3.42	2.61	327	472	484	2.8	4.9	3.3

Table 16.1

	Population (millions)			Average Annual Population Change (percent)			Average Annual Increment to the Population (thousands)			Average Annual Growth of the Labor Force (percent)		
	1960	1990	2025	1965-70	1975-80	1985-90	1965-70	1975-80	1985-90	1960-70	1970-80	1980-90
ASIA	**1,666.8**	**3,108.5**	**4,889.5**	**2.44**	**1.86**	**1.85**	**48,227**	**45,889**	**54,850**	**2.0**	**2.2**	**2.2**
Afghanistan	10.8	16.6	41.1	2.42	0.87	2.63	310	137	408	2.1	1.5	2.7
Bahrain	0.2	0.5	1.0	2.78	4.88	3.65	6	15	17	2.8	8.6	4.7
Bangladesh	51.4	115.6	235.0	2.68	2.83	2.67	1,672	2,328	2,889	1.5	2.1	2.9
Bhutan	0.9	1.5	3.1	1.91	1.70	2.15	19	20	31	1.6	1.8	1.9
China	657.5	1,135.5	1,492.6	2.61	1.43	1.39	20,297	13,773	15,195	2.2	2.5	2.2
Cyprus	0.6	0.7	0.9	1.09	0.65	1.04	7	4	7	0.9	1.3	1.1
India	442.3	853.4	1,445.6	2.28	2.08	2.08	11,951	13,631	16,838	1.5	1.7	2.0
Indonesia	96.2	180.5	263.3	2.33	2.14	1.62	2,648	3,058	2,810	2.0	2.1	2.4
Iran, Islamic Rep	20.3	56.6	122.2	3.30	3.08	3.45	864	1,111	1,792	3.1	3.2	3.3
Iraq	6.8	18.9	50.0	3.19	3.75	3.48	276	454	604	2.7	4.0	3.7
Israel	2.1	4.6	6.9	2.98	2.31	1.58	82	85	70	3.7	2.8	2.3
Japan	94.1	123.5	128.6	1.07	0.93	0.44	1,090	1,057	541	1.8	0.7	0.9
Jordan	1.7	4.3	13.1	3.17	2.34	3.94	67	65	153	2.9	1.0	4.4
Kampuchea, Dem	5.4	8.2	14.0	2.44	(2.07)	2.48	159	(139)	192	2.2	0.7	1.3
Korea, Dom People's Rep	10.5	22.9	39.6	2.76	2.57	2.36	358	435	510	2.3	2.9	2.9
Korea, Rep	25.0	43.6	54.6	2.25	1.55	1.19	679	569	505	3.2	2.6	2.4
Kuwait	0.3	2.1	4.4	9.17	6.24	4.02	55	74	76	7.8	7.4	5.2
Lao People's Dem Rep	2.2	4.1	7.7	2.18	1.16	2.49	56	36	95	2.0	1.3	2.0
Lebanon	1.9	3.0	5.0	2.75	(0.72)	2.11	64	(19)	60	2.4	1.2	2.1
Malaysia	8.1	17.3	27.9	2.66	2.32	2.31	270	301	378	2.7	3.8	2.9
Mongolia	0.9	2.2	5.4	3.09	2.82	3.09	36	44	64	2.5	2.9	2.9
Myanmar	21.7	41.7	72.6	2.29	2.10	2.09	587	676	826	2.0	2.3	1.9
Nepal	9.4	19.1	35.0	2.10	2.67	2.48	229	372	446	1.1	1.8	2.3
Oman	0.5	1.5	4.3	2.71	5.01	3.34	17	44	45	2.2	4.6	3.8
Pakistan	50.0	122.7	267.1	2.79	2.64	3.45	1,712	2,113	3,885	2.1	2.8	2.9
Philippines	27.6	62.4	111.4	3.17	2.53	2.48	1,102	1,150	1,458	2.6	2.5	2.5
Qatar	0.0	0.4	0.9	9.28	5.84	4.15	8	12	14	12.0	8.1	5.7
Saudi Arabia	4.1	14.1	44.8	3.62	5.13	3.96	190	424	507	3.3	5.7	4.0
Singapore	1.6	2.7	3.2	1.97	1.30	1.09	39	30	29	3.0	4.4	1.5
Sri Lanka	9.9	17.2	24.4	2.28	1.71	1.32	270	243	220	2.1	2.3	1.6
Syrian Arab Rep	4.6	12.5	32.3	3.23	3.36	3.57	187	272	409	2.2	3.4	3.6
Thailand	26.4	55.7	80.9	3.08	2.44	1.53	1,021	1,072	820	2.9	2.8	2.3
Turkey	27.5	55.6	89.6	2.51	2.09	1.99	834	883	1,054	1.4	1.7	2.2
United Arab Emirates	0.1	1.6	2.7	8.70	13.97	3.26	16	102	48	14.5	18.8	4.0
Viet Nam	34.7	67.2	118.0	2.17	2.23	2.24	878	1,134	1,422	0.9	2.1	2.8
Yemen Arab Rep	4.0	8.0	23.3	1.47	2.53	3.03	69	143	226	1.2	1.0	2.9
Yemen, People's Dem Rep	1.2	2.5	6.4	2.05	2.36	3.07	29	41	71	1.7	1.7	2.9
EUROPE	**425.1**	**497.7**	**512.3**	**0.67**	**0.45**	**0.23**	**3,049**	**2,144**	**1,113**	**0.6**	**0.7**	**0.6**
Albania	1.6	3.2	5.0	2.68	1.94	1.83	54	49	57	2.3	3.0	2.8
Austria	7.0	7.5	7.0	0.52	0.08	(0.02)	38	6	(1)	(0.8)	0.8	0.6
Belgium	9.2	9.9	9.9	0.40	0.11	0.07	38	11	7	0.3	0.9	0.5
Bulgaria	7.9	9.0	8.9	0.69	0.32	0.11	58	28	10	0.4	0.2	(0.0)
Czechoslovakia	13.7	15.7	17.2	0.24	0.68	0.21	35	102	33	1.2	0.8	0.4
Denmark	4.6	5.1	5.0	0.71	0.25	(0.01)	34	13	0	1.3	1.3	0.5
Finland	4.4	5.0	5.1	0.18	0.29	0.29	8	14	15	0.8	0.8	0.7
France	45.7	56.2	60.4	0.81	0.44	0.36	403	236	201	0.9	0.9	0.8
German Dem Rep	17.2	16.6	16.2	0.06	(0.13)	0.01	9	(22)	1	(0.4)	0.6	0.6
Germany, Fed Rep	55.4	60.5	54.0	0.55	(0.09)	(0.16)	328	(52)	(96)	0.3	0.5	0.3
Greece	8.3	10.0	10.1	0.56	1.28	0.23	48	119	23	0.1	0.7	0.5
Hungary	10.0	10.6	10.2	0.40	0.32	(0.18)	41	34	(18)	1.3	(0.5)	0.1
Iceland	0.2	0.3	0.3	1.22	0.91	0.97	2	2	2	2.5	2.9	1.5
Ireland	2.8	3.7	5.0	0.53	1.18	0.92	16	39	34	0.1	1.1	1.6
Italy	50.2	57.3	54.9	0.65	0.35	0.07	342	199	39	0.1	0.5	0.6
Luxembourg	0.3	0.4	0.3	0.43	0.09	0.04	1	0	0	(0.2)	1.6	0.2
Malta	0.3	0.4	0.4	0.35	2.09	0.49	1	7	2	1.2	2.1	1.0
Netherlands	11.5	14.8	15.1	1.17	0.70	0.37	148	97	54	1.5	1.5	1.2
Norway	3.6	4.2	4.5	0.81	0.39	0.28	31	16	12	1.3	2.0	0.8
Poland	29.6	38.4	45.1	0.64	0.89	0.65	206	310	244	2.0	0.7	0.6
Portugal	8.8	10.3	10.9	(0.19)	1.43	0.25	(16)	135	26	(0.0)	2.5	0.9
Romania	18.4	23.3	25.7	1.35	0.88	0.48	266	191	109	0.6	0.0	0.7
Spain	30.5	39.3	42.5	1.11	1.06	0.38	365	389	146	0.3	0.8	1.1
Sweden	7.5	8.3	8.1	0.78	0.29	(0.03)	62	24	(1)	1.3	1.1	0.4
Switzerland	5.4	6.5	6.1	1.35	(0.25)	0.15	82	(15)	10	1.8	0.3	0.5
United Kingdom	52.4	56.9	57.5	0.47	0.04	0.11	256	21	62	0.5	0.5	0.4
Yugoslavia	18.4	23.8	26.3	0.94	0.87	0.62	187	189	146	0.9	0.8	0.9
U.S.S.R.	**214.3**	**288.0**	**351.5**	**1.01**	**0.82**	**0.78**	**2,404**	**2,139**	**2,209**	**0.6**	**1.6**	**0.7**
OCEANIA	**15.8**	**26.5**	**39.0**	**1.97**	**1.51**	**1.44**	**363**	**333**	**368**	**2.4**	**2.2**	**1.8**
Australia	10.3	16.7	22.7	1.95	1.51	1.22	233	214	197	2.5	2.3	1.7
Fiji	0.4	0.7	1.0	2.29	1.77	1.60	11	11	11	3.5	3.0	2.1
New Zealand	2.4	3.4	4.1	1.41	0.17	0.79	38	5	26	2.2	2.0	1.6
Papua New Guinea	1.9	4.0	8.6	2.00	2.70	2.66	55	78	100	1.8	1.9	2.1
Solomon Islands	0.1	0.3	0.8	2.79	3.01	3.96	22	31	59	X	X	X

Sources: United Nations Population Division and the International Labour Organisation.
Notes: 0 = zero or less than half the unit of measure; X = not available; negative numbers are shown in parentheses.
For additional information, see Sources and Technical Notes.

Table 16.2 Trends in Births, Life Expectancy, Fertility, and Age

	Crude Birth Rate (births per 1,000 population)		Life Expectancy at Birth (years)		Total Fertility Rate		Percentage of Population in Specific Age Groups					
							1970			1990		
	1965-70	1985-90	1965-70	1985-90	1965-70	1985-90	<15	15-65	>65	<15	15-65	>65
WORLD	**33.9**	**27.1**	**54.9**	**61.5**	**5.9**	**3.3**	**37.5**	**57.1**	**5.4**	**32.4**	**59.4**	**8.2**
AFRICA	**47.7**	**44.7**	**44.1**	**51.9**	**6.7**	**6.2**	**44.8**	**52.1**	**3.1**	**45.3**	**51.7**	**3.0**
Algeria	49.8	40.2	51.4	62.5	7.5	6.1	48.4	48.5	4.1	44.4	52.2	3.4
Angola	49.1	47.2	36.0	44.0	6.4	6.4	42.8	57.2	2.9	44.9	52.1	3.0
Benin	49.5	50.5	38.0	46.0	6.9	7.0	43.8	52.1	4.3	47.5	49.7	2.8
Botswana	53.7	47.3	48.8	56.5	6.9	6.3	51.5	45.6	2.5	48.5	48.1	3.4
Burkina Faso	50.9	47.2	39.2	47.2	6.7	6.5	43.4	52.3	2.7	43.8	53.2	3.0
Burundi	46.5	45.7	44.1	48.5	5.8	6.3	41.6	55.9	3.1	45.6	51.1	3.3
Cameroon	42.1	41.6	43.0	52.9	5.8	5.8	41.4	55.9	3.6	43.5	52.6	3.9
Cape Verde	39.5	38.4	50.1	61.5	6.0	5.2	47.5	49.4	5.1	41.6	54.4	4.0
Central African Rep	43.2	44.3	40.0	45.0	5.7	5.9	40.2	56.2	4.0	43.2	53.0	3.8
Chad	45.2	44.2	37.0	45.0	6.1	5.9	41.3	53.6	3.6	42.8	53.6	3.6
Comoros	47.2	45.6	45.0	52.0	6.3	6.2	44.9	51.5	2.9	46.2	51.0	2.8
Congo	45.1	44.4	40.5	48.5	5.9	6.0	42.1	55.0	3.2	44.0	52.6	3.4
Cote d'Ivoire	51.7	50.9	43.0	52.5	7.4	7.4	47.3	49.5	2.3	49.4	48.4	2.2
Djibouti	49.5	47.3	39.0	47.0	6.6	6.6	43.9	53.8	2.3	45.8	51.8	2.4
Egypt	41.8	36.0	49.7	60.6	6.6	4.8	41.4	56.3	4.3	40.9	55.2	3.9
Equatorial Guinea	41.9	42.4	38.0	46.0	5.7	5.7	39.3	56.4	4.5	42.1	53.8	4.1
Ethiopia	48.7	43.7	39.0	41.9	6.7	6.2	44.5	51.0	2.5	44.9	51.6	3.5
Gabon	30.9	38.8	43.0	51.0	4.2	5.0	32.4	61.5	6.0	32.3	61.9	5.8
Gambia, The	49.7	46.8	35.0	37.0	6.5	6.4	41.7	52.3	2.9	44.0	53.1	2.9
Ghana	46.8	44.3	48.0	54.0	6.8	6.4	45.5	51.6	2.7	45.4	51.8	2.8
Guinea	48.5	46.6	35.2	42.2	6.4	6.2	42.5	54.7	2.8	43.7	53.3	3.0
Guinea-Bissau	41.2	40.8	38.0	45.0	5.2	5.4	38.1	57.7	4.2	41.3	54.4	4.3
Kenya	52.2	53.9	48.4	55.3	8.1	8.1	48.2	47.9	3.9	52.1	45.1	2.8
Lesotho	42.7	40.8	45.7	51.3	5.7	5.8	41.3	55.1	3.6	43.1	54.2	2.7
Liberia	45.8	45.0	45.0	51.0	6.4	6.5	43.2	53.7	3.1	45.7	51.1	3.2
Libya	49.5	43.9	50.4	60.8	7.5	6.9	44.9	52.4	2.7	45.8	51.8	2.4
Madagascar	47.0	45.7	43.7	51.5	6.6	6.6	43.3	53.8	2.9	45.1	51.9	3.0
Malawi	53.6	53.0	39.5	47.0	6.9	7.0	46.7	51.0	2.3	46.1	51.2	2.7
Mali	51.6	50.1	37.0	44.0	6.6	6.7	44.5	52.9	2.6	46.6	50.7	2.7
Mauritania	47.3	46.2	38.0	46.0	6.5	6.5	42.8	54.3	2.9	44.6	52.3	3.1
Mauritius	32.2	18.5	61.6	68.2	4.3	1.9	43.8	52.0	2.6	28.4	67.5	4.1
Morocco	48.2	35.3	50.4	60.8	7.1	4.8	47.6	49.4	4.2	40.7	55.7	3.6
Mozambique, People's Rep	46.8	45.0	41.0	47.3	6.5	6.4	43.1	52.3	3.0	44.1	52.7	3.2
Niger	49.4	50.9	37.5	44.5	7.1	7.1	43.6	54.1	4.6	47.3	49.9	2.8
Nigeria	52.3	49.8	42.5	50.5	7.1	7.0	47.0	50.6	2.3	48.4	47.6	4.0
Rwanda	52.4	51.0	44.1	48.5	8.0	8.3	47.3	49.8	2.4	48.9	48.7	2.4
Senegal	46.7	45.7	38.7	45.3	6.7	6.4	43.4	53.5	2.9	44.5	52.5	3.0
Sierra Leone	48.5	48.2	33.5	36.0	6.4	6.5	41.8	55.1	3.1	44.5	52.4	3.1
Somalia	48.2	50.8	39.0	41.9	6.6	6.6	43.7	52.5	3.1	47.6	49.8	2.6
South Africa	38.1	31.7	52.0	55.5	5.9	4.5	41.1	56.2	3.8	37.0	58.8	4.2
Sudan	47.0	44.6	40.9	50.3	6.7	6.4	44.4	53.2	2.7	45.3	51.9	2.8
Swaziland	47.9	46.8	44.4	50.5	6.5	6.5	45.2	51.7	2.8	47.3	49.7	3.0
Tanzania	51.4	50.5	44.1	53.0	6.9	7.1	47.1	49.1	2.4	49.1	48.6	2.3
Togo	44.2	44.9	43.0	52.5	6.2	6.1	43.7	53.7	3.1	45.3	51.5	3.2
Tunisia	41.8	30.3	52.1	63.1	6.8	4.1	46.3	50.9	3.8	37.8	58.2	4.0
Uganda	49.1	50.1	46.0	51.0	6.9	6.9	47.0	50.5	2.6	48.5	49.0	2.5
Zaire	47.0	45.6	44.0	52.0	6.0	6.1	44.3	53.0	2.8	46.2	51.2	2.6
Zambia	48.9	51.2	45.3	53.3	6.7	7.2	46.1	51.4	2.5	49.1	48.6	2.3
Zimbabwe	50.4	41.7	49.0	57.8	7.5	5.8	49.1	33.1	2.7	44.8	52.5	2.7
NORTH & CENTRAL AMERICA	**25.0**	**20.1**	**67.3**	**72.6**	**5.2**	**3.5**	**33.4**	**58.8**	**7.9**	**29.7**	**60.7**	**9.6**
Barbados	23.8	18.5	67.6	73.5	3.5	2.0	37.0	60.0	8.3	25.3	64.5	10.2
Canada	18.4	14.1	72.0	76.3	2.5	1.7	30.2	67.0	7.9	20.9	67.7	11.4
Costa Rica	38.3	28.3	65.6	73.7	5.8	3.3	46.1	51.1	3.2	36.2	59.6	4.2
Cuba	32.0	16.0	68.5	74.0	4.3	1.7	37.1	59.1	5.9	21.8	69.8	8.4
Dominican Rep	44.9	31.3	57.0	64.6	6.7	3.8	47.3	50.1	3.0	37.9	58.7	3.4
El Salvador	45.5	36.3	55.9	67.1	6.6	4.9	46.4	48.0	2.8	44.5	51.8	3.7
Guatemala	45.6	40.8	50.1	62.0	6.6	5.8	45.9	50.7	2.8	45.5	51.3	3.2
Haiti	42.5	34.3	46.2	54.7	6.2	4.7	41.9	55.7	3.8	39.2	56.9	3.9
Honduras	50.1	39.8	50.9	62.6	7.4	5.6	47.2	48.7	2.6	44.6	52.1	3.3
Jamaica	37.3	26.0	66.3	73.8	5.4	2.9	46.9	48.7	5.6	34.4	59.6	6.0
Mexico	44.5	29.0	60.3	67.2	6.7	3.6	46.9	43.3	3.4	37.2	59.0	3.8
Nicaragua	48.4	41.8	51.6	63.3	7.1	5.5	48.3	49.3	2.4	45.8	51.5	2.7
Panama	39.3	26.7	64.3	72.1	5.6	3.1	44.2	51.8	4.1	34.9	60.3	4.8
Trinidad and Tobago	30.3	24.0	65.7	70.2	3.9	2.7	42.1	53.5	4.4	32.0	62.6	5.4
United States	18.0	15.1	70.4	75.0	2.6	1.8	28.3	64.7	9.8	21.5	65.9	12.6
SOUTH AMERICA	**35.7**	**28.8**	**58.4**	**65.5**	**5.2**	**3.6**	**41.1**	**55.5**	**4.0**	**35.4**	**59.6**	**5.0**
Argentina	22.6	21.4	66.0	70.6	3.1	3.0	29.4	67.6	7.0	29.9	61.0	9.1
Bolivia	45.6	42.8	45.1	53.1	6.6	6.1	42.9	53.4	3.3	43.9	52.9	3.2
Brazil	36.4	28.6	57.9	64.9	5.3	3.5	42.3	54.2	3.4	35.2	60.1	4.7
Chile	31.6	23.8	60.6	70.7	4.4	2.7	39.1	57.5	5.1	30.6	63.4	6.0
Colombia	39.6	29.2	58.4	64.8	6.0	3.6	45.4	51.1	3.0	36.2	59.8	4.0
Ecuador	44.5	35.4	56.8	65.4	6.7	4.7	45.3	50.8	3.7	40.6	55.7	3.7
Guyana	35.4	24.8	62.5	69.8	5.3	2.8	47.6	43.5	3.5	34.6	61.2	4.2
Paraguay	39.5	34.8	65.0	66.1	6.4	4.6	46.5	50.6	3.4	40.4	56.0	3.6
Peru	43.6	34.3	51.5	61.4	6.6	4.5	44.0	52.5	3.5	39.2	57.1	3.7
Suriname	40.0	25.9	63.5	69.6	5.9	3.0	48.3	51.7	3.9	34.4	61.1	4.5
Uruguay	20.5	18.9	68.6	71.0	2.8	2.6	27.9	72.1	8.9	26.2	62.6	11.2
Venezuela	40.6	30.7	63.7	69.7	5.9	3.8	45.6	54.4	2.9	38.3	58.0	3.7

Structure, 1965–90

Table 16.2

	Crude Birth Rate (births per 1,000 population)		Life Expectancy at Birth (years)		Total Fertility Rate		Percentage of Population in Specific Age Groups 1970			1990		
	1965-70	1985-90	1965-70	1985-90	1965-70	1985-90	<15	15-65	>65	<15	15-65	>65
ASIA	**38.4**	**27.6**	**53.3**	**61.1**	**5.7**	**3.5**	**40.4**	**59.6**	**4.0**	**32.8**	**62.2**	**5.0**
Afghanistan	53.2	49.3	36.0	39.0	7.1	6.9	42.9	57.1	2.2	42.0	55.2	2.8
Bahrain	43.4	28.2	60.0	70.6	7.0	4.1	46.0	54.0	2.5	32.7	65.3	2.0
Bangladesh	47.5	42.2	43.3	49.6	6.9	5.5	45.4	54.6	3.5	43.9	53.2	2.9
Bhutan	41.8	38.3	40.6	47.9	5.9	5.5	40.3	55.7	3.2	39.7	56.9	3.4
China	36.9	20.5	59.6	69.4	6.0	2.4	39.7	58.1	4.3	26.2	68.0	5.8
Cyprus	21.0	18.6	70.3	74.6	2.8	2.3	31.1	66.4	10.1	25.6	64.1	10.3
India	40.2	32.0	48.0	57.9	5.7	4.3	40.4	56.1	3.7	36.5	59.0	4.5
Indonesia	42.6	27.4	45.1	56.0	5.6	3.3	42.2	54.6	3.1	35.0	61.1	3.9
Iran, Islamic Rep	45.3	42.4	53.2	59.0	7.0	5.6	46.2	50.1	3.1	43.9	52.9	3.2
Iraq	48.8	42.6	53.0	63.9	7.2	6.4	46.6	49.1	2.4	46.4	50.9	2.7
Israel	25.5	21.6	70.8	75.1	3.8	2.9	33.1	56.8	6.7	30.9	60.2	8.9
Japan	17.8	11.4	71.1	77.2	2.0	1.7	24.0	72.9	7.1	18.5	69.8	11.7
Jordan	52.5	45.9	51.7	66.0	8.0	7.2	45.9	51.0	3.1	47.9	49.6	2.5
Kampuchea, Dem	43.9	41.4	45.4	48.4	6.2	4.7	43.2	54.0	2.8	34.9	62.2	2.9
Korea, Dem People's Rep	38.8	28.9	57.6	69.4	5.7	3.6	43.9	49.4	3.5	37.0	59.2	3.8
Korea, Rep	31.9	18.8	57.6	69.4	4.5	2.0	42.0	50.9	3.3	26.5	68.8	4.7
Kuwait	49.7	32.3	64.4	72.7	7.5	4.8	43.4	53.5	1.7	38.7	59.8	1.5
Lao People's Dem Rep	44.4	41.3	40.4	52.0	6.2	5.7	42.3	54.9	2.6	42.6	54.4	3.0
Lebanon	38.8	28.9	62.9	67.2	6.1	3.4	43.9	52.6	4.9	35.3	59.6	5.1
Malaysia	38.5	28.6	59.4	68.6	5.9	3.5	44.6	52.1	3.4	36.2	60.0	3.8
Mongolia	41.9	38.9	58.0	64.5	5.9	5.4	43.8	54.5	3.2	41.8	54.9	3.3
Myanmar	39.1	30.6	49.5	60.0	5.7	4.0	41.2	55.1	3.7	37.2	58.7	4.1
Nepal	45.5	39.6	41.0	47.9	6.2	5.9	41.3	56.1	3.0	42.2	54.7	3.1
Oman	50.0	46.0	43.8	55.4	7.2	7.2	44.1	52.7	2.7	45.8	51.7	2.5
Pakistan	47.8	47.0	46.8	52.1	7.0	6.5	46.3	50.7	3.2	45.7	51.6	2.7
Philippines	40.2	33.2	56.2	63.5	6.0	4.3	45.4	51.9	2.7	40.1	56.5	3.4
Qatar	37.0	30.8	59.0	69.2	7.0	5.6	36.7	60.1	1.7	35.1	63.1	1.8
Saudi Arabia	48.1	42.0	49.9	63.7	7.3	7.2	44.5	52.8	3.2	45.4	52.0	2.6
Singapore	24.9	16.5	67.9	72.8	3.5	1.7	38.7	59.6	3.4	22.8	71.6	5.6
Sri Lanka	31.5	22.5	64.2	70.0	4.7	2.7	41.9	54.9	3.6	32.5	62.3	5.2
Syrian Arab Rep	47.6	44.1	54.0	65.0	7.8	6.8	48.9	47.7	4.4	48.1	49.3	2.6
Thailand	41.8	22.3	56.7	64.2	6.1	2.6	46.2	50.2	3.0	32.7	63.4	3.9
Turkey	39.0	28.4	54.9	64.1	5.6	3.6	41.1	54.5	4.4	34.3	61.4	4.3
United Arab Emirates	38.6	22.6	59.0	69.2	6.8	4.8	34.9	62.1	2.4	31.1	67.2	1.7
Viet Nam	38.3	31.0	47.9	60.8	5.9	4.1	43.8	51.8	4.3	39.2	56.4	4.4
Yemen Arab Rep	48.8	47.9	40.9	50.9	7.0	7.0	43.0	54.6	3.1	48.1	48.7	3.2
Yemen, People's Dem Rep	49.0	47.3	40.9	50.9	7.0	6.7	45.4	50.3	2.6	44.7	52.5	2.8
EUROPE	**17.7**	**13.0**	**70.6**	**74.0**	**2.5**	**1.7**	**24.9**	**72.5**	**11.4**	**19.7**	**66.9**	**13.4**
Albania	34.8	24.0	66.2	72.1	5.1	3.0	42.4	46.2	4.4	32.6	62.1	5.3
Austria	17.0	11.6	69.9	73.9	2.5	1.5	24.5	61.4	14.1	17.6	67.4	15.0
Belgium	15.5	11.7	70.9	74.3	2.3	1.6	23.6	72.0	13.4	18.1	67.2	14.7
Bulgaria	15.8	12.7	70.8	72.6	2.2	1.9	22.8	63.1	9.6	20.0	67.0	13.0
Czechoslovakia	15.5	14.0	70.1	72.0	2.1	2.0	23.2	63.4	11.2	23.3	65.1	11.6
Denmark	16.6	10.7	72.9	75.1	2.2	1.5	23.3	67.1	12.3	17.0	67.5	15.5
Finland	16.3	12.5	69.6	74.6	2.1	1.7	24.6	64.2	9.2	19.3	67.5	13.2
France	17.1	14.0	71.5	75.2	2.6	1.9	24.8	62.9	12.9	20.2	66.0	13.8
German Dem Rep	15.1	12.9	71.3	73.1	2.3	1.7	23.4	67.4	15.5	19.8	67.1	13.1
Germany, Fed Rep	16.6	10.4	70.3	74.5	2.3	1.4	23.2	63.9	13.2	14.9	69.7	15.4
Greece	18.0	11.9	71.0	74.8	2.4	1.7	24.9	61.9	11.1	19.7	66.6	13.7
Hungary	14.3	11.6	69.6	71.3	2.0	1.8	20.8	68.1	11.5	19.9	66.7	13.4
Iceland	22.5	16.8	73.4	77.1	3.2	2.1	32.4	56.1	8.9	25.1	64.5	10.4
Ireland	21.5	18.1	71.1	73.8	3.9	2.5	31.1	60.0	11.2	27.7	62.0	10.3
Italy	18.3	10.8	71.0	75.2	2.5	1.5	24.5	64.3	10.9	17.1	68.7	14.2
Luxembourg	14.5	11.5	69.9	71.9	2.2	1.5	22.0	67.1	12.5	17.1	69.5	13.4
Malta	16.6	14.7	69.4	72.7	2.2	1.9	27.6	59.9	9.0	23.1	66.7	10.2
Netherlands	19.2	11.8	73.6	76.5	2.7	1.5	27.3	63.7	10.2	17.8	69.3	12.9
Norway	17.7	12.4	73.8	76.4	2.7	1.7	24.5	65.3	12.9	18.8	64.8	16.4
Poland	16.6	16.4	69.9	72.4	2.3	2.2	27.0	60.1	8.2	25.2	64.8	10.0
Portugal	21.4	13.5	66.1	73.0	2.9	1.8	28.8	62.0	9.2	21.2	65.9	12.9
Romania	21.3	15.5	68.0	71.1	3.1	2.2	25.9	65.5	8.6	23.4	66.3	10.3
Spain	20.5	12.8	71.6	75.0	2.9	1.7	27.9	62.3	9.8	20.4	66.6	13.0
Sweden	14.8	11.2	74.1	76.8	2.1	1.7	20.8	65.5	13.7	16.5	65.2	18.3
Switzerland	17.7	11.7	72.2	76.5	2.3	1.6	23.3	65.3	11.4	16.4	68.3	15.3
United Kingdom	17.6	13.4	71.4	74.5	2.5	1.8	24.2	62.9	12.9	18.9	65.6	15.5
Yugoslavia	19.8	15.0	66.6	71.7	2.5	2.0	27.5	64.7	7.8	22.9	68.0	9.1
U.S.S.R.	**17.9**	**18.4**	**69.3**	**72.1**	**2.4**	**2.4**	**28.9**	**63.7**	**7.4**	**25.5**	**64.9**	**9.6**
OCEANIA	**24.5**	**20.1**	**64.2**	**69.1**	**3.5**	**2.6**	**32.2**	**60.5**	**7.3**	**26.8**	**64.1**	**9.1**
Australia	19.8	15.0	70.9	75.7	2.9	1.9	28.8	60.0	8.3	22.2	66.8	11.0
Fiji	32.0	27.3	62.7	70.4	4.6	3.2	43.5	55.5	2.4	36.7	59.4	3.9
New Zealand	22.6	15.6	71.3	74.5	3.2	1.9	31.7	59.8	8.5	22.5	66.5	11.0
Papua New Guinea	42.4	38.7	45.1	54.0	6.2	5.7	42.1	54.9	3.0	42.0	55.4	2.6
Solomon Islands	X	X	X	X	X	X	X	X	X	X	X	X

Source: United Nations Population Division.
Notes: X = not available.
For additional information, see Sources and Technical Notes.

Table 16.3 Mortality, Morbidity, and Nutrition, 1980s

	Crude Death Rate (deaths per 1,000 population)		Infant Death Rate (infant deaths per 1,000 live births)		Child Deaths (deaths of children < 5 years old per 1,000 live births)		Maternal Deaths (annual, from pregnancy, per 100,000 live births)	Wasting (percent of children aged 12-23 months)	Stunting (percent of children aged 24-59 months)	Average Calories Available (as percent of need)
	1965-70	1985-90	1965-70	1985-90	1965-70	1985-90	1980-87	1980-87	1980-87	1983-85
WORLD	12	10	103	71	161	105				
AFRICA	22	15	154	106	261	163				
Algeria	17	9	150	74	230	105	130	X	X	113
Angola	28	20	186	137	312	232	X	X	X	X
Benin	29	19	160	110	270	184	1,680	14	X	93
Botswana	18	12	110	67	160	92	300	19	56	93
Burkina Faso	28	19	185	138	320	235	600	17	X	83
Burundi	25	17	140	112	237	191	X	36	52	95
Cameroon	21	16	136	94	230	153	140	2	43	89
Cape Verde	11	10	120	66	176	86	X	X	X	111
Central African Rep	27	20	160	132	270	223	600	X	X	90
Chad	27	20	179	132	302	223	700	X	X	X
Comoros	20	15	115	80	193	127	X	X	X	89
Congo	23	17	110	73	184	115	X	5	27	114
Cote d'Ivoire	23	14	143	96	260	148	X	21	X	106
Djibouti	23	18	166	122	X	X	X	X	X	X
Egypt	18	10	170	85	280	124	80	3	37	130
Equatorial Guinea	26	19	173	127	291	214	X	X	X	X
Ethiopia	25	24	162	154	273	252	X	19	43	X
Gabon	22	16	147	103	250	169	120	X	27	X
Gambia, The	33	21	193	143	350	281	X	X	X	94
Ghana	19	13	117	90	197	145	1,070	28	31	73
Guinea	29	22	192	147	321	249	X	X	X	75
Guinea-Bissau	26	20	173	132	291	223	400	X	X	X
Kenya	20	12	108	72	179	113	170	10	41	93
Lesotho	21	12	140	100	194	135	X	7	23	103
Liberia	21	13	132	87	282	206	X	7	38	101
Libya	17	9	130	82	205	118	80	X	X	153
Madagascar	20	14	195	120	153	90	300	X	X	109
Malawi	26	20	197	150	347	263	250	8	61	105
Mali	27	21	206	169	363	291	X	18	23	76
Mauritania	25	19	173	127	281	214	X	X	X	90
Mauritius	8	5	67	23	93	28	99	20	X	120
Morocco	17	10	138	82	220	118	330	6	12	111
Mozambique, People's Rep	24	19	175	141	294	241	300	X	X	71
Niger	29	21	176	135	296	228	420	26	32	96
Nigeria	22	16	172	105	290	173	1,500	21	X	87
Rwanda	20	17	140	122	237	205	210	23	45	87
Senegal	25	19	168	128	290	222	530	8	27	98
Sierra Leone	34	23	204	154	385	291	450	26	46	80
Somalia	25	20	162	132	273	252	1,100	X	27	89
South Africa	19	10	120	72	168	96	X	X	X	120
Sudan	23	16	156	108	263	175	X	48	63	85
Swaziland	21	13	147	118	220	173	X	X	X	110
Tanzania	21	14	135	106	228	174	370	17	X	100
Togo	21	14	141	94	238	152	84	9	36	96
Tunisia	16	7	138	59	210	99	X	3	45	118
Uganda	19	15	118	103	197	169	300	3	27	98
Zaire	21	14	137	98	232	161	800	11	40	97
Zambia	19	14	115	80	192	127	110	12	41	92
Zimbabwe	17	10	101	72	165	113	150	X	X	88
NORTH & CENTRAL AMERICA	10	9	39	29	55	34				
Barbados	9	8	33	11	55	14	X	X	X	129
Canada	8	7	21	7	25	9	2	X	X	129
Costa Rica	7	4	66	18	88	22	26	3	8	124
Cuba	7	7	49	15	61	18	31	X	X	134
Dominican Rep	13	7	105	65	158	82	56	X	X	109
El Salvador	13	9	112	59	161	84	74	2	54	X
Guatemala	16	9	108	59	193	99	110	3	69	105
Haiti	19	13	172	117	257	170	340	18	52	82
Honduras	16	8	123	69	195	106	82	X	X	98
Jamaica	8	6	45	18	62	23	100	14	9	115
Mexico	10	6	79	47	113	68	92	X	X	135
Nicaragua	15	8	115	62	173	93	65	X	22	X
Panama	8	5	52	23	82	33	90	7	24	105
Trinidad and Tobago	8	6	41	20	50	23	81	10	5	123
United States	10	9	22	10	26	12	9	X	X	138
SOUTH AMERICA	11	8	94	58	130	78				
Argentina	9	9	56	32	68	38	85	X	X	121
Bolivia	20	14	157	110	259	171	480	1	43	88
Brazil	11	8	100	63	139	86	150	X	X	110
Chile	10	6	95	20	112	24	55	1	10	106
Colombia	10	7	74	46	119	68	130	1	21	111
Ecuador	13	8	107	63	156	87	220	X	39	89
Guyana	8	5	56	30	74	37	100	X	X	110
Paraguay	10	7	67	42	105	61	470	X	X	122
Peru	16	9	126	88	200	122	310	11	59	91
Suriname	9	6	55	31	72	37	X	X	X	118
Uruguay	10	10	48	27	54	30	56	14	X	102
Venezuela	8	5	60	36	84	43	65	3	7	103

Table 16.3

	Crude Death Rate (deaths per 1,000 population)		Infant Death Rate (infant deaths per 1,000 live births)		Child Deaths (deaths of children < 5 years old per 1,000 live births)		Maternal Deaths (annual, from pregnancy, per 100,000 live births)	Wasting (percent of children aged 12-23 months)	Stunting (percent of children aged 24-59 months)	Average Calories Available (as percent of need)
	1965-70	1985-90	1965-70	1985-90	1965-70	1985-90	1980-87	1980-87	1980-87	1983-85
ASIA	**14**	**9**	**110**	**73**	**171**	**108**				
Afghanistan	29	23	203	172	357	318	640	X	X	X
Bahrain	10	4	78	26	109	32	X	X	X	X
Bangladesh	21	16	140	119	228	188	600	17	59	84
Bhutan	23	17	164	128	260	196	X	X	X	X
China	11	7	81	32	113	44	44	3	10	109
Cyprus	10	8	29	12	34	16	X	X	X	X
India	18	11	145	99	239	148	500	37	X	98
Indonesia	19	11	124	84	201	117	800	17	X	116
Iran, Islamic Rep	16	8	145	63	224	155	120	23	60	X
Iraq	17	8	111	69	168	94	X	2	14	X
Israel	7	7	25	12	30	16	5	X	X	119
Japan	7	7	16	5	20	8	15	X	X	120
Jordan	16	7	102	44	150	57	X	9	X	X
Kampuchea, Dem	19	17	130	130	193	192	X	X	X	X
Korea, Dem People's Rep	11	5	58	24	76	31	41	X	X	134
Korea, Rep	10	6	58	25	76	31	34	X	X	120
Kuwait	6	3	55	19	73	23	18	2	14	X
Lao People's Dem Rep	19	16	147	110	221	160	X	20	44	X
Lebanon	12	8	52	40	69	49	X	X	X	X
Malaysia	10	6	50	24	72	35	59	12	47	118
Mongolia	11	8	82	45	116	58	140	X	X	116
Myanmar	16	10	110	70	160	85	140	17	X	117
Nepal	24	15	164	128	260	196	850	27	72	93
Oman	23	13	186	100	325	157	X	X	X	X
Pakistan	20	13	145	109	239	165	600	14	1	95
Philippines	11	8	70	45	114	72	80	14	42	102
Qatar	14	4	85	31	120	38	X	X	X	X
Saudi Arabia	19	8	140	71	228	98	X	9	X	128
Singapore	6	6	24	9	31	11	11	9	10	119
Sri Lanka	8	6	61	33	87	43	90	20	31	109
Syrian Arab Rep	15	7	107	48	160	63	280	X	X	129
Thailand	11	7	84	39	118	49	270	8	X	110
Turkey	14	8	153	76	206	92	210	X	X	126
United Arab Emirates	12	4	85	26	120	38	X	X	X	X
Viet Nam	17	10	133	64	197	91	110	7	60	X
Yemen Arab Rep	27	16	186	116	325	196	X	17	69	93
Yemen, People's Dem Rep	25	16	186	120	325	196	100	X	36	95
EUROPE	**10**	**11**	**30**	**13**	**35**	**15**				
Albania	8	6	77	39	107	48	X	X	X	X
Austria	13	12	27	11	31	12	11	X	X	132
Belgium	12	12	23	10	26	12	10	X	X	140
Bulgaria	9	12	31	16	36	19	22	X	X	145
Czechoslovakia	10	12	23	15	27	16	8	X	X	141
Denmark	10	11	16	7	19	9	4	X	X	131
Finland	10	10	15	6	18	7	5	X	X	111
France	11	10	21	8	24	10	13	X	X	132
German Dem Rep	14	13	21	9	26	12	17	X	X	144
Germany, Fed Rep	12	12	23	9	27	11	11	X	X	130
Greece	8	10	42	17	50	16	12	X	X	146
Hungary	11	13	37	20	41	19	28	X	X	134
Iceland	7	7	13	5	17	7	X	X	X	114
Ireland	12	9	23	9	26	11	7	X	X	151
Italy	10	10	33	11	37	12	13	X	X	138
Luxembourg	11	12	21	10	27	10	X	X	X	X
Malta	9	10	28	10	31	13	X	X	X	104
Netherlands	8	9	14	8	17	9	5	X	X	125
Norway	10	11	14	7	18	9	4	X	X	120
Poland	8	10	36	18	41	19	12	X	X	124
Portugal	8	10	61	15	80	20	15	X	X	128
Romania	9	11	52	22	60	28	190	X	X	128
Spain	9	9	33	10	37	11	10	X	X	136
Sweden	10	12	13	6	15	7	4	X	X	113
Switzerland	9	10	17	7	20	8	3	X	X	128
United Kingdom	12	12	19	9	22	11	7	X	X	124
Yugoslavia	9	9	61	25	72	28	27	X	X	142
U.S.S.R.	**8**	**11**	**26**	**24**	**36**	**27**	**X**	**X**	**X**	**133**
OCEANIA	**12**	**9**	**48**	**26**	**87**	**33**				
Australia	9	7	18	8	22	10	11	X	X	126
Fiji	8	5	55	27	73	31	X	X	X	110
New Zealand	9	8	18	11	22	12	20	X	X	129
Papua New Guinea	19	12	130	59	193	84	1,000	X	58	X
Solomon Islands	17	10	116	52	X	X	X	X	X	78

Sources: United Nations Population Division, United Nations Children's Fund, and the Food and Agriculture Organization of the United Nations.
Notes: X = not available.
For additional information, see Sources and Technical Notes.

Table 16.4 Access to Safe Drinking Water, Sanitation, and

	Percentage of Population with Access to:										Numbers of Trained Medical Personnel		
	Safe Drinking Water				Sanitation Services				Health Services 1980-87			Nurses and		
	Urban		Rural		Urban		Rural							
	1980	1985{a}	1980	1985{a}	1980	1985{a}	1980	1985{a}	All	Urban	Rural	Doctors	Midwives	Other
WORLD														
AFRICA														
Algeria	X	85	X	55	X	80	X	40	88	100	80	9,056	X	X
Angola	85	80	10	15	40	27	15	16	30	X	X	481	5,518	1,910
Benin	26	79	15	35	48	60	4	10	18	X	X	238	1,640	522
Botswana	X	100	X	33	X	90	X	25	88	100	85	X	X	X
Burkina Faso	27	43	31	69	38	38	5	5	49	51	48	131	2,899	9,813
Burundi	90	92	20	27	40	90	35	15	61	X	X	216	1,503	X
Cameroon	X	47	X	27	X	25	X	16	41	44	39	8	X	X
Cape Verde	100	100	21	49	34	41	10	X	X	X	X	X	196	X
Central African Rep	X	13	X	X	X	X	X	X	45	X	X	X	X	X
Chad	X	X	X	X	X	X	X	X	X	X	X	X	X	X
Comoros	X	X	X	X	X	90	X	80	X	X	X	31	168	X
Congo	X	42	X	7	X	X	X	X	83	97	70	210	2,746	406
Cote d'Ivoire	X	60	X	77	X	X	X	X	30	61	11	X	X	X
Djibouti	50	50	20	21	43	94	20	50	X	X	X	77	534	X
Egypt	88	100	64	90	X	100	10	65	X	X	X	9,495	12,458	X
Equatorial Guinea	47	X	X	X	99	X	X	X	X	X	X	X	X	X
Ethiopia	16	X	69	X	X	96	X	X	46	X	X	534	1,896	5,907
Gabon	X	X	X	50	X	X	X	X	90	X	X	328	X	3,366
Gambia, The	85	97	X	50	X	X	X	X	X	X	X	X	X	X
Ghana	72	96	33	55	47	63	17	22	60	92	45	817	X	X
Guinea	69	62	2	15	54	X	1	X	X	X	X	X	X	X
Guinea-Bissau	18	19	8	22	21	29	13	18	X	X	X	122	785	X
Kenya	85	X	15	X	89	X	19	X	X	X	X	2,151	17,193	4,581
Lesotho	37	X	11	40	13	22	14	20 b	X	X	X	X	X	X
Liberia	50	100	25	23	X	6	X	2	39	50	30	221	1,152	350
Libya	X	100	X	90	100	100	72	53	X	X	X	5,019	5,565	1,018
Madagascar	80	81	7	17	9	12	X	X	56	X	X	X	X	X
Malawi	77	66	37	49	100	X	81	X	80	X	X	262	1,286	351
Mali	37	48	0	17	79	100	X	X	15	X	X	349	5,223	308
Mauritania	80	73	85	X	5	8	X	X	30	X	X	142	1,230	200
Mauritius	100	100	98	98	100	100	90	95	100	100	100	X	X	X
Morocco	X	73	X	17	X	63	X	13	70	100	50	4,908	22,207	467
Mozambique, People's Rep	X	50	X	12	X	80	X	40	30	X	X	X	X	X
Niger	41	35	32	49	36	X	3	X	41	99	30	160	7,248	6,611
Nigeria	X	100	X	20	X	X	X	5	40	75	30	11,294	74,033	20,150
Rwanda	48	79	55	48	60	77	50	55	27	60	25	163	X	1,550
Senegal	77	79	25	38	100	87	2	X	X	X	X	311	1,393	X
Sierra Leone	50	86	2	20	31	86	6	20	X	X	X	262	2,830	X
Somalia	X	57	X	22	X	44	X	5	27	50	15	325	3,416	X
South Africa	X	X	X	X	X	X	X	X	X	X	X	X	X	X
Sudan	100	90	31	20	73	40	X	5	51	90	40	2,095	12,986	X
Swaziland	X	100	X	7	X	100	X	25	X	X	X	33	477	160
Tanzania	X	80	X	38	X	90	X	78	76	99	72	X	X	X
Togo	70	100	31	41	24	31	10	9	61	X	X	229	1,973	X
Tunisia	100	100	17	52	100	84	X	16	90	100	80	3,453	9,353	11,831
Uganda	X	45	X	12	X	40	X	10	61	90	57	X	X	X
Zaire	X	54	X	20	X	X	X	9	26	40	17	X	X	X
Zambia	X	76	X	41	X	76	X	34	75	X	X	880	5,655	2,773
Zimbabwe	X	100	X	14	X	100	X	14	71	100	62	X	X	3,238
NORTH & CENTRAL AMERICA														
Barbados	99	100	98	100	X	100	X	100	X	X	X	225	1,134	X
Canada	X	100	X	100	X	X	X	X	X	X	X	48,860	85,539	X
Costa Rica	100	100	82	82	99	100	84	88	80	100	63	2,539	5,400	X
Cuba	X	X	X	X	X	X	X	X	X	X	X	18,850	35,062	X
Dominican Rep	85	72	34	24	25	72	4	59	80	X	X	3,555	5,184	X
El Salvador	67	76	40	47	48	89	26	35	56	80	40	1,664	5,038	1,214
Guatemala	90	89	18	39	45	73	20	42	34	47	25	3,544	9,093	X
Haiti	51	59	8	32	42	42	10	14	70	80	70	810	2,537	102
Honduras	93	51	40	49	49	22	26	38	73	85	65	2,800	6,300	614
Jamaica	50	99	46	93	12	92	2	90	X	X	X	1,115	4,675	X
Mexico	90	95	40	50	77	77	12	15	45	X	X	X	87,398	3,207
Nicaragua	67	77	6	13	34	35	X	16	83	100	60	2,110	5,917	250
Panama	100	100	62	64	83	99	59	61	80	95	64	2,167	5,475	410
Trinidad and Tobago	100	100	93	93	96	100	88	100	X	X	X	1,213	4,521	X
United States	X	X	X	X	X	X	X	X	X	X	X	501,200	3,212,700	X
SOUTH AMERICA														
Argentina	61	63	17	17	80	76	35	35	71	80	21	80,100	30,505	X
Bolivia	69	81	10	27	37	51	4	22	63	90	36	4,032	1,066	X
Brazil	83	86	51	53	X	33	1	2	X	X	X	122,818	110,052	X
Chile	100	97	17	22	100	79	10	21	X	X	X	9,684	32,150	X
Colombia	93	100	73	76	93	96	4	13	60	X	X	23,520	44,520	X
Ecuador	79	83	20	33	73	79	17	34	62	90	30	11,033	14,794	X
Guyana	100	100	60	60	73	100	80	87	89	X	X	125	887	X
Paraguay	39	49	9	8	95	66	80	40	61	90	38	2,453	3,584	195
Peru	68	73	18	17	57	67	0	13	X	X	17	18,200	14,900	X
Suriname	100	100	79	94	100	100	79	100	X	X	X	306	1,400	X
Uruguay	96	95	2	27	59	59	6	59	80	X	X	5,756	3,000	2,300
Venezuela	93	88	53	65	60	57	12	5	X	X	X	24,083	15,214	4,342

Health Services, 1980s

Table 16.4

	Safe Drinking Water				Sanitation Services				Health Services 1980 87			Numbers of Trained Medical Personnel		
	Urban		Rural		Urban		Rural						Nurses and	
	1980	1985{a}	1980	1985{a}	1980	1985{a}	1980	1985{a}	All	Urban	Rural	Doctors	Midwives	Other
ASIA														
Afghanistan	28	38	8	17	X	5	X	X	29	80	17	2,957	2,135	329
Bahrain	X	100	X	100	X	100	X	100	X	X	X	518	1,148	19
Bangladesh	26	25	40	66	21	20	1	6	45	X	X	14,944	11,197	X
Bhutan	50	100	5	24	X	100	X	7	19	X	X	52	164	242
China	X	X	X	X	X	X	X	X	X	X	X	926,603	759,485	1,784,425
Cyprus	X	100	X	100	X	100	X	100	X	X	X	911	2,165	725
India	77	79	31	85	27	40	1	4	X	X	X	297,228	429,315	X
Indonesia	35	41	19	37	29	32	21	38	75	X	X	16,698	122,945	37,230
Iran, Islamic Rep	82	90	50	60	96	90	43	24	78	95	60	16,918	43,291	2,488
Iraq	X	95	X	85	X	95	X	70	93	97	78	9,442	9,931	1,465
Israel	X	100	X	97	X	99	X	95	X	X	X	11,895	26,895	17,010
Japan	X	X	X	X	X	X	X	X	X	X	X	181,101	651,660	X
Jordan	100	100	65	88	94	91	34	91	97	98	95	2,958	2,596	623
Kampuchea, Dem	X	X	X	X	X	X	X	X	53	80	50	X	X	X
Korea, Dem People's Rep	X	100	X	100	X	100	X	100	X	X	X	45,120	X	X
Korea, Rep	86	90	61	48	100	100	100	100	93	97	86	35,657	70,783	X
Kuwait	86	100	100	100	100	100	100	100	100	X	X	2,804	8,831	1,134
Lao People's Dem Rep	28	X	20	X	13	X	4	X	X	X	X	551	6,753	2,088
Lebanon	X	100	X	100	X	94	X	18	X	X	X	3,953	X	X
Malaysia	90	100	49	66	100	100	55	67	X	X	X	4,938	12,841	2,152
Mongolia	X	100	X	100	X	100	X	100	X	X	X	3,881	8,083	15,384
Myanmar	38	37	15	27	38	35	15	26	33	100	11	19,931	41,590	X
Nepal	83	77	7	24	16	54	1	1	X	X	X	497	1,707	X
Oman	X	90	X	55	16	88	1	25	91	100	90	1,240	3,460	80
Pakistan	72	84	20	28	42	56	2	5	55	99	35	34,850	20,295	2,050
Philippines	49	81	43	68	81	76	67	66	X	X	X	8,132	19,880	X
Qatar	76	100	43	100	X	100	X	100	X	X	X	646	1,672	95
Saudi Arabia	92	100	87	68	81	100	50	33	97	100	88	17,544	37,670	1,291
Singapore	100	100	NA	NA	80	99	NA	NA	100	100	X	1,086	4,967	890
Sri Lanka	65	82	18	35	80	69	63	41	93	X	X	1,914	11,346	X
Syrian Arab Rep	98	91	54	68	74	72	28	55	75	92	60	8,593	12,550	2,487
Thailand	65	57	63	78	64	81	41	57	70	X	X	8,058	62,505	X
Turkey	95	100	62	70	56	95	X	90	X	X	X	36,427	48,841	X
United Arab Emirates	95	100	81	100	93	100	22	77	90	X	X	1,278	3,328	97
Viet Nam	X	70	32	39	X	X	55	X	80	100	75	19,861	101,448	43,763
Yemen Arab Rep	100	90	18	30	60	66	X	X	30	75	24	1,234	2,965	X
Yemen, People's Dem Rep	85	86	25	35	70	70	15	30	30	X	X	4,942	2,022	X
EUROPE														
Albania	X	100	X	95	X	100	X	100	X	X	X	2,641	13,372	3,110
Austria	X	100	X	100	X	100	X	100	X	X	X	19,451	27,655	X
Belgium	X	100	X	100	X	100	X	100	X	X	X	29,776	X	X
Bulgaria	X	100	X	96	X	100	X	100	X	X	X	24,718	57,500	X
Czechoslovakia	X	100	X	100	X	100	X	100	X	X	X	55,871	106,968	X
Denmark	X	100	X	100	X	100	X	100	X	X	X	12,806	X	X
Finland	X	99	X	90	X	100	X	100	X	X	X	11,071	82,951	X
France	X	100	X	100	X	100	X	100	X	X	X	173,116	X	89,276
German Dem Rep	X	100	X	100	X	100	X	100	X	X	X	37,943	116,600	X
Germany, Fed Rep	X	100	X	100	X	95	X	83	X	X	X	153,895	269,301	X
Greece	X	100	X	95	X	100	X	95	X	X	X	28,212	21,811	8,379
Hungary	X	100	X	95	X	100	X	100	X	X	X	34,758	61,422	X
Iceland	X	100	X	100	X	100	X	100	X	X	X	545	2,724	293
Ireland	X	100	X	100	X	100	X	100	X	X	X	5,180	X	1,131
Italy	X	100	X	100	X	100	X	100	X	X	X	245,116	X	3,697
Luxembourg	X	100	X	100	X	100	X	100	X	X	X	663	102	X
Malta	X	100	X	100	X	100	X	100	X	X	X	413	3,187	453
Netherlands	X	100	X	100	X	100	X	100	X	X	X	32,193	971	X
Norway	X	100	X	100	X	100	X	100	X	X	X	9,176	72,448	X
Poland	X	94	X	82	X	100	X	100	X	X	X	73,199	198,934	X
Portugal	X	97	X	90	X	100	X	95	X	X	X	24,629	X	X
Romania	X	100	X	90	X	100	X	95	X	X	X	40,050	X	X
Spain	X	100	X	100	X	100	X	100	X	X	X	121,500	149,312	X
Sweden	X	100	X	100	X	100	X	100	X	X	X	X	X	X
Switzerland	X	100	X	100	X	100	X	100	X	X	X	9,298	X	3,117
United Kingdom	X	100	X	100	X	100	X	100	X	X	X	92,172	182,897	17,472
Yugoslavia	X	100	X	65	X	78	X	46	X	X	X	42,365	91,253	50,036
U.S.S.R.	X	100	X	100	X	100	X	100	X	X	X	1,170,000	X	X
OCEANIA														
Australia	X	X	X	X	X	X	X	X	X	X	X	36,610	139,434	X
Fiji	94	X	66	X	85	X	60	X	X	X	X	325	1,342	X
New Zealand	X	X	X	X	X	100	X	X	X	X	X	5,747	40,950	X
Papua New Guinea	55	95	10	15	96	99	3	35	X	X	X	269	3,941	301
Solomon Islands	96	91	45	60	80	90	21	X	X	X	X	38	301	X

Sources: World Health Organization and the United Nations Population Division.
Note: a. Data are from 1985 unless otherwise indicated in the Sources and Technical Notes. b. Sewerage only.
0 = zero or less than half of one percent; X = not available; NA = not applicable.
For additional information, see Sources and Technical Notes.

Table 16.5 Fertility, Education, and Child Health, 1980—87

	Adult Female Literacy (percent)		Adult Male Literacy (percent)		Average Number of Children by Mother's Years of Education (1985)		Couples Using Contraception (percent)	Births Attended By Trained Personnel (percent)	ORS Use{a} (litres per 100 diarrhea episodes)	Low-Birth-Weight Infants (percent)	Percent of 1-Year-Olds Fully Immunized, 1986-87, Against			
	1970	1985	1970	1985	none	> 6 years		1983-87	1986	1982-87	TB	DPT	Polio	Measles
WORLD														
AFRICA														
Algeria	11	37	39	63	X	X	X	X	8	9	95	66	66	59
Angola	7	X	16	49	X	X	X	15	20	17	29	10	16	55
Benin	8	16	23	37	7.4	4.3	9.2	34	7	10	67	52	52	38
Botswana	44	69	37	73	X	X	27.8	52	19	8	99	86	88	91
Burkina Faso	3	6	13	21	X	X	X	X	37	18	67	34	34	68
Burundi	10	26	29	43	X	X	8.7	12	12	14	89	73	76	58
Cameroon	19	55	47	68	6.4	5.2	2.4	X	3	13	77	45	43	44
Cape Verde	X	X	X	X	X	X	X	X	X	X	X	X	X	X
Central African Rep	6	29	26	53	X	X	X	X	23	15	53	24	24	30
Chad	2	11	20	40	X	X	X	X	7	11	40	12	12	33
Comoros	X	X	X	X	X	X	X	X	X	X	X	X	X	X
Congo	19	55	50	71	X	X	X	X	8	12	86	71	71	69
Cote d'Ivoire	10	31	26	53	7.4	5.8	2.9	20	17	14	53	71	71	85
Djibouti	X	X	X	X	X	X	X	X	X	X	X	X	X	X
Egypt	20	30	50	59	X	X	29.7	24	24	7	72	82	88	76
Equatorial Guinea	X	X	X	X	X	X	X	X	X	X	X	X	X	X
Ethiopia	1	X	8	X	X	X	X	58	9	X	28	16	15	13
Gabon	22	53	43	70	X	X	X	92	27	16	79	48	48	55
Gambia, The	X	X	X	X	X	X	X	X	X	X	X	X	X	X
Ghana	18	43	43	64	6.8	5.5	9.5	73	26	17	71	37	34	51
Guinea	7	17	21	40	X	X	X	X	4	18	46	15	8	43
Guinea-Bissau	6	17	13	46	X	X	X	16	5	20	82	47	48	60
Kenya	29	49	44	70	8.3	7.3	17.0	X	21	13	86	75	75	60
Lesotho	74	84	49	62	6.2	4.8	5.3	28	25	10	84	77	77	79
Liberia	8	23	27	47	X	X	6.5	89	X	X	68	28	28	55
Libya	13	50	60	81	X	X	X	76	49	5	77	62	62	50
Madagascar	43	62	56	74	X	X	X	62	8	10	42	30	24	10
Malawi	18	31	42	52	X	X	7.0	59	39	10	92	55	50	53
Mali	4	11	11	23	X	X	5.0	27	3	17	29	8	8	11
Mauritania	X	X	X	X	X	X	0.8	23	1	10	91	32	61	69
Mauritius	59	77	77	89	X	X	75.4	90	18	9	87	85	85	68
Morocco	10	22	34	45	6.4	4.2	35.9	X	11	9	87	78	78	76
Mozambique, People's Rep	14	22	29	55	X	X	X	28	54	15	59	51	38	46
Niger	2	9	6	19	X	X	X	47	X	20	28	5	4	27
Nigeria	14	31	35	54	6.6	4.2	4.8	X	2	25	41	20	21	31
Rwanda	21	33	43	61	X	X	10.1	X	X	17	85	78	80	63
Senegal	5	19	18	37	7.3	4.5	11.7	X	1	10	92	53	53	70
Sierra Leone	8	21	18	38	X	X	X	25	46	14	73	30	30	50
Somalia	1	6	5	18	X	X	X	2	50	X	33	25	25	29
South Africa	X	X	X	X	X	X	48.0	X	X	12	X	X	X	X
Sudan	6	14	28	33	6.5	3.4	4.6	20	40	15	46	29	29	22
Swaziland	X	X	X	X	X	X	X	X	X	X	X	X	X	X
Tanzania	18	88	48	93	X	X	X	74	42	14	95	81	80	78
Togo	7	28	27	53	X	X	X	15	23	20	66	41	40	48
Tunisia	17	41	44	68	X	X	41.1	60	3	7	94	89	89	79
Uganda	30	45	52	70	X	X	X	X	21	10	74	39	40	48
Zaire	22	45	61	79	X	X	X	X	19	X	54	36	36	39
Zambia	37	67	66	84	X	X	X	X	32	14	92	66	61	58
Zimbabwe	47	67	63	81	X	X	38.4	69	5	15	86	77	77	73
NORTH & CENTRAL AMERICA														
Barbados	X	X	X	X	X	X	46.5	X	X	X	X	X	X	X
Canada	X	X	X	X	X	X	73.1	99	X	6	X	80	80	85
Costa Rica	87	93	88	94	5	2.7	69.5	93	49	9	81	91	89	43
Cuba	87	96	86	96	X	X	60.0	X	166	8	96	87	86	99
Dominican Rep	65	77	69	78	7	3	50.0	57	21	16	51	80	79	71
El Salvador	53	69	61	75	X	X	47.3	35	180	15	55	53	57	48
Guatemala	37	47	51	63	X	X	25.0	19	17	10	34	16	18	24
Haiti	17	35	26	40	6	2.8	6.9	20	11	17	45	20	28	23
Honduras	50	58	55	61	X	X	34.9	50	48	20	66	58	61	57
Jamaica	97	93	96	90	6.2	4.8	51.4	89	18	8	92	81	82	62
Mexico	69	88	78	92	8.1	3.3	53.0	X	14	15	71	62	97	54
Nicaragua	57	X	58	X	X	X	27.0	X	202	15	93	43	85	44
Panama	81	88	81	89	7	3	58.2	83	18	8	89	73	74	78
Trinidad and Tobago	89	95	95	97	4.6	3.2	51.6	90	2	X	X	79	80	60
United States	99	X	99	X	X	X	68.0	100	X	7	X	37	24	82
SOUTH AMERICA														
Argentina	92	95	94	96	X	X	X	X	12	6	91	75	85	81
Bolivia	46	65	68	84	X	X	26.0	36	59	15	31	24	28	33
Brazil	63	76	69	79	X	X	66.0	73	31	8	68	57	90	55
Chile	88	96	90	97	X	X	X	97	X	7	97	93	95	92
Colombia	76	87	79	89	7	2.6	65.0	51	22	15	80	58	82	59
Ecuador	68	80	75	85	7.8	2.7	44.2	27	34	10	85	51	51	46
Guyana	89	95	94	97	6.6	4.8	31.4	93	23	11	69	67	77	52
Paraguay	75	85	84	91	8.2	2.9	44.8	22	21	6	66	58	93	56
Peru	60	78	81	91	7.3	3.3	45.8	55	7	9	61	42	45	35
Suriname	X	X	X	X	X	X	X	X	X	X	X	X	X	X
Uruguay	93	94	93	93	X	X	X	X	17	8	98	70	70	99
Venezuela	71	85	79	88	7	2.6	49.3	82	73	9	86	54	64	57

Table 16.5

	Adult Female Literacy (percent)		Adult Male Literacy (percent)		Average Number of Children by Mother's Years of Education (1985)		Couples Using Contraception (percent)	Births Attended By Trained Personnel (percent) 1983-87	ORS Use{a} (litres per 100 diarrhea episodes) 1986	Low-Birth-Weight Infants (percent) 1982-87	Percent of 1-Year-Olds Fully Immunized, 1986-87, Against			
	1970	1985	1970	1985	none	> 6 years					TB	DPT	Polio	Measles
ASIA														
Afghanistan	2	8	13	39	X	X	1.6	X	22	19	27	25	25	31
Bahrain	X	X	X	X	X	X	X	X	X	X	X	X	X	X
Bangladesh	12	22	36	43	6.1	5	25.2	X	25	31	14	9	8	6
Bhutan	X	X	X	X	X	X	X	3	20	X	38	27	27	23
China	X	56	X	82	X	X	74.0	X	X	6	85	75	77	77
Cyprus	X	X	X	X	X	X	X	X	X	X	X	X	X	X
India	20	29	47	57	X	X	34.0	33	18	30	46	58	50	17
Indonesia	42	65	66	83	X	X	48.0	43	14	14	82	69	70	61
Iran, Islamic Rep	17	39	40	62	X	X	X	82	25	9	56	74	74	76
Iraq	18	87	50	90	X	X	14.5	50	26	9	99	76	76	69
Israel	83	93	93	97	X	X	X	99	X	7	X	92	93	88
Japan	99	X	99	X	X	X	64.3	100	X	5	85	83	95	73
Jordan	29	63	64	87	9.3	4.9	26.5	75	23	7	2	89	89	87
Kampuchea, Dem	23	65	X	85	X	X	X	47	32	X	54	37	35	54
Korea, Dem People's Rep	X	X	X	X	X	X	X	99	X	X	69	62	70	35
Korea, Rep	81	88	94	96	5.7	3.4	70.4	X	X	9	47	76	80	89
Kuwait	42	63	65	76	X	X	X	99	23	7	4	94	94	95
Lao People's Dem Rep	28	76	37	92	X	X	X	X	72	39	60	28	10	33
Lebanon	58	69	79	86	X	X	53.0	45	15	10	X	91	91	81
Malaysia	48	66	71	81	5.3	3.2	51.4	82	17	9	99	59	62	20
Mongolia	74	86	87	93	X	X	X	99	X	10	53	79	74	61
Myanmar	57	X	85	X	X	X	X	97	12	16	45	23	13	14
Nepal	3	12	23	39	X	X	13.8	10	23	X	78	46	40	22
Oman	X	12	X	47	X	X	X	60	74	14	95	77	77	78
Pakistan	11	19	30	40	6.5	3.1	7.6	24	80	25	72	62	62	53
Philippines	80	85	83	86	5.4	3.8	45.0	X	10	18	92	73	73	68
Qatar	X	X	X	X	X	X	X	X	X	X	X	X	X	X
Saudi Arabia	2	12	15	35	X	X	X	78	97	6	93	89	89	80
Singapore	55	79	82	93	X	X	74.2	100	X	7	92	98	97	94
Sri Lanka	69	83	85	91	X	X	62.0	87	24	28	61	61	62	47
Syrian Arab Rep	20	43	60	76	8.8	4.1	19.8	37	12	9	81	70	70	73
Thailand	72	88	86	94	X	X	66.0	33	26	12	97	80	80	60
Turkey	35	62	69	86	5.9	2.1	51.0	78	42	7	34	71	70	50
United Arab Emirates	7	38	24	58	X	X	X	96	35	X	78	75	75	56
Viet Nam	X	80	X	88	X	X	20.0	99	14	18	59	51	54	42
Yemen Arab Rep	1	3	9	27	8.6	5.4 b	1.0	12	25	9	28	14	14	15
Yemen, People's Dem Rep	9	25	31	59	X	X	X	10	103	13	41	25	25	35
EUROPE														
Albania	X	X	X	X	X	X	X	X	X	7	92	96	94	96
Austria	X	X	X	X	X	X	71.4	X	X	6	90	90	90	60
Belgium	99	X	99	X	X	X	81.0	100	X	5	90	95	99	90
Bulgaria	89	X	94	X	X	X	76.0	100	X	6	99	99	99	99
Czechoslovakia	X	X	X	X	X	X	95.0	100	X	6	99	99	98	98
Denmark	X	X	X	X	X	X	63.0	X	X	6	85	89	94	X
Finland	X	X	X	X	X	X	80.0	X	X	4	90	94	78	81
France	98	X	99	X	X	X	78.7	X	X	5	96	97	97	55
German Dem Rep	X	X	X	X	X	X	X	X	X	6	99	93	94	98
Germany, Fed Rep	X	X	X	X	X	X	77.9	X	X	5	30	30	80	30
Greece	76	88	93	97	X	X	X	X	X	6	X	82	97	81
Hungary	98	X	98	X	X	X	73.1	99	X	10	99	99	99	99
Iceland	X	X	X	X	X	X	X	100	X	3	X	99	99	100
Ireland	X	X	X	X	X	X	X	X	X	4	80	45	90	63
Italy	93	96	95	98	X	X	78.0	X	X	7	30	88	81	21
Luxembourg	X	X	X	X	X	X	X	X	X	X	X	X	X	X
Malta	X	X	X	X	X	X	X	X	X	X	X	X	X	X
Netherlands	X	X	X	X	X	X	76.0	X	X	4	X	95	95	96
Norway	X	X	X	X	X	X	71.0	100	X	4	90	85	90	87
Poland	97	X	98	X	X	X	75.0	X	X	8	95	97	98	91
Portugal	65	80	78	89	3.5	1.8	66.3	X	X	8	71	96	78	66
Romania	91	X	96	X	X	X	58.0	99	X	6	55	90	90	90
Spain	87	92	93	97	X	X	59.4	96	X	X	X	88	80	00
Sweden	X	X	X	X	X	X	78.1	100	X	4	17	99	98	94
Switzerland	X	X	X	X	X	X	71.2	100	X	5	X	90	95	60
United Kingdom	X	X	X	X	X	X	83.0	98	X	7	96	67	85	71
Yugoslavia	76	86	92	97	X	X	55.0	X	X	7	84	90	89	92
U.S.S.R.	97	X	98	X	X	X	X	100	X	6	93	85	99	95
OCEANIA														
Australia	X	X	X	X	X	X	X	99	X	6	X	X	X	68
Fiji	X	X	X	X	X	X	41.0	X	X	X	X	X	X	X
New Zealand	X	X	X	X	X	X	69.5	99	X	5	20	72	74	67
Papua New Guinea	24	35	39	55	X	X	X	34	8	25	70	41	41	34
Solomon Islands	X	X	X	X	X	X	X	X	X	X	X	X	X	X

Sources: United Nations Children's Fund and the United Nations Population Division.
Notes: a. Oral Rehydration Salts. b. Attended some school.
            0 = zero or less than half the unit of measure; X = not available.
            For additional information, see Sources and Technical Notes.

Table 16.6   Selected Causes of Death, 1980s

			Deaths per 100,000 Population											
	All		Infectious/ Parasitic		Malignant Neoplasms		Circulatory		Respiratory		Digestive		Injury	
	Total	Male	Female	Male	Female	Male	Female	Male	Female	Male	Female	Male	Female	Male	Female

	Total	Male	Female	Male	Female	Male	Female	Male	Female	Male	Female	Male	Female	Male	Female
WORLD															
AFRICA															
Mauritius	883.1	1,128.2	679.7	24.4	13.3	75.5	70.2	535.0	323.9	109.8	53.2	63.4	15.9	73.2	32.8
NORTH & CENTRAL AMERICA															
Barbados	596.2	707.8	511.8	16.1	16.0	112.4	111.6	310.6	204.8	41.6	18.2	18.7	13.8	58.9	15.1
Canada	504.5	660.3	379.0	3.7	2.5	169.8	111.5	263.7	144.4	53.9	24.3	23.7	14.6	68.7	25.8
Costa Rica	603.0	733.7	494.5	24.3	13.8	157.4	115.4	241.0	164.7	78.7	56.4	32.5	24.2	78.6	20.1
Cuba	577.8	647.1	506.3	10.2	8.4	127.4	94.8	262.3	220.0	55.8	47.2	23.1	18.8	92.3	45.8
Guatemala	1,023.5	1,114.5	933.5	227.0	205.2	47.9	65.8	128.4	111.0	177.8	152.1	45.9	27.8	103.1	20.9
Trinidad and Tobago	899.9	1,049.0	776.1	22.7	18.1	108.3	94.8	462.3	348.6	80.8	50.1	53.9	23.4	93.3	30.8
United States	559.3	727.7	426.1	11.7	6.4	163.6	109.9	298.3	173.8	57.3	29.0	26.5	15.9	82.8	28.1
SOUTH AMERICA															
Argentina	687.0	854.4	541.8	26.6	18.7	150.5	97.6	377.4	248.0	53.0	30.0	43.6	23.0	68.5	23.9
Chile	685.2	867.2	538.0	29.9	18.3	141.2	112.4	230.1	162.8	99.6	57.8	76.6	33.4	121.0	32.8
Guyana	792.3	964.8	636.2	28.0	15.4	54.9	51.6	353.7	219.5	61.4	41.9	122.4	57.3	86.2	24.7
Suriname	864.2	1,108.2	684.8	56.7	28.4	81.7	87.4	329.5	198.3	87.2	46.1	62.5	34.0	162.5	49.0
Uruguay	677.0	848.4	529.9	21.5	15.0	201.6	119.8	287.9	197.0	61.5	28.2	32.1	19.5	74.0	26.9
ASIA															
Bahrain	772.2	822.0	707.2	13.6	8.0	84.6	51.5	308.3	216.3	58.9	33.3	17.3	19.4	49.0	16.8
Israel	566.2	653.7	488.7	12.3	10.7	117.4	99.7	274.9	202.2	45.2	30.0	17.9	12.1	55.6	30.0
Japan	402.1	528.4	304.9	8.7	4.5	150.8	77.8	173.8	115.9	55.8	24.7	27.5	13.0	54.6	21.4
Korea, Rep	695.1	1,007.4	487.0	38.6	15.0	149.1	64.5	321.5	168.9	47.9	24.9	96.8	28.1	90.1	31.9
Kuwait	615.6	674.9	526.3	20.1	16.1	83.5	65.9	294.9	223.9	56.1	38.5	15.8	12.8	60.7	19.3
Singapore	633.2	789.0	508.9	29.0	14.4	187.3	110.3	266.6	194.2	149.9	72.7	22.0	10.6	48.0	20.2
Sri Lanka	811.0	925.2	687.2	66.9	45.9	38.2	40.7	156.0	84.7	53.6	44.2	25.3	10.9	126.6	46.0
EUROPE															
Austria	560.4	738.7	432.9	3.9	1.4	174.5	111.0	324.4	201.0	38.9	16.1	45.5	20.1	88.1	31.8
Belgium	590.4	775.5	448.1	5.8	3.5	204.8	111.0	264.9	159.1	73.0	23.4	26.9	16.5	74.4	33.7
Bulgaria	761.3	925.1	609.6	9.2	5.3	133.9	83.9	500.7	367.8	75.4	42.0	35.7	14.6	79.8	25.1
Czechoslovakia	789.3	1,041.7	594.6	4.0	2.7	226.8	120.4	506.3	311.2	79.8	38.9	46.8	20.8	88.6	36.5
Denmark	575.9	729.3	453.9	5.3	2.9	181.8	138.9	303.2	167.4	51.8	27.3	26.7	15.7	71.3	36.2
Finland	577.0	806.1	410.7	6.7	3.7	160.0	93.1	391.1	200.8	59.0	22.2	22.0	11.8	103.9	29.5
France	524.6	726.7	367.4	8.7	5.3	208.7	91.0	209.8	115.9	46.2	20.1	44.2	20.8	91.0	37.7
German Dem Rep	671.6	872.3	540.2	4.2	2.3	164.7	102.7	427.2	285.9	59.2	22.1	X	X	X	X
Germany, Fed Rep	533.5	720.1	408.1	5.6	3.1	182.5	111.8	308.9	176.0	46.5	16.9	37.0	18.7	54.4	23.3
Greece	505.3	612.9	410.9	4.6	2.9	143.5	79.1	254.7	193.6	31.3	20.7	22.4	11.3	53.9	21.2
Hungary	827.3	1,095.0	617.9	11.0	4.0	235.5	130.2	497.6	308.3	55.2	22.2	73.4	32.2	130.2	50.8
Iceland	463.8	565.3	377.2	3.6	0.3	136.9	130.3	258.4	135.8	44.0	37.4	11.6	9.8	66.0	24.9
Ireland	669.0	849.2	520.5	7.4	3.5	174.4	126.0	400.3	229.3	115.4	67.1	20.3	13.9	56.5	21.1
Italy	549.1	723.0	414.5	3.7	2.0	193.9	100.5	279.2	177.8	54.7	20.7	49.1	21.3	54.4	20.8
Luxembourg	619.8	814.3	472.2	4.8	6.1	228.2	116.2	323.9	208.3	53.7	17.5	41.0	23.3	81.3	35.5
Malta	611.8	733.5	509.1	3.9	4.2	143.6	93.3	442.1	310.7	43.8	24.4	19.4	13.7	23.9	8.3
Netherlands	508.3	682.2	377.2	4.0	2.8	200.2	108.8	271.6	139.3	54.8	21.0	19.4	13.7	38.9	19.7
Norway	510.0	677.9	372.6	4.9	3.0	147.5	101.3	907.7	148.5	45.8	28.2	18.3	10.8	66.7	24.5
Poland	765.0	1,034.4	562.0	12.5	5.2	200.2	107.3	485.9	284.6	58.9	19.9	33.4	17.3	96.7	26.1
Portugal	595.8	771.5	455.9	8.8	4.2	140.2	82.7	275.6	190.2	54.6	25.5	50.1	18.8	62.4	29.4
Romania	829.6	986.2	692.7	12.5	6.0	125.3	82.6	496.4	408.4	126.3	80.5	55.3	28.6	94.9	32.4
Spain	497.3	647.3	375.7	8.7	4.9	158.2	81.2	241.2	164.8	59.0	24.8	44.4	18.7	54.8	17.5
Sweden	481.7	625.2	363.6	4.2	3.6	127.5	97.8	308.3	162.6	44.5	24.1	17.6	9.1	62.4	23.8
Switzerland	458.4	614.3	337.6	7.2	3.7	173.5	99.4	233.5	128.0	35.1	13.9	21.2	11.2	77.9	32.2
United Kingdom	559.9	716.1	441.6	3.8	2.7	182.8	127.5	324.3	181.3	71.6	36.8	18.9	15.0	40.4	17.3
Yugoslavia	749.0	916.7	611.0	15.8	11.5	148.8	87.2	399.5	304.1	59.5	37.0	40.8	18.4	76.8	25.3
U.S.S.R.	780.1	1,068.3	599.5	X	X	X	X	X	X	X	X	X	X	X	X
OCEANIA															
Australia	510.8	660.2	388.3	3.8	2.0	162.6	103.4	286.6	171.7	51.4	20.8	22.2	12.8	63.6	24.4
New Zealand	596.4	761.9	460.6	5.3	3.3	172.9	120.6	330.7	190.3	78.6	43.2	19.3	12.4	79.7	31.3

Source: World Health Organization.
Note: For additional information, see Sources and Technical Notes.

Sources and Technical Notes

Table 16.1 Size and Growth of Population and Labor Force, 1960–2025

Sources: United Nations Population Division, *World Population Prospects 1988* (U.N., New York, 1989); International Labour Organisation (ILO), *Economically Active Population: Estimates 1950–1980,* *Projections 1985—2025* (ILO, Geneva, 1986).

Population refers to the midyear population. Most data are estimates based on population censuses and surveys. All projections are for the medium-case scenario (see the following discussion). The average annual growth rate takes into account the effects of international migration. Data on several small countries and other enti-

ties not shown in these tables are often included in world and regional tables.

Many of the numbers in Tables 16.1–16.3 are estimated using demographic models based on a country's population size, age and sex distribution, fertility and mortality rates by age and sex groups, growth rates of both urban and rural population, and the levels of internal and international migration.

Recent population censuses and surveys are used to calculate or estimate these parameters. The United Nations Population Division Department of International Economic and Social Affairs (DIESA) compiles and evaluates census and survey results from all countries. These data are adjusted for overenumeration and underenumeration of certain age and sex groups (infants, female children, young males), misreporting of age and sex distributions, changes in definitions, and so forth, when necessary. These adjustments incorporate data from civil registrations, population surveys, earlier censuses, and, when necessary, population models based on information from socioeconomically similar countries. (Because the figures have been adjusted, they are not strictly comparable to the official statistics compiled by the United Nations Statistical Office and published in the *Demographic Yearbook*.)

After any adjustment, the data are scaled to 1980. Similar estimates are made for each five-year period between 1950 and 1980. Historical data are used when deemed accurate, with adjustments and scaling. For many developing countries, accurate historical data do not exist. The Population Division estimates for these countries use available information and demographic models.

Projections are based on estimates of the 1980 population. Age- and sex-specific mortality rates are applied to the 1980 population to determine the number of survivors at the end of each five-year period. Births are projected by applying age-specific fertility rates to the projected female population. The births are given an assumed sex ratio, and the appropriate age- and sex-specific survival rates are applied. Future migration rates are also estimated on an age- and sex-specific basis. Combining future fertility, mortality, and migration rates yields the projected size and composition of the population.

Assumptions about future mortality, fertility, and migration rates are made on a country-by-country basis and, when possible, are based on historical trends. Four scenarios of population growth (high, medium, low, and constant) are created using different assumptions about these rates. The medium-case scenario, for example, assumes medium levels of fertility, mortality, and migration, assumptions that may vary among the countries.

The labor force includes all people who produce economic goods and services. It includes all employed people (employers, the self-employed, salaried employees, wage earners, unpaid family workers, members of producer cooperatives, and members of the armed forces), and the unemployed (experienced workers and those looking for work for the first time).

The ILO determines the average *annual growth of the labor force* by multiplying the activity rates of age/sex groups (the economically active fraction of an age/sex group) by the number of people in those groups. Estimates of activity rates are based on information from national censuses and labor force surveys. ILO adjusts national labor force statistics when necessary to conform to international definitions.

Table 16.2 Trends in Births, Life Expectancy, Fertility, and Age Structure, 1965–90

Source: United Nations Population Division, *World Population Prospects 1988* (U.N., New York, 1989).

The *crude birth rate* is derived by dividing the number of live births in a given year by the midyear population. This ratio is then multiplied by 1,000.

Life expectancy at birth is the average number of years that a newborn baby is expected to live if the age-specific mortality rates effective at the year of birth apply throughout his or her lifetime.

The *total fertility rate* is an estimate of the number of children that an average woman would have if current age-specific fertility rates remained constant during her reproductive years.

The age structure shows the *percentage of the population in specific age groups:* 0–14, 15–65 and over 65 years.

For additional details, refer to the sources or to the Technical Note for Table 16.1. The United Nations includes in regional and world totals small countries and other entities not shown here.

Table 16.3 Mortality, Morbidity, and Nutrition, 1980s

Sources: Crude death rate and infant mortality data: United Nations Population Division, *World Population Prospects 1988* (U.N., New York, 1989); Child and maternal mortality, wasting, stunting, and access to health services: United Nations Children's Fund (UNICEF), *State of the World's Children 1989* (UNICEF, New York, 1989); Calorie supply as percentage of requirements: Food and Agriculture Organization of the United Nations (FAO), unpublished data, September 1986.

The *crude death rate* is derived by dividing the number of deaths in a given year by the midyear population, and multiplying by 1,000.

The *infant death rate* is derived by dividing the number of babies who die before their first birthday in a given year by the number of live births in that year, and multiplying by 1,000.

Child deaths are derived by dividing the number of children under five years of age who die in a given year by the number of live births in that year, and multiplying by 1,000. These data are provided to UNICEF by the United Nations Population Division and the United Nations Statistical Office.

Maternal deaths are the number of deaths from pregnancy- or childbirth-related causes per 100,000 live births. A maternal death is currently defined by the World Health Organization (WHO) as the death of a woman while pregnant or within 42 days of termination of pregnancy from any cause related to or aggra-vated by the pregnancy, including abortion. Most official maternal mortality rates are underestimated because causes of death are often incorrectly classified or unavailable. In some countries, over 60 percent of women's deaths are registered without a specified cause. Maternal mortality is highest in women aged 10–15 years, women over 40 years, and in women with five or more children. Data are provided to UNICEF by WHO and refer to a single year between 1980 and 1987.

Wasting indicates current acute malnutrition and refers to the percentage of children between the ages of 12 and 23 months whose weight-for-height is less than 77 percent of the median weight-for-height of the reference population of the U.S. National Center for Health Statistics (NCHS). *Stunting*, an indicator of chronic undernutrition, refers to the percentage of children between the ages of 24 and 59 months whose height-for-age is less than 77 percent of the median. NCHS, among others, has found that healthy children under the age of five years do not differ appreciably in weight or height. WHO has accepted the NCHS weight-for-age and weight-for-height standards. Children with low weight-for-age are at a high risk of mortality. Data on wasting and stunting, provided to UNICEF by WHO, refer to a single year between 1980 and 1987. It should be noted that many countries do not report data to UNICEF or WHO.

The *average calories available as a percentage of need* are calories from all food sources: domestic production, international trade, stock drawdowns, and foreign aid. The quantity of food available for human consumption, as estimated by FAO, is that which reaches the consumer. The amount actually consumed may be lower than the figures shown, depending on how much is lost in home storage, preparation, and cooking, and how much is fed to pets and domestic animals or discarded.

Table 16.4 Access to Safe Drinking Water, Sanitation, and Health Services, 1980s

Sources. Drinking water and sanitation: World Health Organization (WHO), *The International Drinking Water Supply and Sanitation Decade: Review of Mid-Decade Progress (as at December 1985)* (WHO, Geneva, September 1987); WHO, *The International Drinking Water Supply and Sanitation Decade: Review of National Baseline Data: December 1980* (WHO, Geneva, 1984); WHO, *Global Strategy for Health for All. Monitoring 1988-1989. Detailed analysis of global indicators* (WHO, Geneva, May 1989); and unpublished data (WHO, Geneva, September 1986). Urban and rural fractions of total population: United Nations Population Division, *The Prospects of World Urbanization, Revised as of 1984–85* (U.N., New York, 1987). Numbers of health care personnel: WHO, *1988 World Health Statistics Annual* (WHO, Geneva, 1988).

WHO collected data on drinking water and sanitation from national governments in 1980, 1983, and 1985, using questionnaires completed by public health officials, WHO experts, and Resident Representatives of the United Nations Development Programme. Responses to at least one questionnaire were received from 114 developing countries, representing about 70 percent of the developing world population. Data are shown also for 14 developed countries.

WHO released updated information in 1989. Most new data, shown under "1985," were gathered in 1985. Data for 49 countries were gathered during 1986–88, and information from Lebanon in 1980. For several countries in Africa, dates were not shown.

Urban and rural population were defined by each national government.

WHO defines reasonable access to *safe drinking water* in an urban area as access to piped water or a public standpipe within 200 meters of a dwelling or housing unit. In rural areas, reasonable access implies that a family member need not spend a disproportionate part of the day fetching water. "Safe" drinking water includes treated surface water and untreated water from protected springs, boreholes, and sanitary wells.

Urban areas with access to *sanitation services* are defined as urban populations served by connections to public sewers or household systems such as pit privies, pour-flush latrines, septic tanks, communal toilets, and other such facilities. Rural populations with access were defined as those with adequate disposal such as pit privies, pour-flush latrines, and so forth. Application of these definitions may vary, and comparisons can therefore be misleading.

The population with access to *health services* is defined by UNICEF as the percentage of the population that can reach appropriate local health services by the usual local means of transport in no more than one hour.

The data on *numbers of trained medical personnel* are the latest available to WHO regional offices at the beginning of 1988. Most are from 1983–86; however, some go back to 1977. Comparisons should be made with care, as categories and definitions vary among countries.

Health-care personnel have been combined into three categories:
■ *doctors*: all physicians or surgeons;
■ *nurses and midwives*: all those registered nurses and others in categories in which the term "nurse" or "nursing" appears; all midwives, birth attendants, and others in categories in which the term "midwife" appears;
■ *other*: all others directly involved in diagnosis, treatment, and prevention of disease (e.g., dentists, medical assistants, acupuncturists), and all other reported categories (e.g., pharmacists, laboratory technicians, and hospital administrators).

Access to health personnel can vary substantially within a country. The degree of access in individual countries can be partly inferred from other health data (e.g., infant mortality, immunizations) presented in this chapter.

Table 16.5 Fertility, Education, and Child Health, 1980–87

Sources: Number of children by mother's years of education: United Nations Population Division, "Education in Fertility," *Fertility Behavior in the Context of Development: Evidence from the World Fertility Survey,* Population Studies 100 (U.N., New York, 1987); Contraceptive prevalence: United Nations Population Division, *Recent Levels and Trends of Contraceptive Use as Assessed in 1987* (U.N., New York, 1988); Contraceptive prevalence data for Malawi, Mali, Guatemala, Mexico, Brazil, Colombia, Indonesia, the Philippines, Thailand, and Yemen: United Nations Population Division, *Levels and Trends of Contraceptive Use as Assessed in 1988* (U.N., New York, 1989); Contraceptive prevalence data for Cuba, India, and Viet Nam: National Family Planning and Child Survival Study Project, unpublished data (Center for Population and Family Health, Columbia University, New York, March 1988); Adult literacy and health indicators: United Nations Children's Fund (UNICEF), *State of the World's Children 1989* (UNICEF, New York, 1989); Canada–DPT, polio, measles immunization: Health and Welfare Canada, personal communication, May 1989; Iceland–low birth weight, attended births, immunizations: Ministry of Health and Social Services, personal communication, May 1989; New Zealand—measles immunization (data from 1987): Department of Health, personal communication, May 1989; Switzerland—attended births: Federal Office of Public Health, personal communication, April 1989.

Adult female and male literacy rates refer to the percentage of persons over the age of 15 who can read and write. UNESCO recommends defining as illiterate a person who cannot both read with understanding and write a short and simple statement on his or her everyday life. This concept is widely accepted, but its interpretation and application vary. Actual definitions of adult literacy are not strictly comparable among countries.

The *average number of children by mother's years of education* is shown for women who have completed less than one year of school and those with seven or more years of education. In general, the latter marry nearly four years later, have approximately 25 percent higher contraceptive use, and breast-feed their babies for a shorter period. Education is also associated with lower maternal and child mortality. (For this reason and others, education may not always have a net negative effect on numbers of children who are born or who survive.)

Contraceptive prevalence is the percentage of *couples using contraception* when the woman is of childbearing age. The data were obtained from nationally representative sample surveys conducted over several years of married or cohabiting women between the ages of 15 and 49 years. The ages of women in some surveys varied slightly from this range. These surveys were conducted as part of the World Fertility Survey (WFS), contraceptive prevalence surveys (CPSs), and independent national surveys on fertility and family planning.

The percentage of *births attended by trained personnel* includes all health personnel accepted by national authorities as part of the health system. Personnel included vary by country. Some countries include traditional birthing assistants and midwives; others, only doctors.

ORS use refers to use of oral rehydration salts for children to combat diarrheal disease leading to dehydration or malnutrition.

The percentage of *low-birth-weight infants* refers to all babies weighing 2,500 grams or less at birth. WHO has adopted the standard that healthy babies should weigh more than 2,500 grams at birth. These data refer to a single year between 1982 and 1987.

Immunization data show the *percentage of 1-year-olds fully immunized against: TB* (tuberculosis); *DPT* (diptheria, pertussis [whooping cough], and tetanus); *polio;* and *measles.* Immunization data are supplied by WHO and UNICEF field offices. Data for some countries are outside the range of years or ages indicated.

Table 16.6 Selected Causes of Death, 1980s

Source: World Health Organization (WHO), *1988 World Health Statistics* (WHO, Geneva, 1988).

The data on *deaths per 100,000 population* show underlying causes of death in official statistics transmitted to WHO by appropriate national authorities. Relatively few countries compile these data and report them to WHO. Figures for provisional causes of death are used occasionally; final statistics are substituted as soon as possible. Data are standardized by age, thereby controlling for the effect of differences in countries' age structures. The present table uses the "world" population figures; WHO also reports "European" population, which has a lower proportion in age groups up to the age of 30 and higher numbers thereafter.

The data for *malignant neoplasms* include, but are not limited to, those of the stomach, trachea, bronchus, lungs, and female breast. *Circulatory disease* includes, among others, ischemic heart disease and cerebrovascular disease. Chronic and unspecified bronchitis, emphysema, and asthma are among the diseases in the *respiratory* column; *digestive* diseases include chronic liver disease and cirrhosis. Motor vehicle accidents, suicide, and self-inflicted wounds are among the causes of death in the injury column.

17. Land Cover and Settlements

The data in this chapter show the nature of and changes in the relationship of people to land. Information on country size, population density, land use patterns over time, urban populations, labor force, and population movements is included.

Data in Table 17.1 show that the growth in cropland has slowed, rising only 2.7 percent between 1975–77 and 1985–87. There was a greater percentage increase in many of the countries of Africa, Asia, Latin America, and Oceania, however, partly because of increasing population pressure. On the other hand, the amount of land devoted globally to permanent pasture remained relatively static (0.2 percent decline), although it increased dramatically in Latin America.

Table 17.2 shows that urbanization continues to grow in both developed and developing countries, though the rates are generally higher in developing countries. Some of this growth is caused by rural migration, but much results from high endogenous urban growth rates. People seek improvements in their lives when they move to urban areas. Many find them; others cannot satisfy even the most basic needs. Increasing urbanization can present additional, significant hazards to life and health and to a fragile social and economic fabric. The foremost of these dangers is probably disease caused by crowding, poor sanitation, and poverty.

In 32 African and Asian nations, the urban percentage of the total population has risen by at least 50 percent from 1975 to 1990. It has increased by an average of 5 percent annually in the past 30 years in all of Africa, 3.7 percent in South America, and 3.2 percent in Asia. Annual urban growth rates are now 10 percent or more in Botswana, Cape Verde, Swaziland, Tanzania, and the United Arab Emirates.

In many developing countries, a significant portion of the population now lives in urban areas. For example, 33 percent of the Egyptian people live in 21 cities with populations over 100,000, and almost half the country's total population is in urban areas. In other countries, however, the population remains heavily rural despite the existence of large numbers of major cities. China has 17 cities with populations of at least 2 million, and 234 cities with populations of more than 100,000; however, residents of urban areas represent only 21.4 percent of China's total population. In South America, 76 percent of the population is urban, ranging from a high of 91 percent in Venezuela to a low of 35 percent in Guyana. Indeed, South America is the most urbanized of the world's major regions.

Changes in the labor force have paralleled those in the urban/rural mix of the population. In 1960, 60 percent of the world's workers were agricultural. By 1980, only slightly over half were agricultural workers, and the proportion in agriculture has declined in all regions, despite increasing cropland.

The worldwide increase from 18 to 21 percent employed in industry obscures changes that range from a 10-percent increase in the U.S.S.R. and smaller increases in Africa and Asia, to no change in Europe and reductions in North and Central America. In western economies, in which heavy industry often has been modernized or abandoned, and from which many industries have moved to less expensive labor markets, the portion of the work force in the service sector has risen significantly–up 14 percent in Europe. Indeed, the proportion of the population in the service sector has risen everywhere, with large increases in Oceania and South America.

Table 17.3 presents information on migrations. Some of the urban population rise results from movements of people from rural areas to cities, as well as the natural growth of urban populations. Some of these migrants from rural to urban areas are escaping deterioration of the natural environment or limited land resources in their native areas. Because of these deprivations, the migrants' basic needs, such as food, safe water, and shelter, cannot be met, or they cannot earn their livelihoods; and many of these people are searching for a better life. Some migrants are "refugees" fleeing internal or international armed conflicts that endanger their lives; some are seeking political freedom. In 1988, 14.4 million refugees left their own countries, seeking sanctuary from the vicissitudes of war and banditry. Many African and Asian countries not only lost large numbers of people as refugees, but also received significant refugee populations, further stressing already strained resource bases. In Africa and in Asia, significant movements of "internally displaced refugees" reflect both environmental and political deterioration.

Regardless of the legal, physical, and financial barriers to immigration, a surprisingly high proportion of many countries' populations is foreign born. The presence of immigrants often benefits the host country, because they may supply needed skills or take jobs that indigenous people reject. However, in developing countries, the burden on public services, already strained by movements of their own nationals, may become heavier when immigrants arrive.

Table 17.1 Land Area and Use, 1975–87

	Land Area (000 ha)	Population Density 1989 (per 1,000 ha)	Cropland 1985-87	Cropland % Change Since 1975-77	Permanent Pasture 1985-87	Permanent Pasture % Change Since 1975-77	Forest and Woodland 1985-87	Forest and Woodland % Change Since 1975-77	Other Land 1985-87	Other Land % Change Since 1975-77	Wilderness Area Total	Wilderness Area % of Total Land Area
WORLD	13,076,536	398	1,473,059	2.7	3,215,463	(0.2)	4,074,427	(2.1)	4,313,835	1.3	5,088,731	39
AFRICA	2,963,627	212	184,964	4.6	787,934	(0.5)	689,254	(4.0)	1,301,475	1.9	917,767	31
Algeria	238,174	103	7,528	(0.9)	31,132	(14.2)	4,594	11.5	194,920	2.5	140,424	59
Angola	124,670	78	3,517	0.5	29,000	0.0	53,220	(1.6)	38,933	2.3	27,050	22
Benin	11,062	415	1,839	3.3	442	0.0	3,670	(12.0)	5,111	9.5	1,209	11
Botswana	56,673	22	1,360	0.7	44,000	0.2	962	0.0	10,351	(0.8)	31,255	54
Burkina Faso	27,380	320	3,087	19.4	10,000	0.0	6,840	(8.1)	7,453	1.3	750	3
Burundi	2,565	2,065	1,329	5.3	911	6.3	65	9.6	260	(32.8)	X	X
Cameroon	46,540	235	6,983	7.2	8,300	0.0	24,980	(4.2)	6,277	11.2	1,320	3
Cape Verde	403	913	40	0.0	25	0.0	1	0.0	337	0.0	X	X
Central African Rep	62,298	46	1,997	5.1	3,000	0.0	35,840	(0.3)	21,461	(0.0)	20,917	34
Chad	125,920	44	3,188	4.0	45,000	0.0	13,050	(5.7)	64,682	1.0	61,254	48
Comoros	223	2,256	98	8.5	15	0.0	35	0.0	75	(9.2)	X	X
Congo	34,150	57	678	2.7	10,000	0.0	21,240	(0.9)	2,232	9.1	11,837	35
Cote d'Ivoire	31,800	380	3,613	22.4	3,000	0.0	6,880	(42.1)	18,307	31.2	4,268	13
Djibouti	2,318	170	X	X	200	0.0	6	0.0	2,112	0.0	X	X
Egypt	99,545	530	2,525	(7.5)	0	X	31	0.0	96,989	0.2	42,540	43
Equatorial Guinea	2,805	153	230	0.0	104	0.0	1,295	0.0	1,176	0.0	X	X
Ethiopia	110,100	415	13,930	1.4	45,100	(1.1)	27,500	(3.5)	23,570	5.9	19,716	19
Gabon	25,767	44	452	13.0	4,700	(1.4)	20,000	0.0	615	2.4	7,333	27
Gambia, The	1,000	835	167	10.1	90	0.0	180	(24.7)	563	8.4	X	X
Ghana	23,002	633	2,850	5.5	3,410	(2.8)	8,350	(7.7)	8,392	8.4	X	X
Guinea	24,586	273	1,577	0.4	3,000	0.0	10,060	(8.5)	9,949	10.3	X	X
Guinea-Bissau	2,812	344	330	15.8	1,080	0.0	1,070	0.0	332	(11.9)	X	X
Kenya	56,697	425	2,387	5.6	3,740	(1.0)	3,680	(7.5)	46,890	0.4	11,221	19
Lesotho	3,035	568	307	(9.4)	2,000	0.0	0	0.0	728	4.5	2,133	70
Liberia	9,632	257	371	0.9	240	0.0	2,103	0.0	6,918	(0.0)	1,420	13
Libya	175,954	25	2,136	3.7	13,300	9.0	660	16.1	159,858	(0.8)	65,497	37
Madagascar	58,154	200	3,057	8.5	34,000	0.0	14,900	(9.7)	6,197	28.1	691	1
Malawi	9,408	866	2,376	4.3	1,840	0.0	4,410	(13.2)	782	272.2	781	8
Mali	122,019	74	2,075	4.6	30,000	0.0	8,560	(4.5)	81,384	0.4	58,814	47
Mauritania	102,522	19	198	0.0	39,250	0.0	15,000	(0.9)	48,074	0.3	71,370	69
Mauritius	185	5,892	107	0.3	7	0.0	58	0.0	13	(2.5)	X	X
Morocco	44,630	549	8,443	9.0	20,900	7.7	5,200	0.2	10,087	(17.9)	X	X
Mozambique, People's Rep	78,409	194	3,090	0.3	44,000	0.0	14,970	(6.0)	16,349	6.2	6,130	8
Niger	126,670	54	3,537	32.0	9,253	(8.4)	2,540	(19.1)	111,340	0.5	65,633	55
Nigeria	91,077	1,199	31,252	3.8	20,970	0.8	14,600	(17.0)	24,255	7.5	1,526	2
Rwanda	2,495	2,801	1,113	18.6	410	(27.3)	503	(5.5)	469	1.8	X	X
Senegal	19,253	372	5,225	3.5	5,700	0.0	5,939	(4.2)	2,389	3.7	1,586	8
Sierra Leone	7,162	565	1,793	9.3	2,204	0.0	2,083	(2.3)	1,082	(8.7)	X	X
Somalia	62,734	117	930	3.0	28,850	0.0	8,850	(5.3)	24,104	2.0	10,460	17
South Africa	122,104	282	13,169	(1.7)	81,378	(0.2)	4,515	8.8	23,042	0.2	X	X
Sudan	237,600	103	12,478	1.8	56,000	0.0	47,080	(6.2)	122,042	2.4	79,377	32
Swaziland	1,720	444	164	0.8	1,152	(3.7)	105	0.3	299	16.8	X	X
Tanzania	88,604	297	5,212	2.9	35,000	0.0	42,545	(2.7)	5,847	21.6	7,053	8
Togo	5,439	616	1,429	1.0	200	0.0	1,400	(26.3)	2,410	25.3	X	X
Tunisia	15,536	514	4,766	(3.6)	3,035	8.3	560	9.3	7,175	(1.4)	1,901	12
Uganda	19,955	892	6,670	21.4	5,000	0.0	5,760	(7.9)	2,525	(21.3)	530	2
Zaire	226,760	154	6,647	8.2	9,221	0.0	175,630	(1.9)	35,262	8.7	11,763	5
Zambia	74,072	110	5,195	3.8	35,000	0.0	29,290	(3.0)	4,587	18.3	15,075	20
Zimbabwe	38,667	244	2,757	9.1	4,856	0.0	19,930	0.0	11,124	(2.0)	X	X
NORTH & CENTRAL AMERICA	2,137,796	197	273,829	2.1	367,020	2.5	684,544	(2.2)	812,403	0.5	900,665	42
Barbados	43	6,023	33	0.0	4	0.0	0	0.0	6	0.0	X	X
Canada	922,097	29	46,010	5.4	31,500	32.4	352,000	5.9	492,587	(5.7)	640,587	65
Costa Rica	5,106	576	525	6.1	2,293	34.1	1,640	(22.9)	648	(16.4)	X	X
Cuba	11,086	923	3,305	5.9	2,752	3.0	2,739	13.2	2,290	(20.3)	X	X
Dominican Rep	4,838	1,451	1,473	13.2	2,092	0.0	623	(3.1)	650	(18.9)	X	X
El Salvador	2,072	2,478	733	8.9	610	0.0	106	(35.4)	623	(1.0)	X	X
Guatemala	10,843	824	1,848	10.2	1,360	7.9	4,070	(16.4)	3,565	17.4	X	X
Haiti	2,756	2,316	905	4.4	496	(7.6)	52	(16.1)	1,303	1.0	X	X
Honduras	11,189	445	1,783	5.9	2,520	8.3	3,580	(18.5)	3,306	18.5	1,126	10
Jamaica	1,083	2,293	269	2.2	195	(8.6)	189	(5.0)	430	5.6	X	X
Mexico	190,869	454	24,703	3.0	74,499	0.0	44,620	(11.9)	47,047	12.6	3,050	2
Nicaragua	11,875	315	1,268	3.1	5,200	11.8	3,820	(22.6)	1,587	50.0	1,521	10
Panama	7,599	312	572	4.5	1,310	8.3	3,990	(7.0)	1,727	11.3	X	X
Trinidad and Tobago	513	2,462	119	3.2	11	0.0	224	(4.3)	159	4.1	X	X
United States	916,660	270	189,915	0.9	241,467	(0.2)	265,188	(8.8)	220,090	14.2	44,058	5
SOUTH AMERICA	1,753,473	166	141,221	14.1	472,777	4.0	904,628	(4.5)	234,847	3.1	422,270	24
Argentina	273,669	117	35,950	3.2	142,600	(0.7)	59,600	(1.1)	35,519	1.6	14,976	5
Bolivia	108,439	66	3,398	3.0	26,800	(1.2)	55,830	(1.3)	22,411	4.5	178,100	16
Brazil	845,651	174	76,717	22.7	167,000	6.4	560,420	(4.2)	41,514	0.3	202,061	24
Chile	74,880	173	5,553	4.0	11,900	1.7	8,680	0.0	48,747	(0.8)	23,086	31
Colombia	103,870	300	5,299	3.2	39,804	7.3	51,507	(5.5)	7,260	1.5	15,156	13
Ecuador	27,684	379	2,594	1.4	4,900	61.5	12,098	(19.6)	8,092	14.9	X	X
Guyana	19,685	52	495	21.3	1,230	17.0	16,369	(7.7)	1,591	223.2	12,204	57
Paraguay	39,730	105	2,176	71.2	19,152	26.0	16,564	(20.4)	1,839	(25.2)	7,726	19
Peru	128,000	170	3,710	12.8	27,120	0.0	69,400	(3.5)	27,770	8.1	36,660	29
Suriname	16,147	25	65	49.2	20	22.9	14,860	(0.3)	1,203	2.1	11,080	68
Uruguay	17,481	178	1,444	(0.0)	13,544	(0.6)	666	7.6	1,827	2.3	X	X
Venezuela	88,205	218	3,815	6.0	17,500	3.4	31,335	(8.5)	35,555	6.3	29,742	33

Table 17.1

	Land Area (000 ha)	Population Density 1989 (per 1,000 ha)	Cropland 1985-87	Percentage Change Since 1975-77	Permanent Pasture 1985-87	Percentage Change Since 1975-77	Forest and Woodland 1985-87	Percentage Change Since 1975-77	Other Land 1985-87	Percentage Change Since 1975-77	Wilderness Area Total	Percentage of Total Land Area
ASIA	2,678,653	1,139	451,030	0.8	678,546	(1.2)	539,890	(1.5)	1,009,375	1.3	372,454	14
Afghanistan	65,209	241	8,054	0.1	30,000	0.0	1,900	0.0	25,255	(0.0)	8,740	14
Bahrain	68	7,324	2	0.0	4	0.0	0	0.0	62	0.0	X	X
Bangladesh	13,391	8,404	9,154	0.3	600	0.0	2,126	(3.4)	1,511	3.0	X	X
Bhutan	4,700	316	102	19.0	218	2.0	3,295	2.5	1,085	(8.6)	1,179	25
China	932,641	1,201	97,674	(2.9)	5	0.0	123	0.0	638	0.2	210,776	22
Cyprus	924	751	158	(0.8)	11,978	(5.4)	67,129	1.0	49,210	(1.6)	X	X
India	297,319	2,811	169,002	0.5	11,817	(2.5)	121,494	(0.4)	26,740	(3.0)	1,161	0
Indonesia	181,157	981	21,107	8.1	44,000	0.0	18,020	0.1	86,750	1.3	11,761	6
Iran, Islamic Rep	163,600	336	14,830	(6.9)	4,000	0.0	1,893	(1.6)	32,394	(0.4)	15,685	10
Iraq	43,737	418	5,450	3.0	818	0.0	110	(5.2)	673	(1.5)	6,477	15
Israel	2,033	2,219	432	3.9	626	23.5	25,105	0.4	7,188	1.3	X	X
Japan	37,652	3,265	4,733	(6.0)	791	0.1	70	14.8	7,618	(0.4)	X	X
Jordan	8,893	461	414	5.5	580	0.0	13,372	0.0	644	(1.5)	X	X
Kampuchea, Dem	17,652	456	3,056	0.3	50	0.0	8,970	0.0	639	(24.3)	X	X
Korea, Dem People's Rep	12,041	1,862	2,382	9.4	83	150.0	6,518	(1.4)	1,129	13.8	X	X
Korea, Rep	9,873	4,366	2,143	(4.2)	134	0.0	2	0.0	1,642	(0.2)	X	X
Kuwait	1,782	1,131	4	266.7	800	0.0	13,100	(7.1)	8,280	13.0	X	X
Lao People's Dem Rep	23,080	172	900	5.9	10	0.0	80	(11.1)	633	6.7	437	2
Lebanon	1,023	2,832	300	(9.0)	27	0.0	19,820	(10.8)	8,633	35.6	X	X
Malaysia	32,855	516	4,375	3.1	123,338	(5.0)	15,178	(0.3)	16,652	57.9	2,844	9
Mongolia	156,500	14	1,332	38.9	362	0.0	32,272	0.3	23,051	(0.9)	24,131	15
Myanmar	65,754	621	10,067	0.7	319,080	0.0	116,848	1.0	399,039	0.4	2,547	4
Nepal	13,680	1,366	2,326	0.0	1,984	9.0	2,308	(0.5)	7,061	(2.1)	X	X
Oman	21,246	67	47	24.6	1,000	0.0	0	0.0	20,199	(0.0)	4,769	18
Pakistan	77,088	1,541	20,690	4.0	5,000	0.0	3,150	10.0	48,248	(2.2)	2,737	3
Philippines	29,817	2,043	7,920	6.5	1,180	26.2	11,150	(16.0)	9,567	17.1	X	X
Qatar	1,100	322	4	100.0	50	0.0	0	0.0	1,046	(0.2)	X	X
Saudi Arabia	214,969	63	1,178	6.7	85,000	0.0	1,200	(21.1)	127,591	0.2	67,889	28
Singapore	61	43,836	4	(50.0)	0	0.0	3	0.0	54	8.0	X	X
Sri Lanka	6,474	2,625	1,883	(1.0)	439	0.0	1,747	(2.2)	2,405	2.5	X	X
Syrian Arab Rep	18,406	655	5,627	1.3	8,299	(3.1)	524	16.2	3,956	2.9	X	X
Thailand	51,089	1,075	19,920	17.1	740	32.1	14,662	(18.4)	15,956	2.6	2,809	6
Turkey	76,963	709	27,658	(0.4)	8,800	(13.7)	20,199	0.2	20,307	7.8	X	X
United Arab Emirates	8,360	185	18	61.8	200	0.0	3	50.0	8,139	(0.1)	1,938	26
Viet Nam	32,536	2,019	6,494	1.5	310	14.0	13,000	(3.7)	12,732	3.0	X	X
Yemen Arab Rep	19,500	398	1,355	0.3	7,000	0.0	1,600	0.0	9,545	(0.0)	2,067	11
Yemen, People's Dem Rep	33,297	72	119	9.2	9,065	0.0	1,540	(6.1)	22,573	0.4	9,639	34
EUROPE	472,960	1,050	139,908	(1.0)	83,979	(3.4)	157,164	1.3	91,968	2.5	18,941	4
Albania	2,740	1,164	713	4.9	398	(2.5)	1,043	3.1	586	(8.4)	0	0
Austria	8,273	906	1,516	(6.3)	1,983	(5.9)	3,207	(1.6)	1,567	21.9	0	0
Belgium	3,282	3,026	815	(8.5)	703	(10.4)	698	(0.6)	1,066	17.9	0	0
Bulgaria	11,055	814	4,133	(4.5)	2,034	13.5	3,867	1.6	1,021	(9.5)	0	0
Czechoslovakia	12,538	1,247	5,144	(2.2)	1,642	(5.2)	4,597	1.9	1,156	9.6	0	0
Denmark	4,237	1,208	2,606	(2.0)	214	(22.0)	493	1.4	924	12.9	0	0
Finland	30,461	163	2,404	(2.3)	130	(19.4)	23,222	(0.2)	4,704	2.9	2,939	9
France	55,010	1,018	19,232	2.1	12,074	(8.9)	14,642	0.5	9,063	8.6	0	0
German Dem Rep	10,524	1,582	4,955	(0.7)	1,252	(4.1)	2,977	0.8	1,362	(0.7)	0	0
Germany, Fed Rep	24,428	2,482	7,464	(1.2)	4,528	(7.7)	7,328	0.7	5,108	9.4	0	0
Greece	13,085	767	3,940	2.0	5,255	0.0	2,620	0.1	1,270	(6.2)	0	0
Hungary	9,234	1,145	5,290	(3.2)	1,234	(4.3)	1,659	6.4	1,051	13.8	0	0
Iceland	10,025	25	8	0.0	2,274	(0.2)	120	0.0	7,623	0.0	2,975	29
Ireland	6,889	535	1,008	(16.7)	4,681	3.8	333	9.8	867	(0.1)	0	0
Italy	29,406	1,948	12,130	(1.6)	4,966	(4.2)	6,625	4.9	5,685	1.9	0	0
Luxembourg	259	1,417	X	X	0	0.0	0	0.0	0	0.0	0	0
Malta	32	10,938	13	(2.5)	0	0.0	0	0.0	19	1.8	0	0
Netherlands	3,392	4,334	907	6.8	1,108	(9.3)	300	(0.8)	1,077	6.4	0	0
Norway	30,683	137	856	7.6	99	(2.6)	8,330	0.0	21,399	(0.3)	5,627	17
Poland	30,446	1,066	14,753	(2.0)	4,060	(1.2)	8,730	1.2	2,904	8.6	0	0
Portugal	9,195	1,116	2,757	(3.4)	530	0.0	3,641	0.0	2,267	4.5	0	0
Romania	23,034	1,006	10,649	1.2	4,402	(0.8)	6,340	0.2	1,644	(6.7)	0	0
Spain	49,944	785	20,420	(1.3)	10,270	(6.4)	15,671	3.3	3,582	15.0	0	0
Sweden	41,162	203	2,969	(1.1)	568	(21.8)	28,015	0.6	9,610	0.3	2,315	3
Switzerland	3,977	1,638	412	4.1	1,609	(1.0)	1,052	0.0	904	(0.0)	0	0
United Kingdom	24,160	2,354	7,031	0.8	11,573	0.0	2,298	12.6	3,292	(8.5)	0	0
Yugoslavia	25,540	928	7,774	(2.8)	6,360	0.4	9,344	2.7	2,061	(2.6)	0	0
U.S.S.R.	2,227,200	128	232,473	(0.0)	373,667	0.2	942,667	1.9	678,393	(2.6)	752,022	34
OCEANIA	788,660	33	49,633	14.0	451,539	(3.8)	156,280	(6.7)	185,375	14.3	260,346	30
Australia	761,793	22	47,885	14.1	437,136	(4.0)	106,000	(9.6)	170,772	15.9	229,431	33
Fiji	1,827	404	240	3.7	60	(7.7)	1,185	0.0	342	(1.1)	0	0
New Zealand	26,867	125	518	21.2	13,857	0.7	7,213	2.4	5,279	(6.3)	3,723	14
Papua New Guinea	45,286	86	385	8.9	87	(18.5)	38,270	(0.5)	6,545	3.0	3,903	8
Solomon Islands	2,754	115	56	10.6	39	0.0	2,560	0.0	144	(3.6)	0	0

Sources: Food and Agriculture Organization of the United Nations, United Nations Population Division, and Sierra Club.
Notes: 0 = zero or less than half the unit of measure; X = not available; negative numbers are shown in parentheses.
  For additional information, see Sources and Technical Notes.

Table 17.2 Urban and Rural Populations, Settlements, and

	Urban Population as a Percentage of Total [a]			Average Annual Population Change 1960-90 (percent)		Percentage of Population in Urban Areas by Size of Area (000) [a]						Year of Data [b]	Total Labor Force 1985 (000)	Percentage of Labor Force in					
														Agriculture		Industry		Services	
	1960	1975	1990	Urban	Rural	50-100	100-250	250-500	500-1,000	1,000-2,000	2,000+			1960	1980	1960	1980	1960	1980
WORLD	34.2	38.5	42.7	2.6	1.4								2,163,644	60	51	18	21	22	28
AFRICA	18.3	25.3	34.5	5.0	2.1								213,792	78	69	8	12	14	19
Algeria	30.4	40.3	44.7	4.2	2.1	X	3.5	1.7	2.5	8.0	0.0	77	4,834	67	31	12	27	21	42
Angola	10.4	17.8	28.3	5.9	1.7	X	0.0	5.4	0.0	0.0	0.0	70	3,719	69	74	12	10	19	17
Benin	9.5	21.5	42.0	7.7	1.0	X	3.6	0.0	0.0	0.0	0.0	81	1,964	54	70	9	7	37	23
Botswana	1.8	12.0	23.6	12.6	2.5	X	9.7	0.0	0.0	0.0	0.0	87	381	92	70	3	13	5	17
Burkina Faso	4.7	6.3	9.0	4.6	2.2	X	2.5	4.4	0.0	0.0	0.0	83	3,765	92	87	5	4	3	9
Burundi	2.2	3.0	7.3	6.3	1.9	X	4.6	0.0	0.0	0.0	0.0	87	2,520	90	93	3	2	7	5
Cameroon	13.9	26.9	49.4	6.8	0.6	X	2.3	0.0	6.6	10.4	0.0	86	3,958	87	70	5	8	8	22
Cape Verde	6.7	30.2	61.5	10.0	(0.8)	17.7	0.0	0.0	0.0	0.0	0.0	80	121	70	52	12	23	18	26
Central African Rep	22.7	34.2	46.6	4.5	0.8	X	0.0	18.4	0.0	0.0	0.0	84	1,282	94	72	2	6	4	21
Chad	7.0	15.2	33.3	7.5	1.0	X	3.6	0.0	0.0	0.0	0.0	72	1,790	95	83	2	5	3	12
Comoros	4.1	21.3	27.6	9.7	2.0	X	0.0	0.0	0.0	0.0	0.0	X	204	89	83	4	6	7	11
Congo	33.0	35.8	42.2	3.3	1.9	X	17.1	34.3	0.0	0.0	0.0	84	710	52	62	17	12	31	26
Cote d'Ivoire	19.3	32.2	46.6	7.2	2.7	X	0.0	2.8	0.0	14.5	0.0	75	4,053	89	65	2	8	9	27
Djibouti	49.6	68.5	80.7	7.3	2.2	17.0	0.0	0.0	0.0	0.0	0.0	70	X	X	X	X	X	X	X
Egypt	37.9	43.5	48.8	3.3	1.8	2.6	4.0	4.8	1.5	4.0	19.1	86	12,837	58	46	12	20	30	34
Equatorial Guinea	25.5	46.6	64.5	5.1	(0.6)	X	0.0	0.0	0.0	0.0	0.0	83	169	82	66	6	11	12	23
Ethiopia	6.4	9.5	12.9	4.6	2.0	X	0.3	0.7	0.0	3.7	0.0	87	19,182	88	80	5	8	7	12
Gabon	17.4	30.6	45.7	6.3	1.5	X	0.0	0.0	0.0	0.0	0.0	67	518	85	75	7	11	8	14
Gambia, The	12.4	16.6	22.5	5.1	2.6	12.2	17.1	0.0	0.0	0.0	0.0	80	307	89	84	5	7	7	9
Ghana	23.3	29.8	33.0	3.9	2.2	1.9	1.2	2.5	5.4	0.0	0.0	70	4,963	64	56	14	18	22	26
Guinea	9.9	16.3	25.6	5.4	1.5	X	3.2	0.0	0.0	0.0	0.0	67	2,846	88	81	6	9	6	10
Guinea-Bissau	13.6	20.8	30.8	4.8	1.3	X	12.3	0.0	0.0	0.0	0.0	79	427	87	82	2	4	11	14
Kenya	7.4	12.9	23.6	8.8	4.0	1.2	0.7	2.1	0.0	5.6	0.0	79	8,389	86	81	5	7	9	12
Lesotho	3.4	10.8	20.3	8.7	1.7	X	0.0	0.0	0.0	0.0	0.0	72	730	93	86	2	4	5	10
Liberia	18.6	30.4	44.0	6.1	1.8	X	0.0	19.2	0.0	0.0	0.0	84	808	80	74	10	9	10	16
Libya	22.7	46.8	70.2	8.1	0.9	X	2.9	7.8	15.3	0.0	0.0	73	904	53	18	17	29	30	53
Madagascar	10.6	16.3	25.0	5.7	2.2	X	0.0	3.8	0.0	0.0	0.0	71	4,510	93	81	2	6	5	13
Malawi	4.4	7.7	14.8	7.2	2.5	X	3.4	4.8	0.0	0.0	0.0	87	3,074	92	83	3	7	5	9
Mali	11.1	16.2	19.2	4.3	2.0	1.5	0.0	5.2	0.0	0.0	0.0	76	2,598	94	86	3	2	3	13
Mauritania	6.7	19.6	42.1	8.9	0.8	X	0.0	0.0	0.0	0.0	0.0	76	590	91	69	3	9	6	22
Mauritius	33.2	43.6	42.3	2.5	1.2	17.4	13.0	0.0	0.0	0.0	0.0	86	390	40	28	26	24	35	48
Morocco	29.3	37.8	48.5	4.3	1.5	X	6.3	10.7	8.9	0.0	11.0	81	6,676	62	46	14	25	24	29
Mozambique	3.7	8.6	26.8	9.5	1.6	X	1.3	1.8	6.3	0.0	0.0	86	7,671	81	85	18	7	11	8
Niger	5.8	10.6	19.5	6.9	2.1	X	3.7	0.0	0.0	0.0	0.0	77	3,203	95	91	1	2	4	7
Nigeria	14.4	23.4	35.2	6.5	2.4	X	3.7	1.7	0.9	1.1	0.0	75	36,568	71	68	10	12	17	20
Rwanda	2.4	4.0	7.7	7.4	3.1	X	1.9	0.0	0.0	0.0	0.0	78	3,063	95	93	1	3	4	4
Senegal	31.9	34.2	38.4	3.6	2.7	3.3	3.5	0.0	12.4	0.0	0.0	76	2,897	84	81	5	6	11	13
Sierra Leone	13.0	21.1	32.2	5.2	1.2	X	0.0	13.0	0.0	0.0	0.0	85	1,352	78	70	12	14	10	16
Somalia	17.3	25.6	36.4	5.8	2.3	X	4.9	0.0	0.0	0.0	0.0	72	1,999	88	76	4	8	8	16
South Africa	46.6	50.5	58.9	3.2	1.5	X	8.3	0.0	16.1	14.1	0.0	85	10,831	32	17	30	35	38	49
Sudan	10.3	18.9	22.0	5.4	2.3	X	4.0	2.1	2.6	0.0	0.0	80	6,991	86	71	6	8	8	21
Swaziland	3.9	14.0	33.1	10.6	1.8	X	0.0	0.0	0.0	0.0	0.0	76	273	89	74	4	9	7	17
Tanzania	4.7	10.1	32.8	10.3	2.2	X	3.2	1.1	0.0	4.9	0.0	85	10,913	89	86	4	5	7	10
Togo	9.8	15.8	25.7	6.1	2.1	X	5.0	0.0	0.0	0.0	0.0	70	1,244	80	73	8	10	12	17
Tunisia	36.0	47.6	54.3	3.6	1.1	X	3.3	0.0	8.4	0.0	0.0	84	2,224	56	35	18	36	26	29
Uganda	5.1	8.3	10.4	6.0	3.3	X	0.0	2.1	0.0	0.0	0.0	69	7,054	89	86	4	4	7	10
Zaire	22.3	32.2	39.5	4.7	1.9	X	3.2	1.9	3.5	1.8	8.9	84	11,666	83	72	9	13	8	16
Zambia	17.2	36.3	55.6	7.5	1.2	X	8.4	9.0	8.0	0.0	0.0	80	2,242	79	73	7	10	14	17
Zimbabwe	12.6	19.4	27.6	5.9	2.5	X	2.3	4.9	7.8	0.0	0.0	83	3,410	69	73	11	11	20	17
NORTH & CENTRAL AMERICA	63.2	67.1	71.0	1.9	0.7								176,065	18	12	32	29	50	58
Barbados	35.4	38.6	44.7	1.2	(0.1)	34.8	0.0	0.0	0.0	0.0	0.0	80	127	26	10	27	21	47	69
Canada	68.9	75.6	76.4	1.7	0.4	7.9	9.8	11.0	17.5	5.4	24.3	86	12,723	13	5	34	29	52	65
Costa Rica	36.6	42.2	53.6	4.3	2.0	0.0	0.0	15.2	0.0	0.0	0.0	70	904	51	31	19	23	30	46
Cuba	54.9	64.2	74.9	2.3	(0.7)	5.3	9.9	6.3	0.0	0.0	20.3	86	3,987	39	24	22	29	39	48
Dominican Rep	30.2	45.3	60.4	5.1	0.8	6.1	3.9	0.0	13.1	0.0	0.0	70	1,862	67	46	12	15	21	39
El Salvador	38.3	40.4	44.4	2.9	2.1	6.3	0.0	6.1	0.0	0.0	0.0	71	1,832	62	43	17	19	21	37
Guatemala	33.0	37.1	42.0	3.7	2.4	X	0.0	0.0	9.5	0.0	0.0	81	2,261	67	57	14	17	19	26
Haiti	15.6	22.1	30.3	4.2	1.3	X	0.0	0.0	11.2	0.0	0.0	84	2,822	80	70	6	8	14	22
Honduras	22.7	32.3	43.6	5.6	2.2	4.9	2.4	9.1	13.7	0.0	0.0	85	1,303	70	61	11	16	19	23
Jamaica	33.8	44.1	52.3	3.0	0.4	0.0	0.0	0.0	22.4	0.0	0.0	82	1,095	39	31	25	16	36	52
Mexico	50.8	62.8	72.6	4.1	0.9	3.0	6.3	8.7	5.3	1.7	24.4	80	26,080	55	37	20	29	25	35
Nicaragua	39.6	50.3	59.8	4.7	1.8	8.0	0.0	0.0	18.6	0.0	0.0	79	993	62	47	16	16	22	38
Panama	41.2	49.1	54.8	3.5	1.6	X	10.2	19.9	0.0	0.0	0.0	86	760	51	32	14	18	35	50
Trinidad and Tobago	22.5	48.4	69.1	5.3	(1.7)	34.1	0.0	0.0	0.0	0.0	0.0	82	450	22	10	34	39	44	51
United States	70.0	73.6	74.0	1.3	0.6	X	8.2	8.7	10.6	10.5	39.0	86	116,800	7	4	36	31	57	66
SOUTH AMERICA	51.7	64.5	76.1	3.7	(0.0)								93,776	44	29	22	26	33	45
Argentina	73.6	80.6	86.2	2.0	(0.7)	6.1	4.8	5.4	6.1	6.8	35.1	80	10,884	20	13	36	34	44	53
Bolivia	39.3	41.5	51.4	3.5	1.8	2.2	4.6	11.9	15.6	0.0	0.0	85	1,987	61	46	18	20	21	34
Brazil	44.9	61.8	76.9	4.3	(0.5)	3.5	0.0	0.0	0.0	0.0	0.0	85	49,642	52	31	15	27	33	42
Chile	67.8	78.3	85.6	2.6	(0.9)	9.4	15.5	6.8	34.1	0.0	0.0	85	4,276	30	17	20	25	50	58
Colombia	48.2	60.8	70.3	3.7	0.5	2.7	10.8	4.5	5.0	9.8	14.5	85	9,195	51	34	19	24	30	42
Ecuador	34.4	42.4	56.9	4.8	1.6	X	7.5	0.0	9.8	13.6	0.0	86	2,839	57	39	19	20	24	42
Guyana	29.0	29.6	34.6	2.6	1.8	X	19.6	0.0	0.0	0.0	0.0	76	337	38	27	27	26	35	47
Paraguay	35.6	39.0	47.5	4.0	2.3	3.1	0.0	0.0	19.5	0.0	0.0	82	1,223	56	49	19	21	25	31
Peru	46.3	61.4	70.2	4.2	0.7	7.3	5.7	6.6	5.4	0.0	25.4	85	6,204	52	40	20	18	27	42
Suriname	47.3	44.8	47.5	1.1	1.1	X	48.6	0.0	0.0	0.0	0.0	64	117	30	20	22	20	48	60
Uruguay	80.1	83.0	85.5	0.9	(0.4)	6.3	0.0	0.0	0.0	41.4	0.0	85	1,171	21	16	30	29	49	55
Venezuela	66.6	77.8	90.5	4.3	(1.0)	6.7	6.0	9.0	9.1	14.0	18.8	87	5,871	35	16	22	28	43	56

Labor Force, 1960–90 — Table 17.2

	Urban Population as a Percentage of Total(a)			Average Annual Population Change 1960-90 (percent)		Percentage of Population in Urban Areas by Size of Area (000){a}						Year of Data{b}	Total Labor Force 1985 (000)	Percentage of Labor Force in Agriculture		Industry		Services	
	1960	1975	1990	Urban	Rural	50-100	100-250	260-500	500-1,000	1,000-2,000	2,000+			1960	1980	1960	1980	1960	1980
ASIA	21.5	25.3	29.9	3.2	1.7								1,299,138	75	66	10	15	15	19
Afghanistan	8.0	13.1	21.7	4.9	0.9	X	3.1	0.0	0.0	7.9	0.0	86	4,971	85	X	6	X	9	X
Bahrain	78.6	79.3	82.9	4.2	3.3	20.6	25.2	0.0	0.0	0.0	0.0	81	181	14	3	45	35	42	62
Bangladesh	5.1	9.1	13.6	6.2	2.4	1.1	1.4	0.0	0.0	1.4	3.4	81	28,845	87	75	3	6	10	19
Bhutan	2.5	3.5	5.3	4.5	1.8	X	0.0	0.0	0.0	0.0	0.0	77	632	95	92	2	3	3	5
China	19.0	20.2	21.4	2.2	1.7	X	1.2	2.0	3.2	3.1	12.3	82	617,906	X	74	X	14	X	12
Cyprus	35.6	43.4	52.8	2.0	(0.4)	38.0	0.0	0.0	0.0	0.0	0.0	86	312	42	26	27	34	31	40
India	18.0	21.5	28.0	3.7	1.8	1.6	2.5	1.8	2.7	0.7	4.7	81	293,194	74	70	11	13	15	17
Indonesia	14.6	19.4	28.8	4.5	1.5	X	1.5	1.7	1.2	2.3	5.1	80	63,430	75	57	8	13	17	30
Iran, Islamic Rep	33.6	45.7	54.9	5.2	2.2	3.4	8.7	4.6	8.1	2.5	12.8	82	13,023	54	36	23	33	23	31
Iraq	42.9	61.4	74.2	5.4	0.7	X	4.1	4.1	0.0	12.5	0.0	70	4,259	53	30	18	22	29	48
Israel	77.0	86.6	91.6	3.2	(0.8)	X	22.6	20.1	0.0	38.0	0.0	86	1,610	14	6	35	32	51	62
Japan	62.5	75.7	77.0	1.6	(0.7)	12.0	16.1	16.2	4.9	14.2	0.0	86	59,772	33	11	30	34	37	55
Jordan	42.7	55.4	68.1	4.7	1.1	8.6	4.0	7.9	23.1	0.0	0.0	85	799	44	10	26	26	30	64
Kampuchea, Dem	10.3	10.3	11.6	1.8	1.4	X	0.0	5.4	0.0	0.0	0.0	62	3,602	X	X	X	X	X	X
Korea, Dem People's Rep	65.8	X	X	X	X	X	X	X	X	X	X	X	9,084	62	43	23	30	15	27
Korea, Rep	0.6	0.7	1.1	3.9	1.9	3.2	8.1	7.7	5.4	3.4	36.8	85	16,790	66	36	9	27	25	37
Kuwait	72.3	83.8	95.6	8.0	0.6	X	13.5	0.0	0.0	0.0	0.0	75	677	1	2	34	32	65	67
Lao People's Dem Rep	7.9	11.4	18.6	5.1	1.7	X	3.2	0.0	0.0	0.0	0.0	66	2,014	83	76	4	7	13	17
Lebanon	39.6	68.6	83.7	4.1	(2.8)	X	4.8	0.0	35.2	0.0	0.0	70	769	38	X	23	X	39	X
Malaysia	25.2	30.5	42.3	4.3	1.7	X	11.3	1.9	0.0	0.0	0.0	86	6,171	63	42	12	19	25	39
Mongolia	35.7	48.7	51.2	4.2	2.0	X	0.0	27.0	0.0	0.0	0.0	87	894	70	40	13	21	17	39
Myanmar	19.3	23.9	24.6	3.0	2.0	X	2.3	0.0	1.4	0.0	6.6	83	16,699	X	53	X	19	X	28
Nepal	3.1	4.8	9.6	6.3	2.2	1.1	1.4	0.0	0.0	0.0	0.0	81	6,870	93	93	2	1	5	7
Oman	3.5	6.1	10.6	7.9	3.7	X	0.0	0.0	0.0	0.0	0.0	60	361	X	50	X	22	X	28
Pakistan	22.1	26.4	32.0	4.3	2.6	X	1.8	0.9	3.5	1.1	8.1	81	29,801	61	55	18	16	21	30
Philippines	30.3	35.6	42.4	3.9	2.1	9.6	8.5	5.3	3.4	4.6	12.3	86	19,874	61	52	15	16	24	33
Qatar	72.4	83.6	89.5	8.0	3.8	X	69.0	0.0	0.0	0.0	0.0	86	146	18	3	24	28	59	69
Saudi Arabia	29.7	58.7	77.3	7.6	0.4	X	5.5	3.2	10.6	0.0	0.0	74	3,405	71	48	10	14	19	37
Singapore	100.0	100.0	100.0	1.7	0.0	X	0.0	0.0	0.0	0.0	102.8	87	1,226	8	2	23	38	69	61
Sri Lanka	17.9	22.0	21.4	2.5	1.7	X	3.4	0.0	4.2	0.0	0.0	86	5,920	57	53	13	14	30	33
Syrian Arab Rep	36.8	45.4	51.8	4.6	2.5	1.2	7.6	4.1	0.0	23.9	0.0	87	2,596	54	32	19	32	27	36
Thailand	12.5	15.2	22.6	4.6	2.1	1.3	1.0	0.0	0.0	0.0	9.1	80	26,657	84	71	4	10	12	19
Turkey	29.7	41.6	48.4	4.1	1.3	6.6	6.5	5.7	2.8	3.0	15.6	85	21,385	79	58	11	17	10	25
United Arab Emirates	40.0	79.8	77.8	12.5	6.5	X	35.4	20.0	0.0	0.0	0.0	80	683	X	5	X	38	X	57
Viet Nam	14.7	18.8	21.9	3.6	1.9	X	5.0	0.5	0.0	2.1	10.0	79	28,755	81	68	5	12	14	21
Yemen Arab Rep	3.4	11.0	25.0	9.3	1.5	X	2.0	0.0	0.0	0.0	0.0	75	1,676	70	69	15	9	15	22
Yemen, People's Dem Rep	28.0	34.3	43.3	3.9	1.6	X	0.0	12.7	0.0	0.0	0.0	X	558	83	41	7	18	10	41
EUROPE	60.9	68.8	73.1	1.1	(0.7)								226,373	28	14	39	39	33	47
Albania	30.6	32.8	35.3	2.9	2.1	X	7.3	0.0	0.0	0.0	0.0	86	1,398	71	56	18	26	11	18
Austria	49.9	53.1	57.7	0.7	(0.4)	1.0	5.0	14.6	0.0	19.7	0.0	86	3,504	24	9	46	41	30	50
Belgium	92.5	94.6	96.9	0.4	(2.6)	7.3	9.4	4.9	9.9	0.0	0.0	86	4,092	8	3	41	36	44	61
Bulgaria	38.6	57.5	70.3	2.5	(1.9)	X	10.6	7.1	0.0	12.3	0.0	85	4,483	57	18	25	45	18	37
Czechoslovakia	46.9	58.7	68.6	1.7	(1.3)	6.0	3.9	71.4	0.0	7.6	0.0	85	8,181	26	13	46	49	28	37
Denmark	73.7	82.1	86.4	0.9	(1.8)	2.5	8.7	0.0	0.0	26.4	0.0	86	2,784	18	7	37	32	45	61
Finland	38.1	55.1	67.9	2.3	(1.8)	23.8	0.0	10.5	19.6	0.0	0.0	86	2,488	36	12	31	35	33	53
France	62.4	73.0	74.1	1.3	(0.6)	6.5	10.7	8.8	3.8	4.1	15.6	82	24,639	22	9	39	35	39	56
German Dem Rep	72.3	75.2	77.0	0.1	(0.9)	7.8	9.3	3.6	6.4	7.3	0.0	85	9,518	18	11	48	50	34	39
Germany, Fed Rep	77.4	83.1	86.4	0.7	(1.4)	7.0	10.7	5.7	8.8	7.8	0.0	86	29,403	14	6	48	44	38	50
Greece	42.9	55.3	62.6	1.9	(0.8)	X	3.7	0.0	7.1	0.0	30.6	81	3,780	56	31	20	29	24	40
Hungary	40.0	50.1	60.3	1.6	(1.2)	7.5	105.2	0.0	0.0	0.0	19.5	86	5,215	39	18	34	44	27	38
Iceland	80.3	86.8	90.5	1.6	(1.2)	40.5	?	0.0	0.0	0.0	0.0	85	127	25	10	36	37	39	53
Ireland	45.8	53.6	59.1	1.8	(0.0)	2.1	4.2	0.0	25.4	0.0	0.0	81	1,367	36	19	25	34	39	48
Italy	59.4	65.6	68.6	0.9	(0.4)	9.4	8.9	4.4	2.5	6.7	4.9	81	22,763	31	12	40	41	29	48
Luxembourg	62.1	73.7	84.3	1.6	(2.4)	21.0	0.0	0.0	0.0	0.0	0.0	85	155	15	5	44	35	41	60
Malta	70.0	80.7	87.1	1.0	(2.5)	X	0.0	0.0	0.0	0.0	0.0	80	139	10	5	41	42	50	53
Netherlands	85.0	88.4	88.5	1.0	(0.0)	X	14.6	4.6	14.9	7.1	0.0	86	5,861	11	6	43	32	46	63
Norway	32.1	68.2	74.4	3.4	(2.7)	6.0	11.5	10.9	0.0	0.0	0.0	87	2,039	20	8	37	29	44	62
Poland	47.9	55.2	63.2	1.8	(0.3)	7.5	11.8	6.5	7.5	4.5	0.0	86	19,221	48	29	29	39	23	33
Portugal	22.5	27.8	33.3	1.8	0.0	4.1	0.0	3.2	7.9	0.0	0.0	81	4,563	44	26	29	37	27	38
Romania	34.2	46.2	50.4	2.1	(0.2)	3.2	0.0	0.6	0.0	8.6	0.0	85	11,418	65	31	15	44	20	26
Spain	56.6	69.6	78.4	2.0	(1.5)	8.4	14.5	8.5	6.5	4.4	0.0	86	13,723	42	17	31	37	27	46
Sweden	72.6	82.7	84.0	0.9	(1.4)	11.5	13.2	5.5	8.5	17.4	0.0	87	4,237	14	6	45	33	41	62
Switzerland	51.0	55.7	59.6	1.2	0.0	4.8	0.0	20.4	13.1	0.0	0.0	86	3,173	11	6	50	39	39	55
United Kingdom	85.7	89.8	92.5	0.5	(1.9)	1.2	31.8	15.4	3.5	1.8	12.1	85	27,432	4	3	48	38	48	59
Yugoslavia	27.9	38.5	50.2	2.9	(0.4)	3.7	6.2	3.1	2.8	4.7	0.0	81	10,484	63	32	18	33	19	34
U.S.S.R.	48.8	60.0	67.5	2.1	(0.5)	X	9.0	9.0	7.5	8.1	6.5	86	143,289	42	20	29	39	29	41
OCEANIA	66.3	71.8	70.9	2.1	1.2								11,211	27	20	32	28	41	52
Australia	80.6	85.9	85.5	1.8	0.6	2.7	5.6	4.4	6.3	13.7	40.2	85	7,364	11	7	40	32	49	61
Fiji	29.7	36.7	44.0	3.5	1.4	17.1	20.4	0.0	0.0	0.0	0.0	86	231	60	46	17	17	23	37
New Zealand	76.0	82.8	84.2	1.5	(0.2)	11.3	6.3	18.8	25.0	0.0	0.0	87	1,458	15	11	37	33	48	56
Papua New Guinea	2.7	11.9	15.8	8.7	2.0	2.2	4.0	0.0	0.0	0.0	0.0	86	1,685	89	76	4	10	7	14
Solomon Islands	8.4	17.5	21.8	5.8	1.9	X	0.0	0.0	0.0	0.0	0.0	76	X	X	X	X	X	X	X

Sources: United Nations Statistical Office, International Labour Organisation, and The World Bank.

Notes:
a. Urban population totals do not agree because definitions vary; moreover, data are of varying ages and from different sources.
b. Year shown is that of most recent city-size data; 1985 total country population figures were used to calculate percentage.
0 = zero or less than half the unit of measurement; X = not available; negative numbers are shown in parentheses. Totals may not add due to rounding.
For additional information, see Sources and Technical Notes.

Table 17.3 Population Movements

	Foreign-Born Population (c. 1980)			Government Perception of Significant Migration		Immi-gration	1988 Refugees in Need of Protection and/or Assistance		Permanent Refugee Settlement	"Refugee"-like Populations	Internally Displaced "Refugees"
	Number	Percent	Sex-Ratio (M/F)	+	-	1984	Country of Asylum	Country of Origin	1975-87	1988	1988
WORLD							14,421,800		X		
AFRICA							4,088,260		X		
Algeria	X	X	X			X	167,000	0	X	X	X
Angola	X	X	X		-	X	95,700	395,700	X	X	975,000
Benin	41,000	1.2	1.10			X	3,000	0	X	X	X
Botswana	16,000	1.7	1.30			X	2,700	0	X	X	X
Burkina Faso	111,000	2.1	0.91		-	X	200	0	X	X	X
Burundi	83,000	2.1	1.04			X	76,000	186,600	X	X	X
Cameroon	218,000	3.1	1.23			X	4,700	0	X	X	X
Cape Verde	X	X	X			X	0	0	X	X	X
Central African Rep	45,000	2.5	0.96			X	3,000	0	X	X	X
Chad	X	X	X		-	X	0	41,300	X	X	225,000
Comoros	45,000	4.1	0.88	+	-	X	0	0	X	X	X
Congo	X	4.1	X			X	2,100	0	X	X	X
Cote d'Ivoire	1,474,000	22.0	1.46	+		X	800	0	X	X	X
Djibouti	X	X	X	+		X	2,000	0	X	X	X
Egypt	115,000	0.3	1.42			X	5,660	0	X	60,000	X
Equatorial Guinea	X	X	X	+		X	0	0	X	X	X
Ethiopia	X	X	X			X	700,500	1,036,200	X	X	1,100,000
Gabon	X	X	X	+	-	X	100	0	X	X	X
Gambia, The	54,000	11.1	1.57	+		X	0	0	X	X	X
Ghana	562,000	6.6	1.36	+		X	100	500	X	X	X
Guinea	X	X	X		-	X	0	0	X	X	X
Guinea-Bissau	13,000	1.7	1.00			X	0	5,000	X	X	X
Kenya	157,000	1.0	1.10			X	10,600	0	X	X	X
Lesotho	X	X	X		-	X	4,000	0	X	X	X
Liberia	59,000	4.0	1.51	+		X	200	0	X	X	X
Libya	197,000	8.8	2.13	+		X	0	0	X	24,000	X
Madagascar	X	X	X			X	0	0	X	X	X
Malawi	289,000	5.2	0.95		-	X	630,000	0	X	X	X
Mali	146,000	2.3	1.01		-	X	0	0	X	X	X
Mauritania	28,000	2.1	1.4			X	0	0	X	X	X
Mauritius	14,000	1.7	1.17		-	X	0	0	X	X	X
Morocco	62,000	0.3	X		-	X	800	0	X	X	X
Mozambique, People's Rep	X	X	X		-	X	400	1,147,000	X	X	1,750,000
Niger	X	X	X			X	0	0	X	X	X
Nigeria	X	X	X			X	5,000	0	X	X	X
Rwanda	42,000	0.9	1.04			X	20,600	217,800	X	X	X
Senegal	X	X	X		-	X	5,200	0	X	X	X
Sierra Leone	79,000	3.0	X			X	100	0	X	X	X
Somalia	X	X	X	+	-	X	365,000	350,000	X	X	600,000
South Africa	963,000	3.9	1.94	+	-	X	180,000	24,900	X	X	3,570,000
Sudan	248,000	1.8	1.02	+		X	693,600	355,000	X	X	2,600,000
Swaziland	26,000	5.3	0.90		-	X	70,700	0	X	X	X
Tanzania	416,000	2.4	1.12			X	266,200	0	X	X	X
Togo	60,000	3.1	X			X	500	0	X	X	X
Tunisia	42,000	0.6	X		-	X	100	0	X	X	X
Uganda	X	X	X	+		X	125,000	9,200	X	X	225,000
Zaire	X	X	X	+		X	325,700	53,700	X	X	X
Zambia	231,000	4.1	1.10			X	149,000	0	X	X	X
Zimbabwe	X	X	X	+		X	171,500	0	X	X	X
NORTH & CENTRAL AMERICA							263,440		X		
Barbados	19,000	7.6	0.76		-	X	0	0	X	X	X
Canada	3,867,000	16.1	0.98	+		173,476	0	0	223,637	130,000	X
Costa Rica	22,000	1.2	1.14	+		X	38,700	0	X	175,000	X
Cuba	128,000	1.5	2.45			X	2,000	200	X	X	X
Dominican Rep	32,000	0.9	1.91	+	-	X	6,000	0	X	375,000	X
El Salvador	32,000	0.9	0.97		-	X	400	152,500	X	X	X
Guatemala	40,000	0.7	0.78		-	X	2,100	43,280	X	X	X
Haiti	11,000	0.3	1.03		-	X	0	8,000	X	X	X
Honduras	X	X	X	+	-	X	38,540	0	X	125,000	X
Jamaica	X	X	X		-	X	0	0	X	X	X
Mexico	269,000	0.4	1.00		-	X	162,600	0	X	X	X
Nicaragua	22,000	1.2	1.09		-	X	7,600	54,760	X	X	X
Panama	47,000	2.6	1.17		-	X	1,400	0	X	X	X
Trinidad and Tobago	60,000	6.5	0.94		-	X	0	0	X	X	X
United States	14,080,000	6.2	0.88	+		545,722	0	0	1,166,188	1,662,000	X
SOUTH AMERICA							16,410		X		
Argentina	1,912,000	6.8	1.00		-	72,000 a	4,900	0	X	X	X
Bolivia	54,000	1.2	1.10		-	X	300	0	X	X	X
Brazil	1,811,000	1.0	1.17			X	250	0	X	X	X
Chile	84,000	0.8	1.04			X	450	5,100	X	X	X
Colombia	83,000	0.4	1.08		-	X	410	0	X	X	X
Ecuador	57,000	0.9	X	+		X	700	0	X	X	X
Guyana	6,000	0.8	1.25		-	X	0	0	X	X	X
Paraguay	168,000	5.6	1.11			X	0	0	X	X	X
Peru	67,000	0.4	1.09			X	700	0	X	X	X
Suriname	X	X	X			X	0	8,000	X	X	X
Uruguay	132,000	4.7	0.96		-	X	200	0	X	X	X
Venezuela	1,039,000	7.2	1.10	+		32,000 a	500	0	X	X	X

Table 17.3

	Foreign-Born Population (c. 1980)			Government Perception of Significant Migration		Immi-gration	1988 Refugees in Need of Protection and/or Assistance		Permanent Refugee Settlement	"Refugee"-like Populations	Internally Displaced "Refugees"
	Number	Percent	Sex-Ratio (M/F)	+	-	1984	Country of Asylum	Country of Origin	1975-87	1988	1988
ASIA						9,990,890			X		
Afghanistan	X	X	X		-	X	0	5,927,180	X	X	2,000,000
Bahrain	112,000	32.0	3.09	+		X	0	0	X	X	X
Bangladesh	759,000	1.1	X			X	0	48,500	X	X	X
Bhutan	X	X	X	+		X	0	0	X	X	X
China	X	X	X			X	0	112,000	X	X	X
Cyprus	X	X	X		-	251	0	0	X	X	265,000
India	7,938,000	1.2	1.12			X	246,820	0	X	X	6,000
Indonesia	125,000	0.1	X			X	2,310	0	X	X	X
Iran, Islamic Rep	179,000	0.5	1.19	+	-	X	2,807,000	348,900	X	X	1,000,000
Iraq	X	X	X			X	75,000	508,200	X	20,000	750,000
Israel	1,422,000	42.5	0.92	+	-	X	844,700	0	X	X	X
Japan	669,000	0.6	1.06			X	520	0	X	X	X
Jordan	X	X	X	+		X	870,490	0	X	X	X
Kampuchea, Dem	X	X	X			X	0	354,190	X	X	X
Korea, Dem People's Rep	X	X	X			X	0	0	X	X	X
Korea, Rep	549,000	1.5	1.22		-	X	120	0	X	X	X
Kuwait	576,000	42.4	2.00	+		X	0	0	X	350,000	X
Lao People's Dem Rep	X	X	X			X	0	78,890	X	X	X
Lebanon	X	X	X			X	294,080	0	X	X	650,000
Malaysia	750,000	5.7	1.14			X	102,880	0	X	X	X
Mongolia	X	X	X			X	0	0	X	X	X
Myanmar	X	X	X			X	0	20,800	X	X	6,000
Nepal	234,000	1.6	0.44	+	-	X	12,000	0	X	X	X
Oman	X	X	X	+		X	0	0	X	X	X
Pakistan	X	X	X	+	-	X	3,594,600	0	X	X	X
Philippines	426,000	0.9	1.70		-	X	20,920	90,000	X	X	200,000
Qatar	X	X	X	+		X	0	0	X	X	X
Saudi Arabia	791,000	11.8	2.01	+		X	0	0	X	70,000	X
Singapore	527,000	21.8	1.02	+		X	290	0	X	X	X
Sri Lanka	48,000	0.3	1.43	+	-	X	0	91,500	X	X	450,000
Syrian Arab Rep	210,000	3.3	1.07		-	X	265,220	0	X	X	X
Thailand	273,000	0.6	1.28	+		X	439,860	0	X	X	X
Turkey	868,000	1.9	1.01		-	X	301,200 b	0	X	X	X
United Arab Emirates	356	63.9	3.75	+		X	0	0	X	X	X
Viet Nam	X	X	X		-	X	25,000	74,400	X	X	X
Yemen Arab Rep	X	X	X	+	-	X	62,000	0	X	X	X
Yemen, People's Dem Rep	X	X	X		-	X	0	55,000	X	X	X
EUROPE						54,800			X		
Albania	X	X	X			X	0	0	X	X	X
Austria	291,000	3.9	X	+		X	15,700	0	16,716	X	X
Belgium	879,000	8.9	1.20			47,002	0	0	X	X	X
Bulgaria	X	X	X			X	0	b	X	X	X
Czechoslovakia	74,000	0.5	0.68			1,451	0	2,700	X	X	X
Denmark	160,000	3.1	0.86			17,949	0	0	26,537	X	X
Finland	39,000	0.8	0.88			11,686	0	0	X	X	X
France	6,001,000	11.1	1.09	+		51,425	0	0	169,452	54,500	X
German Dem Rep	X	X	X			X	0	0	X	X	X
Germany, Fed Rep	4,535,000	7.4	1.36	+		457,093	0	0	71,348	103,000	X
Greece	127,000	1.3	1.14	+		X	4,600	0	X	X	X
Hungary	X	X	X			318	10,000	5,000	X	X	X
Iceland	6,000	2.6	0.81			1,939	0	0	X	X	X
Ireland	232,000	6.8	X		-	15,400	0	0	X	X	X
Italy	937,000	1.7	0.75	+		77,002	13,000	0	X	X	X
Luxembourg	87,000	23.8	0.94	+		X	0	0	X	X	X
Malta	X	X	X		-	X	0	0	X	X	X
Netherlands	552,000	3.8	1.32	+	-	58,627	0	0	15,115	14,000	X
Norway	82,000	2.0	1.10			19,688	0	0	9,141	X	X
Poland	2,087,000	6.4	0.85			1,559	0	3,900	X	X	X
Portugal	284,000	2.9	0.91			X	900	0	X	X	X
Romania	X	X	X			X	0	12,200	X	X	X
Spain	365,000	1.1	0.92		-	19,135	9,200	0	30,571	X	X
Sweden	422,000	5.1	1.06	+		29,245	0	0	66,753	19,000	X
Switzerland	1,064,000	16.7	0.93	+		58,617	0	0	19,986	17,000	X
United Kingdom	3,390,000	6.3	0.98	+		201,100	0	0	X	8,000	X
Yugoslavia	296,000	1.3	X		-	96	1,400	0	X	X	X
U.S.S.R.	X	X	X			X	0	0	X	X	156,000
OCEANIA						8,000			X		
Australia	3,004,000	20.6	1.07	+		101,980	0	0	150,859	X	X
Fiji	15,000	2.5	X			X	0	0	X	X	X
New Zealand	464,000	14.8	1.04		-	31,330	0	0	9,255	X	X
Papua New Guinea	1,000	0.0	1.07			X	8,000	0	X	X	X
Solomon Islands	5,000	2.5	1.17			X	0	0	X	X	X

Sources: United Nations, U.S. Committee on Refugees, World Bank, and Council of European Statisticians.

Notes: There were a total of 81,400 Namibian and 2,273,090 Palestinian refugees needing protection and/or assistance during 1988. 25,260 Vietnamese sought asylum in Hong Kong, 440 in Macao, and 180 on the Island of Taiwan. Belize provided asylum to 4,100 Central Americans; French Guiana hosted 8,000 Surinamese. The Gaza Strip and the West Bank are counted as controlled by Israel. There were an additional 62,610 refugees whose origin is not defined.
a. Estimates for 1970-75. b. There could be an additional 250,000 ethnic Turks from Bulgaria taking refuge in Turkey.
+ = significant migration into a country; - = significant migration out of a country; 0 = zero or less than half of the unit of measure;
X = not available, negative numbers are shown in parentheses; for additional information see Sources and Technical Notes.

Sources and Technical Notes

Table 17.1 Land Area and Use, 1975–87

Sources: Land area and use: Food and Agriculture Organization of the United Nations (FAO), unpublished data (FAO, Rome, June 1989); Population density: calculated from FAO and United Nations Population Division, *World Population Prospects 1988* (U.N., New York, 1989); Wilderness area: J. Michael McCloskey and Heather Spalding, "A Reconnaissance-Level Inventory of the Amount of Wilderness Remaining in the World," *Ambio*, Vol.18, No.4 (1989), and (regional and global totals) The Sierra Club, unpublished data, January 1988.

Land area and *land use* data are provided to FAO by national governments in response to annual questionnaires. FAO also compiles data from national agricultural censuses. When official information is lacking, FAO prepares its own estimates or relies on unofficial data. Several countries use definitions of total area and land use that differ from those used in this chapter. Refer to the sources for details.

FAO often adjusts the definitions of land use categories and sometimes substantially revises earlier data. For example, in 1985, FAO began to exclude from the cropland category land used for shifting cultivation but currently lying fallow. In addition, FAO's 1986 data on the area of permanent pasture in Africa were 19 percent lower than the 1983–85 average. Because these changes reflect data-reporting procedures in addition to actual land-use changes, apparent trends should be interpreted with caution.

Land use data are periodically revised and may change significantly from year to year. For the most recent land use statistics, see the latest *FAO Production Yearbook*. *Land area* data are for 1987. They exclude major inland water bodies, national claims to the continental shelf, and Exclusive Economic Zones. (See Chapter 23, "Oceans and Coasts," Table 23.1.) Antarctica is excluded from the world total.

The *population density* and *land use* figures for the world refer to the six inhabited continents. Population density was derived by using the population figures for 1989 published by the U.N. Population Division and 1987 land area data from FAO. Although the population figures were published in 1989, actual censuses and estimates were made in prior years. For additional information on population and methodology, see the Technical Notes to Table 16.1, in Chapter 16.

Cropland includes land under temporary and permanent crops, temporary meadows, market and kitchen gardens, and temporary fallow. Permanent cropland is land under crops that do not need to be replanted after each harvest, such as cocoa, coffee, rubber, fruit, and vines. (It excludes land used to grow trees for wood or timber.)

Permanent pasture is land used five or more years for forage, including natural crops and cultivated crops. This category is difficult for countries to assess because it includes wild land used for pasture. In addition, few countries regularly report data on permanent pasture. As a result, the absence of a change in permanent pasture (e.g., zero percent change for many African and Asian countries) may not reflect actual conditions, but indicate differences in land classification and data reporting. Grassland not used for forage is included under *other land*.

Forest and woodland includes land under natural or planted stands of trees, as well as logged-over areas that will be reforested in the near future.

Other land includes uncultivated land, grassland not used for pasture, built-on areas, wetlands, wasteland, and roads.

Wilderness area refers to lands showing no evidence of development, such as settlements, roads, buildings, airports, railroads, pipelines, powerlines, and reservoirs. The data were derived from 65 detailed, aeronautical, navigational maps published in the early and mid-1980s by the U.S. Defense Mapping Agency at scales of 1:2,000,000 and 1:1,000,000. The maps show human constructs in remote areas to provide orienting landmarks for navigators. Although the maps do not always show agricultural development or logging, these activities usually occur near roads and settlements. The minimum unit of wilderness surveyed was 4,000 square kilometers because it was impossible to identify smaller wilderness areas from these maps.

Wilderness areas include areas classified as forest and woodlands or other land by FAO.

Table 17.2 Urban and Rural Populations, Settlements, and Labor Force, 1960–90

Sources: Urban population as percentage of total: United Nations Population Division, *World Population Prospects* (U.N., New York, 1989); Areas with 50,000—100,000 population: U.N., Department of International Economic and Social Affairs, Statistical Office, *Compendium of Human Settlement Statistics 1983* (U.N., New York, 1985); Areas with population of 100,000 to 2 million or more: U.N. Statistical Office, *Demographic Yearbook 1987* (U.N., New York, 1989); Total labor force: International Labour Organisation (ILO), *Economically Active Population. Estimates: 1950–1980; Projections: 1985–2025* (ILO, Geneva,1986); Labor force by sector: The World Bank, *World Development Report 1984* and *World Development Report 1988* (The World Bank, Washington, D.C., 1984 (1965 data) and 1988 (1980 data)).

Urban population as percentage of total is defined as the portion of the total population residing in urban areas. The rest of the population is defined as rural. Definitions of urban vary from country to country. For a list of individual country definitions, see the sources. *Average annual population change* was calculated by the *World Resources Report*. For additional information on methods of data collection and estimation, refer to the Technical Note for Table 16.1, in Chapter 16, "Population and Health."

The *percentage of population in urban areas by size of area* was calculated using figures for populations of cities and urban aggregations reported in the *Demographic Yearbook* and total national population projections for 1990 in *World Population Prospects*. In the *Demographic Yearbook*, the United Nations reports populations of cities and urban agglomerations whose populations are 100,000 or larger, as well as the populations of capital cities (regardless of size). Where the source contained data for both a city and the urban agglomeration of which it is a part, we used the latter, unless the city-only data were considerably more recent.

For some countries, all of the figures for urban population sizes are several years old. For many other countries, recent figures were available for some cities but not for others. The differences may be as much as 10 years or more. In these cases, the year of the most recent data is shown (*year of data*).Therefore, the figures for the percentage of the population that lives in cities of varying sizes, which were calculated using the city-size data described and 1985 national population figures, are low for many countries–and probably extremely low for some. Moreover, some countries that had no urban aggregations of 50,000 or of 100,000 or more in the data year now have cities of those sizes.

The United Nations defines a "city proper" as "a locality with legally fixed boundaries and an administratively recognized urban status which is usually characterized by some form of local government," and an "urban agglomeration" as "comprising the city or town proper and also the suburban fringe or thickly settled territory lying outside of, but adjacent to, the city boundaries....For some countries or areas, the data relate to entire administrative divisions known, for example, as shi or municipos which are composed of a populated centre and adjoining territory, some of which may contain other quite separate urban localities or be distinctly rural in character." For additional information, refer to the source.

The data on cities of *50,000–100,000* population are poor for Africa (only 10 countries report data) and Asia (11 countries reporting), although these data are fairly complete for most of the rest of the world. A few of these data refer to cities with populations ranging from under 50,000 to 100,000. Note that if one adds the percentages of population by size of urban areas, the total may not correspond to the total percentage of population that is urban. These two sets of data represent potentially differing definitions of "urban" and are from different time periods. The data on the percentage of the population in cities of certain sizes include urban agglomerations that may not correspond with other definitions of urban. Estimates of the proportion of population that is urban–and of the size of urban areas–also could have been made by different observers, with differing purposes. All of these factors, especially year of estimate and definitional variation, may give rise to the observed discrepancies.

All people who work or are seeking work to produce economic goods and services comprise the *total labor force*, which includes employed people and the unemployed (both experienced workers who are without work and those looking for work for the first time). The data for total labor force, as well as *percentage of labor force in agriculture*, *industry*, and *services*, take into account information on the economically active population obtained from national censuses of population, labor force sample surveys, and other surveys conducted through 1985. Estimates are based on mid year, medium variant population figures (see Chapter 16, "Population and Health," for further information on population projections).

Table 17.3 Population Movements

Sources: Foreign born population, sex ratio (male/female) of foreign born population, significant immigration and emigration: United Nations Department of International Economic and Social Affairs (DIESA), *World Population Trends and Policies: 1987 Monitoring Report* (U.N., New York, 1988). Immigration: John J.Kelly, "Improving the Comparability of International Migration Statistics: Contributions by the Conference of European Statisticians from 1971 to Date," *International Migration Review*, Vol.21, No.4 (1988), pp.1017-1037. Immigration for Argentina and Venezuela: Gurushri Swamy, *Population and International Migration* (World Bank Staff Working Papers Number 689, Washington, D.C., 1985). Refugees in need of protection and/or assistance, permanent refugee settlement, "refugee-like" populations, and internally displaced "refugees": U.S. Committee for Refugees, *World Refugee Survey–1988 in Review* (U.S. Committee for Refugees, Washington, D.C., 1989).

The *number* and *percentage* of the *foreign-born population* is one of the few measures of international migration commonly available from national censuses. These particular data represent the most recent available census or alien registration data and date from between 1970 and 1984, although the majority of countries (58%) date from 1980–84. No attempt has been made to project these data to the present. While the percentage of the population that is foreign born gives some indication of the relative impact

of immigration on a receiving country, the absolute impact can be even more striking. Each of these numbers summarizes this century's civil and international wars, political developments, legal and illegal immigration, labor migration, refugee movements, relative economic conditions, and perceptions of areas of opportunity. One of the powers of using data on place of birth or nationality is that it includes all movements of people even when they are otherwise not easily categorized (e.g., the large movements that resulted from the partition of India).

The *sex-ratio* gives a picture of the participation of each sex in these international movements. It is derived by dividing the number of males by the number of females. A sex-ratio of 1.0 indicates that the sexes are represented equally. While for many countries the number of male migrants is greater than the number of female migrants, in global terms the sexes are just about equally represented. In the country with the largest foreign-born population, the United States, females make up 53 percent of the foreign-born population.

The United Nations maintains a data base on government perceptions of migration and its desirability. These data report government perceptions concerning whether immigration or emigration is significant in their countries. In *government perception of significant migration*, a plus sign indicates a perception of significant migration into the country, and a minus sign indicates a perception of significant migration out of the country. Blanks indicate that the government does not feel that either is significant. Some countries report both significant immigration and emigration. In the absence of detailed data on numerical importance of current international migration, these qualitative assessments provide data on its political importance and on the perception of reality.

Migration data are limited, in part, because of differences in definitions. The data (primarily from the Economic Council of Europe) on *migration* in 1984 have been adjusted, where necessary, to correspond to internationally recommended definitions of long-term migrants, although only countries of destination are shown.

The definition of a refugee is limited and specifically includes *refugees in need of protection and/or assistance*. The 1951 United Nations Convention classifies a person in flight from a country as a "refugee" if the flight is based on "a well-founded fear of being persecuted for reasons of race, religion, nationality, membership [in] a particular social group, or political opinion." The U.S. Committee for Refugees reports that

this definition has been somewhat broadened, at least in Africa, to include "external aggression, occupation, foreign domination, or events seriously disturbing to the public order. "This table includes all the refugees registered with the United Nations High Commissioner on Refugees (UNHCR) and with the United Nations Relief and Works Administration (UNRWA) in the Middle East. The U.S. Committee on Refugees has also included in these data those who clearly would be registered as refugees if the country of refuge had asked UNHCR to assist them. Therefore, for example, the 328,500 people from Kampuchea and 20,600 from Myanmar encamped in Thailand are counted as refugees even if unregistered by the UNHCR. Each country's refugee population is shown (*countries of asylum*) as well as each country's contribution to that population (*countries of origin*). The country of asylum is not the country of ultimate settlement.

Several outcomes are possible for refugees, including repatriation, continuing status as a refugee in the country of asylum, and permanent resettlement in a third country. The data on permanently resettled refugees (*permanent refugee settlement*) show the cumulative totals, where available, for the period 1975–87. These people are not included in the totals on refugees in need of assistance and protection. While developing countries are the countries of asylum for most of the world's refugees, few are able to extend permanent status to their charges.

Some people (*1988 "refugee"-like populations*) fall outside of the narrow definition of refugee, but nonetheless fear persecution or death if they were to return to their own countries (and thus are not merely economic migrants, even if their countries of residence consider them such). Because they are undocumented, estimates of their numbers vary. In any case, they clearly fall outside of national and international protections afforded refugees.

The primary legalistic definition of a refugee includes movement across an international border. Millions of people, however, flee from place to place within their own countries for the same reasons (such as fear of persecution) that motivate international refugees. In this table, estimates of *internally displaced "refugees"* include people forcibly resettled for reasons that would have made them refugees if they had crossed a national border. These movements are fueled primarily by armed conflict, and these people are ineligible for much of the assistance given refugees. Estimates of displaced people vary and are unavailable in most parts of the world.

18. Food and Agriculture

The success of the agricultural and pastoral season is the essential resource question for the vast majority of humankind. Only after provision is made for individual survival can people afford to plan for sustainable management of other resources. A woman growing maize in Kenya, a smallholder planting rice in Indonesia, and a farmer producing wheat in the United States—all put varying pressures on the resource base and, depending on the agricultural production system, may contribute to desertification, deforestation, or nitrate pollution of groundwater.

The world's agricultural production of food and other commodities has increased during the past decade and outpaced population growth. (See Table 18.1.) Regional and national data show that this increase did not occur everywhere. South America's per capita food production remained stagnant, while North and Central America saw general declines, as did Africa. Only in Asia, Europe, and the U.S.S.R. has food production grown faster than the population. Despite a 23 percent increase in cereal production, Africa's per capita food output has dropped by 8 percent in the past decade. A large percentage of the growth in European and North American cereal production has been used to feed livestock.

Yield figures demonstrate great differences in the intensity of agricultural activity. On a global average, one hectare yielded about 2.5 metric tons of cereals and 12.5 metric tons of roots and tubers. South American and Asian numbers range close to this average. It took Africa roughly twice and Europe half the area to produce these quantities.

Differences in yields are partly caused by differences in agricultural inputs. (See Table 18.2.) European agriculture applies two and a half times more fertilizer per hectare of cropland than the global average, while South America uses one third and Africa one fifth of the world average. Low rates of fertilizer applications are caused by lack of financial resources, low crop prices compared with fertilizer prices, and poor distribution structures. Other inputs show similar distributions.

Two thirds of the world's agricultural machinery is concentrated in North America and Europe, which farm one third of the world's cropland. Africa, in contrast, with 10 percent of the world's crop area, owns only 2 percent of the world's tractors, many of them out of operation because of high fuel prices and a lack of spare parts.

Outside of mechanized agricultural systems, livestock continue to supply traction power for cultivation and harvest and are an important source for food, raw materials, fertilizer, and energy. Table 18.3 shows that the number of cattle in Europe and North America declined in the past decade but increased in Mexico, many Central American nations, Africa, Asia, and South America.

The world's agricultural production systems are integrated in the world market for commodities and the flows of food aid. (See Table 18.4.) Many countries depend on agricultural trade to earn foreign exchange, but a steady decline in world commodity prices (see Chapter 15, "Basic Economic Indicators," Table 15.3) means increasing exports or decreasing export earnings. The same decreases in commodity prices, however, have made food imports more affordable for nations with agricultural deficits. The trade in food products has grown decisively in the past decade. Europe is now a net exporter of cereals, with the U.S.S.R. and Japan absorbing the greatest quantities of the world's trade in cereals.

Food aid often links agricultural areas producing subsidized surpluses with countries in need. The United States, Canada, and the European Community are the world's largest donors of cereals. Cape Verde, Jamaica, Mauritania, El Salvador, and Djibouti lead the world with the greatest cereal receipts per capita. All five countries have witnessed growing cereal receipts per capita over the past 10 years. Asia's agricultural success is underlined by a 32 percent decline in cereal receipts during the same period.

The inherent fertility of soils and the climatic regime together determine which crops can be grown, what inputs are required, and what outputs are possible. The physical and chemical characteristics of soils also determine the land's susceptibility to environmental degradation and constrain the choice of agricultural methods.

Table 18.5 presents country data on the extent of soils with physical and chemical fertility limitations for agriculture. Many of these constraints can be overcome by increased quantities of nutrients and stringent soil conservation methods. Soil capabilities vary widely among regions and within countries. For example, Africa contains large areas of soils with no inherent fertility limitations. In most of these lands, however, a shortage of moisture prevents rainfed agriculture, as can be seen in Table 18.6.

Table 18.1 Food and Agriculture Production, 1976-88

| | Index of Agricultural Production (1979-81 = 100) | | | | Index of Food Production (1979-81 = 100) | | | | Average Production Cereals | | Average Yields Cereals | | Average Yields Roots and Tubers | |
| | Total | | Per Capita | | Total | | Per Capita | | (000 metric tons) | Percent Change Since | Kilograms Per Hectare | Percent Change Since | Kilograms Per Hectare | Percent Change Since |
	1976-78	1986-88	1976-78	1986-88	1976-78	1986-88	1976-78	1986-88	1986-88	1976-78	1986-88	1976-78	1986-88	1976-78
WORLD	94	116	99	102	94	116	99	102	1,803,327	19	2,561	22	12,479	3
AFRICA	95	116	103	94	94	116	103	95	84,225	23	1,148	12	7,163	10
Algeria	90	121	99	98	90	121	99	98	2,082	25	756	34	8,446	27
Angola	103	102	114	85	98	104	109	87	364	(26)	377	(45)	4,182	9
Benin	93	142	102	115	94	136	101	110	491	44	810	10	8,425	9
Botswana	110	93	123	73	111	93	123	73	33	(57)	407	(7)	5,385	6
Burkina Faso	90	144	97	121	91	141	98	118	1,882	72	715	34	7,898	18
Burundi	93	122	99	100	98	123	104	101	489	49	1,193	5	7,427	(4)
Cameroon	98	116	106	96	99	116	108	96	907	6	963	13	2,402	7
Cape Verde	87	136	90	115	87	137	90	116	14	175	522	4	2,843	(36)
Central African Rep	95	102	102	87	94	102	100	86	122	29	814	48	5,109	58
Chad	98	121	104	103	94	120	100	102	725	16	625	20	5,334	25
Comoros	93	121	103	97	93	121	102	97	23	33	1,092	(7)	3,182	(7)
Congo	92	114	99	95	92	114	99	95	10	(46)	676	15	7,142	18
Cote d'Ivoire	85	122	97	91	85	127	96	95	1,084	34	857	14	5,672	29
Djibouti	X	X	X	X	X	X	X	X	X	X	X	X	X	X
Egypt	94	134	102	111	96	143	104	118	8,881	12	4,661	18	21,303	31
Equatorial Guinea	X	X	X	X	X	X	X	X	X	X	X	X	2,381	(12)
Ethiopia	89	105	95	92	88	106	94	94	5,824	28	1,236	22	3,013	(9)
Gabon	91	110	105	84	91	110	105	83	11	18	1,464	(2)	6,581	11
Gambia, The	114	121	125	99	115	123	127	100	105	73	1,188	24	3,000	(5)
Ghana	97	134	103	105	97	135	103	106	960	54	931	12	6,779	6
Guinea	98	109	105	93	98	109	105	93	622	15	819	(1)	7,184	1
Guinea-Bissau	95	152	110	132	95	152	110	132	222	182	942	42	6,154	14
Kenya	100	127	112	96	103	123	115	92	3,084	1	1,575	4	9,226	20
Lesotho	105	100	113	82	107	98	115	80	161	(27)	659	(44)	15,000	14
Liberia	94	117	103	93	94	120	103	96	288	16	1,213	(2)	7,180	9
Libya	77	119	89	88	77	119	89	88	282	14	660	67	7,098	24
Madagascar	97	117	107	94	97	117	106	94	2,364	11	1,757	2	6,419	5
Malawi	95	106	104	85	95	103	104	83	1,390	(7)	1,103	(9)	3,278	(28)
Mali	91	125	97	103	91	124	97	102	1,959	69	997	26	8,938	(3)
Mauritania	90	110	97	91	90	110	97	91	111	194	638	47	1,916	45
Mauritius	109	117	116	105	111	117	117	105	5	218	3,932	46	19,916	28
Morocco	96	138	103	115	96	138	103	115	6,733	51	1,282	39	14,818	22
Mozambique, People's Rep	97	101	106	85	97	101	106	85	514	(29)	587	(24)	5,843	21
Niger	86	109	93	88	86	109	93	88	1,917	31	420	(2)	7,302	(10)
Nigeria	89	126	99	100	88	127	98	100	12,062	77	1,230	43	10,551	6
Rwanda	86	103	95	82	86	99	95	79	284	13	1,109	0	7,216	(12)
Senegal	110	134	122	113	110	135	122	113	955	26	786	15	4,237	8
Sierra Leone	99	110	106	93	101	110	108	93	527	(17)	1,333	(7)	3,344	(21)
Somalia	96	127	112	99	96	127	112	99	541	110	753	76	10,557	(5)
South Africa	91	101	97	86	91	101	97	87	11,140	(3)	1,641	18	11,876	(2)
Sudan	93	113	102	91	90	112	100	90	3,606	24	521	(24)	2,573	(24)
Swaziland	85	125	94	100	86	127	94	101	97	2	1,428	(7)	1,889	(34)
Tanzania	91	116	101	89	90	116	99	90	3,904	51	1,345	34	6,920	(18)
Togo	89	110	96	89	90	106	97	86	411	62	805	(1)	8,548	(24)
Tunisia	98	120	106	101	98	120	106	101	967	6	797	21	11,574	21
Uganda	111	105	123	83	111	104	122	82	1,040	(33)	1,259	6	5,686	29
Zaire	96	120	105	96	95	119	105	96	1,145	48	846	15	7,288	8
Zambia	116	118	129	89	117	118	130	89	1,309	(21)	1,748	19	3,577	3
Zimbabwe	103	119	113	96	107	110	118	90	2,463	14	1,326	(6)	4,981	26
NORTH & CENTRAL AMERICA	93	100	98	91	94	100	98	92	343,453	3	3,600	11	19,836	7
Barbados	89	80	90	78	89	80	90	78	2	0	2,500	(4)	8,260	(20)
Canada	100	113	104	106	101	113	104	106	47,985	12	2,238	(1)	25,729	15
Costa Rica	94	112	103	92	97	108	106	89	312	6	2,284	13	7,103	(11)
Cuba	90	109	92	106	89	109	91	105	606	10	2,559	8	6,485	7
Dominican Rep	98	107	106	90	99	109	106	92	565	39	3,519	33	5,920	(1)
El Salvador	89	84	94	77	96	100	101	91	688	13	1,708	11	14,162	39
Guatemala	95	106	103	86	94	118	102	97	1,400	40	1,665	20	5,488	43
Haiti	94	106	99	93	95	108	101	95	399	1	1,144	16	4,110	(7)
Honduras	85	110	94	86	88	108	98	85	559	11	1,476	54	7,456	121
Jamaica	101	113	105	102	101	113	105	102	7	(40)	1,447	(24)	12,426	25
Mexico	90	111	97	94	89	112	96	95	22,686	28	2,280	26	14,009	11
Nicaragua	115	82	124	64	111	87	120	68	541	60	1,839	61	12,091	161
Panama	92	117	98	101	93	115	99	99	296	27	1,643	30	9,030	10
Trinidad and Tobago	140	75	148	67	141	76	149	68	9	(61)	2,501	(16)	9,386	(22)
United States	94	97	98	91	94	98	97	91	267,364	0	4,383	16	31,610	12
SOUTH AMERICA	92	116	98	100	93	117	99	100	80,232	26	2,034	22	11,562	7
Argentina	96	107	101	97	95	107	100	97	23,548	(4)	2,457	13	17,358	36
Bolivia	96	114	104	94	95	114	103	95	808	39	1,322	19	5,459	(12)
Brazil	88	123	95	106	91	126	98	108	41,341	44	1,801	32	12,190	5
Chile	89	117	94	104	89	118	94	104	2,764	70	3,370	90	14,083	44
Colombia	91	113	97	97	91	116	97	99	3,406	14	2,554	11	11,300	9
Ecuador	95	119	103	97	94	118	103	96	967	57	1,458	7	6,430	(35)
Guyana	99	85	105	74	99	85	105	74	239	(15)	2,644	14	7,023	3
Paraguay	87	133	96	106	87	134	96	108	1,275	171	1,671	20	15,575	10
Peru	102	116	110	97	105	119	114	99	2,179	41	2,446	30	8,138	15
Suriname	80	108	79	100	80	108	79	100	291	45	3,922	1	6,427	7
Uruguay	96	110	98	105	97	108	99	102	1,049	16	2,105	70	5,773	18
Venezuela	92	114	101	94	91	113	100	93	2,351	89	1,999	13	8,220	9

Table 18.1

| | Index of Agricultural Production (1979-81 = 100) | | | | Index of Food Production (1979-81 = 100) | | | | Average Production Cereals | | Average Yields Cereals | | Average Yields Roots and Tubers | |
| | Total | | Per Capita | | Total | | Per Capita | | (000 metric tons) | Percent Change Since | Kilograms Per Hectare | Percent Change Since | Kilograms Per Hectare | Percent Change Since |
	1976-78	1986-88	1976-78	1986-88	1976-78	1986-88	1976-78	1986-88	1986-88	1976-78	1986-88	1976-78	1986-88	1976-78
ASIA	**91**	**129**	**96**	**113**	**91**	**128**	**96**	**113**	**782,291**	**33**	**2,572**	**33**	**13,996**	**2**
Afghanistan	X	X	X	X	X	X	X	X	4,496	3	1,334	3	15,833	12
Bahrain	X	X	X	X	X	X	X	X	X	X	X	X	16,225	(31)
Bangladesh	94	111	103	92	94	111	102	92	23,812	24	2,177	16	10,398	2
Bhutan	93	134	98	118	93	134	98	117	193	31	1,591	10	8,371	25
China	87	139	90	128	87	138	90	126	355,480	39	3,945	51	15,375	(2)
Cyprus	99	96	101	88	99	96	101	88	112	29	2,061	36	24,055	4
India	97	125	103	107	97	126	103	108	165,125	23	1,627	25	14,417	17
Indonesia	84	134	89	118	84	135	90	119	46,443	69	3,573	45	11,038	26
Iran, Islamic Rep	93	120	103	92	92	120	101	92	12,351	43	1,308	11	16,341	12
Iraq	106	134	118	104	106	134	118	105	2,259	27	909	0	15,911	34
Israel	101	110	109	97	107	116	114	102	268	14	2,295	32	39,003	17
Japan	108	103	111	99	108	105	111	101	14,735	(11)	5,671	(1)	25,934	13
Jordan	86	149	97	114	87	151	98	116	96	26	921	126	22,381	39
Kampuchea, Dem	139	173	135	145	136	169	133	141	2,049	17	1,243	6	7,902	4
Korea, Dem People's Rep	91	129	98	109	91	129	98	109	11,528	49	4,570	20	13,299	4
Korea, Rep	110	107	115	97	108	109	113	98	8,672	(9)	5,685	14	22,334	31
Kuwait	X	X	X	X	X	X	X	X	3	5,910	5,723	152	18,800	22
Lao People's Dem Rep	71	138	74	117	71	138	73	118	1,252	73	2,071	66	10,060	3
Lebanon	X	X	X	X	X	X	X	X	24	(59)	1,874	81	21,016	174
Malaysia	89	141	95	120	83	157	89	134	1,733	(5)	2,670	2	9,379	6
Mongolia	97	114	105	94	98	118	106	97	797	110	1,255	62	10,491	37
Myanmar	87	141	93	121	87	142	93	123	14,700	47	2,810	50	9,590	80
Nepal	97	117	105	98	98	119	106	99	4,455	22	1,600	(3)	5,660	5
Oman	X	X	X	X	X	X	X	X	1	(51)	1,100	4	13,440	(9) a
Pakistan	87	140	94	108	88	136	96	104	19,390	32	1,733	19	9,871	(12)
Philippines	89	107	96	89	89	107	97	89	13,185	31	1,880	28	6,424	8
Qatar	X	X	X	X	X	X	X	X	2	381	3,120	15	8,568	(12)
Saudi Arabia	118	312	137	231	119	315	137	233	2,858	876	4,035	462	15,836	94 a
Singapore	107	92	110	86	106	93	110	86	X	X	X	X	11,190	(3)
Sri Lanka	87	102	92	91	83	102	87	91	2,444	48	2,973	49	9,072	50
Syrian Arab Rep	81	119	89	93	78	119	86	93	3,507	50	1,223	41	16,976	28
Thailand	88	115	94	101	89	115	96	101	23,544	31	2,052	13	13,639	(5)
Turkey	95	115	101	97	94	116	100	98	29,875	23	2,179	21	21,440	39
United Arab Emirates	X	X	X	X	X	X	X	X	5	1,731	3,866	(67) a	10,591	(4)
Viet Nam	85	135	91	115	85	135	90	115	16,058	41	2,647	34	5,911	(14)
Yemen Arab Rep	100	121	108	99	100	121	108	99	713	(11)	859	2	16,724	46
Yemen, People's Dem Rep	101	107	109	88	100	106	107	87	120	9	1,727	12	14,984	12
EUROPE	**93**	**108**	**94**	**106**	**93**	**108**	**94**	**106**	**293,126**	**25**	**4,280**	**29**	**21,982**	**16**
Albania	91	109	97	95	93	109	99	95	1,010	14	2,911	16	9,085	20
Austria	95	105	95	105	95	105	95	105	4,971	14	4,919	17	26,791	13
Belgium	91	114	91	113	91	114	91	114	2,211	13	5,881	35	41,747	20
Bulgaria	93	102	94	101	92	104	93	103	7,878	0	3,818	8	10,302	(9)
Czechoslovakia	98	121	100	119	98	121	100	119	11,481	13	4,602	20	19,072	10
Denmark	93	120	94	119	93	120	94	119	7,756	12	4,976	31	33,782	45
Finland	102	106	103	103	102	106	103	103	2,945	(11)	2,486	(5)	16,820	(0)
France	89	104	91	100	89	104	90	100	53,177	36	5,689	41	33,281	46
German Dem Rep	91	117	91	117	91	116	91	117	10,900	22	4,427	26	24,991	58
Germany, Fed Rep	95	112	95	113	95	112	95	113	25,495	18	5,366	32	34,439	27
Greece	92	104	95	100	91	100	94	97	5,351	40	3,683	48	16,683	13
Hungary	93	107	94	108	93	108	94	108	14,368	16	5,028	23	17,077	23
Iceland	97	100	100	93	97	101	100	94	X	X	X	X	11,000	5
Ireland	98	110	102	104	98	110	102	104	2,045	26	5,646	30	22,766	(12)
Italy	91	102	92	101	91	102	92	101	18,193	12	3,861	18	18,769	13
Luxembourg	X	X	X	X	X	X	X	X	X	X	X	X	X	X
Malta	102	107	109	111	102	107	110	111	10	78	3,834	37	6,842	(14)
Netherlands	90	117	92	113	90	117	92	113	1,198	(1)	6,641	30	42,644	24
Norway	93	109	94	106	93	109	94	106	1,226	21	3,688	10	22,665	(6)
Poland	103	112	106	106	103	112	106	106	25,200	22	3,017	15	18,926	(0)
Portugal	96	108	100	103	95	109	99	103	1,580	25	1,600	64	8,296	(7)
Romania	97	121	100	117	97	121	99	117	31,000	62	5,000	65	24,062	63
Spain	91	115	94	111	92	115	94	111	20,247	41	2,810	34	17,172	21
Sweden	97	98	97	97	97	98	97	97	5,311	(4)	3,870	11	31,830	19
Switzerland	95	109	95	106	95	109	95	106	1,004	30	5,471	23	35,917	10
United Kingdom	89	108	89	107	89	108	89	107	22,383	42	5,651	34	37,230	32
Yugoslavia	97	106	99	101	97	106	99	101	16,163	4	3,872	14	8,196	(10)
U.S.S.R.	**105**	**117**	**108**	**109**	**105**	**118**	**108**	**111**	**196,906**	**(7)**	**1,806**	**6**	**11,995**	**(0)**
OCEANIA	**100**	**111**	**105**	**99**	**101**	**108**	**106**	**97**	**23,095**	**16**	**1,557**	**13**	**11,040**	**9**
Australia	100	111	105	101	101	106	105	97	22,023	16	1,510	13	27,046	28
Fiji	79	98	83	86	79	98	83	86	28	48	2,181	9	8,452	6
New Zealand	98	111	98	105	101	114	101	109	1,032	18	4,586	14	29,218	15
Papua New Guinea	92	116	100	97	94	117	102	97	3	(10)	1,546	5	6,967	0
Solomon Islands	79	110	80	85	79	110	87	85	4	(25)	4,000	32	16,057	19

Source: Food and Agriculture Organization of the United Nations.
Notes: a. Two years of data.
0 = zero or less than half of the unit of measure; X = not available; negative numbers are shown in parentheses.
For additional information, see Sources and Technical Notes.

Table 18.2 Agricultural Inputs, 1975-87

	Cropland		Irrigated Land as a Percentage of Cropland		Average Annual Fertilizer Use (kilograms per hectare of cropland)		Average Annual Pesticide Use (metric tons active ingredient)		Tractors		Harvesters	
	Total (000 hectares) 1987	Hectares Per Capita 1989	1975-77	1985-87	1975-77	1985-87	1975-77	1982-84	Annual Average Number 1985-87	Percent Change Since 1975-77	Annual Average Number 1985-87	Percent Change Since 1975-77
WORLD	1,473,699	0.28	14	15	67	91	X	X	25,252,192	31	3,953,208	28
AFRICA	185,424	0.30	5	6	14	19	X	X	535,798	29	51,846	42
Algeria	7,540	0.31	3	5	19	37	16,457	21,400	81,373	95	8,073	107
Angola	3,550	0.36	X	X	4	4	X	X	10,263	8	0	X
Benin	1,840	0.40	0	0	1	6	X	X	118	29	0	X
Botswana	1,360	1.10	0	0	2	0	X	X	2,157	14	84	37
Burkina Faso	3,140	0.36	0	0	3 a	5	X	X	121	35	0	X
Burundi	1,332	0.25	4	5	1 b	2	22	59	53	379	0	X
Cameroon	6,995	0.64	0	0	3	7	X	X	920	197	0	X
Cape Verde	40	0.11	5	5	2	3 a	X	X	16	66	0	X
Central African Rep	2,005	0.71	X	X	1	1 b	X	X	185	44	13	60
Chad	3,205	0.58	0	0	2	2	X	X	162	17	17	11
Comoros	98	0.19	X	X	0	0	X	X	X	X	X	X
Congo	679	0.35	0	1	4 a	5	X	X	687	5	45	73
Cote d'Ivoire	3,640	0.30	1	2	14	9	X	X	3,350	41	53	112
Djibouti	X	X	X	X	0	0	X	X	7	25	0	X
Egypt	2,560	0.05	100	100	188	347	26,970	19,567	44,000	100	2,250	12
Equatorial Guinea	230	0.53	X	X	0	0	X	X	99	3	0	X
Ethiopia	13,930	0.30	1	1	2	4	600 b	993	3,900	6	150	11
Gabon	452	0.40	X	X	1	4	X	X	1,373	31	0	X
Gambia, The	170	0.20	7	7	9	21	X	101 a	43	(7)	4	44
Ghana	2,870	0.20	0	0	10	4	X	X	3,800	17	427	110
Guinea	1,577	0.24	4	4	1	0	X	X	180	80	0	X
Guinea-Bissau	335	0.35	X	X	1 a	1 b	X	X	48	40	0	X
Kenya	2,420	0.10	2	2	22	46	935 a	1,307	8,536	41	520	22
Lesotho	320	0.19	X	X	5	12	X	X	1,650	45	31	31
Liberia	371	0.15	1	1	15	5	1,223	310	318	22	0	X
Libya	2,145	0.49	10	11	21	24	2,610 b	2,017	29,567	98	0	X
Madagascar	3,067	0.26	18	28	3	4	X	1,630	2,800	14	138	41
Malawi	2,377	0.29	1	1	10	17	X	X	1,350	34	0	X
Mali	2,076	0.23	6	9	5	15	X	683	833	8	47	28
Mauritania	199	0.10	6	6	7	7	11 a	X	315	43	0	X
Mauritius	107	0.10	14	16	225	268	753 b,c,d	981 b,c,d	344	13	0	X
Morocco	8,462	0.35	14	15	23	36	2,225 a	3,350	32,000	45	3,187	14
Mozambique, People's Rep	3,090	0.20	1	3	4	2	X	X	5,750	4	0	X
Niger	3,540	0.51	1	1	1 b	1	451	159 a	172	117	0	X
Nigeria	31,335	0.29	3	3	2	10	X	4,000 a	10,533	37	0	X
Rwanda	1,120	0.16	0	0	0	1	X	X	85	7	0	X
Senegal	5,225	0.73	3	3	9	4	X	X	467	22	147	24
Sierra Leone	1,801	0.45	1	2	1	2 a	X	X	483	200	6	240
Somalia	933	0.13	11	12	4 b	3	X	X	2,043	50	0	X
South Africa	13,169	0.38	8	9	61	61	19,292	11,053	182,767	3	32,000	42
Sudan	12,478	0.51	14	15	6	4	X	X	19,000	110	1,190	25
Swaziland	164	0.21	34	38	61	55	16 b	X	3,385	43	0	X
Tanzania	5,230	0.20	1	3	6	9	2,992 c,d	5,733 c,d	18,550	2	0	X
Togo	1,431	0.43	0	0	2	7	X	X	320	167	0	X
Tunisia	4,680	0.59	3	5	10	21	X	1,330	26,100	7	2,587	9
Uganda	6,705	0.38	0	0	0	0	X	23 c	3,800	109	13	41
Zaire	6,690	0.19	0	0	2	1 b	X	X	2,277	52	0	X
Zambia	5,208	0.64	0	0	12	16	X	X	4,410	5	278	12
Zimbabwe	2,769	0.29	3	7	48	56	865 a	207 b	20,333	6	583	16
NORTH & CENTRAL AMERICA	273,853	0.65	9	9	84	83	X	X	5,667,780	(4)	831,109	0
Barbados	33	0.13	X	X	127	103	X	X	592	18	0	X
Canada	45,990	1.75	1	2	33	49	26,928	54,767	728,091	15	157,945	(2)
Costa Rica	526	0.18	8	21	127	166	3,027	3,667	6,250	10	1,090	18
Cuba	3,320	0.32	21	26	118	192	7,817	9,567	62,462	11	4,212	71
Dominican Rep	1,475	0.21	11	14	51	47	1,961	3,297	2,270	10	0	X
El Salvador	733	0.14	6	15	149	111	1,310	2,838 b	3,397	14	370	37
Guatemala	1,865	0.21	4	4	50	62	4,627	5,117	4,120	10	2,900	21
Haiti	905	0.14	8	8	3 a	3	156 d	X	572	27	0	X
Honduras	1,785	0.36	5	5	14	19	940	859	3,370	16	0	X
Jamaica	269	0.11	12	13	60	66	861	1,420	2,997	16	0	X
Mexico	24,705	0.28	20	21	45	73	19,148	27,630	160,000	60	17,700	36
Nicaragua	1,268	0.34	6	7	31	49	2,943	2,003	2,453	84	0	X
Panama	575	0.24	4	5	43	58	1,542	2,393	6,150	46	1,850	91
Trinidad and Tobago	120	0.10	17	19	60	52	X	X	2,603	21	0	X
United States	189,915	0.77	9	10	102	93	459,400	373,333	4,676,000	(8)	645,000	(0)
SOUTH AMERICA	141,972	0.49	5	6	28	39	X	X	1,169,119	82	110,302	21
Argentina	35,750	1.12	4	5	2	4	7,448	14,313	206,000	14	46,500	13
Bolivia	3,399	0.48	4	5	1	2	612	833	790	13	278	42
Brazil	77,500	0.53	2	3	41	49	59,292	46,698	775,000	148	42,000	31
Chile	5,580	0.43	23	23	20	46	1,838	1,800	37,843	10	8,500	9
Colombia	5,318	0.17	6	9	49	81	19,344	16,100	33,813	36	2,417	28
Ecuador	2,646	0.25	20	21	26	34	5,445	3,110	8,000	52	700	29
Guyana	495	0.48	30	26	26	32	705	658	3,560	5	419	4
Paraguay	2,176	0.52	4	3	1	5	2,957	3,423	9,900	175	0	X
Peru	3,725	0.17	35	33	38	43	2,370	2,753	14,933	19	0	X
Suriname	68	0.17	79	88	85	178	974 a	1,720 b	1,703	38	134	18
Uruguay	1,444	0.47	4	7	43	41	1,390	1,517	32,603	3	4,617	(5)
Venezuela	3,865	0.20	8	9	44	143	6,923	8,143	44,667	43	4,733	92

Table 18.2

	Cropland		Irrigated Land as a Percentage of Cropland		Average Annual Fertilizer Use (kilograms per hectare of cropland)		Average Annual Pesticide Use (metric tons active ingredient)		Tractors		Harvesters	
	Total (000 hectares) 1987	Hectares Per Capita 1989	1975-77	1985-87	1975-77	1985-87	1975-77	1982-84	Annual Average Number 1985-87	Percent Change Since 1975-77	Annual Average Number 1985-87	Percent Change Since 1975-77
ASIA	450,920	0.15	28	31	42	93	X	X	4,790,874	128	1,248,829	113
Afghanistan	8,054	0.51	31	33	6	9	1,000 b	605 a	770	8	0	X
Bahrain	2	0.00	50	50	11 b	567	X	X	X	X	X	X
Bangladesh	9,164	0.08	15	23	29	68	X	234	4,950	52	0	X
Bhutan	103	0.07	X	X	1	1	X	X	X	X	X	X
China	96,976	0.09	43	46	74	195	150,467	159,267	876,591	110	33,107	125
Cyprus	157	0.23	19	19	87	123	X	X	13,405	35	535	67
India	168,990	0.20	20	25	22	52	52,506	53,087	651,424	153	2,839	265
Indonesia	21,220	0.12	26	34	27	100	18,687	16,344	12,511	37	16,471	30
Iran, Islamic Rep	14,830	0.27	36	39	23	63	X	X	111,667	123	2,883	11
Iraq	5,450	0.30	30	32	8	36	X	X	40,305	101	2,977	(37)
Israel	438	0.10	45	64	183	219	600	847	25,697	22	317	7
Japan	4,708	0.04	63	61	398	431	33,960	32,000	1,863,857	123	1,153,540	118
Jordan	414	0.10	9	11	16	34	X	X	4,840	24	340	70
Kampuchea, Dem	3,056	0.38	3	3	0	0	1,593	833	1,358	1	20	0
Korea, Dem People's Rep	2,392	0.11	44	48	244	328	4,000 b	X	72,667	179	0	X
Korea, Rep	2,143	0.05	48	58	334	395	4,675	12,273	16,140	1,856	15,825	23,402
Kuwait	4	0.00	100	27	100 b	139	X	X	91	361	0	X
Lao People's Dem Rep	901	0.23	6	13	0	2 a	X	X	800	100	0	X
Lebanon	301	0.10	26	29	74	81	X	X	3,000	0	90	0
Malaysia	4,380	0.26	7	8	68	154	X	9,730 b	11,567	83	0	X
Mongolia	1,335	0.62	3	3	6	16	X	X	11,267	31	2,728	19
Myanmar	10,060	0.25	10	11	5	18	3,721	15,300	10,282	28	38	84
Nepal	2,339	0.13	12	28	6	20	X	X	2,843	80	0	X
Oman	48	0.03	92	87	12	96	X	X	127	67	27	158
Pakistan	20,760	0.17	70	77	32	81	2,120	1,856	170,000	250	737	83
Philippines	7,930	0.13	14	18	34	50	3,547 a	4,415	19,767	53	587	59
Qatar	4	0.01	X	X	133	160	X	X	86	109	0	X
Saudi Arabia	1,180	0.09	34	36	7	337	X	X	1,750	108	563	88
Singapore	3	0.00	X	X	375	1,391	X	X	56	61	0	X
Sri Lanka	1,887	0.11	25	30	49	106	X	697	27,958	76	5	56
Syrian Arab Rep	5,630	0.47	10	12	15	42	X	4,892 a,c	47,977	164	2,890	37
Thailand	20,050	0.37	15	20	14	26	13,120	22,289	130,333	347	0	X
Turkey	27,927	0.51	7	8	40	59	X	9,000 b	609,623	117	12,289	(16)
United Arab Emirates	19	0.01	44	27	85	153	X	X	X	X	X	X
Viet Nam	6,470	0.10	18	28	58	61	1,693	883 a	41,500	344	0	X
Yemen Arab Rep	1,360	0.18	17	18	3	10	325 c,d	1,614 a,c,d	2,180	66	0	X
Yemen, People's Dem Rep	119	0.05	48	48	7	13	X	X	3,021	56	15	25
EUROPE	140,100	0.28	9	12	207	228	X	X	9,912,318	33	838,235	4
Albania	714	0.22	50	57	104	133	4,510	5,183	10,587	8	1,421	18
Austria	1,510	0.20	0	0	219	220	3,449	4,548	326,060	10	30,314	(2)
Belgium	818	0.08	0	0	511	517	8,847	13,263	122,350	13	8,872	(13)
Bulgaria	4,131	0.46	26	30	160	195	28,287	32,400	54,327	(17)	7,985	(24)
Czechoslovakia	5,134	0.33	3	4	317	324	13,967	14,970	138,625	(2)	20,459	3
Denmark	2,600	0.51	8	16	248	241	4,998	7,729	166,612	(11)	34,183	(19)
Finland	2,411	0.49	2	3	186	216	1,768	2,639	240,000	25	47,000	15
France	19,459	0.35	4	6	266	301	83,017	98,733	1,523,896	11	150,405	(0)
German Dem Rep	4,934	0.30	3	3	354	333	11,900	14,133	161,351	17	17,470	43
Germany, Fed Rep	7,476	0.12	4	4	436	425	23,693	29,836	1,478,858	2	150,000	(13)
Greece	3,940	0.39	24	29	126	172	30,570	29,240	181,000	76	6,499	25
Hungary	5,289	0.50	5	3	270	258	26,267	27,595	54,257	(11)	11,554	(18)
Iceland	8	0.03	X	X	3,578	3,061	3 e,c	5 e,c	13,167	13	17	11
Ireland	983	0.27	X	X	423	655	1,721	2,250	160,000	34	5,163	(1)
Italy	12,167	0.21	22	25	127	178	83,724	98,496	1,270,569	47	42,097	47
Luxembourg	X	X	X	X	0	0	X	X	X	X	X	X
Malta	13	0.04	8	8	30	49	X	X	447	25	10	11
Netherlands	924	0.06	52	59	751	748	6,593	9,670	190,133	19	5,711	(13)
Norway	856	0.20	6	11	290	273	1,494	1,508	151,601	38	17,884	21
Poland	14,739	0.39	1	1	240	232	11,360	15,277	985,885	126	61,449	155
Portugal	2,755	0.27	22	23	89	96	24,375	16,016	77,552	54	4,703	16
Romania	10,686	0.46	16	30	110	180	99,297	17,237	184,489	43	52,352	21
Spain	20,425	0.52	14	16	72	92	55,267	71,533	860,670	64	46,590	14
Sweden	2,953	0.35	2	4	176	138	5,454	5,736	183,943	(2)	47,186	(5)
Switzerland	412	0.06	6	6	380	429	1,945	1,699	106,438	25	4,464	(13)
United Kingdom	6,988	0.12	1	2	275	364	25,137	34,147	521,846	7	54,509	(9)
Yugoslavia	7,766	0.33	2	2	94	131	19,091	31,567	951,324	267	9,937	(10)
U.S.S.R.	232,570	0.81	7	9	76	114	348,767	535,400	2,755,333	15	812,000	18
OCEANIA	48,860	1.87	4	4	34	34	X	X	420,976	(2)	60,888	(4)
Australia	47,105	2.85	4	4	22	26	60,638	65,200	332,000	0	57,100	(3)
Fiji	240	0.33	0	0	52	79	X	X	4,450	31	0	X
New Zealand	522	0.16	37	51	1,250	740	1,651	1,793	81,414	(11)	3,326	(26)
Papua New Guinea	386	0.10	X	X	20	30	X	X	1,163	0	450	38
Solomon Islands	57	0.18	X	X	0	0	X	X	X	X	X	X

Sources: Food and Agriculture Organization of the United Nations, United Nations Industrial Development Organization, United Nations Population Division, and country sources.

Notes: a. Two years of data. b. One year of data. c. May not be active ingredient. d. Imports of pesticides. e. Sales of pesticides.
0 = zero or less than half of the unit of measure; X = not available; negative numbers are shown in parentheses.
For additional information, see Sources and Technical Notes.

Table 18.3 Livestock Populations, 1976-88

	Cattle			Sheep and Goats		Pigs		Equines		Buffaloes and Camels		Chickens	
	Annual Average (000) 1986-88	Percent Change Since 1976-78	Cattle per Capita 1986-88	Annual Average (000) 1986-88	Percent Change Since 1976-78	Annual Average (000) 1986-88	Percent Change Since 1976-78	Annual Average (000) 1986-88	Percent Change Since 1976-78	Annual Average (000) 1986-88	Percent Change Since 1976-78	Annual Average (millions) 1986-88	Percent Change Since 1976-78
WORLD	1,268,516	5	0.3	1,661,928	12	828,818	17	121,717	4	156,603	17	9,665	56
AFRICA	179,654	11	0.3	362,046	15	12,830	42	17,507	3	16,573	16	797	52
Algeria	1,147	2	0.0	17,860	42	5	23	816	(0)	128	(9)	22	32
Angola	3,390	21	0.4	1,230	9	475	30	6	0	0	X	6	17
Benin	903	24	0.2	1,859	7	622	55	6	10	0	X	23	152
Botswana	2,350	(20)	2.0	1,265	75	9	(29)	173	32	0	X	1	48
Burkina Faso	2,890	11	0.3	7,934	87	500	217	270	0	5	(2)	21	101
Burundi	399	(49)	0.1	1,072	20	72	104	0	X	0	X	4	32
Cameroon	4,362	51	0.4	5,368	19	1,289	33	62	20	0	X	14	55
Cape Verde	12	(0)	0.0	81	27	69	175	9	(3)	0	X	0	275
Central African Rep	2,224	118	0.8	1,247	42	371	106	0	X	0	X	3	79
Chad	4,021	6	0.8	4,413	(5)	12	77	405	(3)	504	28	4	26
Comoros	85	12	0.2	105	14	0	X	4	30	0	X	0	48
Congo	70	9	0.0	249	49	47	1	0	X	0	X	1	45
Cote d'Ivoire	923	68	0.1	3,000	41	447	61	2	0	0	X	16	33
Djibouti	68	109	0.2	912	2	0	X	8	35	56	74	0	X
Egypt	1,892	(15)	0.0	2,759	(21)	15	1	1,920	18	2,620	7	30	11
Equatorial Guinea	5	11	0.0	42	9	5	9	0	X	0	X	0	64
Ethiopia	30,333	18	0.7	40,467	1	19	14	7,015	3	1,037	8	57	9
Gabon	9	174	0.0	145	(11)	153	18	0	X	0	X	2	53
Gambia, The	298	3	0.4	394	35	12	33	4	1	0	X	0	38
Ghana	1,245	60	0.1	5,092	31	680	81	29	0	0	X	11	(11)
Guinea	1,800	13	0.3	920	15	49	41	4	0	0	X	13	125
Guinea-Bissau	338	24	0.4	412	32	289	30	4	8	0	X	1	99
Kenya	9,433	(3)	0.4	15,467	61	100	56	2	0	787	32	22	32
Lesotho	522	(0)	0.3	2,450	51	69	(11)	244	27	0	X	1	21
Liberia	42	17	0.0	475	32	137	44	0	X	0	X	4	83
Libya	212	14	0.1	6,625	15	0	X	105	46	182	154	34	632
Madagascar	10,550	18	1.0	1,633	(9)	1,391	131	1	(48)	0	X	20	46
Malawi	1,000	39	0.1	1,147	31	198	4	1	30	0	X	8	7
Mali	4,668	11	0.5	10,833	(4)	59	67	612	16	241	19	18	47
Mauritania	1,217	5	0.7	7,167	2	0	X	166	(5)	810	13	4	23
Mauritius	31	19	0.0	100	32	11	97	0	0	0	0	2	61
Morocco	3,110	(12)	0.1	20,707	3	8	2	1,468	(25)	59	(72)	36	73
Mozambique, People's Rep	1,350	0	0.1	487	18	155	48	20	0	0	X	21	41
Niger	3,433	21	0.5	10,950	23	37	31	804	26	416	27	15	80
Nigeria	12,190	6	0.1	39,133	13	1,300	37	950	0	18	6	177	87
Rwanda	641	1	0.1	1,415	43	90	7	0	X	0	X	1	40
Senegal	2,212	(9)	0.3	4,800	78	448	171	418	(1)	8	25	11	62
Sierra Leone	330	1	0.1	507	39	49	62	0	X	0	X	5	51
Somalia	5,090	29	0.7	32,667	22	10	23	49	12	6,527	21	3	22
South Africa	11,790	(9)	0.4	35,490	(4)	1,453	4	454	1	0	X	36	33
Sudan	22,347	43	1.0	32,954	21	0	X	674	(3)	2,898	15	30	26
Swaziland	648	2	0.9	344	21	19	1	16	1	0	X	1	38
Tanzania	13,063	10	0.5	11,116	25	182	25	171	6	0	X	25	67
Togo	285	27	0.1	1,780	16	296	14	4	88	0	X	2	(3)
Tunisia	607	(29)	0.1	6,943	2	4	34	349	7	182	5	17	36
Uganda	4,338	(14)	0.3	4,602	34	387	130	17	4	0	X	17	25
Zaire	1,373	20	0.0	3,853	15	771	13	0	X	0	X	19	52
Zambia	2,768	33	0.4	417	26	198	(4)	1	40	0	X	14	(28)
Zimbabwe	5,575	(12)	0.6	2,167	(15)	180	(25)	123	12	0	X	10	12
NORTH & CENTRAL AMERICA	166,630	(9)	0.4	33,214	(1)	84,521	2	26,762	6	9	13	1,732	35
Barbados	18	(12)	0.1	89	18	49	17	5	0	0	X	1	104
Canada	11,866	(17)	0.5	723	22	10,408	59	344	(3)	0	X	104	16
Costa Rica	2,288	19	0.8	16	76	228	6	126	5	0	X	5	(21)
Cuba	5,004	(9)	0.5	491	12	2,433	53	755	(11)	0	X	26	26
Dominican Rep	2,092	10	0.3	606	49	389	(46)	582	40	0	X	19	140
El Salvador	1,094	(12)	0.2	19	13	424	(12)	118	6	0	X	4	4
Guatemala	2,150	54	0.3	736	16	867	37	154	2	0	X	15	29
Haiti	1,473	70	0.2	1,243	9	783	(56)	728	8	0	X	12	154
Honduras	2,829	52	0.6	34	13	577	10	260	3	0	X	8	84
Jamaica	290	5	0.1	440	27	247	26	37	(11)	0	X	5	13
Mexico	31,160	20	0.4	16,215	6	17,873	21	12,463	(3)	0	X	231	56
Nicaragua	1,837	(33)	0.5	9	4	748	8	317	(1)	0	X	5	31
Panama	1,447	5	0.6	7	15	240	23	175	4	0	X	7	53
Trinidad and Tobago	78	5	0.1	62	17	84	49	5	(5)	9	13	8	16
United States	102,154	(17)	0.4	12,037	(15)	48,773	(9)	10,645	20	0	X	1,266	33
SOUTH AMERICA	256,446	12	0.9	130,026	7	53,671	5	21,020	8	1,056	200	836	54
Argentina	51,982	(12)	1.7	32,281	(15)	4,067	8	3,288	(5)	0	X	52	62
Bolivia	5,377	44	0.8	11,817	6	1,697	31	1,003	(15)	0	X	11	40
Brazil	134,027	27	0.9	30,636	21	32,573	(6)	9,044	14	1,056	200	520	54
Chile	3,282	(4)	0.3	6,870	8	1,270	36	528	7	0	X	20	19
Colombia	23,957	(1)	0.8	3,547	28	2,512	34	3,200	21	0	X	37	39
Ecuador	3,885	45	0.4	2,086	(14)	4,167	42	736	27	0	X	49	147
Guyana	207	(17)	0.2	197	13	183	43	4	(7)	0	X	15	37
Paraguay	7,435	30	1.9	540	9	1,808	56	367	(0)	0	X	15	50
Peru	3,947	(5)	0.2	14,940	(11)	2,291	12	1,365	2	0	X	50	33
Suriname	72	152	0.2	10	0	19	8	0	20	0	X	5	23
Uruguay	9,864	(3)	3.2	24,599	54	210	(52)	475	(6)	0	X	7	33
Venezuela	12,390	29	0.7	1,808	12	2,864	47	1,007	(0)	0	X	56	71

Table 18.3

	Cattle			Sheep and Goats		Pigs		Equines		Buffaloes and Camels		Chickens	
	Annual Average (000) 1986-88	Percent Change Since 1976-78	Cattle per Capita 1986-88	Annual Average (000) 1986-88	Percent Change Since 1976-78	Annual Average (000) 1986-88	Percent Change Since 1976-78	Annual Average (000) 1986-88	Percent Change Since 1976-78	Annual Average (000) 1986-88	Percent Change Since 1976-78	Annual Average (millions) 1986-88	Percent Change Since 1976-78
ASIA	384,949	11	0.1	612,954	15	406,771	20	43,701	10	138,017	16	3,855	106
Afghanistan	3,710	0	0.3	21,263	(8)	0	X	1,748	3	277	(8)	7	12
Bahrain	6	25	0.0	23	39	0	X	0	X	1	30	1	114
Bangladesh	22,568	(12)	0.2	11,726	23	0	X	45	4	1,900	45	76	35
Bhutan	398	43	0.3	56	118	62	16	43	15	6	43	0	40
China	70,767	31	0.1	167,711	4	339,060	17	26,681	17	21,067	16	1,783	143
Cyprus	43	158	0.1	545	14	231	57	9	(56)	0	X	2	22
India	197,543	9	0.2	157,665	35	10,267	31	2,351	16	75,093	19	225	51
Indonesia	6,478	2	0.0	17,674	59	6,120	108	713	12	2,977	30	395	265
Iran, Islamic Rep	8,350	13	0.2	48,107	2	0	(100)	2,239	0	257	0	105	68
Iraq	1,586	(12)	0.1	10,569	(13)	0	X	483	(12)	199	(37)	75	382
Israel	319	4	0.1	402	10	130	52	11	(2)	10	(5)	23	(3)
Japan	4,701	22	0.0	73	(24)	11,380	40	22	(29)	0	X	338	32
Jordan	30	(4)	0.0	1,576	41	0	0	25	(29)	15	1	58	130
Kampuchea, Dem	1,811	87	0.2	2	7	1,388	363	14	29	679	51	6	54
Korea, Dem People's Rep	1,200	37	0.1	646	33	3,017	68	47	23	0	X	19	8
Korea, Rep	2,712	69	0.1	238	(1)	3,494	104	3	(60)	0	X	56	83
Kuwait	25	140	0.0	283	5	0	X	3	X	8	50	25	378
Lao People's Dem Rep	588	69	0.2	73	128	1,510	103	42	59	999	55	8	101
Lebanon	51	(8)	0.0	604	83	21	31	16	(8)	0	(19)	11	150
Malaysia	617	30	0.0	443	7	2,203	42	5	1	214	(26)	57	24
Mongolia	2,471	3	1.2	17,588	(5)	85	439	2,012	(8)	552	(10)	0	64
Myanmar	9,888	28	0.3	1,430	84	2,968	56	145	29	2,181	25	33	90
Nepal	6,374	(5)	0.4	5,892	13	467	35	0	X	2,893	(27)	10	29
Oman	134	0	0.1	923	225	0	X	24	(5)	81	297	2	131
Pakistan	16,952	14	0.2	58,543	32	0	X	3,455	26	14,640	26	136	271
Philippines	1,754	(0)	0.0	2,134	51	7,298	15	300	(3)	2,913	2	55	11
Qatar	8	(7)	0.0	192	149	0	X	1	(33)	20	93	1	401
Saudi Arabia	323	(2)	0.0	10,757	156	0	X	119	6	415	66	67	517
Singapore	0	(62)	0.0	2	(10)	479	(58)	0	X	0	(83)	8	(41)
Sri Lanka	1,803	9	0.1	541	(1)	94	151	1	(7)	1,007	23	8	53
Syrian Arab Rep	713	12	0.1	13,576	71	1	14	265	(20)	7	(36)	13	52
Thailand	4,949	13	0.1	155	139	4,223	6	20	(24)	6,085	3	83	45
Turkey	12,150	(14)	0.2	53,217	(12)	10	(31)	2,030	(22)	543	(49)	59	33
United Arab Emirates	48	149	0.0	1,222	249	0	X	0	30	117	149	9	3,678
Viet Nam	2,747	75	0.0	438	110	11,885	32	135	(3)	2,707	21	68	18
Yemen Arab Rep	1,027	24	0.1	4,298	21	0	X	523	(16)	61	(1)	20	620
Yemen, People's Dem Rep	96	7	0.0	2,335	16	0	X	170	19	81	(24)	2	31
EUROPE	128,009	(4)	0.3	152,423	15	187,594	16	5,974	(26)	371	(15)	1,252	9
Albania	651	24	0.2	2,370	30	212	51	116	(0)	2	15	5	60
Austria	2,626	4	0.4	134	(38)	3,891	4	44	7	0	X	14	7
Belgium	3,027	0	0.3	183	72	5,745	17	25	(52)	0	X	32	2
Bulgaria	1,678	(2)	0.2	9,834	(4)	3,999	12	486	0	27	(59)	39	4
Czechoslovakia	5,061	9	0.3	1,140	25	6,906	(1)	42	(27)	0	X	46	11
Denmark	2,360	(24)	0.5	91	60	9,228	14	29	(51)	0	X	14	(3)
Finland	1,495	(16)	0.3	69	(37)	1,270	6	37	15	0	X	7	(26)
France	22,335	(7)	0.4	11,685	(3)	12,279	7	330	(24)	0	X	188	12
German Dem Rep	5,784	5	0.3	2,651	37	12,763	11	105	54	0	X	51	5
Germany, Fed Rep	15,273	5	0.3	1,409	23	24,152	17	363	2	0	X	72	(19)
Greece	776	(30)	0.1	14,084	10	1,130	42	323	(43)	1	(74)	30	3
Hungary	1,718	(10)	0.2	2,396	2	8,391	11	98	(36)	0	X	60	3
Iceland	71	15	0.3	770	(12)	13	78	57	18	0	X	0	46
Ireland	5,662	(9)	1.6	3,768	45	978	2	77	(33)	0	X	7	(11)
Italy	8,840	3	0.2	12,599	34	9,277	2	393	(26)	103	29	114	11
Luxembourg	X	X	X	X	X	X	X	X	X	X	X	X	X
Malta	14	(3)	0.0	10	(34)	95	280	2	(30)	0	X	1	3
Netherlands	4,855	5	0.3	1,014	20	14,019	68	64	(16)	0	X	96	35
Norway	953	1	0.2	2,391	30	768	10	16	(22)	0	X	4	7
Poland	10,588	(10)	0.3	4,712	20	19,033	(6)	1,155	(43)	0	X	59	(25)
Portugal	1,343	13	0.1	5,902	19	2,937	46	294	(2)	0	X	18	5
Romania	7,001	16	0.3	19,629	34	14,751	54	719	20	200	(4)	131	68
Spain	5,023	11	0.1	20,595	16	15,370	69	497	(37)	0	X	53	0
Sweden	1,679	(11)	0.2	402	3	2,297	(13)	57	(1)	0	X	11	(3)
Switzerland	1,866	(7)	0.3	440	(3)	1,944	(6)	51	6	0	X	6	(6)
United Kingdom	12,340	(10)	0.2	26,165	14	7,936	(0)	180	24	0	X	119	(6)
Yugoslavia	4,982	(12)	0.2	7,904	2	8,201	6	414	(52)	29	(57)	73	39
U.S.S.R.	121,195	9	0.4	147,738	1	78,225	23	6,163	(6)	578	(10)	1,121	47
OCEANIA	31,634	(22)	1.2	223,527	13	5,205	18	590	6	0	(100)	71	30
Australia	22,950	(27)	1.4	156,792	13	2,628	19	405	(3)	X	X	55	27
Fiji	159	0	0.2	59	8	29	57	41	18	X	X	2	99
New Zealand	8,113	(7)	2.5	66,592	12	430	(7)	100	55	X	X	9	16
Papua New Guinea	105	(20)	0.0	19	156	1,643	20	1	50	X	X	3	149
Solomon Islands	13	(45)	0.0	0	8	51	21	0	X	X	X	0	7

Sources: Food and Agriculture Organization of the United Nations and United Nations Population Division.
Notes: 0 = zero or less than half of the unit of measure; X = not available; negative numbers are shown in parentheses.
For additional information, see Sources and Technical Notes.

Table 18.4 Food Trade and Aid, 1975-87

| | Average Annual Net Trade in Food | | | | | | Average Annual Donations or Receipts of Food Aid | | | | | |
| | Cereals (000 metric tons) | | Oils (metric tons) | | Pulses (metric tons) | | Cereals (000 metric tons) | | Cereals Kg Per Capita | | Oils (metric tons) | Milk (metric tons) |
	1975-77	1985-87	1975-77	1985-87	1975-77	1985-87	1975-77	1985-87	1975-77	1985-87	1985-87	1985-87
WORLD												
AFRICA	9,563	25,130	613,067	2,162,654	(124,732)	178,133	2,832	6,587	7	11	158,900	121,899
Algeria	1,873	4,581	187,103	357,292	39,089	102,265	13	3	1	0	320	1,185
Angola	130	241	14,533	29,348	3,180	16,254	11	68	2	8	3,723	4,107
Benin	37	62	(14,289)	(14,902)	(280) a	250	9	13	3	3	825	1,292
Botswana	26	141	(877)	(5,417)	(933)	5,236	5	44	6	39	3,858	3,409
Burkina Faso	33	136	716	12,453	(6,988)	(2,620)	29	89	5	11	5,018	6,069
Burundi	10	15	866	2,508	186 a	X	5	8	1	2	393	675
Cameroon	86	204	(3,626)	(6,116)	(275)	(32)	5	10	1	1	570	82
Cape Verde	42	70	344	2,306	3,126	2,721	29	53	101	158	626	957
Central African Rep	10	32	359	263	6	165 b	1	9	1	4	100	117
Chad	13	64	397 b	X	6 a	X	28	91	7	18	3,343	3,032
Comoros	12	24	X	283	X	13 a	4	8	12	17	185	233
Congo	41	94	2,757	8,816	52	134	3	1	2	1	65	9
Cote d'Ivoire	150	582	(92,665)	(137,735)	1,423	195	0	0	0	0	21	40
Djibouti	19	50	1,266	4,373	X	783	1	17	4	47	137	205
Egypt	4,319	8,825	403,831	719,725	116,108	48,525	1,628	1,909	44	39	13,545	14,087
Equatorial Guinea	2	8	X	X	X	X	0	5	0	11	78	80
Ethiopia	115	767	1,331	35,817	(86,016)	163	79	746	3	17	26,289	20,775
Gabon	49	58	2,874	3,593	48	37	0	0	0	0	0	1
Gambia, The	27	80	(17,011)	(4,652)	2 a	X	8	18	14	23	433	936
Ghana	117	163	17,404	19,847	2,986	143	57	86	6	6	3,955	1,832
Guinea	61	165	3,543	4,700	X	X	27	66	6	11	161	691
Guinea-Bissau	23	28	193	2,930	104	47	16	19	24	21	248	273
Kenya	(40)	67	47,013	113,506	(15,674)	(21,992)	9	195	1	9	828	571
Lesotho	75	118	1,467	3,000	(4,549)	4,300	17	48	14	30	2,834	3,281
Liberia	52	114	(3,447)	2,168	125	200	1	33	1	15	150	0
Libya	582	1,442	35,953	69,483	6,904	7,433	0	0	0	0	0	0
Madagascar	89	180	14,264	14,301	(15,516)	(3,403)	8	92	1	9	9,862	2,007
Malawi	28	(45)	3,077	5,751	(6,882)	(26,454)	1	7	0	1	87	181
Mali	38	194	(2,100)	1,167	(751)	X	24	142	4	17	1,317	1,858
Mauritania	131	222	3,786	18,270	6 b	303	36	101	24	55	3,954	4,587
Mauritius	142	177	16,478	24,977	6,253	9,658	13	10	15	9	9	201
Morocco	1,302	2,005	139,520	232,512	(118,091)	(20,512)	105	424	6	19	22,553	5,400
Mozambique, People's Rep	184	424	(11,350)	34,613	(8,682)	14,554	96	328	9	23	7,246	4,626
Niger	17	133	(1,844)	7,933	(22,516)	(13,067)	50	113	11	18	1,669	3,414
Nigeria	862	1,331	59,815	184,194	1,453	13,385	1	0	0	0	0	37
Rwanda	13	23	4,050	16,762	(46) a	2,447	12	25	3	4	2,009	3,093
Senegal	349	480	(213,709)	(13,114)	278	533	71	109	14	17	793	4,459
Sierra Leone	35	132	(7,154)	3,426	134 a	10 a	6	37	2	10	680	1,367
Somalia	168	311	13,224	23,833	217	4,067	69	182	21	27	16,609	8,289
South Africa	(2,521)	(1,090)	36,158	183,206	(688)	1,198	0	0	0	0	0	0
Sudan	38	644	(11,526)	56,364	(1,697)	20,333	64	881	4	39	10,074	5,426
Swaziland	14	38	600	723	X	X	0	1	1	1	177	170
Tanzania	228	233	19,891	27,188	(13,369)	(8,333)	115	82	7	3	2,545	4,591
Togo	30 a	65	(221)	2,728	(1)	177	11	13	5	4	1,061	1,758
Tunisia	487	1,071	(16,597)	78,109	(15,689)	3,748	126	223	22	30	834	4,532
Uganda	12 a	21	1,689 a	1,183	57 a	650 a	0	18	0	1	2,484	1,135
Zaire	345	369	(55,933)	(22,000)	156 a	2,500 b	21	98	1	3	225	449
Zambia	108	154	19,017	15,195	726	484 a	16	105	3	14	6,637	215
Zimbabwe	(500)	(236)	(477)	22,968	(3,304)	(837)	0	56	0	7	371	166
NORTH & CENTRAL AMERICA	(87,085)	(88,793)	(836,824)	(741,501)	(265,021)	(497,342)	(6,311)	(7,402)	(18)	(18)	(375,143)	(183,734)
Barbados	42	65	4,520	5,286	1,419	782	0	0	1	0	0	18
Canada	(16,362)	(23,508)	(22,041)	(289,626)	(77,520)	(281,738)	(1,031)	(1,133)	(45)	(44)	(49,286)	(5,517)
Costa Rica	80	155	8,337	11,159	615	625	1	112	1	41	98	294
Cuba	1,695	2,179	87,653	131,966	99,650	97,945	0	0	0	0	1,243	1,278
Dominican Rep	271	551	44,571	79,071	4,893	6,300	25	116	5	18	39,049	3,741
El Salvador	120	197	19,270	37,864	4,469	2,392	4	233	1	48	21,834	7,466
Guatemala	129	217	10,400	51,303	3,328	1,187	12	90	2	11	14,990	9,754
Haiti	135	199	23,324	46,905	476	4,000	50	108	10	18	9,344	5,345
Honduras	74	131	11,039	(12,967)	(2,099)	410	17	130	5	29	1,538	5,056
Jamaica	342	380	24,264	19,844	1,560	380	48	253	23	107	1,985	2,996
Mexico	2,512	4,099	57,418	327,348	(52,080)	103,488	0	7	0	0	9,547	30,672
Nicaragua	49	123	(1,306)	36,123	(1,111)	6,894	2	39	1	12	1,744	1,972
Panama	66	106	17,772	26,518	2,896	7,647	2	1	1	0	24	10
Trinidad & Tobago	211	239	14,760	19,629	8,966	12,012	0	0	0	0	0	0
United States	(76,661)	(74,158)	(1,146,930)	(1,248,287)	(266,716)	(466,818)	(5,442)	(7,357)	(25)	(30)	(427,252)	(246,818)
SOUTH AMERICA	(6,173)	(4,228)	(577,887)	(2,131,413)	(66,100)	(88,373)	199	446	1	2	17,383	44,217
Argentina	(11,498)	(14,470)	(346,349)	(1,776,759)	(141,528)	(201,331)	(16)	(40)	(1)	(1)	0	3
Bolivia	221	366	7,131	7,181	272	1,427	14	207	3	32	5,368	7,277
Brazil	1,279	4,975	(585,302)	(868,151)	44,396	57,778	3	8	0	0	18	16,839
Chile	811	323	59,279	65,285	(36,160)	(70,717)	138	13	13	1	470	6,926
Colombia	378	848	64,204	119,796	4,344	44,906	20	4	1	0	240	585
Ecuador	181	317	37,464	34,565	950	24	5	25	1	3	3,052	1,063
Guyana	(30)	(3)	5,682	4,107	3,776	3,933	0	15	0	16	644	965
Paraguay	36	38	(26,292)	(37,072)	X	33	6	3	2	1	60	282
Peru	1,098	1,463	58,533	53,284	(1,349)	8,639	28	211	2	10	7,531	10,198
Suriname	(16)	(71)	3,658	835	1,219	3,183	0	0	0	0	0	0
Uruguay	(195)	(181)	(8,007)	(10,296)	1,208	1,680	0	0	0	0	0	80
Venezuela	1,556	2,161	151,528	274,633	56,527	61,569	0	0	0	0	0	0

Table 18.4

	Average Annual Net Trade in Food						Average Annual Donations or Receipts of Food Aid					
	Cereals (000 metric tons)		Oils (metric tons)		Pulses (metric tons)		Cereals (000 metric tons)		Cereals Kg Per Capita		Oils (metric tons)	Milk (metric tons)
	1975-77	1985-87	1975-77	1985-87	1975-77	1985-87	1975-77	1985-87	1975-77	1985-87	1985-87	1985-87
ASIA	47,053	67,336	(426,647)	(641,534)	35,219	(185,999)	4,284	2,899	2	1	238,158	70,135
Afghanistan	24	77	16,266	47,733	(2,667)	(10,000)	38	108	3	7	0	0
Bahrain	53	74	1,367	7,382	1,238	1,972	0	0	0	0	0	0
Bangladesh	1,544	1,864	89,212	320,734	801 a	13,062	1,155	1,463	15	14	23,305	19
Bhutan	2	20	3,300	4,200	X	X	0	4	0	3	137	277
China	5,555	7,863	172,957	532,662	(49,157)	(292,721)	(48)	172	(0)	0	1,899	3,593
Cyprus	213	432	8,472	14,492	1,652	789	15	1	24	2	0	134
India	5,209	(460)	369,850	1,447,898	6,278	365,967	605	204	1	0	76,738	26,094
Indonesia	2,135	1,587	(421,941)	(795,160)	248	19,822	633	233	5	1	387	4,475
Iran, Islamic Rep	2,010	4,468	272,008	516,955	11,112	10,955	0	10	0	0	0	33
Iraq	897	3,679	92,553	282,052	24,325	76,300	3	0	0	0	0	0
Israel	1,641	1,806	4,217	34,347	13,317	19,950	67	6	19	1	0	38
Japan	20,444	26,857	399,282	458,415	190,580	170,154	(79)	(413)	(1)	(3)	(511)	(231)
Jordan	323	733	10,606	35,308	(1,507)	18,445	120	31	45	9	684	1,189
Kampuchea, Dem	41 a	72	X	X	X	X	0	21	0	3	1,067	0
Korea, Dem People's Rep	53	228 a	3,599	12,433	X	X	10	0	1	0	0	0
Korea, Rep	3,263	7,662	147,800	310,287	7	18,538	483	0	13	0	0	0
Kuwait	205	443	6,698	20,889	6,309	8,542	0	0	0	0	0	0
Lao People's Dem Rep	83	37	146 b	X	14 b	X	16	4	5	1	8	0
Lebanon	585	530	25,210	58,347	10,751	32,500	68	29	24	11	2,045	1,988
Malaysia	984	2,076	(1,462,972)	(4,647,637)	24,719	41,197	0	0	0	0	0	0
Mongolia	71	(24)	1,005	1,540	X	X	0	0	0	0	0	20
Myanmar	(526)	(548)	28,081	27,616	(32,236)	(81,888)	7	0	0	0	0	8
Nepal	(99)	(11)	20,400	10,425	(2,385)	(13,886)	3	13	0	1	436	1,972
Oman	77	233	5,008	14,713	1,999	3,581	0	0	0	0	0	0
Pakistan	256	(20)	277,313	846,701	94 a	60,809	464	417	6	4	127,903	5,850
Philippines	851	1,139	(737,840)	(992,002)	1,546	11,016	66	199	2	4	978	17,117
Qatar	29	92	1,354	6,229	634	2,153	0	0	0	0	0	0
Saudi Arabia	791	7,658	38,922	177,330	14,023	40,528	(3)	(93)	(0)	(8)	0	0
Singapore	593	649	23,156	101,540	8,403	9,153	0	0	0	0	0	0
Sri Lanka	1,163	736	(39,619)	(33,830)	1,808	39,385	348	309	25	19	468	3,500
Syrian Arab Rep	322	1,091	20,611	70,601	(30,821)	(3,758)	85	31	11	3	536	2,154
Thailand	(4,021)	(7,181)	14,373	(15,396)	(114,345)	(233,579)	1	37	0	1	450	120
Turkey	(152)	436	105,407	249,039	(77,263)	(547,576)	(7)	3	(0)	0	74	211
United Arab Emirates	153	455	6,452	45,439	7,628	17,713	0	0	0	0	0	0
Viet Nam	1,085	583	11,645	(4,600)	X	(18,333)	194	38	4	1	869	943
Yemen Arab Rep	224	839	2,821	61,317	505	300	28	58	5	8	147	223
Yemen, People's Dem Rep	116	267	2,905	32,017	1,776	1,933	11	14	7	6	537	409
EUROPE	32,492	(13,025)	1,678,553	1,142,741	521,613	817,072	(574)	(1,125)	(1)	(2)	(14,052)	(3,505)
Albania	17	(13)	4,967	10,600	X	(417)	0	0	0	0	0	0
Austria	113	(770)	99,363	106,506	5,567	5,705	0	(20)	0	(3)	0	(584)
Belgium	3,056	2,161	109,205	(85,339)	31,513	213,460	(48)	(52)	(5)	(5)	0	(45)
Bulgaria	59	971	(19,707)	10,492	(6,605)	1,118	0	0	0	0	0	0
Czechoslovakia	1,460	312	60,291	45,249	6,783	7,333	0	0	0	0	0	0
Denmark	(279)	(1,366)	(7,445)	155,879	(3,236)	(178,517)	(28)	(17)	(6)	(3)	(103)	(1,128)
Finland	(204)	(402)	(779)	(20,095)	2,431	873	(35)	(22)	(7)	(4)	(1,324)	(1,167)
France	(11,203)	(25,998)	374,010	273,276	55,417	(444,781)	(168)	(202)	(3)	(4)	0	0
German Dem Rep	3,408	1,706	91,969	48,963	9,972	874	0	0	0	0	0	0
Germany, Fed Rep	5,053	1,591	(92,438)	(200,905)	75,503	526,916	(158)	(242)	(3)	(4)	(5,371)	(1,889)
Greece	636	(247)	(5,307)	(84,813)	3,173	23,755	0	(10)	0	(1)	0	0
Hungary	(1,121)	(1,674)	(33,817)	(198,100)	(43,217)	(109,334)	0	0	0	0	0	0
Iceland	28	24	2,109	2,822	334	320	0	0	0	0	0	0
Ireland	600	156	1,564	35,471	8,035	13,536	(4)	(4)	(1)	(1)	(33)	(1,448)
Italy	6,986	4,506	402,529	416,924	107,138	252,356	(54)	(137)	(1)	(2)	(148)	(209)
Luxembourg	X	X	X	X	X	X	X	X	X	X	X	X
Malta	139	133	3,734	6,250	1,262	1,054	2	0	5	0	0	67
Netherlands	3,528	3,741	106,173	(9,512)	99,298	563,760	(84)	(128)	(6)	(9)	(3,310)	(4,665)
Norway	655	388	(60,020)	(15,857)	6,413	5,256	(10)	(41)	(2)	(10)	(45)	0
Poland	5,280	2,340	54,957	80,801	7,424	(62,593)	0	24	0	1	4,196	10,055
Portugal	1,949	1,871	33,180	(70,440)	15,057	10,827	217	0	23	0	44	504
Romania	(238)	190	(119,076)	(41,440)	(3,771)	(10,735)	1	0	0	0	0	0
Spain	4,418	1,678	(16,606)	(315,678)	60,857	79,344	0	(29)	0	(1)	(6)	0
Sweden	(1,167)	(1,251)	26,725	32,595	3,028	8,529	(91)	(77)	(11)	(9)	(7,757)	(7)
Switzerland	1,326	919	46,412	46,680	5,813	11,598	(34)	(38)	(5)	(6)	(93)	(2,804)
United Kingdom	7,809	(3,196)	511,482	844,076	67,846	(102,278)	(79)	(129)	(1)	(2)	(100)	(185)
Yugoslavia	190	(802)	104,539	67,695	5,574	(890)	0	0	0	0	0	0
U.S.S.R.	12,832	32,118	(177,572)	856,328	(43,416)	(41,070)	0	0	0	0	0	0
OCEANIA	(11,185)	(21,068)	(298,378)	(391,910)	(18,492)	(249,889)	(85)	(392)	(4)	(16)	(1,533)	(694)
Australia	(11,450)	(21,251)	(143,912)	(106,242)	9,071	(202,320)	(94)	(393)	(7)	(25)	(1,537)	(526)
Fiji	62	89	(10,583)	(611)	2,883	4,200	10	0	16	0	0	0
New Zealand	21	(211)	(75,903)	(89,978)	(31,100)	(53,863)	0	0	0	0	0	(168)
Papua New Guinea	96	177	(50,324)	(157,648)	71	103	0	0	0	0	4	0
Solomon Islands	6	17	(2,758)	(14,325)	10	12	0	0	0	2	0	0

Sources: Food and Agriculture Organization of the United Nations and United Nations Population Division.
Notes: a. Two years of data. b. One year of data.
Imports and food aid receipts are shown as positive numbers; exports and food aid donations are represented by negative numbers in parentheses.
0 = zero or less than half of the unit of measure; X = not available; negative numbers are shown in parentheses.
For additional information, see Sources and Technical Notes.

Table 18.5 Physical and Chemical Soil Constraints in Selected

	Total Land Area {a} (000 ha)	No Inherent Soil Constraints (000 ha) {b}	Physical Constraints (000 hectares)				Chemical Constraints (000 hectares)						
			Steep Slopes	Shallow Soils	Poor Drainage	Tillage Problems	Low Nutrient Retention	Aluminum Toxicity Hazard	Phosphorus Fixation Hazard	Low Potassium Reserves	Excess of Soluble Salts	Excess of Sodium	Sulphate Acidity
WORLD													
AFRICA	3,011,330	442,733	260,864	397,812	198,239	111,779	396,629	508,304	205,079	614,926	51,325	4,440	3,690
Algeria	238,174	63,374	24,652	56,208	490	537	83	0	0	80	5,153	0	0
Angola	124,670	3,679	6,008	5,958	16,422	1,028	43,930	36,925	10,859	56,045	252	58	0
Benin	11,062	227	516	995	1,178	246	111	964	0	1,284	10	10	0
Botswana	56,673	3,344	822	1,650	2,897	4,918	27,584	203	37	9,311	4,859	646	0
Burkina Faso	27,380	6,497	1,273	2,739	4,285	2,596	1,948	671	264	696	0	0	0
Burundi	2,565	66	356	108	252	100	0	1,369	1,218	1,945	46	0	0
Cameroon	46,540	1,932	1,774	2,041	4,784	1,232	1,109	33,419	5,334	34,643	34	10	34
Cape Verde	403	43	166	229	0	6	12	0	0	0	0	0	0
Central African Rep	62,298	800	2,222	3,476	5,504	179	13,080	33,182	10,564	43,546	85	0	0
Chad	125,920	30,976	5,879	31,399	7,344	8,453	19,678	2,322	28	4,814	3,640	473	0
Comoros	223	9	76	51	0	0	42	17	86	0	0	0	0
Congo	34,150	0	204	58	10,542	0	11,255	16,348	2,549	20,149	27	0	0
Cote d'Ivoire	31,800	137	1,118	1,912	1,723	107	1,098	20,517	17,512	21,491	0	0	11
Djibouti	2,318	679	13	1,240	23	0	0	0	0	0	101	0	17
Egypt	99,545	20,939	12,236	21,481	1,405	78	0	0	0	0	6,787	85	0
Equatorial Guinea	2,805	21	138	145	593	0	292	1,462	41	1,658	1	0	2
Ethiopia	110,100	14,751	35,017	14,788	622	11,190	2,055	6,153	3,869	12,012	2,001	0	147
Gabon	25,767	0	1,315	443	3,234	0	3,557	13,102	3,062	14,616	174	0	348
Gambia, The	1,000	355	14	29	217	0	0	174	0	174	66	0	99
Ghana	23,002	277	998	2,554	2,167	365	936	6,425	5,132	7,145	18	0	0
Guinea	24,586	369	3,680	10,773	1,395	59	517	8,390	4,155	9,200	101	0	305
Guinea-Bissau	2,812	0	492	818	268	0	168	417	0	671	225	0	528
Kenya	56,697	6,109	8,582	8,823	2,153	2,581	2,059	2,973	1,011	5,677	5,127	0	0
Lesotho	3,035	1	1,910	1,526	226	0	0	0	0	0	0	146	0
Liberia	9,632	174	751	1,004	1,835	0	957	7,653	2,854	7,531	0	0	0
Libya	175,954	45,230	9,626	20,828	274	142	78	0	0	0	2,427	6	0
Madagascar	58,154	2,220	6,460	3,018	3,601	743	4,447	21,118	8,519	23,296	512	0	528
Malawi	9,408	1,097	2,573	1,917	556	167	452	2,700	969	3,400	69	0	0
Mali	122,019	34,752	7,693	18,326	6,385	1,238	17,112	1,600	669	2,551	29	0	0
Mauritania	102,522	28,299	6,558	25,375	238	171	9,598	0	0	0	777	0	131
Mauritius	185	7	44	41	5	8	0	21	16	57	0	0	0
Morocco	44,630	9,059	11,469	10,716	844	1,212	0	0	0	0	2,017	27	0
Mozambique, People's Rep	78,409	4,348	5,655	6,780	2,981	1,952	13,674	15,844	7,611	19,480	1,203	0	168
Namibia	82,329	4,260	8,667	11,303	1,257	5,521	18,789	25	0	5,227	3,478	140	0
Niger	126,670	48,232	7,329	14,690	3,259	1,474	36,128	0	0	0	1,079	57	0
Nigeria	91,077	8,937	6,079	13,108	12,000	1,756	11,647	16,663	5,784	19,068	1,206	636	763
Rwanda	2,495	91	983	218	239	12	0	779	736	1,664	0	0	0
Senegal	19,253	3,120	773	2,707	2,321	416	5,503	843	0	916	243	0	376
Sierra Leone	7,162	180	946	1,552	587	0	388	3,975	66	4,003	63	0	193
Somalia	62,734	3,417	4,078	6,277	1,043	1,825	3,204	790	790	2,003	4,453	755	0
South Africa	122,104	1,273	17,589	23,851	5,682	6,146	25,294	3,070	2,287	7,839	1,158	536	0
Sudan	237,600	74,780	17,129	31,642	19,380	42,336	21,729	12,104	4,539	9,756	1,504	11	31
Swaziland	1,720	0	414	326	238	23	125	418	293	547	0	0	0
Tanzania	88,604	2,981	16,124	6,902	8,389	5,649	7,549	30,356	25,922	41,165	490	0	9
Togo	5,439	276	450	979	525	182	70	364	20	607	0	0	0
Tunisia	15,536	1,684	3,563	3,611	313	523		0			1,349	0	0
Uganda	19,955	592	2,255	2,158	2,480	804	701	11,486	3,606	12,480	26	0	0
Zaire	226,760	1,219	6,267	1,057	40,630	596	64,918	162,158	63,182	169,673	156	0	0
Zambia	74,072	1,786	3,241	5,345	14,112	2,594	18,809	30,025	10,895	32,712	0	800	0
Zimbabwe	38,667	893	2,997	4,984	1,341	2,598	5,985	1,221	644	5,619	349	44	0
CENTRAL AMERICA	273,999	58,592	68,810	44,139	15,417	18,671	689	24,157	15,107	27,335	436	1,934	650
Barbados	43	0	1	3	7	6	0	0	0	6	0	2	0
Belize	2,280	66	373	167	495	402	28	648	266	456	0	20	35
Costa Rica	5,106	84	1,132	306	868	118	0	1,277	1,015	1,477	0	12	35
Cuba	11,086	310	1,108	752	1,564	1,605	186	1,924	1,084	2,778	0	264	143
Dominican Rep	4,838	454	1,514	681	287	563	0	737	61	773	0	0	8
El Salvador	2,072	253	743	223	52	310	0	188	33	470	0	2	2
Guatemala	10,843	469	3,223	1,049	1,349	1,219	17	1,278	742	1,342	0	54	1
Haiti	2,756	263	1,035	419	53	106	0	466	75	793	0	0	10
Honduras	11,189	628	3,843	1,026	970	205	0	3,497	2,642	2,738	0	45	81
Jamaica	1,083	114	357	137	52	161	0	184	11	218	0	0	10
Mexico	190,869	55,092	49,125	37,263	6,660	12,492	458	5,086	3,030	6,978	436	1,413	61
Nicaragua	11,875	493	3,571	909	1,870	1,068	0	4,724	3,154	4,624	0	50	112
Panama	7,599	161	2,152	651	1,075	127	0	3,486	2,479	4,038	0	72	151
Trinidad and Tobago	513	69	113	38	0	0	0	349	349	349	0	0	0
SOUTH AMERICA	1,898,326	207,647	328,981	192,612	212,775	26,976	117,192	842,774	599,694	913,587	24,313	25,138	2,090
Argentina	273,669	129,242	32,985	24,903	31,052	4,266	30	1,385	2,027	4,058	14,613	19,432	0
Bolivia	108,439	8,645	27,382	21,841	15,422	748	2,621	30,582	24,055	27,063	454	1,854	0
Brazil	845,651	22,627	109,234	51,435	98,796	10,373	106,245	580,354	423,041	668,226	3,237	276	1,111
Chile	74,880	4,443	33,369	26,425	3,780	847	370	4,186	0	1,686	2,896	0	0
Colombia	103,870	5,053	26,806	9,369	11,652	1,114	1,122	59,890	41,488	54,043	460	0	0
Ecuador	27,684	2,000	8,560	5,375	1,969	806	40	8,136	5,988	9,099	168	0	0
Guyana	19,685	868	3,855	4,094	1,685	0	1,166	12,024	6,988	12,865	108	0	153
Paraguay	39,730	10,730	0	0	17,475	1,401	895	18,401	16,929	16,929	620	3,576	0
Peru	128,000	13,487	65,030	36,115	12,905	1,069	2,139	54,800	35,798	48,640	798	0	0
Suriname	16,147	378	31	62	1,537	0	955	9,956	9,749	13,458	75	0	212
Uruguay	17,481	6,139	912	4,247	2,791	3,738	0	34	29	29	75	0	37
Venezuela	88,205	3,838	20,817	8,746	13,410	2,614	1,501	54,584	34,108	48,973	785	0	501

Countries

Table 18.5

	Total Land Area {a} (000 ha)	No Inherent Soil Constraints (000 ha) {b}	Physical Constraints (000 hectares)				Chemical Constraints (000 hectares)						
			Steep Slopes	Shallow Soils	Poor Drainage	Tillage Problems	Low Nutrient Retention	Aluminum Toxicity Hazard	Phosphorus Fixation Hazard	Low Potassium Reserves	Excess of Soluble Salts	Excess of Sodium	Sulphate Acidity
ASIA, SOUTHEAST	897,615	33,111	261,439	90,678	108,628	75,797	30,008	241,317	190,538	257,455	17,265	1,160	6,711
Bangladesh	13,391	908	1,538	181	8,198	0	0	2,219	843	1,195	104	0	352
Bhutan	4,700	28	2,347	748	0	0	532	1,807	1,599	1,748	0	0	0
India	297,319	15,184	42,183	33,151	14,725	67,431	8,858	22,491	16,126	32,338	8,505	0	562
Indonesia	181,157	7,525	68,288	11,095	37,008	3,459	11,979	60,812	51,591	72,814	575	2	1,076
Kampuchea, Dem	17,652	780	3,933	661	6,250	605	260	9,555	7,457	9,140	33	354	206
Lao People's Dem Rep	23,080	37	17,463	2,337	1,241	195	2	18,404	16,175	17,699	0	109	0
Malaysia	32,855	283	15,710	735	6,403	358	223	21,183	17,828	20,523	67	0	681
Myanmar	65,754	3,436	31,272	5,880	9,297	1,622	0	33,210	29,647	34,463	710	0	1,247
Nepal	13,680	1,060	5,239	2,909	1,096	0	40	4,167	1,416	1,736	0	0	0
Pakistan	77,088	964	31,451	24,493	1,202	77	6,580	108	108	108	7,035	0	0
Philippines	29,817	197	8,594	1,908	3,519	776	162	13,111	6,227	15,048	0	0	0
Singapore	61	0	0	0	5	0	0	53	48	53	0	298	17
Sri Lanka	6,474	488	838	639	1,510	44	135	1,182	821	1,233	138	0	17
Thailand	51,089	983	17,702	3,127	12,443	707	764	33,103	23,619	30,271	0	363	1,027
Viet Nam	32,536	1,238	14,753	2,814	5,570	523	473	19,600	16,760	18,801	98	34	1,526
ASIA, SOUTHWEST	678,017	45,577	161,378	173,856	6,070	6,421	37,003	4,119	2,090	4,099	49,049	0	0
Afghanistan	65,209	132	28,812	23,889	310	0	4,178	0	0	0	3,970	0	0
Bahrain	68	12	6	12	0	0	0	0	0	0	6	0	0
Iran, Islamic Rep	163,600	3,045	66,071	46,672	2,436	100	43	98	49	49	22,001	0	0
Iraq	43,737	4,080	3,853	6,839	292	3,137	1,548	0	0	1,300	5,286	0	0
Israel	2,033	13	236	557	12	342	0	0	0	0	52	0	0
Jordan	8,893	2,968	1,582	2,741	8	83	384	0	0	0	229	0	0
Kuwait	1,782	21	117	235	14	0	430	0	0	373	128	0	0
Lebanon	1,023	19	524	261	4	133	0	0	0	0	0	0	0
Oman	21,246	2,915	2,374	4,749	200	0	731	0	0	0	1,316	0	0
Qatar	1,100	290	246	493	98	0	64	0	0	0	317	0	0
Saudi Arabia	214,969	20,774	16,873	48,629	686	0	26,499	0	0	336	12,138	0	0
Syrian Arab Rep	18,406	858	2,296	3,735	61	1,083	0	0	0	0	53	0	0
Turkey	76,963	4,462	33,551	20,742	1,424	1,543	0	4,021	2,041	2,041	353	0	0
United Arab Emirates	8,360	199	372	504	525	0	861	0	0	0	1,119	0	0
Yemen, Arab Rep	19,500	2,512	1,972	7,929	0	0	862	0	0	0	601	0	0
Yemen, People's Dem Rep	33,297	3,277	2,493	5,869	0	0	1,403	0	0	0	1,480	0	0

Sources: Food and Agriculture Organization of the United Nations and North Carolina State University.
Notes: Extent of soils without and with different constraint (rows) may not be summed per country or region, because some soils have more than one constraint; columns (extent for any constraint) may be summed over countries.
a. Regional totals differ from those in Table 17.1 because of differences in methodology.
b. Areas with no inherent soil constraints occur in different climates and with different growing periods; some of these prevent rainfed cultivation; see Table 18.6. For additional information, see Sources and Technical Notes.

Table 18.6 Physical and Chemical Soil Constraints By Region and Climatic Classes

	Total Land Area {a} (000 ha)	No Inherent Soil Constraints (000 ha)	Physical Constraints (000 hectares)				Chemical Constraints (000 hectares)						
			Steep Slopes	Shallow Soils	Poor Drainage	Tillage Problems	Low Nutrient Retention	Aluminum Toxicity Hazard	Phosphorus Fixation Hazard	Low Potassium Reserves	Excess of Soluble Salts	Excess of Sodium	Sulphate Acidity
AFRICA	3,011,330	442,733	260,864	397,812	198,239	111,779	396,629	508,304	205,079	614,926	51,325	4,440	3,690
Arid	1,503,665	374,457	126,971	275,757	20,941	34,054	152,652	4,918	3,844	22,354	41,769	2,041	423
Semi-arid	491,100	44,140	46,047	54,312	53,744	52,865	79,346	40,548	21,155	55,914	6,431	1,929	650
Humid	1,006,797	23,569	82,763	64,350	123,554	24,373	164,629	462,805	180,047	535,868	3,063	470	2,617
Cold	9,759	567	5,083	3,185	0	487	0	33	33	790	62	0	0
CENTRAL AMERICA	273,999	58,592	68,810	44,139	15,417	18,671	689	24,157	15,107	27,335	436	1,001	650
Arid	105,296	42,451	27,522	24,247	1,253	2,726	77	110	11	145	388	968	6
Semi-arid	60,398	9,148	15,449	9,566	2,857	6,549	268	3,403	1,435	4,975	14	256	145
Humid	107,592	6,939	25,523	10,253	11,295	9,387	344	20,536	13,609	22,110	34	710	499
Cold	713	54	316	73	12	9	0	108	52	105	0	0	0
SOUTH AMERICA	1,898,326	207,647	328,981	192,612	212,775	26,976	117,192	842,774	599,694	913,587	24,311	25,138	2,090
Arid	249,471	67,551	60,731	53,047	7,936	2,665	2,704	7,887	3,180	8,811	12,389	4,295	21
Semi-arid	330,715	52,346	38,835	35,908	39,776	6,837	9,784	85,539	65,663	104,076	2,437	10,378	118
Humid	1,257,359	81,258	190,060	75,176	164,945	17,474	104,704	749,073	530,851	800,700	7,095	10,465	1,951
Cold	60,781	6,492	39,355	28,481	118	0	0	275	0	0	2,390	0	0
ASIA, SOUTHEAST	897,615	33,111	261,439	90,678	108,628	75,797	30,008	241,317	190,538	257,455	17,265	1,160	6,711
Arid	107,392	946	18,800	16,065	2,467	5,903	14,561	0	0	11	11,054	0	35
Semi-arid	188,274	10,160	19,760	17,998	7,817	56,130	373	9,015	6,639	11,413	4,054	121	65
Humid	554,777	19,858	195,472	36,296	97,789	13,764	14,712	229,720	181,966	244,062	1,982	1,039	6,611
Cold	47,172	2,147	27,407	20,319	555	0	362	2,582	1,933	1,969	175	0	0
ASIA, SOUTHWEST	678,017	45,577	161,378	173,856	6,070	6,421	37,003	4,119	2,090	4,099	49,049	0	0
Arid	462,101	38,839	104,094	129,954	3,544	1,681	36,990	0	0	2,009	47,030	0	0
Semi-arid	79,818	5,882	47,870	39,129	1,224	3,508	6	1,274	545	545	1,538	0	0
Humid	22,431	856	9,414	4,773	1,302	1,232	7	2,845	1,545	1,545	481	0	0
Cold	113,667	0	0	0	0	0	0	0	0	0	0	0	0

Sources: Food and Agriculture Organization of the United Nations and North Carolina State University.
Notes: a. Regional totals differ from those in Table 17.1 because of differences in methodology.
For additional information, see Sources and Technical Notes.

Sources and Technical Notes

Table 18.1 Food and Agricultural Production, 1976–88

Source: Food and Agriculture Organization of the United Nations (FAO), unpublished data (FAO, Rome, July 1989).

Indexes of agricultural and food production portray the disposable output (after deduction for feed and seed) of a country's agriculture sector relative to the base period 1979–81. For a given year and country, the index is calculated as follows: the disposable average output of a commodity in terms of weight or volume during the period of interest is multiplied by the 1979–81 average national producer price per unit. The product of this equation represents the total value of the commodity for that period in terms of the 1979–81 price. The values of all crop and livestock products are totaled to an aggregated value of agricultural production in 1979–81 prices. The ratio of this aggregate for a given year to that for 1979–81 is multiplied by 100 to obtain the index number.

The multiplication of disposable outputs with the 1979–81 unit value eliminates inflationary or deflationary distortion. However, the base period's relative prices among the individual commodities are also preserved. Especially in economies with high inflation, price patterns among agricultural commodities can change decisively over time. To overcome the latter problem, FAO generally shifts the base period every five years.

The continental and world index numbers for a given year are calculated by totaling the disposable outputs of all relevant countries for each agricultural commodity. Each of these aggregates is multiplied with a respective 1979–81 average "international" producer price and summed in a total agricultural output value for that region or the world in terms of 1979–81 prices. The total agricultural output value for a given year is then divided by the "international" 1979–81 output value and multiplied by 100 to obtain the continental and world index numbers. This method avoids distortion caused by the use of international exchange rates.

The agricultural production index includes all crop and livestock products originating in each country. The food production index covers all edible agricultural products that contain nutrients. Coffee and tea have virtually no nutritive value and are thus excluded.

Crop yields (*average yields cereals* and *average yields roots and tubers*) are calculated from production and area data. *Average production cereals* include cereal production for feed and seed. Area refers to the area harvested. Cereals comprise all cereals harvested for dry grain, exclusive of crops cut for hay or harvested green. Roots and tubers cover all root crops grown principally for human consumption; root crops grown principally for feed are excluded.

Most of the data in Tables 18.1–18.5 are supplied by national agriculture ministries in response to annual FAO questionnaires or are derived from agricultural censuses. FAO compiles data from more than 200 country reports and from many other sources and enters them into a computerized data base. FAO fills gaps in the data by preparing its own estimates. As better information becomes available, FAO corrects its estimates and recalculates the entire time series when necessary.

Table 18.2 Agricultural Inputs, 1975–87

Sources: Food and Agriculture Organization of the United Nations (FAO), unpublished data (FAO, Rome, July 1989). Population data: United Nations Population Division, *World Population Prospects 1988* (United Nations, New York, 1989). Pesticide consumption: United Nations Industrial Development Organization (UNIDO) Industrial Statistics and Sectoral Surveys Branch, Policy and Perspectives Division, pesticide data base specifically prepared for UNIDO's study *Global Overview of the Pesticide Industry Sub-Sector* (UNIDO, Sectoral Working Paper PPD.98, Vienna, 2 December 1988). Pesticide consumption for Mauritius, Tanzania, Uganda: Environment Liaison Centre, *Africa Seminar on the Use and Handling of Agricultural and Other Pest Control Chemicals* (October 30 to November 4, 1983, Nairobi, Kenya). Pesticide consumption for Haiti: Data are 1972–74 pesticide imports; U.S. AID, *Draft Environmental Profile of Haiti* (U.S. AID, Washington, D.C., 1979). Pesticide consumption for El Salvador (1975–77 data are for 1974–76; 1982–84 data are for 1979) and Suriname (1975–77 data are for 1979–80; 1982–84 data are for 1981): David K. Burton and Bernard J.R. Philogene, *An Overview of Pesticide Usage in Latin America* (Canadian Wildlife Service Latin American Program, Ottawa, undated). Pesticide consumption for Afghanistan (1975–77 data are for 1980; 1982–84 data are for 1981–82) and the Philippines (1975–77 data are for 1980–81): Regional Network for the Production, Marketing, and Control of Pesticides in Asia and the Pacific (RENPAP), formerly Regional Network for the Production, Marketing, and Control of Pesticides in Asia and the Far East (RENPAF), *RENPAF Gazette: Supply of Pesticides in Nine Countries* (RENPAF, Bangkok, July 1985) and *RENPAP Gazette: Pesticide Data Collection System—Second Report* (RENPAP, Bangkok, October 1988). Pesticide consumption for Malaysia: Data are 1988 estimate by Malaysian Agrochemical Association (MACA) provided by the Ministry of Science, Technology and Environment

(Kuala Lumpur, 1989). Pesticide consumption for Syria (1982–84 data are for 1983–84): L'Office Arabe de Presse et de Documentation, *Rapport Economique Syrien 1983–84* (L'Office Arabe de Presse et de Documentation, Damascus, 1984). Pesticide consumption for Yemen (1975–77 data are for 1972–74; 1982–84 data are for 1975–76): U.S. AID, *Draft Environmental Report on Yemen* (U.S. AID, Washington, D.C., 1982). Pesticide consumption for Iceland: Nordic Council and the Nordic Statistical Secretariat, *Yearbook of Nordic Statistics 1985* (Nordic Council and Nordic Statistical Secretariat, Oslo, 1986).

Cropland refers to land under temporary and permanent crops, temporary meadows, market and kitchen gardens, and temporarily fallow land. Permanent cropland is land under crops that do not need to be replanted after each harvest, such as cocoa, coffee, rubber, fruit trees, and vines. Human population data, used to calculate *hectares per capita*, are for 1989. For trends in cropland area, see Chapter 17, "Land Cover and Settlements," Table 17.1.

Irrigated land as a percentage of cropland refers to areas purposely provided with water, including land flooded by river water for crop production or pasture improvement, whether this area is irrigated several times or only once during the year.

Average annual fertilizer use refers to application of nutrients in terms of nitrogen (N), phosphate (P_2O_5) and potash (K_2O). The fertilizer year is July 1–June 30; data refer to the year beginning in July.

Data on *average annual pesticide use* were compiled by UNIDO. In their study, UNIDO assessed production, trade, and consumption of pesticides for 119 countries and 14 geographical subgroups. The calculations were based on trade statistics of pesticide finished products. These statistics were published by the Statistical Office of the United Nations (UNSO), and consumption and trade data were compiled by the Food and Agriculture Organization of the United Nations (FAO). For one third of the countries, UNIDO had to estimate net weight of active ingredient in the pesticides consumed, because country level data were only quantified for finished products or not available at all. Time series were completed by interpolation and extrapolation or, in some cases, by an econometric need model. Parameters in this model included country-specific factors such as climatic zone, degree of development of agriculture, area under crop production, crop structure, and the frequency of pesticide applications. For additional information, refer to UNIDO's *Global Overview of the Pesticide Industry Sub-Sector*.

Data are expressed in net weight of active ingredients in the pesticides consumed. The active ingredients in a pesticide are those chemicals with pesticidal properties. Active ingredients are often mixed with inert ingredients, which dilute or deliver the active ingredients, in a formulated pesticide. Inert ingredients can exert toxic effects of their own in the environment.

Active ingredients vary widely in potency; information on the ingredients of pesticides is necessary to ensure accurate application and to minimize harmful environmental impacts. For example, 1 metric ton of the modern synthetic pyrethroid insecticide "permethrin" is as potent a pesticide as 3–5 metric tons of carbamate or organophosphate, or 10–30 metric tons of DDT. The data shown in this table do not describe the potency of the active ingredients used. As a result, two countries with similar levels of pesticide consumption may be treating different amounts of land and getting very different results. Increasingly potent pesticides have been developed in recent years; thus, a decline in the amount of active ingredients used may not indicate a reduction in the amount of toxic materials introduced into the environment.

For additional information on pesticide data, see E.J. Tait and A.B. Lane, "Insecticide Production, Distribution and Use: Analyzing National and International Statistics" in Joyce Tait and Banpot Napompeth, eds., *Management of Pests and Pesticides: Farmers' Perceptions and Practices* (Westview Press, Boulder, Colorado, United States, and London, 1987).

Tractors generally refer to wheel and crawler tractors used in agriculture. Garden tractors are excluded. *Harvesters* refer to harvesters and threshers.

Table 18.3 Livestock Populations, 1976–88

Sources: Livestock data: Food and Agriculture Organization of the United Nations (FAO), unpublished data (FAO, Rome, July 1989). Human population data: United Nations Population Division, *World Population Prospects 1988* (United Nations, New York, 1989).

Data on livestock include all animals in the country, regardless of place or purpose of their breeding. Data on livestock numbers are collected by FAO; estimates are made by FAO for countries that either do not report data or only partially report data. Human population data, used to calculate *cattle per capita*, are for 1989. FAO notes that the reported number of *chickens* in some countries does not seem accurate. For some countries, data on chickens cover all poultry. *Equines* includes horses, mules, and asses. For more information on FAO agricultural surveys, see the Technical Note for Table 18.1.

Table 18.4 Food Trade and Aid, 1975–87

Sources: Trade data: Food and Agriculture Organization of the United Nations (FAO), unpublished data, (FAO, Rome, July 1989.

Food aid data: FAO, *Food Aid in Figures, No. 6* (FAO, Rome, 1988). Population data 1976 and 1986: United Nations Population Division, *World Population Prospects: Estimates and Projections as Assessed in 1982* (United Nations, New York, 1985) and United Nations Population Division, *World Population Prospects 1988* (United Nations, New York, 1989).

Figures shown for food trade are *net* imports or exports: exports were subtracted from imports.

Two definitions of trade are used by countries reporting trade data. "Special trade" reports only imports for domestic consumption and exports of domestic goods. "General trade" records total imports and total exports, including re-exports. Trade figures for Czechoslovakia, the German Democratic Republic, Hungary, Poland, Romania, and the U.S.S.R. include goods purchased by the country that are re-exported to a third country without ever entering the purchasing country. For information on the definition used by a particular country, see *FAO Trade Yearbook 1987* (FAO, Rome, 1988).

Average annual donations or receipts of food aid are shown as either positive or negative numbers: Receipts are shown as positive numbers; donations are expressed in negative figures. For some countries that are both recipients and donors of food aid, donations were subtracted from receipts.

Trade in *cereals* includes wheat and wheat flour, rice, barley, maize, rye, and oats. Trade in *oils* includes oils from soybeans, groundnuts (peanuts), olives, cottonseeds, sunflower seeds, rapeseeds, colza, mustard seeds, linseeds, palms, coconuts, palm-kernels, castor beans, and maize, as well as animal oils, fats, and greases. Trade in *pulses* includes all kinds of dried leguminous vegetables, with the exception of vetches and lupins.

Food aid refers to the donation or concessional sale of food commodities. *Cereals* include wheat, rice, coarse grains, bulgur wheat, wheat flour, and the cereal component of blended foods. Cereal donations or receipts *(kilograms per capita)* are the result of dividing the three-year averages by the respective 1976 and 1986 population. *Oils* include vegetable oil and butter oil. *Milk* includes skimmed milk powder and other dairy products (mainly cheese). Regional totals include only listed countries and do not reflect donations by the European Community, the Organization of Petroleum Exporting Countries (OPEC), and the World Food Program.

Food aid data are reported by donor countries and international organizations.

Table 18.5 Physical and Chemical Soil Constraints in Selected Countries

Sources: Data and interpretation: W. Couto, Consultant to Land and Water Development Division, Food and Agriculture Organization of the United Nations (FAO)

(FAO, Rome, November 1989). Method based on Fertility Capability Classification system (FCC) developed by North Carolina State University, P.A. Sanchez, W. Couto, S.W. Buol, "The Fertility Capability Soil Classification System: Interpretation, Applicability and Modification," *Geoderma*, Vol. 27 (1982), pp. 283-309. Agroclimatic data from Land and Water Development Division, FAO, *Report on the Agro-Ecological Zones Project*, Vol. 1, *Methodology and Results for Africa, World Soil Resources Report 48/1* (FAO, Rome, 1978); Vol. 2, *Results for Southwest Asia, World Soil Resources Report 48/2* (FAO, Rome, 1978); Vol. 3, *Methodology and Results for South and Central America, World Soil Resources Report 48/3* (FAO, Rome, 1981); Vol. 4, *Results for Southeast Asia, World Soil Resources Report 48/4*, (FAO, Rome, 1980). Soil data from FAO Geographical Information System, based on the *FAO-UNESCO Soil Map of the World* (FAO and United Nations Educational, Scientific and Cultural Organization, Paris, Vol. 1, 1974; Vol. 3, 1975; Vol. 4, 1971; Vol. 6, 1977; Vol. 7, 1977; Vol. 9, 1979). Land area data: FAO, unpublished data (FAO, Rome, July 1989).

Chemical and physical characteristics of soils are crucial determinants of agronomic management and the productivity of a nation's agriculture. The properties of many soils can restrict crop production, and high yields are possible only if these constraints are overcome. The extent of land with soil constraints is an important indicator of agricultural costs, the potential and success of future expansion, and the comparative advantage of a nation's agricultural production.

Data on the extent of physical and chemical soil constraints were calculated from the electronic version of the FAO-UNESCO Soil Maps of the World. Each mapping unit was interpreted, applying the fertility capability classification (FCC) system developed by North Carolina State University. The conversion to FCC units is based on the soil units data, which integrate the mapping units and their associated regimes.

The FCC system was developed to group soils according to their chemical and physical properties that are relevant for fertility and management purposes. It emphasizes quantifiable topsoil parameters, as well as subsoil properties directly relevant to plant growth. The system interprets three categories: topsoil texture (sandy, loamy, clayey, or organic soils), subsoil texture (sandy, loamy, clayey, and rock or other hard root-restricting layer), and 15 modifiers, identifying chemical and physical properties of the soils. In the past 10 years, the FCC system has proven a meaningful tool for describing fertility limitations to crop yield responses in a variety of soils and crops.

The data in Table 18.5 show some of the major soil constraints to crop production, four *physical constraints* and seven *chemical constraints*, calculated as the total area presenting those constraints on the basis of all mapping units for each country and region. All soils are included, whether occurring as dominant soil, associated soil, or inclusions. Areas of nonsoil, such as glaciers, bare rock, or moving dunes, are excluded. Extent of soils without and with different constraints (rows) may not be summed per country or region, because some soils have more than one constraint. Columns (extent for any one constraint) may be summed over countries.

None of the FCC constraints applies for soils with *no inherent soil constraints*. However, such areas with no fertility limitations occur in different climates and lengths of growing period; part of these prevent rainfed cultivation.

Steep slopes are classified as steeply dissected to mountainous. Dominant slopes are over 30 percent.

Shallow soils are mostly lithosoils and other soils presenting restriction for deep root penetration or mechanized tillage. High priority should be given to erosion control when combined with steep relief.

Soils with *poor drainage* are saturated with water during part of the year and might be prone to waterlogging. These soils require drainage to improve crop growth and generally provide a good soil for rice production.

Tillage problems result from clayey textured topsoils with shrink and swell properties. These soils tend to be hard when dry and sticky when wet. Adequate practices are needed to till them at the correct time. These soils can be highly productive, but cannot be farmed with traditional technology. Improved soil tillage

practices can contribute to expanded agriculture on these soils.

Low nutrient retention occurs in soils with a low ability to retain nutrients, mainly potassium, calcium, and magnesium, against leaching. Heavy applications of these nutrients, as well as nitrogen fertilizers, can overcome this constraint. Fertilizers should be split in more than one application. Subsistence agriculture requires a long fallow period. The reduction or elimination of the fallow period results in rapid drop of fertility.

Aluminum toxicity hazard is prevalent mostly in soils of the subhumid and humid tropics. Plants such as soybeans and corn are sensitive to aluminum-toxicity, and their growth will be affected. This hazard can be overcome by applying large quantities of lime or planting more resistant crops, such as certain varieties of rice, cowpeas, and sorghum.

Soils with *phosphorus fixation hazard* result in crop phosphorus deficiency. These soils can produce a satisfying yield for subsistence farmers; but when used for commercial agriculture, they require high levels of phosphate application and special management practices to increase productivity.

Soils with *low potassium reserves* result in potassium deficiency that constrains crop growth. This can be overcome by application of potassium fertilizers.

An *excess of soluble salts* requires special management to avoid damage to salt-sensitive crops or the use of salt-tolerant species and cultivars. Saline soils have

low potential productivity and need reclamation measures to become productive.

Soils with an *excess of sodium* demand special soil management practices for alkaline soils. Drainage and gypsum application can help to reduce this constraint.

Soils constrained by *sulphate acidity* require plants tolerant to high water tables.

Table 18.6 Physical and Chemical Soil Constraints by Region and Climatic Classes

Sources: Food and Agriculture Organization of the United Nations and North Carolina State University; please refer to Technical Note of Table 18.5 for a detailed listing of sources and definitions of soil constraint classes.

Climatic classes are calculated by length of growing period. Length of growing period is defined as the number of days when both temperature and moisture permit crop growth. Days with mean temperatures of more than 5° C and with soil moisture resulting from rainfall at least equivalent to half potential evapotranspiration are considered favorable to growth. A "normal" growing period includes a humid period, during which precipitation exceeds full potential evapotranspiration. Climatic classes in Table 18.6 refer to length of growing period as follows: *arid* 90 days; *semi-arid* 90–179 days; *humid* 179 days; *cold* = 0 days (mean temperature is 5° C while moisture is available).

19. Forests and Rangelands

The extent of the world's forest has declined steadily as agriculture has expanded around the world. Today about one third of the world's land area is covered by forests. Stretching from the evergreen forests of the moist tropics to the vast boreal forests of the subarctic, forests include a broad variety of ecosystems. Forests provide habitat for millions of plant and animal species, recycle nutrients, protect soils and watersheds, and provide many products and services to people. Some societies have used their forests in a sustainable fashion; others have exploited forest products and lands excessively, leaving only poor remnants of the original stands. Too often a short-term utilitarian view has contributed to the decline in forest cover.

Table 19.1 presents data for the early 1980s on the extent of forests, trends in deforestation, and management status. The rate of global deforestation can be used as one indicator of how fast the world is extinguishing species, destroying watersheds, manipulating microclimates, and affecting the carbon and nutrient cycles.

Since the *1980 Tropical Forest Assessment*, the first global tropical forest resource assessment, completed in 1982 by the Food and Agriculture Organization of the United Nations (FAO) and the United Nations Environment Programme (UNEP), almost a decade has passed with no new comparable data on a global level. A second assessment, the *1990 Forest Assessment*, is currently under way; results are promised in 1992. Over the past seven years, however, FAO has reviewed the 1980 assessment, revised estimates for some countries, and added estimates for some missing countries. Table 19.1 includes these revised numbers. Data on deforestation are missing for Europe, North America, the U.S.S.R., and other areas, because FAO focused on tropical deforestation in the previous assessment. According to the Organisation for Economic Co-operation and Development (OECD), total forest cover is not declining in Europe and Canada. The Forest Service of the U.S. Department of Agriculture reports a slight reduction in U.S. forest area for the past 10 years. Periodic inventories for the U.S.S.R. imply that the total forest area is increasing, but improved survey techniques and inclusion of previous uncharted areas may have inflated these data.

Because data from the 1990 assessment will not be published until 1992, the World Resources Institute obtained recently published estimates on deforestation from independent and national research agencies. (See Chapter 7, "Forests and Rangelands," New Estimates on Tropical Deforestation.) These sources suggest that tropical deforestation has accelerated throughout the 1980s. Table 19.1 shows updated deforestation rates for Brazil, Cameroon, Costa Rica, India, Indonesia, Myanmar, the Philippines, Thailand, and Viet Nam.

Table 19.2 presents data on the harvesting and processing of wood. The total harvest of wood, classified as roundwood production, refers to all wood in the rough and is destined for either fuel or industrial uses (e.g., sawlogs, veneer logs, and pulpwood). Global production of roundwood increased by 22 percent in the past decade, and the production of fuelwood and charcoal grew faster than that of industrial roundwood. Data on the production of fuelwood and charcoal must be viewed with caution. In the case of countries that do not report these data, FAO has estimated consumption per head of population to calculate total fuelwood production.

In Africa, South America, and Asia, most roundwood is used for fuel, whereas in North America, Europe, Oceania, and the U.S.S.R. it is produced for industrial use. This difference can also be observed for the top six roundwood producers: the United States, the U.S.S.R., China, India, Brazil, and Canada, which together remove roughly half of the world's timber from their own territory. Canada, the United States, and the U.S.S.R. harvest most of their wood for industrial purposes and were net exporters of roundwood in 1985–87. China, Brazil and India, each cutting roughly the same quantities of roundwood as Western Europe, used most of their wood for fuel and imported more roundwood than they exported in 1985–87. Both in 1975–77 and in 1985–87, Japan led the world as the largest net importer of roundwood.

The production of processed wood—sawnwood and panels—is concentrated in developed countries and in the wood-rich and most industrialized developing countries. For example, production of panels showed large growth rates in the past decade for China, India, and Brazil. In 1987, Indonesia was the largest producer of wood-based panels in the developing countries, following the United States, the U.S.S.R., Japan, Canada, and the Federal Republic of Germany.

A similar concentration can be observed for the production of paper, where most developing countries remain dependent on imports. One exception in the developing world is China, which was the fifth largest producer of paper in the world in 1985–87.

Table 19.1 Forest Resources, 1980s

| | Extent of Forest and Woodland, 1980s (000 hectares) | | | Average Annual Deforestation, 1980s | | | | | | Average Annual Reforestation 1980s | Managed Closed Forest 1980s | Protected Closed Forest 1980s |
| | | | | Closed Forest | | Open Forest | | Total Forest | | | | |
	Closed	Open	Total	Extent (000 ha)	Percent	Extent (000 ha)	Percent	Extent (000 ha)	Percent	(000 ha)	(000 ha)	(000 ha)
WORLD	2,838,770	1,242,768	4,081,538									
AFRICA	219,811	464,591	684,402	1,359	0.6	2,406	0.5	3,822	0.6	355	2,327	9,434
Algeria	1,518	249	1,767	X	X	X	X	40	2.3	66	X	8
Angola	2,900	50,700	53,600	44	1.5	50	0.1	94	0.2	4	X	X
Benin	47	3,820	3,867	1	2.6	66	1.7	67	1.7	0	X	X
Botswana	0	32,560	32,560	X	X	20	0.1	20	0.1	X	X	0
Burkina Faso	271	4,464	4,735	3	1.1	77	1.7	80	1.7	3	X	X
Burundi	27	14	41	1	2.6	0	2.9	1	2.7	3	X	9
Cameroon	16,500 a	6,800 a	23,300 a	100 b	0.6	90 b	1.3	190 b	0.8	2	X	X
Cape Verde	X	X	X	X	X	X	X	X	X	1	X	X
Central African Rep	3,590	32,300	35,890	5	0.1	50	0.2	55	0.2	X	X	X
Chad	500	13,000	13,500	X	X	80	0.6	80	0.6	0	X	0
Comoros	16	X	16	1	3.1	X	X	1	3.1	0	X	X
Congo	21,340	X	21,340	22	0.1	X	X	22	0.1	0	X	130
Cote d'Ivoire	4,458	5,376	9,834	290	6.5	220	4.1	510	5.2	8	1	648
Djibouti	2	68	70	X	X	X	X	X	X	X	X	X
Egypt	X	X	X	X	X	X	X	X	X	2	X	X
Equatorial Guinea	1,295	X	1,295	3	0.2	X	X	3	0.2	X	X	X
Ethiopia	4,350	22,800	27,150	8	0.2	80	0.4	88	0.3	13	X	X
Gabon	20,500	75	20,575	15	0.1	X	X	15	0.1	1	X	X
Gambia, The	65	150	215	2	3.4	3	2.0	5	2.4	0	X	X
Ghana	1,718	6,975	8,693	22	1.3	50	0.7	72	0.8	2	1,167	397
Guinea	2,050	8,600	10,650	36	1.8	50	0.6	86	0.8	0	X	0
Guinea-Bissau	660	1,445	2,105	17	2.6	40	2.8	57	2.7	0	X	X
Kenya	1,105	1,255	2,360	19	1.7	20	1.6	39	1.7	13	70	405
Lesotho	X	X	X	X	X	X	X	X	X	1	X	X
Liberia	2,000	40	2,040	46	2.3	X	X	46	2.3	3	X	X
Libya	134	56	190	X	X	X	X	X	X	39	X	X
Madagascar	10,300	2,900	13,200	150	1.5	6	0.2	156	1.2	15	X	930
Malawi	186	4,085	4,271	X	X	150	3.7	150	3.5	1	X	146
Mali	500	6,750	7,250	X	X	36	0.5	36	0.5	1	X	X
Mauritania	29	525	554	1	2.4	13	2.4	13	2.4	0	X	X
Mauritius	3	X	3	0	3.3	X	X	0	3.3	0	X	X
Morocco	1,533	1,703	3,236	X	X	X	X	13	0.4	16	421	7
Mozambique, People's Rep	935	14,500	15,435	10	1.1	110	0.8	120	0.8	5	X	25
Niger	100	2,450	2,550	3	2.5	65	2.6	67	2.6	3	X	X
Nigeria	5,950	8,800	14,750	300	5.0	100	1.1	400	2.7	32	0	X
Rwanda	120	110	230	3	2.6	2	1.8	5	2.3	4	X	11
Senegal	220	10,825	11,045	X	X	50	0.5	50	0.5	4	0	63
Sierra Leone	740	1,315	2,055	6	0.8	X	X	6	0.3	0	X	X
Somalia	1,540	7,510	9,050	4	0.2	10	0.1	14	0.1	2	X	X
South Africa	300	X	300	X	X	X	X	X	X	63	10	290
Sudan	650	47,000	47,650	4	0.6	500	1.1	504	1.1	17	50	X
Swaziland	4	70	74	X	X	X	X	0	X	7	X	X
Tanzania	1,440	40,600	42,040	10	0.7	120	0.3	130	0.3	11	0	410
Togo	304	1,380	1,684	2	0.7	10	0.7	12	0.7	1	X	X
Tunisia	186	111	297	X	X	X	X	5	1.7	4	163	X
Uganda	765	5,250	6,015	10	1.3	40	0.8	50	0.8	2	440	45
Zaire	105,750	71,840	177,590	182	0.2	188	0.3	370	0.2	1	X	5,690
Zambia	3,010	26,500	29,510	40	1.3	30	0.1	70	0.2	3	5	220
Zimbabwe	200	19,620	19,820	0	X	80	0.4	80	0.4	6	X	X
NORTH & CENTRAL AMERICA	541,009	261,276	802,285	1,072	0.2	20	0.0	1,251	0.1	2,552	102,884	36,812
Barbados	X	X	X	X	X	X	X	X	X	X	X	X
Canada	264,100	172,300	436,400	X	X	X	X	X	X	720	X	4,870
Costa Rica	1,638	160	1,798	124 c	7.6	X	X	124	6.9	1	X	320
Cuba	1,455	X	1,455	2	0.2	X	X	2	0.1	14	200	X
Dominican Rep	629	X	629	4	0.6	X	X	4	0.6	1	X	X
El Salvador	141	X	141	5	3.2	X	X	5	3.2	0	X	X
Guatemala	4,442	100	4,542	90	2.0	X	X	90	2.0	10	X	62
Haiti	48	X	48	2	3.8	X	X	2	3.8	0	X	X
Honduras	3,797	200	3,997	90	2.3	X	X	90	2.3	X	58	X
Jamaica	67	X	67	2	3.0	X	X	2	3.0	1	X	2
Mexico	46,250	2,100	48,350	595	1.3	20	1.0	615	1.3	28	X	360
Nicaragua	4,496	X	4,496	121	2.7	X	X	121	2.7	1	250	X
Panama	4,165	X	4,165	36	0.9	X	X	36	0.9	1	X	X
Trinidad and Tobago	208	X	208	1	0.4	X	X	1	0.4	1	14	X
United States	209,573 d	86,416 d	295,989 d	X	X	X	X	159 e	0.1	1,775	102,362	31,198
SOUTH AMERICA	653,605	204,520	858,125	9,837	1.5	1,293	0.6	11,180	1.3	760	0	16,761
Argentina	44,500	X	44,500	X	X	X	X	X	X	50	X	2,594
Bolivia	44,010	22,750	66,760	87	0.2	30	0.1	117	0.2	2	X	X
Brazil	357,480	157,000	514,480	8,000 f	2.2	1,050	0.7	9,050	1.8	561	0	4,660
Chile	7,550	X	7,550	X	X	X	X	50	0.7	93	X	845
Colombia	46,400	5,300	51,700	820	1.8	70	1.3	890	1.7	11	X	2,280
Ecuador	14,250	480	14,730	340	2.4	0	X	340	2.3	6	X	350
Guyana	18,475	220	18,695	2	0.0	1	0.2	3	0.0	0	X	12
Paraguay	4,070	15,640	19,710	190	4.7	22	0.1	212	1.1	1	X	90
Peru	69,680	960	70,640	270	0.4	0	X	270	0.4	8	X	850
Suriname	14,830	170	15,000	3	0.0	X	X	3	0.0	0	X	580
Uruguay	490	X	490	X	X	X	X	X	X	6	X	X
Venezuela	31,870	2,000	33,870	125	0.4	120	6.0	245	0.7	24	X	4,500

Table 19.1

| | Extent of Forest and Woodland, 1980s (000 hectares) | | | Average Annual Deforestation, 1980s | | | | | | Average Annual Reforestation 1980s (000 ha) | Managed Closed Forest 1980s (000 ha) | Protected Closed Forest 1980s (000 ha) |
| | | | | Closed Forest | | Open Forest | | Total Forest | | | | |
	Closed	Open	Total	Extent (000 ha)	Percent	Extent (000 ha)	Percent	Extent (000 ha)	Percent			
ASIA	**409,418**	**82,147**	**491,565**	**3,931**	**1.0**	**57**	**0.1**	**4,405**	**0.9**	**5,708**	**48,705**	**19,417**
Afghanistan	810	400	1,210	X	X	X	X	0	X	X	100	X
Bahrain	X	X	X	X	X	X	X	X	X	X	X	X
Bangladesh	927	X	927	8	0.9	X	X	8	0.9	21	795	52
Bhutan	2,100	40	2,140	1	0.1	X	X	1	0.1	1	0	X
China	97,847	17,200	115,047	X	X	X	X	0	X	4,552	X	1,635
Cyprus	153	24	177	X	X	X	X	X	X	X	153	25
India	36,540 g	27,660 g	64,200 g	1,500 h	4.1	X	X	1,500	2.3	173	31,917	6,743
Indonesia	113,895	3,000	116,895	900 i	0.8	20	0.7	920	0.8	164	40	5,430
Iran, Islamic Rep	2,750	1,000	3,750	X	X	X	X	20	0.5	X	400	120
Iraq	70	1,160	1,230	X	X	X	X	X	X	X	X	X
Israel	80	20	100	X	X	X	X	X	X	2	56	7
Japan	23,890	1,390	25,280	X	X	X	X	X	X	240	X	X
Jordan	X	50	50	X	X	X	X	0	X	3	X	X
Kampuchea, Dem	7,548	5,100	12,648	25	0.3	5	0.1	30	0.2	0	X	X
Korea, Dem People's Rep	4,800	X	4,800	X	X	X	X	X	X	200	X	X
Korea, Rep	4,087	X	4,887	X	X	X	X	X	X	84	X	X
Kuwait	X	X	X	X	X	X	X	X	X	X	X	X
Lao People's Dem Rep	8,410	5,215	13,625	100	1.2	30	0.6	130	1.0	2	X	X
Lebanon	X	20	20	X	X	X	X	0	X	X	X	X
Malaysia	20,996	X	20,996	255	1.2	X	X	255	1.2	25	2,499	959
Mongolia	9,528	X	9,528	X	X	X	X	X	X	X	X	X
Myanmar	31,941	X	31,941	677 j	2.1	X	X	677	2.1	0	3,419	299
Nepal	1,941	180	2,121	84	4.3	X	X	84	4.0	5	X	330
Oman	X	X	X	X	X	X	X	X	X	X	X	X
Pakistan	2,185	295	2,480	7	0.3	2	0.7	9	0.4	9	410	15
Philippines	9,510	X	9,510	143 k	1.5	X	X	143	1.5	63	X	690
Qatar	X	X	X	X	X	X	X	X	X	X	X	X
Saudi Arabia	30	170	200	X	X	X	X	0	X	X	X	X
Singapore	X	X	X	X	X	X	X	X	X	X	X	X
Sri Lanka	1,659	X	1,659	58	3.5	X	X	58	3.5	16	X	193
Syrian Arab Rep	60	90	150	X	X	X	X	0	X	X	60	X
Thailand	9,235	6,440	15,675	X	X	X	X	397 l	2.5	31	X	2,220
Turkey	8,856	11,343	20,199	X	X	X	X	X	X	82	8,856	139
United Arab Emirates	X	X	X	X	X	X	X	X	X	X	X	X
Viet Nam	8,770	1,340	10,110	173 m	2.0	X	X	173	1.7	36	X	560
Yemen Arab Rep	X	10	10	X	X	X	X	0	X	X	X	X
Yemen, People's Dem Rep	X	X	X	X	X	X	X	X	X	X	X	X
EUROPE	**137,005**	**21,887**	**158,892**	**X**	**X**	**X**	**X**	**X**	**X**	**1,031**	**74,628**	**1,732**
Albania	1,280	0	1,280	X	X	X	X	X	X	X	X	X
Austria	3,754	0	3,754	X	X	X	X	X	X	21	1,489	0
Belgium	682	80	762	X	X	X	X	X	X	19	272	0
Bulgaria	3,328	400	3,728	X	X	X	X	X	X	50	3,600	100
Czechoslovakia	4,435	145	4,580	X	X	X	X	X	X	37	4,435	X
Denmark	466	18	484	X	X	X	X	X	X	X	330	56
Finland	19,885	3,340	23,225	X	X	X	X	X	X	158	10,578	294
France	13,875	1,200	15,075	X	X	X	X	X	X	51	2,957	92
German Dem Rep	2,700	285	2,985	X	X	X	X	X	X	X	2,697	85
Germany, Fed Rep	6,989	218	7,207	X	X	X	X	X	X	62	3,886	X
Greece	2,512	3,242	5,754	X	X	X	X	X	X	X	1,603	55
Hungary	1,612	25	1,637	X	X	X	X	X	X	19	1,612	41
Iceland	X	100	100	X	X	X	X	X	X	X	X	X
Ireland	347	33	380	X	X	X	X	X	X	9	298	0
Italy	6,363	1,700	8,063	X	X	X	X	X	X	15	699	162
Luxembourg	X	X	X	X	X	X	X	X	X	X	38	0
Malta	X	X	X	X	X	X	X	X	X	X	X	X
Netherlands	294	61	355	X	X	X	X	X	X	2	225	0
Norway	7,635	1,066	8,701	X	X	X	X	X	X	79	1,130	60
Poland	8,588	138	8,726	X	X	X	X	X	X	106	8,099	103
Portugal	2,627	349	2,976	X	X	X	X	X	X	4	X	7
Romania	6,265	410	6,675	X	X	X	X	X	X	X	5,940	X
Spain	6,906	3,905	10,811	X	X	X	X	X	X	92	8,007	40
Sweden	24,400	3,442	27,842	X	X	X	X	X	X	207	14,301	230
Switzerland	935	189	1,124	X	X	X	X	X	X	7	627	7
United Kingdom	2,027	151	2,178	X	X	X	X	X	X	40	1,505	0
Yugoslavia	9,100	1,390	10,490	X	X	X	X	X	X	53	6,300	400
U.S.S.R.	**791,600**	**137,000**	**928,600**	**X**	**X**	**X**	**X**	**X**	**X**	**4,540**	**791,600**	**20,000**
OCEANIA	**86,322**	**71,347**	**157,669**	**25**	**0.0**	**1**	**0.0**	**26**	**0.0**	**117**	**0**	**55**
Australia	41,658	65,085	106,743	X	X	X	X	X	X	62	X	X
Fiji	811	0	811	2	0.2	X	X	2	0.2	9	X	X
New Zealand	7,200	2,300	9,500	X	X	X	X	X	X	43	X	X
Papua New Guinea	34,230	3,945	38,175	22	0.1	1	0.0	23	0.1	2	0	55
Solomon Islands	2,423	17	2,440	1	0.0	X	X	1	0.0	1	X	X

Sources: Food and Agriculture Organization of the United Nations, United Nations Economic Commission for Europe, and country data sources.

Notes: a. Stocks for 1986. b. Annual deforestation for 1976-86. c. Annual deforestation for 1977-83. d. Stocks for 1987. e. Annual deforestation for 1977-87. f. Annual deforestation for 1987; please refer to technical notes for other estimates on deforestation. g. Stocks for 1982. h. Annual deforestation for 1975-82. i. Annual deforestation for 1979-84. j. Annual deforestation for 1975-81. k. Annual deforestation for 1981-88. l. Annual deforestation for 1978-85. m. Annual deforestation for 1976-81. Deforestation of total forest area for some countries may not match, because of missing information in some classes.

Regional and world totals include listed countries with data only.

0 = zero or less than half the unit of measure; X = not available.

For additional information, see Sources and Technical Notes.

Table 19.2 Wood Production and Trade, 1975-87

	Roundwood Production (000 m3)						Processed Wood Production (000 m3)				Paper Production (000 metric tons)		Average Annual Net Trade Roundwood (a) (000 m3)	
	Total		Fuel and Charcoal		Industrial Roundwood		Sawnwood		Panels					
		Change Since (%)		Change Since (%)		Change Since (%)		Change Since (%)		Change Since (%)		Change Since (%)		
	1985-87	1975-77	1985-87	1975-77	1985-87	1975-77	1985-87	1975-77	1985-87	1975-77	1985-87	1975-77	1975-77	1985-87
WORLD	3,255,039	22	1,680,540	28	1,574,499	17	482,975	12	117,289	25	202,438	41		
AFRICA	449,503	33	396,196	35	53,307	21	8,200	39	1,878	63	2,387	89	(5,678)	(3,261)
Algeria	1,945	35	1,709	36	237	32	13	0	50	0	117	289	60	249
Angola	5,020	24	4,006	33	1,013	(3)	5	(95)	7	(70)	15	15	0	0
Benin	4,543	33	4,310	34	233	29	10	11	0	0	0	0	0	0
Botswana	1,225	47	1,149	46	76	48	0	0	0	0	0	0	X	X
Burkina Faso	6,931	25	6,618	25	314	25	1	(25)	0	0	0	0	0	0
Burundi	3,742	28	3,697	27	45	39	3	246	0	0	0	0	X	X
Cameroon	12,165	36	9,391	31	2,774	62	650	107	73	0	5	0	(489)	(533)
Cape Verde	X	X	X	X	X	X	X	X	X	X	X	X	X	X
Central African Rep	3,426	24	2,990	32	436	(13)	54	(32)	5	72	0	0	(112)	(56)
Chad	3,655	25	3,139	25	517	25	1	(17)	0	0	0	0	X	X
Comoros	X	X	X	X	X	X	X	X	X	X	X	X	X	X
Congo	2,525	39	1,634	30	891	60	62	13	58	2	0	0	(142)	(296)
Cote d'Ivoire	11,870	9	8,255	44	3,615	(31)	764	29	236	151	0	0	(2,974)	(1,010)
Djibouti	0	0	0	0	0	0	0	0	0	0	0	0	X	X
Egypt	2,058	29	1,962	29	95	29	0	0	44	8	150	32	101	235
Equatorial Guinea	607	44	447	14	160	445	47	231	10	900	0	0	X	X
Ethiopia	38,927	28	37,114	28	1,813	35	45	(48)	15	7	10	33	0	0
Gabon	3,873	9	2,573	16	1,300	(2)	126	47	228	142	0	0	(1,309)	(1,100)
Gambia, The	856	8	835	6	21	106	1	0	0	0	0	0	X	X
Ghana	9,590	20	8,496	39	1,094	(42)	365	(20)	65	(3)	0	0	(413)	(208)
Guinea	4,351	24	3,737	26	614	17	90	0	2	0	0	0	0	(8)
Guinea-Bissau	561	9	422	5	139	24	16	0	0	0	0	0	0	0
Kenya	33,784	50	32,195	51	1,589	46	181	15	69	459	105	153	(50)	0
Lesotho	539	28	539	28	0	0	0	0	0	0	0	0	X	X
Liberia	5,131	39	4,405	47	726	2	257	56	5	150	0	0	(361)	(293)
Libya	635	15	536	6	99	103	31	151	0	0	6	25	X	X
Madagascar	7,068	34	6,261	32	807	50	234	37	1	0	10	75	(1)	(1)
Malawi	6,725	34	6,414	35	311	11	22	(42)	4	(39)	0	0	0	0
Mali	5,052	30	4,733	30	319	27	6	4	0	0	0	0	X	X
Mauritania	12	33	7	40	5	25	0	0	0	0	0	0	X	X
Mauritius	25	(42)	17	(25)	8	(60)	2	(63)	0	X	0	0	X	X
Morocco	2,063	49	1,292	45	772	56	149	96	105	84	108	87	217	214
Mozambique, People's Rep	15,255	41	14,270	44	985	8	32	(82)	57	898	2	33	(8)	(1)
Niger	4,041	32	3,791	32	249	32	0	0	0	0	0	0	X	X
Nigeria	98,603	44	90,735	41	7,868	89	2,712	186	223	147	64	394	(37)	(59)
Rwanda	5,842	13	5,602	10	240	300	13	465	2	0	0	0	X	X
Senegal	4,099	30	3,539	28	560	37	11	106	0	0	0	0	X	X
Sierra Leone	7,922	18	7,781	19	141	1	13	(44)	0	0	0	0	0	0
Somalia	4,531	40	4,463	40	68	21	14	0	2	0	0	0	2	0
South Africa	18,761	12	7,078	1	11,683	19	1,662	(1)	398	9	1,647	82	(14)	(73)
Sudan	20,099	34	18,206	34	1,893	32	13	(8)	2	(62)	9	60	X	X
Swaziland	2,223	(13)	560	16	1,663	(19)	136	37	8	129	0	0	(217)	(198)
Tanzania	23,892	42	22,398	42	1,495	52	106	90	6	(36)	0	0	(6)	0
Togo	789	31	621	32	168	28	5	0	0	0	0	0	1	0
Tunisia	2,849	26	2,729	26	120	41	3	0	97	260	54	208	49	40
Uganda	12,935	38	11,247	39	1,688	34	23	(32)	6	362	2	62	0	0
Zaire	31,381	34	28,843	34	2,539	34	121	28	53	117	2	33	(46)	(155)
Zambia	9,946	25	9,418	25	528	24	50	16	13	274	2	0	1	9
Zimbabwe	7,391	39	6,003	38	1,388	45	149	19	34	140	77	103	1	(3)
NORTH & CENTRAL AMERICA	723,881	41	159,739	154	564,141	25	158,046	27	40,949	22	82,283	31	(17,626)	(18,918)
Barbados	X	X	X	X	X	X	X	X	X	X	X	X	X	X
Canada	179,536	34	6,623	75	172,913	33	57,201	63	6,253	43	15,255	35	145	(424)
Costa Rica	3,124	(8)	2,617	32	508	(64)	398	(34)	48	(16)	13	76	0	1
Cuba	3,294	41	2,707	44	587	32	109	4	132	X	143	36	X	X
Dominican Rep	982	113	976	116	6	(33)	0	0	0	0	10	7	0	37
El Salvador	4,902	34	4,820	34	82	6	44	27	0	0	16	220	0	0
Guatemala	7,012	24	6,871	32	141	(70)	99	(67)	8	(41)	16	(18)	(2)	(2)
Haiti	6,055	27	5,816	28	239	0	14	0	0	0	0	0	5	0
Honduras	5,505	26	4,683	41	822	(21)	415	(30)	8	(32)	0	0	(22)	(2)
Jamaica	127	114	13	86	114	118	29	(6)	4	(51)	21	X	11	2
Mexico	21,497	22	14,182	31	7,315	7	2,253	6	774	162	2,473	86	(3)	13
Nicaragua	3,675	26	2,795	37	880	0	222	(45)	14	40	0	0	2	0
Panama	2,047	26	1,708	14	339	163	45	(18)	12	11	25	55	2	7
Trinidad and Tobago	62	(37)	22	38	40	(52)	20	(39)	0	0	0	0	2	2
United States	485,760	46	105,756	346	380,005	23	97,178	14	33,698	17	64,312	28	(17,697)	(18,573)
SOUTH AMERICA	313,783	36	221,104	25	92,679	75	25,297	52	3,757	40	7,076	89	(59)	(1,768)
Argentina	11,177	23	5,755	13	5,422	37	1,067	62	398	25	947	65	8	1
Bolivia	1,348	13	1,199	31	149	(46)	94	(28)	5	(44)	1	100	(1)	0
Brazil	237,779	39	171,670	25	66,109	95	17,969	58	2,538	31	4,395	121	23	44
Chile	16,185	48	6,169	17	10,015	77	2,341	84	235	294	400	49	(57)	(1,794)
Colombia	17,526	17	14,853	24	2,673	(11)	721	(23)	113	12	460	68	(7)	0
Ecuador	8,670	48	6,140	48	2,530	46	1,243	59	171	290	41	21	0	0
Guyana	218	18	18	80	200	14	61	(18)	0	0	0	0	(28)	(18)
Paraguay	8,100	62	4,987	27	3,113	189	795	139	92	468	8	743	(2)	0
Peru	7,733	6	6,523	18	1,210	(32)	535	(10)	39	(44)	155	5	4	0
Suriname	205	(31)	14	(49)	191	(29)	67	0	13	(53)	0	0	(11)	(9)
Uruguay	3,283	22	3,026	28	257	(19)	57	(48)	13	(11)	56	70	13	0
Venezuela	1,307	15	684	36	623	(2)	327	(6)	140	76	612	43	20	12

Table 19.2

	Roundwood Production (000 m3)						Processed Wood Production (000 m3)				Paper Production (000 metric tons)		Average Annual Net Trade Roundwood {a} (000 m3)	
	Total		Fuel and Charcoal		Industrial Roundwood		Sawnwood		Panels					
		Change Since (%)		Change Since (%)		Change Since (%)		Change Since (%)		Change Since (%)		Change Since (%)		
	1985-87	1975-77	1985-87	1975-77	1985-87	1975-77	1985-87	1975-77	1985-87	1975-77	1985-87	1975-77	1975-77	1985-87
ASIA	**1,001,413**	**18**	**750,331**	**18**	**251,081**	**19**	**100,830**	**21**	**23,318**	**53**	**40,434**	**81**	**28,630**	**43,144**
Afghanistan	6,734	10	5,163	12	1,572	6	400	6	1	0	0	0	X	X
Bahrain	X	X	X	X	X	X	X	X	X	X	X	X	X	X
Bangladesh	27,849	30	26,999	32	849	(15)	86	(53)	13	(71)	111	113	0	0
Bhutan	3,224	12	2,946	12	278	17	5	0	0	0	0	0	0	(7)
China	269,062	28	174,088	22	94,974	40	26,287	55	1,877	364	9,563	178	898	7,018
Cyprus	80	(37)	23	(11)	57	(44)	60	(1)	24	0	0	0	4	6
India	250,279	23	226,321	22	23,958	41	17,460	133	442	135	1,759	86	(55)	378
Indonesia	158,075	22	129,600	22	28,475	22	7,875	164	5,722	X	640	954	(18,561)	(825)
Iran, Islamic Rep	6,757	2	2,381	4	4,376	0	163	3	54	(57)	78	8	168	117
Iraq	141	21	91	36	50	0	8	0	2	0	28	12	X	X
Israel	118	3	11	(3)	107	4	0	0	146	9	147	38	125	189
Japan	32,650	(7)	573	(30)	32,077	(7)	29,245	(23)	9,084	5	21,356	43	50,846	43,425
Jordan	9	4	5	33	4	(20)	0	0	0	0	13	117	2	28
Kampuchea, Dem	5,423	7	4,856	7	567	8	43	0	2	0	0	0	(6)	0
Korea, Dem People's Rep	4,596	16	3,996	18	600	0	280	0	0	0	80	0	X	X
Korea, Rep	6,713	(29)	4,395	(37)	2,318	(10)	3,575	43	1,247	(34)	2,749	232	6,437	6,582
Kuwait	X	X	X	X	X	X	X	X	X	X	X	X	X	X
Lao People's Dem Rep	4,228	24	3,898	21	330	71	16	(69)	5	154	0	0	(1)	(25)
Lebanon	484	0	459	(2)	25	25	33	(1)	46	(4)	43	(4)	34	16
Malaysia	40,185	27	7,745	26	32,441	27	5,768	15	1,337	58	65	244	(14,684)	(20,480)
Mongolia	2,390	0	1,350	0	1,040	0	470	0	4	0	0	0	(57)	0
Myanmar	19,069	24	16,158	22	2,911	40	496	29	15	29	20	97	(62)	(149)
Nepal	16,127	25	15,567	26	560	0	220	0	0	0	1	(14)	(119)	(126)
Oman	X	X	X	X	X	X	X	X	X	X	X	X	X	X
Pakistan	21,399	38	20,059	33	1,340	187	55	(14)	64	146	84	104	49	27
Philippines	35,822	5	30,027	20	5,794	(46)	1,091	(30)	562	(6)	191	(29)	(3,105)	(582)
Qatar	X	X	X	X	X	X	X	X	X	X	X	X	X	X
Saudi Arabia	X	X	X	X	X	X	X	X	X	X	X	X	X	X
Singapore	X	X	X	X	X	X	X	X	X	X	X	X	X	X
Sri Lanka	8,697	21	8,020	20	678	36	21	(37)	12	(49)	24	26	(1)	(37)
Syrian Arab Rep	48	7	15	36	34	(2)	9	575	27	118	5	81	31	18
Thailand	36,900	17	32,442	23	4,458	(14)	1,027	(39)	311	161	485	124	(211)	24
Turkey	16,235	(54)	10,448	(63)	5,786	(15)	4,923	70	781	97	436	23	36	62
United Arab Emirates	X	X	X	X	X	X	X	X	X	X	X	X	X	X
Viet Nam	25,356	24	22,072	24	3,284	24	354	(36)	40	135	68	34	X	X
Yemen Arab Rep	X	X	X	X	X	X	X	X	X	X	X	X	X	X
Yemen, People's Dem Rep	296	30	296	30	0	0	0	0	0	0	0	0	0	4
EUROPE	**353,415**	**13**	**57,568**	**13**	**295,847**	**13**	**84,473**	**4**	**32,665**	**8**	**57,875**	**37**	**16,933**	**16,310**
Albania	2,330	0	1,608	0	722	0	200	0	12	0	8	(1)	0	X
Austria	13,981	11	1,413	45	12,568	9	5,921	3	1,365	23	2,235	65	1,839	4,205
Belgium	3,376	26	538	149	2,838	15	832	25	2,060	16	841	15	2,274	2,509
Bulgaria	4,581	2	1,769	72	2,812	(19)	1,464	(12)	573	17	451	0	354	309
Czechoslovakia	18,907	10	1,449	(20)	17,458	14	5,226	18	1,008	52	1,262	15	(2,411)	(1,521)
Denmark	2,236	32	402	512	1,834	13	879	14	342	(8)	296	37	(427)	(670)
Finland	41,393	27	3,089	(46)	38,304	42	7,346	24	1,356	1	7,669	75	4,271	4,304
France	39,890	7	10,428	(2)	29,462	10	8,814	(3)	2,542	(12)	5,522	25	3	(3,593)
German Dem Rep	10,841	28	661	45	10,181	27	2,462	9	1,223	36	1,319	11	580	(48)
Germany, Fed Rep	31,583	4	3,702	22	27,880	2	9,700	(4)	6,879	(1)	9,508	56	(537)	(960)
Greece	2,897	4	1,948	(2)	949	19	396	6	393	44	282	32	295	236
Hungary	6,836	22	3,035	24	3,801	21	1,262	14	387	24	511	35	917	(236)
Iceland	X	X	X	X	X	X	X	X	X	X	X	X	X	X
Ireland	1,256	219	46	590	1,210	212	300	317	64	(48)	34	(71)	(29)	(314)
Italy	9,398	41	4,792	52	4,606	31	2,141	8	2,498	(6)	4,700	15	5,775	5,273
Luxembourg	X	X	X	X	X	X	X	X	X	X	X	X	X	X
Malta	X	X	X	X	X	X	X	X	X	X	X	X	X	X
Netherlands	1,118	19	109	190	1,008	11	405	50	89	(50)	2,038	34	387	188
Norway	10,397	19	875	74	9,521	15	2,283	12	683	18	1,589	33	1,390	1,462
Poland	23,663	0	3,905	118	19,758	(2)	6,575	(22)	2,113	16	1,333	0	(934)	(1,500)
Portugal	9,320	19	598	1	8,722	20	1,640	(8)	759	144	639	68	211	(247)
Romania	24,459	16	4,571	(13)	19,888	25	4,425	(2)	1,032	12	801	17	(185)	(47)
Spain	16,596	39	2,844	37	13,752	39	2,483	4	1,909	27	3,105	53	887	665
Sweden	52,507	(1)	4,424	69	48,083	(5)	11,565	4	1,341	(31)	7,390	52	1,852	5,643
Switzerland	4,767	23	883	24	3,884	23	1,676	16	657	11	1,083	56	(142)	266
United Kingdom	5,082	39	151	1	4,931	40	1,782	24	1,115	58	3,935	(1)	408	(105)
Yugoslavia	16,002	11	4,328	7	11,673	13	4,696	26	1,275	27	1,324	67	142	488
U.S.S.R.	**374,857**	**(3)**	**86,800**	**6**	**288,057**	**(5)**	**100,400**	**(11)**	**13,340**	**37**	**10,095**	**14**	**(18,415)**	**(17,677)**
OCEANIA	**38,189**	**23**	**8,802**	**30**	**29,387**	**22**	**5,730**	**(2)**	**1,383**	**34**	**2,288**	**28**	**(5,659)**	**(9,670)**
Australia	19,907	36	2,878	111	17,029	28	3,200	(8)	925	30	1,593	36	(3,620)	(6,912)
Fiji	249	49	37	200	212	37	91	7	9	26	0	0	0	(4)
New Zealand	9,341	(1)	50	(80)	9,291	1	2,275	9	429	47	695	14	(1,263)	(1,033)
Papua New Guinea	7,917	31	5,533	14	2,384	101	117	(18)	19	(13)	0	0	(537)	(1,414)
Solomon Islands	589	30	210	9	379	46	17	103	1	313	0	0	(241)	(297)

Source: Food and Agriculture Organization of the United Nations.
Notes: a. Imports of roundwood are shown as positive numbers; exports are represented by negative numbers.
0 = zero or less than half of the unit of measure; X = not available; negative numbers are shown in parentheses; m3 = cubic meters.
For additional information, see Sources and Technical Notes.

Sources and Technical Notes

Table 19.1 Forest Resources, 1980s

Sources: Developing countries: Food and Agriculture Organization of the United Nations (FAO) Forest Resources Division, *An Interim Report on the State of the Forest Resources in the Developing Countries* (FAO, Rome, 1988). Developed countries, Cyprus, Israel, and Turkey: FAO and United Nations Economic Commission for Europe (ECE), *The Forest Resources of the ECE Region* (ECE, Geneva, 1985).

Deforestation data (1987) for Brazil: Alberto Waingort Setzer, Marcos da Costa Pereira, Alfredo da Costa Pereira, Jr., and Sérgio Alberto de Oliveira Almeida, "Relatório de Atividades do Projeto IBDF-INPE "SEQE"—Ano 1987," unpublished paper (Instituto Nacional de Pesquisas Espaciais (INPE), São José dos Campos, São Paulo, Brazil, May 1988). Deforestation data (1977–83) for Costa Rica: Steven A. Sader and Armond T. Joyce, "Deforestation Rates and Trends in Costa Rica, 1940 to 1983," *Biotropica*, Vol. 20, No. 1, 1988. Stocks (1986) and deforestation data (1976–86) for Cameroon: Joint Interagency Planning and Review Mission for the Forestry Sector, *Cameroon Tropical Forestry Action Plan* (Joint Interagency, Rome, 1988). Stocks (1982) and deforestation data (1975–82) for India: B.B. Vohra, "Confusion on the Forestry Front," unpublished paper (December 1987), based on reconciliation of figures in National Remote Sensing Agency, *Mapping of Forest Cover in India from Satellite Imagery 1972–75 & 1980–82* (Government of India, Hyderabad, 1983). Deforestation data (1979–84) for Indonesia: World Bank Asia Regional Office, *Indonesia: Forests, Land and Water: Issues in Sustainable Development* (August 10, 1988) pp. 2–4. Deforestation data (1975–81) for Myanmar: Kyaw, U.S., "National Report: Burma," in *Proceedings of Ad Hoc FAO/ECE/FINNIDA Meeting of Experts on Forest Resource Assessment, Kotka, Finland, 26–30 October 1987* (Finnish International Development Agency, Helsinki, 1987). Deforestation data (1981–88) for the Philippines: Philippines Forest Management Bureau, Department of Environment and Natural Resources, unpublished mimeographed tables (August 1988). Deforestation data (1978–85) for Thailand: Royal Forestry Department of Thailand, Planning Division, *Forestry Statistics of Thailand, 1986*, Center for Agricultural Statistics, Office of Agricultural Economics, Ministry of Agriculture and Cooperatives. Deforestation data (1976-81) for Viet Nam: Professor Vo Quy, University of Hanoi, "Vietnam's Ecological Situation Today," in *ESCAP Environment News*, Vol. 6, No. 4, October-December 1988, p. 5. Stocks (1987) United States: Waddell, K.L., Oswald, D.D., Powell, D.W., *Forest Statistics of the United States 1987* (Forest Service, U.S. Department of Agriculture, Pacific

Northwest Research Station, Portland, Oregon, United States, 1989). Deforestation data (1977–87) for the United States: personal communication (Forest Service, U.S. Department of Agriculture, December 1989).

Reforestation data for China: State Statistical Bureau, *China: A Statistics Survey in 1985* (New World Press, Beijing, 1985). Reforestation data for Jordan: Library of Congress, Science and Technology Division, *Draft Environmental Report on Jordan* (Library of Congress, Washington, D.C., August 1979). Reforestation data for Yugoslavia: Socijalisticka Federativna Republika Jugoslavija Savenzi Zavod Za Statistiku, *Statisticki Godisnjak Jugoslavija 1983, 1984, 1985* (Savenzi Zavod Za Statistiku, Belgrade, 1984, 1985, 1986).

Extent of forest and woodland refers to natural stands of woody vegetation in which trees predominate. FAO and ECE use slightly different definitions of open and closed forest. FAO defines *closed forest* as land where trees cover a high proportion of the ground and where grass does not form a continuous layer on the forest floor. ECE defines a forest as closed when tree crowns cover more than 20 percent of the area and when the area is used primarily for forestry. *Open forest*, as defined by FAO, consists of mixed forest/grasslands with at least 10 percent tree cover and a continuous grass layer. As defined by ECE, open forests are not used for agricultural purposes; have 5–20 percent of their area covered by tree crowns; have no more than half a hectare covered by groups of trees; or have shrubs or stunted trees covering more than 20 percent of their area.

In both FAO and ECE definitions, "natural" means all stands except plantations and includes stands that have been degraded to some degree by catastrophic fire, logging, agriculture, or acid precipitation. Trees are distinguished from shrubs on the basis of height: a mature tree has a single well-defined stem and is taller than 7 meters, and a mature shrub is usually less than 7 meters tall.

Average annual deforestation refers to the permanent clearing of forest lands for use in shifting cultivation, permanent agriculture, or settlements. As defined here, deforestation does not include other alterations such as selective logging, which can substantially affect forests, forest soil, wildlife and its habitat, and the global carbon cycle. For an analysis of tropical deforestation based on a broader definition of deforestation, see Norman Myers, *The Primary Source* (Norton, New York, 1984).

Average annual reforestation refers to the establishment of plantations for industrial and nonindustrial uses. Reforestation does not include regeneration of old tree crops (through either natural regeneration or forest management), although some countries may report regeneration as reforestation. Many trees are also

planted for nonindustrial uses, such as village wood lots. Reforestation data often exclude this component.

Data for developing countries are based on the 1980 Tropical Forest Resources Assessment, a joint project of FAO and the United Nations Environment Programme (UNEP). The survey assessed the tropical forests of 76 tropical developing countries, covering 97 percent of the total area of developing countries in the tropics. Data for the study were collected from research institutes; correspondence with national forestry services; visits to national forestry, land use, and survey institutions in some of the major forestry countries; visits to FAO regional offices; photographic surveys of all or part of five countries; satellite imagery of all or part of 19 countries; and side-looking airborne radar surveys of four additional countries. Three countries (Myanmar, India, and Peru) prepared their own national reports. In many cases, FAO adjusted data to fit common definitions and to correspond to the baseline year of 1980.

The FAO *1988 Interim Report* expanded the country coverage of the 1980 assessment to 129 developing countries. In that document, FAO evaluated the overall reliability of data on closed forest areas and deforestation rates for the original 76 developing countries. FAO classified their estimates on closed forest areas and deforestation as very good or good for 15 countries or parts of countries (containing 40 percent of the closed forest areas of the 76 tropical countries). These countries or parts of countries are Benin, Brazil (north), Cameroon (south), Colombia, Cote d'Ivoire, The Gambia, Haiti, Liberia, Malaysia (peninsular), Nepal, Paraguay (east), Sierra Leone, Togo, Trinidad and Tobago, and Venezuela. For 40 countries or parts of countries (covering an additional 40 percent of closed forest area), FAO assessed their data on forest cover as very good or good and their data on deforestation as satisfactory or poor. These countries or parts of countries are Angola, Bangladesh, Belize, Bhutan, Bolivia, Burkina Faso, Burundi, Congo (north), Congo (south), Costa Rica, Dominican Republic, El Salvador, French Guiana, Guatemala, Guinea, Guinea Bissau, Guyana, Honduras, India (15%), Jamaica, Kampuchea, Laos, Madagascar, Malaysia (Sabah), Malaysia (Sarawak), Mexico, Mozambique, Myanmar, Namibia (north west), Nigeria, Panama, Papua New Guinea, Peru, Philippines, Senegal, Sri Lanka, Sudan, Thailand, Viet Nam, and Zaire. Estimates on the forest cover and rate of deforestation in the remaining 21 countries or parts of countries (comprising about 20 percent of the total forest area and 29 percent of the total area of open forest) were judged as satisfactory or poor.

Recently published estimates were used in place of deforestation rates for closed forests from the *Interim Report* for Brazil, Costa Rica, Indonesia, Myanmar, the Philippines and Viet Nam.

The deforestation rate for Brazil is based on the analysis of satellite images, which assessed areas burned in the Legal Amazon during the dry season of 1987. The authors of that analysis estimated that at least 40 percent of the land being burned was recently cut forest, indicating a loss of 8 million hectares because of widespread deforestation in 1987. However, 1987 was considered a year of widespread deforestation in the Brazilian Amazon. It was the last year before tax laws that favored those who cleared Amazon land were suspended. A follow-up study by Alberto Setzer of INPE in 1988 showed a drop in the area being deforested in the Legal Amazon to 4.8 million hectares. Another study, conducted by INPE, used LANDSAT Thematic Mapper images and calculated a different deforestation rate for the Legal Amazon. The author concluded that 17.6 million hectares were deforested between 1978 and 1988, with an average annual rate of 1.7 million hectares; Roberto Pereira da Cunha, "Deforestation Estimates Through Remote Sensing: The State of the Art in the Legal Amazon" (INPE, Society of Latin American Remote Sensing Specialists (SELPER), São José dos Campos, São Paulo, Brazil, 1989).

Costa Rica's deforestation rate was obtained using digital analysis of maps based on satellite (LANDSAT) multispectral scanner images for 1977 and 1983.

The World Bank calculated deforestation in Indonesia by estimating the total area lost to shifting cultivation, government-sponsored tree crop projects, transmigration programs, logging, and forest fires.

The deforestation rate for Myanmar is based on aerial photographs, field studies, and satellite imagery.

Philippine deforestation is derived from a comparison of 1980 LANDSAT data with land cover statistics that had been prepared by the Swedish Space Corporation using a remote sensing technique.

The University of Hanoi data came from the Forest Inventory and Planning Institute (FIPI) of the Ministry of Forestry in Hanoi. It estimates that between 1976 and 1981 Viet Nam's area of rich broadleaf forest was reduced by 865,000 hectares. It suggests that the current forest loss is even higher, at about 200,000 hectares annually.

Recently published estimates on the loss of total forest area were used for Thai-

land and Cameroon. The numbers on the extent of forests in Cameroon and India, and on the loss of closed forests in India, were substituted as well. The estimate for the average annual deforestation of Thailand's total forest area was provided by the Royal Forestry Department and is based on LANDSAT imagery. The Cameroon study estimated the 1986 area of closed and open forests by extrapolation from 1975 satellite images. It found that between 1.8 and 2 million hectares had been deforested between 1976 and 1986. India's National Remote Sensing Agency calculated the new deforestation rate and the 1982 estimates for open and closed forests. The Agency used satellite imagery collected from 1980 to 1982 and found that 10.4 million hectares of closed forest were lost during the seven years from 1975 to 1982. Please refer to country sources for additional information.

The FAO/ECE survey covered all types of forests in the 32 member countries of the ECE. Data for this study were drawn from four types of sources: official data supplied in response to questionnaires; estimates by experts in some countries; recent ECE and FAO publications, country reports, and official articles; and estimates by the professional staff conducting the study. Most data refer to the period around 1980, but no attempt was made to adjust the data to a baseline year.

For an evaluation and detailed comparison of forest statistics, see Alan Grainger, "Quantifying Changes in Forest Cover in the Humid Tropics: Overcoming Current Limitations," *Journal of World Forest Resource Management* Vol. 1, pp. 3–23 (1984); and J.M. Melillo, C.A. Palm, R.A. Houghton, et al., "A Comparison of Two Recent Estimates of Disturbance in Tropical Forests," *Environmental Conservation*, Vol. 12, No. 1 (1985).

Table 19.2 Wood Production and Trade, 1975–87

Source: Food and Agriculture Organization of the United Nations (FAO), unpublished data (FAO, Rome, May 1989).

Total roundwood production refers to all wood in the rough, whether destined for industrial or fuelwood uses. All wood felled or otherwise harvested from forests and trees outside the forest with or without bark, round, split, roughly squared, or other forms such as roots and stumps is included.

Fuel and charcoal production covers all rough wood used for cooking, heating, and power production. Wood intended for charcoal production, pit kilns, and portable ovens is included.

Industrial roundwood production comprises all roundwood products other than fuelwood and charcoal: sawlogs, veneer logs, sleepers, pitprops, pulpwood, and other industrial products.

Processed wood production includes sawnwood and panels, as follows.

Sawnwood is wood that has been sawn, planed, or shaped into products such as planks, beams, boards, rafters, or railroad ties. Wood flooring is excluded. Sawnwood generally is thicker than 5 millimeters.

Panels include all wood-based panel commodities such as veneer sheets, plywood, particle board, and compressed or noncompressed fiberboard.

Paper production includes newsprint, printing and writing paper, and other paper and paperboard.

Average annual net trade of *roundwood* is the balance of imports minus exports. Trade in roundwood includes sawlogs and veneer logs, fuelwood, pulpwood, other industrial roundwood, and the roundwood equivalent of trade in charcoal, wood residues, and chips and particles. All trade data refer to both coniferous and nonconiferous wood. Imports are usually on a cost, insurance, freight basis. Exports are generally on a free-on-board basis.

FAO compiles forest products data from responses to annual questionnaires sent to national governments. Data from other sources, such as national statistical yearbooks, are also used. In some cases, FAO prepares its own estimates. FAO continually revises its data using new information; the latest figures are subject to revision.

Statistics on the production of fuelwood and charcoal are lacking for many countries. FAO uses population data and country-specific, per capita consumption figures to estimate fuelwood and charcoal production. Consumption of nonconiferous fuelwood ranges from a low of 0.0016 cubic meters per capita per year in Jordan to a high of 0.9783 cubic meters per capita per year in Benin. Consumption was also estimated for coniferous fuelwood. For both coniferous and nonconiferous fuelwood, the per capita consumption estimates were multiplied by the number of people in the country to determine national totals.

20. Wildlife and Habitat

Population growth, the expansion of agriculture and livestock raising, the building of cities and roads, and pollution are among the many pressures that diminish the quantity and quality of natural habitat remaining in the world. Along with illegal hunting, habitat reduction and its degradation threaten biodiversity and the economic, scientific, and cultural values it represents.

Most countries recognize the need to protect natural habitats, but few agree on the extent of that protection. Table 20.1 shows the area protected for each country. "National Protected Systems" total 3.7 percent of he Earth's land area, but the total area protected in each country varies from a high of 38 percent for Ecuador to less than 1 percent for 55 countries. The number and extent of areas under "International Protection Systems" have risen over the years. However, legal protection alone cannot ensure actual protection. Limited resources, lack of political will, and other factors often limit the amount of enforcement possible. Even when an area is protected, activities just outside its boundaries can have a severe impact on its fragile ecosystems.

The data in Table 20.2 are at least partial measures of pressure on habitat and a country's ability to declare and enforce habitat protection. In Algeria, for example, 15 of the 97 known mammalian species (15.5 percent) are in jeopardy. In contrast, only 32 of Mexico's 439 mammal species (7.3 percent) are similarly categorized. Data on indigenous animals are often hard to assemble, as is indicated by the lack of information on nonmammals in Africa and on Asian species in general. (In this table, swallowtail butterflies, widely distributed and highly visible, represent invertebrate species.) As new information becomes available, it is clear that, throughout the world, the numbers of species reported as known and threatened represent only a fraction of those that exist.

The trade in animals and animal products can have devastating effects on wildlife populations. Parties to the Convention on International Trade in Endangered Species of Wild Flora and Fauna (CITES) pledge to abide by certain regulations and to report trade in designated animals and animal products to the World Conservation Union (formally the International Union for Conservation of Nature and Natural Resources

[IUCN]). Table 20.3 details CITES membership and reporting history, as well as the trade in live primates, cat skins, raw ivory, live parrots, and reptile skins.

These numbers, representing elephants, cats, and large reptiles killed, and parrots and primates trapped (with associated mortality in capture and transport), show primarily the legal international trade, with large gaps in information worldwide. Illegal and domestic commerce would greatly increase these figures and would indicate an even greater reduction in wild populations.

The loss of habitat probably is the principal threat to the survival of endangered plants and animals. Table 20.4 attempts to show some of the changes in habitat that have been caused by human activity. Although generalized maps of habitat change are common and detailed studies of change in specific areas are available, country-level data are difficult to obtain. This table is a first attempt to detail habitat loss globally by country. Subsequent volumes should improve on this coverage and detail. It is undeniable that much critical habitat has already been lost (e.g., 70 percent of Greece's forests, virtually all of the tall grass prairie in the United States, 91 percent of Sudan's moist forests), and that which remains can be degraded or tamed.

Table 20.5 attempts to show the zoo populations of animals that are endangered, vulnerable, rare, or extinct in the wild. Estimating wild populations is difficult, and for many threatened species little is known of their potential for recovery. Fewer than half of the most threatened animals are held and propagated in zoos. The survival of a species represented by small populations in a limited number of zoos is at risk from chance events, disease, diminished genetic diversity, genetic drift, and even the chance variation from the optimal sex ratio.

Propagation in zoos often has the goal of the ultimate reintroduction of these populations into the wild. Returns to the wild are complex and expensive and have had mixed results. They are not even possible unless suitable habitat either survives or is rehabilitated. (See Chapter 8, "Wildlife and Habitat," sections on Biodiversity and Habitat Loss and A Last Resort: Zoos and Botanical Gardens.)

Table 20.1 National and International Protection of Natural

| | National Protection Systems | | | | | International Protection Systems | | | | |
| | All Protected Areas | | Percent of National Land Area Protected | Marine and Coastal Protected Areas Only | | Biosphere Reserves | | Natural World Heritage Sites | Wetlands of International Importance | |
	Number	Area (ha)	Protected	Number	Area (ha)	Number	Area (ha)	Number	Number	Area (ha)
WORLD	5,289	529,081,551	4.0	977	211,405,562	276	149,251,700	78	432	28,473,698
AFRICA	521	101,875,251	3.4	43	9,569,857	40	19,919,499	20	31	3,208,004
Algeria	17	496,687	0.2	1	2,392	1	7,200,000	1	2	8,400
Angola	3	889,700	0.7	2	61,700	0	0	--	--	--
Benin	2	843,500	7.6	0	0	1	880,000	0	--	--
Botswana	9	10,025,000	17.7	NA	NA	0	0	--	--	--
Burkina Faso	7	738,900	2.7	NA	NA	1	16,300	0	--	--
Burundi	0	0	0.0	NA	NA	0	0	0	--	--
Cameroon	12	1,702,200	3.6	1	160,000	3	850,000	1	--	--
Cape Verde	0	0	0.0	0	0	0	0	0	--	--
Central African Rep	7	3,904,000	6.3	NA	NA	2	1,640,200	1	--	--
Chad	1	114,000	0.1	NA	NA	0	0	--	--	--
Comoros	0	0	0.0	0	0	0	0	--	--	--
Congo	10	1,353,100	4.0	1	300,000	2	172,000	0	--	--
Cote d'Ivoire	10	1,958,000	6.2	1	30,000	2	1,480,000	3	--	--
Djibouti	1	10,000	0.5	0	0	0	0	--	--	--
Egypt	9	685,300	0.7	3	62,200	1	1,000	0	2	105,700
Equatorial Guinea	0	0	0.0	0	0	0	0	--	2	105,700
Ethiopia	25	6,872,600	6.2	1	200,000	0	0	1	--	--
Gabon	6	1,753,000	6.8	2	1,058,000	1	15,000	0	3	1,080,000
Gambia, The	0	0	0.0	0	0	0	0	0	--	--
Ghana	8	1,175,075	5.1	0	0	1	7,770	0	1	7,260
Guinea	1	13,000	0.1	0	0	2	133,300	1	--	--
Guinea-Bissau	0	0	0.0	0	0	0	0	--	--	--
Kenya	30	3,094,979	5.4	3	6,500	4	851,359	--	--	--
Lesotho	1	6,805	0.2	NA	NA	0	0	--	--	--
Liberia	1	130,700	1.4	0	0	0	0	--	--	--
Libya	3	155,000	0.1	0	0	0	0	0	--	--
Madagascar	31	1,031,312	1.8	1	1,750	0	0	0	--	--
Malawi	9	1,066,900	11.3	NA	NA	0	0	1	--	--
Mali	6	876,100	0.7	NA	NA	1	771,000	0	3	162,000
Mauritania	2	1,483,000	1.4	0	0	0	0	0	1	1,173,000
Mauritius	1	3,611	2.0	1	3,611	1	3,594	--	--	--
Morocco	10	298,359	0.7	2	13,025	0	0	0	4	10,580
Mozambique, People's Rep	0	0	0.0	0	0	0	0	0	--	--
Niger	4	1,654,240	1.3	NA	NA	0	0	0	1	220,000
Nigeria	4	960,082	1.1	0	0	1	460	0	--	--
Rwanda	2	262,000	10.5	NA	NA	1	15,065	--	--	--
Senegal	9	2,177,259	11.3	4	81,009	3	1,093,756	2	4	99,720
Sierra Leone	3	100,700	1.4	0	0	0	0	--	--	--
Somalia	0	0	0.0	0	0	0	0	--	--	--
South Africa	152	5,801,751	4.8	13	151,947	0	0	--	7	208,044
Sudan	13	8,115,500	3.4	0	0	2	1,900,970	0	--	--
Swaziland	3	39,545	2.3	NA	NA	0	0	--	--	--
Tanzania	20	11,913,075	13.4	0	0	2	2,337,600	4	--	--
Togo	6	463,000	8.5	0	0	0	0	--	--	--
Tunisia	6	44,780	0.3	1	4,030	4	32,425	1	1	12,600
Uganda	18	1,332,029	6.7	0	0	1	220,000	0	--	15,000
Zaire	9	8,827,000	3.9	0	0	3	297,700	4	--	--
Zambia	19	6,359,000	8.6	NA	NA	0	0	0	--	--
Zimbabwe	19	2,760,267	7.1	NA	NA	0	0	1	--	--
NORTH & CENTRAL AMERICA	890	193,908,695	9.1	214	135,780,803	64	93,327,040	21	55	15,001,629
Barbados	0	0	0.0	0	0	0	0	--	--	--
Canada	311	33,885,111	3.7	48	7,105,715	5	842,738	6	30	12,937,549
Costa Rica	25	609,770	12.0	7	194,490	2	728,955	1	--	--
Cuba	15	866,743	7.8	6	226,813	4	323,600	0	--	--
Dominican Rep	13	550,434	11.4	7	269,554	0	0	0	--	--
El Salvador	7	22,151	1.1	0	0	0	0	--	--	--
Guatemala	13	99,083	0.9	3	13,400	0	0	1	--	--
Haiti	2	7,700	0.3	0	0	0	0	0	--	--
Honduras	15	580,369	5.2	1	350,000	1	500,000	1	--	--
Jamaica	0	0	0.0	0	0	0	0	0	--	--
Mexico	47	5,582,625	2.9	11	1,119,301	6	1,288,454	1	1	47,480
Nicaragua	6	43,300	0.4	1	4,000	0	0	0	--	--
Panama	14	1,311,382	17.3	6	897,888	1	597,000	1	--	--
Trinidad and Tobago	6	16,088	3.1	2	3,388	0	0	--	--	--
United States	396	79,039,521	8.6	107	54,317,034	45	19,046,293	11	7	970,090
SOUTH AMERICA	453	80,123,051	4.6	94	24,716,645	23	10,834,241	8	4	216,877
Argentina	69	10,974,795	4.0	7	1,498,648	4	2,004,980	2	--	--
Bolivia	12	4,837,143	4.5	NA	NA	3	435,000	0	--	--
Brazil	160	20,096,133	2.4	20	2,031,530	0	0	1	--	--
Chile	69	11,983,013	16.0	32	10,049,838	7	2,406,633	0	1	4,877
Colombia	35	5,613,965	5.4	9	615,400	3	2,514,375	0	--	--
Ecuador	13	10,619,171	38.4	5	8,975,200	2	1,446,244	2	--	--
Guyana	1	11,655	0.1	0	0	0	0	0	--	--
Paraguay	9	1,120,538	2.8	NA	NA	0	0	0	--	--
Peru	22	5,482,935	4.3	4	709,978	3	2,506,739	3	--	--
Suriname	13	734,800	4.6	5	128,400	0	0	--	1	12,000
Uruguay	7	30,278	0.2	1	3,290	1	200,000	--	1	200,000
Venezuela	43	8,618,625	9.8	11	704,361	0	0	--	1	9,968

Areas, 1989

Table 20.1

	National Protection Systems					International Protection Systems				
	All Protected Areas		Percent of National Land Area Protected	Marine and Coastal Protected Areas Only		Biosphere Reserves		Natural World Heritage Sites	Wetlands of International Importance	
	Number	Area (ha)		Number	Area (ha)	Number	Area (ha)	Number	Number	Area (ha)
ASIA	1,126	61,367,197	2.3	189	13,987,130	34	6,070,242	8	34	1,480,356
Afghanistan	4	142,438	0.2	NA	NA	0	0	0	--	--
Bahrain	0	0	0.0	0	0	0	0	--	--	--
Bangladesh	8	96,787	0.7	3	32,386	0	0	0	--	--
Bhutan	5	876,058	18.6	NA	NA	0	0	--	--	--
China	179	7,903,526	0.8	20	1,183,524	6	1,602,305	1	--	--
Cyprus	3	7,449	0.8	0	0	0	0	0	--	--
India	288	13,170,318	4.4	14	473,802	0	0	5	2	119,373
Indonesia	141	14,067,051	7.8	68	8,940,876	6	1,482,400	--	--	--
Iran, Islamic Rep	30	3,626,271	2.2	3	725,130	9	2,609,731	0	18	1,297,550
Iraq	0	0	0.0	0	0	0	0	0	--	--
Israel	19	236,003	11.6	1	30,900	0	0	--	--	--
Japan	61	2,400,781	6.4	30	637,055	4	116,000	--	2	5,571
Jordan	7	92,900	1.0	0	0	0	0	0	1	7,372
Kampuchea, Dem	0	0	0.0	0	0	0	0	--	--	--
Korea, Dem People's Rep	2	57,890	0.5	0	0	1	132,000	--	--	--
Korea, Rep	17	557,766	5.7	3	284,671	1	37,430	0	--	--
Kuwait	0	0	0.0	0	0	0	0	--	--	--
Lao People's Dem Rep	0	0	0.0	NA	NA	0	0	0	--	--
Lebanon	1	3,500	0.3	0	0	0	0	0	--	--
Malaysia	39	1,101,353	3.4	9	52,225	0	0	0	--	--
Mongolia	13	317,840	0.2	NA	NA	0	0	--	--	--
Myanmar	2	173,070	0.3	0	0	0	0	--	--	--
Nepal	11	958,500	7.0	NA	NA	0	0	2	1	17,500
Oman	2	54,000	0.3	1	1,000	0	0	0	--	--
Pakistan	57	7,583,390	9.8	1	15,540	1	31,355	0	9	20,990
Philippines	32	520,816	1.7	5	30,639	1	23,545	0	--	--
Qatar	0	0	0.0	0	0	0	0	0	--	--
Saudi Arabia	5	807,650	0.4	2	475,000	0	0	0	--	--
Singapore	1	2,715	4.8	0	0	0	0	--	--	--
Sri Lanka	38	739,771	11.4	6	303,484	2	9,376	1	--	--
Syrian Arab Rep	0	0	0.0	0	0	0	0	0	--	--
Thailand	75	4,676,757	9.1	10	625,204	3	26,100	0	--	--
Turkey	15	246,026	0.3	3	113,785	0	0	2	--	--
United Arab Emirates	0	0	0.0	0	0	0	0	--	--	--
Viet Nam	56	858,354	2.6	2	33,743	0	0	0	1	12,000
Yemen Arab Rep	0	0	0.0	0	0	0	0	0	--	--
Yemen, People's Dem Rep	0	0	0.0	0	0	0	0	0	--	--
EUROPE	1,347	31,326,547	6.6	180	7,699,895	81	3,761,665	11	277	2,568,185
Albania	13	54,500	2.0	5	27,500	0	0	--	--	--
Austria	129	1,593,894	19.3	NA	NA	4	27,000	--	5	102,369
Belgium	5	83,920	2.6	0	0	0	0	--	6	9,007
Bulgaria	39	129,125	1.2	0	0	17	25,201	2	4	2,097
Czechoslovakia	66	1,985,974	15.8	NA	NA	4	176,974	--	--	--
Denmark	58	282,028	6.7	3	11,600	1	70,000,000	0	38	734,202
Finland	34	806,450	2.6	0	0	0	0	0	11	101,343
France	73	4,500,922	8.2	27	848,549	6	485,927	1	1	85,000
German Dem Rep	20	70,189	0.7	5	7,532	2	24,960	0	8	46,187
Germany, Fed Rep	86	2,756,757	11.3	9	724,840	1	13,100	0	20	313,600
Greece	61	534,231	4.1	13	84,420	2	8,840	2	11	107,400
Hungary	46	511,149	5.5	0	0	5	128,884	0	13	110,594
Iceland	19	789,050	7.9	5	509,000	0	0	--	1	20,000
Ireland	5	24,151	0.4	0	0	2	8,808	--	14	9,112
Italy	100	1,266,395	4.3	18	211,165	3	3,798	0	45	54,458
Luxembourg	3	64,900	25.1	NA	NA	0	0	0	--	--
Malta	0	0	0.0	0	0	0	0	0	1	6
Netherlands	47	150,914	4.4	10	53,521	1	260,000	--	11	306,348
Norway	65	4,761,838	15.5	12	3,508,478	1	1,555,000	0	14	16,256
Poland	75	2,192,965	7.2	4	73,091	4	25,836	1	5	7,090
Portugal	27	619,707	6.8	8	132,325	1	395	0	2	30,563
Romania	18	152,322	0.7	0	0	3	41,213	--	--	--
Spain	110	2,561,133	5.1	9	75,160	10	614,977	1	3	52,392
Sweden	68	1,706,892	4.1	5	11,640	1	96,500	0	20	271,075
Switzerland	19	120,989	3.0	NA	NA	1	16,870	0	2	1,816
United Kingdom	84	2,569,026	10.6	35	1,194,354	13	44,258	3	40	168,576
Yugoslavia	76	1,035,626	4.1	12	226,720	2	350,000	4	2	18,094
U.S.S.R.	171	20,773,379	0.9	22	4,925,038	19	9,331,366	0	12	2,987,185
OCEANIA	774	39,499,579	4.7	229	14,547,462	12	4,743,223	10	31	3,319,634
Australia	625	36,481,224	4.8	184	13,035,051	12	4,743,223	8	29	3,304,690
Fiji	2	5,342	0.3	1	4,020	0	0	--	--	--
New Zealand	122	2,828,078	10.5	32	1,385,626	0	0	2	2	14,944
Papua New Guinea	3	7,323	0.0	0	0	0	0	--	--	--
Solomon Islands	0	0	0.0	0	0	0	0	--	--	--

Source: World Conservation Monitoring Centre.
Notes: 0 = zero or less than half the unit of measurement; NA = not applicable; -- = country is not a party to the Convention.
For additional information, see Sources and Technical Notes.

Table 20.2 Globally Threatened Animal Species, 1989

	Mammals		Birds		Reptiles		Amphibians		Swallowtail Butterflies	
	Number of Species Known	Number Threatened	Number of Species Known	Number Threatened	Number of Species Known	Number Threatened	Number of Species Known	Number Threatened	Number of Species Known	Number Threatened
WORLD										
AFRICA										
Algeria	97	15	X	4	X	2	X	X	X	X
Angola	275	13	X	12	X	7	X	X	27	1
Benin	187	7	X	X	X	5	X	X	X	X
Botswana	162	8	1,044	3	X	1	X	X	X	X
Burkina Faso	147	6	X	X	X	X	X	X	X	X
Burundi	103	6	X	4	X	1	X	X	15-20	1
Cameroon	297	18	848	16	X	5	X	1	39	2
Cape Verde	9	3	X	4	X	3	X	X	X	X
Central African Rep	208	8	X	1	X	3	X	X	24-29	1
Chad	131	11	X	1	X	3	X	X	7-8	0
Comoros	X	X	X	9	X	1	X	X	3-4	2
Congo	198	12	X	3	X	7	X	X	37-38	1
Cote d'Ivoire	232	23	756	6	X	6	X	1	X	X
Djibouti	22	6	X	1	X	X	X	X	6-7	0
Egypt	105	12	X	2	X	4	X	X	X	X
Equatorial Guinea	182	13	X	3	X	4	X	1	13-21	1
Ethiopia	242	11	847	10	6	3	X	X	15-16	0
Gabon	190	9	X	5	X	7	X	X	25-31	1
Gambia, The	108	5	X	X	X	5	X	X	X	X
Ghana	222	10	721	7	X	7	X	X	X	X
Guinea	188	13	X	4	X	4	X	1	X	X
Guinea-Bissau	109	11	X	2	X	7	X	X	X	X
Kenya	307	12	860	12	106	5	97	4	30	5
Lesotho	33	2	X	4	X	X	X	X	X	X
Liberia	193	14	X	9	X	6	X	1	X	1
Libya	76	9	X	1	X	3	X	X	X	X
Madagascar	X	X	X	29	X	11	X	X	13	3
Malawi	192	5	624	9	X	1	X	X	22	0
Mali	136	11	X	2	X	2	X	X	X	X
Mauritania	61	10	X	2	X	5	X	X	X	X
Mauritius	X	2	X	8	X	6	X	X	2	1
Morocco	108	13	X	3	X	4	X	X	X	X
Mozambique, People's Rep	183	6	X	16	X	5	X	X	16	0
Namibia	161	12	X	6	X	2	X	X	X	X
Niger	131	9	X	1	X	1	X	X	X	X
Nigeria	274	57	831	8	114	9	19	3	X	1
Reunion	X	X	X	2	X	3	X	X	2	1
Rwanda	147	9	X	6	X	2	X	X	18-21	2
Sao Tome	7	0	X	7	X	1	X	X	X	X
Senegal	166	13	X	3	X	8	X	X	X	X
Seychelles	X	X	X	10	X	2	X	3	X	1
Sierra Leone	178	11	X	7	X	6	X	X	12	0
Somalia	173	14	639	6	X	4	X	X	X	X
South Africa	279	20	X	10	X	3	X	X	20	0
Sudan	266	19	X	5	X	3	X	X	X	X
Swaziland	46	3	X	4	X	1	X	X	X	X
Tanzania	310	15	1016	27	X	7	X	X	34	1
Togo	196	6	X	0	X	6	X	X	X	X
Tunisia	77	14	X	2	X	2	X	X	X	X
Uganda	311	10	989	14	X	2	X	X	31-32	2
Western Sahara	15	7	X	3	X	2	X	X	X	X
Zaire	409	24	1086	27	X	6	X	X	48	3
Zambia	228	6	728	8	X	2	X	X	23	0
Zimbabwe	194	7	635	7	X	1	X	X	X	X
AMERICAS										
Argentina	255	26	927	18	204	7	124	1	36-37	1
Bahamas	17	2	218	8	39	18	6	0	5	0
Belize	121	9	504	1	107	8	26	0	X	X
Bermuda	X	1	X	1	X	X	X	X	1	0
Bolivia	267	24	1177	5	180	10	96	0	43-44	2
Brazil	394	42	1567	35	467	19	487	1	74	8
Canada	210	7	426	14	42	1	41	0	18	0
Cayman Islands	X	X	X	3	X	5	X	X	3	0
Chile	90	10	393	6	82	3	38	0	2-3	0
Colombia	358	25	1665	28	383	24	375	0	59	0
Costa Rica	203	10	796	5	218	8	151	1	X	X
Cuba	39	9	286	14	100	10	40	0	13	1
Ecuador{a}	280	21	1447	17	345	36	350	0	64	0
El Salvador	129	7	432	3	92	7	38	0	X	X
French Guiana	142	11	628	3	136	14	89	0	29-31	3
Greenland (Denmark)	26	7	X	2	X	X	X	X	X	X
Guatemala	174	9	666	8	204	10	99	0	30-31	X
Guyana	198	12	728	3	137	14	105	0	30-31	1
Hispaniola	23	3	211	2	134	6	53	0	8	2
Honduras	179	8	672	5	161	9	57	0	X	X
Jamaica	29	2	223	5	38	4	20	0	7	2
Lesser Antilles	37	4	193	29	94	14	15	0	3	0
Mexico	439	32	961	123	717	35	284	4	52	2
Netherlands Antilles	9	0	171	0	22	3	2	0	X	X
Nicaragua	177	9	610	4	162	9	59	0	X	X

Table 20.2

	Mammals		Birds		Reptiles		Amphibians		Swallowtail Butterflies	
	Number of Species Known	Number Threatened	Number of Species Known	Number Threatened	Number of Species Known	Number Threatened	Number of Species Known	Number Threatened	Number of Species Known	Number Threatened
Panama	217	13	920	6	212	10	155	2	X	X
Paraguay	157	14	630	8	110	8	69	0	26-32	0
Peru	359	30	1642	10	297	15	235	0	58-59	2
Puerto Rico	17	1	220	11	46	15	26	1	2-3	0
Suriname	200	11	670	3	131	12	99	0	30-31	1
Tobago	29	2 b	157	6 b	39	8 b	8	0 b	13-14	0 b
Trinidad	85	X	347	X	76	X	15	0	X	X
United States	466	49	1,090	79	368	26	222	8	30-31	1
Uruguay	77	7	367	3	66	9	37	1	7-8	0
Venezuela	305	18	1295	8	246	20	183	0	35-39	1
ASIA										
Afghanistan	X	10	X	2	X	2	X	X	19	1
Bangladesh	X	5	X	4	X	8	X	X	10-15	X
Bhutan	X	12	X	3	X	1	X	X	22-30	2
Brunei	X	6	X	X	X	3	X	X	35-37	1
China	X	30	X	7	X	X	X	X	131-136	7
India	341	29	1,178	135	400	12	181	X	91	2
Indonesia	X	22	X	14	X	11	X	X	121	14
Iran, Islamic Rep	X	9	X	3	X	6	X	X	7-9	0
Iraq	X	8	X	3	X	1	X	X	6-7	0
Japan	186	4	632	35	85	3	58	1	22	0
Korea, Dem Rep	X	8	X	10	X	X	X	X	14	0
Korea, Rep	X	8	X	9	X	X	X	X	14-15	0
Malaysia	X	8	X	7	X	X	X	X	54-56	3
Mongolia	X	6	X	4	X	X	X	X	11	1
Myanmar	300	14	1,000	3	360	8	X	X	70	1
Nepal	X	17	X	2	X	4	X	X	37-38	1
Pakistan	X	13	X	6	X	9	X	X	14	0
Philippines	96	7	541	2	197	1	60	X	49	9
Sri Lanka	X	7	X	2	X	34	X	4	15	1
Turkey	124	2	217	36	X	X	X	1	X	1
Viet Nam	273	X	774	X	180	X	80	X	X	X
EUROPE									11	2
Austria	83	38	201	121	X	X	X	X	X	1
Belgium	X	X	X	X	X	X	X	X	X	0
Czechoslovakia	X	X	390	X	X	X	X	X	X	X
Denmark	49	14	190	41	5	0	14	3	X	0
Finland	62	7	232	14	5	1	5	1	X	1
France	113	59	342	136	36	14	29	18	X	2
Germany, Fed Rep	94	44	305	98	12	9	19	11	X	1
Greece	X	X	X	X	X	X	X	X	X	1
Ireland	31	5	139	33	1	0	3	1	X	0
Italy	97	13	419	60	46	24	28	13	X	2
Luxembourg	60	X	140	54	8	7	16	11	X	0
Netherlands	60	39	257	85	7	6	15	10	X	0
Norway	54	4	220	23	5	1	5	2	X	1
Portugal	56	25	288	113	24	X	17	X	X	0
Spain	135	17	357	22	64	1	24	1	X	1
Sweden	65	10	250	17	6	0	13	5	X	1
Switzerland	86	X	190	74	15	X	20	X	X	1
United Kingdom	77	24	233	35	11	5	14	2	X	0
Yugoslavia	X	X	X	X	X	X	X	X	X	1
U.S.S.R.	357	78	765	80	144	37	34	9	42-44	3
OCEANIA										
Australia	320	43	700	23	550	9	150	6	19	0
Fiji	X	X	X	5	X	5	X	X	1	0
New Caledonia	X	1	X	4	X	X	X	X	3-4	0
New Zealand	69	14	000	16	39	7	5	X	0	0
Papua New Guinea	X	10	X	1	X	7	X	X	37	9
Solomon Islands	X	1	X	3	X	5	X	X	15	3
Vanuatu	X	1	X	1	X	2	X	X	4	0
Western Samoa	X	X	X	1	X	1	X	X	1	0

Sources: World Conservation Monitoring Centre, Organisation for Economic Co-operation and Development, and United Nations Economic and Social Commission for Asia and the Pacific.

Notes: a. Includes the Galapagos Islands. b. Refers to both Trinidad and Tobago.
X = not available.
For additional information, see Sources and Technical Notes.

Table 20.3   CITES-Reported Trade in Wildlife and Wildlife

	CITES Reporting Requirement Met {a} (percent)	Mammals 1986 Live Primates (number)		Mammals 1986 Cat Skins (number)		Mammals 1988 Raw Ivory {b} (kilograms)		Birds 1986 Live Parrots (number)		Reptiles 1986 Reptile Skins {c} (number)	
		Imports	Exports	Imports	Exports	Imports	Exports	Imports	Exports	Imports	Exports
WORLD		51,256	51,256	192,402	192,402	429,549	429,558	696,002	618,539	10,480,798	10,480,798
AFRICA		423	8,879	753	2,272	2,254	130,391	14,255	169,238	65,980	399,256
Algeria	33	1	0	X	X	8	0	1	0	X	X
Angola {d}	NA	X	X	X	X	X	X	0	5	X	X
Benin	0	X	1	0	3	X	X	0	1	0	40,009
Botswana	67	0	34	449	32	38	0	1,083	0	X	5
Burkina Faso {d}	NA	X	X	X	X	X	X	1	1	X	X
Burundi	X	X	X	0	2	X	X	0	34	X	X
Cameroon	100	3	84	3	6	0	2,538	2	10,320	0	86,407
Cape Verde {d}	NA	X	X	X	X	X	X	0	1	X	X
Central African Rep	43	0	1	0	37	0	260	0	1,085	0	37
Chad	NA	X	X	X	X	X	X	0	0	X	X
Comoros {d}	NA	X	X	X	X	X	X	0	0	X	X
Congo	100	0	4	0	2	0	18,806	10	32	0	343
Cote d'Ivoire {d}	NA	0	1	X	X	201	476	4	1,705	0	15
Djibouti {d}	NA	X	X	X	X	0	10,901	0	0	X	X
Egypt	0	2	1	X	X	X	X	1,598	2	65	1
Equatorial Guinea {d}	NA	X	X	X	X	X	X	0	0	X	X
Ethiopia	X	0	1,883	0	1	0	2,160	X	X	0	1
Gabon	X	0	3	X	X	0	13,542	1	4	0	4
Gambia, The	30	X	X	X	0	X	X	0	0	X	X
Ghana	73	0	152	0	2	X	X	0	6,561	0	127
Guinea	17	X	6	0	33	X	X	0	1,990	0	2,323
Guinea-Bissau {d}	NA	X	X	X	X	X	X	0	2	X	X
Kenya	25	2	5,074	1	20	X	X	2	26	259	1
Lesotho {d}	NA	1	0	0	1	X	X	0	0	X	X
Liberia	100	0	1	0	2	X	X	1	9,368	X	X
Libya {d}	NA	6	X	X	X	X	X	21	2	3	X
Madagascar	92	0	23	X	X	X	X	1	9,434	0	676
Malawi	80	X	X	0	2	38	782	0	0	0	684
Mali {d}	NA	X	X	0	500	X	X	21	3,144	0	135,648
Mauritania {d}	NA	X	X	X	X	X	X	X	X	X	X
Mauritius	83	4	400	X	X	X	X	1,060	0	20,000	0
Morocco	18	389	24	X	X	30	0	19	3	92	X
Mozambique, People's Rep	67	X	X	0	3	0	7,302	2	0	0	1
Niger	33	1	0	0	0	X	X	1	7	30,000	50,000
Nigeria	0	1	0	0	42	X	X	0	39	0	31,136
Rwanda	17	X	X	X	X	X	X	0	3	X	X
Senegal	90	4	470	X	X	22	0	12	28,430	0	153
Sierra Leone {d}	NA	0	20	0	1	X	X	0	0	0	65
Somalia	100	0	4	X	X	0	22,638	X	X	X	X
South Africa	92	9	19	208	931	1,914	8,791	8,122	8,050	15,488	1,614
Sudan	25	X	X	0	4	X	X	54	1	0	33,053
Swaziland {d}	NA	X	X	7	0	X	X	1,464	0	X	X
Tanzania	71	0	287	0	342	0	22,581	0	84,228	0	763
Togo	63	0	365	X	X	X	X	0	4,199	0	5,140
Tunisia	100	X	X	16	0	X	X	3	1	71	X
Uganda {d}	NA	X	10	0	2	X	X	1	1	0	1,002
Zaire	55	X	12	0	3	0	11,009	27	108	0	6
Zambia	33	X	X	0	47	0	1,622	0	0	0	2,954
Zimbabwe	67	X	X	6	93	3	6,983	66	450	0	7,086
NORTH & CENTRAL AMERICA		23,588	6,756	38,683	89,590	5,051	126	318,162	26,550	1,388,169	600,161
Barbados {d}	NA	0	439	X	X	X	X	113	0	X	X
Canada	100	1,757	380	21,109	24,104	187	62	9,551	151	41,473	20,251
Costa Rica	67	3	0	2	1	X	X	18	4	0	1
Cuba {d}	NA	12	0	X	X	X	X	1	X	X	X
Dominican Rep	X	4	0	X	X	X	X	128	57	X	X
El Salvador {d}	NA	X	X	X	X	X	X	0	17	0	70,000
Guatemala	86	X	30	0	2	X	X	0	3,628	2	0
Haiti {d}	NA	X	X	X	X	X	X	1	2	19	0
Honduras	100	0	1	0	2	X	X	42	15,816	X	4
Jamaica {d}	NA	X	X	X	X	X	X	4	300	100	X
Mexico {d}	NA	128	3	29	59	61	0	363	221	81,557	5,211
Nicaragua	80	0	13	0	2	X	X	0	2,389	0	113
Panama	78	X	X	X	X	X	X	212	54	0	123,120
Trinidad and Tobago	0	3	X	X	X	X	X	17	2	X	X
United States	83	21,675	5,725	17,543	65,419	4,803	64	305,997	3,853	1,265,018	381,461
SOUTH AMERICA		82	6,135	11	6,150	3	X	2,811	256,634	47,191	1,403,942
Argentina	100	5	6	7	0	X	X	25	177,992	2,100	1,153,967
Bolivia	50	0	361	0	6,124	X	X	0	17	0	97,922
Brazil	33	24	20	2	1	X	X	36	75	23,906	16
Chile	50	15	X	X	X	3	0	532	1,347	X	X
Colombia	50	30	2	1	2	X	X	4	30	5,597	0
Ecuador	67	X	X	0	11	X	X	0	62	X	X
Guyana	60	0	5,305	0	3	X	X	2	30,324	0	52,333
Paraguay	40	3	0	0	1	X	X	0	42	11,398	17,212
Peru	83	0	311	1	4	X	X	296	17,032	127	2,852
Suriname	100	1	120	0	1	X	X	1,912	8,737	X	X
Uruguay	67	2	X	X	X	X	X	0	20,967	363	77
Venezuela	56	2	4	0	3	X	X	0	9	3,700	79,563

Products, 1980s

Table 20.3

	CITES Reporting Requirement Met {a} (percent)	Mammals 1986 Live Primates (number)		Mammals 1986 Cat Skins (number)		Mammals 1988 Raw Ivory {b} (kilograms)		Birds 1986 Live Parrots (number)		Reptiles 1986 Reptile Skins {c} (number)	
		Imports	Exports	Imports	Exports	Imports	Exports	Imports	Exports	Imports	Exports
ASIA		7,921	24,662	16,093	72,969	355,782	269,655	60,365	126,538	4,682,360	6,878,809
Afghanistan	0	X	X	X	X	X	X	X	X	X	X
Bahrain {d}	NA	X	X	X	X	X	X	432	1	X	X
Bangladesh	100	8	0	0	1	X	X	110	9,939	0	20,000
Bhutan {d}	NA	X	X	X	X	X	X	0	500	X	X
China	100	89	1,029	1	68,274	50,384	0	11	500	24	0
Cyprus	42	1	1	1	0	3	0	662	1	101	0
Hong Kong	100	110	30	1,081	85	179,608	135,938	7,017	318	179,215	52,818
India	100	21	2	0	2	3,879	0	162	15,445	0	809
Indonesia	88	12	10,569	X	X	X	X	146	58,832	0	3,081,313
Iran, Islamic Rep	18	X	X	X	X	X	X	1	0	X	X
Iraq {d}	NA	X	X	X	X	X	X	1	0	X	X
Israel	0	107	0	81	4,573	X	X	2,546	0	3,311	25,675
Japan	100	4,795	49	14,321	31	100,985	33,149	27,790	26	802,397	160,181
Jordan	0	X	X	X	X	X	X	375	1	X	X
Kampuchea, Dem {d}	NA	X	X	X	X	X	X	X	X	X	X
Korea, Dem People's Rep {d}	NA	7	0	X	X	X	X	X	X	X	X
Korea, Rep {d}	NA	4	0	339	0	100	0	1	150	46,393	16
Kuwait {d}	NA	9	0	X	X	X	X	1,741	9	2	X
Lao People's Dem Rep {d}	NA	0	95	X	X	X	X	X	7	X	X
Lebanon {d}	NA	X	95	28	0	71	0	5	1	0	100
Malaysia	78	2	8	X	X	2	0	652	15,012	268	3,919
Mongolia {d}	NA	X	X	X	X	X	X	X	X	X	X
Myanmar	X	0	50	X	X	X	X	X	X	X	X
Nepal	75	X	X	X	X	X	X	18	0	X	X
Oman {d}	NA	1	0	X	X	X	X	366	4	X	X
Pakistan	100	4	0	X	X	7	0	329	3	0	2
Philippines	100	0	12,502	X	X	27	0	6	488	110,150	17,559
Qatar {d}	NA	X	X	X	X	X	X	839	2	X	X
Saudi Arabia {d}	NA	51	0	2	1	X	X	2,061	70	265	0
Singapore	X	2	227	0	1	17,722	100,568	6,566	6,756	2,856,468	1,637,061
Sri Lanka	38	17	8	X	X	5	0	348	3	X	X
Syrian Arab Rep {d}	NA	X	X	X	X	X	X	2	4	X	X
Taiwan {d}	NA	2,459	20	17	0	2,955	0	4,819	13,840	623,543	75,789
Thailand	75	207	70	X	X	34	0	359	4,619	46	1,650,317
Turkey {d}	NA	X	X	222	0	X	X	3	1	59,409	500
United Arab Emirates {e}	NA	14	0	0	1	X	X	2,845	7	767	152,750
Viet Nam {d}	NA	X	2	X	X	X	X	X	X	X	X
Yemen Arab Rep {d}	NA	X	0	X	X	X	X	X	X	1	X
Yemen, People's Dem Rep {d}	NA	1	0	X	X	X	X	X	X	X	X
EUROPE		16,063	4,781	136,736	18,335	53,197	28,394	298,973	37,234	4,070,201	1,193,904
Albania {d}	NA	X	X	X	X	X	X	X	X	X	X
Austria	100	131	15	1,262	660	355	0	3,087	43	98,949	14,931
Belgium	100	806	75	227	182	38,730	24,034	20,357	13,126	15,200	12,501
Bulgaria {d}	NA	2	0	X	X	6	0	13	0	X	X
Czechoslovakia {d}	NA	33	11	X	X	X	X	36	1,253	X	X
Denmark	100	48	0	7,976	1,048	X	X	7,982	98	307	5
Finland	64	1	8	220	77	6	0	0	2	2,427	0
France	100	1,870	229	7,701	45	4,486	2,479	18,843	1,053	755,617	177,893
German Dem Rep	55	176	6	1	2,744	X	X	65	3,263	0	28
Germany, Fed Rep	100	538	310	82,240	4,164	1,168	0	60,499	458	42,813	128.651
Greece {d}	NA	1	0	1,897	13	270	0	775	2	14,955	0
Hungary	50	33	1	22	0	X	X	2,371	850	33	0
Iceland {d}	NA	X	X	X	X	X	X	X	X	X	X
Ireland {d}	NA	2	1	1	0	X	X	50	0	X	1
Italy	100	1,150	4	9,505	651	424	39	8,607	12	1,026,928	183,998
Luxembourg	100	0	1	X	X	2	0	1	0	1	0
Malta	X	30	0	42	22	X	X	407	0	825	0
Netherlands	100	2,786	2,797	13	1	1,213	0	27,822	13,809	348,175	10,826
Norway	100	1	1	20	0	11	0	68	4	5	0
Poland {d}	NA	57	4	X	X	X	X	154	0	X	X
Portugal	17	28	0	4	0	X	X	4,217	58	1,102	0
Romania {d}	NA	29	0	X	X	X	X	2	2	X	X
Spain	100	141	17	2,261	3	2,231	0	11,406	207	1,009,054	112,987
Sweden	100	511	116	265	297	15	0	16,454	2,792	19	0
Switzerland	100	268	111	6,657	7,622	58	0	3,582	67	201,955	78,564
United Kingdom	100	5,811	1,048	16,400	806	4,222	1,842	34,520	129	551,281	473,519
Yugoslavia {d}	NA	1,600	0	X	X	X	X	4	0	555	X
U.S.S.R.	73	2,953	2	2	2,855	X	X	225	0	2	0
OCEANIA		127	84	17	0	67	X	304	266	734	25,363
Australia	100	117	65	14	0	67	0	180	244	134	400
Fiji {d}	NA	X	X	X	X	X	X	X	X	X	X
New Zealand	X	9	19	3	0	X	X	124	14	600	X
Papua New Guinea	73	1	0	X	X	X	X	0	8	0	24,963
Solomon Islands {d}	NA	X	X	X	X	X	X	X	X	X	X

Source: World Conservation Monitoring Centre.
Notes: a. Includes all trade reported by members of the Convention on International Trade in Endangered Species of Wild Flora and Fauna (CITES). b. World and Asian totals for raw ivory include 13,195 kg imported into Macao and 995 kg exported from Macao. c. Reptile skins include skins of snakes, lizards, and crocodilians. d. Not a member of CITES. e. Sudan reported exports of 63,679 kg of raw ivory in 1987, none in 1988. f. Withdrew from CITES in 1988.
X = not available; NA = not applicable. For additional information, see Sources and Technical Notes.

Table 20.4   Habitat Loss, 1980s

Habitat Types (areas in square kilometers)

	All Forests — Current sq km	% Lost	Dry Forests — Current sq km	% Lost	Moist Forests — Current sq km	% Lost	Savanna/Grassland — Current sq km	% Lost	Desert/Scrub — Current sq km	% Lost	Wetlands/Marsh — Current sq km	% Lost	Mangroves — Current sq km	% Lost
WORLD	X	X	X	X	X	X	26,153,072	X	X	X	X	50	X	X
AFRICA	X	X	X	38	X	X	19,263,575	X	X	X	X	X	X	X
Algeria	X	X	X	X	X	X	X	X	X	X	7,296	X	X	X
Angola	514,284	45	402,614	45	111,670	48	245,902	17	4,560	20	X	X	1,100	50
Benin	47,420	59	44,060	55	3,360	80	0	0	0	0	0	0	0	0
Botswana	112,926	62	112,926	62	0	0	122,470	53	0	0	23,310	10	0	0
Burkina Faso	49,640	80	49,640	80	0	0	7,680	70	0	0	0	0	0	0
Burundi	1,540	88	1,140	91	30	95	2,460	80	0	0	140	X	0	0
Cameroon	184,678	59	29,490	69	155,188	56	3,760	72	0	0	160	80	4,860	40
Cape Verde	X	X	X	X	X	X	0	0	0	0	0	0	X	X
Central African Rep	279,328	55	146,670	51	132,658	59	0	0	0	0	0	0	0	0
Chad	58,480	80	58,480	80	0	0	119,580	72	0	0	660	90	0	0
Comoros	X	X	X	X	X	X	X	X	X	X	X	X	X	X
Congo	174,420	49	0	0	174,420	49	0	0	0	0	2,900	X	0	0
Cote d'Ivoire	68,630	78	35,620	60	33,010	85	0	0	0	0	320	X	640	60
Djibouti	0	0	0	0	0	70	10,000	50	1,200	20	0	X	90	70
Egypt	X	X	X	X	X	X	X	X	X	X	8,085	X	X	X
Equatorial Guinea	12,850	50	0	0	12,850	50	0	0	0	0	0	0	120	60
Ethiopia	55,700	86	55,700	86	0	0	274,685	61	5,250	30	0	0	0	0
Gabon	172,870	35	0	0	172,450	35	0	0	0	0	0	0	1,150	50
Gambia, The	840	91	720	90	120	95	0	0	0	0	0	0	510	70
Ghana	46,220	80	26,700	71	19,520	86	0	0	0	0	8,532	X	630	70
Guinea	74,400	69	17,990	71	56,410	69	0	0	0	0	5,250	X	1,200	60
Guinea-Bissau	5,120	80	0	0	5,120	80	0	0	0	0	0	0	3,150	70
Kenya	22,738	71	21,298	67	1,440	90	276,816	43	0	0	0	0	930	70
Lesotho	8,510	67	8,510	67	0	0	1,410	70	0	0	400	X	0	0
Liberia	14,240	87	80	20	14,160	87	0	0	0	0	0	0	360	70
Libya	X	X	X	X	X	X	X	X	X	X	X	X	X	X
Madagascar	130,490	75	114,006	62	16,484	84	15,094	78	0	0	1,970	X	1,302	40
Malawi	39,770	56	39,770	56	0	0	0	0	0	0	1,120	60	0	0
Mali	76,700	78	76,700	78	0	0	83,676	80	0	0	20,000	X	0	0
Mauritania	60	90	60	90	0	0	46,100	88	0	0	0	0	0	0
Mauritius	X	X	X	X	X	X	X	X	X	X	X	X	X	X
Morocco	X	X	X	X	X	X	X	X	X	X	333	X	X	X
Mozambique, People's Rep	331,365	57	331,365	57	0	0	6,960	20	0	0	1,710	10	2,760	60
Niger	22,780	80	22,780	80	0	0	109,850	75	0	0	380	80	0	0
Nigeria	210,420	76	143,390	70	67,030	83	4,980	80	0	0	420	80	12,200	50
Rwanda	1,840	80	1,840	80	0	0	1,570	90	0	0	800	X	0	0
Senegal	25,200	82	22,500	80	2,700	90	11,200	80	0	0	15	X	420	40
Sierra Leone	7,440	88	480	40	6,960	89	0	0	0	0	0	0	3,400	50
Somalia	6,420	67	6,420	67	0	0	363,740	40	7,120	4	0	0	540	70
South Africa	204,440	46	204,440	46	0	0	322,572	62	8,800	0	0	0	450	50
Sudan	153,672	74	151,620	73	2,052	91	360,069	68	0	0	111,695	X	0	0
Swaziland	7,720	56	7,720	56	0	0	0	0	0	0	0	0	0	0
Tanzania	361,365	40	358,665	39	2,700	80	143,522	49	0	0	15,454	X	2,120	60
Togo	19,300	65	16,220	57	3,080	83	0	0	0	0	8,683	X	X	X
Tunisia	X	X	X	X	X	X	X	X	X	X	X	X	X	X
Uganda	33,714	79	20,620	67	13,094	86	10,420	71	0	0	14,200	X	0	0
Zaire	832,548	57	91,346	54	741,202	57	54,050	30	0	0	2,150	50	1,250	50
Zambia	446,060	30	446,060	30	0	0	81,750	18	0	0	11,060	10	0	0
Zimbabwe	171,688	56	171,688	56	0	0	0	0	0	0	0	0	0	0
NORTH & CENTRAL AMERICA	X	X	X	X	X	X	X	X	X	X	X	X	X	X
Barbados	X	X	X	X	X	X	X	X	X	X	0	X	X	X
Canada	4,426,060	X	X	X	X	X	276,629	X	X	X	1,270,000	X	X	X
Costa Rica	X	X	X	X	X	X	X	X	X	X	818	X	390	X
Cuba	X	X	X	X	X	X	X	X	X	X	17,465	X	4,000	X
Dominican Rep	X	X	X	X	X	X	X	X	X	X	48,442	X	90	X
El Salvador	X	X	X	X	X	X	X	X	X	X	768	X	450	X
Guatemala	X	60	X	X	X	X	X	X	X	X	2,202	X	500	60
Haiti	X	X	X	X	X	X	X	X	X	X	1,129	X	180	X
Honduras	X	X	X	X	X	X	X	X	X	X	6,490	X	1,450	X
Jamaica	1,841	X	X	X	X	X	X	X	X	X	138	X	70	X
Mexico	384,608	66	X	X	X	X	X	X	1,000,000	X	32,640	X	6,600	X
Nicaragua	X	X	X	X	X	X	X	X	X	X	20,532	X	600	X
Panama	X	X	X	X	X	X	X	X	X	X	6,472	X	4,860	X
Trinidad and Tobago	X	X	X	X	X	X	X	X	X	X	213	X	40	X
United States	2,994,780	26	X	X	X	X	30,000	100	X	X	870,000	54	X	X
SOUTH AMERICA	X	X	X	X	X	X	7,890,629	X	X	X	X	X	X	X
Argentina	360,000	50	X	X	X	X	1,300	X	X	X	61,689	X	X	X
Bolivia	X	X	X	X	X	X	X	X	X	X	24,191	X	X	X
Brazil	X	X	X	X	X	X	X	X	X	X	296,903	X	25,000	X
Chile	X	X	X	X	X	X	X	X	X	X	88,267	X	X	X
Colombia	X	X	X	X	X	X	X	X	X	X	19,281	X	4,400	X
Ecuador	X	X	X	X	27,000	89	X	X	X	X	9,926	X	1,601	X
Guyana	167,322	X	X	X	X	X	X	X	X	X	8,139	X	1,500	X
Paraguay	X	X	X	X	X	X	X	X	X	X	57,236	X	X	X
Peru	X	X	X	X	280	X	X	X	X	X	13,033	X	280	X
Suriname	X	X	X	X	X	X	X	X	X	X	16,250	X	1,150	X
Uruguay	X	X	X	X	X	X	X	X	X	X	6,250	X	X	X
Venezuela	X	X	X	X	X	X	X	X	X	X	145,006	X	X	X

	Habitat Types (areas in square kilometers)													
	Forests						Savanna/ Grassland		Desert/ Scrub		Wetlands/ Marsh		Mangroves	
	All Forests		Dry Forests		Moist Forests									
	Current sq km	% Lost	Current sq km	% Lost	Current sq km	% Lost	Current sq km	% Lost	Current sq km	% Lost	Current sq km	% Lost	Current sq km	% Lost
ASIA	X	X	X	X	X	X	2,678,653	X	X	X	X	X	X	X
Afghanistan	X	X	X	X	X	X	X	X	X	X	400	X	X	X
Bahrain	X	X	X	X	X	X	X	X	X	X	X	X	X	X
Bangladesh	4,820	96	0	0	4,820	96	0	0	0	0	680	96	2,910	73
Bhutan	34,500	34	7,000	30	15,980	35	0	0	0	0	65	X	0	0
China	X	X	129,313	41	X	X	X	X	3,916,800	X	42,000	X	X	X
Cyprus	X	X	X	X	X	X	X	X	X	X	101	X	X	X
India	499,285	78	357,846	81	141,439	56	0	0	85,266	88	9,408	79	1,894	85
Indonesia	X	51	105,029	27	498,996	54	0	0	0	0	118,717	39	21,011	45
Iran, Islamic Rep	X	X	X	X	X	X	X	X	X	X	14,175	X	X	X
Iraq	X	X	X	X	X	X	X	X	X	X	19,205	X	X	X
Israel	X	X	X	X	X	X	X	X	X	X	1,697	X	X	X
Japan	X	X	X	X	X	X	X	X	X	X	2,500	X	4	X
Jordan	X	X	X	X	X	X	X	X	X	X	10	X	X	X
Kampuchea, Dem	38,850	78	16,078	81	22,772	74	0	0	X	X	3,893	45	156	5
Korea, Dem People's Rep	X	X	X	X	X	X	0	0	0	0	1,360	0	X	X
Korea, Rep	X	X	X	X	X	X	X	X	X	X	837	X	X	X
Kuwait	X	X	X	X	X	X	X	X	X	X	X	X	X	X
Lao People's Dem Rep	68,972	68	37,943	67	31,029	75	0	0	0	0	0	0	0	0
Lebanon	X	X	X	X	X	X	X	X	X	X	X	85	X	X
Malaysia	180,077	42	28,524	19	151,553	45	0	0	0	0	22,142	35	7,310	32
Mongolia	X	X	X	X	X	X	X	X	X	X	17,082	X	X	X
Myanmar	241,305	64	120,002	68	121,303	65	12,025	74	286	93	490	98	1,714	58
Nepal	53,813	54	8,820	16	44,993	58	0	0	0	0	2,907	X	0	0
Oman	X	X	X	X	X	X	X	X	X	X	X	X	X	X
Pakistan	7,635	86	1,835	96	5,800	27	0	0	28,108	69	3,200	74	1,540	78
Philippines	63,429	79	10,942	60	52,487	81	0	0	0	0	13,220	X	777	61
Qatar	X	X	X	X	X	X	X	X	X	X	X	X	X	X
Saudi Arabia	X	X	X	X	X	X	1,200,000	X	>705,000	X	X	X	X	X
Singapore	X	X	X	X	X	X	X	X	X	X	2	X	5	X
Sri Lanka	6,095	86	4,464	76	1,631	94	0	0	4,950	75	5,122	X	0	0
Syrian Arab Rep	X	X	X	X	X	X	X	X	X	X	375	X	X	X
Thailand	131,071	73	83,299	78	47,772	57	0	0	0	0	832	96	191	87
Turkey	X	X	X	X	X	X	X	X	X	X	13,910	X	X	X
United Arab Emirates	X	X	X	X	X	X	X	X	X	X	X	X	30	X
Viet Nam	67,584	76	21,048	68	46,536	79	0	0	0	0	260	100	1,468	62
Yemen Arab Rep	X	X	X	X	X	X	X	X	X	X	X	X	X	X
Yemen, People's Dem Rep	X	X	X	X	X	X	X	X	X	X	X	X	X	X
EUROPE	X	X	X	X	X	X	X	X	X	X	X	X	X	X
Albania	X	X	X	X	X	X	X	X	X	X	327	X	X	X
Austria	X	X	X	X	X	X	X	X	X	X	294	X	X	X
Belgium	X	X	X	X	X	X	X	X	X	X	66	X	X	X
Bulgaria	X	X	X	X	X	X	X	X	X	X	145	X	X	X
Czechoslovakia	X	X	X	X	X	X	X	X	X	X	694	X	X	X
Denmark	X	X	X	X	X	X	X	X	X	X	7,159	X	X	X
Finland	X	X	X	X	X	X	X	X	X	X	3,000	X	X	X
France	X	X	X	X	X	X	X	X	X	X	11,714	X	X	X
German Dem Rep	X	X	X	X	X	X	X	X	X	X	3,531	X	X	X
Germany, Fed Rep	X	X	X	X	X	X	X	X	X	X	11,128	X	X	X
Greece	80,940	70	X	X	X	X	X	X	X	X	865	X	X	X
Hungary	X	X	X	X	X	X	X	X	X	X	937	X	X	X
Iceland	X	X	X	X	X	X	X	X	X	X	4,427	X	X	X
Ireland	X	X	X	X	X	X	X	X	X	X	1,150	X	X	X
Italy	X	X	X	X	X	X	X	X	X	X	30,000	94	X	X
Luxembourg	X	X	X	X	X	X	X	X	X	X	X	X	X	X
Malta	X	X	X	X	X	X	X	X	X	X	1	X	X	X
Netherlands	X	X	X	X	X	X	X	X	X	X	3,534	X	X	X
Norway	X	X	X	X	X	X	X	X	X	X	1,516	X	X	X
Poland	X	X	X	X	X	X	X	X	X	X	1,942	X	X	X
Portugal	X	X	X	X	X	X	X	X	X	X	847	X	X	X
Romania	X	X	X	X	X	X	X	X	X	X	4,825	X	X	X
Spain	X	X	X	X	X	X	X	X	X	X	4,450	X	X	X
Sweden	X	X	X	X	X	X	X	X	X	X	20,976	X	X	X
Switzerland	X	X	X	X	X	X	X	X	X	X	1,780	X	X	X
United Kingdom	X	X	X	X	X	X	X	X	X	X	4,464	X	X	X
Yugoslavia	X	X	X	X	X	X	X	X	X	X	888	X	X	X
U.S.S.R.	375,730	38	X	X	X	X	X	X	X	X	28,372	X	X	X
OCEANIA	X	X	X	X	X	X	X	X	X	X	X	X	X	X
Australia	420,000	X	155,000	76	X	X	4,100,000	X	5,500,000	X	17,000	~95	11,617	X
Fiji	X	X	X	X	X	X	X	X	X	X	X	X	197	X
New Zealand	X	X	188,069	69	X	X	65,000	~90	X	X	322,404	90	198	X
Papua New Guinea	X	X	3500	X	236,000	X	28,000	X	X	X	50,000	X	6,000	X
Solomon Islands	X	X	X	X	X	X	X	X	X	X	X	X	X	X

Sources: World Conservation Union and other sources.
Notes: X = not available; 0 = zero or less than half the unit of measure.
For additional information, see Sources and Technical Notes.

Table 20.5 Rare Species of Animals in Zoos, 1980s

	IUCN Status{b}	Number in Wild	Native to	Zoo Census 1986 (1988){a} Number of Zoos	Number of Captive Animals Male	Female	Not Known	Origins (percent) Wild	Captive	Captive Births{c} 1985	1988
MAMMALS											
Horse, Przewalski's wild {d,e}	Ex	0	China, Mongolia	101	283	377	0	0	100	75	29
Addax (e)	E		N. Africa	50	116	255	38	0	100	91	49
Anoa, lowland {d,e}	E		Indonesia	8	14	16	0	0	100	30	1
Anoa, mountain {e}	E		Indonesia	4	8	5	0	54	46	7	0
Ass, African wild {d}	E		N.E. Africa	2	14	21	0	X	X	4	0
Aye-aye	E	3	Madagascar	2	3	4	0	100	0	X	0
Bandicoot, rabbit-eared	E		Australia	3	3	5	0	50	50	1	X
Bear, Baluchistan	E		Iran, Pakistan	X	X	X	X	X	X	X	X
Bettong, brush-tailed {d,e}	E	~300	Australia	13	77	84	5	100	0	42	2
Bilby, greater{e}	E		Australia	3	3	5	0	50	50	X	X
Cat, Pakistan sand {d}	E		Pakistan	4	6	6	0	0	100	1	X
Chimpanzee, West African {e}	E	<17,000	West Africa	8	60	79	10	28	57	X	0
Cougar, Florida	E	>30	United States	2	1	1	0	50	50	X	0
Deer, Bactrian	E		Afghanistan, Nepal	13	16	16	0	9	91	2	3
Deer, Manipur brow-antlered	E		India	13	43	41	10	0	100	4	0
Deer, Persian fallow {d}	E		Iraq, Iran	5	17	41	0	0	100	7	X
Deer, swamp {d}	E		India, Nepal	27	102	192	15	0	100	54	9
Deer, Thailand brow-antlered	E		S.E. Asia	1	1	3	0	0	100	X	1
Drill {d,e}	E	Unknown	Cameroon, Bioko, Nigeria	17	21	30	0	0	100	3	X
Duiker, Jentink's	E		Cote d'Ivoire, Liberia	3	4	5	0	33	67	3	0
Elephant, Asian {e}	E		Asia	81	36	202	0	53	20	0	1
Fox, Rodrigues flying	E		Rodrigues Island	3	12	12	55	13	87	75	27
Fox, Seychelles flying	E		Aldabra Island	1	3	3	0	X	X	X	0
Gazelle, Cuvier's {d}	E		N.W. Africa	5	37	38	0	0	100	75	4
Gazelle, Saudi goitered	E		Arabia	4	27	29	0	0	100	56	8
Gazelle, slender-horned {d}	E		N. Africa	7	28	37	0	X	X	67	29
Gibbon, Javan	E		Indonesia	14	21	21	1	67	33	1	2
Gibbon, Kloss's	E		Indonesia	2	2	0	0	X	X	X	3
Gorilla, eastern lowland {e}	E	3,000-5,000	Zaire	2	5	5	0	50	50	X	0
Gorilla, mountain {e}	E	370-440	Rwanda, Uganda, Zaire	1	1	0	0	X	X	25	X
Lemur, broad-nosed gentle {e}	E		Madagascar	1	1	1	0	100	0	X	0
Lemur, Sclater's {d,e}	E		Madagascar	1	4	1	0	X	X	X	2
Leopard, snow {d,e}	E	~1,000	Asia	71	145	147	0	0	100	34	27
Lion, Asiatic {d,e}	E	~250	India	31	85	98	13	100	0	213	0
Macaque, lion-tailed {d,e}	E	>2,000	S. India	54	157	175	6	0	100	30	23
Markhor, straight-horned	E		Afghanistan, Pakistan	X	X	X	X	X	X	15	X
Marmoset, buffy-headed	E	~50	Brazil	1	2	1	0	X	X	X	X
Marmoset, buffy-tufted-ear	E		Brazil	2	9	6	3	33	67	3	0
Monkey, Central American squirrel	E		Costa Rica, Panama	5	7	9	0	6	94	4	0
Monkey, Nilgiri leaf	E		S. India	6	15	16	0	87	13	0	0
Monkey, red-shanked douc {d}	E		Laos, Viet Nam	8	20	21	0	33	67	3	0
Monkey, Tonkin leaf	E		S.E. Asia, China	7	4	4	0	36	64	3	7
Muntjac, Fea's	E		Myanmar, Thailand	1	1	4	9	60	40	X	X
Orangutan {d,e}	E		Borneo, Sumatra	13	18	30	0	0	100	9	16
Oryx, Arabian {d,e}	E		Asia, Saudi Arabia	20	105	138	5	0	X	248	59
Oryx, scimitar-horned {e}	E		N. Africa	54	191	349	4	0	100	~544	93
Rabbit, volcano	E		Mexico	1	3	3	3	23	77	4	0
Rhinoceros, black {d,e}	E	~3,500	Africa	46	55	75	0	45	55	6	5
Rhinoceros, great Indian {d,e}	E	>1,200	India, Nepal	30	40	26	0	30	70	4	1
Rhinoceros, northern square-lipped {d,e}	E		Zaire, Sudan	5	7	7	0	71	29	X	0
Rhinoceros, Sumatran {d,e}	E	>500	S.E. Asia	3	0	0	13	X	X	X	0
Saki, southern bearded	E		Brazil	1	0	1	0	X	X	X	X
Seal, Hawaiian monk	E	>500	United States	1	2	0	0	0	100	X	X
Sika, Formosan	E		Taiwan	17	100	175	18	X	X	71	23
Solendon, Haitian	E		Hispaniola	1	1	0	0	X	X	X	X
Tamarin, cotton-top {d,e}	E		Colombia	79	407	357	63	0	100	187	19
Tamarin, golden lion {e}	E	>25	Brazil	61	223	229	5	0	100	86	52
Tamarin, golden-headed lion {e}	E	~200	Brazil	5	18	9	0	70	30	22	27
Tamarin, golden-rumped lion	E	~100	Brazil	1	X	X	21	X	X	1	X
Tapir, Malayan {d,e}	E		S.E. Asia	49	57	70	0	30	70	13	3
Tarsir, Philippine	E		Philippines	4	14	15	0	83	17	X	2
Tiger {d,e}	E	~15,000	Asia	X	461	437	0	~5	~95	205	41
Wolf, red {d,e}	E	0-50	United States	8	36	52	1	0	100	16	9
Zebra, Grevy's {d,e}	E		Ethiopia, Kenya	67	125	264	0	21	79	56	33
Anteater, giant {d,e}	V		C. and S. America	44	X	X	107	76	24	2	4
Armadillo, giant	V		South America	2	2	2	0	X	X	X	X
Ass, Asiatic wild {d}	V		Asia	~9	150	248	7	0	100	50	14
Babirusa {d,e}	V		Indonesia	12	23	24	0	0	100	5	6
Banteng {d}	V		S.E. Asia	30	79	126	4	X	X	~209	7
Bat, ghost {e}	V		Australia	1	2	1	2	40	60	X	X
Bear, polar {d,e}	V	>10,000	Arctic	66	70	116	14	30	52	26	6
Bear, spectacled {d,e}	V		S. America	37	52	44	0	27	73	12	5
Bison, European {d}	V	1377	Poland, U.S.S.R.	11	X	X	1,329	0	96	76	4
Bontebok	V		S. Africa	16	25	54	0	0	100	10	6
Camel, wild Bactrian	V		China, Mongolia	41	58	99	2	X	84	82	17
Cat, little spotted {d,e}	V		C. and S. America	5	8	6	0	73	27	0	0
Chamois, Abruzzo	V		Italy	1	3	6	8	23	77	3	X
Cheetah {d,e,f}	V	>1,500	Africa, M.E., Iran, U.S.S.R.	112	235	252	3	42	58	30	35
Chimpanzee, central and eastern {e}	V	<163,000	Equatorial Africa	86	946	1,024	10	18	53	51	64
Chimpanzee, pygmy {d,e}	V	~13,000	Zaire	11	27	32	0	34	66	5	1
Civet, Malagasy	V		Madagascar	1	1	0	0	0	100	X	0
Deer, Calamian	V		Philippines	2	6	4	0	50	50	0	X
Deer, marsh	V		C. and S. America	3	3	5	0	62	38	2	X
Desman, Russian	V		U.S.S.R.	1	1	1	0	X	X	X	X

Table 20.5

	IUCN Status{b}	Number in Wild	Native to	Zoo Census 1986 (1988){a}							
				Number of Zoos	Number of Captive Animals			Origins (percent)		Captive Births{c}	
					Male	Female	Not Known	Wild	Captive	1985	1988
Dhole {e}	V		Asia	11	26	18	0	20	80	6	0
Dog, African wild	V		Africa	55	129	128	8	0	100	34	19
Dog, bush {d}	V		S. America	14	27	39	0	0	100	17	8
Dolphin, Amazon River (Boto)	V		South America	1	1	0	0	100	0	X	0
Dugong	V		Indian and Pacific Oceans	1	1	1	0	X	X	X	X
Duiker, zebra	V		Cote d'Ivoire, Liberia, Sierra Leone	3	4	7	0	55	45	6	0
Echidna, long-beaked	V	~300,000	New Guinea	3	3	3	0	X	X	X	0
Elephant, African	V	~600,000	Africa	76	25	167	0	82	0	2	0
Fox, Mauritian flying	V		Mauritius	1	3	3	0	X	X	2	X
Gaur {d}	V	~1,000	S.E. Asia	22	75	98	0	0	100	173	32
Gazelle, Dama	V		N. Africa	20	44	97	0	0	100	141	34
Gazelle, Dorcas	V		N. Africa, M. East	6	37	59	0	0	100	3	6
Gazelle, mountain	V		Arabia, Middle East	6	39	52	0	X	X	91	1
Gazelle, red-fronted	V		Senegal and Ethiopia	1	5	9	0	36	64	4	X
Gazelle, Speke's	V		Somalia, Ethiopia	6	26	54	0	0	100	80	5
Gibbon, black {d,e}	V		China, S.E. Asia	17	25	24	1	62	38	4	4
Gibbon, Hoolock	V		India, Myanmar	1	0	1	0	100	0	X	0
Gibbon, pileated	V		S.E. Asia, Thailand	14	24	23	0	70	30	2	0
Gorilla, western lowland {d,e}	V	35,000-40,000	Africa	111	221	277	1	59	41	X	1
Guenon, owl-faced (Monkey)	V	Unknown	Rwanda, Uganda, Zaire	10	20	22	7	33	67	7	1
Hippopotamus, pygmy {d,e}	V		W. Africa	69	70	121	0	24	76	15	3
Hyaena, brown {d}	V		S. Africa	14	19	13	0	56	41	X	0
Jaguar {e}	V		Americas	72	95	100	7	X	X	69	12
Lemur, black {e}	V	~200	Madagascar	26	88	86	0	0	100	11	18
Lemur, mongoose {e}	V	~100	Madagascar, Comoros	21	46	34	0	0	100	3	10
Leopard, clouded {d,e}	V		Asia	49	83	77	0	22	78	9	13
Loris, pygmy	V		S.E. Asia	7	21	19	2	47	45	X	5
Macaque, Barbary	V	12,000-23,000	North Africa	39	438	605	54	0	100	174	3
Manatee, Amazonian	V		Brazil	2	2	1	0	X	X	X	X
Manatee, West Indian	V		N., C., and S. America	4	7	8	0	87	13	X	X
Mangabey, collared	V	Unknown	West Africa	19	92	149	0	25	75	38	X
Mangabey, crested	V		Kenya, Tanzania	6	11	14	1	19	28	3	
Mangabey, sooty {d}	V	Unknown	W. Africa	27	62	106	0	19	81	X	0
Mangabey, white-crowned	V		W. Africa	5	9	16	0	76	24	X	2
Margay	V		C. & S. America	34	42	39	0	49	51	2	4
Markhor	V		W. Himalaya	18	56	77	0	0	100	27	X
Marmoset, tassel-eared	V		Brazil	1	1	1	0	X	X	X	0
Monkey, black spider	V		South America	5	64	113	15	60	40	9	3
Monkey, Diana {d}	V	Unknown	West Africa	63	83	103	5	24	76	15	5
Monkey, Geoffroy's spider	V	~20,000	Central America, Mexico	58	94	202	0	X	X	39	16
Monkey, long-haired spider	V		South America	10	6	12	0	62	28	3	0
Monkey, L'hoest's	V	Unknown	Rwanda, Burundi	2	2	3	2	71	29	4	0
Monkey, proboscis	V		Indonesia	5	6	11	0	29	71	X	2
Monkey, Sichuan golden snub-nosed {d}	V		China	16	22	24	0	59	41	X	0
Monkey, woolly	V		South America	29	37	51	0	68	32	3	5
Ocelot {e}	V		N., C. & S. America	30	35	55	0	X	X	22	6
Otter, European {d,e}	V		Europe, Asia	30	54	66	0	37	63	17	3
Otter, giant	V		South America	6	7	4	0	X	X	X	X
Otter, La Plata	V		South America	3	3	2	0	X	X	X	X
Possum, Leadbeater's {e}	V		Australia	3	29	36	1	36	64	18	0
Saki, white-nosed	V		Brazil	1	1	0	0	X	X	X	0
Serow, Formosan	V		Taiwan	2	8	5	0	61	39	2	X
Sifaka, diademed (golden-crowned)	V		Madagascar	1	1	2	0	33	67	X	1
Tahr, Nilgiri	V		India	5	17	26	2	0	100	45	5
Takin, golden	V		China	3	3	3	0	0	100	X	0
Tapir, Central American {d}	V		C. and N.W. South America	5	9	7	0	44	56	1	2
Tapir, mountain	V		N.W. South America	2	5	3	0	37	63	0	1
Tree-kangaroo, Doria's {e}	V		New Guinea	5	5	7	1	87	13	X	0
Tree-kangaroo, Goodfellow's {e}	V		New Guinea	11	11	16	16	50	50	1	1
Uakari, black-headed	V		Brazil, Colombia, Venezuela	2	2	1	0	X	X	X	0
Uakari, red and white	V		Brazil, Colombia, Peru	1	1	1	0	X	X	X	0
Vicuna {d}	V	~80,000	S. America	18	53	44	0	0	100	8	1
Waterbuck, Lechwe {d}	V		S. Africa	20	136	247	2	0	100	49	50
Wolf, grey	V		North Am., M. East, Eurasia	56	159	149	4	X	X	164	32
Wolf, maned {d,e}	V		S. America	51	104	93	7	18	82	29	10
Zebra, Hartmann's mountain {d,e}	V		Angola, Namibia	22	33	84	0	18	82	16	19
Antelope, Hunter's	R		Kenya, Somalia, Tanzania	3	7	4	0	27	73	X	0
Baboon, Gelada	R	<880,000	Ethiopia	25	41	92	1	0	100	13	3
Baboon, Hamadryas	R	Unknown	Ethiopia, Somalia, Saudi Arabia	19	84	119	17	4	87	200	22
Bettong, burrowing	R		Australia	X	X	X	X	X	X	X	0
Deer, Kuhl's	R		Indonesia	2	9	14	10	85	15	293	X
Deer, Bawean	R		Indonesia	2	9	14	10	X	X	3	X
Lemur, Coquerel's dwarf {e}	R		Madagascar	1	10	12	0	18	82	7	9
Marmoset, Goeldi's {d}	R		South America	33	124	118	8	0	100	60	52
Monkey, golden leaf	R		Bhutan, India	2	4	4	0	75	25	X	X
Panda, giant {d,e}	R	~1,000	China	6	X	X	17	89	11	0	0
BIRDS											
Rail, Guam {e}	Ex	0	Oceania	12	36	34	13	X	X	18	13
Amazon, St. Lucia	E	~645	St. Lucia	3	5	5	5	93	7	1	5
Condor, California {e}	E	0	United States	7	31	29	8	X	X	1	0
Crane, whooping	E	~150	Canada, United States	3	22	21	14	35	65	7	5
Eared-pheasant, brown	E		China	5	8	8	0	6	87	25	X
Kestrel, Mauritius	E	~50	Mauritius	X	X	X	29	X	X	16	X
Pheasant, cheer	E		India, Nepal, Pakistan	4	11	7	4	X	90	47	6

Table 20.5   Rare Species of Animals in Zoos, 1980s

(continued)

	IUCN Status{b}	Number in Wild	Native to	Number of Zoos	Male	Female	Not Known	Wild	Captive	1985	1988
				Zoo Census 1986 (1988){a}		Number of Captive Animals	Not	Origins (percent)		Captive Births{c}	
Pheasant, Elliot's	E		China	9	12	10	6	X	92	77	14
Pigeon, Mauritius pink {d}	E	<20	Mauritius	7	33	20	3	11	89	47	X
Barbet, toucan	V		Colombia, Ecuador	X	X	X	X	X	X	1	0
Conure, golden	V		Brazil	28	53	47	54	56	44	25	2
Crane, hooded {d,e}	V	7,000	Asia, U.S.S.R.	23	28	36	12	55	45	8	5
Crane, red-crowned {d,e,g}	V	1,400	Asia, U.S.S.R.	61	103	112	46	24	76	29	6
Crane, Siberian {d}	V	2,700	Asia, U.S.S.R.	7	26	13	3	50	50	3	0
Curassow, blue-billed	V		Colombia	2	8	12	0	16	58	1	0
Duck, white-winged	V	~30	India, S.E. Asia	13	59	56	47	0	100	59	22
Eared-pheasant, white {d}	V		Myanmar, China, India	14	27	27	13	X	X	28	13
Falcon, peregrine	V	>9,000	Global	X	X	X	X	X	X	3	360
Fish-eagle, white-tailed {e}	V	>1,200	Palearctic	56	44	57	40	85	15	4	1
Goose, Hawaiian (Nene)	V	~425	United States	49	114	106	15	3	89	184	37
Goose, ruddy-headed	V	<1,000	X	13	13	15	0	1	96	16	1
Parrot, thick-billed {e}	V		Mexico	18	33	29	9	46	54	4	11
Peacock-pheasant, Palawan	V		Philippines	22	38	28	19	3	92	24	28
Peafowl, green {e}	V	~250	S. Asia	7	9	6	1	0	100	24	4
Pelican, Dalmatian	V	>1,300	W. Palearctic	4	2	1	5	25	75	3	0
Pheasant, Bulwer's	V		Indonesia, Malaysia	4	8	10	0	0	100	7	0
Pheasant, Edwards's	V		Viet Nam	8	10	5	0	X	66	22	1
Pheasant, Mikado	V	~5,000	Taiwan	5	8	10	0	X	94	73	10
Pheasant, Swinhoe's	V	~5,000	Taiwan	16	16	19	1	X	72	141	9
Teal, New Zealand brown	V	>2,200	New Zealand	7	31	29	8	0	100	20	X
Whistling-duck, West Indian	V		Caribbean Islands	13	11	18	3	X	78	159	1
Eagle, harpy	R		S. and C. America	17	20	20	0	100	0	1	9
Lorikeet, Tahiti	R	>1,500	French Polynesia	1	3	6	2	15	81	1	1
Moorhen, Gough	R		United Kingdom	X	X	X	X	X	X	7	X
Parakeet, scarlet-chested	R		Australia	14	17	22	6	X	95	85	5
Parrot, golden-shouldered	R	~250	Australia	3	16	6	0	5	95	9	9
REPTILES											
Alligator, Chinese {d,e,g}	E	~300	China	24	X	X	122	47	53	14	0
Boa, Puerto Rican	E		Puerto Rico	12	30	26	7	35	65	11	14
Boa, Round Island keel-scaled	E		Round Island	1	3	6	8	53	47	8	7
Cobra, central Asian	E		C. Asia	19	13	13	24	84	16	5	0
Crocodile, Cuban	E	~1.000	Cuba	23	13	27	43	50	50	18	1
Crocodile, estuarine	E	~27,000	Asia, Australia, W. Pacific	3	1	2	5	25	12	51	0
Crocodile, Morelet's	E	>2,000	C. America	7	9	10	49	29	71	7	0
Crocodile, Siamese	E	~200	S.E. Asia	4	6	4	2	50	16	3	0
Gecko, Gunther's	E		Round Island	5	12	85	15	8	92	19	0
Gharial	E	~270	S.E. Asia	3	5	10	0	X	73	112	0
Gharial, false	E		S.E. Asia	8	7	11	16	29	16	9	0
Boa, Jamaican	V		Jamaica	20	44	34	63	0	100	44	3
Gila monster, banded	V		Mexico, United States	7	10	4	16	63	37	1	0
Gila monster, reticulate	V		Mexico, United States	59	20	20	35	72	28	5	0
Iguana, Fiji crested	V		Fiji	5	8	6	0	64	36	1	0
Lizard, sail-finned	V		Philippines	9	2	8	7	47	41	14	2
Python, Indian	V		S. and S.E. Asia	5	2	3	1	X	83	88	0
Rattlesnake, ridge-nosed	V		Mexico, United States	3	7	4	2	46	54	9	0
Snake, Eastern indigo	V		United States	14	22	12	19	13	50	14	15
Tortoise, desert	V		Mexico, United States	11	16	17	15	52	10	2	0
Tortoise, Galapagos giant	V		Galapagos	27	66	47	47	52	48	26	0
Tortoise, Hermann's	V		Europe	5	21	21	4	49	42	10	15
Tortoise, radiated {e}	V		Madagascar	42	X	X	361	63	37	17	34
Turtle, yellow-spotted sideneck	V		Brazil	X	X	X	X	X	X	6	2
Viper, Lebetine	V		Greece	9	9	9	0	28	72	9	0
Rattlesnake, Aruba Island {d,e}	R		Aruba	8	32	35	0	13	87	13	0
Skink, Round Island	R		Round Island	3	5	9	76	0	100	26	0
Tortoise, Travancore	R		India, Indonesia	1	0	1	7	X	87	4	2
Tuatara	R	~100,000	New Zealand	6	14	13	16	63	37	9	0
Turtle, bog	R		United States	7	14	14	20	40	60	2	2
Viper, Armenian	R		Iran, Turkey, U.S.S.R.	11	9	12	8	79	21	10	0
Viper, Transcaucasian long-nosed	R		Turkey, U.S.S.R.	6	9	11	2	77	23	7	0
AMPHIBIANS											
Salamander, Texas blind	E		United States	1	0	0	11	0	100	X	X
Toad, Houston	E		United States	1	50	50	0	0	100	X	X
Frog, Goliath	V		Cameroon, Equatorial Guinea	1	0	0	3	X	X	X	X
Salamander, Japanese giant	R		Japan	20	10	11	86	63	37	2	X

Sources: International Union for Conservation of Nature and Natural Resources (IUCN, the World Conservation Union), International Species Information System (ISIS), International Zoo Yearbook, The Peregrine Fund, International Crane Foundation, International Council for Bird Preservation, and American Association of Zoological Parks and Aquariums (AAZPA).

Notes: a. Numbers shown in italics from ISIS. b. Status from the World Conservation Union (IUCN), Ex: Extinct, E: Endangered, V: Vulnerable, R: Rare.
c. Births in 1985 and 1988, less deaths within first 30 days. d. Studbook exists. e. Species Survival Plan(s) (SSP) or equivalent(s).
f. Breeding group of unknown size at Pretoria, South Africa. g. Unknown numbers in Chinese zoos.
0 = zero or less than half the unit of measure; X = not available. For additional information, see Sources and Technical Notes.

Sources and Technical Notes

Table 20.1 National and International Protection of Natural Areas, 1989

Sources: Protected Areas Data Unit of the World Conservation Monitoring Centre (WCMC), unpublished data (WCMC, Cambridge, United Kingdom, July 1988 [marine and coastal protected areas], December 1988 [all protected areas], and December 1989 [areas under "International Protection Systems."]).

National Protection Systems in Table 20.1 combine natural areas in five World Conservation Union management categories (areas are at least 1,000 hectares; access is at least partially restricted):
■ Scientific reserves and strict nature reserves possess outstanding, representative ecosystems. A reserve's size is determined by the area required to ensure the integrity of the site. In many reserves, natural perturbations (e.g., insect epidemics and forest fires) are allowed.
■ National parks and provincial parks are relatively large areas of national or international significance not materially altered by humans. Visitors may use them for recreation and study.
■ Natural monuments and natural landmarks contain unique geological formations, special animals or plants, or unusual habitat.
■ Managed nature reserves and wildlife sanctuaries are protected for specific purposes, such as conservation of a significant plant or animal species. Some areas require management.
■ Protected landscapes and seascapes may be entirely natural or may include cultural landscapes (e.g., scenically attractive agricultural areas). Examples would include coastlines, lake shores, and hilly or mountainous terrain along scenic highways.

Marine and coastal protected areas only refer to all protected areas with littoral, coral, island, marine, or estuarine components. The area given is the whole protected area.

The figures in Table 20.1 do not include locally or provincially protected sites, privately owned areas, or those where consumptive uses of wildlife are permitted. National lists usually include sites that are listed under *International Protection Systems.*

Biosphere Reserves are representative of terrestrial and coastal environments that have been internationally recognized under the Man and the Biosphere Programme of the United Nations Educational, Scientific, and Cultural Organization. They were selected for their value to conservation and are intended to foster the scientific knowledge, skills, and human values to support sustainable development. Each reserve must contain a diverse, natural ecosystem of a specific biogeographical province, large enough to be an effective conservation unit. For further details, refer to M. Udvardy, *A Classification of the Biogeographical Provinces of the World* (IUCN, Morges, Switzerland, 1975), and to *World Resources 1986*, Chapter 6. Each reserve also must include a minimally disturbed core area for conservation and research and may be surrounded by buffer zones where traditional land uses, experimental ecosystem research, and ecosystem rehabilitation may be permitted.

Natural World Heritage Sites are areas of "outstanding universal value." (A World Heritage Site can include structures, but the table includes only "natural" sites.) Any party to the World Heritage Convention may nominate sites that contain examples of a major stage of the earth's evolutionary history; a significant ongoing geological process; a unique or superlative natural phenomenon, formation, or feature; or habitat for threatened species.

Any party to the Convention on Wetlands of International Importance Especially as Waterfowl Habitat (Ramsar, Iran, 1971), who agrees to respect the site's integrity and to establish wetland reserves, can designate *Wetlands of International Importance.*

Because categories overlap, the total number of protected sites is less than the sum of all the categories. Sites in small countries not in the table are included in continental and world totals (i.e., Bahamas, Antarctica, Greenland, the Netherlands Antilles, Taiwan, and Namibia). The United States and Canada share one World Heritage Site, as do Guinea and Côte d'Ivoire. These sites are counted only once in continental and world totals.

Table 20.2 Globally Threatened Animal Species, 1989

Sources: International Union for Conservation of Nature and Natural Resources (IUCN), *The IUCN Mammal Red Data Book, Part I* (IUCN, Gland, Switzerland, 1982); International Council for Bird Preservation (ICBP)/IUCN, *Threatened Birds of Africa and Related Islands* (ICBP/IUCN, Cambridge, United Kingdom, 1985); IUCN, *Threatened Swallowtail Butterflies of the World* (IUCN, Gland, Switzerland, 1985); World Conservation Monitoring Centre (WCMC), series of reports on *Conservation of Biological Diversity* (for some data on Botswana, Côte d'Ivoire, Ethiopia, Guinea-Bissau, Kenya, Nigeria, Senegal, India, Myanmar, and the Philippines) (WCMC, Cambridge, 1988 and 1989); unpublished data (WCMC, Cambridge, United Kingdom, 1988).

Africa (bird species known): Jeffery A. Sayer, Simon Stuart, IUCN, *Environmental Conservation*, Vol. 15, No. 3 (Autumn 1988); Madagascar: IUCN, *Madagascar: An Environmental Profile* (IUCN, Cambridge, 1987); Mexico: Conservation International (CI), *Mexico's Living Endowment: An Overview of Biological Diversity* (CI, n.p., April 1989); Panama (bird species known): James R. Karr, Acting Director, Smithsonian Tropical Research Institute, September 25, 1987 (personal communication); Canada, United States including Caribbean and Pacific islands, Europe, Turkey, Japan, and New Zealand (all species known and threatened), and Australia (birds, reptile, and amphibian species known): Organisation for Economic Co-operation and Development (OECD), *OECD Environmental Data Compendium 1989* (OECD, Paris, 1989); Viet Nam: Vo Quy, "Viet Nam's Ecological Situation Today," *ESCAP Environment News*, Vol. 6, No. 4 (October-December 1988); Czechoslovakia (bird species known): *CSSR Red Data Book*, 1988; U.S.S.R.: A.V. Yablokov, Ostroumov, *Okhrana Zhivoi Prirody* (*The Conservation of Living Nature*), Moscow, 1983.

The World Conservation Union classifies threatened and endangered species in six categories:
■ Endangered. "Taxa in danger of extinction and whose survival is unlikely if the causal factors continue operating."
■ Vulnerable. "Taxa believed likely to move into the Endangered category in the near future if the causal factors continue operating."
■ Rare. "Taxa with world populations that are not at present Endangered or Vulnerable, but are at risk."
■ Indeterminate. "Taxa known to be Endangered, Vulnerable, or Rare but where there is not enough information to say which of the three categories is appropriate."
■ Out of Danger. "Taxa formerly included in one of the above categories, but which are now considered relatively secure because effective conservation measures have been taken or the previous threat to their survival has been removed."
■ Insufficiently Known. "Taxa that are suspected but not definitely known to belong to any of the above categories."

The number of threatened species listed for most countries includes species that are endangered, vulnerable, rare, indeterminate, and insufficiently known, but excludes introduced species. The total number of species includes introductions. The data on mammals exclude cetaceans (whales and porpoises).

Table 20.3 CITES-Reported Trade in Wildlife and Wildlife Products, 1980s

Sources: World Conservation Monitoring Centre (WCMC), unpublished data (WCMC, Cambridge, United Kingdom, August 1989); "Interpretation and Implementation of the Convention: Trade in Ivory from African Elephants," report from the Seventh Meeting of the Conference of the Parties to the Convention on International Trade in Endangered Species of Wild Fauna and Flora (CITES), Lausanne, Switzerland, October 9–20, 1989.

CITES members agree to prohibit commercial international trade in endangered species and to closely monitor trade in species that may become depleted by trade. Species are listed in the appendixes to CITES based on the degree of rarity and of threat from trade. Trade is prohibited for about 675 species in Appendix I and is regulated for at least 27,000 species in Appendix II. Appendix III (seldom used) allows countries to prohibit trade in nationally threatened species. Parties to the Convention are required to submit annual reports, including trade records, to the CITES-United Nations Environment Programme Secretariat in Switzerland. WCMC compiles these data from those reports. Figures refer primarily to legal trade, though illegal trade is included when known.

The *CITES reporting requirement met* column refers to the percentage of years for which a country has submitted an annual report to the CITES Secretariat since it became a party to the Convention, through 1986. Countries that had ratified the CITES Convention by June 30, 1989, are listed as members of CITES.

Live primates (1986) include all captive-bred and wild-caught specimens of all non-human primate species.

Cat skins (1986) include skins of all species of Felidae, excluding a small number of skins reported only by weight or length.

Raw ivory (1988) refers to trade in African elephant ivory, reported by weight. Trade from Singapore comes from a CITES-registered stockpile.

Live parrots (1986) include captive-bred and wild-caught individuals of all psittacine species (parrots, macaws, cockatoos, etc.) except the budgerigar and the cockatiel.

Reptile skins (1986) include whole skins, reported by number, of all crocodilians and many commonly traded lizard and snake species. About 56 percent of the global reptile skins total is lizards, 26 percent is snakes, and 18 percent is crocodilians.

This table shows gross trade. The totals generally overestimate the actual number of specimens traded because the same specimen could be imported and reexported by a number of countries in a single year. However, the impact of international trade on a particular species can be greater than the numbers reported because of mortality (during capture, transit, and quarantine), illegal trade, trade to or from countries that are not CITES members, and omission of domestic trade data.

Table 20.4 Habitat Loss, 1980s

Sources: Except as noted, data are from John T. and Kathy MacKinnon, *Review of the Protected Areas in the Afrotropical Realm* and *Review of the Protected Areas System in the Indo-Malayan Realm* (International Union for Conservation of Nature and Natural Resources (IUCN), Gland, Switzerland, 1986).

All forests: Africa, "potential forest" and extent remaining: K.J. Gregory and D.E. Walling, eds., *Human Activity and Environmental Processes* (John Wiley & Sons, Chichester, United Kingdom, 1987); Canada, "forest regions," 48 percent of total land: D.F.W. Pollard and M.R. McKechnie, *World Conservation Strategy-Canada* (Environment Canada, Ottawa, 1986); Guatemala: "The International News," *Nature Conservancy Magazine*, Vol. 39, No. 3 (1989); Jamaica and Mexico: Robert C. West and John P. Augelli, *Middle America: Its Lands and Peoples* (Prentice-Hall, Englewood Cliffs, N.J., 1966); United States, contiguous states: Michael Williams, "The Death and Rebirth of the American Forest: Clearing and Reversion in the United States," in John F. Richards and Richard P. Tucker, eds., *World Deforestation in the 20th Century* (Duke University Press, Durham, N.C., 1988); Argentina: Instituto Forestal Naccional, *Argentina Forestal. Sintesis de la situacion actual, politica y proyecciones* (IFONES, Corrientes, Argentina, 1988); Guyana, "forested portion": Derek A. Scott and Montserrat Carbonell, compilers, *A Directory of Neotropical Wetlands* (IUCN, World Conservation Monitoring Centre (WCMC), and International Waterfowl Research Bureau, Cambridge and Slimbridge, United Kingdom, 1986); China: Kenneth Ruddle and Wu Chuanjun, eds., *Land Resources of the People's Republic of China* (United Nations University, Tokyo, 1983); Cyprus, "forest cover" above 1,000 feet, and Greece: J.V. Thirgood, *Man and the Mediterranean Forest. A History of Resource Depletion* (Academic Press, London, 1981); Saudi Arabia: A.A. Abol El Rahman, "The Deserts of the Arabian Peninsula," in Michael Evenari, Imanuel Noy Muir, and David Goodall, eds., *Ecosystems of the World. 12A. Hot Deserts and Arid Shrublands* (Elsevier, Amsterdam, 1985); U.S.S.R., European portion: J.T. Richards, "World environmental history and economic development," in William C. Clark and R.E. Munn, eds., *Sustainable Development of the Biosphere* (Cambridge University Press, Cambridge, United Kingdom, 1986); Australia: J.D. Ovington, ed., *Ecosystems of the World. 10. Temperate Broad-Leaved Evergreen Forests* (Elsevier, Amsterdam, 1983); New Zealand, "native bush": C.P. Glasby, "Modification of the Environment in New Zealand," *Ambio*, Vol. XV, No. 5 (1986); Papua New Guinea: Norman Myers, "Threatened Biotas: 'Hot

Spots' in Tropical Forests," *The Environmentalist*, Vol. 8, No. 3 (1988).

Moist Forests: Western Ecuador, lowland forest: Norman Myers, "Threatened Biotas: 'Hot Spots' in Tropical Forests," *The Environmentalist*, Vol. 8, No. 3 (1988); Papua New Guinea: IUCN/United Nations Environment Programme (UNEP), *Review of the Protected Areas System in Oceania* (IUCN, Gland, Switzerland, 1986).

Savanna/grassland: World, Africa, South America, Asia (India and Southeast Asia, percent of land surface): Monica M. Cole, *The Savannas. Biogeography and Geobotany* (Academic Press, London, 1986); Canada, grassland, 3 percent of total land: D.F.W. Pollard and M.R. McKechnie, *World Conservation Strategy—Canada* (Environment Canada, Ottawa, 1986); United States: Edward C. Wolf, *On The Brink of Extinction: Conserving the Diversity of Life*, Worldwatch Paper 78 (Worldwatch Institute, Washington D.C., June 1987); China, "current grazing area": Kenneth Ruddle and Wu Chuanjun, eds., *Land Resources of the People's Republic of China* (United Nations University, Tokyo, 1983); Saudi Arabia, "natural grassland": A.A. Abol El Rahman, "The Deserts of the Arabian Peninsula," in Michael Evenari, Imanuel Noy Muir, and David Goodall, eds., *Ecosystems of the World. 12A. Hot Deserts and Arid Shrublands* (Elsevier, Amsterdam, 1985); Australia: J. Walker and A.N. Gillison, "Australian Savannas," in B.J. Huntley and B.H. Walker, eds., *Ecology of Tropical Savannas* (Springer-Verlag, Berlin, 1982); New Zealand, South Island only: J.T. Richards, "World Environmental History and Economic Development," in William C. Clark and R.E. Munn, eds., *Sustainable Development of the Biosphere* (Cambridge University Press, Cambridge, United Kingdom, 1986); Papua New Guinea: IUCN/UNEP, *Review of the Protected Areas System in Oceania* (IUCN, Gland, Switzerland, 1986).

Desert/scrub: Egypt, western coastal desert: Emily E. Whitehead *et al.*, eds., *Arid Lands Today and Tomorrow*. Proceedings of an International Research and Development Conference (Westview Press, Boulder, Colorado, 1988); Mexico, arid and semi-arid lands in Sonoran, Chihuahuan, Quereton, Hidalguense, Poblan, Guerrerense, Oaxacan, and Yucatec zones: F. Medellin-Leal and A. Gomez-Gonzalez, "Management of Natural Vegetation in the Semi-Arid Ecosystems of Mexico," in B.H. Walker, ed., *Management of Semi-Arid Ecosystems* (Elsevier, Amsterdam, 1979); Argentina, southeastern coastal desert: Daniel H.K. Amiran and Andrew W. Wilson, eds., *Coastal Deserts. Their Natural and Human Environments* (University of Arizona, Tucson, 1973); China, current extent of "desert, snow, rocks": Kenneth Ruddle and Wu Chuanjun, *Land Resources of the People's Republic of China* (United Nations University, Tokyo, 1983); Saudi Arabia: A.A. Abol El Rahman, "The Deserts of the Arabian Peninsula," in Michael Evenari, Imanuel Noy Muir, and David Goodall, eds., *Ecosystems of the World. 12A. Hot Deserts and Arid Shrublands* (Elsevier, Amsterdam, 1985); U.S.S.R.: M.P. Petrov, "Land Use of Semi-Desert in the U.S.S.R.," in B.H. Walker, ed., *Management of Semi-Arid Ecosystems* (Elsevier, Amsterdam, 1979); Australia: A.D. Wilson and R.D. Graetz, "Management of the Semi-Arid and Arid Rangelands of Australia," in B.H. Walker, ed., 1979, *Management of Semi-Arid Ecosystems* (Elsevier, Amsterdam, 1979).

Wetlands/marsh: World: John Zelazny and J. Scott Feieraband, eds., *National Wildlife Federation, Corporate Conservation Council, Proceedings. Wetlands. Increasing Our Wetland Resources* (National Wildlife Federation, Washington, D.C., 1988); Burundi, Congo, Côte d'Ivoire, Liberia, Madagascar, Rwanda, Senegal, Uganda, Guyana, China, Japan, Democratic People's Republic of Korea (all numbers are for peatlands, 1984): Edward Maltby, *Waterlogged Wealth. Why Waste the World's Wet Places?* (International Institute for Environment and Development, London, 1986); Tanzania (Malagarasi, Kilombero, and Rufiji flood plans and swamps), Mali (Niger Central Delta), Sudan (Mochan Swamps, Sudd, Kenamuke Swamp), Ghana (Volta River): Royal Tropical Institute (RTI), Rural Development Program, *A Resource Planning Data Review on Major African Inland Swamp and Flood Plain Ecosystems* (RTI, Amsterdam, 1987); North and Central America except Canada, Palearctic Mexico, and United States, South America except Guyana: Derek A. Scott and Montserrat Carbonell, compilers, *A Directory of Neotropical Wetlands* (IUCN/CMC and International Waterfowl Research Bureau, Cambridge and Slimbridge, United Kingdom, 1986); Canada: National Wetlands Working Group, Canada Committee on Ecological Land Classification, *Wetlands of Canada.* Ecological Land Classification Series No. 24 (Polyscience Publications Inc., n.p., 1988); United States: U.S. Congress, Office of Technology Assessment, *Technologies to Maintain Biological Diversity* (U.S. Government Printing Office, Washington, D.C., 1987); Bhutan, Republic of Korea, Mongolia, Nepal, Philippines, Singapore, Sri Lanka: Derek A. Scott, compiler, *A Directory of Asian Wetlands* (IUCN, Gland, Switzerland, 1989); Japan: Zbigniew Karpowicz, *Wetlands in East Asia—A Preliminary Review and Inventory* (International Council for Bird Preservation, July 1985); wetlands of international importance in Afghanistan, Albania, Algeria, Cyprus, Egypt, European countries, Iran, Iraq, Israel, Jordan, Malta, Morocco, Syria, Tunisia, and Turkey: Erik Carp, compiler, *A Directory of Western Palearctic Wetlands* (UNEP, Nairobi, and IUCN, Gland, Switzerland, 1980); Australia, southern portion: J.F. Richards, "World Environmental History and Economic Development," in William C. Clark and R.E. Munn, eds., *Sustainable Development of the Biosphere* (Cambridge University Press, Cambridge,

United Kingdom, 1986); New Zealand: G.P. Glasby, "Modification of the Environment in New Zealand," *Ambio*, Vol. 15, No. 5 (1986); Papua New Guinea: IUCN/UNEP, *Review of the Protected Areas System in Oceania* (IUCN, Gland, Switzerland, 1986).

Mangroves: North and Central American countries, Colombia, Guyana, Peru, Suriname, Japan, United Arab Emirates, Australia, Fiji, New Zealand: Working Group on Mangrove Ecosystems of the IUCN Commission on Ecology, in cooperation with UNEP and the World Wildlife Fund, *Global Status of Mangrove Ecosystems*, Committee on Ecology Papers No. 3 (IUCN, Gland, Switzerland, 1983); Singapore: Miguel D. Fortes, "Mangrove and Seagrass Beds of East Asia: Habitats Under Stress," *Ambio*, Vol. 17, No. 3 (1988); Papua New Guinea: IUCN/UNEP, *Review of the Protected Areas System in Oceania* (IUCN, Gland, Switzerland, 1986).

Classifications of habitat within biogeographical divisions and vegetation mapping are inexact arts. Several different maps exist of biogeographical divisions. The MacKinnons followed the United Nations Educational, Scientific, and Cultural Organization/AETFAT/United Nations Statistical Organization vegetation map of Africa with classifications by F. White for the Afrotropical Realm. For the Indomalayan Realm, they generally followed the classifications of Miklos Udvardy, the maps by T.C. Whitmore for the Malesian section and Indo-China, and the maps by H.G. Champion and H.K. Seth for the Indian subcontinent. Please see the sources for additional details.

The MacKinnons relied on field investigations, interviews and other personal communications, and published sources. In a few cases (e.g., for China), other sources supplement their data. The categories shown in Table 20.4 are aggregated from the MacKinnons' data. *AFRICA: all forests:* areas under the forest categories below; *dry forests:* upland montane forest/nonforest, dry forest, woodland; *moist forests:* lowland rainforest; *savanna/grassland:* salt-pan vegetation, brushland/thicket, shrubland, grassland, halophytic; *desert/scrub:* desert; *wetlands:* wetland; *mangroves:* mangrove forest/swamp. *ASIA: dry forests:* sub-alpine, dry dipterocarp, mixed deciduous, submontane dry evergreen, forest on limestone, Himalayan dry temperate, subtropical dry evergreen, tropical dry deciduous; *moist forests:* subtropical broadleaved hill, ironwood, lowland rain, tropical moist deciduous, montane wet temperate, tropical semievergreen, subtropical pine, tropical montane evergreen, moist lowland (TL), tropical dry evergreen, heath, monsoon, tropical montane deciduous, forest on ultrabasic; *savanna/grassland:* savanna forest; *desert/scrub:* tropical thorn forest, desert/semi-desert, tropical thorn scrub; *wetlands:* freshwater swamp, peat swamp, seasonal marsh/seasonal salt marsh; *mangroves:* mangrove.

Most other sources use similar nomenclature. In some cases, only a portion of a

given vegetation type (e.g., peatlands, as opposed to wetlands) may be discussed. Some data were not used because they could not be disaggregated to specific countries. The determination of the extent of vegetation types in a country is difficult, and estimates vary significantly.

Some data on current extent of habitat may include restorations, although the vegetation may differ significantly from the original. In addition, the table cannot distinguish pristine habitats from those that are significantly degraded. Though the information in this table is as complete as possible, much is missing; and the data must be considered to be preliminary.

■ *Forests*: estimates for "forests," not further defined, in this category, and aggregated MacKinnon forest data.

■ *Dry forests* and *moist forests*: forest types in the MacKinnon studies and other forests, where the correct category is known. This distinction usually applies to tropical forests.

■ *Savanna/grassland:* excludes areas whose original vegetation is known to have been other than grassland (e.g., cut forests, irrigated desert, drained wetlands).

■ *Wetlands/marsh:* Many estimates of the extent of wetlands are likely to be low. Some include only peatlands; others exclude peatlands, whose inclusion could increase total wetlands, especially in Finland, Ireland, Norway, Poland, and Sweden. Some wetland figures may include lakes, ponds, streams, and areas that are periodically flooded; others note only permanently inundated areas.

■ *Mangroves*: These estimates refer specifically to mangrove forests or swamps.

For additional information, please refer to the sources cited for this table, to other tables in this chapter (especially Table 20.1), and to Chapter 8, "Wildlife and Habitat." Readers are invited to submit data to expand or improve this table.

Table 20.5 Rare Species of Animals in Zoos, 1980s

Sources: Species, status, distribution: International Union for Conservation of Nature and Natural Resources (IUCN), *1988 IUCN Red List of Threatened Animals* (IUCN, Gland, Switzerland, and Cambridge, United Kingdom, 1988). Studbook information: P.J.S. Olney, Pat Ellis, and Benedicte Sommerfelt, eds., *International Zoo Yearbook 27* (The Zoological Society of London, London, 1988). Common names in the list of animals are shown as they appear in the *IUCN Red List*.

Number in wild (mammals): African primates: IUCN, *Threatened Primates of Africa: The IUCN Red Data Book* (IUCN, Gland, Switzerland, 1988); long-beaked echidna, brush-tailed bettong, ghost bat, golden lion tamarin, golden-headed lion tamarin, golden-rumped lion tamarin, red wolf, polar bear, Florida cougar, vicuña, buffy-headed marmoset, Hawaiian monk seal, Chinese alligator, Morelet's crocodile, estuarine crocodile, Cuban crocodile, Siamese crocodile, gharial, tuatara: IUCN, *Mammal Red Data Book, Part I* (IUCN, Gland, Switzerland, 1982); aye-aye, black and mongoose lemurs: Captive Breeding Specialist Group, Species Survival Commission, IUCN, *annual meeting*, Stuttgart, B.R.D., September 1988; tiger (all five species, upper end of range): *AAZPA Newsletter*, Vol. 30, No. 1 (January 1989); lion-tailed macaque: D.A. Lindberg, A.M. Lyles, and N.M. Czekala, "Status and Reproductive Potential of Lion-Tailed Macaque in Captivity," *Zoo Biology*, Supp. 1, 1989; cheetah: Laurie Marker and Stephen J. O'Brien, "Captive Breeding of the Cheetah (*Acinonyx jubatus*) in North American Zoos (1871-1986)," *Zoo Biology*, Vol. 8, No. 1 (1989); buffy-headed marmoset: Russell Mittermeier, *et al.*, "Conservation of primates in the Atlantic forest region of eastern Brazil," *International Zoo Yearbook*, Vol. 22 (Royal Zoological Society, London, 1983); snow leopard: Seneca Zoo Society, *Zoonewsletter* (September 1989); gaur: S.M. Junior *et al.*, "Techniques for Collection and Cryopreservation of Gaur (*Bos Gaurus*) Semen," *AAZPA 1989 Regional Proceedings*; Asiatic lion: S.J. O'Brien *et al.*, "Evidence for African Origins of Founders of the Asiatic Lion Species Survival Plan," *Zoo Biology*, Vol. 6, No. 2 (1987); black, Sumatran, and Indian rhinoceros: Jeffry O. Cohn, "Halting the Rhino's Demise," *Bio Science*, Vol. 38, No. 11 (1988); giant panda: *Discover*, Vol. 10, No. 9 (September 1989); African elephant: World Wildlife Fund, TRAFFIC (U.S.A.), Vol. 9, No. 2 (June 1989); scimitar-horned oryx: IUCN, World Conservation Monitoring Centre (CMC), unpublished data (CMC, Cambridge, United Kingdom, December 1988); banteng: *Tiger Paper*, December 1988; European bison, *AAZPA Newsletter*, Vol. 30, No. 10 (October 1989).

Number in wild (birds, reptiles, amphibians): Dalmatian pelican, white-winged duck, New Zealand brown teal, ruddy-headed goose, white-tailed fish-eagle, peregrine falcon (includes members of five other endangered or rare falcon subspecies), Mikado pheasant, Swinhoe's pheasant, green peafowl, Tahiti lorikeet, golden-shouldered parrot: Warren B. King, *Endangered Birds of the World—The ICBP Red Data Book* (Smithsonian Institution Press, Washington, D.C, 1981); St. Lucia Amazon: "The St. Lucia parrot *Amazona versicolor* 1975–1986: Turning the tide for

a vanishing species," *The Dodo. Journal of the Jersey Wildlife Preservation Trust*, No. 23 (1986); Mauritius pink pigeon: Diane J. Bell and John R.M. Hartley, "An Investigation into Current Captive Breeding Problems with the Pink Pigeon *Nesoenas Mayeri*," *The Dodo. Journal of the Jersey Wildlife Preservation Trust*, No. 24 (1987); Hawaiian goose: J. Michael Scott *et al.*, "Conservation of Hawaii's Vanishing Avifauna," *BioScience*, Vol. 38, No. 4 (1988); Mauritius kestrel: *The Peregrine Fund. World Center for Birds of Prey. Newsletter*, No. 17 (Spring 1989); whooping, Siberian, and red-crowned cranes: International Crane Foundation, Baraboo, Wisconsin, United States (ICF), personal communication, November 1989; hooded crane: Scott R. Swengel and George W. Archibald, "The status of crane breeding in 1986," *Proceedings of the Fifth World Conference on Breeding Endangered Species* (in press, 1989).

Number of zoos, number of captive animals (1986), *captive births* (1985): P.J.S. Olney, Pat Ellis, and Benedicte Sommerfelt, eds., *International Zoological Yearbook 27* (The Zoological Society of London, London, 1988). *Number of zoos* and *number of captive animals* (1988) of species and/or subspecies not listed in International Zoo Yearbook, and 1988 *captive births*: *ISIS Species Distribution Report. Abstract. Mammals, Birds, Reptiles, Amphibians. As of 31 December 1988.* (4 volumes) (International Species Information System, Apple Valley, Minnesota, United States, 1989)(all ISIS data are shown in italics).

Additional *number of zoos, number of captive animals*, or *captive births*: whooping, hooded, red-crowned, and Siberian cranes: ICF, personal communication, November 1989; American peregrine falcon: The Peregrine Fund (PF), Boise, Idaho, United States, personal communication, June 1989; Mauritius kestrel: PF, *The Peregrine Fund. Annual Report October 1, 1987—September 30, 1988* (PF, Boise, Idaho, United States, 1989); red-crowned and Siberian cranes: Scott R. Swengel and George W. Archibald, "The status of crane breeding in 1988," in *Proceedings of the Fifth World Conference on Breeding Endangered Species in Captivity* (in press, 1989); European bison, radiated tortoise, giant anteater, and Chinese alligator: *AAZPA Newsletter*, Vol. 30, No. 10 (October 1989).

Studbooks, usually maintained by zoos, for individual species or subspecies, pro-

vide a history of every captive animal. These data enable mate selection for genetic diversity. Species Survival Plans (SSPs, or their equivalent) are plans for the propagation and preservation of wild animals, considering both genetic diversity and demography. Each SSP attempts to bring together all of the zoos in a country or group of countries that hold the animal.

World Conservation Union Status: All species or subspecies that the World Conservation Union explicitly lists as endangered, vulnerable, or rare, and that are held in zoos are included. (See Sources and Technical Notes for Table 20.2.) If an entire species is not jeopardized, data are shown here only for those subspecies at risk.

Number in the wild: The *Mammal Red Data Book, Threatened Primates of Africa: The IUCN Red Data Book*, and *Endangered Birds of the World—The ICBP Red Data Book* contain the most complete wild population figures. Population estimates are often several years old and must be treated with extreme caution. The wild populations of some species or subspecies have increased because of release of captive-bred animals.

Area native to: The countries in which species and subspecies live in the wild appear as shown in the *Red Data Book*.

Number of zoos, number of captive animals, origins of captive populations, captive births: The number of zoos holding each animal, populations in zoos, and captive births in 1985 were taken from the *International Zoo Yearbook*, when available. Its census data are from 1986, and data on captive births from 1985. It has a wide geographic coverage, with reports from over 500 zoos. Data from the International Species Information System (ISIS) network are used for species not in the *Yearbook*. In the fall of 1989, 347 zoos (housing 106,455 live animals) made up the ISIS network. Approximately two thirds were in North America; and the remainder were in Europe, Asia, and Australasia, with a small number in South America. Figures for origins of captive population from the *Yearbook* and from ISIS are generally similar; discrepancies may be caused by the fact that data are from different years. Numbers are approximate. Numbers of *captive births* are the total births or hatchings minus deaths within the first 30 days of life. This criterion for survival is used by increasing numbers of zoos.

21. Energy, Materials, and Wastes

Energy use on a grand scale fuels modern economies and brings fundamental improvements to the lives of millions on our planet. Differences in energy use, however, contribute to disparities in industrial development among and within countries; and the profligate combustion of fossil fuels bears environmental costs, including growing concentrations of carbon dioxide, urban air pollution, and acidification of water, soils, and vegetation. (See Chapter 24, "Atmosphere and Climate.") Data in the following tables can differ in detail from those of other sources, including the British Petroleum data cited in Chapter 9, primarily because of differences in definitions.

The world's production of commercial energy (see Table 21.1) has grown by 15 percent in the past decade, declining only in 1980–82. Liquid fuels, 42 percent of all commercial production, are still the dominant commercial fuels. Although their global production dropped by 4 percent between 1977 and 1987, it increased in 1979 and again in 1984 and has been on the rise in the past two years. In contrast, gaseous fuels, the least polluting fossil fuels, showed the greatest increase in production during that period, with a temporary decline in 1982.

Globally, 1987 per capita consumption of commercial energy fell back to its 1977 level of 56 gigajoules per capita. The U.S.S.R., Africa, and Asia each witnessed a rise of over 20 percent in per capita consumption of commercial energy. Rates of growth in Africa and Asia, however, mask differences both between and within countries of these regions. In poorer countries, the impoverished depend mainly on fuelwood and other biomass for domestic heating and cooking. Detailed consumption statistics for these noncommercial fuels are not available.

Energy intensity is defined as the amount of energy consumed for each dollar of gross national product. A decline in energy intensity indicates energy efficiency improvements, or a structural shift to a less energy-intensive economy (e.g., a more service-based economy), or a combination. In North America, Japan, and most countries of the European Community—which together use about half the world's commercial energy—energy intensity has declined 15 to 25 percent in the past decade.

About 5 percent of global commercial energy is supplied by primary electricity production. (See Table 21.2.) In 1987, fossil fuel-fired plants accounted for over 60 percent of the world's total production of electricity. About 16 percent of the world's electricity in 1987 came from nuclear sources, showing an increase of 233 percent over the past decade.

The current extent of the world's energy reserves and resources (see Table 21.3) determines future energy mixes, sets priorities in development, and governs both costs and environmental impacts. Coal has the most abundant reserves of all fossil fuels. Three countries, China, the United States, and the U.S.S.R. together control more than 75 percent of the known bituminous coal reserves.

The production and consumption of metals (see Table 21.4) are central to many modern industrial processes. Both production and consumption can create a variety of environmental impacts, such as the creation of large volumes of waste; discharge of pollutants to land, air, and water; and the accumulation of certain metals in the biosphere. The United States, the U.S.S.R., Japan, and countries of the European Community consume the majority of the world's metals. Aluminum, the most abundant metal, has a life index (the ratio between 1988 world reserves and the 1988 world production level) of 224 years. But the known reserves of lead, mercury, tin, and zinc should be exhausted in less than 25 years, assuming 1988 production levels.

Roughly one sixth of the world's countries operate commercial nuclear reactors, but over half of the world's net capacity is installed in the United States, the U.S.S.R., and France. (See Table 21.5.) One of the challenges to the management of present and future nuclear power production is the safe and final disposal of spent fuel. At present, the world's spent fuel is stored in interim facilities, nearly all at the reactor site, with the exception of Finland, West Germany, and Sweden, where a centralized site away from the reactor is used. The United Kingdom leads the world in spent fuel inventories in relation to its land area.

Industrial development creates large quantities of wastes. The data on the generation of wastes in Table 21.6 must be viewed with caution, because the definitions used by countries may vary considerably. The gradual exhaustion of space for waste landfills and growing disposal costs create incentives to ship toxic materials to other countries, legally or clandestinely. Highly publicized trade in toxic wastes in recent years has led to an international convention banning all exports of hazardous materials without prior consent of the importing country. (See Chapter 25, "Policies and Institutions.") Table 21.6 shows some of the legal international trade in toxic wastes.

Table 21.1 Commercial Energy, 1977-87

| | Production (petajoules) | | | | | | | | Consumption | | | | | |
| | Total {a} | | Solid | | Liquid | | Gas | | Total | | Per Capita | | Per Constant 1980 $US of GNP | |
	1987	Change Since (%) 1977	1987	Change Since (%) 1977	1987	Change Since (%) 1977	1987	Change Since (%) 1977	(peta-joules) 1987	Change Since (%) 1977	(giga-joules) 1987	Change Since (%) 1977	(kilo-joules) 1987	Change Since (%) 1977
WORLD	294,526	15	91,091	27	123,175	(4)	66,696	39	282,924	20	56	0	X	X
AFRICA	16,989	7	4,126	88	10,859	(16)	1,833	215	7,353	68	12	20	X	X
Algeria	3,627	45	0	X	2,313	2	1,313	473	975	192	42	110	18,881	(2)
Angola	744	146	0	X	733	148	6	100	24	20	3	0	X	X
Benin	15	X	0	X	15	X	0	X	6	50	1	0	4,341	8
Botswana	X	X	X	X	X	X	X	X	X	X	X	X	X	X
Burkina Faso	0	X	0	X	0	X	0	X	6	100	1	X	3,009	21
Burundi	0	X	0	X	0	X	0	X	3	200	1	X	2,644	134
Cameroon	369	6,050	0	X	360	17,900	0	X	85	270	8	167	7,499	50
Cape Verde	0	X	0	X	0	X	0	X	0	(100)	0	(100)	0	(100)
Central African Rep	0	X	0	X	0	X	0	X	4	100	1	0	4,511	87
Chad	0	X	0	X	0	X	0	X	3	0	1	0	X	X
Comoros	0	X	0	X	0	X	0	X	1	0	2	(33)	6,293	(40)
Congo	265	249	0	X	264	252	0	X	22	1,000	12	1,100	10,327	509
Cote d'Ivoire	41	4,000	0	X	37	X	0	X	70	35	6	(14)	6,753	8
Djibouti	0	X	0	X	0	X	0	X	4	100	11	57	X	X
Egypt	2,132	94	0	X	1,941	89	169	293	991	110	20	67	32,254	23
Equatorial Guinea	0	X	0	X	0	X	0	X	1	0	2	(33)	X	X
Ethiopia	2	100	0	X	0	X	0	X	36	140	1	X	7,704	93
Gabon	335	(29)	0	X	326	(31)	7	(450)	36	80	34	21	9,987	152
Gambia, The	0	X	0	X	0	X	0	X	3	50	4	33	9,898	12
Ghana	17	6	0	X	0	X	0	X	55	10	4	(20)	11,609	(3)
Guinea	1	X	0	X	0	X	0	X	14	17	2	0	7,299	(5)
Guinea-Bissau	0	X	0	X	0	X	0	X	2	100	2	100	15,444	59
Kenya	8	167	0	X	0	X	0	X	66	16	3	(25)	7,637	(24)
Lesotho	X	X	X	X	X	X	X	X	X	X	X	X	X	X
Liberia	1	0	0	X	0	X	0	X	10	(47)	4	(64)	10,812	(39)
Libya	2,179	(50)	0	X	2,007	(53)	173	14	339	220	83	108	16,263	334
Madagascar	1	0	0	X	0	X	0	X	12	(14)	1	(50)	4,131	(12)
Malawi	2	100	0	X	0	X	0	X	8	(11)	1	(50)	5,380	(38)
Mali	1	X	0	X	0	X	0	X	6	0	1	0	2,954	(24)
Mauritania	0	X	0	X	0	X	0	X	42	500	23	360	56,627	387
Mauritius	1	X	0	X	0	X	0	X	17	89	16	60	10,109	27
Morocco	29	0	22	5	1	0	3	0	230	53	10	25	11,004	16
Mozambique, People's Rep	1	(96)	1	(90)	0	X	0	X	14	(56)	1	(67)	9,040	X
Niger	2	X	2	X	0	X	0	X	10	100	2	100	4,621	82
Nigeria	2,756	(38)	4	(50)	2,600	(40)	144	0	495	80	5	25	5,460	97
Rwanda	1	0	0	X	0	X	0	X	6	200	1	X	4,318	89
Senegal	0	X	0	X	0	X	0	X	28	0	4	(20)	7,815	(19)
Sierra Leone	X	X	0	X	0	X	0	X	8	14	2	0	7,184	(0)
Somalia	0	X	0	X	0	X	0	X	12	9	2	0	16,852	(4)
South Africa	3,956	92	3,939	92	0	X	0	X	3,154	44	83	15	37,868	17
Sudan	2	0	0	X	0	X	0	X	43	(14)	2	(33)	6,637	0
Swaziland	X	X	X	X	X	X	X	X	X	X	X	X	X	X
Tanzania	2	0	0	X	0	X	0	X	27	4	1	(50)	4,947	(11)
Togo	0	X	0	X	0	X	0	X	5	(17)	2	(33)	4,767	(26)
Tunisia	226	18	0	X	209	16	17	55	145	63	19	27	13,598	6
Uganda	2	(33)	0	X	0	X	0	X	12	0	1	0	6,024	15
Zaire	76	15	3	(25)	54	15	0	X	61	15	2	0	5,746	12
Zambia	42	(13)	11	(35)	0	X	0	X	55	(17)	7	(46)	16,725	(6)
Zimbabwe	151	50	142	60	0	X	0	X	189	49	21	11	29,279	9
NORTH & CENTRAL AMERICA	75,881	13	20,885	25	29,932	18	20,920	(7)	81,089	0	197	(13)	X	X
Barbados	4	300	0	X	3	200	1	X	11	38	43	34	12,785	11
Canada	9,739	27	1,394	104	3,699	17	3,229	9	7,518	9	291	(2)	23,629	(20)
Costa Rica	10	100	0	X	0	X	0	X	41	11	15	(17)	8,237	(6)
Cuba	39	225	0	X	37	236	1	0	426	17	42	11	X	X
Dominican Rep	3	X	0	X	0	X	0	X	83	19	12	(8)	12,413	(5)
El Salvador	6	100	0	X	0	X	0	X	27	(13)	5	(29)	8,502	1
Guatemala	10	900	0	X	8	X	0	X	41	(15)	5	(38)	5,434	(22)
Haiti	1	0	0	X	0	X	0	X	9	0	1	(50)	6,446	(15)
Honduras	3	50	0	X	0	X	0	X	26	0	6	(25)	9,681	(23)
Jamaica	0	X	0	X	0	X	0	X	74	(25)	31	(34)	29,484	(18)
Mexico	7,317	143	233	69	5,990	169	1,011	76	4,130	66	50	32	20,129	22
Nicaragua	2	X	0	X	0	X	0	X	30	(17)	9	(36)	15,008	11
Panama	7	X	0	X	0	X	0	X	39	X	17	X	9,708	X
Trinidad and Tobago	496	(16)	0	X	338	(32)	158	72	207	7	169	(9)	41,908	1
United States	58,242	4	19,258	21	19,857	2	16,519	(13)	68,079	(3)	280	(12)	20,645	(25)
SOUTH AMERICA	11,705	28	597	143	8,021	9	2,002	92	8,311	34	30	7	X	X
Argentina	1,736	39	9	(31)	967	3	657	143	1,745	31	56	12	33,475	36
Bolivia	139	2	0	X	43	(38)	91	42	61	9	9	(18)	23,830	25
Brazil	2,152	168	138	68	1,228	256	114	185	3,178	36	22	5	11,301	(7)
Chile	197	13	46	21	74	25	34	(37)	345	17	28	0	12,069	(12)
Colombia	1,479	175	397	268	820	167	171	101	717	43	24	14	18,041	(0)
Ecuador	392	(1)	0	X	373	(4)	3	50	182	96	18	50	15,658	63
Guyana	0	X	0	X	0	X	0	X	14	(46)	14	(56)	38,965	(15)
Paraguay	10	900	0	X	0	X	0	X	32	100	8	33	6,155	26
Peru	438	84	4	300	368	88	25	32	343	19	17	(6)	14,595	(8)
Suriname	10	233	0	X	7	X	0	X	14	(50)	36	(54)	20,815	(32)
Uruguay	15	150	0	X	0	X	0	X	59	(23)	19	(30)	6,285	(31)
Venezuela	5,137	(8)	2	(50)	4,141	(18)	908	78	1,616	43	88	6	23,118	44

Table 21.1

	Production (petajoules)								Consumption					
	Total {a}		Solid		Liquid		Gas		Total		Per Capita		Per Constant 1980 $US of GNP	
		Change Since (%)		Change Since (%)		Change Since (%)		Change Since (%)	(peta-joules)	Change Since (%)	(giga-joules)	Change Since (%)	(kilo-joules)	Change Since (%)
	1987	1977	1987	1977	1987	1977	1987	1977	1987	1977	1987	1977	1987	1977
ASIA	73,824	(2)	27,132	65	37,798	(32)	6,623	144	63,353	54	21	24	X	X
Afghanistan	121	21	5	0	0	X	113	22	60	100	4	100	X	X
Bahrain	269	26	0	X	100	(17)	168	83	200	94	430	26	X	X
Bangladesh	144	324	0	X	5	X	137	328	202	138	2	100	12,101	61
Bhutan	0	X	0	X	0	X	0	X	1	X	1	X	X	X
China	25,930	63	19,407	72	5,616	43	547	16	23,469	65	22	47	42,962	(33)
Cyprus	0	X	0	X	0	X	0	X	49	69	72	53	15,516	(6)
India	6,114	94	4,379	73	1,273	199	235	370	6,462	78	8	33	26,037	14
Indonesia	3,863	6	51	629	2,706	(22)	1,080	567	1,382	60	8	33	13,795	(3)
Iran, Islamic Rep	5,464	(56)	36	38	4,783	(59)	622	(16)	1,932	35	38	(7)	X	X
Iraq	4,424	(9)	0	X	4,275	(12)	146	217	367	70	22	22	X	X
Israel	2	(33)	0	X	1	0	2	0	358	58	82	32	13,977	14
Japan	1,440	41	336	(31)	26	0	86	(28)	13,367	7	110	0	9,718	(29)
Jordan	1	X	0	X	1	X	0	X	117	193	31	107	28,432	72
Kampuchea, Dem	0	X	0	X	0	X	0	X	6	500	1	X	X	X
Korea, Dem People's Rep	1,482	22	1,377	19	0	X	0	X	1,698	32	79	3	X	X
Korea, Rep	628	85	467	40	0	X	0	X	2,173	85	52	63	19,903	(9)
Kuwait	2,919	(33)	0	X	2,719	(36)	200	40	501	169	269	66	14,711	162
Lao People's Dem Rep	4	100	0	X	0	X	0	X	4	0	1	0	X	X
Lebanon	2	(33)	0	X	0	X	0	X	109	54	39	50	X	X
Malaysia	1,525	290	0	X	1,030	174	477	3,569	609	132	38	90	19,470	39
Mongolia	88	132	88	132	0	X	0	X	107	88	53	43	X	X
Myanmar	92	35	2	100	42	(25)	44	389	78	44	2	0	10,381	(8)
Nepal	2	100	0	X	0	X	0	X	12	140	1	X	4,401	67
Oman	1,512	113	0	X	1,427	101	85	X	327	1,944	245	1,189	20,992	778
Pakistan	542	121	48	100	85	305	353	96	814	119	7	40	22,117	13
Philippines	70	438	23	283	12	X	0	X	450	(2)	8	(20)	12,487	(19)
Qatar	809	(16)	0	X	624	(31)	184	192	210	173	642	63	59,257	309
Saudi Arabia	9,697	(50)	0	X	8,790	(55)	907	3,679	2,328	531	185	302	X	X
Singapore	0	X	0	X	0	X	0	X	366	5	140	(6)	20,129	(48)
Sri Lanka	8	100	0	X	0	X	0	X	61	56	4	33	11,310	(3)
Syrian Arab Rep	522	34	0	X	509	33	7	600	337	99	30	43	25,493	62
Thailand	343	1,806	75	1,150	89	X	164	X	770	87	14	56	16,359	4
Turkey	823	113	635	158	110	1	11	X	1,536	56	29	21	18,906	7
United Arab Emirates	3,862	(7)	0	X	3,222	(21)	640	1,064	803	703	552	289	36,909	589
Viet Nam	171	(7)	164	(10)	0	X	0	X	217	27	3	0	X	X
Yemen Arab Rep	38	X	0	X	38	X	0	X	40	233	5	150	7,913	60
Yemen, People's Dem Rep	0	X	0	X	0	X	0	X	61	144	27	93	89,931	153
EUROPE	41,672	25	18,893	2	9,186	145	9,296	5	64,362	9	130	5	X	X
Albania	186	59	34	143	126	54	15	7	116	61	38	31	X	X
Austria	246	(19)	31	(24)	45	(40)	42	(57)	883	7	118	8	10,475	(11)
Belgium	271	15	117	(38)	0	X	1	0	1,616	(2)	163	(2)	12,415	(16)
Bulgaria	613	48	542	45	12	140	5	X	1,557	30	173	27	X	X
Czechoslovakia	1,990	3	1,861	(1)	6	20	26	(19)	2,879	3	185	(1)	X	X
Denmark	292	1,290	0	X	193	819	99	X	802	1	157	1	10,847	(14)
Finland	142	129	22	144	0	X	0	X	824	18	167	14	13,317	(16)
France	1,967	36	526	(28)	154	90	137	(54)	6,064	(6)	109	(10)	8,176	(22)
German Dem Rep	2,812	17	2,671	18	2	(33)	92	(17)	3,849	14	231	14	X	X
Germany, Fed Rep	4,438	(4)	3,178	(10)	159	(30)	558	(17)	10,023	1	165	2	11,087	(17)
Greece	316	129	250	91	51	X	5	X	718	34	72	24	16,656	15
Hungary	645	3	253	(15)	116	5	236	8	1,187	9	112	10	48,542	(8)
Iceland	15	67	0	X	0	X	0	X	39	11	157	0	10,055	(24)
Ireland	116	176	49	26	0	X	63	X	367	35	101	22	19,200	17
Italy	887	12	12	9	165	251	556	6	5,991	21	105	18	11,453	(6)
Luxembourg	0	X	0	X	0	X	0	X	120	(18)	326	(19)	15,725	(46)
Malta	0	X	0	X	0	X	0	X	18	64	52	63	12,700	7
Netherlands	2,820	(19)	0	X	196	188	2,611	(23)	3,107	28	213	22	16,689	11
Norway	3,657	280	11	(15)	2,087	268	1,187	865	830	30	199	26	11,653	(10)
Poland	5,287	5	5,108	7	7	(56)	158	(35)	5,320	15	141	5	88,829	X
Portugal	40	(5)	7	17	0	X	0	X	398	37	39	26	14,415	5
Romania	2,676	6	630	58	430	(20)	1,551	6	3,109	16	136	10	X	X
Spain	769	29	424	8	69	97	27	X	2,401	(0)	62	(6)	9,821	(16)
Sweden	504	91	1	X	0	X	0	X	1,225	(11)	147	(12)	8,828	(20)
Switzerland	202	30	0	X	0	X	0	X	726	6	111	4	6,025	(13)
United Kingdom	9,732	55	2,500	(16)	5,182	223	1,837	20	8,522	6	150	6	13,327	(15)
Yugoslavia	1,046	39	666	59	167	1	90	14	1,663	38	71	29	22,084	15
U.S.S.R.	68,457	34	15,522	1	26,175	14	25,295	110	54,724	35	194	24	X	X
OCEANIA	5,997	82	3,937	97	1,203	32	727	160	3,730	30	147	12	X	X
Australia	5,671	83	3,883	100	1,147	30	589	162	3,243	33	201	16	18,727	1
Fiji	1	X	0	X	0	X	0	X	8	(20)	11	(35)	7,032	(26)
New Zealand	323	62	53	(7)	56	93	138	146	373	13	113	7	15,085	1
Papua New Guinea	2	100	0	X	0	X	0	X	34	48	9	13	11,554	16
Solomon Islands	0	X	0	X	0	X	0	X	2	100	7	40	11,331	12

Sources: United Nations Statistical Office, United Nations Population Division, and The World Bank.

Notes:
a. Total includes primary electricity (hydro, nuclear, geothermal), which accounts for about 5 percent of global commercial energy production. The production of primary electricity was assessed at the heat value of electricity (1 kilowatt hour = 3.6 million joules at 100 percent efficiency). See Table 21.2 for electricity production.
1 petajoule = 1,000,000,000,000,000 joules = 947,800,000,000 Btus; 1 gigajoule = 1,000,000,000 joules = 947,800 Btus.
0 = zero or less than half of the unit of measure; X = not available; negative numbers are shown in parentheses; GNP = gross national product.
For additional information, see Sources and Technical Notes.

Table 21.2 Production and Trade of Electricity, 1977-87

	Production (gigawatt-hours)										Trade (gigawatt-hours)			
	Total		Fossil-Fuel Fired		Hydroelectric		Geothermal		Nuclear		Import		Export	
		Change Since (%)		Change Since (%)		Change Since (%)		Change Since (%)		Change Since (%)		Change Since (%)		Change Since (%)
	1987	1977	1987	1977	1987	1977	1987	1977	1987	1977	1987	1977	1987	1977
WORLD	10,467,157	43	6,699,498	27	2,037,585	36	35,461	297	1,694,613	233	244,316	106	236,060	97
AFRICA	250,658	73	202,875	108	43,494	(9)	359	X	3,930	X	2,668	(65)	2,575	(66)
Algeria	13,400	204	13,100	216	300	15	0	X	0	X	120	X	170	X
Angola	1,800	38	465	33	1,335	41	0	X	0	X	0	X	0	X
Benin	5	0	5	0	0	X	0	X	0	X	160	119	0	X
Botswana	X	X	X	X	X	X	X	X	X	X	X	X	X	X
Burkina Faso	125	79	125	79	0	X	0	X	0	X	0	X	0	X
Burundi	54	5,300	2	100	52	X	0	X	0	X	75	159	0	X
Cameroon	2,392	78	67	8	2,325	81	0	X	0	X	0	X	0	X
Cape Verde	28	367	28	367	0	X	0	X	0	X	0	X	0	X
Central African Rep	92	59	18	1,700	74	30	0	X	0	X	0	X	0	X
Chad	51	9	51	9	0	X	0	X	0	X	0	X	0	X
Comoros	14	133	12	100	2	X	0	X	0	X	0	X	0	X
Congo	235	91	2	(97)	233	270	X	X	X	X	53	X	0	X
Cote d'Ivoire	2,200	77	910	(11)	1,290	481	0	X	0	X	0	X	0	X
Djibouti	172	93	172	93	0	X	0	X	0	X	0	X	0	X
Egypt	32,500	131	26,500	428	6,000	(34)	0	X	0	X	0	X	0	X
Equatorial Guinea	17	(15)	15	(17)	2	0	0	X	0	X	0	X	0	X
Ethiopia	810	37	160	(26)	650	73	0	X	0	X	0	X	0	X
Gabon	876	98	201	42	675	124	0	X	0	X	0	X	0	X
Gambia, The	44	42	44	42	0	X	0	X	0	X	0	X	0	X
Ghana	4,758	7	82	55	4,676	6	0	X	0	X	0	X	281	57
Guinea	500	25	333	15	167	52	X	X	X	X	0	X	0	X
Guinea-Bissau	14	17	14	17	0	X	0	X	0	X	0	X	0	X
Kenya	2,629	136	359	(1)	1,911	155	359	X	0	X	176	(35)	0	X
Lesotho	X	X	X	X	X	X	X	X	X	X	X	X	X	X
Liberia	825	(8)	506	(15)	319	6	0	X	0	X	0	X	0	X
Libya	14,260	403	14,260	403	0	X	0	X	0	X	0	X	0	X
Madagascar	504	38	234	26	270	49	0	X	0	X	0	X	0	X
Malawi	578	79	14	(46)	564	90	0	X	0	X	0	X	1	(67)
Mali	204	158	42	8	162	305	0	X	0	X	0	X	0	X
Mauritania	120	38	95	9	25	X	0	X	0	X	0	X	0	X
Mauritius	488	27	348	6	140	150	0	X	0	X	0	X	0	X
Morocco	7,120	90	6,500	171	620	(54)	0	X	0	X	0	X	X	X
Mozambique, People's Rep	500	(90)	440	(2)	60	(99)	0	X	0	X	330	95	0	(100)
Niger	157	80	157	80	0	X	0	X	0	X	135	207	0	X
Nigeria	9,905	141	7,695	592	2,210	(26)	0	X	0	X	0	X	100	127
Rwanda	174	15	4	0	170	16	0	X	0	X	8	300	3	X
Senegal	752	43	752	43	0	X	0	X	0	X	0	X	0	X
Sierra Leone	196	3	196	3	0	X	0	X	0	X	0	X	0	X
Somalia	255	311	255	311	0	X	0	X	0	X	0	X	0	X
South Africa	122,465	63	117,790	61	745	(64)	0	X	3,930	X	0	(100)	300	69
Sudan	1,055	17	539	17	516	17	0	X	0	X	0	X	0	X
Swaziland	X	X	X	X	X	X	X	X	X	X	X	X	X	X
Tanzania	874	27	264	58	610	17	0	X	0	X	0	X	0	X
Togo	40	(38)	36	(36)	4	(50)	0	X	0	X	238	118	0	X
Tunisia	4,549	164	4,436	162	113	277	0	X	0	X	0	X	3	X
Uganda	655	(10)	11	(15)	644	(10)	0	X	0	X	0	X	107	(61)
Zaire	5,295	28	139	85	5,156	27	0	X	0	X	3	(70)	110	83
Zambia	8,479	(2)	38	(73)	8,441	(1)	0	X	0	X	20	300	1,500	(42)
Zimbabwe	7,645	95	5,150	1,010	2,495	(28)	0	X	0	X	1,350	(48)	0	(100)
NORTH & CENTRAL AMERICA	3,341,495	27	2,190,118	17	598,957	28	19,889	337	532,531	92	50,949	122	54,576	140
Barbados	425	61	425	61	0	X	0	X	0	X	0	X	0	X
Canada	496,335	52	102,752	33	316,322	42	0	X	77,261	191	3,471	29	47,427	138
Costa Rica	2,930	60	50	(89)	2,880	111	0	X	0	X	175	X	100	X
Cuba	13,594	76	13,550	78	44	(40)	0	X	0	X	0	X	0	X
Dominican Rep	5,296	136	4,346	99	950	1,567	0	X	0	X	0	X	0	X
El Salvador	1,900	46	130	(68)	1,030	96	740	97	0	X	0	X	0	X
Guatemala	1,770	23	1,090	(8)	680	160	0	X	0	X	0	X	0	X
Haiti	450	109	130	155	320	95	0	X	0	X	0	X	0	X
Honduras	1,085	59	205	(4)	880	88	0	X	0	X	160	1,500	2	(89)
Jamaica	2,385	0	2,260	(0)	125	15	0	X	0	X	0	X	0	X
Mexico	104,791	99	82,024	149	18,435	(4)	4,332	632	0	X	117	117	2,042	3,827
Nicaragua	1,063	(7)	495	(51)	268	99	300	X	0	X	200	953	10	0
Panama	2,902	X	870	X	2,032	X	0	X	X	X	0	X	0	X
Trinidad and Tobago	3,315	110	3,315	110	0	X	0	X	0	X	0	X	0	X
United States	2,685,627	21	1,961,145	13	254,695	14	14,517	305	455,270	81	46,826	132	4,995	82
SOUTH AMERICA	390,861	96	89,442	50	293,983	113	0	X	7,436	354	17,005	10,206	23	(92)
Argentina	52,165	61	23,791	(5)	21,909	280	0	X	6,465	295	180	122	7	250
Bolivia	1,520	21	391	11	1,129	24	0	X	0	X	2	X	0	X
Brazil	202,287	101	15,770	115	185,546	98	0	X	971	X	16,813	88,389	10	(93)
Chile	15,636	60	3,489	7	12,147	87	0	X	0	X	0	(100)	0	X
Colombia	35,368	130	9,810	103	25,558	143	0	X	0	X	0	(100)	0	(100)
Ecuador	5,668	151	1,094	(29)	4,574	534	0	X	0	X	0	X	0	X
Guyana	385	(11)	380	(12)	5	X	0	X	0	X	0	X	0	X
Paraguay	2,825	440	5	(96)	2,820	588	0	X	0	X	10	X	X	X
Peru	14,195	65	3,145	21	11,050	83	0	X	0	X	0	X	0	X
Suriname	1,330	(6)	395	(33)	935	13	0	X	0	X	0	X	0	X
Uruguay	4,526	57	316	(76)	4,210	169	0	X	0	X	0	(100)	6	X
Venezuela	54,704	133	30,604	150	24,100	115	0	X	0	X	0	(100)	0	(100)

Table 21.2

	Production (gigawatt-hours)										Trade (gigawatt-hours)			
	Total		Fossil-Fuel Fired		Hydroelectric		Geothermal		Nuclear		Import		Export	
		Change Since (%)		Change Since (%)		Change Since (%)		Change Since (%)		Change Since (%)		Change Since (%)		Change Since (%)
	1987	1977	1987	1977	1987	1977	1987	1977	1987	1977	1987	1977	1987	1977
ASIA	2,034,571	84	1,403,797	64	356,248	64	6,400	1,003	268,126	677	1,216	66	3,062	675
Afghanistan	1,257	57	493	94	764	40	0	X	0	X	0	X	0	X
Bahrain	3,020	172	3,020	172	0	X	0	X	0	X	0	X	0	X
Bangladesh	5,895	205	5,365	258	530	21	0	X	0	X	0	X	0	X
Bhutan	21	31	13	44	8	14	0	X	0	X	10	233	0	X
China	497,267	123	397,260	126	100,007	110	0	X	0	X	0	X	X	X
Cyprus	1,512	76	1,512	76	0	X	0	X	0	X	0	X	0	X
India	217,500	120	154,182	163	57,918	52	0	X	5,400	138	16	167	30	(3)
Indonesia	34,810	244	27,310	229	7,290	301	210	X	0	X	0	X	0	X
Iran, Islamic Rep	37,910	111	31,510	125	6,400	60	0	X	0	X	0	X	0	X
Iraq	22,860	256	22,250	280	610	6	0	X	0	X	0	X	0	X
Israel	17,491	57	17,491	57	0	X	0	X	0	X	0	X	359	166
Japan	698,970	31	423,490	(0)	84,070	10	1,600	176	189,810	500	0	X	0	X
Jordan	3,486	480	3,467	477	19	X	X	X	0	X	X	X	364	X
Kampuchea, Dem	X	X	X	X	X	X	X	X	X	X	0	X	0	X
Korea, Dem People's Rep	50,200	67	21,100	69	29,100	66	0	X	0	X	0	X	0	X
Korea, Rep	80,250	186	35,592	34	5,344	284	0	X	39,314	55,272	0	X	0	X
Kuwait	18,400	185	18,400	185	0	X	0	X	0	X	0	X	0	X
Lao People's Dem Rep	1,100	144	50	178	1,050	143	0	X	0	X	20	233	755	327
Lebanon	4,600	188	3,990	399	610	(24)	0	X	0	X	40	X	0	X
Malaysia	17,387	131	12,477	85	4,910	538	0	X	0	X	0	X	23	X
Mongolia	3,153	165	3,153	565	0	(100)	0	X	0	X	70	X	0	X
Myanmar	2,279	107	1,158	5	1,121	X	0	X	0	X	0	(100)	0	X
Nepal	538	243	26	44	512	268	0	X	0	X	32	10	21	250
Oman	3,793	590	3,793	590	0	X	0	X	0	X	0	X	0	X
Pakistan	33,475	208	17,723	236	15,250	194	0	X	502	19	0	X	0	X
Philippines	23,852	58	14,100	9	5,220	147	4,532	X	0	X	0	X	0	X
Qatar	4,420	331	4,420	331	0	X	0	X	0	X	0	X	0	X
Saudi Arabia	37,100	411	37,100	411	0	X	0	X	0	X	0	X	0	(100)
Singapore	11,814	131	11,814	131	0	X	0	X	0	X	0	X	0	X
Sri Lanka	2,707	110	530	1,028	2,177	76	0	X	0	X	0	X	0	X
Syrian Arab Rep	7,161	251	5,661	1,959	1,500	(15)	0	X	0	X	X	X	130	X
Thailand	29,992	157	25,917	208	4,075	25	0	X	0	X	416	135	18	200
Turkey	44,353	116	25,677	114	18,618	117	58	X	0	X	572	16	0	X
United Arab Emirates	13,100	343	13,100	343	0	X	0	X	0	X	0	X	0	X
Viet Nam	5,300	53	3,300	15	2,000	233	0	X	0	X	0	X	0	X
Yemen Arab Rep	718	755	718	755	0	X	0	X	0	X	0	X	0	X
Yemen, People's Dem Rep	465	100	465	100	0	X	0	X	0	X	0	X	0	X
EUROPE	2,620,577	33	1,427,274	5	490,834	10	6,863	173	695,606	332	172,178	100	140,824	85
Albania	3,840	34	490	(27)	3,350	52	0	X	0	X	0	X	650	1,200
Austria	50,220	35	14,640	14	35,580	45	0	X	0	X	3,997	66	9,606	51
Belgium	62,375	32	19,925	(43)	477	5	6	X	41,967	252	5,660	9	7,778	18
Bulgaria	43,470	46	28,497	40	2,538	(28)	0	X	12,435	111	5,326	32	952	35
Czechoslovakia	85,825	29	58,707	(5)	4,904	12	0	X	22,214	19,386	11,931	122	8,510	202
Denmark	29,398	31	29,186	30	29	32	183	X	0	X	4,172	65	1,758	(7)
Finland	53,464	69	20,024	17	13,794	15	0	X	19,646	683	6,104	339	507	1
France	356,200	70	37,000	(68)	67,900	(11)	0	X	251,300	1,301	8,700	(28)	38,400	428
German Dem Rep	114,180	24	101,244	18	1,726	38	0	X	11,210	115	7,451	181	3,664	70
Germany, Fed Rep	415,812	24	265,033	(6)	18,932	16	0	X	131,847	266	22,177	29	18,370	64
Greece	30,087	58	27,305	60	2,779	45	3	X	0	X	977	1,358	362	1,348
Hungary	29,749	27	18,594	(20)	169	14	0	X	10,986	X	12,610	133	1,997	107
Iceland	4,210	61	5	(93)	3,957	57	248	1,450	0	X	0	X	0	X
Ireland	12,636	40	11,520	39	1,116	49	0	X	0	X	0	X	0	X
Italy	198,292	20	155,627	44	39,505	(23)	2,986	19	174	(95)	24,818	342	1,672	(41)
Luxembourg	573	(50)	473	(55)	100	18	0	X	0	X	4,003	47	440	122
Malta	944	126	944	126	0	X	0	X	0	X	0	X	0	X
Netherlands	68,411	17	64,852	19	1	X	2	X	3,556	(4)	3,644	270	21	(91)
Norway	103,810	44	514	124	103,296	44	0	X	0	X	2,932	11	3,311	111
Poland	145,832	33	141,779	33	4,053	69	0	X	0	X	10,422	235	8,703	182
Portugal	20,101	40	10,949	187	9,151	(8)	1	X	0	X	3,700	871	675	(27)
Romania	73,090	22	60,500	20	12,590	66	0	X	0	X	3,100	78	0	(100)
Spain	133,168	44	63,908	37	27,999	(29)	0	X	41,261	500	3,171	126	4,704	101
Sweden	146,625	63	6,810	(59)	72,233	35	6	X	67,576	239	2,169	(35)	6,185	10
Switzerland	56,976	27	958	(49)	34,317	(3)	0	X	21,701	181	10,859	96	20,314	29
United Kingdom	300,247	7	240,974	1	4,035	3	0	X	55,238	38	11,672	X	37	X
Yugoslavia	80,792	66	46,616	92	26,253	8	3,428	X	4,495	X	2,583	660	2,208	47
U.S.S.R.	1,664,924	45	1,258,115	30	219,825	50	0	X	186,984	437	300	(66)	35,000	180
OCEANIA	164,071	52	127,877	64	34,244	19	1,950	53	0	X	0	X	0	X
Australia	132,172	60	117,926	71	13,406	(2)	840	X	0	X	0	X	0	X
Fiji	430	54	80	(71)	350	X	0	X	0	X	0	X	0	X
New Zealand	27,030	27	6,210	15	19,710	35	1,110	(13)	0	X	0	X	0	X
Papua New Guinea	1,797	72	1,359	80	438	52	0	X	0	X	0	X	0	X
Solomon Islands	30	67	30	67	0	X	0	X	0	X	0	X	0	X

Source: United Nations Statistical Office.
Notes: 0 = zero or less than half of the unit of measure; X = not available; negative numbers are shown in parentheses.
For additional information, see Sources and Technical Notes.

Table 21.3 Reserves and Resources of Commercial Energy

	Bituminous Coal (million metric tons) 1987		Lignite and Subbituminous Coal (million metric tons) 1987		Crude Oil (million t)	Natural Gas (billion m3)	Uranium (metric tons)		Hydroelectric (megawatts)	
	Proved Reserves in Place	Proved Recoverable Reserves	Proved Reserves in Place	Proved Recoverable Reserves	Proved Recoverable Reserves 1987	Proved Recoverable Reserves 1987	Recoverable at Less Than $80 per kg 1987	Recoverable at $80-130 per kg 1987	Technical Potential	Installed Capacity 1987
WORLD	1,696,519	1,075,473	749,747	522,507	123,559	109,326	1,676,820	679,125	X	584,977
AFRICA	133,861	62,631	1,518	279	8,033	7,249	639,410	138,950	X	18,029
Algeria	X	43	X	X	1,593	3,000	26,000	X	287	285
Angola	X	X	X	X	156	50	X	X	17,220 a	400
Benin	X	X	X	X	X	X	X	X	500	0
Botswana	7,000	3,500	X	X	X	X	X	X	1	X
Burkina Faso	X	X	X	X	X	X	X	X	200	0
Burundi	X	X	X	X	X	X	X	X	289 a	12
Cameroon	X	X	X	X	71	110	X	X	23,000 a	528
Cape Verde	X	X	X	X	X	X	X	X	X	0
Central African Rep	X	X	4	4	X	X	8,000	8,000	2,000	22
Chad	X	X	X	X	X	X	X	X	30	0
Comoros	X	X	X	X	X	X	X	X	10	1
Congo	X	X	X	X	98	70	X	X	10,000 a	120
Cote d'Ivoire	X	X	X	X	16	100	X	X	3,000	885
Djibouti	X	X	X	X	X	X	X	X	X	0
Egypt	25	13	X	40	600	290	X	X	3,210	2,700
Equatorial Guinea	X	X	X	X	X	24	X	X	2,000	1
Ethiopia	X	X	23	11	X	24	X	X	4,000 b	230
Gabon	X	X	X	X	130	17	14,000	4,650	6,500 a	125
Gambia, The	X	X	X	X	X	X	X	X	X	0
Ghana	X	X	X	X	3	X	X	X	2,000	1,072
Guinea	X	X	X	X	X	X	X	X	5,000	47
Guinea-Bissau	X	X	X	X	X	X	X	X	60	0
Kenya	X	X	X	X	X	X	X	X	841 a	354
Lesotho	X	X	X	X	X	X	X	X	450	X
Liberia	X	X	X	X	X	X	X	X	2,000	81
Libya	X	X	X	X	2,865	728	X	X	X	0
Madagascar	1,000	X	75	X	X	X	X	X	7,800	45
Malawi	25	12	X	X	X	X	X	X	900	146
Mali	X	X	X	X	X	X	X	X	2,000	45
Mauritania	X	X	X	X	X	X	X	X	X	0
Mauritius	X	X	X	X	X	X	X	X	65	59
Morocco	134	45	44	X	X	2	X	X	2,453	619
Mozambique, People's Rep	X	240	X	X	X	65	X	X	15,000	1,523
Niger	X	70	X	X	X	X	170,710	2,200	235	0
Nigeria	X	21	338	169	2,200	2,380	X	X	12,400	1,900
Rwanda	X	X	X	X	X	40	X	X	600	56
Senegal	X	X	X	X	X	X	X	X	500	0
Sierra Leone	X	X	X	X	X	X	X	X	1,300	2
Somalia	X	X	X	X	X	6	X	6,600	50	0
South Africa	121,218	55,333	X	X	X	28	324,800	101,500	X	572 c
Sudan	X	X	X	X	41	85	X	X	380 a	225
Swaziland	2,020	1,820	X	X	X	X	X	X	600	X
Tanzania	304	200	X	X	X	116	X	X	4,000 a	259
Togo	X	X	X	X	X	X	X	X	270	4
Tunisia	X	X	X	X	245	85	X	X	65	64
Uganda	X	X	X	X	X	X	X	X	1,200	156
Zaire	600	600	X	X	15	1	1,800	X	120,000	2,486
Zambia	X	X	69	55	X	X	X	X	12,000	2,245
Zimbabwe	1,535	734	965	X	X	X	X	X	3,500 a	633
NORTH & CENTRAL AMERICA	233,097	118,055	220,598	106,038	13,132	10,708	269,500	368,240	X	154,418
Barbados	X	X	X	X	1	X	X	X	X	0
Canada	5,585	3,831	15,205	3,135	960	2,730	148,000	95,000	118,596 a	56,848
Costa Rica	X	X	27	X	X	X	X	X	9,472 a	736
Cuba	X	X	X	X	X	X	X	X	X	49
Dominican Rep	X	X	X	X	X	X	X	X	503 a	165
El Salvador	X	X	X	X	X	X	X	X	664 a	233
Guatemala	X	X	X	X	6	X	X	X	8,674 a	445
Haiti	X	X	13	X	X	X	X	X	152	70
Honduras	X	X	21	X	X	X	X	X	4,800 a	130
Jamaica	X	X	X	X	X	X	X	X	67 a	25
Mexico	1,569	1,252	793	634	7,703	2,119	4,500	3,240	34,400 a	7,780
Nicaragua	X	X	X	X	X	X	X	X	4,106	103
Panama	X	X	X	X	X	X	X	X	3,031	551
Trinidad and Tobago	X	X	X	X	77	294	X	X	X	0
United States	225,943	112,972	204,539	102,269	4,385	5,565	117,000	270,000	183,287	87,192
SOUTH AMERICA	17,245	11,074	7,999	2,648	8,973	4,658	172,200	4,165	X	67,083
Argentina	X	X	195	130	308	670	9,200	2,600	37,208 d	6,591
Bolivia	X	X	X	X	22	143	X	X	18,000	295
Brazil	X	X	3,276	1,245	361	105	163,000	X	150,322 a	40,106
Chile	79	31	4,500	1,150	40	120	X	45	26,487 a	2,279
Colombia	16,524	9,666	X	X	216	110	X	X	83,640 a	4,675
Ecuador	X	X	28	23	157	12	X	X	36,000 a	917
Guyana	X	X	X	X	X	X	X	X	12,620 a	2
Paraguay	X	X	X	X	X	X	X	X	4,585 a	3,340
Peru	X	960	X	100	75	18	X	1,520	60,000	2,150
Suriname	X	X	X	X	X	X	X	X	2,334	189
Uruguay	X	X	X	X	X	X	X	X	2,000 a	1,039
Venezuela	642	417	X	X	7,794	3,480	X	X	37,186 a	5,500

Table 21.3

	Bituminous Coal (million metric tons) 1987		Lignite and Subbituminous Coal (million metric tons) 1987		Crude Oil (million t)	Natural Gas (billion m3)	Uranium (metric tons)		Hydroelectric (megawatts)	
	Proved Reserves in Place	Proved Recoverable Reserves	Proved Reserves in Place	Proved Recoverable Reserves	Proved Recoverable Reserves 1987	Proved Recoverable Reserves 1987	Recoverable at Less Than $80 per kg 1987	Recoverable at $80-130 per kg 1987	Technical Potential	Installed Capacity 1987
ASIA	807,793	674,352	146,927	131,344	82,323	37,536	34,580	21,460	X	109,270
Afghanistan	112	66	X	X	X	64	X	X	25,000	281
Bahrain	X	X	X	X	17	195	X	X	X	0
Bangladesh	1,054	X	X	X	0	360	X	X	800	197
Bhutan	X	X	X	X	X	X	X	X	X	3
China	650,220	610,800	120,220	120,100	2,451	895	X	X	436,197 d	28,000
Cyprus	X	X	X	X	X	X	X	X	4,700 b	0
India	129,154	60,648	2,100	1,900	657	500	34,580	10,960	41,000 a	17,003
Indonesia	X	1,000	X	2,000	1,142	2,068	X	X	141,800 a	1,600
Iran, Islamic Rep	3,754	193	2,295	X	13,048	13,864	X	X	11,200 a	1,804
Iraq	X	X	X	X	13,600	745	X	X	X	100
Israel	X	X	X	X	0	0	X	X	X	0
Japan	8,348	856	175	17	6	29	X	6,600	26,840 a	34,650
Jordan	X	X	X	X	1	0	X	X	17 a	11
Kampuchea, Dem	X	X	X	X	X	X	X	X	10,000	10
Korea, Dem People's Rep	2,000	300	300	300	X	X	X	X	X	4,600
Korea, Rep	238	158	X	X	X	X	X	X	2,000	2,236
Kuwait	X	X	X	X	12,700	1,050	X	X	X	0
Lao People's Dem Rep	X	X	X	X	X	X	X	X	28,000	200
Lebanon	X	X	X	X	X	X	X	X	200 a	246
Malaysia	15	4	X	X	434	1,462	X	X	11,846 a,e	1,090
Mongolia	12,000	X	12,000	X	X	X	X	X	32,000 a	0
Myanmar	5	2	X	X	8	268	X	X	28,800 a	258
Nepal	X	X	X	X	X	X	X	X	X	161
Oman	X	X	X	X	550	270	X	X	X	0
Pakistan	X	X	145	102	13	635	X	X	3,106 a	2,901
Philippines	X	X	170	82	2	X	X	X	6,598 a	2,153
Qatar	X	X	X	X	430	4,440	X	X	X	0
Saudi Arabia	X	X	X	X	22,712	3,963	X	X	X	0
Singapore	X	X	X	X	X	X	X	X	X	0
Sri Lanka	X	X	X	X	X	X	X	X	1,253 d	801
Syrian Arab Rep	X	X	X	X	235	144	X	X	1,282	827
Thailand	X	X	1,663	914	13	105	X	X	3,428 a	2,256
Turkey	593	175	7,847	5,929	37	25	X	3,900	43,000 a	5,004
United Arab Emirates	X	X	X	X	13,340	5,765	X	X	X	0
Viet Nam	300	150	12	X	X	X	X	X	18,000	320
Yemen Arab Rep	X	X	X	X	X	X	X	X	X	0
Yemen, People's Dem Rep	X	X	X	X	X	X	X	X	X	0
EUROPE	308,251	59,992	163,272	99,508	2,797	6,656	91,130	90,310	X	161,485
Albania	X	X	15	X	27	7	X	X	2,800 a	680
Austria	X	X	350	65	11	12	X	X	11,360 d	10,575
Belgium	715	410	X	X	X	X	X	X	X	1,328
Bulgaria	36	30	4,418	3,700	2	5	X	X	5,282 f	1,975
Czechoslovakia	5,400	1,870	6,100	3,500	4	13	X	X	2,165 d	2,890
Denmark	X	X	63	X	62	125	X	27,000 g	14 a	10
Finland	X	X	X	X	X	X	X	1,500	3,620 a	2,586
France	790	213	181	45	30	34	50,030	11,760	20,395 d	24,100
German Dem Rep	X	X	47,000	21,000	1	190	X	X	X	1,844
Germany, Fed Rep	44,000	23,919	55,000	35,150	36	179	800	4,000	4,200 a	6,760
Greece	X	X	5,312	3,000	3	4	400	X	4,140 a	2,137
Hungary	1,407	596	8,306	3,865	39	125	X	X	900 a	46
Iceland	X	X	X	X	X	X	X	X	12,800 a	756
Ireland	7	5	12	9	X	26	X	X	X	512
Italy	X	X	75	39	91	300	4,800	X	13,000 a	17,879
Luxembourg	X	X	X	X	X	X	X	X	X	1,132
Malta	X	X	X	X	X	X	X	X	X	0
Netherlands	1,406	497	X	X	26	1,770	X	X	100 d	2
Norway	X	X	38	10	1,543	2,773	X	X	34,000 a	25,394
Poland	63,800	28,700	13,000	11,700	2	130	X	X	2,400 d	1,976
Portugal	8	3	38	33	X	X	7,100	1,400	7,184 d	3,173
Romania	70	X	3,900	X	175	235	X	X	7,600 a	4,640
Spain	532	379	700	391	0	14	26,000	7,650	12,440 a	14,453
Sweden	X	X	4	1	X	X	2,000	37,000	10,000 a	16,700
Switzerland	X	X	X	X	X	X	X	X	8,200 a	11,510
United Kingdom	190,000	3,300	1,000	500	710	630	X	X	1,120 d	1,409
Yugoslavia	80	70	17,760	16,500	30	84	X	X	13,600 a	7,000
U.S.S.R.	130,000	104,000	157,000	137,000	8,000	41,080	X	X	766,200 d	62,695
OCEANIA	66,273	45,369	52,433	45,690	301	1,439	470,000	56,000	X	11,997
Australia	66,220	45,340	50,600	45,600	279	1,281	470,000	56,000	5,050 d	7,029
Fiji	X	X	X	X	X	X	X	X	103 a	80
New Zealand	49	27	1,833	90	22	130	X	X	12,182 d	4,648
Papua New Guinea	X	X	X	X	X	28	X	X	19,600 a	140
Solomon Islands	X	X	X	X	X	X	X	X	X	0

Sources: World Energy Conference, The World Bank, and United Nations Statistical Office.

Notes: Global and regional totals include countries not listed.

All data on exploitable and theoretical hydropotential assume use of 5,000 hours per year as representative for all hydropower (57 percent load factor).

a. Exploitable potential at large-scale sites (over 1 megawatt) only. b. Exploitable potential at small-scale site (under 1 megawatt) only. c. Figure for South Africa includes Botswana, Lesotho, Swaziland, and Namibia. d. Exploitable potential. e. Refers to Peninsular Malaysia only. f. Theoretical potential. g. Includes Greenland.

0 = zero or less than half of the unit of measure; X = not available; t = metric tons; m3 = cubic meters; billion = one thousand million.

For additional information, see Sources and Technical Notes.

Table 21.4 Production, Consumption, and Reserves of Selected

	Annual Production (000 metric tons)					Annual Consumption (000 metric tons)			
	1973	1978	1983	1988		1973	1978	1983	1988
ALUMINUM {a}									
Australia	17,595.1	24,293.0	24,372.0	35,000.0	United States	5,076.7	4,978.1	4,221.0	4,612.4
Guinea	3,048.0	10,456.0	12,421.0	15,600.0	Japan	1,574.1	1,656.1	1,820.8	2,123.2
Brazil	849.4	1,160.0	7,199.0	8,750.0	U.S.S.R.	1,480.0	1,830.0	1,850.0	1,800.0
Jamaica	13,599.2	11,777.0	7,683.0	7,408.0	Germany, Fed Rep	855.7	952.3	1,085.0	1,232.6
U.S.S.R.	4,267.2	4,600.0	4,600.0	4,600.0	France	450.1	532.7	613.4	650.0
Suriname	7,110.0	5,188.0	3,400.0	3,394.0	China	230.0	560.0	600.0	600.0
China	762.0	1,500.0	1,600.0	3,200.0	Italy	336.0	404.0	430.0	581.0
Yugoslavia	2,167.0	2,565.0	3,500.0	3,034.0	United Kingdom	491.5	402.2	323.4	427.4
India	1,292.4	1,663.0	1,923.0	3,011.0	Canada	331.8	338.8	248.0	421.6
Hungary	2,599.9	2,899.0	2,917.0	2,906.0	India	148.5	224.0	218.5	337.0
Ten Countries Total	**53,290.1**	**66,101.0**	**69,615.0**	**86,903.0**	**Ten Countries Total**	**10,974.4**	**11,878.2**	**11,410.1**	**12,785.2**
World Total	**70,351.9**	**79,851.0**	**78,634.0**	**97,144.3**	**World Total**	**13,600.6**	**15,325.8**	**15,352.0**	**17,683.3**
Bauxite, World Reserves 1988 (000 metric tons)				21,800,000	World Reserves Life Index (years)				224
Bauxite, World Reserve Base 1988 (000 metric tons)				232,000,000	World Reserve Base Life Index (years)				2,388
CADMIUM									
U.S.S.R.	2.5	2.5	3.0	3.0	Japan	1.5	1.1	1.9	4.8
Japan	3.2	2.3	2.2	2.6	United States	5.7	4.5	4.1	4.2
United States	3.4	1.5	1.1	1.9	U.S.S.R.	1.9	2.5	2.5	2.6
Canada	1.4	1.0	1.5	1.7	Belgium	1.4	1.5	2.3	2.1
Belgium	1.1	1.1	1.3	1.5	United Kingdom	1.6	1.4	1.4	1.5
Germany, Fed Rep	1.2	1.1	1.1	1.2	France	1.2	1.1	1.0	1.1
Mexico	0.2	0.8	0.6	0.9	Germany, Fed Rep	2.2	1.8	1.5	1.0
Australia	0.7	0.7	1.1	0.9	China	X	X	0.3	0.4
China	0.1	0.2	0.3	0.7	Brazil	0.1	0.1	0.3	0.4
Finland	0.2	0.6	0.6	0.7	Italy	0.4	0.5	0.4	0.4
Ten Countries Total	**14.0**	**11.8**	**12.7**	**14.9**	**Ten Countries Total**	**15.9**	**14.5**	**15.8**	**18.5**
World Total	**17.2**	**15.7**	**17.4**	**19.8**	**World Total**	**17.8**	**16.9**	**18.3**	**21.2**
World Reserves 1988 (000 metric tons)				535	World Reserves Life Index (years)				X b
World Reserve Base 1988 (000 metric tons)				970	World Reserve Base Life Index (years)				X b
COPPER									
Chile	735.4	1,035.5	1,257.5	1,472.0	United States	2,221.2	2,193.1	1,803.9	2,268.4
United States	1,558.5	1,357.6	1,038.1	1,419.7	Japan	1,201.8	1,241.4	1,216.3	1,330.7
Canada	824.0	659.4	653.0	756.5	U.S.S.R.	1,100.0	1,330.0	1,300.0	1,250.0
U.S.S.R.	700.4	865.0	570.0	640.0	Germany, Fed Rep	727.2	780.0	737.0	797.5
Zaire	488.6	423.8	536.5	530.0	China	283.0	367.0	380.0	465.0
Poland	155.0	321.0	402.3	440.0	Italy	300.0	344.0	325.0	445.0
Zambia	706.6	643.0	541.0	400.0	France	407.8	319.0	390.0	408.9
China	99.8	200.0	175.0	300.0	United Kingdom	541.2	501.6	358.0	327.7
Peru	202.7	366.4	318.8	298.3	Belgium	164.4	289.5	258.2	317.8
Mexico	80.5	87.2	196.0	280.2	Korea, Rep	23.8	73.5	152.3	266.3
Ten Countries Total	**5,551.4**	**5,958.9**	**5,688.2**	**6,536.8**	**Ten Countries Total**	**6,970.4**	**7,439.1**	**6,920.7**	**7,877.3**
World Total	**7,116.9**	**7,632.9**	**7,661.8**	**8,453.4**	**World Total**	**8,763.3**	**9,449.2**	**9,102.7**	**10,659.8**
World Reserves 1988 (000 metric tons)				350,000	World Reserves Life Index (years)				41
World Reserve Base 1988 (000 metric tons)				560,000	World Reserve Base Life Index (years)				66
LEAD									
Australia	402.8	400.3	480.6	475.0	United States	1,093.2	1,398.7	1,134.8	1,201.0
U.S.S.R.	471.7	520.0	435.0	440.0	U.S.S.R.	600.0	760.0	805.0	790.0
United States	547.1	529.7	465.6	394.0	Japan	267.3	266.5	359.6	406.5
Canada	387.8	319.8	251.5	368.4	Germany, Fed Rep	293.7	335.8	318.3	373.5
China	99.8	145.0	160.0	255.0	United Kingdom	282.2	336.5	292.9	302.5
Mexico	179.3	170.6	184.3	170.2	China	170.0	210.0	215.0	250.0
Peru	183.4	182.7	207.4	149.0	Italy	180.2	251.0	234.0	246.0
Korea, Dem People's Rep	90.7	105.0	75.0	110.0	France	213.7	211.7	196.1	215.6
Yugoslavia	119.3	124.5	114.0	100.0	Korea, Rep	X	27.5	41.2	146.0
Bulgaria	105.0	117.0	95.0	97.0	Yugoslavia	69.6	82.0	133.3	128.9
Ten Countries Total	**2,586.9**	**2,614.6**	**2,468.4**	**2,558.6**	**Ten Countries Total**	**3,169.9**	**3,879.7**	**3,730.2**	**4,060.0**
World Total	**3,487.0**	**3,478.7**	**3,358.3**	**3,381.3**	**World Total**	**4,441.6**	**5,372.4**	**5,248.1**	**5,665.4**
World Reserves 1988 (000 metric tons)				75,000	World Reserves Life Index (years)				22
World Reserve Base 1988 (000 metric tons)				125,000	World Reserve Base Life Index (years)				37
MERCURY									
U.S.S.R.	1.8	2.1	2.2	2.3	United States	1.9	2.0	1.7	X
Spain	2.1	1.0	1.4	1.5	U.S.S.R.	0.8	1.8	X	X
United States	0.1	0.8	0.9	X	China	0.5	0.5	X	X
Algeria	0.5	1.1	0.3	0.7	India	0.2	0.4	0.2	X
China	0.9	0.7	0.7	0.7	Italy	0.6	0.4	0.1	X
Czechoslovakia	0.2	0.2	0.1	0.2	United Kingdom	0.7	0.4	0.3	X
Finland	0.0	0.0	0.1	0.1	Germany, Fed Rep	0.8	0.4	0.3	X
Mexico	0.7	0.1	0.2	0.1	Romania	0.1	0.3	X	X
Turkey	0.3	0.2	0.2	0.1	Japan	0.6	0.2	0.3	X
Yugoslavia	0.5	X	0.1	0.1	South Africa	0.1	0.1	X	X
Ten Countries Total	**7.2**	**6.2**	**6.2**	**5.7**	**Ten Countries Total**	**6.2**	**6.5**	**2.8**	**X**
World Total	**9.3**	**6.3**	**6.2**	**5.7**	**World Total**	**7.9**	**7.7**	**7.4**	**X**
World Reserves 1988 (000 metric tons)				128	World Reserves Life Index (years)				22
World Reserve Base 1988 (000 metric tons)				241	World Reserve Base Life Index (years)				42

Metals, 1973-88

Table 21.4

NICKEL

Annual Production (000 metric tons)

	1973	1978	1983	1988
Canada	244.0	138.3	128.1	199.0
U.S.S.R.	135.2	147.9	169.6	190.0
New Caledonia	107.4	66.1	46.2	67.7
Australia	40.6	82.4	76.6	62.4
Indonesia	20.8	31.9	49.4	53.0
Cuba	36.5	37.0	37.6	44.0
South Africa	19.4	22.5	20.5	34.8
Dominican Rep	30.1	14.3	19.6	29.3
China	X	10.0	13.0	27.0
Botswana	0.5	16.0	18.2	26.5
Ten Countries Total	**634.5**	**566.4**	**578.8**	**733.6**
World Total	**710.0**	**660.4**	**674.7**	**835.1**

World Reserves 1988 (000 metric tons): 54,000
World Reserve Base 1988 (000 metric tons): 120,600

Annual Consumption (000 metric tons)

	1973	1978	1983	1988
Japan	111.2	99.0	114.8	161.7
United States	179.4	163.9	136.9	141.2
U.S.S.R.	100.0	127.0	145.0	130.0
Germany, Fed Rep	54.8	67.4	63.0	89.4
France	29.6	35.5	32.9	39.6
Italy	20.4	24.5	22.5	28.6
China	18.0	19.0	18.0	27.5
United Kingdom	31.5	32.0	21.8	26.7
Sweden	26.8	20.0	16.4	17.1
Spain	5.3	8.8	8.3	17.0
Ten Countries Total	**577.0**	**597.1**	**579.6**	**678.8**
World Total	**649.4**	**699.6**	**687.4**	**851.7**

World Reserves Life Index (years): 65
World Reserve Base Life Index (years): 144

TIN

Annual Production (000 metric tons)

	1973	1978	1983	1988
Brazil	5.4	7.0	13.3	43.7
Indonesia	22.3	27.4	26.6	30.6
Malaysia	72.3	62.7	41.4	28.9
China	20.0	14.0	15.0	25.0
U.S.S.R.	29.0	34.0	22.0	16.0
Thailand	20.9	30.2	19.9	14.2
Bolivia	30.3	30.9	25.3	10.5
Australia	10.8	11.9	9.3	7.2
United Kingdom	3.8	2.8	4.0	4.5
Peru	0.2	0.0	2.8	4.4
Ten Countries Total	**215.1**	**221.6**	**179.5**	**185.0**
World Total	**233.8**	**241.4**	**196.6**	**200.8**

World Reserves 1988 (000 metric tons): 4,260
World Reserve Base 1988 (000 metric tons): 4,280

Annual Consumption (000 metric tons)

	1973	1978	1983	1988
United States	61.1	50.4	35.6	37.6
Japan	38.8	29.5	30.4	32.2
U.S.S.R.	19.0	24.0	29.0	30.0
Germany, Fed Rep	16.9	15.0	14.2	19.4
China	14.0	54.7	13.0	14.0
United Kingdom	18.4	13.9	10.2	10.2
France	11.7	9.9	7.6	7.8
Korea, Rep	1.0	2.1	2.6	7.3
Brazil	3.9	5.8	4.3	6.7
Italy	8.4	6.8	4.1	6.0
Ten Countries Total	**193.2**	**212.1**	**151.0**	**171.2**
World Total	**254.3**	**232.8**	**205.6**	**235.3**

World Reserves Life Index (years): 21
World Reserve Base Life Index (years): 21

ZINC

Annual Production (000 metric tons)

	1973	1978	1983	1988
Canada	1,226.6	1,066.9	1,069.7	1,351.7
U.S.S.R.	671.3	770.0	805.0	810.0
Australia	480.5	473.3	699.0	765.7
China	99.8	160.0	160.0	527.0
Peru	411.9	457.5	576.4	488.8
Mexico	271.4	244.9	266.3	262.2
United States	434.4	302.7	296.7	256.4
Spain	87.3	146.8	167.7	255.5
Korea, People's Rep	159.7	145.0	140.0	225.0
Sweden	118.5	162.8	204.2	186.9
Ten Countries Total	**3,961.3**	**3,929.9**	**4,385.0**	**5,129.1**
World Total	**5,709.4**	**5,928.2**	**6,367.6**	**6,976.7**

World Reserves 1988 (000 metric tons): 147,000
World Reserve Base 1988 (000 metric tons): 295,000

Annual Consumption (000 metric tons)

	1973	1978	1983	1988
United States	1,363.9	1,021.0	933.0	1,110.3
U.S.S.R.	840.0	990.0	1,050.0	1,080.0
Japan	814.9	732.5	770.8	774.2
Germany, Fed Rep	438.2	391.0	405.2	445.6
China	190.0	185.0	290.0	385.0
France	290.4	281.7	270.5	290.0
Italy	220.0	221.0	208.0	250.0
United Kingdom	305.4	247.6	177.2	192.5
Belgium	180.1	139.8	165.7	174.8
Korea, Rep	23.1	58.0	113.2	173.0
Ten Countries Total	**4,666.0**	**4,267.6**	**4,383.6**	**4,875.4**
World Total	**6,267.1**	**6,192.8**	**6,272.8**	**7,114.5**

World Reserves Life Index (years): 21
World Reserve Base Life Index (years): 42

IRON ORE

Annual Production (000 metric tons)

	1973	1978	1983	1988
U.S.S.R.	216,094.1	246,239.8	245,189.2	251,000.0
Brazil	50,503.3	84,981.3	88,712.0	145,040.0
China	55,880.0	70,002.4	71,120.0	105,000.0
Australia	84,823.8	83,130.1	71,034.7	96,084.0
United States	89,071.7	82,888.3	38,163.0	57,515.0
India	35,561.0	38,835.6	38,798.0	52,322.0
Canada	50,210.7	41,748.5	33,493.5	38,742.0
South Africa	10,954.5	24,205.2	16,604.5	25,248.0
Sweden	34,725.9	21,485.4	14,264.6	20,440.0
Venezuela	23,108.9	13,514.8	9,715.0	18,789.0
Ten Countries Total	**650,933.9**	**707,031.4**	**627,094.5**	**810,180.0**
World Total	**845,660.5**	**848,724.7**	**740,166.2**	**916,431.0**

World Reserves 1988 (000 metric tons): 153,416,000
World Reserve Base 1988 (000 metric tons): 216,408,000

Annual Consumption (000 metric tons)

	1973		1978	1983	1988	
U.S.S.R.	181,583.0	c	198,160.0	202,384.0	204,000.0	d
China	61,581.0	c	125,812.0	118,045.0	169,300.0	d
Japan	144,639.0		127,086.0	117,845.0	118,353.0	d
United States	130,314.0		111,901.0	62,203.0	61,051.0	d
Germany, Fed Rep	58,825.0		47,369.0	41,829.0	44,210.0	d
Brazil	20,600.0	c	18,001.0	18,734.0	37,411.0	d
France	39,868.0		34,169.0	24,074.0	22,762.0	d
Italy	15,803.0		17,926.0	16,213.0	19,781.0	d
Poland	13,304.0		20,509.0	17,307.0	18,167.0	d
Czechoslovakia	15,134.0		17,993.0 e	18,046.0	17,832.0	d
Ten Countries Total	**681,651.0**		**718,926.0**	**636,680.0**	**712,867.0**	**d**
World Total	**845,660.5**		**848,724.7**	**740,166.2**	**952,896.0**	**d**

World Reserves Life Index (years): 167
World Reserve Base Life Index (years): 236

STEEL, CRUDE

Annual Production (000 metric tons)

	1973	1978	1983	1988
U.S.S.R.	131,461.4	151,438.0	152,516.6	163,000.0
Japan	119,324.0	102,106.3	97,180.2	105,681.0
United States	136,804.9	124,314.5	76,762.7	90,650.0
China	27,216.0	31,780.1	39,953.1	59,200.0
Germany, Fed Rep	49,521.3	41,254.0	35,729.2	41,023.0
Brazil	7,149.6	12,107.5	14,660.4	24,536.0
Italy	20,995.3	24,283.0	21,673.9	23,668.0
France	25,264.6	22,841.5	17,623.3	19,003.0
Korea, Rep	1,157.6	3,138.9	11,915.2	19,113.0
Poland	14,057.1	19,251.7	16,236.2	17,000.0
Ten Countries Total	**532,951.9**	**532,515.5**	**484,250.7**	**562,874.0**
World Total	**697,703.0**	**709,933.0**	**662,794.0**	**777,784.0**

Annual Consumption (000 metric tons)

	1973		1978	1983	1988	
U.S.S.R.	137,554.0	b	153,436.0	157,578.0	163,199.0	d
United States	144,120.0	b	145,013.0	94,011.0	101,642.0	d
China	25,980.0	b	42,511.0	51,909.0	78,092.0	d
Japan	78,976.0	b	66,652.0	65,614.0	75,757.0	d
Germany, Fed Rep	34,860.0	b	33,294.0	30,242.0	29,100.0	d
Italy	23,651.0	b	19,601.0	18,815.0	23,428.0	d
Poland	17,379.0	b	19,511.0	14,812.0	15,752.0	d
India	8,700.0	b	11,131.0	12,385.0	15,536.0	d
Korea, Rep	3,256.0	b	7,010.0	8,620.0	15,070.0	d
United Kingdom	24,268.0	b	19,509.0	14,200.0	15,021.0	d
Ten Countries Total	**498,744.0**	**b**	**517,668.0**	**468,186.0**	**532,597.0**	**d**
World Total	**704,392.0**	**b**	**720,897.0**	**668,272.0**	**740,527.0**	**d**

Sources: U.S. Bureau of Mines, World Bureau of Metal Statistics (Ware, United Kingdom), and other sources.
Notes: World reserves life index equals 1988 world reserves divided by 1988 world production.
World reserve base life index equals 1988 world reserve base divided by 1988 world production.
a. Production refers to bauxite, consumption data to aluminum. b. A life index would be misleading because production data include secondary metal.
c. Data refer to 1974. d. Data refer to 1987. e. Data refer to 1979.
0 = zero or less than half the unit of measure; X = not available. For additional information, see Sources and Technical Notes.

Table 21.5 Nuclear Power and Waste, 1970-89

| | Number of Commercial Reactors Operable 1970 | Number of Commercial Reactors (as of December 31, 1988) | | | | | Number of Research Reactors 1989 {a} | Net Capacity of Commercial Reactors (Megawatts) | | | Reactor Years of Experience (to December 31, 1988) | | Spent Fuel Inventories | | |
		Opera-ble	Under Cons-truction	Shut-down	Sus-pended	Can-celed		Opera-ble 1970	Opera-ble 1988	Under Cons-truction 1988	Years	Months	(cumulative tons of heavy metal) 1970	1988	(kg per 000 ha) 1988
WORLD	66	429	105	37	16	38	324	15,471	310,822	84,871	5,040	9	X	X	X
AFRICA	0	2	0	0	0	0	5	0	1,842	0	8	3	0	100	X
Algeria	0	0	0	0	0	0	1	0	0	0	0	0	0	0	0.0
Egypt	0	0	0	0	0	0	1	0	0	0	0	0	0	0	0.0
Libya	0	0	0	0	0	0	1	0	0	0	0	0	0	0	0.0
South Africa	0	2	0	0	0	0	1	0	1,842	0	8	3	0	100	0.8
Zaire	0	0	0	0	0	0	1	0	0	0	0	0	0	0	0.0
NORTH & CENTRAL AMERICA	14	126	15	14	8	33	110	6,433	107,458	13,337	1,467	10	151	28,606	X
Canada	1	18	4	3	0	0	14	22	12,185	3,524	206	0	96	11,000	11.9
Cuba	0	0	2	0	0	0	0	0	0	816	0	0	0	0	0.0
Jamaica	0	0	0	0	0	0	1	0	0	0	0	0	0	0	0.0
Mexico	0	0	2	0	0	0	3	0	0	1,308	0	0	0	0	0.0
United States	13	108	7	11	8	33	92	6,411	95,273	7,689	1,261	10	55	17,606	19.2
SOUTH AMERICA	0	3	2	0	0	0	15	0	1,561	1,937	27	4	0	0	X
Argentina	0	2	1	0	0	0	5	0	935	692	20	7	0	X	X
Brazil	0	1	1	0	0	0	4	0	626	1,245	6	9	0	X	X
Chile	0	0	0	0	0	0	2	0	0	0	0	0	0	0	0.0
Colombia	0	0	0	0	0	0	1	0	0	0	0	0	0	0	0.0
Peru	0	0	0	0	0	0	1	0	0	0	0	0	0	0	0.0
Venezuela	0	0	0	0	0	0	2	0	0	0	0	0	0	0	0.0
ASIA	6	59	26	1	1	2	57	1,658	40,726	18,131	564	4	0	7,200	X
Bangladesh	0	0	0	0	0	0	1	0	0	0	0	0	0	0	0.0
China	0	0	3	0	0	0	9	0	0	2,148	0	0	0	0	0.0
China, Taiwan	0	6	0	0	0	0	5	0	4,924	0	44	1	0	900	254.2
India	2	6	8	0	0	0	5	400	1,154	1,760	72	8	X	X	X
Indonesia	0	0	0	0	0	0	3	0	0	0	0	0	0	0	0.0
Iran, Islamic Rep	0	0	2	0	0	2	1	0	0	2,392	0	0	0	0	0.0
Iraq	0	0	0	0	0	0	2	0	0	0	0	0	0	0	0.0
Israel	0	0	0	0	0	0	2	0	0	0	0	0	0	0	0.0
Japan	4	38	12	1	0	0	18	1,258	28,253	10,931	394	0	X	5,600	148.7
Korea, Dem People's Rep	0	0	0	0	0	0	1	0	0	0	0	0	0	0	0.0
Korea, Rep	0	8	1	0	0	0	3	0	6,270	900	36	4	0	700	71.3
Malaysia	0	0	0	0	0	0	1	0	0	0	0	0	0	0	0.0
Pakistan	0	1	0	0	0	0	1	0	125	0	17	3	0	X	X
Philippines	0	0	0	0	1	0	1	0	0	0	0	0	0	0	0.0
Thailand	0	0	0	0	0	0	1	0	0	0	0	0	0	0	0.0
Turkey	0	0	0	0	0	0	2	0	0	0	0	0	0	0	0.0
Viet Nam	0	0	0	0	0	0	1	0	0	0	0	0	0	0	0.0
EUROPE	35	183	36	19	7	1	107 b	5,942	125,402	30,236	2,285	10	0	55,100	X
Austria	0	0	0	0	0	1	3	0	0	0	0	0	0	0	0.0
Belgium	0	7	0	1	0	0	5	0	5,480	0	86	7	0	700	84.6
Bulgaria	0	5	2	0	0	0	1	0	2,585	1,906	43	8	0	X	X
Czechoslovakia	0	8	8	1	0	0	3	0	3,264	5,120	44	1	0	X	X
Denmark	0	0	0	0	0	0	2	0	0	0	0	0	0	0	0.0
Finland	0	4	0	0	0	0	1	0	2,310	0	39	4	0	400	13.1
France	3	55	9	5	0	0	20	1,190	52,588	12,245	488	1	X	12,700	230.9
German Dem Rep	1	5	6	0	0	0	5	70	1,694	3,432	72	5	X	X	X
Germany, Fed Rep	1	23	2	6	0	0	21	328	21,491	1,520	279	3	X	3,300	135.1
Greece	0	0	0	0	0	0	2	0	0	0	0	0	0	0	0.0
Hungary	0	4	0	0	0	0	3	0	1,645	0	14	2	0	X	X
Italy	2	2	0	2	3	0	6	397	1,120	0	77	10	X	1,400	47.6
Netherlands	1	2	0	0	0	0	2	52	508	0	35	9	X	200	59.0
Norway	0	0	0	0	0	0	2	0	0	0	0	0	0	0	0.0
Poland	0	0	2	0	0	0	3	0	0	880	0	0	0	0	0.0
Portugal	0	0	0	0	0	0	1	0	0	0	0	0	0	0	0.0
Romania	0	0	5	0	0	0	2	0	0	3,300	0	0	0	0	0.0
Spain	1	10	0	0	4	0	0	153	7,519	0	82	7	X	2,800	56.1
Sweden	0	12	0	1	0	0	2	0	9,693	0	135	2	0	1,900	46.2
Switzerland	1	5	0	0	0	0	4	350	2,952	0	68	10	X	700	176.0
United Kingdom	25	40	2	3	0	0	15	3,402	11,921	1,833	810	10	X	30,900	1,279.0
Yugoslavia	0	1	0	0	0	0	3	0	632	0	7	3	0	100	3.9
U.S.S.R.	11	56	26	3	0	2	28	1,438	33,823	21,230	687	2	X	X	X
OCEANIA	0	0	0	0	0	0	2	0	0	0	0	0	0	0	X
Australia	0	0	0	0	0	0	2	0	0	0	0	0	0	0	0.0

Sources: International Atomic Energy Agency, U.S. Department of Energy, and Food and Agriculture Organization of the United Nations.
Notes:
a. Data as of September 1989.
b. The total includes one research reactor in operation under the Commission of European Communities (CEC).
0 = zero or less than half of the unit of measure; X = not available.
For additional information, see Sources and Technical Notes.

Table 21.6 Waste Generation and Trade in Selected Countries

	Annual Municipal Waste Generation					Annual Industrial Waste Generation				Annual Hazardous and Special Waste Generation				
	Total (000 metric tons)			Per Capita (kg) 1985	Per Unit Area (t per km2) 1985	Year of Estimate	Total (000 t)	Per Million $US of Industrial Gross Domestic Product (t)	Per Unit Area (t per km2)	Year of Estimate	Total (000 t per year)	Per Unit Area (t per km2)	Imports (000 t)	Exports (000 t)
	1975	1980	1985											
AFRICA														
Guinea	X	X	X	X	X	X	X	X	X	X	X	X	15.0	X
Nigeria	X	X	X	X	X	X	X	X	X	X	X	X	4.0	X
South Africa	X	X	X	X	X	X	X	X	X	X	X	X	0.0	X
Zimbabwe	X	X	X	X	X	X	X	X	X	X	X	X	6.9 a	X
AMERICAS														
Brazil	X	X	X	X	X	X	X	X	X	X	X	X	40.0	X
Canada	X	12,600	16,000	630	1.7	1980	61,000	730	6.6	1980	3,290	0.4	130.0	65.0
Costa Rica	X	X	534 b	216 b	10.5 b	X	X	X	X	X	X	X	X	X
Haiti	X	X	X	X	X	X	X	X	X	X	X	X	4.5	X
Mexico	X	X	32	0	0.0	1986	192	4	0.1	X	X	X	7.0	X
United States	140,000	160,000	178,000 b	762 b	19.4 b	1985	628,000	513	68.5	1985	265,000	28.9	45.3	203.4
Venezuela	X	X	X	X	X	X	X	X	X	X	X	X	2.0	X
ASIA														
Cyprus	X	X	X	X	X	1985	56	X	6.0	X	X	X	X	X
Hong Kong	X	X	X	X	X	1987	6	1	6.1	X	X	X	X	X
India	X	X	X	X	X	X	X	X	X	1980	35,722	12.0	X	X
Israel	X	X	1,400	331	68.9	X	X	X	X	X	30	1.5	X	X
Japan	38,074	41,511	41,530	344	110.3	1985	312,000	573	828.6	1986	666	1.8	3.0 a	X
Korea, Rep	X	X	15,746 b	396 b	159.5 b	1981	7,030	274	71.2	1981	180	1.8	X	X
Lebanon	X	X	X	X	X	X	X	X	X	X	X	X	2.4	X
Malaysia	X	X	X	X	X	X	X	X	X	1985	419 c	1.3 d	X	X
Singapore	X	1,082	1,498	585	2,455.7	X	X	X	X	X	X	X	X	X
EUROPE														
Austria	1,407 e	1,673 f	1,727 b	231 b	20.9 b	1983	13,258	510	160.3	1983	200	2.4	0.3	3.4
Belgium	2,900	3,082	X	313 g	93.9 g	1980	8,000	186	243.8	1980	915	27.9	914.1 h	13.2
Bulgaria	X	X	6,773	756	61.3	X	X	X	X	X	X	X	X	X
Czechoslovakia	X	X	X	X	X	1982	80,910	X	645.3	X	X	X	X	X
Denmark	X	2,046	2,161	422	51.0	1985	1,317	95	31.1	1985	125	3.0	X	20.0 a
Finland	X	X	2,000	408	6.6	1985	15,000	841	49.2	1985	124	0.4	X	2.8 a
France	X	14,000	15,000	272	27.3	1984	50,000	301	90.9	1984	2,000	3.6	95.9	25.0 h
German Dem Rep	X	X	X	X	X	X	X	X	X	X	X	X	814.3	X
Germany, Fed Rep	20,423 i	21,417	19,387 j	317 j	79.4 j	1984	55,932	198	229.0	1985	5,000	20.5	75.0 h	1,695.6 h
Greece	X	2,500	X	259 g	19.1 g	1980	3,904	378	29.8	X	X	X	X	X
Hungary	X	X	7,000	657	75.8	1985	21,146	2,509	229.0	1984	7,081	76.7	1.5	X
Iceland	X	X	93	386	0.9	1985	105	X	1.0	X	X	X	28.6	X
Ireland	555	640	1,100 j	309 j	16.0 j	1984	1,580	346	22.9	1984	20	0.3	X	20.0 j
Italy	14,095	14,041	15,000	263	51.0	1980	35,000	207	119.0	1980	2,000	6.8	X	22.8
Luxembourg	119	128	131	357	50.7	1985	135	X	52.2	1985	4	1.5	X	4.0 a
Netherlands	X	6,565 k	6,510	449	191.9	1986	3,942	66	116.2	1986	1,500	44.2	320.0 a	250.0 a
Norway	1,700	1,700	1,970	474	6.4	1980	2,186	93	7.1	1980	120	0.4	X	0.3
Poland	X	X	7,900	212	25.9	1985	274,885	X	902.8	X	X	X	X	1.4
Portugal	X	1,948	2,246	221	24.4	1980	11,200	1,110	121.8	1986	1,049	11.4	X	X
Romania	X	X	X	X	X	X	X	X	X	X	X	X	4.0 a	X
Spain	X	8,028 l	10,600	275	21.2	1986	5,108	60	10.2	1987	1,708	3.4	X	2.6
Sweden	2,400	2,510	2,650	317	6.4	1980	4,000	102	9.7	1980	500	1.2	X	15.0 m
Switzerland	1,900	2,240	2,500	386	62.9	X	X	X	X	1987	120	3.0	7.1	68.0 n
United Kingdom	16,036 o	15,816	17,737 p	313 p	73.4 p	1984	50,000	327	207.0	1986	3,900	16.1	82.5	X
Yugoslavia	X	X	X	X	X	X	X	X	X	X	X	X	2.4	X
U.S.S.R.	X	X	X	X	X	1985	306,311	X	13.8	X	X	X	X	X
OCEANIA														
Australia	X	10,000	X	679 g	1.3 g	1980	20,000	386	2.6	1980	300	0.0	X	0.7 a
New Zealand	1,150	2,106 q	X	606 q	7.8 n	1982	300	38	1.1	1982	45	0.2	X	0.1

Sources: Organisation for Economic Co-operation and Development, United Nations Environment Programme, and other sources.
Notes: a. Refers to 1983. b. Refers to 1983. c. Thousand cubic meters per year. d. Cubic meters per year. e. Refers to 1973. f. Refers to 1979. g. Refers to 1980.
h. Refers to 1985. i. Refers to 1977. j. Refers to 1984. k. Refers to 1981. l. Refers to 1978. m. Refers to 1980. n. Refers to 1986. o. Refers to 1976.
p. Refers to 1987. q. Refers to 1982.
0 = zero or less than half the unit of measure; X = not available; t = metric tons; km2 = square kilometers.
For additional information, see Sources and Technical Notes.

Sources and Technical Notes

Table 21.1 Commercial Energy, 1977–87

Sources: Energy: United Nations Statistical Office, *Energy Statistics Yearbook 1980* and *1987* (United Nations, New York, 1982 and 1989). Gross national product: The World Bank, unpublished data (The World Bank, Washington, D.C., April 1989).

Energy data are compiled by the United Nations Statistical Office (UNSO) primarily from responses to questionnaires sent to national governments, supplemented by official national statistical publications and data from intergovernmental organizations. When official numbers are not available, UNSO prepares estimates based on the professional and commercial literature.

Total production of commercially traded fuels includes the production of solid, liquid, and gaseous fuels and the production of primary electricity. Electricity production data are shown in Table 21.2. *Solid fuels* include bituminous coal, lignite, peat, and oil shale burned directly. *Liquid fuels* include crude petroleum and natural gas liquids. *Gas* includes natural gas and other petroleum gases. Fuelwood, bagasse, charcoal, and all forms of solar energy are excluded from production figures, even when traded commercially.

Consumption refers to "apparent consumption" and is defined as domestic production plus net imports, minus net stock increases, minus aircraft and marine bunkers. *Total consumption* includes energy from solids, liquids, gases, and primary electricity.

All the production data and the total consumption data are in petajoules (10^{15} joules). One petajoule is the same as 0.0009478 Quads (10^{15} British Thermal Units) and is the equivalent of 163,400 "U.N. standard" barrels of oil or 34,140 "U.N. standard" metric tons of coal. The heat content of various fuels has been converted to coal-equivalent and then petajoule-equivalent values using country- and year-specific conversion factors. For example, a metric ton of bituminous coal produced in Argentina has an energy value of 0.843 metric tons of standard coal equivalent (7 million kilocalories). A metric ton of bituminous coal produced in Turkey has an energy value of 0.871 metric tons of standard coal equivalent. The original national production data for bituminous coal were multiplied by these conversion factors and then by 29.3076×10^{-6} to yield petajoule equivalents. Other fuels were converted to coal equivalent and petajoule-equivalent terms in a similar manner.

South Africa refers to the South Africa Customs Union: South Africa, Botswana, Lesotho, Swaziland, and Namibia.

Table 21.2 Production and Trade of Electricity, 1977–87

Source: United Nations Statistical Office, *Energy Statistics Yearbook 1980* and *1987* (United Nations, New York, 1982 and 1989).

Electricity *production* data generally refer to gross production. Data for the Dominican Republic, Finland, France (including Monaco), Iceland, Mexico, Switzerland, the United States, Zambia, and Zimbabwe refer to net production. Gross production is the amount of electricity produced by a generating station before consumption by station auxiliaries and transformer losses within the station are deducted. Net production is the amount of electricity remaining after these deductions. Typically, net production is 5–10 percent less than gross production. Energy production from pumped storage is not included in gross or net electricity generation.

A gigawatt-hour of electricity is 10^9 watt-hours. A gigawatt-hour is the equivalent of 3,412 million British Thermal Units or 3.6×10^{12} joules.

Electricity production includes both public and self-producer power plants. Public power plants produce electricity for many users. They may be operated by private, cooperative, or governmental organizations. Self-producer power plants are operated by organizations or companies to produce electricity for internal applications, such as factory operations.

Table 21.3 Reserves and Resources of Commercial Energy

Sources: World Energy Conference (WEC), *1989 Survey of Energy Resources* (WEC, London, 1989). Hydroelectric technical potential: The World Bank, *A Survey of the Future Role of Hydroelectric Power in 100 Developing Countries* (The World Bank, Washington, D.C., 1984). Hydroelectric installed capacity: United Nations Statistical Office, *Energy Statistics Yearbook 1987* (United Nations, New York, 1989). United States (hydroelectric potential and installed capacity): U.S. Federal Energy Regulatory Commission (FERC), unpublished data (FERC, Washington, D.C., 1988).

Energy resource estimates are based on geological, economic, and technical criteria. Resources are first graded according to the degree of confidence in the extent and location of the resource, based on available geological information. Judgments on the technical and economic feasibility of exploiting the resource are then incorporated into the assessment.

Proved reserves in place represent the total resource that is known to exist in specific locations and in specific qualities. *Proved recoverable reserves* are the fraction of proved reserves in place that can be extracted with existing technology under present and expected economic conditions. Additional energy resources, comprising those that are currently subeconomic, are not shown.

Bituminous coal includes anthracite. Anthracite is probably only a small fraction of the total (3–4 percent), but it is impossible to calculate the exact amount of anthracite included in the figures.

In the *lignite and subbituminous coal* aggregate, lignite accounts for 67 percent of the global proved reserves in place and 75 percent of the global proved recoverable reserves.

Crude oil includes natural gas liquids and reservoir gas recovered in liquid form in surface separators or plant facilities.

Uranium data refer to known uranium deposits of a size and quality that could be recovered within specified production cost ranges (under $80 per kilogram and $80–130 per kilogram) using currently proven mining and processing technology.

Hydroelectric technical potential refers to the annual energy potential of all sites where it is physically possible to construct dams, with no consideration of economic return or of adverse impacts of site development. Data for 69 countries refer to "exploitable potential," the annual energy that could be generated by hydroelectric plants within the limits of current technology and under present and expected local economic conditions. Data for one country refer to "theoretical potential," the annual energy potentially available in the country if all natural water flows were turbined to sea level or to the water level of a country's border with 100 percent mechanical efficiency. Theoretical potential is estimated from national precipitation and water runoff data and is not limited by technical or economic criteria. Exploitable potential, the most conservative estimate of a country's hydroelectric resources, is shown when available; theoretical potential is shown only when technical and exploitable potential figures are not available. All three types of estimates include sites where hydroelectric generating plants are currently in place. All data on exploitable and theoretical hydroelectric potential assume use of 5,000 hours per year as representative for all hydropower (57 percent load factor). Data on technical potential assigned site-specific load factors.

Installed capacity refers to the combined generating capacity of hydroelectric plants installed in the country as of December 31, 1987.

Table 21.4 Production, Consumption, and Reserves of Selected Metals, 1973–88

Sources: Production data for 1973, 1978, and 1983: U.S. Bureau of Mines (U.S. BOM), *Minerals Yearbook 1975, 1980,* and *1986* (U.S. Government Printing Office, Washington, D.C., 1975, 1980, and 1986). Production data for 1988: U.S. BOM, unpublished data (U.S. BOM, July 1989).

Consumption data for aluminum, cadmium, copper, lead, nickel, tin, and zinc: World Bureau of Metal Statistics, *World Metal Statistics* (World Bureau of Metal Statistics, Ware, United Kingdom, February 1979, December 1980, and August 1989). Consumption data for mercury: Roskill Information Services Ltd., *Roskill's Metals Databook, 5th Edition 1984* (Roskill, London, March 1984; Roskill Information Services Ltd., *Statistical Supplement to The Economics of Mercury, 4th Edition 1978* (Roskill, London, 1980); Roskill Information Services Ltd., *The Economics of Mercury, 5th Edition 1984* (Roskill, London, 1984). Consumption data for iron ore: United Nations Economic Commission for Europe (ECE), *Annual Bulletin of Steel Statistics for Europe 1976, 1982, 1986,* and *1987* (United Nations, New York, 1977, 1983, 1987, and 1988); Organisation for Economic Co-operation and Development (OECD), *The Iron and Steel Industry in 1974, 1978, 1983,* and *1987* (OECD, Paris, 1976, 1980, 1985, and 1988); International Iron and Steel Institute, *Steel Statistical Yearbook 1984* and *1988* (International Iron and Steel Institute, Brussels, 1984 and 1988). Consumption data for crude steel: International Iron and Steel Institute, *Steel Statistical Yearbook 1984* and *1988* (International Iron and Steel Institute, Brussels, 1984 and 1988).

Reserves and reserve base data: U.S. Bureau of Mines (U.S. BOM), *Mineral Commodity Summaries 1989* (U.S. Government Printing Office, Washington, D.C., 1989).

The U.S. BOM prepares trade, consumption, and other data on commodities for the United States as well as for all other countries of the world (depending on availability of reliable data) based on information from government mineral and statistical agencies, the United Nations, and U.S. and foreign technical and trade literature.

The World Bureau of Metal Statistics publishes consumption data for the metals presented, excluding mercury, iron, and steel. Data on the metals included were supplied by metal companies, government agencies, trade groups, and statistical bureaus. Obviously incorrect data have been revised, but most data were compiled and reported without adjustment or retrospective revisions.

The countries listed represent the top 10 producers and the top 10 consumers of each material in 1988.

The *annual production* data are the metal content of the ore mined for *copper, lead, mercury, nickel, tin,* and *zinc. Aluminum* (bauxite) and *iron ore* production are expressed in gross weight of ore mined (marketable product). Iron ore production refers to iron ore, iron ore concentrates, and iron ore agglomerates (sinter and pellets). *Cadmium* is the production of the refined metal. *Crude steel* production is defined as the total of usable ingots, continuously cast semifinished products, and liquid steel for castings. The United Nations definition of crude steel is the equivalent of the term "raw steel" as used by the United States.

Annual consumption of metal refers to the domestic use of refined metals. These metals include metals refined from either primary (raw) or secondary (recovered) materials. Metal used in a product that is then exported is considered consumed by the producing country rather than by the importing country. Data on *mercury* consumption must be viewed with caution; they include estimates on consumption of secondary materials, which are generally not reported. Consumption of *iron ore* was calculated by adding net imports to the quantities of iron ore and concentrates reported as delivered to consuming industries. Data for Brazil, China, and the U.S.S.R. are calculated as apparent consumption, the net of production plus imports minus exports. Such a consumption number makes no allowance for stock inventories. Because different countries report different grades of iron ore, consumption data are not strictly comparable among countries. Because world consumption of iron ore is roughly equal to world production, world production data were used for the world consumption totals. Worldwide stock inventories are assumed to be negligible. *Crude steel* consumption is calculated as apparent consumption. The International Iron and Steel Institute converted imports and exports into crude steel equivalent by using a factor of $1.3/(1 + 0.175c)$, where c is the domestic proportion of crude steel that is continuously cast. Such an adjustment avoids distortion of the export or import share relative to domestic production.

The *world reserve base life index* and the *world reserves life index* are expressed in years remaining. They were computed by dividing the 1988 world reserve base and world reserves by the respective world production rate of 1988. The underlying assumption is a constant world production at the 1988 level and capacity.

The reserve base is the portion of the mineral resource that meets grade, quality, thickness, and depth criteria defined by current mining and production practices. It includes both measured and indicated reserves and refers to those resources that are both currently economic and marginally economic, as well as some of those that are currently subeconomic.

Mineral reserves are the part of the reserve base that could be economically extracted or produced at the time of the assessment. Reserves do not signify that extraction facilities are in place and operative.

Table 21.5 Nuclear Power and Waste, 1970–89

Sources: International Atomic Energy Agency (IAEA), *Nuclear Power Reactors in the World*, April 1989 ed. (IAEA, Vienna, 1989). Research reactors: International Atomic Energy Agency (IAEA), unpublished data (IAEA, Vienna, September 1989). Spent fuel data: U.S. Department of Energy (DOE), Energy Information Administration (EIA), *World Nuclear Fuel Cycle Requirements* (U.S. DOE/EIA, Washington, D.C., 1989) and U.S. DOE/EIA unpublished data (U.S DOE/EIA, Washington, D.C., 1988).

The *number of commercial reactors* includes *operable* reactors, which produce electricity for the commercial electricity grid, although not necessarily at full power.

Reactors *under construction* are those for which major placing of concrete, usually for the base mat of the reactor building, is done. The time period covered is up to December 31, 1988. *Shut-down reactors* are those officially shut down by the owner and taken out of operation permanently. Reactors *suspended* are plants for which construction work has been suspended but not canceled. The number *canceled* refers to power plants that were permanently canceled while under construction. The period covered for the categories "suspended" and "canceled" is 1988 only.

The *number of research reactors* includes all research, test, and training reactors, as well as critical assemblies. In some cases, prototype power reactors are included. Otherwise, the category might be called "nonpower reactors" as these reactors are sometimes referred to in the United States.

Under *net capacity of commercial reactors*, the electricity requirements of generating plants (usually about 5–10 percent of gross generation) have been deducted. Capacity planned or *under construction* refers to the total additional capacity that would be possible if all the reactors planned or under construction were completed.

Reactor years of experience is the operation experience accumulated by operating and shut-down reactors. The reactor-years are counted from the date of first grid connection up to December 31, 1988, for operating reactors and up to time of shut down for shut-down reactors.

Spent fuel inventories are expressed as cumulative totals up to the years given and are net of reprocessing. Heavy metal refers to the actinide elements (uranium, plutonium, etc.) contained in the spent fuel.

Table 21.6 Waste Generation and Trade in Selected Countries

Sources: Generation of municipal, industrial, hazardous, and special waste in Organisation for Economic Co-operation and

Development (OECD) countries: OECD, *Environmental Data Compendium 1989* (OECD, Paris, 1989). Bulgaria, Cyprus, Czechoslovakia, Hungary, Iceland, Poland, and the U.S.S.R.: United Nations Statistical Commission and Economic Commission for Europe (ECE), *Environment Statistics in Europe and North America* (United Nations, New York, 1987). Hong Kong, India, and Malaysia: United Nations Environment Programme (UNEP), *Environmental Data Report* (Basil Blackwell, Oxford, 1989). Costa Rica: Gary Hartshorn, Lynne Hartshorn, Augustin Atmella, et al. *Costa Rica: Country Environmental Profile* (Tropical Science Center and United States Agency for International Development, San Jose, 1982). Mexico: I. Fernando Ortiz Monasterio, *Manejo de los Desechos Industriales Peligrosos en Mexico* (Universo Veintiuno, Mexico, 1987). Israel: Uri Marinov, State of Israel Environmental Protection Service, personal communication (Jerusalem, Israel, 1988). Republic of Korea: Soo-Saeng Han, *The State of the Environment in Korea* (Office of Environment, Republic of Korea, 1983). Singapore: Ministry of the Environment, Singapore, *Annual Report '85* (Ministry of the Environment, Singapore, 1985).

Imports and exports of hazardous waste: United Nations Environment Programme (UNEP), *Environmental Data Report* (Basil Blackwell, Oxford, 1989). Exports for Australia, Denmark, Finland, France, Ireland, Luxembourg, Sweden, and Switzerland and imports for West Germany: OECD, *Environmental Data Compendium 1989* (OECD, Paris, 1989). Imports for Japan and Romania, and imports and exports for the Netherlands: Jim Vallette, *The International Trade in Wastes: A Greenpeace Inventory*, 4th ed. (Greenpeace, Washington, 1989).

Industrial gross domestic product: The World Bank, *World Development Report 1982–1988* (Oxford University Press, New York, 1982–1988). Geographical area: United Nations Food and Agricultural Organization (FAO), unpublished data (FAO, Rome, July 1989). Population data: United Nations Population Division, *World Population Prospects: Estimates and Projections as Assessed in 1982* (United Nations, New York, 1985); United Nations Population Division, *World Population Prospects 1988* (United Nations, New York, 1989).

Waste data were collected by various means and are not strictly comparable among countries. OECD collects data using questionnaires completed by government representatives. Refer to the

Technical Note for Table 15.1 in Chapter 15, "Basic Economic Indicators," for details concerning gross domestic product. Area data exclude inland bodies of water.

Annual municipal waste generation refers to the trash collected from households, commercial establishments, and small industries. *Per capita* and *per unit area* generation of municipal waste are for the most recent year available. *Annual industrial waste generation* refers to both chemical and nonchemical materials. Amounts depend on the definition of waste used in a country, the levels of industrial production, and the types of technology used. Waste generation *per million $US of industrial gross domestic product* is the portion of gross domestic product (GDP) contributed by industry. *Annual hazardous and special waste generation* refers to waste known to contain potentially harmful substances. Definitions of hazardous waste vary among countries. Nuclear wastes are not included in this table. Refer to Table 21.5 for data on spent fuel from nuclear reactors.

The reported data for *imports* and *exports* of hazardous waste are preliminary estimates that may be subject to some variation. Because the quantity of wastes crossing borders is difficult to determine and because many transfrontier movements take place illegally, only documented shipments are listed. Recent media reports on uncompleted or attempted shipments of hazardous materials indicate that the volume and extent of the toxic trade, especially to developing countries, is far greater than portrayed in this table. Thus the data reflect only the inadequacy of current worldwide notification and documentation procedures. Most of the entries for Europe are by classification from the exporting country. If not otherwise noted, data refer to 1987.

Country-specific information is presented alphabetically, as follows:
■ Australia—industrial waste includes hazardous and special waste.
■ Austria—industrial waste includes hazardous waste; exports include 400 metric tons of waste delivered for incineration.
■ Belgium—imports of hazardous waste include 14,400 metric tons of waste delivered for incineration and 893,000 metric tons from the Federal Republic of Germany for sea disposal. Actual exports of hazardous waste may be less than the reported value. The Federal Republic of Germany issued a permit for incineration of 6,650 metric tons of wastes from Belgium,

the Federal Republic of Germany, France, and Italy. The relative percentage of this total is not known. All 6,650 metric tons are allocated to Belgium's export figure.
■ Bulgaria—municipal waste refers to household waste only.
■ Canada—hazardous waste is measured in wet weight.
■ Czechoslovakia—industrial waste data are obtained from infrequent surveys; coverage is confined to national industrial enterprises.
■ Denmark—includes only hazardous waste that has been disposed of legally.
■ Finland—municipal waste includes only waste originating in households.
■ France—hazardous waste is the amount of toxic or hazardous waste. Special wastes, totaling 18 million metric tons per year, are not included.
■ Hong Kong—industrial waste includes nonchemical waste only.
■ Hungary—municipal waste weight is estimated from volume data. The hazardous waste figure is based on enterprise declarations.
■ Iceland—municipal waste refers to nonindustrial solid waste.
■ Italy—the hazardous and special waste export figure is greater than the listed total. Large quantities of wastes have been exported to Africa and Southwest Asia, but no specific data are available.
■ Netherlands—industrial waste refers to chemical waste. It includes wastes generated only by enterprises employing 10 or more people and includes office and canteen wastes; hazardous waste refers to "notifiable" wastes only. Most of the waste imports are halogenated solvents.
■ New Zealand—industrial waste is nonchemical waste only.
■ Norway—industrial waste includes chemical waste only; export data are for halogenated substances only.
■ United Kingdom—municipal, industrial, and hazardous waste generation are for England and Wales only. Hazardous waste refers to notifiable wastes.
■ United States—the 1983 municipal waste figure is an OECD estimate. Industrial waste includes wastewaters that meet the U.S. definition of solid waste.

For a more detailed discussion of hazardous waste data collection and reports on hazardous waste management in 12 countries, see William S. Forester and John H. Skinner, *International Perspectives on Hazardous Waste Management* (Academic Press, London, 1987).

22. Freshwater

The world's supply of freshwater is limited and, like many other resources, is not equally distributed among the peoples of the world. But clean freshwater is essential for health, agriculture, and industry. Its control and distribution are ancient functions of government.

Other than through desalinization, governments can control access to three sources of water. The renewable freshwater resource easiest to regulate is net precipitation (precipitation minus evapotranspiration). (See Table 22.1.) Governments routinely regulate the use, quality, and control of runoff or groundwater resulting from precipitation within its boundaries. But precipitation can vary dramatically from year to year even in normal times, and during a period of climatic change this variability might become more extreme. Many countries, particularly those in Southwest Asia and northern Africa, do not receive adequate precipitation to meet their needs and instead rely on a second source, river flows from neighboring countries, to supply the difference.

The quality and quantity of runoff and groundwater from neighboring countries is less accessible to a government's control. Governments resort to international agreements (such as that between Sudan and Egypt regulating the flow of the Nile) to protect their water supplies.

A third—and nonrenewable source—of freshwater are ancient aquifers. Whether closed or simply drawn down faster than they are recharged, the fossil waters of these aquifers can be tapped to fuel intensive irrigated agriculture and urban life (e.g., the Ogallala Aquifer in the United States and the Nubian Aquifer in Libya). Once these aquifers are depleted, the agriculture, industry, and settlements they support must either find new sources or be abandoned.

The amount of freshwater per capita that is withdrawn (taken for use) from surface and subsurface sources is not correlated with either a country's economic wealth or the size of its internal renewable water resources. Many arid countries, such as Afghanistan, Sudan, Egypt, People's Democratic Republic of Yemen, the Islamic Republic of Iran, and Iraq, annually withdraw over 1,000 cubic meters per capita; most of this water is used in irrigation. Countries that use almost all or even more than their total renewable freshwater resources include: Egypt, Libya, Cyprus, Israel, Qatar, Saudi Arabia, United Arab Emirates, the People's Democratic Republic of Yemen, and Malta. Some temperate and developed countries also use large amounts of freshwater. The United States, for example, uses 2,162 cubic meters per capita annually; its total use is higher than any other country. Canada, Bulgaria, the Netherlands, Portugal, Romania, the U.S.S.R., and Australia use over 1,000 cubic meters per capita.

Streams, rivers, and lakes are used for power generation, fishing, recreation, and transportation. They also provide and nourish habitat for wildlife and affect the microclimates of whole ecosystems. But the flows of these streams and rivers, as well as lake and groundwater resources, can be depleted as water is withdrawn for agriculture, industry, and domestic use. Evaporation from artificial lakes or impoundments can add a substantial fraction to this depletion, especially in arid countries. For most of the countries of the world, including some developed North American and European countries, agriculture uses the greatest share of freshwater resources. Industrial uses and ordinary domestic uses take large shares of withdrawals in wetter countries (where irrigation is uncommon) and in the highly industrialized countries of the world. These uses tend to change the character of freshwater that is ultimately returned for reuse farther downstream. Often this water is polluted chemically, biologically, or thermally. (See Chapter 10, "Freshwater.")

There is growing concern over the quality of freshwater available to the world's peoples. Table 22.2 shows some water quality data from selected monitoring stations that report to the only global water monitoring entity, the United Nation's Environment Programme's Global Environment Monitoring System's Water Quality Monitoring Project (GEMS/WATER). GEMS water quality data are spotty, both spatially and temporally. Not all stations measure or report on all water quality parameters, or even the same water quality parameters from year to year. Lake and groundwater stations tend to report better quality of water than do rivers because they are usually used or constructed primarily for drinking water.

Heavily polluted rivers (such as the Lerma in Mexico, the Sabarmati in India, and the Espierre in Belgium) show low levels of dissolved oxygen, high levels of biochemical oxygen demand, and high concentrations of faecal coliforms, the sentinel organisms used to represent the presence of pathogens. Most people in the world depend on surface water and shallow groundwater. Improving the quality of these waters is a major public health effort.

Table 22.1   Freshwater Resources and Withdrawal

	Annual Internal Renewable Water Resources 1990		Annual River Flows		Annual Withdrawal				Sectoral Withdrawal (percent)		
	Total (cubic km)	Per Capita (000 cubic meters)	From Other Countries (cubic km)	To Other Countries (cubic km)	Year of Data	Total (cubic km)	Percentage of Water Resources {a}	Per Capita (cubic meters)	Domestic	Industry	Agriculture
WORLD	40,673.00 b	7.69			1987 b	3296	8	660	8	23	69
AFRICA	4,184.00 b	6.46			1987 b	144	3	244	7	5	88
Algeria	18.90	0.75	0.20	0.70	1980	3.00	16	161	22	4	74
Angola	158.00 b	15.77	X	X	1987 b	0.48	0	43	14	10	76
Benin	26.00	5.48	X	X	1987 b	0.11	0	26	28	14	58
Botswana	1.00	0.78	17.00	X	1980	0.09	1	98	5	10	85
Burkina Faso	28.00 b	3.11	X	X	1987 b	0.15	1	20	28	5	67
Burundi	3.60 b	0.66	X	X	1987 b	0.10	3	20	36	0	64
Cameroon	208.00	18.50	X	X	1987 b	0.40	0	30	46	19	35
Cape Verde	0.20	0.53	0.00	0.00	1972	0.04	20	148	9	2	89
Central African Rep	141.00 b	48.40	X	X	1987 b	0.07	0	27	21	5	74
Chad	38.40 b	6.76	X	X	1987 b	0.18	0	35	16	2	82
Comoros	1.02 b	1.97	0.00	0.00	1987 b	0.01	1	15	48	5	47
Congo	181.00 b	90.77	621.00	X	1987 b	0.04	0	20	62	27	11
Cote d'Ivoire	74.00	5.87	X	X	1987 b	0.71	1	68	22	11	67
Djibouti	0.30	0.74	0.00	X	1973 b	0.01	2	28	28	21	51
Egypt	1.80	0.03	56.50	0.00	1985	56.40	97	1,202	7 c	5 c	88 c
Equatorial Guinea	30.00 b	68.18	X	X	1987 b	0.01	0	11	81	13	6
Ethiopia	110.00	2.35	X	X	1987 b	2.21	2	48	11	3	86
Gabon	164.00 b	140.05	X	X	1987 b	0.06	0	51	72	22	6
Gambia, The	3.00	3.50	19.00	X	1982	0.02	0	33	7	2	91
Ghana	53.00	3.53	X	X	1970	0.30	1	35	35	13	52
Guinea	226.00 b	32.87	X	X	1987 b	0.74	0	115	10	3	87
Guinea-Bissau	31.00 b	31.41	X	X	1987 b	0.01	0	18	31	6	63
Kenya	14.80	0.59	X	X	1987 b	1.09	7	48	27	11	62
Lesotho	4.00 b	2.25	X	X	1987 b	0.05	1	34	22	22	56
Liberia	232.00	90.84	X	X	1987 b	0.13	0	54	27	13	60
Libya	0.70	0.15	0.00	0.00	1985	2.62	374	262	15	10	75
Madagascar	40.00	3.34	0.00	0.00	1984	16.30	41	1,675	1	0	99
Malawi	9.00 b	1.07	X	X	1987 b	0.16	2	22	34	17	49
Mali	62.00 b	6.62	X	X	1987 b	1.36	2	159	2	1	97
Mauritania	0.40	0.20	7.00	X	1978	0.73	10	473	12	4	84
Mauritius	2.20	1.99	0.00	0.00	1974	0.36	16	415	16	7	77
Morocco	30.00	1.19	0.00	0.30	1985	11.00	37	501	6 c	3 c	91 c
Mozambique, People's Rep	58.00 b	3.70	X	X	1987 b	0.76	1	53	24	10	66
Niger	14.00	1.97	30.00	X	1987 b	0.29	1	44	21	5	74
Nigeria	261.00 b	2.31	47.00	X	1987 b	3.63	1	44	31	15	54
Rwanda	6.30 b	0.87	X	X	1987 b	0.15	2	23	24	8	68
Senegal	23.20 b	3.15	12.00	X	1987 b	1.36	4	201	5	3	92
Sierra Leone	160.00 b	38.54	X	X	1987 b	0.37	0	99	7	4	89
Somalia	11.50	1.52	0.00	X	1987 b	0.81	7	167	3	0	97
South Africa	50.00	1.42	X	X	1970	9.20	18	404	16	17	67
Sudan	30.00	1.19	100.00	56.50	1977	18.60	14	1,089	1	0	99
Swaziland	6.96 b	8.82	X	X	1987 b	0.29	4	414	5	2	93
Tanzania	76.00 b	2.78	X	X	1970	0.48	1	36	21	5	74
Togo	11.50	3.33	X	X	1987 b	0.09	1	40	62	13	25
Tunisia	3.75	0.46	0.60	0.00	1985	2.30	53	325	13	7	80
Uganda	66.00 b	3.58	X	X	1970	0.20	0	20	32	8	60
Zaire	1,019.00 b	28.31	X	X	1987 b	0.70	0	22	58	25	17
Zambia	96.00 b	11.35	X	X	1970	0.36	0	86	63	11	26
Zimbabwe	23.00 b	2.37	X	X	1987 b	1.22	5	129	14	7	79
NORTH & CENTRAL AMERICA	6,945.00 b	16.26			1987 b	697	10	1,692	9	42	49
Barbados	0.05	0.20	0.00	0.00	1962	0.03	51	117	52	41	7
Canada	2,901.00	109.37	X	X	1980	36.15	1	1,501	18	70	12
Costa Rica	95.00	31.51	X	X	1970	1.35	1	779	4	7	89
Cuba	34.50	3.34	0.00	0.00	1975	8.10	23	868	9	2	89
Dominican Rep	20.00	2.79	X	X	1987 b	2.97	15	453	5	6	89
El Salvador	18.95	3.61	X	X	1975	1.00	5	241	7	4	89
Guatemala	116.00	12.61	X	X	1970	0.73	1	139	9	17	74
Haiti	11.00	1.69	X	X	1987 b	0.04	0	46	24	8	68
Honduras	102.00	19.85	X	X	1970	1.34	1	508	4	5	91
Jamaica	8.30	3.29	0.00	0.00	1975	0.32	4	157	7	7	86
Mexico	357.40	4.03	X	X	1975	54.20	15	901	6	8	86
Nicaragua	175.00	45.21	X	X	1975	0.89	1	370	25	21	54
Panama	144.00	59.55	X	X	1975	1.30	1	744	12	11	77
Trinidad and Tobago	5.10 b	3.98	0.00	0.00	1975	0.15	3	149	27	38	35
United States	2,478.00	9.94	X	X	1985	467.00	19	2,162	12 c	46 c	42 c
SOUTH AMERICA	10,377.00 b	34.96			1987 b	133	1	476	18	23	59
Argentina	694.00	21.47	300.00	X	1976	27.60	3	1,059	9	18	73
Bolivia	300.00 b	41.02	X	X	1987 b	1.24	0	184	10	5	85
Brazil	5,190.00	34.52	1760.00	X	1987 b	35.04	1	212	43	17	40
Chile	468.00 b	35.53	X	X	1975	16.80	4	1,625	6	5	89
Colombia	1,070.00	33.63	X	X	1987 b	5.34	0	179	41	16	43
Ecuador	314.00	29.12	X	X	1987 b	5.56	2	561	7	3	90
Guyana	241.00 b	231.73	X	X	1971	5.40	2	7,616	1	0	99
Paraguay	94.00 b	21.98	220.00	X	1987 b	0.43	0	111	15	7	78
Peru	40.00	1.79	X	X	1987 b	6.10	15	294	19	9	72
Suriname	200.00 b	496.28	X	X	1987 b	0.46	0	1,181	6	5	89
Uruguay	59.00 b	18.86	65.00	X	1965	0.65	1	241	6	3	91
Venezuela	856.00	43.37	461.00	X	1970	4.10	0	387	43	11	46

Table 22.1

	Annual Internal Renewable Water Resources		Annual River Flows		Annual Withdrawal				Sectoral Withdrawal (percent)		
	Total (cubic km)	1990 Per Capita (000 cubic meters)	From Other Countries (cubic km)	To Other Countries (cubic km)	Year of Data	Total (cubic km)	Percentage of Water Resources {a}	Per Capita (cubic meters)	Domestic	Industry	Agriculture
ASIA	**10,485.00**	**3.37**			**1987 b**	**1,531.00**	**15**	**526**	**6**	**8**	**86**
Afghanistan	50.00	3.02	X	X	1987 b	26.11	52	1,436	1	0	99
Bahrain	0.00	0.00	X	X	1975	0.20	X	735	60	36	4
Bangladesh	1,357.00	11.74	1000.00	X	1987 b	22.50	1	211	3	1	96
Bhutan	95.00 b	62.66	X	X	1987 b	0.02	0	15	36	10	54
China	2,800.00	2.47	0.00	X	1980	460.00	16	462	6	7	87
Cyprus	0.90	1.28	0.00	0.00	1985	0.54	60	807	7 c	2 c	91 c
India	1,850.00	2.17	235.00	X	1975	380.00	18	612	3	4	93
Indonesia	2,530.00	14.02	X	X	1987 b	16.59	1	96	13	11	76
Iran, Islamic Rep	117.50	2.08	X	X	1975	45.40	39	1,362	4	9	87
Iraq	34.00	1.80	66.00	X	1970	42.80	43	4,575	3	5	92
Israel	1.70	0.37	0.45	0.00	1986	1.90	88	447	16 c	5 c	79 c
Japan	547.00	4.43	0.00	0.00	1980	107.80	20	923	17	33	50
Jordan	0.70	0.16	0.40	X	1975	0.45	41	173	29	6	65
Kampuchea, Dem	88.10	10.68	410.00	X	1987 b	0.52	0	69	5	1	94
Korea, Dem People's Rep	67.00 b	2.92	X	X	1987 b	14.16	21	1,649	11	16	73
Korea, Rep	63.00	1.45	X	X	1976	10.70	17	298	11	14	75
Kuwait	0.00	0.00	0.00	X	1974	0.01	X	10	64	32	4
Lao People's Dem Rep	270.00	66.32	X	X	1987 b	0.99	0	228	8	10	82
Lebanon	4.80	1.62	0.00	0.86	1975	0.75	16	271	11	4	85
Malaysia	456.00	26.30	X	X	1975	9.42	2	765	23	30	47
Mongolia	24.60	11.05	X	X	1987 b	0.55	2	272	11	27	62
Myanmar	1,082.00	25.96	X	X	1987 b	3.96	0	103	7	3	90
Nepal	170.00	8.88	X	X	1987 b	2.68	2	155	4	1	95
Oman	2.00	1.36	0.00	X	1975	0.43	22	561	3	3	94
Pakistan	298.00	2.43	170.00	X	1975	153.40	33	2,053	1	1	98
Philippines	323.00	5.18	0.00	0.00	1975	29.50	9	693	18	21	61
Qatar	0.02	0.06	0.00	X	1975	0.04	174	234	36	26	38
Saudi Arabia	2.20	0.16	0.00	X	1975	2.33	106	321	45	8	47
Singapore	0.60	0.22	0.00	0.00	1975	0.19	32	84	45	51	4
Sri Lanka	43.20	2.51	0.00	0.00	1970	6.30	15	503	2	2	96
Syrian Arab Rep	7.60	0.61	27.90	30.00	1976	3.34	9	449	7	10	83
Thailand	110.00	1.97	69.00	X	1987 b	31.90	18	599	4	6	90
Turkey	196.00	3.52	7.00	69.00	1985	15.60	8	317	24 c	19 c	57 c
United Arab Emirates	0.30	0.19	0.00	X	1980	0.42	140	429	11	9	80
Viet Nam	376.00 b	5.60	X	X	1987 b	5.07	1	81	13	9	78
Yemen Arab Rep	1.00	0.12	0.00	X	1987 b	1.47	147	X	4	2	94
Yemen, People's Dem Rep	1.50	0.60	0.00	X	1975	1.93	129	1,167	5	2	93
EUROPE	**2,321.00 b**	**4.66**			**1987 b**	**359**	**15**	**726**	**13**	**54**	**33**
Albania	10.00	3.08	11.30	X	1970	0.20	1	94	6	18	76
Austria	56.30	7.51	34.00	X	1980	3.13	3	417	19	73	8
Belgium	8.40	0.85	4.10	X	1980	9.03	72	917	11	85	4
Bulgaria	18.00	2.00	187.00	X	1980	14.18	7	1,600	7	38	55
Czechoslovakia	28.00	1.79	62.60	X	1980	5.80	6	379	23	68	9
Denmark	11.00	2.15	2.00	X	1977	1.40	11	277	30	27	43
Finland	110.00	22.11	3.00	X	1980	3.70	3	774	12	85	3
France	170.00	3.03	15.00	20.50	1984	33.30	18	606	16	69	15
German Dem Rep	17.00	1.02	17.00	X	1980	9.13	27	545	14	68	18
Germany, Fed Rep	79.00	1.30	82.00	X	1981	41.40	26	671	10	70	20
Greece	45.15	4.49	13.50	3.00	1980	7.00	12	726	8	29	63
Hungary	6.00	0.57	109.00	X	1980	5.38	5	502	9	55	36
Iceland	170.00	671.94	0.00	0.00	1987 b	0.09	0	349	31	63	6
Ireland	50.00	13.44	0.00	X	1972	0.40	1	135	16	74	10
Italy	179.40	3.13	7.60	0.00	1981	46.35	25	811	14	27	59
Luxembourg	1.00	2.72	4.00	X	1976	0.06	1	166	42	45	13
Malta	0.03	0.07	0.00	0.00	1978	0.02	92	68	76	8	16
Netherlands	10.00	0.68	80.00	X	1980	14.20	16	1,004	5	61	34
Norway	405.00	96.15	8.00	X	1980	2.00	0	489	20	72	8
Poland	49.40	1.29	6.80	X	1980	16.80	30	472	16	60	24
Portugal	34.00	3.31	31.60	X	1980	10.50	16	1,062	15	37	48
Romania	37.00	1.59	171.00	X	1980	25.40	12	1,144	8	33	59
Spain	110.30	2.80	1.00	17.00	1985	26.30	24	682	12	26	62
Sweden	176.00	21.11	4.00	X	1980	3.98	2	479	36	55	9
Switzerland	42.50	6.52	7.50	X	1985	3.20	6	502	23	73	4
United Kingdom	120.00	2.11	0.00	X	1980	28.35	24	507	20	77	3
Yugoslavia	150.00	6.29	115.00	200.00	1980	8.77	3	393	16	72	12
U.S.S.R.	**4,384.00**	**15.22**	**300.00**	**X**	**1980**	**353.00**	**8**	**1,330**	**6**	**29**	**65**
OCEANIA	**2,011.00 b**	**75.96**			**1987 b**	**23**	**1**	**907**	**18**	**16**	**76**
Australia	343.00	20.48	0.00	0.00	1975	17.80	5	1,306	65	2	33
Fiji	28.55 b	38.12	0.00	0.00	1987 b	0.03	0	37	20	20	60
New Zealand	397.00	117.49	0.00	0.00	1980	1.20	0	379	46	10	44
Papua New Guinea	801.00 b	199.70	X	X	1987 b	0.10	0	25	29	22	49
Solomon Islands	44.70 b	149.00	0.00	0.00	1987 b		0	18	40	20	40

Sources: Bureau of Geological and Mining Research, National Geological Survey, France; U.S. Geological Survey; and Institute of Geography, National Academy of Sciences, U.S.S.R.

Notes:
a. Water resources include both internal renewable resources and river flows from other countries.
b. Estimated by the Institute of Geography, U.S.S.R.
c. Sectoral percentages date from the year of other withdrawal data.
0 = zero or less than half the unit of measure, X = not available.
For additional information, see Sources and Technical Notes.

Table 22.2   Water Quality at Selected GEMS/WATER Stations,

	Median Dissolved Oxygen (mg/l)			Median Biochemical Oxygen Demand (BOD) (mg/l)			Median pH			Median Faecal Coliforms (no./100 ml)			Median Dissolved Mercury (microgram/l)			Median Dissolved Lead (mg/l)		
	1979-81	1982-84	1985-87	1979-81	1982-84	1985-87	1979-81	1982-84	1985-87	1979-81	1982-84	1985-87	1979-81	1982-84	1985-87	1979-81	1982-84	1985-87
RIVERS																		
AFRICA																		
KENYA-Thika	X	X	8.8	X	X	3.0	X	7.5	7.1	X	X	X	X	X	X	X	X	X
Nairobi	X	X	2.3	X	X	30.0	X	7.3	7.6	X	X	X	X	X	X	X	X	X
SUDAN-Blue Nile at Khartoum	7.0	7.7	8.5	2.6	2.1	3.1	7.8	8.1	7.9	2	7	X	X	X	X	0.000	0.000	0.000
TUNISIA-Sources de Zaghouan Galerie	9.4	9.0	X	3.3	X	X	7.5	7.5	X	2	7	X	X	X	X	X	X	X
TANZANIA-Ruvu Mlandizi	7.3	7.1	X	X	0.0	X	7.9	8.1	8.0	170	X	X	X	X	X	X	X	X
NORTH & CENTRAL AMERICA																		
CANADA-St. Lawrence	X	X	10.3	X	X	X	8.1	7.9	8.2	X	X	X	X	X	X	X	X	X
GUATEMALA-Pixcaya	7.0	6.8	X	20.0	3.9	X	7.9	8.1	X	1275	24000	X	X	X	X	X	X	X
MEXICO-Colorado	7.9	8.3	8.8	6.0	3.6	1.3	8.0	7.8	8.0	240	122	23	X	X	X	X	X	X
Blanco	4.9	4.7	3.4	9.8	14.3	6.5	7.6	8.0	7.7	19500	40000	40000	X	X	X	X	X	X
Lerma	0.1	0.1	0.6	51.4	61.7	18.9	8.5	8.2	7.6	180000	100000	5965	X	X	X	X	X	X
PANAMA-San Felix	8.2	8.1	8.0	2.0	2.0	2.0	7.8	7.9	7.8	883	925	460	X	X	X	X	0.145	X
Aguas Claras	7.9	8.0	8.3	2.0	2.0	X	7.6	7.7	7.5	224	162	130	X	X	X	X	0.150	X
UNITED STATES-Mississippi	8.1	8.1	X	X	X	X	X	X	X	299	925	X	0.10	0.25	X	0.002	0.002	X
Sacramento	9.4	10.6	9.6	X	X	X	X	X	X	37	50	X	0.05	0.10	0.10	0.000	0.002	X
Hudson	9.6	11.6	X	X	X	X	X	X	X	990	680	410	0.10	0.15	0.15	0.002	0.003	0.001
SOUTH AMERICA																		
ARGENTINA-Parana Rosario	7.5	5.9	7.3	2.5	1.1	2.1	X	7.0	7.4	X	X	4300	X	X	X	X	X	X
Rio de la Plata, Buenos Aires	7.6	7.4	7.6	0.9	1.1	1.0	7.4	7.2	7.3	620	310	230	X	X	X	X	X	X
Paraguay	6.9	4.9	X	0.8	0.9	X	7.3	7.4	X	493	614	X	X	X	X	X	X	X
BRAZIL-Guandu (Tomada d'Agua)	7.9	7.8	7.7	1.2	0.8	1.2	7.0	6.8	6.8	2	4900	4	X	0.10	X	X	0.023	X
Paraiba do Sul (Barra Mansa)	X	7.6	7.6	X	1.2	1.6	7.0	6.9	6.9	3	13000	4900	X	0.10	X	X	0.024	X
Jacui (JA 042)	X	8.1	7.7	X	1.0	1.0	X	6.9	7.0	X	330	230	X	0.00	0.00	X	0.019	0.013
CHILE-Mapocho en Los Almendros	11.4	12.4	10.6	1.0	0.9	0.8	7.5	7.3	6.9	2	2	2	X	X	X	X	X	X
Maipo	11.4	14.0	12.8	1.0	1.2	0.8	8.1	8.1	8.0	330	855	1100	X	X	X	X	X	X
COLOMBIA-Cauca Juanchito	X	5.1	5.4	X	2.2	2.2	X	7.1	7.1	X	X	X	X	X	X	X	X	X
ECUADOR-Daule	X	7.0	X	X	1.2	X	7.9	7.2	X	515	2400	X	X	X	X	X	X	X
San Pedro	8.1	8.0	7.6	10.0	2.3	3.2	8.0	7.8	8.0	16000	11600	80190	X	X	X	X	X	X
PERU-Rimac	7.7	8.2	X	X	X	X	8.0	7.9	X	1100	X	X	X	X	X	X	0.220	X
URUGUAY-Santa Lucia	X	9.0	7.0	X	2.0	1.2	X	7.4	7.3	X	0	0	X	0.00	X	X	0.000	X
Uruguay Salto	X	8.9	8.0	X	1.2	0.8	X	6.9	7.3	X	20	23	X	0.00	X	X	0.000	X
Rio de la Plata, Colonia	X	8.8	8.8	X	1.0	0.8	X	7.1	7.5	X	30	190	X	0.00	X	X	0.000	X
ASIA																		
BANGLADESH-Brahmaputra	6.5	7.8	X	2.4	2.7	X	7.2	7.2	X	2300	1800	X	X	X	X	X	X	X
Lower Ganges (Padha)	6.3	6.6	X	3.8	1.6	X	7.5	7.3	X	1900	1800	X	X	X	X	X	X	X
Surma	6.3	7.0	X	3.0	2.7	X	7.2	7.0	X	1800	2600	X	X	X	X	X	X	X
CHINA-Changjiang	8.8	8.1	8.6	0.5	0.7	0.8	7.7	8.1	7.9	240	350	380	0.20	0.20	0.20	0.005	0.005	0.005
Huanghe	9.9	9.7	9.6	1.7	1.5	1.7	8.3	8.2	8.1	893	1153	920	0.20	0.20	0.20	0.001	0.002	0.002
Zhujiang	7.4	7.7	7.9	0.5	0.6	0.6	8.0	8.0	8.1	505	600	405	0.10	0.10	0.10	0.005	0.002	0.002
INDIA-Sabarmati near Dharoi	9.4	9.4	9.0	1.8	2.0	1.7	8.3	8.0	8.2	270	220	79	X	X	X	X	X	X
Sabarmati at Ahmedabad	0.0	0.0	0.0	59.0	62.5	84.5	7.7	7.8	7.8	5400000	3500000	1700000	X	X	X	X	X	X
Mahi near Sevalia	8.9	9.0	9.1	0.0	2.0	2.2	8.4	8.3	8.5	2800	2400	750	X	X	X	X	X	X
Mahi at Vasad, Baroda	8.2	8.7	8.5	2.2	1.6	2.0	8.4	8.3	8.6	3100	11000	590	X	X	X	X	X	X
INDONESIA-Citarum	7.1	X	3.2	2.5	X	6.4	7.7	X	7.1	150000	X	175000	0.00	X	0.67	0.021	X	0.008
IRAN-Shur at Tehran	8.9	9.2	X	1.3	1.4	X	7.8	8.1	X	X	X	X	X	X	X	X	X	X
Zayandeh at Isfahan	8.4	8.1	X	1.6	1.4	X	7.9	7.8	X	1100	1100	X	X	X	X	X	X	X
Sefid Rud Downstream	8.8	9.9	X	0.9	1.0	X	8.0	8.1	X	310	600	X	X	X	X	X	X	X
JAPAN-Shinano at Zuiun Bridge	10.4	10.3	10.3	1.6	1.5	1.6	7.1	7.2	7.2	320	320	245	2.50	0.50	0.50	0.014	X	0.020
Kiso at Shimo-ochiai	11.1	11.1	11.0	0.8	0.9	0.6	7.1	7.1	7.1	490	270	490	0.00	X	X	0.002	0.002	0.002
Kiso at Inuyama	10.9	10.7	10.3	1.0	1.1	0.9	7.1	7.1	7.1	500	330	665	0.00	X	X	0.002	0.002	0.002
Yodo at Hirakata	8.7	8.6	8.5	3.1	3.0	3.5	7.4	7.4	7.3	69500	59000	230000	0.50	X	X	0.010	X	X
KOREA, REP-Han	X	9.9	10.9	X	1.0	1.6	7.4	7.5	7.4	X	5	8	X	X	0.02	X	X	X
MALAYSIA-Klang	2.8	3.1	2.7	5.7	5.5	7.0	7.0	7.0	6.7	X	X	525000	X	0.00	X	X	X	0.000
Gombak	9.6	X	X	0.4	0.7	X	7.0	7.2	X	3500	10000	X	X	X	X	X	X	X
Muda	7.2	7.3	6.9	2.0	1.0	1.0	6.7	7.0	7.1	1800	1800	3000	X	X	X	X	X	X
PAKISTAN-Ravi (upstream Lahore)	7.2	7.0	6.6	1.0	1.0	2.0	7.3	7.4	7.3	426	333	227	0.00	X	X	0.000	X	X
Ravi (downstream Lahore)	6.9	7.0	5.4	1.0	1.0	8.0	7.4	7.5	7.4	387	567	500	0.00	X	X	0.000	X	X
Indus at Kotri	7.5	7.5	6.0	4.5	6.0	4.0	7.5	7.4	7.5	100	110	109	0.00	X	X	0.000	X	X
PHILIPPINES-Pampanga	8.3	6.7	X	1.4	3.3	X	7.9	8.0	X	7900	1700	X	0.05	0.00	X	0.010	X	X
Cagayan	7.8	7.9	7.8	0.5	0.5	1.0	7.9	X	X	4900	X	X	0.11	X	X	X	X	X
THAILAND-Chao Phrya	6.3	6.6	6.1	1.1	1.1	1.0	7.3	7.3	7.5	780	1400	700	0.00	0.11	X	0.000	0.020	0.000
Prasak	6.1	7.4	6.5	1.0	1.2	1.2	7.4	7.5	7.5	330	500	675	0.00	0.02	X	0.000	0.030	X
TURKEY-Cark Suyu, Beskopruler	9.1	10.3	9.2	1.5	1.3	1.4	8.0	7.9	8.0	X	X	64	X	X	X	X	X	X
Sakarya, Adatepe	9.1	8.7	9.0	0.5	1.9	2.0	7.6	8.0	8.0	X	X	10000	X	X	X	X	0.018	X
Porsuk	8.9	9.3	9.2	1.2	1.2	1.3	8.1	8.0	8.1	X	X	360	X	X	X	X	X	X
EUROPE																		
BELGIUM-Lys	2.7	4.9	2.2	6.5	4.3	4.5	7.5	7.5	7.5	800	1000	61500	X	X	X	X	X	X
Espierre	0.5	0.8	0.0	195.0	96.0	184.0	6.4	7.0	7.3	13000	53000	X	X	X	X	X	X	X
FINLAND-Tornionjoki	12.6	12.3	12.1	X	X	X	6.8	6.8	6.8	X	X	X	0.00	X	X	0.000	X	X
Kymjoki	11.4	10.2	11.0	X	X	X	6.6	6.6	6.5	X	X	X	0.00	X	X	0.001	X	X
Kalkkinen	12.3	11.7	11.8	X	X	X	7.0	7.0	7.0	X	X	X	0.05	0.10	X	0.001	0.000	X
HUNGARY-Tisza at Szolnok	9.4	9.6	8.6	2.7	2.5	2.4	7.4	7.6	7.7	330	86	137	0.00	0.20	X	0.000	0.045	X
Danube at Budapest	10.2	10.4	10.0	3.6	5.8	6.0	7.8	7.7	7.9	3500	3500	3000	0.00	0.20	X	0.000	0.030	X
NETHERLANDS-Rhine (frontier)	8.1	8.0	X	3.0	3.0	X	7.6	7.6	X	18000	8000	X	X	X	X	X	X	X
Maas (frontier)	10.5	8.7	X	3.0	3.0	X	7.8	7.7	X	17000	11000	X	X	X	X	X	X	X
NORWAY-Glama at Askim	10.0	X	X	X	X	X	6.9	6.9	6.8	41	28	45	X	X	X	0.001	0.001	X
PORTUGAL-Tejo at Santarem	8.9	8.5	8.5	2.7	1.8	1.5	7.2	7.5	7.4	920	4500	3850	0.20	X	X	0.050	0.045	X
SPAIN-Mino en Pte Mayor Oroza	9.1	9.6	X	1.0	1.2	X	7.2	6.8	X	6000	X	X	X	X	X	0.003	X	X
Guadiana en Pte Palmas	5.3	5.1	X	1.1	2.3	X	7.7	7.6	X	15	100	X	X	X	X	0.003	0.003	X
Ebro en Mendavia	9.2	9.3	X	3.8	3.8	X	7.7	7.6	X	2000	20000	X	X	X	X	0.002	X	X
UNITED KINGDOM-Thames	10.0	9.9	10.1	X	X	X	8.1	8.0	7.9	X	X	X	X	X	X	X	X	X
Exe	11.4	11.0	12.0	1.4	1.8	X	7.3	7.5	7.6	1950	2000	550	X	X	X	X	X	X
Trent	10.1	9.8	10.2	X	X	X	7.6	7.8	7.9	X	X	X	X	X	X	0.010	0.003	0.002

1979–87 — Table 22.2

	Median Dissolved Oxygen (mg/l)			Median Biochemical Oxygen Demand (BOD) (mg/l)			Median pH			Median Faecal Coliforms (no./100 ml)			Median Dissolved Mercury (microgram/l)			Median Dissolved Lead (mg/l)		
	1979-81	1982-84	1985-87	1979-81	1982-84	1985-87	1979-81	1982-84	1985-87	1979-81	1982-84	1985-87	1979-81	1982-84	1985-87	1979-81	1982-84	1985-87
RIVERS (continued)																		
OCEANIA																		
AUSTRALIA-Murray at Mannum	7.3	7.6	7.5	X	X	X	7.9	7.9	7.8	35	130	80	X	X	X	X	X	X
FIJI-Waimanu	7.5	7.8	7.9	1.0	0.7	0.9	6.9	7.0	7.1	0	1800	1875	X	X	X	X	X	X
NEW ZEALAND-Waikato at Taupo Gates	9.4	9.7	9.8	0.4	0.4	0.5	8.0	8.0	7.9	1	1	0	0.10	X	X	0.005	X	X
Waikato at Mercer Bridge	8.9	9.4	9.2	1.3	1.4	1.1	7.4	7.5	7.3	700	350	250	0.10	X	X	0.005	X	X
LAKES																		
AFRICA																		
KENYA-Lake Victoria, Kisumu	6.5	7.7	X	X	X	X	7.7	8.2	8.2	X	X	X	X	X	X	X	X	X
SUDAN-Jebel Aulia Reservoir	7.8	7.8	9.5	2.3	1.7	3.4	8.0	8.2	8.0	X	X	X	X	X	X	0.000	0.000	0.000
TUNISIA-Reservoir Beni Mtir	9.0	8.3	X	8.0	X	X	7.0	7.0	X	0	2400	X	X	X	X	X	X	X
TANZANIA-Lake Victoria South Port	8.2	X	X	X	X	X	7.2	7.4	7.6	200	X	X	X	X	X	X	X	X
NORTH & CENTRAL AMERICA																		
CANADA-Lake Ontario, Mid-lake	12.8	13.0	12.8	X	X	X	8.2	7.8	8.1	0	0	X	X	X	X	X	X	X
GUATEMALA-Amatitlan	7.3	4.6	X	14.7	7.0	X	8.5	8.2	X	97	43	X	X	X	X	X	X	X
MEXICO-Lago de Chapala	7.0	6.9	7.3	1.5	1.2	1.4	8.8	8.9	8.7	3	4	4	X	X	X	X	X	X
Presa de la Amistad	7.8	X	8.5	1.6	1.2	1.1	8.1	8.2	8.4	X	X	X	X	X	X	X	X	X
PANAMA-Lagomadden Station 001	6.0	6.5	5.9	X	X	X	7.4	7.4	7.7	4	8	7	X	X	X	X	X	X
SOUTH AMERICA																		
ARGENTINA-Embalse Salto Grande	8.6	X	6.8	X	X	X	7.5	X	8.0	X	X	1	X	X	X	X	X	X
BOLIVIA-Water Supply Station 001	6.3	6.3	X	1.0	1.0	X	7.3	7.1	X	0	0	X	X	X	X	X	X	X
BRAZIL-Reservatorio de Guarapiranga	X	7.6	7.5	X	1.0	2.0	X	6.8	6.8	X	95	30	X	0.00	X	X	0.001	X
Reservatorio de Promissao	X	8.1	8.7	X	1.5	1.5	X	6.9	7.5	X	50	105	X	0.00	X	X	0.001	X
Rio Paraguacu Pedra do Cavalo	X	6.6	6.4	X	1.2	1.6	X	7.4	7.3	X	23	17	X	0.20	0.20	X	0.050	0.050
ASIA																		
BANGLADESH-Kaptai Lake	6.0	7.1	X	2.0	1.6	X	7.3	7.4	X	40	30	X	X	X	X	X	X	X
CHINA-Lake Tai	9.8	9.1	10.0	X	0.8	0.9	7.8	7.9	7.7	5	4	2	0.59	0.23	0.05	0.002	0.001	0.002
HONG KONG-Plover Cove Station	8.4	8.5	8.0	X	X	X	7.4	7.7	8.0	0	2	0	X	X	X	X	X	X
IRAN-Dariush Kabir Reservoir and Dam	11.5	11.5	X	1.8	0.9	X	8.3	8.0	X	2	2	X	X	X	X	X	X	X
JAPAN-Sagami Reservoir	10.8	10.1	9.7	1.4	1.7	1.4	7.7	7.4	7.4	130	225	170	X	X	X	X	X	X
Lake Biwa off Miidera	10.7	10.5	11.0	1.9	1.7	1.7	8.0	8.1	7.9	240	93	43	0.50	X	X	0.010	X	X
PAKISTAN-Kalri Lake	7.2	7.2	6.7	X	X	X	8.1	8.2	8.3	1650	1900	1800	X	X	X	X	X	X
PHILIPPINES-La Mesa Reservoir	7.8	7.7	6.3	0.3	0.4	0.8	7.7	7.4	7.4	X	X	X	X	X	X	X	X	X
THAILAND-Sirikit Reservoir	6.2	7.0	7.3	1.0	1.0	1.0	7.9	7.4	7.8	20	32	200	0.02	0.44	X	X	0.015	X
EUROPE																		
FINLAND-Paajarvi Lake Station 95	9.6	10.2	9.4	X	X	X	6.9	6.9	6.9	X	X	X	0.05	X	X	X	X	X
Yli-Kitka Lake Station 144	10.1	9.4	9.8	X	X	X	7.1	7.1	6.8	X	X	X	0.00	X	X	0.021	0.000	X
HUNGARY-Lake Balaton	10.1	11.0	10.0	3.0	2.8	2.3	8.3	8.4	8.4	X	X	X	X	X	0.00	X	X	X
NETHERLANDS-Ijsselmeer	12.3	12.0	X	3.9	3.0	X	8.5	8.5	X	X	2	X	X	X	X	X	X	X
PORTUGAL-Castel de Bode Reservoir	9.5	9.0	8.5	2.3	1.1	0.8	7.1	6.9	6.9	0	1	0	0.20	0.20	X	0.050	0.100	X
SWEDEN-Lake Vanern	11.7	11.5	11.5	X	X	X	7.2	7.2	7.2	X	X	X	X	X	X	X	X	X
UNITED KINGDOM-Rutland Reservoir	10.8	11.6	X	1.8	X	X	8.3	8.3	X	X	X	X	X	X	X	X	X	X
Tunstall Reservoir	X	10.7	10.4	X	X	X	X	7.1	6.9	X	X	X	X	0.10	X	X	0.010	0.003
OCEANIA																		
AUSTRALIA-Mount Bold Reservoir	8.4	8.0	X	X	1.0	1.0	7.7	7.7	7.8	12	8	5	X	X	X	X	X	X
Lake Burragorang	8.0	7.6	8.4	X	X	X	7.6	7.5	7.5	2	X	X	0.20	0.50	0.50	0.602	1.000	1.900
GROUNDWATER																		
AFRICA																		
TUNISIA-Medjerdah Sloughia	7.0	5.8	X	1.9	X	X	7.5	7.5	X	0	0	X	X	X	X	X	X	X
Djouggar	8.9	9.5	X	2.9	X	X	7.5	7.5	X	0	15	X	X	X	X	X	X	X
Cherichira	8.9	8.0	X	4.6	X	X	7.5	7.0	X	0	2	X	X	X	X	X	X	X
Aintahouna	6.0	8.5	X	3.0	X	X	7.0	7.0	X	0	0	X	X	X	X	X	X	X
TANZANIA-Makutopora Basin, Dodoma	3.0	X	X	X	X	X	8.1	8.1	8.3	X	X	X	X	X	X	X	X	X
NORTH & CENTRAL AMERICA																		
MEXICO-Pozo Hacienda Tahdzibichen	3.9	3.8	4.8	2.4	2.9	2.4	7.7	7.6	7.2	400	430	40	X	X	X	X	X	X
Pozo en Aguascalientes	X	X	X	X	0.1	0.4	7.4	7.4	7.5	X	3	3	X	X	X	X	X	X
Pozo en la Region Lagunera	X	X	X	X	X	X	7.6	7.7	7.5	X	X	X	X	X	X	X	X	X
SOUTH AMERICA																		
ARGENTINA-Salta Dto. Anta (Tolloche)	8.4	9.1	X	X	3.9	X	8.4	8.6	X	X	0	X	X	X	X	X	X	X
CHILE-Pozo en Panamericana 1377	X	3.8	4.6	0.6	0.5	X	7.2	7.4	7.1	2	2	X	X	X	X	X	X	X
URUGUAY-Acuifero Rivera	X	8.0	7.3	X	0.7	X	X	6.2	6.2	X	0	0	X	X	X	X	X	X
ASIA																		
BANGLADESH-Ground Water	X	1.8	X	X	7.0	X	7.4	7.4	X	0	0	X	X	X	X	X	X	X
Ground Water from Muladi	X	1.9	X	X	0.8	X	7.7	7.0	X	0	0	X	X	X	X	X	X	X
INDIA-Well at Tarvai	3.6	3.1	4.0	1.7	1.2	2.0	7.4	7.3	7.2	150	7	7	X	X	X	X	X	X
Well at Eluru near Alwaye	3.0	3.9	4.6	1.6	0.9	1.3	6.6	6.6	7.0	140	92	240	X	X	X	X	X	X
Well at Peddavoora	5.2	4.6	3.2	2.0	1.2	1.5	7.7	7.6	7.5	210	11	7	X	X	X	X	X	X
INDONESIA-Well Near Cibeureum	2.0	X	3.5	X	X	0.4	7.5	7.3	7.6	3	X	0	0.00	X	0.44	0.021	X	0.007
IRAN-Well No. 1, Tehran	8.4	8.2	X	X	X	X	7.2	7.4	X	X	X	X	X	X	X	X	X	X
Well No. 5 in Shiraz	8.7	X	X	1.2	0.6	X	7.9	7.7	X	540	49	X	X	X	X	X	X	X
JAPAN-Urawa Purification Plant	6.2	6.4	6.2	X	X	X	7.5	7.6	7.4	0	0	0	X	X	X	X	X	X
Suginami Filtration Plant	7.0	7.3	8.1	0.7	0.7	X	6.2	6.1	6.1	0	0	1	X	X	X	X	X	X
TURKEY-Well in Eskisehir Plain	8.9	8.6	8.2	X	X	X	7.7	7.6	7.5	0	0	0	X	X	X	X	X	X
EUROPE																		
HUNGARY-Deep Well in Pest Country	0.0	X	0.1	X	X	X	8.0	X	X	X	X	X	X	X	X	X	X	X
SWEDEN-Station 34014 at Drangsmark	5.0	4.5	6.5	X	X	X	6.5	6.5	6.4	X	X	X	X	X	X	X	X	X
Station 70013 at Odskolt	9.8	9.2	9.9	X	X	X	5.8	5.7	5.7	X	X	X	X	X	X	X	X	X
Station 23008 at Tarnsjo	11.0	10.7	11.8	X	X	X	7.3	7.3	7.3	X	X	X	X	X	X	X	X	X

Source: Global Environmental Monitoring System.
Notes: 0 = zero or less than half the unit of measure; X = not available.
For additional information, see Sources and Technical Notes.

Sources and Technical Notes

Table 22.1 Freshwater Resources and Withdrawal

Sources: Water resources and withdrawal data as footnoted: J. Forkasiewicz and J. Margat, *Tableau Mondial de Données Nationales d'Economie de l'Eau, Ressources et Utilisation* (Departement Hydrogéologie, Orléans, France, 1980). Data for Algeria, Egypt, Libya, Morocco, Tunisia, Cyprus, Israel, Lebanon, Syrian Arab Republic, Turkey, Albania, France, Greece, Italy, Malta, Spain, and Yugoslavia: J. Margat, Bureau de Recherches Géologiques et Minières, Orléans, France, April 1988 (personal communication). Resource data as footnoted and withdrawal data: Alexander V. Belyaev, Institute of Geography, U.S.S.R. National Academy of Sciences, Moscow, September 1989 and January 1990 (personal communication). Withdrawal and Sectoral Use data for the United States: W.B. Solley, C.F. Merk, and R.R. Pierce, "Estimated Use of Water in the United States, in 1985," *U.S. Geological Survey Circular*, No. 1004 (U.S. Geological Survey, Reston, Virginia, United States, 1988). Population: United Nations Population Division, *World Population Prospects 1988* (United Nations, New York, 1989).

Margat compiles water resources and withdrawal data from published documents, including national, United Nations, and professional literature. Data for small countries and countries in arid and semi-arid zones are less reliable than are those for larger and wetter countries.

Belyaev compiles data on water resources and withdrawals from the world's literature and estimates resources and consumption from models using other data, such as area under irrigated agriculture, livestock populations, and precipitation, when necessary.

Annual internal renewable water resources refers to the average annual flow of rivers and aquifers generated from endogenous precipitation. Caution should be used when comparing different countries because these estimates are based on differing sources and dates. These annual averages also disguise large seasonal, interannual, and long-term variations. When data for *annual river flows to and from other countries* are not shown, the internal renewable water resources figure *may* include these flows.

Water is withdrawn when it is taken from a surface or underground source and conveyed to the place of use. *Annual withdrawal as a percentage of water resources* refers to *total* water withdrawal, not counting evaporative losses from storage basins, as a percentage of internal renewable water resources and river flows from other countries. *Per capita annual internal renewable water resources* data were created using 1990 population estimates. *Per capita annual withdrawal* fig-

ures were calculated using national population data for the year of data shown for withdrawal.

Sectoral withdrawal is classified in three categories: *domestic* (drinking water, homes, commercial establishments, public services [e.g., hospitals], and municipal use or provision), *industry* (including water withdrawn to cool thermoelectric plants), and agriculture (irrigation and livestock).

Withdrawal data are based on both national reports and models using estimates from other data (e.g., numbers of cattle, percentage of irrigated land and crop mix, etc.). Thus these data should be used with care. Totals may not add due to rounding error.

Table 22.2 Water Quality at Selected GEMS/WATER Stations, 1979–87

Source: Global Environment Monitoring System, Water Monitoring Project (GEMS/WATER), provided by the Canada Centre for Inland Waters, unpublished data (Burlington, Ontario, Canada, December 1989). Water quality guidelines: GEMS, *Assessment of Freshwater Quality* (United Nations Environment Programme (UNEP) and the World Health Organization (WHO), prepared by the Monitoring and Assessment Research Centre, London, 1988); WHO, *Guidelines for Drinking-Water Quality, Volume 1, Recommendations* (WHO, Geneva, 1984).

UNEP and WHO collaborate within GEMS to assemble and monitor water quality data on a global basis. The GEMS/WATER system includes 240 river, 43 lake, and 60 groundwater monitoring stations.

GEMS water quality data are available from 1979 to present. Data shown in this table comprise a subset of the 50 indicators of water quality that can be reported within the GEMS system. Not all stations collect all data, and the frequency and physical accuracy of measurement vary between stations. The median value was selected to highlight general trends and to minimize the importance of radical outliers that might be due to measurement error or relatively short-term pollution or degradation episodes. Three-year periods were used to minimize seasonal and interannual variability and, again, to emphasize general trends, if any.

Median dissolved oxygen is a critical factor in the health or potential health of aquatic organisms. In general, for life, growth, and reproduction, values of 2 mg/l are required for scavenger fish and 4 mg/l or more for game and other sensitive fish. Lower values of dissolved oxygen would indicate poor stocks of fish and other oxygen-dependent organisms.

Median biochemical oxygen demand (BOD) stands for the biodegradability of

the total organic matter dissolved or suspended in the water under study. Along with dissolved oxygen, this is the most commonly reported water quality indicator within the GEMS system. Sewage or other organic pollutants, especially in warm waters already low in dissolved oxygen, could lead to severe oxygen depletion with adverse impacts on aquatic life. While not a direct measure of pathogenic organisms, BOD is a surrogate for the potential health effects of untreated water. Rivers can be said to be seriously polluted if they have a BOD of 6.5 mg/l or more, that is, if more than 6.5 mg of oxygen would be required to oxidize the organic matter in a liter of water.

The *median pH* value is a measure of the acid-base equilibrium of an aqueous system. The pH of most natural water sources varies between 6.5 and 8.5, which is the suggested guideline for drinking water. Highly acidified waters have values in the range of 4.5–5.5. The acidification of fresh waters, as a consequence of acid precipitation, was first observed in Sweden and Norway. Acidification has immediate impacts on aquatic life, and it increases corrosion in water collection and distribution systems. Low pH values for water are associated with increased corrosion and increases in the dissolved concentrations of heavy metals including lead and cadmium.

Median faecal coliforms are most commonly associated with the faeces of animals and humans. This measure is used as a sentinel indicator for the presence or potential presence of myriad other pathogenic organisms that are more difficult to observe and measure. The absence, or near absence, of faecal coliforms means the absence of most other pathogens. The exception is that in minimally treated water that destroys coliform organisms, the cyst forms of *Giardia* or *Amoeba*, *inter alia*, can survive. Water for human consumption should usually contain zero faecal coliforms per 100 ml sample.

Median dissolved mercury concentrations are of importance for both direct human consumption of water and human consumption of fish and shellfish that have concentrated mercury in their tissues. The latter is the primary route for human health effects. Drinking water should not contain mercury in all forms that exceeds 0.001 mg/l.

Median dissolved lead is naturally present in much of the world's water supplies at about 1–10 micrograms per liter. The discharges from sewage and wastewater treatment plants are usually enriched in heavy metals, including lead. The presence of large amounts of lead is usually a sign of industrial pollution. Children, infants, fetuses in utero, and pregnant women are at most risk from lead poisoning. Water should not contain more than 0.05 mg/l of lead.

23. Oceans and Coasts

Oceans are vital to support life on earth. They modify climatic regimes, are a reservoir of dissolved gases such as carbon dioxide, and provide habitat and food. People, in turn, are pivotal in maintaining the life and health of the oceans. People determine the rate of economic development in coastal zones, the quantities of food harvested from the seas, and the amounts of waste dumped into global waters. The tables in this chapter present data on some of these interactions between humans and oceans.

The impact of human activities is most dramatic in coastal waters and semi-enclosed seas. Table 23.1 shows coastal pressures caused by demographic trends, maritime trade, and offshore oil and gas exploration.

Increasing numbers of people are settling along the world's coastline. If the trend toward greater urbanization persists, it will multiply impacts on fragile and productive coastal ecosystems. For most coastal nations, growth will continue without appropriate coastal zone management, aggravated by inadequate resources to provide housing, sanitation, and employment.

Maritime trade connects the economies of the world. Used to ship vital raw materials such as crude petroleum and other petroleum products, ores, and cereals, as well as manufactured goods, maritime trade requires intensive investment in infrastructure and creates areas at high-risk of accidents and environmental damage in fragile coastal zones. (See Chapter 11, "Oceans and Coasts," Box 11.1.) Europe as a region ships the largest volume of goods and generates heavy traffic in semi-enclosed seas such as the North Sea and the Mediterranean.

The extraction of offshore oil and gas contributes to the economic value of the coastal zone, but creates waste and poses the risk of major oil spills. Roughly one fourth of the world's oil production comes from offshore areas. World offshore crude oil output increased by 27 percent in the past decade, with the greatest production increases in Angola, Brazil, Mexico, and the North Sea oil-producing countries. The world's offshore gas production grew by 19 percent in the past decade, showing an increase in all regions except Africa and North and Central America.

The continued health of the oceans and the long-term livelihood of the millions employed in fishing depend on harvests that do not exceed the sustainable yield. But many regional fisheries in fishing areas such as the Northwest Pacific and the Southern Ocean show signs of drastic overfishing. Table 23.2 presents data on fisheries by country, and Table 23.3 lists fisheries by oceans and their estimated sustainable yields.

Globally, the annual marine catch increased by 30 percent over the past decade with the U.S.S.R. and Japan together landing one fourth of the world's total.

Approximately 14 percent of the world fish catch comes from inland waters and the total freshwater catch, including aquaculture, is increasing. In Asia, the region with the world's largest share of aquaculture production, fish raised from aquaculture has contributed decisively to an increased supply of fish used for human consumption. Aquaculture competes with humans for land and clean water, and the nutrients generated by fish food and wastes degrade water quality.

The statistics in Table 23.4 illustrate that the Antarctic fisheries turned from one species to another during the past 18 years as each harvested species has declined because of overfishing. The catch of Antarctic cod, for example, has dropped drastically, while the Antarctic icefish has become the dominant species of finfish caught. In 1987, krill represented 78 percent of the total catch in the waters of the Antarctic, a drastic rise from a mere 3 percent in 1970–74.

Data on the discharge of pollutants into coastal waters and measurements of marine pollutants have been collected in a variety of coastal areas. Few countries, however, have prepared reliable estimates, and the methods of data collection vary widely. The North Atlantic Ocean and Arctic Ocean covered by the 1972 Convention for the Prevention of Marine Pollution by Dumping from Ships and Aircraft (Oslo Convention) is one area with a time-series of data with relatively consistent measurement methods. Table 23.5 depicts trends in waste dumping and the resulting input of heavy metals.

The total of industrial wastes dumped in the Oslo Convention area has declined, primarily because of a reduction in the dumping of liquids, sludges, containerized wastes, and phosphogypsum wastes. Wastes from the production of titanium dioxide, as well as fly ash and colliery wastes, have remained more or less constant. The overall input of heavy metals from these sources has decreased. The Oslo Convention figures, which are restricted to a relatively small regional area, represent the only current global data collection on ocean dumping.

Table 23.1 Coastal Areas and Resources

	Length of Coastline (kilometers)	Maritime Area (000 square kilometers) Shelf to 200-m Depth	Maritime Area Exclusive Economic Zone	Percentage of Urban Population in Large Coastal Cities	Average Annual Volume of Goods Loaded and Unloaded 1983-85 (000 metric tons) Petroleum Crude	Petroleum Product	Dry Cargo	Offshore Oil and Gas Resources Annual Production Oil (000 metric tons) 1978	1988	Gas (million cubic meters) 1978	1988	Proven Reserves Oil (million t) 1988	Gas (billion m3) 1988
WORLD	594,008	21,427	115,484	X	X	X	X	571,741	723,457	267,042	318,822	34,307	29,650
AFRICA	37,908	1,326	11,981	X	326,806	46,620	258,392	56,504	100,614	3,025	2,787	4,650	3,818
Algeria	1,183	13.7	137.2	74	24,589 a	27,211	16,923	0	0	0	0	0	0
Angola	1,600	66.9	605.7	100	8,018	772	1,525	4,712	22,454	X	405	326	44
Benin	121	X	27.1	100	0	278	728	X	X	X	X	101	X
Cameroon	402	10.6	15.4	54	6,146	1,121	3,299	X	9,317	0	0	551	35
Cape Verde	965	X	789.4	0	0	123	266	0	0	0	0	0	0
Comoros	340	X	249.0	0	0	4	99	0	0	0	0	0	0
Congo	169	8.9	24.7	33	5,147	376	3,255	2,612	6,657	X	X	548	62
Cote d'Ivoire	515	10.3	104.6	84	1,755	1,081	6,551	X	769	X	14	35	10
Djibouti	314	X	6.2	0	0	830	517	0	0	0	0	0	0
Egypt	2,450	37.4	173.5	23	139,746	2,451	24,614	19,721	29,448	744	1,189	381	125
Equatorial Guinea	296	X	283.2	0	0	24 a	128	0	0	0	0	0	0
Ethiopia	1,094	47.7	75.8	0	750	434	1,198	0	0	0	0	0	0
Gabon	885	46.0	213.6	0	6,147 b	302	726	6,872	5,392	6	X	408	X
Gambia, The	80	X	19.5	100	0	32	222	0	0	0	0	0	0
Ghana	539	20.9	218.1	72	1,059 c	266	2,329	249	11	X	X	4	X
Guinea	346	38.4	71.0	100	0	224	10,684	0	0	0	0	0	0
Guinea-Bissau	274	X	150.5	100	0	12	135	0	0	0	0	0	0
Kenya	536	14.4	118.0	25	1,953	224	3,162	0	0	0	0	0	0
Liberia	579	19.6	229.7	100	500	88	16,114	0	0	0	0	0	0
Libya	1,770	83.7	338.1	100	46,663 a	4,235	6,468	X	996	X	X	109	42
Madagascar	4,828	180.4	1,292.0	0	0 b	348	763	0	0	0	0	0	0
Mauritania	754	44.2	154.3	100	0 b	108	9,037	0	0	0	0	0	0
Mauritius	177	91.6	1,183.0	100	0	236	1,494	0	0	0	0	0	0
Morocco	1,835	62.1	278.1	59	4,557 c	138 c	31,192	0	0	0	0	0	0
Mozambique, People's Rep	2,470	104.3	562.0	86	405	165 c	4,852	0	0	0	0	0	0
Namibia	1,489	X	X	0	X	X	X	0	0	0	0	0	0
Nigeria	853	46.3	210.9	20	56,580 b	1,425	12,284	19,142	23,486	2,274	1,179	1,999	3,413
Reunion	201	X	X	100	0	226	1,208	0	0	0	0	0	0
Senegal	531	31.6	205.7	78	317	281 c	4,400	X	X	X	X	41	X
Seychelles	491	X	1,349.3	0	0	86	116	0	0	0	0	0	0
Sierra Leone	402	26.4	155.7	100	161	14 d	1,721	0	0	0	0	0	0
Somalia	3,025	60.7	782.8	100	176	38	690	0	0	0	0	0	0
South Africa	2,881	143.4	1,553.4	32	16,080	88 c	73,688	X	X	X	X	X	30
Sudan	853	22.3	91.6	10	1,161	342	2,321	0	0	0	0	0	0
Tanzania	1,424	41.2	223.2	68	580	636	1,995	X	X	X	X	X	57
Togo	56	1.0	2.1	100	0 b	191 a	995	0	0	0	0	0	0
Tunisia	1,143	50.8	85.7	100	3,866 b	1,403	11,347	2,252	1,410	X	X	38	X
Zaire	37	1.0	1.0	3	1,426 b	782 c	1,237	944	674	X	X	109	X
NORTH & CENTRAL AMERICA	183,950	5,632	18,759	X	287,431	132,656	711,001	66,695	145,043	150,941	123,193	6,625	3,710
Antigua and Barbuda	153	X	X	X	X	61 c	82	0	0	0	0	0	0
Bahamas	3,542	85.7	759.2	100	19,035	7,266	3,370	0	0	0	0	0	0
Barbados	97	0.3	167.3	0	137	150 a	573	0	0	0	0	0	0
Belize	386	X	X	0	0	63 c	211	0	0	0	0	0	0
Bermuda	103	X	X	0	0 b	358	209	0	0	0	0	0	0
Canada	90,908	2,903.4	2,939.4	15	8,959	6,565	180,439	X	3	X	X	279	300
Cayman Islands	160	X	X	0	1,179	32	96	0	0	0	0	0	0
Costa Rica	1,290	15.8	258.9	0	476	200	2,401	0	0	0	0	0	0
Cuba	3,735	X	362.8	76	5,100	4,350	16,916	0	0	0	0	0	0
Dominica	148	X	20.0	0	0	4	78	0	0	0	0	0	0
Dominican Rep	1,288	18.2	268.8	77	1,559	653	3,464	0	0	0	0	0	0
El Salvador	307	17.8	91.9	0	614	65 a	1,099	0	0	0	0	0	0
Greenland	44,087	X	X	0	0	173 c	386	0	0	0	0	0	0
Grenada	121	X	27.0	0	0	19	58	0	0	0	0	0	0
Guadeloupe	306	X	X	0	0 b	330 c	1,145	0	0	0	0	0	0
Guatemala	400	12.3	99.1	0	540	349	3,906	0	0	0	0	0	0
Haiti	1,771	10.6	160.5	100	0	111	899	0	0	0	0	0	0
Honduras	820	53.5	200.9	9	364	272	1,947	0	0	0	0	0	0
Jamaica	1,022	40.1	297.6	100	1,052	1,029	7,959	0	0	0	0	0	0
Martinique	290	2.4	X	0	349	361	858	0	0	0	0	0	0
Mexico	9,330	442.1	2,851.2	2	62,905	5,725	15,534	2,002	82,979	1,115	11,360	5,168	1,303
Nicaragua	910	72.7	159.8	0	482	152	1,115	0	0	0	0	0	0
Panama	2,490	57.3	306.5	66	1,447	622	1,201	0	0	0	0	0	0
Trinidad and Tobago	362	29.2	76.8	0	8,535	3,477	5,198	8,742	5,837	4,429	3,872	79	256
United States	19,924	1,870.7	9,711.4	41	149,781 c	83,683	458,637	55,950	56,224	145,396	107,961	1,099	1,851
SOUTH AMERICA	30,663	1,985	10,125	X	96,816	40,756	256,624	57,338	73,583	681	14,373	2,392	1,114
Argentina	4,989	796.4	1,164.5	58	X	3,871	36,583	X	X	X	X	34	5
Brazil	7,491	768.6	3,168.4	30	32,039 c	5,426	150,779	1,939	18,725	681	4,693	639	116
Chile	6,435	27.4	2,288.2	86	1,913	124	14,682	X	450	X	934	71	65
Colombia	2,414	67.9	603.2	14	1,511 c	3,052	9,563	X	X	X	2,533	10	40
Ecuador	2,237	47.0	1,159.0	55	9,500 b	1,198	2,634	X	X	X	X	10	30
French Guiana	378	X	X	X	0	120	141	0	0	0	0	0	0
Guyana	459	50.1	130.3	100	0	446	1,586	0	0	0	0	0	0
Peru	2,414	82.7	1,026.9	73	1,351	1,101	11,818	1,440	5,156	X	X	30	3
Suriname	386	X	101.2	100	0	700 c	6,757	0	0	0	0	0	0
Uruguay	660	56.6	119.3	100	1,193	78 a	1,127	0	0	0	0	0	0
Venezuela	2,800	88.1	363.8	19	49,157 b	24,666	20,944	53,958	49,252	X	6,213	1,599	854

Table 23.1

	Length of Coastline (kilometers)	Maritime Area (000 square kilometers) Shelf to 200-m Depth	Maritime Area Exclusive Economic Zone	Percentage of Urban Population in Large Coastal Cities	Average Annual Volume of Goods Loaded and Unloaded 1983-85 (000 metric tons) Petroleum Crude	Product	Dry Cargo	Offshore Oil and Gas Resources Annual Production Oil (000 metric tons) 1978	1988	Gas (million cubic meters) 1978	1988	Proven Reserves Oil (million t) 1988	Gas (billion m3) 1988
ASIA	163,609	6,768.6	20,258.5		878,838	228,114	1,073,957	287,919	213,338	23,169	55,194	16,247	10,342
Bahrain	161	5.1	5.1	100	0	12,025	4,348	11,952	12,706	X	X	177	X
Bangladesh	580	54.9	76.8	32	1,076	691	6,192	0	0	0	0	0	0
Brunei	161	X	X	0	8,393 b	5,593	841	9,512	4,471	10,213	8,414	177	218
China	14,500	869.8	1,355.8	X	24,650	5,108	80,619	100	145	X	X	X	28
Cyprus	648	6.5	99.4	41	512	486	3,420	0	0	0	0	0	0
Hong Kong	733	X	X	100	0 b	5,088	43,074	0	0	0	0	0	0
India	12,700	452.1	2,014.9	27	19,638	4,094 c	43,995	3,075	30,787	36	4,476	381	411
Indonesia	54,716	2,776.9	5,408.6	77	55,810	22,897	31,112	27,153	22,093	5,799	8,580	544	793
Iran, Islamic Rep	3,180	107.0	155.7	4	87,405	4,293	9,312	32,594	16,085	X	X	3,907	793
Iraq	58	0.7	0.7	0	X	X	X	0	0	0	0	0	0
Israel	273	4.5	23.3	59	6,890	657	15,870	0	0	0	0	0	0
Japan	13,685	480.5	3,861.1	52	194,724	51,623	432,882	125	62	620	475	1	5
Jordan	26	X	0.7	0	0	0	13,104	0	0	0	0	0	0
Kampuchea, Dem	443	X	55.6	0	0	0 b	110	0	0	0	0	0	0
Korea, Dem People's Rep	2,495	X	129.6	X	2,200	1,007	1,610	0	0	0	0	0	0
Korea, Rep	2,413	244.6	X	29	26,424	6,491	92,706	0	0	0	0	0	0
Kuwait	499	12.0	12.0	100	28,357 b	20,250	7,959	0	0	0	0	0	0
Lebanon	225	4.5	22.6	100	X	391 c	2,070	0	0	0	0	0	0
Macao	40	X	X	100	0	207	588	0	0	0	0	0	0
Malaysia	4,675	373.5	475.6	62	18,433	8,767	37,406	11,205	13,081	X	13,645	313	1,399
Maldives	644	X	959.1	0	0	5	85	0	0	0	0	0	0
Myanmar	3,060	229.5	509.5	77	0	54 d	1,642	X	X	0	0	0	110
Oman	2,092	61.1	561.7	0	19,522	44 c	3,664	X	X	X	X	X	45
Pakistan	1,046	58.3	318.5	34	4,363 c	2,129	10,881	0	0	0	0	0	0
Philippines	22,540	178.4	1,786.0	64	7,542 c	590	23,189	X	297	0	0	3	X
Qatar	563	24.0	24.0	100	14,193	654	2,412	12,949	9,462	X	1,018	299	4,399
Saudi Arabia	2,510	77.9	186.2	31	182,788	25,120	40,544	130,546	73,768	X	5,445	9,357	1,416
Singapore	193	0.3	0.3	100	31,440	32,009	37,840	0	0	0	0	0	0
Sri Lanka	1,340	26.8	517.4	84	1,621	497	5,167	0	0	0	0	0	0
Syrian Arab Rep	193	X	10.3	0	12,001	2,645	4,704	0	0	0	0	0	0
Thailand	3,219	257.6	85.8	94	7,215	2,605	25,401	X	959	X	5,685	54	227
Turkey	7,200	50.4	236.6	48	51,742	3,014	26,030	0	0	0	0	0	0
United Arab Emirates	1,448	59.3	59.3	100	51,735	4,475	9,011	48,709	29,423	6,502	7,248	802	498
Viet Nam	3,444	327.9	722.1	52	0	163	1,258	0	0	0	0	231	X
Yemen, Arab Rep	523	24.7	33.9	0	0	241 c	2,463	0	0	0	0	0	0
Yemen, People's Dem Rep	1,383	X	550.3	100	2,500 c	1,925	1,124	0	0	0	0	0	0
EUROPE	69,643	1,952	21,451	X	514,526	254,024	1,144,191	72,672	159,668	69,472	98,041	3,604	4,595
Albania	418	5.5	12.3	0	0	129	1,628	0	0	0	0	0	0
Belgium	64	2.7	2.7	20	3,384	19,102	93,832	0	0	0	0	0	0
Bulgaria	354	12.3	32.9	18	12,500	644	16,001	0	0	0	0	0	0
Denmark	3,379	68.6	1,464.2	94	6,847	6,995	27,909	423	2,910	X	1,768	46	96
Finland	1,126	98.1	98.1	68	10,335 c	4,822	35,116	0	0	0	0	0	0
France	3,427	147.8	10,263.1	19	70,815	30,059	102,594	0	0	0	0	0	0
German Dem Rep	901	X	9.6	6	0 b	3,295	21,022	0	0	0	0	0	0
Germany, Fed Rep	1,488	40.8	40.8	13	21,868	12,190	94,922	X	533	X	75	1	5
Greece	13,676	24.7	505.1	80	10,154 a	2,681	20,904	X	1,206	X	93	99	X
Iceland	4,988	133.8	866.9	100	0	505	1,644	0	0	0	0	0	0
Ireland	1,448	125.9	380.3	100	2,250 c	2,525	12,287	X	X	19	3,070	X	40
Italy	4,996	144.1	552.1	38	97,828	39,090	88,698	215	3,105	393	1,067	68	96
Malta	140	13.0	66.2	0	2 d	403 c	1,278	0	0	0	0	0	0
Netherlands	451	84.7	84.7	52	77,383	43,731	197,922	X	926	5,479	16,446	224	300
Norway	5,832	102.9	2,024.8	85	16,018	6,626	41,302	17,752	45,170	14,213	28,945	1,469	2,871
Poland	491	28.5	28.5	10	743	2,131	46,999	0	0	0	0	0	0
Portugal	1,693	39.1	1,774.2	71	7,271	2,727	13,611	0	0	0	0	0	0
Romania	225	24.4	31.9	0	12,548 a	5,096	20,713	0	0	0	0	0	0
Spain	4,964	170.5	1,219.4	54	42,382	14,865	82,751	996	1,505	8,787	857	39	23
Sweden	3,218	155.3	155.3	77	14,079	14,976	62,697	0	0	0	0	0	0
United Kingdom	12,429	492.2	1,785.3	39	99,424	38,603	137,760	53,286	104,315	40,581	45,720	1,651	1,150
Yugoslavia	3,935	06.7	52.5	8	7,501	2,584	22,508	0	0	0	0	5	15
U.S.S.R.	46,670	1,249.5	4,490.3	12	75,217	48,997	110,144	9,500	10,010	10,998	13,748	571	4,531
OCEANIA	61,565	2,514	28,420	X	7,987	13,020	228,682	20,655	21,169	8,755	11,486	219	1,539
Australia	25,760	2,269.2	4,496.3	98	6,564	9,059	204,426	20,655	20,120	8,755	8,685	171	1,048
Cook Islands	120	X	1,830.0	0	0	9	27	0	0	0	0	0	0
French Polynesia	2,525	X	5,030.0	0	0	226	310	0	0	0	0	0	0
Fiji	1,129	2.1	1,135.3	100	0	567	812	0	0	0	0	0	0
Kiribati	1,143	X	3,550.0	0	0	6	32	0	0	0	0	0	0
New Caledonia	2,254	X	1,740.0	0	0	355	1,811	0	0	0	0	0	0
New Zealand	15,134	242.8	4,833.2	94	1,295 c	1,697	14,094	X	1,050	X	2,801	7	142
Niue	64	X	390.0	0	X	X	X	0	0	0	0	0	0
Papua New Guinea	5,152	X	2,366.6	100	0 b	587 c	3,000	X	X	X	X	41	350
Solomon Islands	5,313	X	1,340.0	0	0	45	523	0	0	0	0	0	0
Tonga	419	X	700.0	0	0	15	58	0	0	0	0	0	0
Tuvalu	24	X	328.2	0	X	X	X	0	0	0	0	0	0
Vanuatu	2,528	X	680.0	0	0	19	97	0	0	0	0	0	0

Sources: United Nations Statistical Office, United Nations Office for Ocean Affairs and the Law of the Sea, and Offshore Magazine.
Notes: a. Two years of data. b. Goods loaded. c. Goods unloaded. d. One year of data.
0 = zero or less than half the unit of measure; X = not available; t = metric tons; billion = thousand million; m3 = cubic meters.
For additional information, see Sources and Technical Notes.

Table 23.2 Marine and Freshwater Catches and Aquaculture

| | Average Annual Marine Catch | | Average Annual Freshwater Catch | | Average Annual Aquaculture Production (000 metric tons) | | | | | | | | | | | |
	(000 metric tons) 1985-87	Percent Change Since 1975-77	(000 metric tons) 1985-87	Percent Change Since 1975-77	Freshwater Fishes 1984-86	1987	Diadromous Fishes 1984-86	1987	Marine Fishes 1984-86	1987	Crustaceans 1984-86	1987	Molluscs 1984-86	1987	Other 1984-86	1987
WORLD	78,955.2	30	11,388.4	63	X	X	X	X	X	X	X	X	X	X	X	X
AFRICA	3,049.4	8	1,646.1	17	X	X	X	X	X	X	X	X	X	X	X	X
Algeria	68.7	77	0.1	X	0.0	0.0	X	X	X	X	0.0	0.0	0.0	0.0	X	X
Angola	63.4	(44)	8.0	0	0.0 a	0.0	X	X	X	X	X	X	X	X	X	X
Benin	9.1	82	30.7	50	0.0	0.0	X	X	X	X	X	X	X	X	X	X
Botswana	X	X	1.7	19	X	X	X	X	X	X	X	X	X	X	X	X
Burkina Faso	X	X	7.0	24	0.1	0.0	X	X	X	X	X	X	X	X	X	X
Burundi	X	X	5.7	(68)	0.0	0.0	X	X	X	X	X	X	X	X	X	X
Cameroon	64.2	65	20.0	0	0.2	0.1	X	X	X	X	X	X	X	X	X	X
Cape Verde	7.9	65	0.0	X	X	X	X	X	X	X	X	X	X	X	X	X
Central African Rep	X	X	13.0	24	0.3	0.1	X	X	X	X	X	X	X	X	X	X
Chad	X	X	111.7	2	X	X	X	X	X	X	X	X	X	X	X	X
Comoros	5.2	38	0.0	X	X	X	X	X	X	X	X	X	X	X	X	X
Congo	17.3	7	13.0	63	0.1	0.1	X	X	X	X	X	X	X	X	X	X
Cote d'Ivoire	78.2	9	27.9	489	0.6	0.8	X	X	X	X	X	X	X	X	X	X
Djibouti	0.4	62	0.0	X	X	X	X	X	X	X	X	X	X	X	X	X
Egypt	41.7	46	189.9	150	40.8	51.3	X	X	X	X	X	X	X	X	X	X
Equatorial Guinea	3.6	(10)	0.4	X	X	X	X	X	X	X	X	X	X	X	X	X
Ethiopia	0.5	(55)	3.5	81	X	X	X	X	X	X	X	X	X	X	X	X
Gabon	19.0	236	1.9	373	0.0	0.0	X	X	X	X	X	X	X	X	X	X
Gambia, The	10.1	(15)	2.7	228	0.0	0.0	X	X	X	X	X	X	X	X	X	X
Ghana	272.8	29	50.0	22	0.4	0.4	X	X	X	X	X	X	X	X	X	X
Guinea	28.0	186	2.0	100	0.0	0.0	X	X	X	X	X	X	X	X	X	X
Guinea-Bissau	3.6	20	0.0	X	X	X	X	X	X	X	X	X	X	X	X	X
Kenya	6.4	48	112.6	244	0.1	0.1	0.1	0.1	X	X	0.0	0.0	X	X	X	X
Lesotho	X	X	0.0	(35)	0.0	0.0	0.0	0.0	X	X	X	X	X	X	X	X
Liberia	11.4	87	4.0	0	0.1	0.1	X	X	X	X	X	X	X	X	X	X
Libya	7.9	116	0.0	X	X	X	X	X	X	X	X	X	X	X	X	X
Madagascar	17.6	25	46.0	11	0.2	0.2	0.0 a	0.0	X	X	X	X	X	X	X	X
Malawi	X	X	74.5	4	0.1	0.1	X	X	X	X	0.0	0.0	X	X	X	X
Mali	X	X	58.9	(39)	0.0	0.0	X	X	X	X	X	X	X	X	X	X
Mauritania	94.2	289	6.0	(44)	X	X	X	X	X	X	X	X	X	X	X	X
Mauritius	14.4	102	0.0	11,100	0.0	0.0	X	X	0.0	0.0	0.0	0.0	0.0	0.0	X	X
Morocco	519.6	103	1.3	181	0.0	0.0	0.0	0.0	X	X	X	X	0.2	0.1	X	X
Mozambique, People's Rep	33.4	41	1.3	(73)	0.0 b	0.0	X	X	X	X	X	X	X	X	X	X
Niger	X	X	2.3	(68)	0.0	0.0	X	X	X b	X	X	X	X	X	X	X
Nigeria	153.8	(38)	99.2	(59)	6.3	5.7	X	X	0.3 b	0.3	X	X	X	X	X	X
Rwanda	X	X	1.3	14	0.0	0.1	X	X	X	X	X	X	X	X	X	X
Senegal	265.4	(16)	15.0	X	0.0	0.0	X	X	X	X	0.0	0.0	0.0	0.0	X	X
Sierra Leone	36.8	(36)	16.2	1,326	0.0	0.0	X	X	X	X	X	X	X	X	X	X
Somalia	16.7	76	0.0	X	X	X	X	X	X	X	X	X	X	X	X	X
South Africa	709.8	22	0.8	700	X	X	X	X	0.0 a	0.0	X	X	0.1 a	0.2	X	X
Sudan	0.9	27	23.8	4	0.0	0.0	X	X	X	X	X	X	X	X	X	X
Swaziland	X	X	0.0	149	0.0	0.0	X	X	X	X	X	X	X	X	X	X
Tanzania	44.9	(9)	263.1	37	0.0	0.0	X	X	X	X	X	X	X	X	X	X
Togo	14.5	53	0.7	(63)	0.0	0.0	X	X	X	X	0.0	0.1	0.1	0.1	X	X
Tunisia	93.6	91	0.0	X	X	X	X	X	X	X	X	X	X	X	X	X
Uganda	X	X	186.1	(0)	0.1	0.0	X	X	X	X	X	X	X	X	X	X
Zaire	2.0	(75)	158.1	51	0.3	0.7	X	X	X	X	X	X	X	X	X	X
Zambia	X	X	68.0	23	0.5	1.0	X	X	X	X	X	X	X	X	X	X
Zimbabwe	X	X	17.5	304	0.0	0.0	0.1	0.1	X	X	0.0	0.0	X	X	X	X
NORTH & CENTRAL AMERICA	8,306.2	68	276.0	111	X	X	X	X	X	X	X	X	X	X	X	X
Barbados	3.9	(6)	0.0	X	X	X	X	X	X	X	X	X	X	X	X	X
Canada	1,416.2	33	43.4	1	X	X	2.4	3.8	0.1 b	X	X	X	6.9	8.3	X	X
Costa Rica	20.1	51	0.3	426	0.1	0.1	0.0	0.0	X	X	0.0	0.2	0.0	0.0	X	X
Cuba	209.2	21	17.1	848	15.0	16.0	X	X	X	X	X	0.3	1.1	1.1	X	X
Dominican Rep	16.9	217	1.7	232	0.1	0.2	X	X	X	X	0.2	0.3	X	X	X	X
El Salvador	14.9	140	2.2	81	0.0	X	X	X	0.6 a	0.5	0.5	0.5	X	X	X	X
Guatemala	2.2	(31)	0.2	(60)	0.1	0.1	X	X	X	X	0.3	0.5	X	X	X	X
Haiti	7.6	101	0.3	0	X	X	X	X	X	X	X	X	X	X	X	X
Honduras	12.1	151	0.2	14	0.1	0.3	X	X	X	X	0.6	1.8	X	X	X	X
Jamaica	9.2	(9)	1.3	X	0.9	1.5	X	X	X	X	0.0	0.0	0.0	0.0	X	X
Mexico	1,181.3	146	135.4	1,681	7.5	7.4	0.3	0.4	X	X	0.4	0.6	40.5	50.7	X	X
Nicaragua	3.8	(62)	0.1	(74)	0.0	0.0	X	X	X	X	X	X	X	X	X	X
Panama	192.7	7	0.5	X	0.3	0.7	X	X	X	X	3.4	2.8	X	X	X	X
Trinidad and Tobago	3.0	(31)	0.0	X	X	X	X	X	X	X	X	X	X	X	X	X
United States	5,075.0	76	73.3	(2)	144.5	180.9	52.4	82.0	X	X	34.1	44.5	136.9	130.5	X	X
SOUTH AMERICA	12,275.2	104	340.1	42	X	X	X	X	X	X	X	X	X	X	X	X
Argentina	453.4	65	8.7	(27)	X	X	0.2	0.3	X	X	X	X	0.0 a	0.0	X	X
Bolivia	X	X	4.8	272	X	0.0	X	0.0	X	X	X	X	X	X	X	X
Brazil	607.5	9	213.7	32	25.7	16.0	X	X	X	X	1.4	1.5	0.1	0.1	X	X
Chile	5,062.6	322	0.9	X	X	X	1.3	2.8	X	X	X	X	1.7	2.0	5.7	9.2
Colombia	24.4	5	45.8	1	0.4	0.7	0.4	0.4	0.0 a	0.0	1.5 a	2.2	X	X	X	X
Ecuador	922.4	189	0.9	X	X	0.0	X	X	X	X	31.5	73.0	X	0.0	X	X
Guyana	40.6	86	0.8	X	0.0	0.0	X	X	X	X	0.0	0.0	X	X	X	X
Paraguay	X	X	10.2	251	0.0	0.0	X	X	X	X	X	X	X	X	X	X
Peru	4,745.6	38	32.3	279	0.1 c	0.4	0.6	0.5	0.0 b	0.0	1.6	3.1	2.2	0.3	X	X
Suriname	4.2	(23)	0.1	(43)	X	X	X	X	X	X	0.0	0.0	X	X	X	X
Uruguay	138.4	287	0.8	239	0.1	0.3	0.3	0.3	X	X	X	0.0	0.2	0.0	X	X
Venezuela	272.8	91	21.2	185	0.1	0.3	0.3	0.3	X	X	X	X	X	X	X	X

Table 23.2

	Average Annual Marine Catch		Average Annual Freshwater Catch		Average Annual Aquaculture Production (000 metric tons)											
	(000 metric tons)	Percent Change Since	(000 metric tons)	Percent Change Since	Freshwater Fishes		Diadromous Fishes		Marine Fishes		Crustaceans		Molluscs		Other	
	1985-87	1975-77	1985-87	1975-77	1984-86	1987	1984-86	1987	1984-86	1987	1984-86	1987	1984-86	1987	1984-86	1987
ASIA	32,303.6	35	7,725.4	92	X	X	X	X	X	X	X	X	X	X	X	X
Afghanistan	X	X	1.5	0	X	X	X	X	X	X	X	X	X	X	X	X
Bahrain	7.9	127	0.0	X	X	X	X	X	X	X	X	X	X	X	X	X
Bangladesh	209.3	106	584.9	8	115.4	143.1	X	X	X	X	13.2	22.1	X	X	X	X
Bhutan	X	X	1.0	0	X	X	X	X	X	X	X	X	X	X	X	X
China	4,626.7	41	3,415.0	220	2,325.7	3,379.5	X	X	14.0	29.5	50.0	156.0	417.3	711.1	1,604.2	1,324.5
Cyprus	2.5	137	0.1	65	0.0	0.0	0.1	0.0	0.0	0.0	X	X	X	X	X	X
India	1,710.9	19	1,169.0	43	732.0	732.0	X	X	X	X	12.0	12.0	2.3	2.3	X	X
Indonesia	1,860.8	73	609.7	51	164.4	182.8	94.5	104.4	4.1	4.3	37.7	42.6	X	X	62.7	60.0
Iran, Islamic Rep	112.7	83	27.2	275	X	X	X	X	X	X	X	X	X	X	X	X
Iraq	5.2	(38)	15.7	(8)	4.5	4.5	X	X	X	X	X	X	X	X	X	X
Israel	9.6	(0)	13.8	(9)	11.4	13.1	0.2	0.4	0.7	0.6	0.0	0.0	X	X	X	X
Japan	11,532.1	18	210.0	4	25.2	25.8	75.8	81.8	187.7	208.1	2.3	3.0	361.2	405.2	575.2	502.2
Jordan	0.1	38	0.0	X	X	0.1	X	X	X	X	X	X	X	X	X	X
Kampuchea, Dem	6.3	(41)	63.0	(15)	1.6	1.6	X	X	X	X	X	X	X	X	X	X
Korea, Dem People's Rep	1,596.7	50	103.3	87	X	X	X	X	X	X	11.0	11.0	88.0	88.0	620.0	620.0
Korea, Rep	2,821.3	40	55.3	236	1.7	5.1	1.0	3.1	1.6	1.8	0.1	0.2	364.5	449.3	452.5	417.3
Kuwait	8.3	69	0.0	X	X	X	X	X	X	X	X	X	X	X	X	X
Lao People's Dem Rep	X	X	20.0	0	2.5	2.5	X	X	X	X	X	X	X	X	X	X
Lebanon	1.5	(19)	0.1	0	X	X	0.3	0.4	X	X	X	X	X	X	X	X
Malaysia	609.3	14	9.2	275	8.0	9.1	0.4	0.8	0.1	0.1	0.2	0.3	51.3	45.7	X	X
Mongolia	X	X	0.4	31	X	X	X	X	X	X	X	X	X	X	X	X
Myanmar	524.3	43	147.7	10	4.9	5.5	X	X	X	X	X	X	X	X	X	X
Nepal	X	X	9.7	252	3.1	5.4	X	X	X	X	X	X	X	X	X	X
Oman	104.2	(33)	0.0	X	X	X	X	X	X	X	X	0.0	X	X	X	X
Pakistan	333.7	73	83.6	182	10.0	10.0	X	X	X	X	X	X	X	X	X	X
Philippines	1,378.2	15	545.1	115	48.2	83.7	204.0	197.5	X	X	31.0	36.9	33.8	22.0	164.6	220.8
Qatar	2.4	1	0.0	X	X	X	X	X	X	X	X	X	X	X	X	X
Saudi Arabia	44.9	93	0.0	X	0.0 b	0.0	X	X	0.0 b	0.0	0.0 b	0.0	X	X	X	X
Singapore	19.4	24	0.2	(66)	X	X	0.2	0.2	0.1	0.2	0.3	0.5	0.6	1.0	X	X
Sri Lanka	147.6	23	34.9	169	0.3	0.0	X	X	X	X	0.2	0.4	X	X	X	X
Syrian Arab Rep	0.7	(35)	4.2	137	1.5	1.6	0.1	0.1	X	X	X	X	X	X	X	X
Thailand	2,135.4	29	173.4	21	69.7	78.3	0.7	1.3	0.4	0.6	19.1	35.5	35.4	35.9	0.0 a	0.0
Turkey	551.0	332	44.5	139	1.8	2.1	0.8	1.0	0.0 b	0.0	X	X	X	X	X	X
United Arab Emirates	79.1	21	0.0	X	X	0.0	X	X	0.0	0.0	0.0 a	X	X	X	X	X
Viet Nam	593.1	46	241.6	38	232.0	230.0	X	X	X	X	18.3	28.6	X	X	1.4	1.7
Yemen Arab Rep	21.3	32	0.0	X	X	X	X	X	X	X	X	X	X	X	X	X
Yemen, People's Dem Rep	49.3	(10)	0.0	X	X	X	X	X	X	X	X	X	X	X	X	X
EUROPE	12,298.0	(4)	449.2	53	X	X	X	X	X	X	X	X	X	X	X	X
Albania	9.1	127	3.2	X	0.3	0.3	0.0	0.0	0.0 b	0.0	X	X	1.5	1.3	X	X
Austria	X	X	4.6	99	1.1	1.2	3.0	3.0	X	X	X	X	X	X	X	X
Belgium	41.1	(11)	0.4	X	0.1	0.2	0.3	0.4	X	X	X	X	X	X	X	X
Bulgaria	93.7	(36)	12.9	57	11.2	11.0	1.2	1.5	X	X	X	X	0.0	0.2	X	X
Czechoslovakia	X	X	20.7	19	19.1	19.6	1.1	1.1	X	X	X	X	X	X	X	X
Denmark	1,746.5	(4)	22.3	45	X	X	23.9	25.9	X	X	X	X	X	X	X	X
Finland	124.8	36	32.8	45	X	X	10.2	12.7	X	X	X	X	X	X	X	X
France	819.6	7	34.8	X	7.8 a	7.4	25.6	31.0	0.1	0.2	0.0	0.0	179.3	187.3	0.0 a	X
German Dem Rep	180.3	(34)	20.7	41	13.4	13.2	6.3	7.7	X	X	X	X	X	X	X	X
Germany, Fed Rep	185.8	(57)	24.1	58	6.0	5.4	14.0	14.5	X	X	X	X	37.5	29.4	X	X
Greece	115.1	23	9.7	12	0.2	0.3	1.9	1.9	0.1 b	0.1	X	X	0.2	0.4	X	X
Hungary	X	X	36.6	13	17.5	17.5	0.2 c	0.4	X	X	X	X	X	X	X	X
Iceland	1,656.3	48	0.6	21	X	X	0.2	0.6	X	X	X	X	X	X	X	X
Ireland	235.2	155	0.2	X	X	X	1.4	2.9	X	X	X	X	12.0	15.3	X	X
Italy	514.7	36	51.4	143	2.3	2.0	26.4	33.5	4.4	4.0	X	X	47.3	48.9	X	X
Luxembourg	X	X	0.0	X	X	X	X	X	X	X	X	X	X	X	X	X
Malta	1.5	2	0.0	X	X	X	X	X	X	X	X	X	X	X	X	X
Netherlands	460.3	47	4.4	21	0.1	0.3	0.2	0.4	X	X	X	X	88.4	99.1	X	X
Norway	1,981.8	(36)	0.4	(12)	X	X	36.8	56.2	X	0.0	X	X	0.2 b	0.1	X	X
Poland	636.8	(10)	29.7	21	18.0	18.9	1.6	1.6	X	X	X	X	X	X	X	X
Portugal	264.0	6	1.4	1,287	X	X	1.0	1.2	0.7	0.8	X	X	5.0	7.0	X	X
Romania	194.0	122	63.7	26	38.0	38.0	X	X	X	X	0.0	0.0	X	X	X	X
Spain	1,409.5	(2)	27.4	61	13.3	12.1	0.5 a	0.4	0.1 c	3.3	244.6	249.2	X	X
Sweden	219.8	12	3.3	(68)	0.0 a	X	2.9	4.8	X	X	0.0	0.0	0.7	2.6	X	X
Switzerland	X	X	4.6	18	X	X	0.3	0.3	X	X	X	X	X	X	X	X
United Kingdom	893.6	(10)	13.9	X	0.1	0.1	19.0	26.5	X	X	X	0.0	0.3	2.1	X	X
Yugoslavia	52.3	53	25.6	4	13.3	13.0	0.2	0.3	0.0	0.1	X	X	2.0	2.6	X	X
U.S.S.R.	10,040.5	12	940.3	13	244.8	261.9	22.9	27.0	0.1	0.0	X	X	0.1	0.2	3.5	3.5
OCEANIA	682.3	144	11.1	405	X	X	X	X	X	X	X	X	X	X	X	X
Australia	177.2	54	2.4	62	0.2 b	0.2	0.9 c	1.5	X	X	0.0 c	0.0	9.3	9.4	0.1	0.1
Fiji	26.5	375	3.4	557	0.0	0.0	X	X	X	0.0	0.0	0.0	0.0	0.0	0.5 a	1.4
New Zealand	360.0	409	0.2	86	X	X	0.3	0.9	X	X	X	X	12.8	17.8	X	X
Papua New Guinea	10.7	(71)	5.0	5,841	X	X	X	X	X	X	X	X	X	X	X	X
Solomon Islands	48.0	118	0.0	X	X	X	X	X	X	X	0.0	0.0	X	X	X	X

Source: Food and Agriculture Organization of the United Nations.
Notes: Total of aquaculture production is included in the country totals for marine and freshwater catches (except United States).
a. Two years of data. b. One year of data. c. Number may not match total of 1987 because of missing data in some classes.
0 = zero or less than half of the unit of measure; X = not available; negative numbers are shown in parentheses.
For additional information, see Sources and Technical Notes.

Table 23.3 Annual Catch and Sustainable Yield of Regional Marine Fisheries

	Average Annual Catch{a} (000 metric tons)		Estimated Sustainable Yield{b} (000 metric tons per year)		Average Annual Catch{a} (000 metric tons)		Estimated Sustainable Yield{b} (000 metric tons per year)
	1975-77	1985-87			1975-77	1985-87	
ALL MARINE FISHERIES	60,948	78,955	62,250-95,950				
Atlantic, Northwest				**Atlantic, Northeast**			
Total catch	3,372	2,930	3,400-4,300	Total catch	12,585	10,690	10,100-12,300
Cods, hakes, haddocks	867	881		Cods, hakes, haddocks	4,143	3,580	
Herrings, sardines, anchovies	613	503		Herrings, sardines, anchovies	1,923	1,747	
Clams, cockles, arkshells	226	418		Redfishes, basses, congers	1,162	1,275	
Redfishes, basses, congers	251	209		Jacks, mullets, sauries	3,217	1,740	
Flounders, halibuts, soles	236	212		Mackerels, snoeks, cutlassfishes	757	580	
All others	1,179	708		All others	1,381	1,768	
Atlantic, Western Central				**Atlantic, Eastern Central**			
Total catch	1,510	2,168	3,200-5,100	Total catch	3,613	3,047	2,900-3,700
Herrings, sardines, anchovies	650	992		Herrings, sardines, anchovies	1,386	1,180	
Miscellaneous marine fishes	134	275		Jacks, mullets, sauries	648	405	
Shrimps, prawns	169	186		Miscellaneous marine fishes	371	353	
Oysters	176	194		Redfishes, basses, congers	340	298	
Redfishes, basses, congers	150	115		Tunas, bonitos, billfishes	243	246	
All others	232	405		All others	626	565	
Atlantic, Southwest				**Atlantic, Southeast**			
Total catch	893	1,847	2,600-3,900	Total catch	2,656	2,303	2,500-3,100
Squids, cuttlefishes, octopuses	6	456		Herrings, sardines, anchovies	994	837	
Cods, hakes, haddocks	236	593		Jacks, mullets, sauries	612	645	
Redfishes, basses, congers	215	277		Cods, hakes, haddocks	686	510	
Herrings, sardines, anchovies	172	143		Miscellaneous marine fishes	63	68	
Shrimps, prawns	45	76		Redfishes, basses, congers	99	78	
All others	218	301		All others	202	166	
Indian Ocean, Western				**Indian Ocean, Eastern**			
Total catch	2,051	2,605	2,700-4,200	Total catch	1,190	2,277	1,500-2,200
Redfishes, basses, congers	426	491		Miscellaneous marine fishes	607	983	
Tunas, bonitos, billfishes	158	446		Redfishes, basses, congers	129	214	
Miscellaneous marine fishes	344	524		Herrings, sardines, anchovies	119	191	
Herrings, sardines, anchovies	339	379		Jacks, mullets, sauries	58	127	
Shrimps, prawns	242	240		Shads	3	112	
All others	543	524		All others	274	649	
Pacific, Northwest				**Pacific, Northeast**			
Total catch	17,668	25,187	13,500-16,500	Total catch	2,142	3,156	2,600-3,200
Herrings, sardines, anchovies	1,888	5,831		Cods, hakes, haddocks	1,228	1,933	
Cods, hakes, haddocks	3,950	5,282		Flounders, halibuts, soles	182	368	
Miscellaneous marine fishes	3,634	4,716		Salmons, trouts, smelts	182	383	
Mackerels, snoeks, cutlassfishes	2,176	1,869		Redfishes, basses, congers	177	154	
Redfishes, basses, congers	1,470	1,399		Herrings, sardines, anchovies	112	78	
All others	4,549	6,091		All others	262	240	
Pacific, Western Central				**Pacific, Eastern Central**			
Total catch	5,447	6,369	5,800-7,800	Total catch	1,506	2,623	2,200-3,000
Miscellaneous marine fishes	1,812	1,657		Herrings, sardines, anchovies	742	1,350	
Tunas, bonitos, billfishes	621	1,184		Tunas, bonitos, billfishes	436	534	
Jacks, mullets, sauries	729	840		Miscellaneous marine fishes	97	246	
Herrings, sardines, anchovies	570	683		Shrimps, prawns	75	141	
Redfishes, basses, congers	544	637		Mackerels, snoeks, cutlassfishes	4	161	
All others	1,170	1,368		All others	152	191	
Pacific, Southwest				**Pacific, Southeast**			
Total catch	387	745	1,200-2,000	Total catch	4,698	10,628	3,700-10,300
Cods, hakes, haddocks	98	169		Herrings, sardines, anchovies	3,590	7,663	
Jacks, mullets, sauries	42	124		Jacks, mullets, sauries	532	2,294	
Redfishes, basses, congers	58	148		Cods, hakes, haddocks	161	173	
Miscellaneous marine fishes	21	72		Redfishes, basses, congers	49	100	
Squids, cuttlefishes, octopuses	32	112		Mackerels, snoeks, cutlassfishes	108	66	
All others	136	120		All others	258	332	
Mediterranean and Black Sea				**Antarctic**			
Total catch	1,225	1,966	1,090-1,410	Total catch	172	416	X
Herrings, sardines, anchovies	584	852		Krill, planktonic crustaceans	55	338	
Jacks, mullets, sauries	95	216		Redfishes, basses, congers	108	73	
Redfishes, basses, congers	117	150		Jacks, mullets, sauries	1	1	
Miscellaneous marine fishes	113	130		Miscellaneous marine fishes	8	3	
Mussels	22	96		Miscellaneous marine molluscs	0	0	
All others	294	521		All others	1	0	

Source: Food and Agriculture Organization of the United Nations.
Notes: a. Catch includes all fish, crustaceans, and molluscs harvested in marine fishing areas. b. Estimated sustainable yields refers to marine fish, crustaceans, and cephalopods. It excludes ocean pelagic species (about 3-5 percent of total potential) and molluscs. Figures may not add because of rounding.
0 = zero or less than half the unit of measure; X = not available.
For additional information, see Sources and Technical Notes.

Table 23.4 Antarctic Fisheries, 1970-87

Antarctic Marine Area	Species	Main Fishing Countries, 1987	Average Annual Catch												State of Exploitation
			1970-74		1975-79		1980-84		1985		1986		1987		
			(000 metric tons)	Percent of Total	(000 metric tons)	Percent of Total	(000 metric tons)	Percent of Total	(000 metric tons)	Percent of Total	(000 metric tons)	Percent of Total	(000 metric tons)	Percent of Total	
Southern Atlantic	Grey rockcod	U.S.S.R.	0	0	2	1	0	0	1	0	0	0	0	0	Unknown
	Humped rockcod	Poland, U.S.S.R.	0	0	8	4	7	2	8	4	2	0	3	1	Heavily exploited
	Marbled rockcod	Poland, U.S.S.R.	83	90	7	4	11	3	2	1	0	0	0	0	Severely depleted
	Patagonian rockcod	U.S.S.R.	0	0	3	2	18	5	12	5	16	3	9	2	Unknown
	Mackerel icefish	U.S.S.R., Poland	3	3	62	33	63	17	16	7	14	3	72	17	Heavily exploited
	Other fishes	FRG, Poland, U.S.S.R.	1	1	16	8	17	5	8	4	4	1	5	1	
	Total fish catch		87	95	98	52	116	31	47	21	36	8	89	21	
	Krill	U.S.S.R., Japan, Chile	5	5	92	48	253	69	181	79	426	92	326	79	Lightly exploited
	Total catch		92	100	190	100	369	100	228	100	462	100	415	100	
Southern Indian	Grey rockcod	U.S.S.R.	22	23	11	14	7	6	7	23	3	8	3	7	Heavily exploited
	Marbled rockcod	U.S.S.R.	47	49	13	16	5	4	2	6	1	3	0	0	Severely exploited
	Mackerel icefish	U.S.S.R., France	24	25	28	35	10	8	8	26	17	45	4	10	Heavily exploited
	Other fishes	U.S.S.R.	3	3	1	1	1	1	8	26	1	3	3	7	
	Total fish catch		96	100	53	66	23	18	25	81	22	58	10	24	
	Krill	U.S.S.R., Japan	0	0	27	34	104	82	6	19	16	42	31	76	Lightly exploited
	Total catch		96	100	80	100	127	100	31	100	38	100	41	100	
Southern Pacific	Other fishes		0	0	0	0	0	0	0	0	0	0	0	0	
	Total fish catch		0	0	0	0	0	0	0	0	0	0	0	0	
	Krill	Japan	0	0	1	100	5	100	5	100	4	100	2	100	Lightly exploited
	Total catch		0	100	1	100	5	100	5	100	4	100	2	100	
TOTAL ANTARCTIC			188		271		501		264		504		458		

Source: Food and Agriculture Organization of the United Nations.
Notes: 0 = zero or less than half the unit of measure; FRG = Federal Republic of Germany. For additional information, see Sources and Technical Notes.

Table 23.5 Waste and Metals Dumped in Oslo Convention Area

Total Input (000 metric tons)	Metals	1976	1977	1978	1979	1980	1981	1982	1983	1984	1985	1986
Industrial wastes		7,202.7	7,501.0	8,500.9	9,900.3	9,100.8	8,700.1	7,840.1	7,400.2	5,500.0	5,900.0	5,840.0
Liquid wastes and sludges		1,400.0	1,600.0	1,700.0	1,700.0	1,500.0	1,100.0	940.0	1,100.0	1,100.0	1,000.0	1,100.0
	Cadmium (t)	0.1	0.2	0.2	0.2	0.1	0.1	0.0	0.0	0.0	0.1	0.0
	Copper (t)	14.0	3.1	2.8	13.0	6.6	11.0	5.5	4.4	2.1	2.4	2.9
	Lead (t)	2.1	9.8	10.6	9.3	1.7	2.5	1.8	1.8	0.3	1.9	1.8
	Mercury (t)	0.1	0.1	0.1	0.1	0.1	0.2	0.0	0.0	0.0	0.0	0.0
	Zinc (t)	460.0	530.0	560.0	390.0	27.0	31.0	72.0	19.0	14.0	14.0	6.8
"Deep water" solid wastes		2.7	1.0	0.9	0.3	0.8	0.1	0.1	0.2	0.0	0.0	0.0
TiO2 production wastes		1,900.0	2,100.0	2,200.0	2,800.0	2,300.0	2,700.0	2,400.0	2,500.0	2,400.0	2,400.0	2,200.0
	Cadmium (t)	X	X	X	X	0.2	0.1	0.1	0.1	0.4	0.0	0.1
	Copper (t)	X	X	X	X	2.5	3.7	1.1	0.6	1.3	0.7	0.6
	Lead (t)	X	X	X	X	6.8	8.5	4.1	3.5	4.8	3.4	3.7
	Mercury (t)	X	X	X	X	0.0	0.0	0.0	0.0	0.0	0.0	0.0
	Zinc (t)	X	X	X	X	73.0	84.0	49.0	80.0	80.0	82.0	72.0
Phosphogypsum wastes		1,700.0	2,000.0	2,600.0	3,100.0	3,300.0	2,900.0	2,600.0	1,600.0	1,600.0	800.0	640.0
	Cadmium (t)	9.6	3.9	4.6	5.2	4.6	3.9	1.9	1.1	2.1	0.9	1.8
	Copper (t)	21.0	6.6	7.6	7.6	3.9	3.5	3.0	2.6	3.7	1.7	1.5
	Lead (t)	15.0	5.6	5.8	7.0	10.0	5.9	4.6	1.3	1.8	0.1	0.3
	Mercury (t)	0.3	0.1	0.1	0.2	0.3	0.2	0.2	0.2	0.2	0.1	0.0
	Zinc (t)	73.0	22.0	28.0	22.0	17.0	14.0	6.7	5.2	12.0	3.0	3.5
Fly ash and colliery wastes		2,200.0	1,800.0	2,000.0	2,300.0	2,000.0	2,000.0	1,900.0	2,200.0	400.0	1,700.0	1,900.0
	Cadmium (t)	X	X	X	0.1	0.1	0.3	0.4	0.3	0.0	0.1	0.2
	Copper (t)	77.0	66.0	200.0	230.0	180.0	190.0	190.0	220.0	43.0	160.0	200.0
	Lead (t)	25.0	19.0	220.0	250.0	240.0	250.0	240.0	280.0	48.0	210.0	250.0
	Mercury (t)	X	X	X	0.3	0.3	0.3	0.4	0.3	0.0	0.2	0.2
	Zinc (t)	45.0	44.0	420.0	490.0	440.0	450.0	460.0	540.0	110.0	400.0	490.0
Sewage sludge		7,600.0	8,200.0	8,700.0	8,800.0	9,200.0	8,600.0	8,200.0	7,400.0	7,600.0	7,700.0	8,200.0
	Cadmium (t)	11.0	12.0	9.1	9.0	10.0	6.9	5.6	4.1	4.1	3.7	3.5
	Copper (t)	260.0	280.0	250.0	260.0	240.0	220.0	220.0	170.0	160.0	150.0	150.0
	Lead (t)	150.0	170.0	220.0	210.0	190.0	170.0	180.0	160.0	160.0	170.0	170.0
	Mercury (t)	4.2	4.0	3.0	2.9	3.8	2.6	1.6	1.6	1.0	1.3	1.1
	Zinc (t)	790.0	810.0	920.0	1,000.0	730.0	690.0	500.0	460.0	510.0	330.0	490.0
Dredged materials		72,000.0	77,000.0	85,000.0	65,000.0	110,000.0	110,000.0	100,000.0	150,000.0	71,000.0	100,000.0	91,000.0
	Cadmium (t)	42.0	50.0	63.0	43.0	72.0	63.0	53.0	73.0	34.0	31.0	22.0
	Copper (t)	1,300.0	1,100.0	1,400.0	1,300.0	1,900.0	1,700.0	1,600.0	1,600.0	1,300.0	1,500.0	1,300.0
	Lead (t)	2,700.0	2,300.0	2,700.0	2,700.0	4,000.0	3,800.0	3,400.0	3,600.0	2,700.0	2,800.0	2,600.0
	Mercury (t)	19.0	16.0	43.0	34.0	40.0	35.0	31.0	34.0	27.0	28.0	40.0
	Zinc (t)	9,800.0	9,700.0	11,000.0	10,000.0	15,000.0	16,000.0	14,000.0	13,000.0	9,700.0	10,000.0	11,000.0
TOTAL METALS	CADMIUM (t)	62.7	66.1	76.9	57.5	87.0	74.3	61.0	78.6	40.6	35.9	27.6
	COPPER (t)	1,672.0	1,455.7	1,860.4	1,810.6	2,333.0	2,128.2	2,019.6	1,997.6	1,510.1	1,814.8	1,655.0
	LEAD (t)	2,892.1	2,504.4	3,156.4	3,176.3	4,448.5	4,236.9	3,830.5	4,046.6	2,914.9	3,185.4	3,025.8
	MERCURY (t)	23.6	20.2	46.2	37.5	44.5	38.4	33.2	36.1	28.2	29.6	41.4
	ZINC (t)	11,168.0	11,106.0	12,928.0	11,902.0	16,287.0	17,269.0	15,087.7	14,104.2	10,426.0	10,829.0	12,062.3

Source: Oslo Commission.
Notes: TiO2 = Titanium dioxide.
0 = zero or less than half of the unit of measure; X = not available; t = metric tons. For additional information, see Sources and Technical Notes.

Sources and Technical Notes

Table 23.1 Coastal Areas and Resources

Sources: Length of marine coastline: United Nations Office of Ocean Affairs and the Law of the Sea, unpublished data (United Nations, New York, June 1989); U.S. Central Intelligence Agency, *The World Factbook 1988* (U.S. Government Printing Office, Washington, D.C., 1988). Shelf area to 200-meter depth: John P. Albers, M. Devereux Carter, Allen L. Clark, et al., *Summary Petroleum and Selected Mineral Statistics For 120 Countries, Including Offshore Areas* (U.S. Government Printing Office, Geological Survey Professional Paper 817, Washington, D.C., 1973). Exclusive economic zone: United Nations Office of Ocean Affairs and the Law of the Sea, unpublished data (United Nations, New York, June 1989); French Polynesia and New Caledonia: Anthony Bergin, "Fisheries Surveillance in the South Pacific," *Ocean & Shoreline Management*, Vol. 11 (1988), p. 468.

Coastal population: United Nations Population Division, *Demographic Yearbook 1987* (United Nations, New York, 1989); *The Times Atlas of the World* (John Bartholomew & Son Limited and Times Book Limited, London, 1988); U.S. Defense Mapping Agency, Hydrographic/Topographic Center, *World Port Index* (U.S. Government Printing Office, Washington, D.C., 1984).

Volume of goods loaded and unloaded: United Nations Statistical Office, Department of International Economic and Social Affairs, *Statistical Yearbook 1985/86* (United Nations, New York, 1988).

Offshore oil and gas resources: *Offshore Magazine* (PennWell Publishing Company, Tulsa, Oklahoma, United States, June 20, 1983; and June 1989); 1978 offshore gas production for Spain and the United States: *1988 Energy Statistics Sourcebook* (PennWell Publishing Company, Tulsa, Oklahoma, United States, 1988).

The United Nations Office for Ocean Affairs and the Law of the Sea compiles information concerning coastal claims from the following sources: the U.N. Legislative Series; official gazettes; communications to the Secretary General; legal journals; and other publications. National claims to maritime zones fall into five categories: territorial sea, contiguous zone, exclusive economic zone (EEZ), exclusive fishing zone, and continental shelf. The extent of the continental *shelf to 200-meter depth* and the *exclusive economic zone* for those countries with marine coastline are presented in Table 23.1. Only the potential and not the actual established area of the EEZ are shown. At present, half of the world's countries have established a full EEZ. Please refer to *World Resources 1988–89*, Table 22.1, for a listing of national claims to territorial seas, EEZs, and continental shelf.

An EEZ may be established by a nation out to 200 nautical miles to claim all the resources within the zone, including fish and all other living resources; minerals; and energy from wind, waves, and tides. Nations may also claim rights to regulate scientific exploration, protect the marine environment, and establish marine terminals and artificial islands. The EEZ data shown do not reflect the decisions of some countries, such as those in the European Community, to collectively manage EEZs in some areas. When countries' EEZs overlap—such as those of the United States and Cuba, which both have 200-mile EEZs, yet are only 90 miles apart—they must agree on a maritime boundary between them, often a halfway point.

The shelf area to the 200-meter isobath represents one indicator of potential offshore oil and gas resources because of sedimentation from continental areas. Other indicators include geology and geography. Questions of accessibility and water depth affect the economics of exploration and production and may constrain operations in water in depths greater than 200 meters. Significant deep-water operations currently take place in the North Sea.

The percentage of urban population in large coastal cities was calculated using data on cities and urban agglomerations whose populations are 100,000 or larger, reported in the *Demographic Yearbook*. The total population living in urban centers with more than 100,000 people located at the coast was divided by the total population living in urban centers with more than 100,000 people. Thus, 100 percent means that all inhabitants of urban centers of 100,000 or larger are living at the coast. "Coastal centers" were defined by *The Times Atlas* and usually are no more than 20 kilometers inland. Major ports located farther inland, such as Hamburg, are included as well. Data on urban agglomeration were preferred over that of component cities. The data are the most recent country-level estimates available, but range from 1970 to 1986. The percentage of urban population in large coastal cities can be used as an indicator of population and development pressures exerted on coastal resources.

The United Nations Statistical Office based its estimates of *average annual volume of goods loaded and unloaded* in maritime transport mostly on information available in external trade statistics. *Petroleum* products exclude bunkers and those products not generally carried by tanker, namely: paraffin wax, petroleum coke, asphalt, and lubricating oil, which are included with the data for *dry cargo*.

Offshore Magazine annually queries national governments for statistics on *offshore oil and gas resources*. These data are supplemented with figures from oil- and gas-producing companies, expert sources, and published literature. National govern-

ments often have difficulty providing offshore gas production figures; the data are more frequently obtained from alternate sources. Figures for offshore *oil* and *gas* production in Middle Eastern countries are particularly difficult to obtain and, as a result, are less reliable.

The regional total for Asian offshore oil production includes production within the Neutral Zone, an area of disputed sovereignty that lies between Saudi Arabia and Kuwait. Profits from production in the Neutral Zone are shared by the two countries.

Proven reserves of offshore crude oil and gas represent the fraction of total resources that can be recovered in the future, given the present, and expected, economic conditions and existing technological limits. Please refer to Table 21.3 for data on both offshore and onshore reserves.

Table 23.2 Marine and Freshwater Catches and Aquaculture

Sources: Fishcatch: Food and Agriculture Organization of the United Nations (FAO), *Yearbook of Fishery Statistics 1984* and *1987* (FAO, Rome, 1986 and 1989). Aquaculture: Food and Agriculture Organization of the United Nations (FAO), Fisheries Department, *Aquaculture Production (1984–1987)* (FAO, Rome, November 1989).

Average Annual Marine and *freshwater catch* data refer to marine and freshwater fish killed, caught, trapped, collected, bred, or cultivated for commercial, industrial, and subsistence use. Crustaceans, molluscs, and miscellaneous aquatic animals are included. Statistics for mariculture, aquaculture and other kinds of fish farming are included in the country totals. Quantities taken in recreational activities are excluded. Figures are the national totals averaged over a three-year period; they include fish caught by a country's fleet anywhere in the world. Catches of freshwater species caught in seas with low salinity are included in the statistics of the appropriate marine area. Catches of diadromous (migratory between salt and freshwater) species are shown either in the marine or inland area where caught.

Data are represented as nominal catch, which is the landings converted to a live-weight basis, that is, weight when caught. Landings for some countries are identical to catches.

International fishery data are continually revised. The *Yearbook of Fishery Statistics 1987*, the latest edition, contains FAO's most up-to-date published figures.

Data are provided annually to the FAO Fisheries Department by national fishery offices and regional fishery commissions. Some countries' data are only provisional for the latest year; for other countries, no data are available. If no new data are submitted, FAO uses the previous year's figures or makes estimates based on other information.

Years refer to calendar years except for Antarctic fisheries data, which are for split years (July 1–June 30). Data for Antarctic fisheries are given for the calendar year in which the split year ends (see Table 23.4 for trends in Antarctic fisheries).

Average annual aquaculture production refers to the farming of aquatic organisms, including fish, molluscs, crustaceans, and aquatic plants. Farming is defined as the active intervention in the rearing process of aquatic organisms, such as regular stocking, feeding, and protection from predators. For statistical purposes, aquaculture includes only aquatic organisms that are harvested by an individual or corporate body that has owned them throughout their rearing period. Aquatic organisms that are exploitable by the public as a common property resource, with or without appropriate licenses, are the harvest of fisheries.

FAO's global collection of aquaculture statistics by questionnaire was begun in 1984; today, these data are a regular feature of the annual FAO survey of world fishery statistics.

FAO's 840 "species items" are summarized in six categories. *Freshwater fishes* include carps, barbels, tilapias and other freshwater fishes. *Diadromous fishes* include, among others, sturgeons, river eels, salmons, trouts, and smelts. *Marine fishes* include a variety of species such as flounders, halibuts, and redfishes. *Crustaceans* include, among others, freshwater crustaceans, crabs, lobsters, shrimps, and prawns. *Molluscs* include freshwater molluscs, oysters, mussels, scallops, clams, and squids. *Other* includes frogs, turtles, and aquatic plants. Data on whales and other mammals are excluded from this table. For a detailed listing of species, please refer to the most recent *FAO Yearbook of Fishery Statistics*.

Table 23.3 Annual Catch and Sustainable Yield of Regional Marine Fisheries

Sources: Marine fishery production: Food and Agriculture Organization of the United Nations (FAO), *Yearbook of Fishery Statistics 1981* and *1987* (FAO, Rome, 1983 and 1989). Estimated fishery potential: M.A. Robinson, *Trends and Prospects in World Fisheries* (FAO, Fisheries Department, Rome, 1984).

FAO divides the world's oceans into 19 marine statistical areas and organizes *average annual catch* data by 840 "species items," species groups separated at the family, genus, or species level. The species groups shown in Table 23.3 are FAO groupings, which include species similar to those named. For example, the group designated "herrings, sardines, and anchovies" also includes menhadens, pilchards,

and bonefish. The species groups listed represent the top five categories caught in each fishing zone in 1987. Years shown are three-year averages. (Refer to the Technical Note for Table 23.2 for the definition of nominal fish catch and additional information on FAO's fishery data base.)

Data on *estimated sustainable yield* are FAO estimates of marine fisheries' biologically realizable potential. These estimates refer to the maximum harvest that can be sustained by a fishery year after year, given average environmental conditions. An assumed level of incidental take (catching one species while fishing for another) is subtracted from estimates of potential. The figures exclude the potential harvest from culturing marine fish. Maximum sustainable yield estimates are not strictly comparable to the catch data shown because they exclude molluscs. Estimates of oceanic pelagic species (about 3–5 percent of the total marine potential) are unavailable at the regional level and are also excluded.

Table 23.4 Antarctic Fisheries, 1970–87

Source: Food and Agriculture Organization of the United Nations (FAO), *Review of the State of World Fishery Resources* (FAO, Fisheries Circular No. 710, Revision 6, Rome, 1989).

The Antarctic fisheries comprise three FAO fishery zones (*Antarctic marine areas*). "Southern Atlantic" refers to FAO Region 48, an area bounded by 70° west longitude and 30° east longitude, and extending north from the Antarctic coast to 50° south and 60° south latitudes. "Southern Indian" refers to FAO Region 58, an area bounded by 30° east longitude and 150° east longitude, and extending north from the Antarctic coast to 45° south and 55° south latitudes. "Southern Pacific" refers to FAO Region 88, an area bounded by 150° east longitude and 70° west longitude, and extending north from the Antarctic coast to 60° south latitude. *Main fishing countries* for these areas are listed for 1987.

Most of the Antarctic *average annual catch* is composed of krill and two endemic families of finfish: Antarctic cod (Nototheniidae) and icefish (Channichthyidae). Catches of other finfish are very low. Over 90 percent of the total 1987 Antarctic catch came from waters around the islands of South Georgia and South Orkney in Region 48 (Southern Atlantic). Data are for split years (July 1–June 30) and are given for the calendar year in which the split year ends. Refer to the Technical Note for Table 23.2 for the definition of nominal fish catch and additional information on FAO's fishery data base.

The scientific body responsible for the Antarctic area is the Commission for the Conservation of Antarctic Marine Living Resources. It collects data on catch and effort, assesses the fish stock, and recommends management measures.

Table 23.5 Waste and Metals Dumped in Oslo Convention Area

Source: Oslo Commission, *Thirteenth Annual Report* (Oslo Commission, London, 1989).

Ocean dumping data are reported to the Oslo Commission by 13 western European countries (Belgium, Denmark, Finland, France, Federal Republic of Germany, Iceland, Ireland, Netherlands, Norway, Portugal, Spain, Sweden, and the United Kingdom) who are contracting parties to the 1972 Convention for the Protection of Marine Pollution by Dumping from Ships and Aircraft (Oslo Convention). The Oslo Convention applies to dumping in the North Atlantic and Arctic Ocean bounded by 36° north latitude, 42° west longitude, and 51° east longitude. Dumping elsewhere in the world, as well as in internal waters, such as harbors and estuaries, lies outside the limits of the Oslo Convention area. It is covered under the 1972 Convention on the Prevention of Marine Pollution by Dumping of Wastes and Other Matter (London Dumping Convention).

The Oslo Convention bans the dumping of certain dangerous substances, except when they occur in wastes in trace quantities, because of their toxicity, persistence, and bioaccumulation. These include crude oil, mercury, cadmium, and high-level radioactive wastes. Other hazardous substances such as arsenic, lead, and copper may be dumped but require special permits.

The Oslo Commission secretariat records permits for incineration of chemical wastes at sea and dumping permits for three types of waste: industrial waste, sewage sludge, and dredged materials.

Industrial wastes in Table 23.5 have been divided into five waste categories:

Liquid wastes and sludges include liquid waste products from a variety of industrial processes such as caustic washing from oil refining, sugar refining wastes, and brine from herbicide manufacture.

"Deep water" solid wastes refer to the disposal of containerized solid waste at a United Kingdom deep water dumpsite in the Atlantic.

TiO₂ production wastes refer to titanium dioxide, one of the top 20 (by quantity) inorganic chemicals currently produced in the world. Titanium dioxide is used primarily as a white pigment in the manufacture of paints, plastic, paper, rubber, and artificial fibers. Depending on the selected raw materials and technology, titanium dioxide production generates waste products such as copperas (red mud), spent acid (sulfuric acid), and metal chlorides. Under unfavorable hydrodynamic conditions, the dumping of titanium dioxide wastes can harm marine flora and fauna. Belgium, the Federal Republic of Germany, Italy, and the Netherlands have decided that future discharges of TiO_2 will be prohibited.

Phosphogypsum wastes, a byproduct from the manufacture of phosphoric acid, contain residual free acid (phosphoric acid) and a number of impurities originating from phosphate rock, such as fluorides, heavy metals, and radium. The disposal of phosphogypsum to the seas can result in a reduction of pH in the sea water immediately adjacent to the discharge area and a build-up of metals in the water column and in marine organisms, especially where there is no regular and significant tidal flow.

Fly ash and colliery wastes include two kinds of wastes. Fly ash, the refuse from coal-fired power stations, consists of large, fused lumps of clinker and a considerable amount of very fine powder. It is composed mainly of silicon dioxide with oxides of aluminum and iron, and a variety of metals are present in trace concentrations. Colliery wastes are predominantly made up of natural stone, but include small fragments of coal and thus have a high metal content. Dumping grounds receiving fly ash or colliery wastes have shown extreme impoverishment of the marine fauna because of the smothering effect of these wastes.

Sewage sludge contains much of the solid material separated from liquid effluent during the treatment of municipal wastes. It can contain both toxic pollutants, such as metals and organic chemicals, as well as nontoxic pollutants, such as fecal bacteria, viruses, and nutrients.

Dredged materials include sediments from harbors, estuaries, and navigation channels that are dumped at sea. Dredged materials do not usually contain high levels of contaminants, but dredge spoils from harbors tend to contain higher levels than dredge spoils from other sources. Disposal of dredged materials varies considerably from year to year, depending on the frequency of maintenance dredging.

The effects of *metals* in the marine environment and the resulting impacts on humans are both wide ranging and complex. Transition metals such as copper and zinc may be toxic in high concentration, but are essential nutrients at low concentration. Heavy metals such as cadmium, lead, and mercury may also be required in low concentration for cell metabolism, but are toxic at slightly higher concentrations. Heavy metals tend to build up in sediment and living organisms. Shellfish are especially efficient at concentrating metals from the water and sediment in which they live. For a detailed discussion of potential harmful effects of cadmium and lead in the marine environment, please refer to Joint Group of Experts on the Scientific Aspects of Marine Pollution (GESAMP), *Reports and Studies No. 22, Review of Potentially Harmful Substances—Cadmium, Lead and Tin* (World Health Organization, Geneva, 1985).

The quality of the data presented on metal inputs varies over time and among waste categories. Analytical methods, measuring the concentration of metals in waste, have improved in the past decade. Data analysis is seriously compromised when the sample's metal content is less than the instrument's detection limit. In such cases, the instrument's lowest detection limit is reported as the sample's metal content. Different instruments, measuring low metal concentrations, can give different results, reflecting only differences in the quality of the instruments instead of differences in the concentration of metals. Measurements at the beginning of these time series also tend to be based on less frequent analyses and less accurate instruments than do those taken more recently.

Trend data for metal input from fly ash, colliery waste, phosphogypsum wastes, and titanium dioxide production wastes are closely related to the tonnage of wastes dumped. Their metal content depends on the raw materials used, which tend to fluctuate only between certain narrow limits.

Large errors can occur in estimating the concentration of heavy metals in dredged materials, especially when a poor estimate, based on an inadequate sampling technique, is multiplied by large quantities of dredged materials. Furthermore, some dredged materials originate from sea channels and thus do not represent new inputs.

A large proportion of trace metals measured in solid wastes is actually bound to a mineral matrix. Analytical techniques produce overestimates of the quantities of metals directly available to the marine environment.

24. Atmosphere and Climate

The earth's atmosphere is not a limitless sink for human wastes. Emissions from fossil fuel production and consumption, deforestation, agriculture, industry, and other activities that serve humankind continue to increase and accumulate. These emissions could change our planet's climate, destroy the protective stratospheric ozone layer, acidify rainfall, and directly affect the health of people, plants, and animals. This chapter provides data that give some context to potential risk as well as each country's contribution to that risk. (See Chapter 2, "Climate Change: A Global Concern," for further information.)

Table 24.1 details how much each country contributes to the greenhouse effect through the emission of carbon dioxide (CO_2, shown as the amount of elemental carbon it contains), methane (CH_4), and chlorofluorocarbons (CFCs). The industrialized countries emit most of the CO_2 from the burning of fossil fuels, especially the United States, which ranks first, and the U.S.S.R., which ranks second. But data on the destruction of forests, a surprising range of human economic activities that produce CH_4, and the emission of CFCs show that even developing countries can emit large quantities of greenhouse gases to the atmosphere. Brazil, for example, ranked third in the world in 1987 as a producer of CO_2 because of massive deforestation.

The data in Table 24.1 show emissions from the mix of fossil fuels for each country, contributions from cement production and land use change, CH_4 emissions by source, and an estimate of CFC use. CO_2, CH_4, and CFCs make up about 86 percent of the contribution to current greenhouse heating. Other important greenhouse gases include nitrous oxide, the emissions of which are poorly understood, and tropospheric ozone, produced by reactions involving other pollutants.

Both CO_2 and CH_4 have large sources and sinks that are independent of human activities and which had been in reasonable balance until the past few hundred years. Emissions resulting from human activities have upset this balance leading to increases in the concentrations of both CO_2 and CH_4. (See Table 24.3.) The increases in concentrations total less than the amount emitted by human activity, so sinks for both CO_2 and CH_4 must have also increased.

The total emission of CO_2 caused by human activities in 1987 was about 8.6 billion metric tons (expressed in terms of the total carbon content of the CO_2), but the total in the atmosphere increased by only about 3.7 billion metric tons. For CH_4 the ratio was even larger. Methane emissions in 1987 were about 255 million metric tons, but the atmosphere had a net increase of only about 43 million metric tons. Table 24.2 attributes these atmospheric increases to countries in proportion to the fraction of the total CO_2 and CH_4 emissions that can be assigned to each.

Each country's share of increases of CH_4 and CFC in the atmosphere can be expressed in terms of the amount of CO_2 that would produce the same greenhouse heating effect. These equivalents, together with their share of increases in CO_2 concentrations, are each country's share of the increasing greenhouse effect.

The United States, U.S.S.R., and Brazil are also the largest contributors to the annual increase in the greenhouse effect, China is fourth because of equal effects from of CO_2 and CH_4, and India is fifth, primarily because of its emissions of CH_4. Japan is the sixth largest contributor.

Table 24.3 shows the increasing load over time of atmospheric pollutants that potentially threaten our climate and the stratospheric ozone layer. These trends, if unbroken, foreshadow even higher concentrations that could lead to climate change and increased ultraviolet radiation reaching the earth from the sun. Increases in atmospheric CO_2 concentrations shown in Table 24.3 reflect increasing emissions shown in Table 24.4. Total annual emissions of CO_2 from the burning of fossil fuels and cement production have increased over threefold since 1950.

Other emissions more directly affect human health, but they are difficult to estimate or are not measured in many places. Emission of these pollutants is concentrated in urban environments. Table 24.5 shows reported levels of sulfur dioxide, suspended particulate matter, and smoke in selected cities in the Global Environment Monitoring System (GEMS) of the United Nations Environment Programme. Data on most cities, even those with notorious air pollution such as Mexico City and Los Angeles, are not available within GEMS. Available data point to the special risk for people living in the developing countries of Asia in large cities such as Shenyang and Calcutta. In contrast, even São Paulo and Santiago appear healthful.

Table 24.6 shows estimates of the total amount of selected air pollutants dumped into the atmosphere by selected countries. Of special interest are declines, among those reporting, in industrialized countries of sulfur dioxide and oxides of nitrogen, both causes of acid rain. These declines are due in part to the deliberate conversion to lower sulfur fuels as well as to pollution controls in industry and transport.

Table 24.1 Sources of Current Greenhouse Gas Emissions

| | Anthropogenic Additions to the Carbon Dioxide Flux (000 metric tons carbon) | | | | | | | Anthropogenic Additions to the Methane Flux c.1987 (000 metric tons of methane) | | | | | | CFC Use Per Capita 1986 (kg) |
| | 1987 | | | | | c.1987 Land Use Change | Per Capita (metric tons) | Solid Waste | Livestock | Hard Coals | Wet Rice | Pipeline Leakage | Per Capita (metric tons) | |
	Cement	Solid	Liquid	Gas	Flaring									
WORLD	140,000	2,300,000	2,300,000	900,000	50,000	2,800,000	1.7	44,000	76,000	16,000	66,000	53,000	0.05	0.2
AFRICA	6,600	74,000	64,000	16,000	11,000	390,000	0.9	1,800	9,000	910	1,700	4,500	0.03	0.0
Algeria	880	820	5,800	9,000	2,800	X	0.8	52	150	0	X	3,700	0.17	0.1
Angola	48	0	450	81	660	5,500	0.7	20	120	X	6	X	0.02	X
Benin	41	0	99	0	0	2,500	0.6	9	42	X	1	X	0.01	X
Botswana	0	420	0	0	0	700	1.0	2	110	X	X	X	0.10	X
Burkina Faso	0	0	120	0	0	4,200	0.5	16	120	X	11	X	0.02	X
Burundi	0	4	41	0	0	(2)	0.0	11	25	X	2	X	0.01	X
Cameroon	0	1	1,600	0	0	34,000	3.4	22	150	0	8	X	0.02	X
Cape Verde	0	0	9	0	0	X	0.0	1	1	X	X	X	0.00	X
Central African Rep	0	0	71	0	0	3,500	1.3	6	58	X	4	X	0.03	X
Chad	0	0	56	0	0	4,200	0.8	12	170	X	7	X	0.04	X
Comoros	0	0	13	0	0	X	0.0	1	3	X	5	X	0.03	X
Congo	8	0	420	1	46	3,200	2.0	3	4	X	1	X	0.00	0.0
Cote d'Ivoire	89	0	1,300	0	0	100,000	9.1	23	41	X	110	X	0.03	0.1
Djibouti	0	0	72	0	0	X	0.2	1	9	X	X	X	0.03	X
Egypt	1,400	800	16,000	2,300	0	X	0.4	110	200	X	230	460	0.02	0.1 a
Equatorial Guinea	0	0	19	0	0	250	0.6	1	0	X	X	X	0.00	X
Ethiopia	34	0	700	0	0	7,800	0.2	84	1,200	X	X	X	0.03	X
Gabon	19	0	660	90	610	1,800	3.0	3	1	X	X	X	0.00	0.0
Gambia, The	0	0	49	0	0	200	0.3	1	12	X	4	X	0.03	X
Ghana	37	2	780	0	0	7,500	0.6	31	49	X	17	X	0.01	0.1
Guinea	0	0	260	0	0	8,700	1.4	13	69	X	200	X	0.06	X
Guinea-Bissau	0	0	33	0	0	3,000	3.3	2	9	X	52	X	0.08	X
Kenya	180	67	1,100	0	0	1,600	0.1	46	530	X	8	X	0.03	0.0
Lesotho	X	X	X	X	X	X	X	X	X	X	X	X	0.03	X
Liberia	12	0	170	0	0	7,500	3.3	5	4	X	47	X	0.06	0.1
Libya	370	1	4,600	1,900	420	X	1.8	9	47	X	X	360	0.10	X
Madagascar	5	10	220	0	0	23,000	2.1	23	380	X	430	X	0.10	X
Malawi	10	19	110	0	0	16,000	2.1	16	36	X	9	50	0.02	X
Mali	3	0	100	0	0	2,100	0.3	18	300	X	38	X	0.05	X
Mauritania	0	4	860	0	0	X	0.5	4	130	X	2	X	0.08	X
Mauritius	0	52	270	0	0	X	0.3	2	2	X	X	X	0.00	X
Morocco	520	1,200	3,800	45	0	X	0.2	53	220	4	1	X	0.01	0.0
Mozambique, People's Rep	61	46	210	0	0	7,000	0.5	33	49	X	38	X	0.01	X
Niger	5	48	150	0	0	1,600	0.3	14	210	0	6	X	0.04	X
Nigeria	480	79	6,900	1,900	6,200	58,000	0.7	220	610	1	150	X	0.01	0.1
Rwanda	0	0	99	0	0	290	0.1	14	30	X	1	X	0.01	X
Senegal	51	0	580	0	0	2,900	0.5	15	97	X	16	X	0.02	0.1
Sierra Leone	0	0	150	0	0	990	0.3	8	14	X	70	X	0.05	X
Somalia	0	0	260	0	0	990	0.2	13	580	X	1	X	0.09	X
South Africa	1,700	67,000	9,300	0	0	X	2.3	630	980	880	0	X	0.08	0.1
Sudan	27	0	870	0	0	27,000	1.2	50	1,000	X	1	X	0.05	X
Swaziland	0	120	0	0	0	X	0.2	2	23	X	X	X	0.03	X
Tanzania	41	3	520	0	0	4,800	0.2	50	560	0	120	X	0.03	X
Togo	50	0	75	0	0	690	0.3	7	17	X	4	X	0.01	0.0
Tunisia	460	66	2,300	400	28	X	0.4	17	64	X	X	X	0.01	0.1
Uganda	3	0	190	0	0	2,200	0.1	36	200	X	10	X	0.02	X
Zaire	54	210	700	0	0	35,000	1.1	79	52	1	120	X	0.01	0.0
Zambia	51	280	410	0	0	4,200	0.6	14	86	3	4	X	0.01	X
Zimbabwe	0	3,600	590	0	0	4,100	0.9	19	200	20	X	X	0.03	0.1
NORTH & CENTRAL AMERICA	16,000	470,000	680,000	280,000	6,300	90,000	3.8	18,000	10,000	3,900	700	27,000	0.15	0.6
Barbados	27	0	210	12	0	X	1.0	1	1	X	X	X	0.01	0.3
Canada	1,700	26,000	52,000	29,000	1,400	0	4.3	1,700	760	150	X	7,800	0.40	0.8
Costa Rica	71	0	660	0	0	15,000	5.7	6	92	X	6	X	0.04	0.1
Cuba	480	120	8,600	12	0	49	0.9	24	240	X	83	X	0.03	0.1
Dominican Rep	150	0	1,600	0	0	99	0.3	14	77	X	48	X	0.02	0.1
El Salvador	83	0	470	0	0	180	0.1	13	35	X	2	X	0.01	0.1
Guatemala	180	0	770	0	0	10,000	1.3	19	97	X	2	X	0.01	0.0
Haiti	27	0	160	0	0	26	0.0	12	61	X	5	X	0.02	0.0
Honduras	54	0	460	0	0	9,800	2.2	9	89	X	3	X	0.02	0.0 a
Jamaica	34	0	1,600	0	0	58	0.7	5	14	X	0	X	0.01	0.1
Mexico	2,700	5,700	58,000	14,000	1,200	32,000	1.4	180	1,500	42	22	4,500	0.08	0.1 a
Nicaragua	14	0	560	0	0	17,000	4.9	8	75	X	11	X	0.03	0.1
Panama	48	4	700	0	0	5,400	2.7	5	55	X	13	X	0.05	0.1
Trinidad and Tobago	43	0	740	2,100	1,900	87	4.0	3	3	X	0	X	0.01	0.3
United States	9,800	430,000	540,000	240,000	1,800	6,000	5.0	16,000	7,000	3,700	510	15,000	0.17	0.8 a
SOUTH AMERICA	6,700	16,000	92,000	27,000	3,700	240,000	5.8	620	13,000	100	850	2,300	0.07	0.1
Argentina	860	1,000	17,000	10,000	1,300	X	1.0	70	3,100	2	18	540	0.12	0.1
Bolivia	41	0	840	160	78	6,800	1.2	15	220	X	14	50	0.05	0.0
Brazil	3,500	10,000	38,000	1,500	540	1,200,000 b	9.1	320	7,500	37	490	510	0.08	0.1 a
Chile	200	1,300	5,100	450	98	X	0.6	29	180	8	6	130	0.03	0.1
Colombia	810	3,400	7,200	2,300	240	120,000	4.6	68	890	53	180	110	0.04	0.1
Ecuador	270	0	3,600	36	220	39,000	4.4	22	140	X	21	X	0.02	0.1
Guyana	0	0	280	0	0	340	0.6	2	12	X	14	X	0.06	0.0
Paraguay	20	0	430	0	0	7,400	2.0	8	190	X	6	X	0.06	0.0
Peru	270	160	5,600	340	94	45,000	2.5	45	190	1	36	X	0.02	0.0
Suriname	7	8	330	0	0	350	1.8	1	2	X	35	X	0.11	0.1
Uruguay	55	1	890	0	0	X	0.3	8	460	X	13	X	0.17	0.1
Venezuela	730	130	12,000	12,000	1,100	18,000	2.4	40	450	0	19	980	0.08	0.1

Table 24.1

	Anthropogenic Additions to the Carbon Dioxide Flux (000 metric tons carbon)							Anthropogenic Additions to the Methane Flux c.1987 (000 metric tons of methane)						CFC Use Per Capita 1986 (kg)
	1987					c.1987 Land Use Change	Per Capita (metric tons)	Solid Waste	Livestock	Hard Coals	Wet Rice	Pipeline Leakage	Per Capita (metric tons)	
	Cement	Solid	Liquid	Gas	Flaring									
ASIA	60,000	770,000	480,000	90,000	19,000	870,000	0.8	8,700	23,000	5,500	62,000	8,300	0.04	0.0
Afghanistan	14	120	550	320	92	X	0.1	42	270	1	100	X	0.03	X
Bahrain	0	0	2,100	2,300	0	X	9.4	1	0	X	X	70	0.15	0.2
Bangladesh	42	34	1,500	1,800	0	1,900	0.1	230	1,400	X	4,600	X	0.06	0.0
Bhutan	0	1	8	0	0	220	0.2	3	13	X	18	X	0.03	X
China	24,000	480,000	84,000	7,300	810	X	0.6	2,500	4,400	4,200	18,000	X	0.03	0.0 a
Cyprus	120	110	870	0	0	X	1.6	2	6	X	X	X	0.01	X
India	5,000	110,000	35,000	3,200	1,700	140,000	0.4	1,800	10,000	830	18,000	180	0.04	0.0 a
Indonesia	1,600	2,200	21,000	3,400	6,700	220,000	1.5	380	430	9	4,900	400	0.04	0.0 a
Iran, Islamic Rep	1,700	1,100	26,000	8,400	2,500	X	0.8	100	540	4	250	1,100	0.04	0.1
Iraq	1,400	1	8,600	520	2,700	X	0.8	36	130	X	25	X	0.01	0.1
Israel	280	2,500	5,300	21	0	X	1.9	120	22	X	X	X	0.03	0.7
Japan	9,500	75,000	140,000	23,000	24	X	2.1	2,400	260	80	1,200	X	0.03	0.5 a
Jordan	310	0	2,400	0	0	X	0.4	6	10	X	X	X	0.00	X
Kampuchea, Dem	0	0	120	0	0	4,800	0.6	17	83	X	840	X	0.13	X
Korea, Dem People's Rep	1,100	36,000	2,800	0	0	X	1.9	47	42	200	430	X	0.04	X
Korea, Rep	3,500	23,000	20,000	1,200	0	X	1.1	96	83	120	630	X	0.02	0.1 a
Kuwait	140	0	4,500	4,100	360	X	4.9	4	8	X	X	580	0.32	0.6 a
Lao People's Dem Rep	0	0	56	0	0	85,000	22.4	10	66	X	240	X	0.09	X
Lebanon	120	0	2,200	0	0	X	0.8	5	4	X	X	X	0.00	X
Malaysia	390	360	7,700	2,300	610	38,000	3.1	36	37	X	320	50	0.03	0.1 a
Mongolia	27	1,800	670	0	0	X	1.2	4	250	3	X	X	0.12	X
Myanmar	53	79	730	600	27	150,000	4.0	89	450	0	2,300	X	0.08	0.0
Nepal	14	62	150	0	0	6,700	0.4	38	490	X	660	X	0.07	X
Oman	0	0	4,500	1,100	320	X	4.4	3	14	X	X	640	0.49	X
Pakistan	930	1,800	6,700	4,700	420	770	0.1	220	1,500	10	970	X	0.03	0.0
Philippines	400	1,200	8,200	0	0	68,000	1.3	130	230	6	1,800	X	0.04	0.0
Qatar	41	0	580	2,500	0	X	9.5	1	1	X	X	480	1.47	X
Saudi Arabia	1,300	0	32,000	12,000	1,000	X	3.7	26	59	X	X	4,800	0.39	0.3
Singapore	210	11	7,600	0	0	X	3.0	6	2	X	X	X	0.00	0.8
Sri Lanka	81	0	1,000	0	0	1,700	0.2	37	110	X	340	X	0.03	X
Syrian Arab Rep	570	1	6,700	99	190	X	0.7	23	100	X	X	X	0.01	X
Thailand	1,200	2,100	10,000	2,200	0	94,000	2.1	120	470	X	4,500	X	0.10	0.0 a
Turkey	3,000	18,000	16,000	150	0	X	0.7	110	980	18	26	X	0.02	0.1
United Arab Emirates	340	0	4,700	7,100	1,900	X	9.6	3	8	X	X	X	0.01	0.9
Viet Nam	210	3,700	1,200	0	0	58,000	1.0	140	220	28	2,800	X	0.06	X
Yemen Arab Rep	100	0	810	0	0	X	0.1	15	57	X	X	X	0.01	X
Yemen, People's Dem Rep	0	0	1,500	0	0	X	0.6	5	22	X	X	X	0.01	X
EUROPE	33,000	550,000	440,000	170,000	5,100	0	2.4	9,200	9,200	2,300	220	6,300	0.05	0.6
Albania	120	1,000	1,300	200	0	0	0.9	47	47	X	2	X	0.03	X
Austria	620	3,900	7,700	2,600	0	0	2.0	150	160	X	X	X	0.04	0.7 a
Belgium	790	8,900	13,000	4,100	0	0	2.7	210	140	28	X	X	0.04	0.9 a
Bulgaria	770	18,000	12,000	2,900	0	0	3.7	X	200	1	9	X	0.02	0.1
Czechoslovakia	1,400	46,000	13,000	5,200	0	0	4.2	250	300	130	X	X	0.04	0.1
Denmark	270	8,500	7,300	810	6	0	3.3	100	170	X	X	X	0.05	0.9 a
Finland	220	4,800	8,800	820	0	0	3.0	97	96	X	X	X	0.04	0.7
France	3,200	21,000	56,000	14,000	0	0	1.7	1,100	1,500	82	7	1,500	0.08	0.9 a
German Dem Rep	1,600	71,000	13,000	4,300	0	0	5.4	300	360	X	X	X	0.04	0.7
Germany, Fed Rep	3,400	79,000	73,000	26,000	0	0	3.0	1,200	920	440	X	X	0.04	0.9 a
Greece	1,800	7,100	7,200	70	0	0	1.6	200	140	X	10	X	0.03	0.9 a
Hungary	560	8,800	6,200	5,300	0	0	2.0	170	150	12	6	220	0.05	0.1
Iceland	15	130	420	0	0	0	2.0	5	11	X	X	X	0.06	0.4
Ireland	190	3,800	2,900	840	0	0	2.1	70	360	0	X	X	0.12	0.9 a
Italy	4,900	15,000	63,000	19,000	0	0	1.8	1,100	590	X	100	20	0.03	0.9 a
Luxembourg	42	1,000	990	210	0	0	6.1	7	41	X	X	X	0.13	0.9 a
Malta	0	130	250	0	0	0	1.1	8	1	X	X	X	0.02	X
Netherlands	420	7,300	7,300	21,000	45	0	2.5	290	350	X	X	2,200	0.19	0.9 a
Norway	230	840	6,900	840	3,500	0	2.9	82	76	3	X	510	0.16	0.2 a
Poland	2,200	110,000	11,000	5,500	0	0	3.4	600	800	960	X	X	0.06	0.2
Portugal	790	1,900	5,800	0	0	0	0.8	200	100	1	17	X	0.03	0.9 a
Romania	1,900	21,000	13,000	22,000	460	0	2.5	370	560	44	23	X	0.04	0.0
Spain	3,200	17,000	25,000	1,600	38	0	1.2	770	450	80	42	X	0.03	0.9 a
Sweden	300	3,000	12,000	150	0	0	1.9	170	110	0	X	X	0.03	0.4 a
Switzerland	540	460	9,000	870	0	0	1.7	130	120	X	X	X	0.04	0.9
United Kingdom	1,800	71,000	52,000	31,000	1,100	0	2.8	1,100	950	540	X	1,800	0.08	0.9 a
Yugoslavia	1,200	19,000	11,000	3,200	0	0	1.5	460	420	2	5	X	0.04	0.2
U.S.S.R.	19,000	370,000	340,000	300,000	5,100	0	3.7	4,400	8,100	2,600	320	3,700	0.07	0.4 a
OCEANIA	960	36,000	26,000	9,800	27	2,700	3.1	1,100	2,900	680	59	1,100	0.24	0.6
Australia	810	35,000	21,000	7,900	0	0	4.0	1,000	1,900	670	52	770	0.27	0.8 a
Fiji	13	12	130	0	0	X	0.2	2	7	X	6	X	0.02	0.1
New Zealand	120	1,200	2,600	1,900	27	0	1.8	64	960	11	X	370	0.43	0.6
Papua New Guinea	0	1	640	0	0	2,700	0.9	9	6	X	X	X	0.00	X
Solomon Islands	0	0	37	0	0	X	0.1	1	1	X	X	1	0.01	X

Sources: Carbon Dioxide Information Analysis Center; R.A. Houghton, R.D. Boone, J.R. Fruci, et al.; World Resources Institute; Other sources.

Notes: a. Reported; see Technical Notes. b. Brazil land use change data are due to rampant deforestation activity in 1987. A 1988 deforestation estimate using similar methodology would result in total carbon emissions from land use change of 800,000,000 metric tons. A Brazilian Space Agency (INPE) estimate of average annual deforestation from 1978 to 1988 would lead to an estimate of average carbon emissions totaling 380,000,000 metric tons.
0 = zero or less than half of the unit of measure; X = not available; negative numbers are shown in parentheses.
Regional totals can include countries not shown on this list. For additional information, see Chapters 2, 7, 19, and the Sources and Technical Notes.

Table 24.2 Net Additions to the Greenhouse Heating Effect

	Attributed Atmospheric Concentration Increases (000 metric tons)							
	Carbon Dioxide Emissions (000 t. carbon)			Methane Emissions		1986 CFC Use		Net Total
	Fossil Fuels and Cement 1987	Annual Land Use Change	Net Annual Atmospheric Increase	Net Annual Atmospheric Increase	Equivalent Carbon Dioxide Heating Effect (000 t. carbon)	Net Annual Atmospheric Increase	Equivalent Carbon Dioxide Heating Effect (000 t. carbon)	Atmospheric Increase (000 t. carbon)
WORLD	**2,500,000**	**1,200,000**	**3,700,000**	**43,000**	**800,000**	**771.5**	**1,400,000**	**5,900,000**
AFRICA	**75,000**	**170,000**	**240,000**	**3,000**	**57,000**	**23.8**	**42,000**	**340,000**
Algeria	8,400	X	8,400	650	12,000	2.3	4,100	25,000
Angola	540	2,400	2,900	25	470	0.0	X	3,400
Benin	61	1,100	1,100	9	160	0.0	X	1,300
Botswana	190	310	490	20	370	0.0	X	860
Burkina Faso	54	1,800	1,900	25	470	0.0	X	2,400
Burundi	20	(1)	19	6	120	0.0	X	140
Cameroon	720	15,000	16,000	31	580	0.0	X	16,000
Cape Verde	4	X	4	0	5	0.0	X	9
Central African Rep	31	1,500	1,600	11	210	0.0	X	1,800
Chad	24	1,800	1,900	31	590	0.0	X	2,400
Comoros	6	X	6	2	31	0.0	X	37
Congo	210	1,400	1,600	1	26	0.0	X	1,600
Cote d'Ivoire	630	43,000	44,000	29	550	1.1	2,000	47,000
Djibouti	31	X	31	2	32	0.0	X	64
Egypt	9,000	X	9,000	170	3,100	2.9 a	5,100	17,000
Equatorial Guinea	8	110	120	0	4	0.0	X	120
Ethiopia	320	3,400	3,700	220	4,000	0.0	X	7,800
Gabon	600	780	1,400	1	12	0.0	X	1,400
Gambia, The	22	85	110	3	53	0.0	X	160
Ghana	360	3,300	3,600	16	300	1.4	2,400	6,300
Guinea	110	3,800	3,900	47	890	0.0	X	4,800
Guinea-Bissau	14	1,300	1,300	11	200	0.0	X	1,500
Kenya	610	680	1,300	99	1,800	0.0	X	3,100
Lesotho	X	X	X	6	120	0.0	X	X
Liberia	81	3,300	3,400	9	180	0.2	410	4,000
Libya	3,200	X	3,200	70	1,300	0.0	X	4,500
Madagascar	100	10,000	10,000	140	2,600	0.0	X	13,000
Malawi	58	6,800	6,900	19	350	0.0	X	7,300
Mali	46	910	960	59	1,100	0.0	X	2,100
Mauritania	380	X	380	23	430	0.0	X	810
Mauritius	140	X	140	1	15	0.0	X	160
Morocco	2,400	X	2,400	46	860	0.0	X	3,300
Mozambique, People's Rep	140	3,000	3,200	20	380	0.0	X	3,600
Niger	89	700	790	38	720	0.0	X	1,500
Nigeria	6,800	25,000	32,000	160	3,100	10.2	18,000	53,000
Rwanda	43	130	170	8	140	0.0	X	310
Senegal	270	1,300	1,500	21	400	0.7	1,200	3,100
Sierra Leone	66	430	500	16	290	0.0	X	790
Somalia	110	430	550	100	1,900	0.0	X	2,400
South Africa	34,000	X	34,000	420	7,800	3.3	5,800	47,000
Sudan	390	12,000	12,000	180	3,300	0.0	X	15,000
Swaziland	53	X	53	4	78	0.0	X	130
Tanzania	250	2,100	2,300	120	2,300	0.0	X	4,600
Togo	54	300	360	5	87	0.0	X	440
Tunisia	1,400	X	1,400	14	250	0.8	1,300	3,000
Uganda	86	950	1,000	42	780	0.0	X	1,800
Zaire	420	15,000	16,000	42	790	0.0	X	16,000
Zambia	320	1,800	2,100	18	340	0.0	X	2,500
Zimbabwe	1,800	1,800	3,600	40	760	0.9	1,500	5,900
NORTH & CENTRAL AMERICA	**630,000**	**39,000**	**670,000**	**10,000**	**190,000**	**227.2**	**400,000**	**1,300,000**
Barbados	110	X	110	0	5	0.1	130	250
Canada	48,000	X	48,000	1,700	33,000	20.7	36,000	120,000
Costa Rica	320	6,600	7,000	17	330	0.3	490	7,800
Cuba	4,000	21	4,000	59	1,100	1.0	1,800	6,900
Dominican Rep	770	43	810	23	440	0.7	1,200	2,400
El Salvador	240	79	320	8	160	0.5	860	1,300
Guatemala	410	4,300	4,800	20	370	0.0	X	5,100
Haiti	82	12	93	13	250	0.0	X	340
Honduras	220	4,300	4,500	17	320	0.2 a	350	5,200
Jamaica	700	25	720	3	61	0.2	420	1,200
Mexico	35,000	14,000	49,000	1,100	20,000	5.2 a	9,100	78,000
Nicaragua	250	7,300	7,500	16	290	0.4	610	8,400
Panama	330	2,400	2,700	12	230	0.2	400	3,300
Trinidad and Tobago	2,100	38	2,100	1	20	0.4	640	2,800
United States	530,000	2,600	540,000	7,100	130,000	197.4 a	350,000	1,000,000
SOUTH AMERICA	**64,000**	**640,000**	**710,000**	**2,900**	**54,000**	**19.4**	**34,000**	**800,000**
Argentina	13,000	X	13,000	630	12,000	3.1	5,500	31,000
Bolivia	490	2,900	3,400	51	950	0.0	X	4,400
Brazil	23,000	540,000 b	560,000	1,500	28,000	8.9 a	16,000	610,000
Chile	3,100	X	3,100	59	1,100	1.3	2,200	6,400
Colombia	6,100	54,000	60,000	220	4,100	3.0	5,200	69,000
Ecuador	1,800	17,000	19,000	31	570	1.0	1,700	21,000
Guyana	120	150	270	5	89	0.0	X	360
Paraguay	200	3,200	3,400	34	640	0.0	X	4,100
Peru	2,800	19,000	22,000	47	870	0.0	X	23,000
Suriname	150	150	300	6	120	0.0	68	490
Uruguay	410	X	410	81	1,500	0.3	540	2,500
Venezuela	12,000	7,700	19,000	250	4,700	1.8	3,200	27,000

Table 24.2

	Attributed Atmospheric Concentration Increases (000 metric tons)							
	Carbon Dioxide Emissions (000 t. carbon)			Methane Emissions		1986 CFC Use		Net Total
	Fossil Fuels and Cement 1987	Annual Land Use Change	Net Annual Atmospheric Increase	Net Annual Atmospheric Increase	Equivalent Carbon Dioxide Heating Effect (000 t. carbon)	Net Annual Atmospheric Increase	Equivalent Carbon Dioxide Heating Effect (000 t. carbon)	Net Total Atmospheric Increase (000 t. carbon)
ASIA	**610,000**	**380,000**	**990,000**	**18,000**	**340,000**	**111.2**	**190,000**	**1,500,000**
Afghanistan	480	X	480	70	1,300	0.0	X	1,800
Bahrain	1,900	X	1,900	12	220	0.1	160	2,300
Bangladesh	1,500	850	2,300	1,100	20,000	0.0	X	22,000
Bhutan	4	97	100	6	110	0.0	X	210
China	260,000	X	260,000	4,800	90,000	18.0 a	32,000	380,000
Cyprus	480	X	480	1	24	0.0	X	500
India	67,000	61,000	130,000	5,200	98,000	0.4 a	700	230,000
Indonesia	15,000	95,000	110,000	1,000	19,000	5.4 a	9,500	140,000
Iran, Islamic Rep	17,000	X	17,000	340	6,400	5.1	9,000	33,000
Iraq	5,700	X	5,700	32	590	1.7	3,000	9,300
Israel	3,500	X	3,500	23	440	3.1	5,400	9,300
Japan	110,000	X	110,000	650	12,000	57.5 a	100,000	220,000
Jordan	1,200	X	1,200	3	51	0.0	X	1,200
Kampuchea, Dem	52	2,100	2,100	160	3,000	0.0	X	5,100
Korea, Dem People's Rep	18,000	X	18,000	120	2,300	0.0	X	20,000
Korea, Rep	21,000	X	21,000	160	2,900	3.1 a	5,400	29,000
Kuwait	4,000	X	4,000	100	1,900	1.0 a	1,800	7,600
Lao People's Dem Rep	24	37,000	37,000	53	1,000	0.0	X	38,000
Lebanon	1,000	X	1,000	2	30	0.0	X	1,000
Malaysia	5,000	17,000	22,000	74	1,400	1.4 a	2,500	26,000
Mongolia	1,100	X	1,100	42	790	0.0	X	1,900
Myanmar	650	67,000	68,000	480	9,000	0.0	X	77,000
Nepal	98	2,900	3,000	200	3,800	0.0	X	6,800
Oman	2,600	X	2,600	110	2,100	0.0	X	4,700
Pakistan	6,400	330	6,700	460	8,600	0.0	X	15,000
Philippines	4,300	29,000	34,000	360	6,700	0.0	X	40,000
Qatar	1,400	X	1,400	81	1,500	0.0	X	2,900
Saudi Arabia	20,000	X	20,000	820	15,000	3.8	6,600	42,000
Singapore	3,400	X	3,400	1	26	2.1	3,700	7,100
Sri Lanka	480	720	1,200	82	1,500	0.0	X	2,700
Syrian Arab Rep	3,300	X	3,300	21	400	0.0	X	3,700
Thailand	6,800	41,000	48,000	850	16,000	2.0 a	3,500	67,000
Turkey	16,000	X	16,000	190	3,600	5.3	9,200	29,000
United Arab Emirates	6,100	X	6,100	2	35	1.3	2,300	8,400
Viet Nam	2,200	25,000	28,000	540	10,000	0.0	X	38,000
Yemen Arab Rep	400	X	400	12	230	0.0	X	630
Yemen, People's Dem Rep	630	X	630	5	87	0.0	X	720
EUROPE	**520,000**	**X**	**520,000**	**4,600**	**85,000**	**274.8**	**480,000**	**1,100,000**
Albania	1,200	X	1,200	16	300	0.0	X	1,500
Austria	6,500	X	6,500	51	960	5.2 a	9,100	17,000
Belgium	12,000	X	12,000	64	1,200	7.0 a	12,000	25,000
Bulgaria	15,000	X	15,000	35	660	0.9	1,600	17,000
Czechoslovakia	29,000	X	29,000	110	2,200	1.6	2,700	33,000
Denmark	7,400	X	7,400	46	860	3.6 a	6,300	15,000
Finland	6,400	X	6,400	33	610	3.5	6,100	13,000
France	41,000	X	41,000	710	13,000	39.3 a	69,000	120,000
German Dem Rep	39,000	X	39,000	110	2,100	11.6	20,000	62,000
Germany, Fed Rep	79,000	X	79,000	430	8,000	42.9 a	75,000	160,000
Greece	7,000	X	7,000	59	1,100	7.1 a	12,000	20,000
Hungary	9,100	X	9,100	94	1,800	1.1	1,900	13,000
Iceland	210	X	210	3	51	0.1	170	440
Ireland	3,400	X	3,400	72	1,300	2.6 a	4,500	9,200
Italy	45,000	X	45,000	310	5,800	40.4 a	71,000	120,000
Luxembourg	990	X	990	8	150	0.3 a	450	1,600
Malta	170	X	170	1	27	0.0	X	200
Netherlands	16,000	X	16,000	470	8,800	10.3 a	18,000	43,000
Norway	5,300	X	5,300	110	2,100	0.7 a	1,200	8,700
Poland	56,000	X	56,000	400	7,400	7.5	13,000	76,000
Portugal	3,700	X	3,700	55	1,000	7.2 a	13,000	17,000
Romania	25,000	X	25,000	170	3,100	0.0	X	28,000
Spain	21,000	X	21,000	230	4,200	27.5 a	48,000	73,000
Sweden	6,900	X	6,900	47	870	3.6 a	6,300	14,000
Switzerland	4,700	X	4,700	42	790	5.9	10,000	16,000
United Kingdom	69,000	X	69,000	740	14,000	40.3 a	71,000	150,000
Yugoslavia	15,000	X	15,000	150	2,800	4.7	8,200	26,000
U.S.S.R.	**450,000**	**X**	**450,000**	**3,200**	**60,000**	**101.0 a**	**180,000**	**690,000**
OCEANIA	**32,000**	**1,200**	**33,000**	**990**	**19,000**	**14.1**	**25,000**	**76,000**
Australia	28,000	X	28,000	750	14,000	12.0 a	21,000	63,000
Fiji	66	X	66	2	45	0.1	130	240
New Zealand	2,500	X	2,500	240	4,400	2.0	3,500	10,000
Papua New Guinea	280	1,200	1,400	2	45	0.0	X	1,500
Solomon Islands	16	X	16	0	8	0.0	X	24

Sources: Carbon Dioxide Information Analysis Center; R.A. Houghton, R.D. Boone, J.R. Fruci, et al.; World Resources Institute; Other sources.
Notes: a. Reported. b. 1987 may have been an anomalously high year. An estimate for 1988, using a similar methodology, would have added 340,000,000 metric tons to the carbon flux from land use change. A Brazilian Space Agency (INPE) study of average annual deforestation from 1978 to 1988 would show an an addition of 161,000,000 metric tons.
0 = zero or less than half of the unit measure; X = not available; negative numbers are shown in parentheses; t. = metric ton.
For further information see Chapters 2, 7, 19, and the Sources and Technical Notes.

Table 24.3 Atmospheric Concentrations of Greenhouse and Ozone-Depleting Gases, 1959—88

Year	(parts per million) Carbon Dioxide (CO2)	Carbon tetra-chloride (CCl4)	Methyl chloro-form (CH3CCl3)	CFC-11 (CCl3F)	CFC-12 (CCl2F2)	CFC-22 (CHClF2)	CFC-113 (C2Cl3F3)	Total Gaseous Chlorine	Nitrous Oxide (N2O)	Methane (CH4)	Carbon Monoxide (CO)
			(parts per trillion)						(parts per billion)		
Preindustrial	c.280.0	0	0	0	0	0	0	0	c.285.0	c.700	X
1959	315.8	X	X	X	X	X	X	X	X	X	X
1960	316.8	X	X	X	X	X	X	X	X	X	X
1961	317.5	X	X	X	X	X	X	X	X	X	X
1962	318.3	X	X	X	X	X	X	X	X	1,354	X
1963	318.8	X	X	X	X	X	X	X	X	X	X
1964	X	X	X	X	X	X	X	X	X	X	X
1965	319.9	X	X	X	X	X	X	X	X	1,386	X
1966	321.2	X	X	X	X	X	X	X	X	1,338	X
1967	322.0	X	X	X	X	X	X	X	X	1,480	X
1968	322.8	X	X	X	X	X	X	X	X	1,373	X
1969	323.9	X	X	X	X	X	X	X	X	1,385	X
1970	325.3	X	X	X	X	X	X	X	X	1,431	X
1971	326.2	X	X	X	X	X	X	X	X	1,436	X
1972	327.3	X	X	X	X	X	X	X	X	1,500	X
1973	329.5	X	X	X	X	X	X	X	X	1,624	X
1974	330.1	X	X	X	X	X	X	X	X	1,596	X
1975	331.0	104	70	120	200	X	X	1,202	291.4	1,541	X
1976	332.0	106	78	133	217	X	X	1,290	293.3	1,490	X
1977	333.7	115	86	148	239	X	X	1,416	294.6	1,471	X
1978	335.3	123	94	159	266	X	X	1,544	296.4	1,531	X
1979	336.7	116	112	167	283	46	X	1,621	296.3	1,545	X
1980	338.5	121	126	179	307	52	X	1,755	297.6	1,554	X
1981	339.8	122	127	185	315	59	X	1,797	298.5	1,569	72
1982	341.0	121	133	193	330	64	X	1,863	301.0	1,591	72
1983	342.6	126	144	205	350	71	24	1,983	300.9	1,615	70
1984	344.3	130	150	213	366	76	27	2,072	300.4	1,629	73
1985	345.7	130	158	223	384	85	31	2,163	301.5	1,643	75
1986	347.0	127	169	232	404	98	35	X	302.5	1,656	75
1987	348.7	133	168	247	421	105	41	X	304.5	1,673	X
1988	351.3	133	182	263	439	X	52	X	306.3	1,697	X

Sources: Scripps Institute of Oceanography and Oregon Graduate Center.
Notes: X = not available; c. = circa. All estimates are parts by volume. For additional information, see Sources and Technical Notes.

Table 24.4 World Carbon Dioxide Emissions from Fossil Fuels and Cement Manufacture

Global Emissions of Carbon Dioxide to the Atmosphere
Fossil Fuel Consumption, Gas Flaring, and Cement Manufacturing
(millions of metric tons of carbon)

Year	Total	Solid	Liquid	Gas	Gas Flaring	Cement
1950	1,638	1,078	423	97	23	18
1951	1,775	1,137	479	115	24	20
1952	1,803	1,127	504	124	26	22
1953	1,848	1,132	533	131	27	24
1954	1,871	1,123	557	138	27	27
1955	2,050	1,215	625	150	31	30
1956	2,185	1,281	679	161	32	32
1957	2,278	1,317	714	178	35	34
1958	2,338	1,344	732	192	35	36
1959	2,471	1,390	790	214	36	40
1960	2,586	1,419	850	235	39	43
1961	2,602	1,356	905	254	42	45
1962	2,708	1,358	981	277	44	49
1963	2,855	1,404	1,053	300	47	51
1964	3,016	1,442	1,138	328	51	57
1965	3,154	1,468	1,221	351	55	59
1966	3,314	1,485	1,325	380	60	63
1967	3,420	1,455	1,424	410	66	65
1968	3,596	1,456	1,552	445	73	70
1969	3,809	1,494	1,674	487	80	74
1970	4,090	1,571	1,838	515	87	78
1971	4,241	1,571	1,946	553	88	84
1972	4,409	1,587	2,056	582	95	89
1973	4,647	1,594	2,240	607	110	95
1974	4,655	1,591	2,244	615	108	96
1975	4,628	1,686	2,131	620	95	95
1976	4,894	1,723	2,313	644	111	103
1977	5,033	1,786	2,390	645	105	108
1978	5,082	1,802	2,383	673	107	116
1979	5,365	1,899	2,535	713	100	119
1980	5,263	1,921	2,409	724	89	120
1981	5,129	1,930	2,272	734	72	121
1982	5,097	1,993	2,181	732	70	121
1983	5,088	1,998	2,166	735	63	125
1984	5,263	2,088	2,194	796	57	128
1985	5,382	2,196	2,175	826	56	130
1986	5,565	2,250	2,281	845	54	136
1987	5,650	2,276	2,287	895	50	140

Source: Carbon Dioxide Information Analysis Center.
Note: Totals differ from the sum of other columns due to rounding error. For additional information, see Sources and Technical Notes.

Table 24.5 Air Pollution in Selected Cities

	City	Sulfur Dioxide				Gravimetrically Determined Suspended Particulate Matter				Smoke			
		Site Years	Number of days over 150 ug/cubic meter			Site Years	Number of days over 230 ug/cubic meter			Site Years	Number of days over 150 ug/cubic meter		
			Min.	Avg.	Max.		Min.	Avg.	Max.		Min.	Avg.	Max.
NORTH AMERICA													
Canada	Hamilton	8	0	3	7	10	0	8	14	X	X	X	X
	Montreal	10	0	10	32	15	0	0	6	X	X	X	X
	Toronto	9	0	1	3	14	0	1	7	X	X	X	X
	Vancouver	5	0	0	0	12	0	0	7	X	X	X	X
United States	Birmingham	X	X	X	X	9	0	7	28	X	X	X	X
	Chattanooga	X	X	X	X	16	0	1	17	X	X	X	X
	Chicago	4	0	1	2	7	0	6	14	X	X	X	X
	Fairfield	X	X	X	X	5	0	0	0	X	X	X	X
	Houston	3	0	0	0	7	0	0	0	X	X	X	X
	New York	12	1	8	22	12	0	0	0	X	X	X	X
	St. Louis	3	1	3	8	X	X	X	X	X	X	X	X
SOUTH AMERICA													
Brazil	Rio de Janeiro	X	X	X	X	6	0	11	35	X	X	X	X
	Sao Paulo	11	0	12	32	X	X	X	X	11	16	31	52
Chile	Santiago	9	0	19	55	X	X	X	X	9	11	102	299
Colombia	Cali	1	0	0	0	X	X	X	X	X	X	X	X
	Medellin	3	0	0	0	3	0	0	0	X	X	X	X
Venezuela	Caracas	8	0	0	0	X	X	X	X	8	0	0	0
ASIA													
China	Beijing	8	0	68	157	8	145	272	338	X	X	X	X
	Guangzhou	12	0	30	74	10	7	123	283	X	X	X	X
	Shanghai	10	0	16	32	10	19	133	277	X	X	X	X
	Shenyang	7	43	146	236	13	117	219	347	X	X	X	X
	Xian	7	4	71	114	10	189	273	327	X	X	X	X
Hong Kong	Hong Kong	10	0	15	74	X	X	X	X	11	0	3	18
India	Bombay	13	0	3	32	12	23	100	207	X	X	X	X
	Calcutta	8	0	25	85	8	189	268	330	X	X	X	X
	Delhi	12	0	6	49	12	212	294	338	X	X	X	X
Indonesia	Jakarta	X	X	X	X	7	4	173	268	X	X	X	X
Iran	Tehran	15	6	104	163	15	8	174	347	15	12	122	249
Israel	Tel Aviv	9	0	3	24	X	X	X	X	X	X	X	X
Japan	Osaka	20	0	0	0	20	0	0	2	X	X	X	X
	Tokyo	15	0	0	0	15	0	2	4	X	X	X	X
Malaysia	Kuala Lumpur	1	0	0	0	5	10	37	59	X	X	X	X
Philippines	Manila	4	3	24	60	7	0	14	225	X	X	X	X
Korea, Rep	Seoul	6	5	87	121	X	X	X	X	X	X	X	X
Thailand	Bangkok	3	0	0	0	12	5	97	209	X	X	X	X
EUROPE													
Belgium	Brussels	13	0	12	32	X	X	X	X	13	0	0	2
Denmark	Copenhagen	3	0	0	0	6	0	0	1	6	0	0	0
Finland	Helsinki	8	0	2	7	11	0	19	75	X	X	X	X
France	Gourdon	4	27	46	64	X	X	X	X	9	0	3	7
German, Fed Rep	Frankfurt	6	8	20	38	3	0	0	0	X	X	X	X
	Munich	3	0	0	1	X	X	X	X	X	X	X	X
Greece	Athens	3	1	9	15	X	X	X	X	X	X	X	X
Ireland	Dublin	6	0	1	3	X	X	X	X	6	0	6	15
Italy	Milan	8	6	29	167	X	X	X	X	X	X	X	X
Netherlands	Amsterdam	10	0	1	5	X	X	X	X	X	X	X	X
Poland	Warsaw	13	3	10	19	X	X	X	X	14	4	17	33
	Wroclaw	15	1	8	22	X	X	X	X	15	9	30	73
Portugal	Lisbon	X	X	X	X	7	4	12	28	X	X	X	X
Spain	Madrid	7	0	35	95	X	X	X	X	4	4	60	126
United Kingdom	Glasgow	5	4	14	21	X	X	X	X	5	2	6	8
	London	6	0	7	17	X	X	X	X	6	0	0	0
Yugoslavia	Zagreb	15	3	30	80	15	13	34	57	X	X	X	X
OCEANIA													
Australia	Melbourne	13	0	0	0	4	0	0	0	X	X	X	X
	Sydney	12	0	2	11	10	0	3	19	X	X	X	X
New Zealand	Auckland	12	0	0	0	X	X	X	X	12	0	0	0
	Christchurch	12	0	0	2	X	X	X	X	12	0	8	25

Sources: Global Environment Monitoring System (GEMS).
Notes: ug = microgram; Min. = minimum; Avg. = average; Max. = maximum; 0 = zero or less than half the unit of measure; X = not available.
For additional information, see Sources and Technical Notes.

Table 24.6 Emissions of Pollutants in Selected Countries

	Sulphur Dioxide			Suspended Particulate Matter			Oxides of Nitrogen			Carbon Monoxide			Lead			
Estimated Annual Emissions (thousands of metric tons)	1973-75	1979-81	1982-84	1973-75	1979-81	1982-84	1973-75	1979-81	1982-84	1973-75	1979-81	1982-84	1973-75	1979-81	1982-84	1985
NORTH AMERICA																
Canada	5,880 a	4,610 a	3,760 a	2,080 a	1,870 a	X	1,750 a	1,910 a	1,870 a	10,700 a	9,930 a	X	14.5	X	11.5	0.5
United States	25,600	23,330	21,100	10,400	8,470	6,900	19,200	20,670	19,500	81,200	76,030	69,230	147.0	78.4	46.9	21.1
CENTRAL AMERICA																
Mexico	X	X	X	X	X	X	X	X	X	X	X	X	X	19.6	8.4	X
ASIA																
China	X	14,210	12,920	X	16,200	13,740 b	X	4,400	4,130	X	X	X	X	X	X	X
India	1,610	X	X	X	X	X	X	X	X	X	X	X	X	X	X	X
Israel	230 b	290 b	2 b	16 b	20 b	19 b	95 b	110 b	125 b	320 b	360 b	420 b	X	X	X	X
Japan	2,620	1,640	1,610	X	X	X	1,800	1,340	1,420	X	X	X	X	X	X	X
Kuwait	X	X	450	X	X	180	X	X	100	X	X	620	X	X	X	X
Thailand	X	120 c	310	X	40 c	230	X	30 c	130	X	120 c	X	X	X	3.0	X
Turkey	X	X	X	X	X	X	X	X	X	X	3,710	X	X	X	X	X
EUROPE																
Austria	X	330	140	X	50	55	X	200	210	X	1,130 d	1,070 d	X	X	X	X
Belgium	1,000	860	700	175	140	125	380	390	340	1,150	1,250	1,000	X	X	1.4	X
Bulgaria	X	1,030	1,140	X	X	X	X	X	150	X	X	X	X	X	X	X
Czechoslovakia	X	3,100	X	X	X	X	X	1,200	X	X	X	X	X	X	X	X
Denmark	430	460	410	X	X	X	200	240	250	X	600 e	250 e	X	X	X	X
Finland	540	570	360	X	X	X	160	280	250	X	600	X	X	X	X	X
France	3,760	3,410	2,305	295	265	210	1,675	1,855	1,730	X	6,550 d	6,330 d	X	X	X	X
German Dem Rep	X	4,000	4,000	X	X	X	X	X	X	X	X	X	X	X	X	X
Germany, Fed Rep	3,600	3,300	2,750	950 d	750 d	650 d	2,600	3,100	3,000	11,700	9,000	7,400	X	X	X	X
Greece	270	400	360	155 d	170 d	185 d	105	130	150	X	X	X	X	X	X	X
Hungary	X	1,635	1,460	X	550	500	X	290	300	X	1,370	1,400	0.6 f	X	X	X
Iceland	X	6	6	X	X	X	X	10	10	X	X	X	X	X	X	X
Ireland	X	X	X	80	90	100	X	70	60	390	490	480	0.7	0.8	0.5	0.4
Italy	X	3,210	2,230	X	435 d	410 d	X	1,510	1,530	X	5,480 d	5,420 d	X	X	X	X
Luxembourg	X	20	10	X	X	X	X	20	20	X	X	X	X	X	X	X
Netherlands	320 g	390 g	260 g	30 g	40 g	35 g	390 g	470 g	450 g	1,460 g	1,170 g	990 g	X	1.4	1.3	X
Norway	155	145	110	30 h	30 h	25 h	150	170	180	540	66	595	X	0.6	0.4	X
Poland	2,080	2,600	3,700	2,230	2,120	3,350	90	X	1,770	590	X	3,300	X	X	X	X
Portugal	180	260	305	75	120	X	105	210	190	460	525	X	X	X	X	X
Romania	X	200	X	X	X	X	X	X	X	X	X	X	X	X	X	X
Spain	X	2,670	X	X	1,520	X	620	810	X	X	3,780	X	X	X	X	X
Sweden	690	480	285	170	X	40	310	320	295	1,400	1,390	1,600	1.5 f	1.1 f	0.7 f	0.9 f
Switzerland	110	120	100	30	30	20	160	200	210	740	710	640	X	1.0	X	0.5
United Kingdom	5,430	4,740	3,750	450 i	300 i	230 i	1,870	1,900	1,770	4,820	5,090	5,180	7.9 f	7.2 f	7.0 f	6.5 f
Yugoslavia	X	820	X	X	X	X	X	X	X	X	X	X	X	X	X	X
OCEANIA																
New Zealand	X	90 d	X	X	X	X	X	X	X	X	X	X	X	X	X	X

Source: Global Environment Monitoring System (GEMS).

Notes: a. Excludes forest fires. b. Fossil fuel combustion only. c. Bangkok only. d. Excludes industrial processes. e. Mobile sources only. f. Gasoline (petrol) engined road vehicles only. g. Fossil fuel combustion in domestic stoves and automobiles only. h. Fossil fuels and wood. i. Smoke from coal combustion.
0 = zero or less than half the unit measure; X = not available.
For additional information, see Sources and Technical Notes.

Sources and Technical Notes

Table 24.1 Sources of Current Greenhouse Gas Emissions

Sources: Cement, liquid fuel, solid fuel, gas, and gas flaring additions to the carbon dioxide flux: Carbon Dioxide Information Analysis Center (CDIAC), Environmental Sciences Division, Oak Ridge National Laboratory (CDIAC, unpublished data, Oak Ridge, Tennessee, July 1988). Land use change: R.A. Houghton, R.D. Boone, J.R. Fruci, et al., "The Flux of Carbon from Terrestrial Ecosystems to the Atmosphere in 1980 Due to Changes in Land Use: Geographic Distribution of the Global Flux," Tellus Vol. 39B, No. 1–2 (1987), pp. 122–139; and, World Resources Institute (WRI), unpublished analysis based on recent reassessments of rates of deforestation in eight countries. (See Table 19.1, Sources and Technical Notes.) Methane from municipal solid waste: Jean Lerner, personal communication (National Aeronautics and Space Administration [NASA] Goddard Space Flight Center, Institute for Space Studies, May 1989). Methane from livestock: Jean Lerner, Elaine Mathews, and Inez Fung, "Methane Emissions From Animals: A Global High-Resolution Data Base," Global Biogeochemical Cycles, Vol. 2, No. 2 (June 1988), pp. 139–156; and Jean Lerner (NASA Goddard Space Flight Center, Institute for Space Studies, New York, May 1989), personal communication. Methane from coal mining and from wet rice agriculture: WRI estimate. Methane from pipeline leakage: U.S. Department of Energy (DOE), Energy Information Administration, International Energy Annual 1986 (DOE, Washington, D.C., 1986). Chloroflurocarbon (CFC) use per capita: WRI estimate of relative per capita consumption based on information in the U.S. Environmental Protection Agency (EPA), Stratospheric Protection Program, Office of Program Development, Office of Air and Radiation, Appendices to Regulatory Impact Analysis: Protection of Stratospheric Ozone (Washington, D.C., August 1988), Vol. 2, Part 2, Appendix K, pp. K-2-4–K-2-6 , and data on average per capita CFC consumption in broad classes, Alliance for Responsible CFC Use, unpublished data (Alliance for Responsible CFC Use, Arlington, Virginia, 1989).

Carbon dioxide (CO_2), methane (CH_4), CFC-11, and CFC-12 are the four most important greenhouse gases. This table shows sources of their current annual emissions, combining CFC-11 and CFC-12. Nitrous oxide and tropospheric ozone are also important to the greenhouse effect but less well studied and more difficult to estimate. Although ozone in the troposphere and nitrous oxide emissions are associated with industrial economies, ozone has an average lifetime measured in hours and is a product of particular chemical processes associated with temperature inversions above heavily polluted cities. Nitrous oxide emissions by country have proven difficult to estimate in part because significant emissions from developing countries are poorly understood.

This table includes data on *anthropogenic additions to the carbon dioxide flux* (*cement* manufacture, *solid* fuels, *liquid* fuel, *gas* fuels, *gas flaring*, and *land use change*). CDIAC annually calculates emissions of CO_2 from the manufacture of cement and the burning of fossil fuels for most of the countries of the world. (See Technical Notes to Table 24.1.) Country-level estimates are not as accurate as the time trends they describe. Estimates for country emissions do not include "bunker fuels" used in international transport. The world totals, however, include bunker fuels. Carbon releases from *land use change* are based on the work of R.A. Houghton, R.D. Boone, and J.R. Fruci, *et al.* They estimated the world flux of carbon in 1980 from deforestation, reforestation, logging, and changes in agricultural area for most of the world's tropical countries. Estimates were made using rates of decay and regrowth for 15 ecosystems, including 9 ecosystems in the tropics and 6 land uses. The estimates explicitly include shifting cultivation and the diversion of forest fallow to permanent clearing. They are also consistent, based on a sound methodology, and global. They are the most complete estimates available, but subject to modification should better data become available. Please consult the source for more detail.

WRI has modified the estimates from Houghton, Boone, and Fruci, *et al.*, using new data on deforestation in eight countries (Brazil, Cameroon, Costa Rica, India, Indonesia, Myanmar, the Philippines, and Viet Nam; see Table 19.1). Their carbon flux estimates for closed forests were used to scale new carbon estimates in proportion to the change in deforestation estimates.

WRI also subtracted the weight of carbon contained in sawlogs and veneer logs (Food and Agriculture Organization of the United Nations, FAO, *1987 Forest Products Yearbook*, FAO, Rome, 1989) produced in each tropical country from carbon releases calculated from land use change. Carbon was estimated as making up 45 percent of the weight of these wood products. This step was taken to approximate the carbon sequestered from the global carbon cycle by the production of durable wooden goods in each country. This is only an estimate because portions of other forest products are also sequestered (e.g., books in libraries, pit props, utility poles); and portions of saw and ve-

neer logs are consumed (e.g., wastewood, disposal of plywood sheets used in concrete form building, etc.). This should lead to a small underestimate of total CO_2 emissions because it includes logs from areas not counted as deforested.

Other estimates exist of deforestation, of forest-specific measures of biomass per hectare, and of the percentage of carbon per unit of forest biomass. In fact, controversy continues regarding the exact parameters to use in these calculations. The parameters used for this calculation, however, were based on consistent definitions and common data sources. Even if slightly lower values were used for deforestation and biomass per area, the magnitude of carbon emissions would remain about the same. These estimates then are a good first approximation to current (i.e., 1985–90, average-annual) emissions that result from land use changes. There is some suggestion that northern temperate and boreal forest areas are net sinks for atmospheric carbon, although this too is controversial.

Methane emissions from municipal *solid waste* were calculated by multiplying the 1986 population by per capita emission coefficients developed for each country by H.G. Bingemer and P.J. Crutzen in "The Production of CH_4 from Solid Wastes," *Journal of Geophysical Research*, Vol. 92, No. (D2) (1987), pp. 2181-2187. R.J. Cicerone and R.S. Oremland ("Biogeochemical Aspects of Atmospheric Methane," *Global Biogeochemical Cycles*, Vol. 2, No. 4, December 1988, pp. 299–327) suggest a likely range for annual world emissions from landfills at 30-70 million metric tons.

Methane emissions from domestic *livestock* were calculated using FAO statistics on animal populations and published estimates of methane emissions from each animal. The animals studied included cattle and dairy cows, water buffalo, sheep, goats, camels, pigs, horses, and caribou. P.J. Crutzen, I. Aselmann, and W. Seiler ("Methane Production by Domestic Animals, Wild Ruminants, Other Herbivorous Fauna, and Humans," *Tellus*, Vol. 38B, 1986, pp. 271–284) produced estimates of animal methane production based on energy intake under several different management methods for several different feeding regimes. These differing emission coefficients were then assigned to each country, based on the specifics of that country's animal husbandry practices and the nature and quality of feed available. Cicerone and Oremland's *Aspects of Atmospheric Methane* shows a likely range of 65–100 million metric tons for emissions from enteric fermentation in domestic animals.

Methane from *hard coal* mining was estimated at 0.5 percent (EPA, *Policy Options for Stabilizing Global Climate: Draft Report to Congress*, 1:II-28, Washington, D.C., 1989) of the mass of hard coals (anthracite, bituminous, and subbituminous) mined (United Nations, *Energy Statistics Yearbook 1986*, New York, 1988) in each country. Methane trapped within the rock is released by mining, and it is one of the

hazards of underground coal mining. Cicerone and Oremland (*Aspects of Atmospheric Methane*) show a likely range of 25–45 million metric tons of methane emitted annually in the course of mining coal.

Methane from the practice of *wet rice* agriculture was calculated from the area of rice production (as reported by the FAO, *FAO Production Yearbook 1984*, FAO, Rome, 1985) and estimates of emission by A. Holzappel-Pshorn and W. Seiler ("Methane Emission During a Cultivation Period from an Italian Rice Paddy," *Journal of Geophysical Research*, Vol. 91, No. D11, October 1986, pp. 803–811, 814) that totaled 54 grams per square meter over a 140-day growing season. The areas of rice production, reported in the *FAO Production Yearbook*, were modified by deducting areas of upland (dryland) rice production based on country-specific or region-specific ratios of dry- to wet-rice production (Dana G. Dalrymple, *Development and Spread of High-Yielding Rice Varieties in Developing Countries*, Bureau of Science and Technology, Agency for International Development, Washington, D.C., 1986). Other estimates of the methane flux from rice paddies exist. The one used here was based on a technique that captured methane produced anaerobically before the growth of the rice plant as well as capturing the bulk of methane production that is transported through the rice plant throughout the growing period. Growing periods vary by rice cultivar, as do requirements for fertilizers and pesticides that could, conceivably, influence methanogenesis. In the tropics, with modern varieties, sufficient fertilizer, and adequate water, two crops are possible. Thus, this Italian example may underestimate tropical methanogenesis.

The cultivation of rice, however, uses common techniques in both temperate and tropical climes—even if the cultivars are not so well adapted. The preparation of the impoundments wherein wet rice is grown—the creation of a hardpan overlain by soft anaerobic muck—creates similar environmental and chemical regimes wherever it occurs. Nonetheless, variations in water quality, soils, ambient temperature, precision of water control, and presence of cultivated algae or fish could also affect the total flux of methane.

Wet rice agriculture is practiced under four main water regimes: irrigated (52.8 percent of the total rice area), rainfed (similar to irrigated, 22.6 percent of the total), deep water (often dry in the early part of the season, may be planted to floating rice, 8.2 percent of the world's rice area), and tidal (3.4 percent of the area). Cicerone and Oremland (*Aspects of Atmospheric Methane*) suggest a likely range of 60–170 million metric tons for methane emissions associated with wet rice agriculture.

Methane from natural gas *pipeline leakage* was converted from billions of cubic feet to cubic meters (1:0.028), cubic meters of CH_4 to grams (1:653, at standard temperature and pressure). There is reason to believe that pipeline leaks in the

U.S.S.R. are grossly understated—although U.S.S.R. natural gas volume is sometimes mistakenly overstated—but other estimates are non-existent. Cicerone and Oremland (*Aspects of Atmospheric Methane*) suggest a likely range of 25-50 million metric tons of methane emitted because of leaks associated with natural gas drilling, venting, and transmission leaks.

The only other major anthropogenic source of methane, unaccounted for in this table, is emissions consequent to the burning of biomass. Extensive biomass burning, especially in the tropics, is believed to release large amounts of methane. Cicerone and Oremland (*Aspects of Atmospheric Methane*) put the likely range of those emissions at 50–100 million metric tons.

Other natural sources of methane include wetlands, methane hydrate destabilization in permafrost, termites, freshwater lakes, oceans, and enteric emissions from other animals. Natural sources account for an estimated 25 percent of all methane emissions. Cicerone and Oremland (*Aspects of Atmospheric Methane*) show likely ranges of methane emissions at 100–200 million metric tons from natural wetlands, 10–100 million metric tons from termites, 5–25 million metric tons from the oceans, 1–25 million metric tons from fresh waters, and possible current releases of 5 million metric tons (potentially rising to 100 million metric tons if temperatures increase in the high arctic) from methane hydrate destabilization.

WRI has estimated total *CFC use per capita* (of CFC-11 and CFC-12) rounded to the nearest tenth of a kilogram for many countries. It used data on 1986 per capita production/use from 18 countries and the European Community (EC) to peg consumption in other similar countries. This estimate was based in part on the general level of total CFC (including CFC-113 and CFC-22) consumption (i.e., less than 0.3 kg, 0.3–0.5 kg, and over 0.5 kg, from the Alliance for Responsible CFC Use) and other relevant information. These data are, therefore, a mix of reported and estimated numbers. Nonetheless, they probably reflect relative consumption, and describe absolute consumption reasonably well. Consumption data for the EC were simply distributed among its members. Thus, all EC members are tied as the highest per capita consumers of CFCs. In fact, they probably differ dramatically among themselves. For example, Portugal's per capita consumption of CFC-11 and CFC-12 is probably close to 0.2 kg instead of 0.9 kg, but the latter figure is based on the EC as a whole.

Table 24.2 Net Additions to the Greenhouse Heating Effect

Sources: Carbon emissions from fossil fuels and cements: Carbon Dioxide Information Analysis Center (CDIAC), Environmental Sciences Division, Oak Ridge National Laboratory, unpublished data (CDIAC, Oak Ridge, Tennessee, July 1989). Carbon emissions from land use change: R.A. Houghton, R.D. Boone, J.R. Fruci, *et al.*, "The Flux of Carbon from Terrestrial Ecosystems to the Atmosphere in 1980 due to Changes in Land Use: Geographic Distribution of the Global Flux," *Tellus* Vol. 39B, No. 1–2 (1987), pp. 122–139; World Resources Institute (WRI), unpublished analysis. Methane emissions: derived from an analysis and compilation of data on anthropogenic methane sources. (See Technical Notes for Table 24.1.) Equivalent carbon heating effects: WRI estimates. Annual Chlorofluorocarbon (CFC) use: U.S. Environmental Protection Agency (EPA), Stratospheric Protection Program, Office of Program Development, Office of Air and Radiation, *Appendices to Regulatory Impact Analysis: Protection of Stratospheric Ozone* (EPA, Washington, D.C., August 1988); WRI analyses.

Only a fraction of carbon dioxide (CO_2) emitted during a year remains in the atmosphere to increase greenhouse heating. This amount totals approximately 3.7 billion metric tons of carbon. This table apportions this total carbon increase to each country, based on its share of the gross emissions of carbon dioxide.

CDIAC annually estimates *carbon dioxide* emissions from the burning of fossil fuels and the manufacture of cement for most of the countries of the world. (See Technical Notes to Tables 24.4 and 24.1.) These country-level estimates are not as accurate as the time trends they describe. These country totals include gas flaring.

The calculation of carbon releases from *annual land use change* is based on the work of R.A. Houghton, R.D. Boone, and, J.R. Fruci, *et al.* (See the Technical Notes for Table 24.1 for further information.)

The atmospheric concentration of *methane* increases by about 43 million metric tons per year, much less than that portion of total human-induced emissions estimated in Table 24.1, 255 million metric tons. In this table, *the net annual atmospheric increase* in methane concentrations is apportioned among countries, in proportion to each country's contribution to the total anthropogenic emissions documented in Table 24.1. An *equivalent carbon dioxide heating effect* of this methane increase is then calculated (see following discussion) to allow summing of greenhouse gas emission effects within a country.

Methane emissions were summed from individual estimates (see Table 24.1 and its Technical Notes) of anthropogenic emissions owing to the decomposition of solid waste, animal emissions from enteric fermentation (livestock flatulence), hard coal mining, wet rice agriculture, and leakages from natural gas pipelines.

CFC use is based primarily on reports on CFC-11 and CFC-12 that were submitted to a United Nations Environment Programme (UNEP) workshop on CFC production in Rome, May 1986. (See Technical Notes for Table 24.1.) Estimates of CFC production by country are difficult to obtain. They are held in confidence by UNEP if reported as required for signatories to the Montreal Protocol to Protect Stratospheric Ozone. Other potential sources of these data are major manufacturers of CFCs who hold these data as proprietary. The use or production of CFCs, given current levels of recycling, almost invariably leads to emissions at some later date. In fact, current production figures are not much greater than current emission estimates. Annual production or use is, therefore, a measure of CFCs to be emitted at a later date. In this table, however, responsibility for emissions is counted in the current year of production.

Use and production are confounded here, and in the available data. The U.S. EPA estimated 1985 and 1986 world production of CFC-11 and CFC-12 and also listed specific production/use figures for 18 countries and the 12 countries of the European Community. These specific country production/use figures are reported here. Other country figures are estimated from a WRI analysis of per capita consumption. (See Table 24.1.) Few data are available from the U.S.S.R. An estimate of 101,000 metric tons reported by WRI in *World Resources 1988–89* (WRI, Washington, D.C., 1988), p. 343, is the most recent estimate available. World production/use of CFC-11 and CFC-12 in 1986, as enumerated in this table, totals 771,500 thousand metric tons, compared with an estimate of 785,200 metric for 1985 and an estimate of 850,000 metric tons for 1986, both reported by the U.S. EPA.

To facilitate comparisons, net annual increases of methane in the atmosphere and CFC emissions are shown as the equivalent amount of carbon which, if released as carbon dioxide, would have the same heating effect. *Equivalent carbon dioxide effect* is calculated from Table 2-1 in the U.S. EPA's *Policy Options for Stabilizing Global Climate* (U.S. EPA, Washington, D.C., 1989), which was adapted from the work of V. Ramanathan of the University of Chicago.

This table lists for each greenhouse gas the "radiative forcing for a uniform increase in trace gases from current levels" in degrees Celsius per part per billion by volume. They are estimates of the increase in global temperature for an increase of one additional part per billion in the concentration of each greenhouse gas in the atmosphere. These estimates explicitly exclude feedback effects. The figure for methane (CH_4) is the lower bound of a range of possible heating effects, and thus probably sets a lower limit on the influence of methane sources considered here on global warming. A total carbon heating coefficient was calculated. Because of limited production data, estimates of CFC-11

Gas	Molecular Weight	Radiative Forcing °C/ppb*	CO₂ Heating Coefficient
CO₂	43.999	0.000004	1.0
CH₄	16.043	0.0001	68.6
CFC-11	137.368	0.07	
CFC-12	120.914	0.08	6,414.3

*ppb = parts per billion (one thousand million) by volume.

Note: Estimates of CO₂ equivalence were converted to carbon emission equivalents for presentation by dividing by 3.664 (the ratio of the molecular weight of CO₂ to the atomic weight of carbon).

and CFC-12 were combined and average molecular weights and radiative forcing coefficients were calculated (these data do not include the influence of other CFCs or the halons). One metric ton of methane provides the same radiative forcing as about 70 metric tons of CO₂. One metric ton of CFC-11 and -12 provides the same forcing as 6,400 metric tons of CO₂.

Table 24.3 Atmospheric Concentrations of Greenhouse and Ozone-Depleting Gases, 1959–88

Sources: Carbon dioxide: Charles D. Keeling, R.B. Bacastow, A. F. Carter, *et al.*, "A Three-Dimensional Model of Atmospheric CO₂ Transport Based on Observed Winds: 1. Observational Data and Preliminary Analysis," *Aspects of Climate Variability in the Pacific and the Western Americas*, American Geophysical Union (AGU) Monograph No. 55 (AGU, Washington, D.C., 1989), pp. 165–236 Other gases: R.A. Rasmussen and M.A.K. Khalil, "Atmospheric Trace Gases: Trends and Distributions over the Last Decade," *Science*, Vol. 232, pp. 1623–1624. Concentrations after 1985 of CCl₄, CH₃CCl₃, CCl₃F (CFC-11), CCl₂F₂ (CFC-12), and N₂O; M.A.K. Khalil and R.A. Rasmussen, unpublished data (Oregon Graduate Center, Beaverton, United States, September 1989). C₂Cl₃F₃ (CFC-113); M.A.K. Khalil and R.A. Rasmussen, unpublished data (Oregon Graduate Center, Beaverton, United States, September 1989). Methane data (CH₄), 1979–88; M.A.K. Khalil, R.A. Rasmussen, "Atmospheric Methane: Recent Global Trends," in preparation (1989). Methane data, 1962-78; M.A.K. Khalil, R.A. Rasmussen, and M.J. Shearer, "Trends of Atmospheric Methane During the 1960's and 70's," *Journal of Geophysical Research*, Vol. 94, No. D15, December 1989, pp. 18,279–18,288.

The trace gases listed in Tables 24.3–24.2 affect stratospheric ozone or contribute to the greenhouse effect or both. For further details concerning these processes, refer to Chapter 2, "Climate Change: A Global Concern."

Carbon dioxide (CO₂) accounts for about half the increase in the greenhouse effect and is emitted to the atmosphere by natural and anthropogenic processes. See the Technical Note for Table 24.4 for further details.

Atmospheric carbon dioxide concentrations are monitored at many sites worldwide; the data presented here are from Mauna Loa, Hawaii (19.53° North latitude, 155.58° West longitude). Trends at Mauna Loa reflect global trends, although carbon dioxide concentrations differ significantly among monitoring sites at any given time. The average annual concentration at the South Pole in 1988, for example, was 2.4 parts per million lower than at Mauna Loa.

Annual means disguise large daily and seasonal variations in carbon dioxide concentrations. The seasonal variation is caused by photosynthetic plants storing larger amounts of carbon, from carbon dioxide, during the summer than in the winter. Some annual mean figures were derived from interpolated data.

Data are revised to correct for drift in instrument calibration, hardware changes, and perturbations to "background" conditions. Additional details concerning data collection, revisions, and analysis are contained in C.D. Keeling, *et al.*, "Measurement of the Concentration of Carbon Dioxide at Mauna Loa Observatory, Hawaii," *Carbon Dioxide Review: 1982*, W.C. Clark, ed. (Oxford University Press, New York, 1982).

Carbon tetrachloride (CCl₄) is an intermediate product in the production of CFC-11 and CFC-12. It is also used in other chemical and pharmaceutical applications and for grain fumigation. Compared with other gases, CCl₄ makes a small contribution to the greenhouse effect and to stratospheric ozone depletion.

Methyl chloroform (CH₃CCl₃) is used primarily as an industrial degreasing agent and as a solvent for paints and adhesives. Its contribution to the greenhouse effect and to stratospheric ozone depletion is also small.

CFC-11 (CCl₃F), CFC-12 (CCl₂F₂), CFC-22 (CHClF₂) and CFC-113 (C₂Cl₃F₃) are potent depletors of stratospheric ozone. Together, their cumulative impact may equal one fourth of the greenhouse contribution of carbon dioxide.

Total Gaseous Chlorine is calculated by multiplying the number of chlorine atoms in each of the chlorine-containing gases (carbon tetrachloride, methyl chloroform, and the CFCs) by the concentration of that gas.

Nitrous oxide (N₂O) is emitted by aerobic decomposition of organic matter in oceans and soils, by bacteria, by combustion of fossil fuels and biomass (fuelwood and cleared forests), and by the use of nitrogen fertilizers. N₂O is an important depletor of stratospheric ozone; present levels may contribute one twelfth the amount contributed by carbon dioxides toward the greenhouse effect.

Methane (CH₄) is emitted through the release of natural gas and as one of the products of anaerobic respiration. Sources of anaerobic respiration include the soils of moist forests, wetlands, bogs, tundra, and lakes. Emission sources associated with human activities include livestock management (enteric fermentation in ruminants,

i.e., animal flatulence), anaerobic respiration in the soils associated with wet rice agriculture, and combustion of fossil fuels and biomass (fuelwood and cleared forests). Methane acts to increase ozone in the troposphere and lower stratosphere; its cumulative greenhouse impact is currently thought to be one third that of carbon dioxide, but on a molecule-for-molecule basis its effect, ignoring any feedback or involvement in any atmospheric processes, is 20–30 times that of CO₂.

Carbon monoxide (CO) is emitted by motor traffic, other fossil fuel combustion, and slash-and-burn agriculture. Increasing levels of carbon monoxide can lead to an increase in tropospheric ozone and a buildup of other trace gases, particularly methane, in the atmosphere.

Data for all gases except carbon dioxide and carbon monoxide are from January values monitored at Cape Meares, Oregon (45° north latitude, 124° west longitude). Although gas concentrations at any given time vary among monitoring sites, these data reflect global trends. Data for carbon monoxide were taken from several sites and averaged to reflect global concentrations and trends.

Table 24.4 World Carbon Dioxide Emissions from Fossil Fuels and Cement Manufacture

Source: Carbon Dioxide Information Analysis Center (CDIAC), Environmental Sciences Division, Oak Ridge National Laboratory. Oak Ridge, Tennessee, testimony by Gregg Marland and Tom Boden before the U.S. Senate Committee on Energy and Natural Resources, July 26, 1989, and personal communication.

CDIAC calculates world emissions from data on the net apparent consumption of fossil fuels (based on the World Energy Data Set maintained by the United Nations Statistical Office), and from data on world cement manufacture (based on the Cement Manufacturing Data Set maintained by the U.S. Bureau of Mines). Emissions are calculated using global average fuel chemistry and usage.

Estimates of world emissions are probably within 10 percent of actual emissions. Individual country estimates (see Table 24.1) could depart more severely from reality. CDIAC points out that the time trends from a consistent and uniform time series "should be more accurate than the individual values."

Emissions of carbon dioxide are usually calculated and reported in terms of its content of elemental *carbon*. The actual mass of carbon dioxide can be calculated by multiplying the carbon mass by 3.664.

Total emissions consist of the sum of the carbon in CO₂ produced during the consumption of *solid*, *liquid*, and *gas* fuels (primarily but not exclusively coals, petroleum products, and natural gas), produced during *gas flaring* (the practice of burning off gas released in the process of petroleum extraction, a practice that is declining) and the production of *cement* (by calcining calcium carbonate to produce calcium oxide, 0.136 metric tons of carbon are released as CO_2 for each ton of cement production).

Combustion of different fossil fuels releases carbon dioxide at different rates for the same energy production. Burning oil releases about 1.5 times the amount of carbon dioxide released from burning natural gas; coal combustion releases about twice the carbon dioxide of natural gas.

It was assumed that approximately 1 percent of the coal used by industry and power plants was not burned, and an additional few percent were converted to non-oxidizing uses. Other oxidative reactions of coal are assumed to be of negligible importance in carbon budget modeling. Carbon emissions from gas flaring and cement production are also included. These two sources emit about 3 percent of the carbon emitted by fossil fuel combustion. Fossil fuel emissions include those released from "bunker fuels" in international transport and are thus not ascribable to particular countries. See the Technical Notes for Table 24.1 for further information.

Table 24.5 Air Pollution in Selected Cities

Source: World Health Organization (WHO)/United Nations Environment Programme (UNEP), and the Global Environment Monitoring System (GEMS)/AIR Monitoring Project, Monitoring and Assessment Research Centre (MARC), *Assessment of Urban Air Quality* (MARC, London, 1988).

Air quality in selected cities is given for the number of *site years* of observation (number of sites multiplied by the number of years of operation) for *sulphur dioxide* (SO_2), *suspended particulate matter* (SPM), and *smoke*. These data are presented for the *minimum*, the *average*, and the *maximum* number of days that the pollutant exceeded WHO guidelines for all years of observation (site year). See the Technical Notes for Table 24.6 for some additional details.

WHO recommends that SO_2 exposure should not exceed *150 micrograms per cubic meter* on more than seven days a year. Many reporting cities exceed this level on an average basis. This is of particular concern for young children and people at risk of respiratory illness. Exposure, along with acute respiratory illness, could lead to chronic respiratory illness later in life. GEMS estimates that over 600 million people live in urban areas where SO_2 levels exceed WHO guidelines.

The health effects of SPM are in part dependent on the biological and chemical makeup and activity of the particles. Heavy metal particles, or hydrocarbons

condensed onto dust particles, can be especially toxic. There are two commonly used methods to measure SPM: high-volume gravimetric sampling and smoke shade methods. Gravimetric sampling determines the mass of particulates in a given volume of air. Smoke shade methods relate the reflectance of a stain left on filter paper that has had ambient air drawn through it to the concentration of particulate in the air. Smoke shade data cannot be used interchangeably with gravimetrically determined mass measurements because the smoke shade measurement is predominantly an indication of dark material in the air, which may not be proportional to the total weight of suspended matter. High-volume data may be twice as large as concurrent smoke shade results. In this table, *gravimetrically determined suspended particulate matter* measurements are shown and compared with the WHO guideline of *230 micrograms per cubic meter*. The *smoke* shade method is also shown and compared with its guideline of *150 micrograms per cubic meter*.

On average, most cities shown exceed the WHO guidelines on more than the specified seven days a year. GEMS estimates that over 1.25 billion people live in cities with unacceptable levels of SPM. See source for additional details.

Table 24.6 Emissions of Pollutants in Selected Countries

Source: World Health Organization (WHO)/United Nations Environment Programme (UNEP), and the Global Environment Monitoring System (GEMS)/AIR Monitoring Project, Monitoring and Assessment Research Centre (MARC), *Assessment of Urban Air Quality* (MARC, London, 1988).

These data should be used carefully. Because different methods and procedures may have been used in each country, the best comparative data may be time trends within a country.

Sulfur dioxide (SO_2) is created by both natural and anthropogenic activities. Anthropogenic sources include fossil fuel combustion and industrial activities. High levels of sulfur dioxide and *suspended particulate matter* (SPM) may cause respiratory problems among adults and children and may also result in illness of the lower respiratory tract, primarily in children. Anthropogenic emissions of SO_2 are estimated to range from 160–180 million metric tons per year worldwide. In the atmosphere, SO_2 oxidizes and, with moisture, becomes sulphuric acid. This acid precipitation, made more acidic by the simultaneous addition of nitric acid, has effects far distant from its source and is responsible for declines in forests in

North America and Europe, negative effects on soils and crops, and the destruction of architectural treasures.

Total emissions of SO_2 increase at the rate of about 4 percent annually. These estimates are calculated from data on fossil fuel production, their sulphur content, and mean emission rates. Not included in this table is the U.S.S.R., which in 1979 was estimated to emit 28 million metric tons of SO_2. When the U.S.S.R. estimate is combined with almost 68 million tons listed in this table, over 50 percent of world emissions are accounted for. SO_2 emissions have declined recently in several developed countries and China.

SPM arises from numerous anthropogenic and natural sources. Among the anthropogenic sources are combustion, industrial and agricultural practices, and the formation of sulfates from SO_2 emissions. SPM emissions total perhaps 300 million metric tons (50 percent are sulphates) of which just 27 million are accounted for in this table. SPM emissions are difficult to estimate. Health effects are dependent on the particular chemical and biological properties of the individual particles.

Oxides of nitrogen (NO_X) are important pollutants. All oxides of nitrogen contribute to acid rain, in the form of nitric acid; and NO_X is a precursor to tropospheric ozone that plagues many urban areas. Oxides of nitrogen are difficult to estimate, but total world production is estimated at about 75 million metric tons.

Carbon monoxide (CO), is formed both naturally and from industrial processes, including the incomplete combustion of fossil and other carbon-bearing fuels. Emissions from automobiles are the most important source, especially in urban environments. Estimates of world anthropogenic emissions range from 300 to 1,600 million metric tons annually. Carbon monoxide interferes with oxygen uptake in the blood, producing chronic anoxia leading to illness or, in the case of massive and acute poisoning, even death.

Lead (Pb) emissions could total up to 450,000 metric tons, of which almost all are anthropogenic. Alkyl lead, an anti-knock additive to ordinary gasoline, accounts for 60 percent of global emissions and up to 90 percent in individual countries. Children are especially vulnerable to lead poisoning, which affects heme biosynthesis and the nervous system.

GEMS coordinates the gathering of environmental information by national governments. MARC derived these particular data from reports published by individual countries on the state of the environment, or of specific pollutants, or submitted to GEMS in response to a questionnaire. See the source for additional details.

25. Policies and Institutions

Two basic indicators of a country's commitment to environmental protection are its participation in relevant international agreements and the collection and dissemination of environmental information.

Table 25.1 presents information on participation in critical treaties and other international agreements. The most recent treaty cited is the Convention on the Control of Transboundary Movements of Hazardous Wastes and Their Disposal, adopted in Basel, Switzerland, in March 1989, and signed by 43 countries (41 are included in Table 25.1), including the European Community (EC) as an entity, as of January 25, 1990. To enter into force, 20 countries must ratify, accept, approve, or accede to it; only Jordan had ratified it by the end of January 1990. The treaty is intended to restrict international traffic of hazardous wastes. Before shipping hazardous wastes, exporting countries must have written consent from the importing countries as well as the countries the shipment will cross. The exporting country must manage the shipment in an "environmentally sound" manner. Each country is empowered to prohibit imports of hazardous wastes.

Fifty-eight countries including the EC have ratified, accepted, approved, or acceded to the Vienna Convention for Protection of the Ozone Layer. The purpose of this agreement is to protect human health and the environment by taking measures to control activities that produce adverse effects. The agreement entered into force on September 22, 1988. By January 22, 1990, 54 countries and the EC have ratified, accepted, approved, or acceded to the Montreal *Protocol on Substances that Deplete the Ozone Layer* (CFC control in Table 25.1), which requires 50 percent reductions in the production of chlorofluorocarbons by 1999. The Protocol entered into force on January 1, 1989.

Additional countries have become party to older agreements shown in Table 25.1. The agreements are as follows:

■ *Nuclear Test Ban (1963)*—to prohibit atmospheric and underwater nuclear weapons tests.
■ *Wetlands (Ramsar) (1971)*—to stem the encroachment on and loss of wetlands.
■ *Biological and Toxin Weapons (1972)*—to prohibit acquisition and retention of biological agents and toxins.

■ *World Cultural and Natural Heritage (1972)*—to protect cultural and natural heritage sites of outstanding value.
■ *Ocean Dumping (1972)*—to control pollution of the seas by dumping.
■ *Endangered Species (CITES)(1973)*—to protect endangered species from over-exploitation.
■ *Ship Pollution (1978)*—to eliminate international pollution by oil and other harmful substances.
■ *Migratory Species (1979)*—to protect wild animal species that cross international borders.
■ *Law of the Sea (1982)*—to establish a comprehensive legal regime for the seas and oceans.

Individually, these agreements are limited. Together they raise governments' perception of the validity of international action to protect the environment. They form a web of precedent and commitment that can expand in the future.

The sources of environmental and natural resource information listed in Table 25.2 are believed to be comprehensive assessments of natural resource and environmental conditions on national, regional, and global levels. They summarize available information on the condition of the natural resource base, highlight major environmental problems, document trends, and often suggest policies for resource management.

It is a sign of growing environmental awareness that almost all countries have become members of the United Nations Environment Programme's INFOTERRA system, which encourages and facilitates sharing information on environmental problems and their causes, impacts, and solutions.

Much environmental information is collected by aid donors, or at the request or instigation of international organizations. This information is often merely an adjunct to action plans (e.g., the Tropical Forestry Action Plan, the National Conservation Strategy) or describes merely a subset of the total environment (National Conservation Profile). But, ultimately, countries must collect and analyze their own information to plan the sustainable use of their resources. Of special interest is the still rare production of national state-of-the-environment reports and national compendia of environmental data that describe the condition and trends of important environmental measures. Global and regional sources of information provide both country information and a ready comparison of each country's performance.

Table 25.1 Participation in Global Conventions Protecting the

	Nuclear Test Ban 1963	Wetlands (Ramsar) 1971	Biological and Toxin Weapons 1972	World Heritage 1972	Ocean Dumping 1972	Endangered Species (CITES) 1973	Ship Pollution (MARPOL) 1978	Migratory Species 1979	Law of the Sea 1982	Ozone Layer 1985	CFC Control 1987	Hazardous Wastes Movement 1989	Regional Seas (UNEP)
WORLD													
AFRICA													
Algeria	S	CP		CP		CP	CP		S				M*,ML*,MSP*
Angola									S				
Benin	CP		CP	CP		CP		CP	S				WCA
Botswana	CP		S			CP			S				
Burkina Faso	S			CP					S	CP	CP		
Burundi	S		S	CP		CP			S				
Cameroon	S			CP		CP		CP	CP	CP	CP		WCA*
Cape Verde	CP		CP	CP	CP				CP				
Central African Rep	CP		S	CP		CP		S	S	CP			
Chad	CP				S	CP		S	S				
Comoros									S				
Congo			CP	CP		CP			S			S	WCA
Cote d'Ivoire	CP		S	CP	CP		CP	S	CP				WCA*
Djibouti									S				
Egypt	CP	CP	S	CP		CP	CP	CP	CP	CP	CP		M*,ML*,MSP*
Equatorial Guinea									S	CP			
Ethiopia	S		CP	CP		CP			S				
Gabon	CP	CP	S	CP	CP	CP	CP		S				WCA
Gambia, The	CP		S	CP		CP			CP				WCA*
Ghana	CP		CP	CP		CP		CP	CP	CP	CP		WCA
Guinea				CP		CP			CP				WCA*
Guinea-Bissau	CP		CP						CP				
Kenya	CP		CP		CP	CP			CP	CP	CP		
Lesotho			CP		S	S			S				
Liberia	CP		S		S	CP	CP		S				WCA
Libya	CP		CP	CP	CP				S				M*,ML
Madagascar	CP		S	CP		CP		S	S				EA
Malawi	CP	CP	S	CP		CP			S				
Mali	S	CP	S	CP				CP	CP				
Mauritania	CP	CP		CP					S				WCA
Mauritius	CP		CP			CP			S				
Morocco	CP	CP	S	CP	CP	CP	S		S	S	S		M*,ML*,MSP
Mozambique, People's Rep				CP		CP			S				
Niger	CP	CP	CP	CP		CP		CP	CP				
Nigeria	CP		CP	CP	CP	CP		CP	CP	CP	CP		WCA*
Rwanda	CP		CP			CP			S				
Senegal	CP	CP	CP	CP	S	CP		CP	CP			S	WCA*
Sierra Leone	CP		CP						S				
Somalia	S		S		S	CP		CP	CP				EA,R*
South Africa	CP	CP	CP	CP	CP	CP	CP		S	CP	CP		
Sudan	CP			CP		CP			CP				R*
Swaziland	CP								S				
Tanzania	CP		S	CP		CP			CP				
Togo	CP		CP		S	CP		S	CP	CP		S	WCA*
Tunisia	CP	CP	CP	CP	CP	CP	CP	CP	CP	CP	CP	CP	M*,ML*,MSP*
Uganda	CP	CP		CP		CP		S	S	CP	CP		
Zaire	CP		CP	CP	CP	CP			CP				
Zambia	CP			CP		CP			CP				
Zimbabwe				CP		CP			S				
NORTH & CENTRAL AMERICA													
Barbados			CP						S				C*
Canada	CP	CP	CP	CP	CP	CP			S	CP	CP	S	
Costa Rica	CP		CP	CP	CP	CP			S				
Cuba			CP	CP	CP				CP				
Dominican Rep	CP		CP	CP	CP	CP			S				
El Salvador	CP		S						S				
Guatemala	CP		CP	CP	CP	CP			S	CP	CP	S	C
Haiti	S		S	CP	CP				S			S	
Honduras	CP		CP	CP	CP	CP			S				C
Jamaica	S		CP	CP				S	CP				C*
Mexico	CP	CP	CP	CP	CP	CP	S		CP	CP	CP	S	C*
Nicaragua	CP		CP	CP		CP			S				C
Panama	CP		CP	CP	CP	CP	CP	CP	S	CP	CP	S	SEP*, C*
Trinidad and Tobago	CP					CP			CP	CP	CP		C*
United States	CP	CP	CP	CP	CP	CP	CP			CP	CP		C*,SP
SOUTH AMERICA													
Argentina	CP		CP	CP	CP	CP			S	S	S	S	
Bolivia	CP		CP	CP	S	CP			S				
Brazil	CP		CP	CP	CP	CP	CP		CP				
Chile	CP	CP	CP	CP	CP	CP		CP	S	S	S		SEP*
Colombia	CP		CP	CP	S	CP	CP		S			S	SEP*,C
Ecuador	CP		CP	CP		CP						S	SEP*
Guyana			S	CP		CP			S				
Paraguay	S		CP	CP		CP		CP	CP				
Peru	CP		CP	CP		CP	CP	CP	S	CP			SEP*
Suriname		CP			CP	CP	CP	CP	S				
Uruguay	CP	CP	CP		S	CP	CP		S	CP		S	
Venezuela	CP		CP		S	CP				CP	CP	S	C*

Environment, 1989

Table 25.1

	Nuclear Test Ban 1963	Wetlands (Ramsar) 1971	Biological and Toxin Weapons 1972	World Heritage 1972	Ocean Dumping 1972	Endangered Species (CITES) 1973	Ship Pollution (MARPOL) 1978	Migratory Species 1979	Law of the Sea 1982	Ozone Layer 1985	CFC Control 1987	Hazardous Wastes Movement 1989	Regional Seas (UNEP)
ASIA													
Afghanistan	CP		CP	CP	CP	CP			S			S	
Bahrain			CP						CP			S	P*
Bangladesh	CP		CP	CP		CP			S				
Bhutan	CP		CP						S				
China	CP		CP	CP	CP	CP	CP		S	CP			
Cyprus	CP		CP	CP		CP	CP		CP			S	M*,ML*, MSP*
India	CP	CP	CP	CP		CP	CP	CP	S				
Indonesia	CP		S			CP			CP		S		
Iran, Islamic Rep	CP	CP	CP	CP		CP			S				P*
Iraq	CP		S	CP					CP				P*
Israel	CP					CP	CP	CP	S	CP	S	S	M*,ML,MSP*
Japan	CP	CP	CP		CP	CP	CP		S	CP	CP		
Jordan	CP	CP	CP	CP	CP	CP				CP	CP	CP	R
Kampuchea, Dem			CP			S			S				
Korea, Dem People's Rep			CP				CP		S				
Korea, Rep	CP		CP	CP			CP		S				
Kuwait	CP		CP		S	S			CP			S	P*
Lao People's Dem Rep	CP		CP	CP					S				
Lebanon	CP		CP	CP	S		CP		S			S	M*,ML
Malaysia	CP		S	CP		CP			S	CP	CP		
Mongolia	CP		CP						S				
Myanmar	CP		S				CP		S				
Nepal	CP	CP	S	CP	S	CP			S				
Oman				CP	CP	CP	CP		CP				P*
Pakistan	CP	CP	CP	CP	CP	CP		CP	S				
Philippines	CP		CP	CP	CP	CP		S	CP		S	S	
Qatar			CP	CP					S				P*
Saudi Arabia			CP	CP					S			S	P*,R*
Singapore	CP		CP			CP			S	CP	CP		
Sri Lanka	CP		CP	CP		CP		S	S	CP	CP		
Syrian Arab Rep	CP		S	CP		CP				CP	CP	S	M*
Thailand	CP		CP	CP		CP			S	CP	CP		
Turkey	CP		CP	CP								S	M*,ML*,MSP*
United Arab Emirates			S		CP				S	CP	CP	S	P*
Viet Nam			CP	CP		S			S				
Yemen Arab Rep	S		CP	CP					S				R*
Yemen, People's Dem Rep	CP		CP	CP					CP				R
EUROPE													
Albania													
Austria	CP	CP	CP			CP	CP		S	CP	CP		
Belgium	CP	CP	CP		CP	CP	CP		S	CP	CP	S	
Bulgaria	CP	CP	CP	CP			CP		S				
Czechoslovakia	CP		CP				CP		S				
Denmark	CP	CP	CP	CP	CP	CP	CP	CP	S	CP	CP	S	
Finland	CP	CP	CP	CP	CP	CP	CP	CP	S	CP	CP	S	
France		CP	CP	CP	CP	CP	CP	S	S	CP	CP	S	*,MSP*,C*,EA*,SP
German Dem Rep	CP	CP	CP	CP	CP	CP	CP		S	CP	CP		
Germany, Fed Rep	CP	CP	CP	CP	CP	CP	CP	CP		CP	CP	S	
Greece	CP	CP	CP	CP	CP		CP	S	S	CP	CP	S	M*,ML*,MSP*
Hungary	CP	CP	CP	CP	CP	CP	CP	CP	S	CP	CP	S	
Iceland	CP	CP	CP		CP		CP		CP	CP	CP		
Ireland	CP	CP	CP		CP	S		CP	S	CP	CP	S	
Italy	CP	CP	CP	CP	CP	CP	CP	CP	S	CP	CP	S	M*,ML*,MSP*
Luxembourg	CP		CP	CP	S	CP		CP	S	CP	CP	S	
Malta	CP	CP	CP	CP		CP			S	CP	CP		M*,ML*,MSP*
Netherlands	CP	CP	CP		CP	CP	CP	CP	S	CP	CP	S	C*
Norway	CP	CP	CP	CP	CP	CP	CP	CP	S	CP	CP	S	
Poland	CP	CP	CP	CP	CP	S	CP		S				
Portugal	S	CP	CP	CP	CP	CP	CP	CP	S	CP	CP	S	
Romania	CP		CP						S				
Spain	CP	CP	CP	CP	CP	CP	CP	CP	S	CP	CP	S	M*,ML*,MSP*
Sweden	CP	CP	CP	CP	CP	CP	CP	CP	S	CP	CP	S	
Switzerland	CP	CP	CP		CP	CP	CP		S	CP	CP	S	
United Kingdom	CP	CP	CP	CP	CP	CP	CP	CP		CP	CP	S	C*,SP
Yugoslavia	CP	CP	CP	CP	CP		CP			CP			M*,MSP*
U.S.S.R.	CP	CP	CP	CP	CP	CP	CP		S	CP	CP		
OCEANIA													
Australia	CP	CP	CP	CP	CP	CP	CP		S	CP	CP		SP*
Fiji	CP		CP						CP	CP	CP		SP*
New Zealand	CP	CP	CP	CP	CP	CP			S	CP	CP	S	SP
Papua New Guinea	CP		CP			CP	CP		S				SP*
Solomon Islands			CP		CP	CP			S				SP*

Sources: United Nations Treaties Section, United Nations Environment Programme, World Conservation Union, and the U.S. Department of State.

Notes: Regional Seas letter codes (M, ML, etc.) indicate signature of specific Regional Seas conventions; see Sources and Technical Notes for further detail.
CP = contracting party (has ratified or taken equivalent action); S = signatory; * = ratification (or equivalent action) of Regional Seas convention.
Some small countries, signatories or contracting parties to the conventions and protocols listed, are not included in this table.
For formal titles of the conventions and protocols listed, and for additional information, see Sources and Technical Notes.

Table 25.2 Sources of Environmental and Natural Resource

		Sources of National Environmental Information {a}						
	INFOTERRA Member	State of the Environment Report {a}	National Conservation Profile {a}	National Conservation Strategy {a}	U.S. AID I {a}	U.S. AID II {a}	Environmental Statistical Compendium {a}	Tropical Forestry Action Plan {a}
AFRICA								
Algeria	Yes					N	N	
Angola	Yes							
Benin	Yes							
Botswana	Yes	1986 {b}	D	D				
Burkina Faso	Yes				1980 {b}	1982 {b}		IP
Burundi	Yes				1981 {b}			
Cameroon	Yes		1989 {c}		1981 {b}			(D) 1987 {b}
Cape Verde	Yes				1980 {b}			
Central African Rep	Yes		IP {c}					
Chad	Yes			IP				
Comoros	Yes							
Congo	Yes		IP {c}					IP
Cote d'Ivoire	Yes		D	ND				
Egypt	Yes				1988 {b}			
Ethiopia	Yes		D {c}	IP				IP
Gabon	Yes		IP {c}					
Gambia, The	Yes				1981 {b}			
Ghana	Yes		1988 {b}, IP {c}		1980 {b}			1986 {b}
Guinea	Yes		IP {c}		1983 {b}			(D) 1986
Guinea-Bissau	Yes		D	IP				
Kenya	Yes	1987 {b}	1988 {b}	IP				
Lesotho	Yes			IP	1982 {b}			IP
Liberia	Yes				1988 {b}			
Libya	Yes							
Madagascar	Yes	1987 {b}	D {b}	1984				
Malawi	Yes			IP	1982 {b}			
Mali	Yes			IP	1980 {b}			IP
Mauritania	Yes			IP	1979 {b}	1989 {b}		IP
Mauritius	Yes							
Morocco	Yes				1980 {b}			
Mozambique, People's Rep	Yes							
Niger	Yes				1980 {b}			
Nigeria	Yes		1988 {b}	IP				
Rwanda	Yes				1981 {b}	1987 {b}		
Sao Tome and Principe	Yes		IP {c}					
Senegal	Yes		D {c}	ND	1980 {b}			IP
Seychelles	Yes			ND				
Sierra Leone	No			ND				IP
Somalia	Yes			IP	1979 {b}			IP
South Africa	No			1980				
Sudan	Yes				1982 {b}	1983 {b}		
Swaziland	No				1980 {b}			
Tanzania	Yes		1988 {b}	IP				(D) 1988 {b}
Togo	Yes			IP				IP
Tunisia	Yes				1980 {b}			
Uganda	Yes		1988 {b}	ND	1982 {b}			
Zaire	Yes		1988 {b}, IP {c}	ND	1979 {b}	1981 {b}		IP
Zambia	Yes			1985 {b}	1982 {b}			
Zimbabwe	Yes			1987 {b}	1982 {b}			
NORTH & CENTRAL AMERICA								
Barbados	Yes			IP	1982 {b}			
Bahamas	Yes							
Belize	Yes			ND	1982	1984 {b}		IP
Canada	Yes	1986 {b}		ND {b}			1986 {b}	
Costa Rica	Yes	1988 {b}		1989	1981 {b}	1982 {b}		
Cuba	No							IP
Dominican Rep	No				1981 {b}			1987 {b}
El Salvador	Yes			IP	1982 {b}	1985 {b,c}		
Guatemala	Yes			IP	1981 {b}	1984 {b,c}		
Haiti	Yes				1979 {b}	1985 {b}		IP
Honduras	Yes				1981 {b}	1982 {b,c}		1987 {b}
Jamaica	Yes			IP	1982 {b}	1987 {b}		IP
Mexico	Yes	1986 {b,c}	1988 {b}					1988 {b}
Nicaragua	No			IP	1981 {b}			(D){b}
Panama	Yes			IP	1980 {b}	1980 {b,c}	1985	(D) {b}
St. Lucia	Yes			1987				
Trinidad and Tobago	No	IP						
United States	Yes	1986{c}, 1987{c}		IP			1979{b}, 1983{b}	
SOUTH AMERICA								
Argentina	Yes							1988 {b}
Bolivia	Yes				1979 {b}	1988 {b}		1989 {b}
Brazil	Yes		1988 {b}					
Chile	Yes	1985 {c}						
Colombia	Yes		1988 {b}	IP				(D) 1988 {b}
Ecuador	Yes		1988 {b}		1979 {b}	1983 {b,c}		(D) 1988 {b}
Guyana	Yes					1982 {b}		(D) 1988 {b}
Paraguay	No	1985				1982		

Information, 1989

Table 25.2

	INFOTERRA Member	State of the Environment Report {a}	National Conservation Profile {a}	National Conservation Strategy {a}	U.S. AID I {a}	U.S. AID II {a}	Environmental Statistical Compendium {a}	Tropical Forestry Action Plan {a}
				Sources of National Environmental Information {a}				
Peru	Yes		1988 {b}	IP	1979 {b}	1986 {b}		1987 {b}
Uruguay	Yes							
Venezuela	Yes							
ASIA								
Bahrain	Yes	1988						
Bangladesh	Yes			IP	1988{b}			
China	Yes	1984 {b}					N.D.	
Cyprus	Yes	1987						
Hong Kong	Yes	1985{b}, 1988						
India	Yes	1981{b}, 1985{b}	1989 {b}	IP	1988{b}			
Indonesia	Yes	1989		IP	1987		1983 {b,c}	IP
Iran, Islamic Rep	Yes							
Iraq	Yes							
Israel	Yes	1979 {b}						
Japan	Yes	1987 {b}					1980	
Jordan	Yes			IP	1979 {b}			
Korea, Rep	Yes	1983						
Kuwait	Yes	1987						
Laos	No		1988 {b}	IP				IP
Lebanon	Yes							
Malaysia	Yes	1983/1987 {b}	1988 {b}	IP				IP
Mongolia	Yes							
Myanmar	No		1989 {b}		1982 {b}			
Nepal	Yes			1988*	1979 {b}			
Oman	Yes		1988 {b}	IP	1981 {b}			
Pakistan	Yes			1987	1981 {b}	1986/1988 {c}	1984 {b}	
Philippines	Yes	1986 {b}	1988	IP	1980 {b}		1979	
Qatar	Yes	1987 {b}						
Saudi Arabia	Yes	1984						
Singapore	Yes	1985 {b}			1988 {b}			
Sri Lanka	Yes	1978		1989	1978 {b}	1988 {b}		
Syrian Arab Rep	Yes				1981 {b}			
Thailand	Yes			IP	1979 {b}	1987 {b}		
Turkey	Yes	1981 {b}						
United Arab Emirates	Yes							
Viet Nam	Yes			1985 {b}				IP
Yemen Arab Rep	Yes				1980	1982		
Yemen, People's Dem Rep	Yes							
EUROPE								
Austria	Yes						1985 {b,c}	
Belgium	Yes	1979						
Bulgaria	Yes							
Czechoslovakia	Yes							
Denmark	Yes	1982						
Finland	Yes	1985					1987 {b,c}	
France	Yes	1985 {b}		IP			1986	
German Dem Rep	Yes							
Germany, Fed Rep	Yes						1985{b}, 1986 {b,c}	
Greece	Yes							
Hungary	Yes	1986 {b,c}					1981	
Iceland	Yes	1986 {b}						
Ireland	Yes	1985 {b}						
Italy	Yes	1987 {c}		ND			1984 {c}	
Luxembourg	Yes	1984 {c}						
Malta	Yes							
Netherlands	Yes	1985 {b}					1987 {b}	
Norway	Yes			IP			1983 {b,c}	
Poland	Yes						1985 {c}	
Portugal	Yes							
Romania	Yes							
Spain	Yes	1977		IP				
Sweden	Yes	1984 {b}					1985 {b,c}	
Switzerland	Yes			IP				
United Kingdom	Yes			1983			1986 {b}	
Yugoslavia	Yes	1983		IP			1985 {c}	
USSR	**Yes**							
OCEANIA								
Australia	Yes	1987 {b}		1983			1985 {b}	
Fiji	Yes			IP				ND
New Zealand	Yes	1988		1985				
Papua New Guinea	Yes							IP
Samoa	Yes							
Vanuatu	Yes			IP				

Table 25.2 Sources of Environmental and Natural Resource Information, 1989 (continued)

Sources of Global and Regional Environmental Information

World:

Lester R. Brown, et al., *State of the World 1990* (W.W. Norton, New York, 1990). {b}

Martin W. Holdgate, Mohammed Kassas, and Gilbert F. White, *World Environment 1972-82* (Tycooly, Dublin, 1982). {b}

United Nations Environment Programme (UNEP), *The State of the Environment 1987* (UNEP, Nairobi, 1987). {b}

United Nations Environment Programme, *Environmental Data Report* (Basil Blackwell, Oxford, 1989). {b}

World Commission on Environment and Development, *Our Common Future* (Oxford University Press, Oxford, 1987). {b}

World Bank, *World Development Report* (World Bank, Oxford University Press, New York, 1989). {b}

World Resources Institute (WRI), *A Directory of Country Environment Studies* (WRI, Washington, D.C., 1990). {b}

World Resources Institute, International Institute for Environment and Development, *World Resources 1988-89* (Basic Books, New York, 1988). {b}

Africa:

Bureau for Africa, U.S. Agency for International Development (U.S. AID), *Natural Resources and Environmental Concerns In Sub-Saharan Africa* (U.S. AID, Washington, D.C., 1986). {b}

L.O. Lewis and L. Berry, *African Environments and Resources* (Unwin Hyman, Boston, 1988).

International Union for Conservation of Nature and Natural Resources (IUCN), *The IUCN Sahel Report: A Long-Term Strategy for Environmental Rehabilitation* (IUCN, Gland, Switzerland, 1986). {b}

Program for International Development, *Renewable Resource Trends in East Africa* (Program for International Development, Clark University, Worcester, Massachusetts, 1984). {b}

Food and Agriculture Organization of the United Nations (FAO), *Natural Resources and the Human Environment for Food and Agriculture in Africa* (FAO, Rome, 1986). {b}

Latin America:

Conservando el Patrimonio Natural de la Region Neotropical, Eric Cardich, ed. (Gland, Switzerland, 1986). {b}

Marc J. Dourojeanni, *Renewable Natural Resources of Latin America and the Caribbean, Situation and Trends* (World Wildlife Fund, Washington, D.C., 1982). {b}

Inter-American Development Bank (IDB), *Natural Resources in Latin America* (IDB, Washington, D.C., 1983). {b}

H. Jeffrey Leonard, *Natural Resources and Economic Development in Central America* (International Institute for Environment and Development, Washington, D.C., 1987). {b}

Jorge Morello, *Perfil Ecologico de Sudamerica* (Instituto de Cooperacion Iberoamericana, Barcelona, 1984). {b,c}

Europe and North America, Other Developed Countries:

Commission of the European Communities (CEC), *The State of the Environment in the European Community 1986* (CEC, Luxembourg, 1986). {b}

United Nations Statistical Commission and Economic Commission for Europe, *Environment Statistics in Europe and North America* (United Nations, New York, 1987). {b}

Eurostat, *Statistics Related to the Environment* (Eurostat, Luxembourg, 1987). {b}

Docter-Institute for Environmental Studies-Milan and the Commission of the European Communities, *European Environmental Yearbook 1987* (Docter International U.K., London, 1987). {b}

Organisation for Economic Co-operation and Development (OECD), *OECD Environmental Data Compendium 1989* (OECD, Paris, 1989). {b}

Organisation for Economic Co-operation and Development (OECD), *State of the Environment 1985* (OECD, Paris, 1985). {b}

Asia and Oceania:

A.L. Dahl and L.L Baumgart, *The State of the Environment in the South Pacific* (United Nations Environment Programme, Geneva, 1983). {b}

United Nations Economic and Social Commission for Asia and the Pacific (ESCAP), *State of the Environment in Asia and the Pacific,* Vols. 1 and 2 (ESCAP, Bangkok, 1985). {b}

Sources: World Resources Institute, International Institute for Environment and Development, World Conservation Union, U.S. Agency for International Development, World Conservation Monitoring Unit, and the United Nations Environment Programme.

Notes: INFOTERRA: member of INFOTERRA environmental monitoring network; U.S. AID I and U.S. AID II: U.S. Agency for International Development, Environmental Profile, Phase I or Phase II; a. Publication date of most recent edition; multiple dates indicate different reports. b. Copy held in World Resources Institute library. c. Not available in English. Some Statistical Reports have English table headings. D = Draft; IP = in preparation; ND = published, no date. For additional information, see Sources and Technical Notes.

Sources and Technical Notes

Table 25.1 Participation in Global Conventions Protecting the Environment, 1989

Sources: United Nations Environment Programme (UNEP), "Environmental Law in the United Nations Environment Programme" (UNEP, Nairobi, 1985); UNEP Governing Council, "Register of International Treaties and Other Agreements in the Field of the Environment, Supplement 1" (UNEP, Nairobi, April 1987); UNEP Governing Council, "Register of International Treaties and Other Agreements in the Field of the Environment, 1989" (UNEP, Nairobi, 1989); UNEP, "Status of Regional Agreements Negotiated in the Framework of the Regional Seas Programme, Rev. 1" (UNEP, Nairobi, February 1988); U.S. Department of State, *Treaties in Force* (U.S. Department of State, Washington, D.C., 1987); U.S. Department of State, unpublished data (U.S. Department of State, Washington, D.C., March 1988 and June 1989); Treaties Section, United Nations Secretariat (U.N.), unpublished data (U.N., New York, March 1988 and June 1989); International Union for Conservation of Nature and Natural Resources (IUCN), Environmental Law Centre, unpublished data (IUCN, Bonn, May 1989); UNEP, Environmental Law and Machinery Unit, unpublished data (UNEP, Nairobi, February 1990).

A country becomes a signatory of a treaty when a person given authority by the national government signs it. Unless otherwise provided in the treaty, a signatory is under no duty to perform the obligations stipulated before the treaty comes into force for the country. A country's signature indicates a commitment to undertake domestic action to ratify, accept, approve, or accede to the treaty. A country is a contracting party when the treaty comes into force with respect to the country. Typically this occurs when the country has ratified the treaty or otherwise adopted the provisions of the treaty as national law and when a prescribed number of countries indicates their consent to be bound by the treaty and register their instruments of ratification, acceptance, approval, or accession with the treaty's depositary (which may be a national government, a United Nations organization, or another international organization; some treaties have multiple depositaries).

The complete titles of the conventions and treaties summarized in Table 25.1, and their places and dates of adoption, are as follows:

■ *Nuclear Test Ban*: Treaty Banning Nuclear Weapon Tests in the Atmosphere, in Outer Space, and Under Water (Moscow, 1963); to prohibit atmospheric and underwater nuclear weapons tests and other nuclear explosions and prohibit tests in any other environment if radioactive debris would be present outside the territory of the country conducting the test.

■ *Wetlands (Ramsar)*: Convention on Wetlands of International Importance Especially as Waterfowl Habitat (Ramsar, 1971); to stem the encroachment on and loss of wetlands, by establishing a List of Wetlands of International Importance, and providing that parties will establish wetland nature reserves and consider their international responsibilities for migratory waterfowl.

■ *World Heritage*: Convention Concerning the Protection of the World Cultural and Natural Heritage (Paris, 1972); to establish a system to protect cultural and natural heritage sites of outstanding value.

■ *Ocean Dumping*: Convention on the Prevention of Marine Pollution by Dumping of Wastes and Other Matter (London, Mexico City, Moscow, Washington, D.C., 1972); to control pollution of the seas by dumping, by prohibiting the dumping of certain materials and regulating ocean disposal of others, authorizing regional agreements, and establishing a mechanism for assessing liability and settling disputes.

■ *Biological and Toxin Weapons*: Convention on the Prohibition of the Development, Production, and Stockpiling of Bacteriological (Biological) and Toxin Weapons, and on their Destruction (London, Moscow, Washington, D.C., 1972); to prohibit acquisition and retention of biological agents and toxins that are not justified for peaceful purposes and of the means of delivering them for hostile purposes or armed conflict.

■ *Endangered Species (CITES)*: Convention on International Trade in Endangered Species of Wild Fauna and Flora (CITES) (Washington, D.C., 1973); to protect endangered species from over-exploitation by tightly controlling trade in live or dead animals and in animal parts through a system of permits.

■ *Ship Pollution (MARPOL)*: Protocol of 1978 Relating to the International Convention for the Prevention of Pollution from Ships, 1973 (London, 1978); a modification of the 1973 convention to eliminate international pollution by oil and other harmful substances and to minimize accidental discharge of such substances.

■ *Migratory Species*: Convention on the Conservation of Migratory Species of Wild Animals (Bonn, 1979); to protect wild animal species that cross international borders, by promoting international agreements.

■ *Law of the Sea*: United Nations Convention on the Law of the Sea (Montego Bay, 1982); to establish a comprehensive legal regime for the seas and oceans, establish rules for environmental standards and enforcement provisions, and develop international rules and national legislation to prevent and control marine pollution.

■ *Ozone Layer*: Vienna Convention for the Protection of the Ozone Layer (Vienna, Austria, 1985); to protect human health and the environment by conducting research on ozone layer modification and its effects and on alternative substances and technologies, monitoring the ozone layer, and taking measures to control activities that produce adverse effects.

■ *CFC Control*: Protocol on Substances That Deplete the Ozone Layer (Montreal, 1987); to require a 50 percent reduction in production of chlorofluorocarbons (CFCs) by 1999, with allowances for consumption increases in developing countries.

■ *Hazardous Waste Movement*: Convention on the Control of Transboundary Movements of Hazardous Wastes and their Disposal (Basel, 1989); to restrict and control the international traffic in hazardous wastes.

■ *Regional Seas:* A series of conventions and their associated protocols addressing region-specific marine-related environmental issues. Areas of seas or ocean zones included under Regional Sea conventions are the Mediterranean, west and central Africa, eastern Africa, the Red Sea and the Gulf of Aden, the Caribbean, the South Pacific, the Southeast Pacific, and the Kuwait region.

Some of the symbols used to indicate participation in Regional Sea conventions denote several related conventions and protocols. An asterisk (*) follows the convention abbreviation if a country has ratified at least one of the conventions or protocols. The full titles of Regional Seas conventions, their date of adoption, and the abbreviations used in the table are listed below.

M. Convention for the Protection of the Mediterranean Sea against Pollution (1976). Protocol for the Prevention of Pollution of the Mediterranean Sea by Dumping from Ships and Aircraft (1976). Protocol Concerning Co-operation in Combating Pollution of the Mediterranean Sea by Oil and Other Harmful Substances in Cases of Emergency (1976).

ML. Protocol for the Protection of the Mediterranean Sea against Pollution from Land-Based Sources (1980).

MSP. Protocol Concerning Mediterranean Specially Protected Areas (1982).

WCA. Convention for Co-operation in the Protection and Development of the Marine and Coastal Environment of the West and Central African Region (1981). Protocol Concerning Co-operation in Combating Pollution in Cases of Emergency (1981).

EA. Convention for the Protection, Management and Development of the Marine and Coastal Environment of the Eastern

African Region (1985). Protocol Concerning Protected Areas and Wild Fauna and Flora in the Eastern African Region (1985). Protocol Concerning Co-operation in Combating Marine Pollution in Cases of Emergency in the Eastern African Region (1985).

R. Regional Convention for the Conservation of the Red Sea and Gulf of Aden (1982). Protocol Concerning Regional Co-operation in Combating Pollution by Oil and Other Harmful Substances in Cases of Emergency (1982).

C. Convention for the Protection and Development of the Marine Environment of the Wider Caribbean Region (1983). Protocol Concerning Co-operation in Combating Oil Spills in the Wider Caribbean Region (1983).

SEP. Convention for the Protection of the Marine Environment and Coastal Area of the Southeast Pacific (1981). Agreement on Regional Co-operation in Combating Pollution of the Southeast Pacific by Oil and Other Harmful Substances in Cases of Emergency (1981). Supplementary Protocol to the Agreement on Regional Co-operation in Combating Pollution of the Southeast Pacific by Oil and Other Harmful Substances in Cases of Emergency (1983). Protocol for the Protection of the Southeast Pacific against Pollution from Land-Based Sources (1983).

SP. Convention for the Protection of the Natural Resources and Environment of the South Pacific Region (1986).

P. Kuwait Regional Convention for Co-operation on the Protection of the Marine Environment from Pollution (1978). Protocol Concerning Regional Co- operation in Combating Pollution by Oil and Other Harmful Substances in Cases of Emergency (1978).

The United Nations Convention on the Law of the Sea and the Convention on the Control of Transboundary Movements of Hazardous Wastes and their Disposal have not yet entered into force. The Vienna Convention for the Protection of the Ozone Layer entered into force on September 22, 1988, after the required 20 countries ratified it. The Protocol on Substances that Deplete the Ozone Layer entered into force January 1, 1989, when the requirement of ratification by at least 11 countries, accounting for at least two thirds of 1986 estimated world chlorofluorocarbon (CFC) consumption, had been met. The Convention on the Control of Transboundary Movements of Hazardous Wastes and their Disposal will

enter into force when 20 countries have ratified it.

The European Community has signed the Convention on the Conservation of Migratory Species of Wild Animals, the United Nations Convention on the Law of the Sea, the Vienna Convention for the Protection of the Ozone Layer, and the Protocol on Substances That Deplete the Ozone Layer. It has also signed conventions on three regional seas as well as associated protocols.

The Eastern African and South Pacific Regional Seas conventions and their protocols have not yet entered into force.

Information on the number of Natural World Heritage Sites and Wetlands of International Importance is contained in Chapter 20, "Wildlife and Habitat," Table 20.1. For information on treaty terms, refer to the sources.

Table 25.2 Sources of Environmental and Natural Resource Information, 1989

Source: Compiled by the World Resources Institute. *INFOTERRA*, the International Referral System for Sources of Environmental Information, is a network of national information centers established by the United Nations Environment Programme (UNEP) for the exchange of environmental information. Each member country compiles a register of institutions willing to share expertise in environmentally related areas, such as atmosphere and climate, energy, food and agriculture, plant and animal wildlife, and pollution. An international directory is developed from the national registers; the directory passes on queries to experts who can answer them. In 1985, the network answered more than 10,500 queries, over half of which came from developing countries.

State of the Environment Reports are published by government agencies, multilateral organizations, universities, and nongovernmental organizations. They analyze the condition and management of a country's natural resources and document its progress or failure in sustaining its natural resource base.

National Conservation Profiles are published by the World Conservation Monitoring Center in support of the conservation of biological diversity. Typical profiles report the state of the flora and fauna, the extent of exploitation, and the nature of existing and proposed conservation activities.

National Conservation Strategy reports are prepared by some countries that have adopted a National Conservation Strategy, endorsed by the national government. Some countries that have not officially adopted a National Conservation Strategy may have published a draft document for discussion. For more detailed information on the status of National Conservation Strategies, see past issues of the *IUCN Bulletin Supplement*. The U.S. Agency for International Development (U.S. AID) sponsors the production of two series of environmental profiles. *Phase I* profiles are compiled from published literature. *Phase II* profiles are based on more extensive field studies, often written in collaboration with government institutions or local nongovernmental organizations. Phase II profiles include analyses of the laws, policies, and institutions that affect the environment and natural resource management in the country, and often propose strategies to redress problems. Phase II profiles are comparable to state of the environment reports in scope and detail.

Environmental Statistical Compendium reports national environmental statistical data primarily through graphs and tables, and contains little analysis. The United Nations Environment Programme State of Environment Unit publishes reports entitled *The State of the Environment: National Reports. A UNEP Series*. Reports have been published so far on Qatar, Kenya, and Bahrain, and are in preparation for Ecuador, Haiti, Indonesia, Morocco, Nepal, Thailand, and Zimbabwe. Other reports in this column were published by the countries independently or under the auspices of an agency such as U.S. AID.

Tropical Forestry Action Plans (TFAPs) are developed by individual countries, with the coordination of the Food and Agriculture Organization of the United Nations, to ensure the development and rational use of tropical forests and related resources. They are also intended to coordinate and guide the activities of the international donor community. TFAPs focus on five priority areas: forestry in land use, forest-based industrial development, fuelwood and energy, conservation of tropical forest ecosystems, and institutions.

The brief bibliography of Regional Sources of Environmental Information that follows the table includes general statistical and analytical publications for each region. It includes neither specialized reports nor journal articles.

World Map

World Map

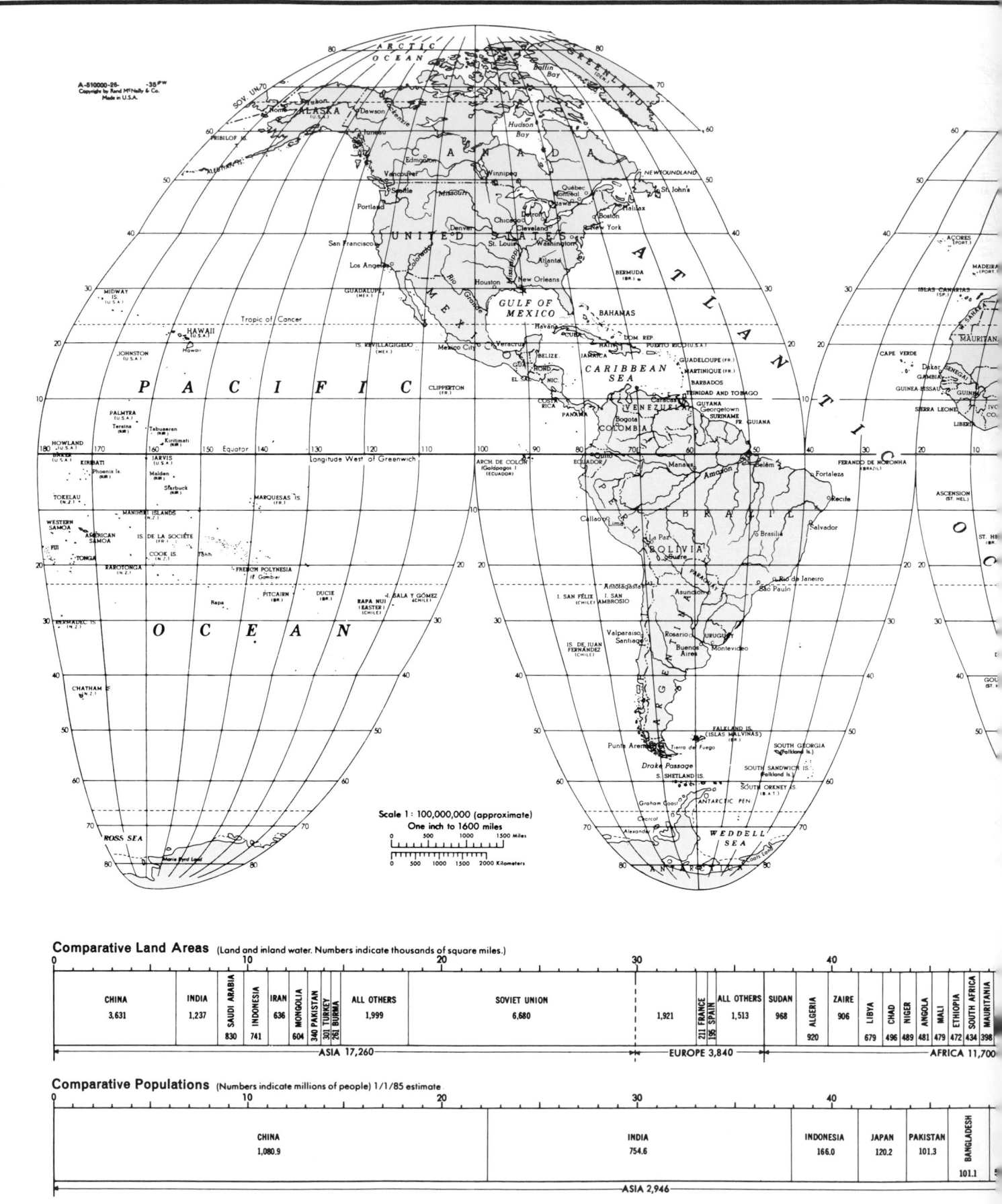

Scale 1 : 100,000,000 (approximate)
One inch to 1600 miles

Comparative Land Areas (Land and inland water. Numbers indicate thousands of square miles.)

0	10	20	30	40

| CHINA 3,631 | INDIA 1,237 | SAUDI ARABIA 830 | INDONESIA 741 | IRAN 636 | MONGOLIA 604 | PAKISTAN 340 | TURKEY 301 | BURMA 261 | ALL OTHERS 1,999 | SOVIET UNION 6,680 | 1,921 | FRANCE 211 | SPAIN 195 | ALL OTHERS 1,513 | SUDAN 968 | ALGERIA 920 | ZAIRE 906 | LIBYA 679 | CHAD 496 | NIGER 489 | ANGOLA 481 | MALI 479 | ETHIOPIA 472 | SOUTH AFRICA 434 | MAURITANIA 398 |

ASIA 17,260 — EUROPE 3,840 — AFRICA 11,700

Comparative Populations (Numbers indicate millions of people) 1/1/85 estimate.

0	10	20	30	40

| CHINA 1,080.9 | INDIA 754.6 | INDONESIA 166.0 | JAPAN 120.2 | PAKISTAN 101.3 | BANGLADESH 101.1 |

ASIA 2,946

World Resources 1990–91

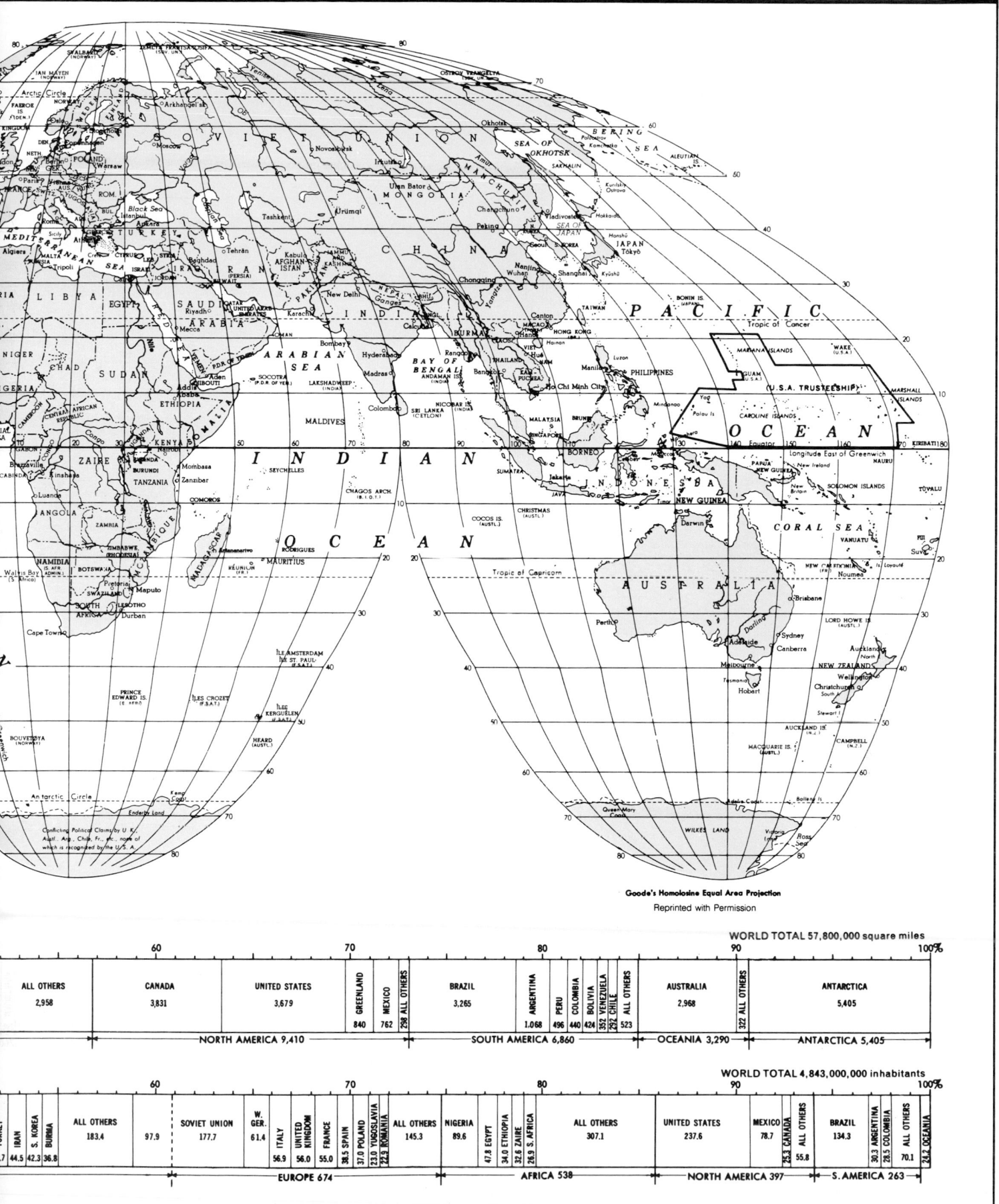

Goode's Homolosine Equal Area Projection

Reprinted with Permission

WORLD TOTAL 57,800,000 square miles

	60		70			80				90			100%

ALL OTHERS 2,958	CANADA 3,831	UNITED STATES 3,679	GREENLAND 840	MEXICO 762	ALL OTHERS 298	BRAZIL 3,265	ARGENTINA 1,068	PERU 496	COLOMBIA 440	BOLIVIA 424	VENEZUELA 382	CHILE 292	ALL OTHERS 523	AUSTRALIA 2,968	ALL OTHERS 322	ANTARCTICA 5,405

NORTH AMERICA 9,410 — SOUTH AMERICA 6,860 — OCEANIA 3,290 — ANTARCTICA 5,405

WORLD TOTAL 4,843,000,000 inhabitants

	60		70			80			90			100%

IRAN 44.5	S. KOREA 42.3	BURMA 36.8	ALL OTHERS 183.4	97.9	SOVIET UNION 177.7	W. GER. 61.4	ITALY 56.9	UNITED KINGDOM 56.0	FRANCE 55.0	SPAIN 38.5	POLAND 37.0	YUGOSLAVIA 23.0	ROMANIA 22.9	ALL OTHERS 145.3	NIGERIA 89.6	EGYPT 47.8	ETHIOPIA 34.0	ZAIRE 32.6	S. AFRICA 26.9	ALL OTHERS 307.1	UNITED STATES 237.6	MEXICO 78.7	CANADA 25.3	ALL OTHERS 55.8	BRAZIL 134.3	ARGENTINA 30.3	COLOMBIA 28.5	ALL OTHERS 70.1	OCEANIA 24.7

EUROPE 674 — AFRICA 538 — NORTH AMERICA 397 — S. AMERICA 263

Index

Numbers in italics refer to pages with data tables or technical notes.

Index

Index

Index

Index

Index

rates of (by country), *270-71, 274-75*
service delivery and, 37
trend toward, 2, 8, 66-58
Uruguay, urbanization and, 66

Vector Control Research Center (India),
59
Vector mosquitoes, 57
Vehicles
see also Automobiles; Transportation
emissions controls for, 40, 209, 210, 212
fuel-efficient prototype, 150, 151-52
growth in worldwide fleet of, 147-48
impact of growth in, 148
pollution control measures for, 209
worldwide use of fuel in light, 152
Venezuela, energy trade by, 143
Viet Nam, carbon release by, 109
Viterito, Arthur, 80

Waldsterben, 205
Ward, James, 177
Waste
dumped in Oslo Convention area, *341,
343-44*
generation of and trade in (selected
countries), *325, 327-28*
pollution in Latin America by, 39-42
radioactive (by country), *324, 327*
toxic, from coca production, 44
transboundary dumping of hazardous,
190
Wastewater
composition and treatment of, 162-63
pathogens, 183
priority of treatment technique
development for, 174
treatment in Latin America, 41-42
Water and Sanitation for Health Project
(WASH)—U.S.AID, 70
Western Europe

see also Europe; individual countries
energy consumption and production by
countries in, 142, 143-44
energy efficiency results in, 25
public opinion on environmental issues
in, 9-10
Wetlands
see also Habitats
aquatic habitats in, 127-28
Whales, 192-193
Wildlands and Human Needs Program,
World Wildlife Fund, 129
Wildlife
conservation technique modification,
134
habitat, 121-37, *299-314*
trade in wildlife products and, *304-05,
311-12*
Wind turbines, 27
Withdrawals, freshwater (by country),
330-31, 334
Women
African role of, 92
AIDS infected, 60
environmental concerns of, 10
reforestation work performed by, 110
of reproductive age, 53
Wood. *See* Forests
World Bank
African food production prospects
report by, 92
banking projects supported by, 78
environmental concerns in projects of,
9
experience with sub-Saharan projects
by, 91-92
support for freshwater and sanitation
projects, 71-72
sustainable projects by, 203
Tropical Forestry Action Plan strategy
developed by, 108-9
the Cubatao project loan by, 41
Third World water and sanitation
access estimate by, 69-70, 71, 73
World Climate Conferences, 5
World Congress on National Parks
(Indonesia), 128

World Conservation Union, coral reefs
estimate by, 128
World Health Organization (WHO)
AIDS data and work by, 60-62
farm aid agency study by, 71
malaria estimates and work by, 56-59
pollution standards of, 40, 162-64
sanitation service estimate by, 69-70
urban population estimate by, 202
water quality studies by, 164
World Meteorological Organization, 30
climate change and policy responses
study done by, 5
World Resources Institute (WRI)
energy efficiency and economic growth
studies in developing countries by,
25, 147
global warming policy
recommendations by, 111
natural resources accounting method
developed by, 234, 237
Tropical Forestry Action Plan strategy
developed by, 108-09
World Wildlife Fund (WWF)
analysis of Japanese timber trade by,
118
elephant numbers in Africa estimate by,
135

Yields
climate change effects on agricultural,
111
of marine fisheries, 180, *338- 343*

Zaire
acid rain in, 118
ivory smuggling in, 135
Zoos
biodiversity maintenance in, 129
rare animal species protected in,
308-10, 313-14

The World Resources Institute (WRI) is a research and policy institute helping governments, the private sector, environmental and development organizations, and others address a fundamental question: How can societies meet human needs and nurture economic growth while preserving the natural resources and environmental integrity on which life and economic vitality ultimately depend?

Through its policy studies, WRI aims to generate accurate information about global resources and environmental conditions, analyze emerging issues, and develop creative yet workable policy responses. In seeking to deepen public understanding, it publishes a variety of reports and papers, undertakes briefings, seminars and conferences, and offers material for use in the press and on the air.

In developing countries, WRI provides technical support, policy analysis, and other services for governments and nongovernmental organizations that are trying to manage natural resources sustainably.

A central task of WRI is to build bridges between scholarship and action, bringing the insights of scientific research, economic analysis, and practical experience to the attention of policymakers and other leaders around the world.

WRI's projects are now directed at two principal concerns:

■ The effects of natural resources deterioration on economic development and on the alleviation of poverty and hunger in developing countries; and
■ The new generation of globally important environmental and resource problems that threaten the economic and environmental interests of the United States and many other countries.

WRI is an independent, not-for-profit corporation that receives its financial support from private foundations, governmental and intergovernmental institutions, private corporations, and interested individuals.

WRI is currently carrying out the following policy research programs: Forests and Biodiversity, Economics and Institutions; Climate, Energy, and Pollution; Resource and Environmental Information; and Special Initiatives in Institutions and Technology.

WRI's Center for International Development and Environment provides services for developing countries in the sustainable management of natural resources. With an overarching objective of capacity building, these services include policy advice, technical program support, training, data management, and information dissemination.

World Resources Institute

1709 New York Avenue, N.W.
Washington, D.C. 20006, U.S.A.

WRI's Board of Directors:
Matthew Nimetz, *Chairman*
John E. Cantlon, *Vice Chairman*
John H. Adams
Robert O. Anderson
Robert O. Blake
John E. Bryson
Ward B. Chamberlin
Richard M. Clarke
Edwin C. Cohen
Louisa C. Duemling
Alice F. Emerson
John Firor
José Goldemberg
Michio Hashimoto
Cynthia R. Helms
Curtis A. Hessler
Martin Holdgate
James A. Joseph
Thomas E. Lovejoy
Alan R. McFarland, Jr.
Robert S. McNamara
Scott McVay
Paulo Nogueira-Neto
Thomas R. Odhiambo
Ruth Patrick
Alfred M. Rankin, Jr.
James Gustave Speth
Maurice F. Strong
M.S. Swaminathan
Mostafa K. Tolba
Russell E. Train
Alvaro Umaña
George M. Woodwell

Officers:
James Gustave Speth, *President*
Mohamed T. El-Ashry, *Vice President for Research and Policy Affairs*
J. Alan Brewster, *Vice President for Administration and Finance*
Wallace D. Bowman, *Secretary and Treasurer*
Jessica T. Matthews, *Vice President*

The United Nations Environment Programme (UNEP) was established in 1972 and given by the United Nations General Assembly a broad and challenging mandate to stimulate, coordinate, and provide policy guidance for sound environmental action throughout the world. Initial impetus for UNEP's formation came out of the largely nongovernmental and antipollution lobby in industrialized countries. This interest in pollutants remains, but right from the early years, as perceptions of environmental problems broadened to encompass those arising from the misuse and abuse of renewable natural resources, the promotion of environmentally sound or sustainable development became a main purpose of UNEP.

From the global headquarters in Nairobi, Kenya, and seven regional and liaison offices worldwide, UNEP's staff of some 200 scientists, administrators, and information specialists carry out UNEP's program, which is laid down and revised every two years by a Governing Council of representatives from its 58 member states. These members are elected on a staggered basis for three years by the United Nations General Assembly.

Broadly, this program aims to stimulate research into major environmental problems, promote environmentally sound management at both national and international levels by encouraging the application of the research results, and make such actions and findings known to the public—from scientists and policymakers to industrialists and school children.

By the terms of its mandate, UNEP runs its program in cooperation with numerous other United Nations agencies, governments, intergovernmental organizations, and nongovernmental organizations. It focuses on climate change, pollution, water resources, desertification control, forests, oceans and regional seas, biological diversity, human settlements, renewable sources of energy, environmentally sound management of industry, toxic chemicals, and international environmental lawmaking.

The essential base for environmentally sound management is provided by UNEP's work on the monitoring and assessment of the state and trends of the global environment. This is carried out in conjunction with agency partners, through the Global Environment Monitoring System (GEMS). The Global Resource Information Database (GRID), an element of GEMS, stores and analyzes geographically referenced environmental and resource data, and provides the essential link between monitoring and assessment and sound environmental management by putting information in forms useful to planners and managers. GEMS, the Geneva-based International Register of Potentially Toxic Chemicals, and INFOTERRA provide both the international community and individual countries and organizations with vital environmental information they need to take action.

United Nations Environment Programme

P.O. Box 30552
Nairobi, Kenya

Executive Director
Mostafa K. Tolba

Deputy Executive Director
William H. Mansfield III

Regional and Liaison Offices
Latin America and the Caribbean:
UNEP Regional Office for Latin America and Caribbean
Edificio de Naciones Unidas
Presidente Mazaryk 29
Apartado Postal 6-718
Mexico 5, D.F., Mexico

West Asia:
UNEP Regional Office for West Asia
1083 Road No. 425
Jufair 342
P.O. Box 26814
Manama, Bahrain

Asia and the Pacific:
(UNEP Regional Office for Asia and the Pacific)
United Nations Building
Rajadamnern Avenue
Bangkok 10200, Thailand

Europe:
UNEP Regional Office for Europe
Palais des Nations
CH-1211 Geneva 10, Switzerland

Africa:
UNEP Regional Office for Africa
UNEP Headquarters
P.O. Box 30552
Nairobi, Kenya

New York:
UNEP Liaison Office
UNDC Two Building
Room 0803
Two, United Nations Plaza
New York, New York 10017, U.S.A.

Washington:
UNEP Liaison Office
Ground Floor
1889 F Street, N.W.
Washington, D.C. 20006, U.S.A

The United Nations Development Programme (UNDP) is the world's largest multilateral source of grant funding for development cooperation. It was created in 1965 through a merger of two predecessor programs for United Nations technical cooperation. Its funds, which total $1.3 billion for 1990, come from the yearly voluntary contributions of member states of the United Nations or its affiliated agencies. A 48-nation Governing Council composed of both developed and developing countries approves major programs and policy decisions.

Through a network of offices in 113 developing countries, and in cooperation with over 30 international and regional agencies, UNDP works with 152 governments to promote higher standards of living, faster economic growth, and environmentally sound development. Currently, it is providing financial and technical support for over 6,000 projects designed to build governments' management capacities, train human resources, and transfer technology. These projects cover such fields as agriculture, forestry, land reclamation, water supply, environmental sanitation, energy, meteorology, industry, education, transport, communications, public administration, health, housing, trade, and development finance. Currently, projects valued at approximately $500 million are targeted on activities concerned with environmental aspects of development.

All UNDP-supported activities emphasize the permanent enhancement of self-reliant, sustainable development. Projects are therefore designed to:
■ Survey, assess, and promote the effective management of natural resources; industrial, commercial and export potentials; and other development assets.
■ Stimulate capital investments to help realize these possibilities.
■ Train people in a wide range of vocational and professional skills.
■ Transfer appropriate technologies that respect and enhance the environment and stimulate the growth of local technological capabilities.
■ Foster economic and social development, with particular emphasis on meeting the needs of the poorest segments of the population.

In each developing country, UNDP also plays the chief coordinating role for operational development activities undertaken by the whole United Nations system. Globally, UNDP has been assigned numerous coordinating roles—from administering special-purpose funds such as those entrusted to the United Nations Sudano-Sahelian Office, to chairing the interagency steering committee of the International Drinking Water and Supply and Sanitation Decade. It also focuses on bringing women more fully into the process, forstering participatory grassroots development, and encouraging entrepreneurship.

United Nations Development Programme

1 U.N. Plaza
New York, New York 10017

Administrator
William H. Draper III

Associate Administrator
Andrew J. Joseph

Regional Bureau for Africa
Assistant Administrator and Director
Pierre-Claver Damiba

Regional Bureau for Asia and the Pacific
Assistant Administrator and Director
Krishan G. Singh

Regional Bureau for Arab States and Europe
Assistant Administrator and Director
Mohammed A. Nour

Regional Bureau for Latin America and the Caribbean
Assistant Administrator and Director
Augusto Ramirez-Ocampo

Bureau for Special Activities
Assistant Administrator and Director
Aldo Ajello

Office for Project Service
Assistant Administrator and Director
Bernt A. Bernander